Ecology and Conservation of Fishes

Ecology and Conservation of Fishes

HAROLD M. TYUS

CRC Press
Taylor & Francis Group
Boca Raton   London   New York

CRC Press is an imprint of the
Taylor & Francis Group, an **informa** business

Cover: Center photo of two biologists and an endangered pallid sturgeon is given courtesy of Steven Krentz, U.S. Fish and Wildlife Service. The remaining eight images are from various sources that are acknowledged elsewhere in this book.

CRC Press
Taylor & Francis Group
6000 Broken Sound Parkway NW, Suite 300
Boca Raton, FL 33487-2742

© 2012 by Taylor & Francis Group, LLC
CRC Press is an imprint of Taylor & Francis Group, an Informa business

No claim to original U.S. Government works

Printed in the United States of America on acid-free paper
Version Date: 2011908

International Standard Book Number: 978-1-4398-5854-7 (Hardback)

This book contains information obtained from authentic and highly regarded sources. Reasonable efforts have been made to publish reliable data and information, but the author and publisher cannot assume responsibility for the validity of all materials or the consequences of their use. The authors and publishers have attempted to trace the copyright holders of all material reproduced in this publication and apologize to copyright holders if permission to publish in this form has not been obtained. If any copyright material has not been acknowledged please write and let us know so we may rectify in any future reprint.

Except as permitted under U.S. Copyright Law, no part of this book may be reprinted, reproduced, transmitted, or utilized in any form by any electronic, mechanical, or other means, now known or hereafter invented, including photocopying, microfilming, and recording, or in any information storage or retrieval system, without written permission from the publishers.

For permission to photocopy or use material electronically from this work, please access www.copyright.com (http://www.copyright.com/) or contact the Copyright Clearance Center, Inc. (CCC), 222 Rosewood Drive, Danvers, MA 01923, 978-750-8400. CCC is a not-for-profit organization that provides licenses and registration for a variety of users. For organizations that have been granted a photocopy license by the CCC, a separate system of payment has been arranged.

Trademark Notice: Product or corporate names may be trademarks or registered trademarks, and are used only for identification and explanation without intent to infringe.

Library of Congress Cataloging-in-Publication Data

Tyus, Harold M.
  Ecology and conservation of fishes / Harold M. Tyus.
      p. cm.
  Includes bibliographical references and index.
  ISBN 978-1-4398-5854-7 (hardcover : alk. paper)
  1. Fishes--Ecology. 2. Fishes--Conservation. I. Title.

QL639.8.T98 2012
597--dc23   2011036045

Visit the Taylor & Francis Web site at
http://www.taylorandfrancis.com

and the CRC Press Web site at
http://www.crcpress.com

This book is dedicated to W. L. Minckley, William M. Lewis Jr., and Ptychocheilus lucius, who gave me the insight, fortitude, and means to accomplish this work.

Contents

Preface xix
Acknowledgments xxi
Use and Features of the Book xxiii
Author Biography xxv
A Lexicon of Greek and Latin Word Roots Used in this Text xxvii

Part I
Introduction

Chapter 1
Ecology of Fishes: Content and Scope 3

History of Ecology 3
Fish Ecology Explored 4
What Is a Fish? 5
Fish: The First Vertebrate 7
Summary 7

Part II
Evolutionary Ecology of Fishes

Chapter 2
Aquatic Evolution, Origins, and Affinities 11

Aquatic Evolution 11
Evolutionary Ecology 11
Origins and Affinities of Fishes 12
Paleoecology of Fishes 15
Summary 16

Chapter 3
Aquatic Environment 17

Aquatic Ecosystems 17
Properties of Water 17
Seawater 21
Fish in Water: Where Is the Gravity? 21
Summary 22

Part III
Fish Diversity

Chapter 4
Diversity 1: Chordates to Sharks 27

Introduction 27
From Chordate to Vertebrate 27
Agnathans: Hagfishes and Lampreys 28

  Conodonts and Ostracoderms 31
  Early Gnathostomes 32
  Chondrichthyes 34
  Summary 37

Chapter 5
Diversity 2: Teleostomes to Bony Fishes 39

  Radiation of Teleostomes 39
  Acanthodians: Spiny Ones 40
  Sarcopterygians: Lobe-Fin Fishes 40
    Coelacanths 41
    Lungfishes 42
  Actinopterygians: Ray-Fins 43
    Relict Bony Fishes 45
  Summary 46

Chapter 6
Diversity 3: Teleosts 49

  Diversity and Adaptation 49
  Lower Teleosts 50
    Bonytongues 50
    Eels 51
    Herrings, Sardines, Menhaden, and Anchovies 52
    Minnows, Suckers, Characins, and Catfishes 53
    Whitefish, Arctic Grayling, Trout, Salmon, and Pike 54
  Cods and Anglerfishes 55
  Higher Teleosts 56
    Perciformes 56
    Flatfishes and Tetraodonts 57
  Summary 59

Chapter 7
Radiations, Extinctions, and Biodiversity 61

  Life on Earth Has Not Been Easy 61
  Fish Extinctions and a Few Questions 63
    Abiotic Change 63
    Biotic Factors 64
    Pseudoextinction 64
    Persistence of Survivors 65
    Enhanced Radiations 65
    Lessons from Long-Term Survivors 65
  Ecological Concepts 66
  Case Study: Fishes of Fossil Lake 66
  Summary 68

Part IV
Freshwater Ecosystems

Chapter 8
Zoogeography of Fishes 71
- Patterns and Species Diversity 71
- Factors Affecting Distribution 71
- Adaptation 72
- Continental Movement 73
- Fishes of Zoogeographic Regions 75
- Vicariance Biogeography 77
- Pleistocene Glaciation 78
- The Future 79
- Ecological Concepts 79
- Summary 80

Chapter 9
Lotic Systems: Flowing Water and the Terrestrial Environment 83
- A Drop of Rain 83
- Flowing Water 83
- Characteristics of Streams 85
  - Water Is (Almost) Always Moving 85
  - Channel Complexity 86
  - Riparian and Floodplain Features 86
  - Hydrology 87
  - Trophic Status and Energy Transport 88
- Ecological Concepts 89
- Summary 91

Chapter 10
Coldwater Streams 93
- Structure and Function 93
- Coldwater Fishes 95
- Constraints on Trout 97
- Case Study: Greenback Cutthroat Trout 100
- Summary 102

Chapter 11
Fishes of Warmwater Streams and Rivers 103
- A Warmwater Fish Viewpoint 103
- The Stream Connected 103
- Stream Fishes 106
- Large River Fish Faunas 108
- Case Study: The North American Paddlefish 109
- Summary 112

Chapter 12
Lentic Systems: Standing Water 115

    The Drop Is Stored (Temporarily) 115
    Standing Water Ecosystems 116
    Characteristics of Lakes 117
    Structure 120
    Function 120
    Fish in Lakes 120
    Ecological Concepts 122
    Case Study: Lake Baikal 123
    Summary 124

Chapter 13
Fishes of Temperate and Tropical Great Lakes 125

    General 125
    Fishes of Temperate Lakes 125
        The Laurentian Great Lakes: A History of Change 126
    Fishes of Tropical Lakes 133
    Case Study: Cichlids of East African Great Lakes 134
        The Lakes 134
        Cichlids 135
        The Nile Perch Arrives 136
    Summary 139

Chapter 14
Artificial Lakes and Groundwater Reservoirs 141

    Artificial Lakes: Reservoirs 141
    Structure and Function 142
    Ecological Concepts 145
    Cumulative Effects 146
    Fish and Reservoirs 147
        Impoundments 147
        Large Reservoirs 148
    Groundwater 149
    Case Study: Death Valley and Devils Hole 150
        Devils Hole Pupfish 151
        The Fight to Save the Fish 152
    Summary 153

Part V
Estuarine and Marine Ecosystems

Chapter 15
Estuaries and Coastal Zone 157

    Coastal Zone 157
    What Are Estuaries? 157

CONTENTS

- Drowned River Estuaries 159
  - Structure 159
  - Function 160
- Estuarine Fishes 163
  - Plankton-Based Systems 166
  - Oyster Reefs 166
- Ecological Concepts 167
- Case Study: Alewives as Migrating Subsystems 168
  - Background 168
  - Significance 168
  - Alewife Run at Mattamuskeet 169
  - A Study of Alewives 170
  - Tragedy Strikes and a Lesson Learned 172
- Summary 173

Chapter 16
Marine Environments, Intertidal Fishes, and Sharks 175

- Oceanography and Marine Ecology 175
- Intertidal Zone: Structure and Function 179
  - Beaches 179
  - Rocky Shorelines 180
- Marine Fishes 181
  - Perspectives 181
  - Diversity Scrutinized 181
  - Declining Abundance 182
- Ecological Concepts 183
- Case Study: The Ultimate Marine Predator 183
  - How Do Ecologists View Sharks? 183
  - Sharks as Human Predators 184
  - Humans as Shark Predators 184
  - Shark Swimming as Multitasking 185
- Summary 186

Chapter 17
Neritic Province and Fisheries 189

- Inshore Ocean in Perspective 189
  - Structure 189
  - Function 190
- Pelagic Systems 190
- Benthic Systems 190
  - Soft Substrates 191
  - Seagrass Flats 191
  - Rocky Substrates 192
  - Kelp Forests 192
  - Coral Reefs 192
- Neritic Fishes and the Temperate Zone 194
- Marine Commercial Fisheries 196
- Case Study: Cod and Northwest Atlantic Groundfishery 196
  - America or Codland? 196

    New Technology: Dragging, Bycatch and Bykill 198
    Fisheries of the Grand Banks 199
    Why Did the Fishery Collapse? 200
Summary 202

Chapter 18
Oceanic Province and Epipelagic Fishes 203

Province 203
    Structure 203
    Function 204
Epipelagic Zone 204
    Conditions 204
    Fishes 205
Upwellings 208
Case Study: Peruvian Anchoveta 209
Summary 211

Chapter 19
Deep Sea: Twilight to the Abyss 213

Features of the Deep Sea and Its Fishes 213
    Structure 214
    Function 215
Mesopelagic Zone 215
    Conditions 215
    Fishes 216
Abyss 218
Deep Benthic and Benthopelagic 220
Seamounts 221
Fish Adaptations in the Deep Sea 222
    In General 222
    Bioluminescence 222
    Buoyancy 223
    Sensory 223
Deepwater Fisheries 224
Case Study: Deep-Sea Anglerfish 224
Ecological Concepts 227
Summary 228

Part VI
Fish Adaptation

Chapter 20
Fitness, Morphology, and Ecophysiology 231

Adaptation and Fitness 231
Fish Morphology and Ecophysiology 232
    Integrated Fish Response 232
    Locomotion, Shape, and Function 233

CONTENTS

 Physicochemical Adaptation with Organs 235
  Dealing with Temperature: Warm Muscle 235
  Obtaining Oxygen from Water and Air 237
  Buoyancy and the Swim Bladder 238
  Osmoregulation 239
 Morphology and Sensory Systems 240
  Feeding Morphology 240
   Body Shape 241
   Mouth 241
   Teeth 242
   Gill Rakers 242
   Gut 242
  Sensory Systems 242
   Vision 243
   Hearing 243
   Chemosensory 243
   Mechanoreception 244
   Electricity 244
   Magnetism 244
 Summary 244

Chapter 21
Energy, Metabolism, and Growth 247

 Energy Budgets 247
 Fish Energetics 251
 Metabolism 252
  In Perspective 252
  Energy Source and Quality 252
  Metabolic Output and Rate 253
 Growth and Aging 254
 Metabolic Stress 257
 Stress in Fishes 258
 Case Study: Measuring Growth and Age in Hard Tissues 258
 Summary 260

Chapter 22
Adaptation, Niche, and Species Interactions 263

 Adaptation 263
 The Niche 265
 Niche Overlap and Response 268
 Species Interactions 269
  Intraspecific Competition 269
  Interspecific Interactions 270
 Summary 273

Chapter 23
Populations, Growth, and Regulation 275

 Fish Populations in General 275
 Present Status of Fish Populations 275

Population Characteristics 276
Population Growth 278
Population Regulation 280
Equilibrium/Nonequilibrium 283
    Nonequilibrium (Density Independence) 283
    Equilibrium (Density Dependence) 283
    Complexity of Regulation 284
Carrying Capacity Problem 284
Commercial Exploitation 285
    Overfishing 285
    Fallacy of Maximum Sustained Yield 286
    Fishery-Induced Depensation 287
    A Look at Fisheries Yield Models 287
Summary 288

Chapter 24
Instinct, Learning, and Social Behavior 289

Why the Interest in Behavior? 289
Instinctive Behavior and Innate Mechanisms 289
    Orientation 290
    Kineses, Reflexes, and Taxes 290
Biological Clock (Biorhythms) 291
Cognition and Learning 292
Nonreproductive Social Behavior 293
    Communication 293
    Spacing Behavior 294
    Multispecies Groups 295
Cooperation 296
Machiavellian Intelligence 296
How to Study Behavior 297
Case Study: Behavioral Interactions 298
Summary 301

Chapter 25
Trophic Concept and Feeding 303

Trophic Concept 303
Trophic Cascade 304
    Effect: Bottom-Up and Top-Down 304
    Fishery-Induced Trophic Cascades 305
Feeding Adaptations 306
    Trophic Categories 306
        Detritivores 306
        Scavengers 307
        Herbivores 307
        Omnivores 308
        Carnivores 309
    Resource Sharing and Trophic Adaptability 311
Food and Selectivity 312

    Foraging Behavior and Theory 313
        Optimal Foraging 313
        Predatory Behavior and Prey Response 314
        Prey Defense 315
        Basic Prey Model 316
        Patch Model 316
    Case Study: Prey Response—A Matter of Humps? 318
    Summary 321

Chapter 26
Reproductive Ecology and Life History Patterns 323

    Reproductive Process 323
    Life History Patterns 324
        Timing of Reproduction and Environmental Cues 324
        Sex and Mating 325
        Sexual Selection 326
        Alternative Breeding Tactics 326
        Spawning Site Selection 327
        Parental Care of Eggs and Young 327
    Reproductive Effort and Energy Allocation 327
        To the Young 327
        To the Reproductive Adults 329
    Two Life History Strategies 330
        Semelparity 330
        Iteroparity 330
    Reproductive Tradeoffs: r and K Selection and a 3-D Continuum 331
    Case Study: Timing of Spawning 332
    Summary 334

Chapter 27
Migration 335

    Fish Move, Disperse, and Migrate 335
    Finding the Way Back—Homing 336
        Home Stream Concept 336
        A Few Terms 337
        Orientation Mechanisms 338
    Spawning Migrations 340
        Oceanodromy 340
        Diadromy 340
        Potamodromy 341
    Examples and Descriptions 341
        Anadromous Migrations and Homing of Pacific Salmon 341
        Catadromous Migrations of Anguillid Eels 343
        Oceanadromous Migrations of Atlantic Herring 344
        Potamodromous Migrations 346
    Case Study: Migration of Colorado Pikeminnow 346
    Summary 350

Chapter 28
Larval Fish 353

- Introduction and Importance 353
- Reproduction and Early Life 355
- Description and Taxonomy 356
- Larval Ontogeny 357
- Larval Fish Ecology 359
  - Ecological Interactions 359
  - Habitat Selection 359
  - Feeding 360
  - Predation 361
  - Why Larvae? 363
- The Niche Revisited 365
- Fisheries Ecology and Recruitment Concepts 366
  - Historic Concepts 366
  - Recent Concepts 367
  - Marine versus Freshwater Environments 370
- Case Study—Larval Fish Movement 370
- Summary 373

Part VII
Applied Ecology: The Human Factor

Chapter 29
Exploitation and Fisheries Management 377

- Introduction 377
- Historic Perspective 378
  - Exploitation 378
  - Natural Resource Conservation 381
- Management Practices 382
  - In General 382
  - Fisheries 383
    - Regulation 383
    - Habitat Management 384
    - Manipulation of Organisms 384
- Fisheries: Practices and Problems 384
  - A Scientific Approach 384
  - Freshwater Fisheries 385
  - Estuarine Fisheries 388
  - Marine Coastal Fisheries 388
  - Offshore Marine Fisheries 389
- Concepts of Sustainability 391
- Ecosystem Approach 392
- Case Study: Fish Salvage at Tracy 394
  - Operation and Change 394
  - Demands on Operation 398
    - Endangered Species Acts 398
    - Invasion of Nonnative Species 398
    - Smart Predaceous Fish 398

    Increasing Demand for Pumping Water 398
    Entrainment of Early Life Stages 399
    Predation at Release Sites 399
    Stress 399
Future of the Facility 399
Summary 399

Chapter 30
Conservation of Fishes I: Crisis and a Response 401

Introduction 401
Biodiversity 401
Biodiversity Crisis 402
Why and How Are Species Going Extinct? 406
How Many Fish Do We Need? 409
Species Problem 410
A Response: The New Conservation 411
Endangered Fish Recovery? 412
Case Study: Can Science Save the Salmon? 413
    Declining Pacific Salmon 413
    Four "H"s 415
        Harvest 415
        Hatcheries 416
        Habitat 417
        Hydropower 419
    Breaching the Lower Snake River Dams 419
    Can Science Save the Salmon? 421
Summary 421

Chapter 31
Conservation of Fishes II: Understanding the Decline 423

Five Causes 423
Physical Habitat Alteration 424
    Effects 424
        Lakes 424
        Streams 425
        Estuaries 425
        Marine Systems 426
    Issues 426
Introduced Species 429
    Effects 429
        Lakes 429
        Streams 429
        Estuaries 430
        Marine Systems 430
    Issues 431
Overfishing 434
    Effects 434
        Lakes, Ponds, and Springs 434
        Rivers and Streams 435

        Estuaries 435
            Marine Systems 435
        Issues 435
    Hybridization 436
        Effects 436
            Lakes 436
            Rivers and Streams 436
            Estuarine and Marine Systems 437
        Issues 437
    Water Pollution 438
        Effects 438
            Lakes 438
            Rivers and Streams 438
            Estuaries 439
            Marine Systems 439
        Issues 439
    Are All Suspects Guilty? 441
    Case Study: Chesapeake Bay—An Ecological Disaster 441
        Background 441
        Oyster Reefs 442
        Loss of Planktivores 444
        Ecological Disaster 444
    Summary 446

Chapter 32
Changes and the Future 447

    Introduction 447
    Interesting Times 447
    Global Climate 449
        Changes in Progress 449
        Warming/Cooling: Evidence and Tipping Points 452
    GCC: Effects on Fish and Habitat 454
        Direct and Indirect Effects 454
        Ecology and Fish Production 454
        Invasive Species 456
    Fish and Fisheries in the Future: Bad News and Good Prospects 457
    Welcome to the Twenty-First Century 459
    Summary 460

Appendix 463

    Introduction 463
    Instructions 464
    Agnathans 465
    Gnathostomes (Jaws in the Mouth) 466
    Lower Teleosts 468
    Higher Teleosts 471

Glossary 475
Literature Cited 483
Taxonomic Index 517
Subject Index 521

Preface

My ichthyology and fishery science professors assured their classes that ocean fisheries would supply adequate food for an increasing human population (e.g., Daniel and Minot 1954). But this was not to be true. Instead, major marine fisheries have collapsed, and there are worldwide declines in freshwater fish populations. Where impacts have been due to overfishing without major physical habitat destruction, recovery has been possible. Where habitat destruction also has occurred, short-term recovery has been lacking.

The cause of freshwater fish declines and losses is no mystery. It is primarily due to the effects of an ever-increasing human population on fragile and isolated systems. These effects have reduced the distribution and abundance of fish populations by widespread changes that happen too rapidly for natural selection to respond.

How pervasive are continuing anthropogenic effects on the environment? Some, if not most, climatologists believe that human interference has so altered planetary function that the relatively stable period that began after the last ice age (i.e., Holocene) has ended. A new era of climate change is in progress, and it has been named the Anthropocene, a term attributed to Dutch scientist Paul Cruzen (Pearce 2007). Disturbed by the lack of success of endangered fish recovery programs, I wondered how we could better prepare for additional perturbation. I decided that better education across disciplines was needed.

The concept of this book took shape after I joined the research faculty at the University of Colorado, where I developed and taught an ecology of fishes course for the Department of Ecology and Evolutionary Biology. The class also was heavily supported by environmental sciences students. This book is based on that course and on student input. It has a functional approach with an applied orientation to promote better understanding and use of ecological concepts. Case studies are used as lessons in reality. Also, by student request, I have worked to make the book more interesting by including personal anecdotes and experiences.

There is a need for more applied ecologists in fishery management and conservation, and better training is needed to infuse management with practical application of ecological theory. The message: There are few (perhaps no) "natural" ecosystems remaining, nongame fishes are in decline, invasive species are rampant, endangered fishes are not being recovered, human impacts continue, and solutions are difficult. The prognosis: Not good. Fishery scientists and managers must become more effective at understanding and dealing with resource issues. If not, fish species, communities, and entire ecosystems will continue to decline as habitats change and species are lost. I hope this book will help.

Acknowledgments

This book is a product of 15 years of experience in teaching the Ecology of Fishes class (EBIO 4460/5460) at the University of Colorado at Boulder, and I thank the faculty at the Department of Ecology and Evolutionary Biology for giving me the opportunity. I also acknowledge the aid provided by the Center for Limnology, Cooperative Institute for Environmental Sciences (CIRES), which provided me with assistance, encouragement, and additional resources. My efforts were greatly aided by the many excellent texts cited herein and by undergraduate and graduate research activities at the University of Colorado and the University of Denver. I use the American Fisheries Society's Committee on Names of Fishes (Nelson et al. 2004) as the authority for names of fishes and Nelson (2006) as the authority for phylogenetic systematics. I am grateful for the use of images provided by Jon Miller and J. H. McCutchan, Jr., illustrations provided in U.S. Government publications, and permission to use copyrighted material from the American Fisheries Society and other publishers. In addition, J. H. McCutchan, Jr. assisted with photography and computer applications. Many of the examples and case studies I use in this book were obtained during my association with North Carolina State University, the Army Corps of Engineers, the U.S. Fish and Wildlife Service, and the University of Colorado. I am indebted to reviewers who provided their encouragement and suggestions for improving various portions of the draft manuscript: W. R. Courtenay, Jr., E. A. Frimpong, L. A. Hawkins, J. F. Kitchell, W. M. Lewis Jr., J. H. McCutchan Jr., J. Miller, N. J. Nikirk, D. E. Portz, W. M. Robinson, C. L. Roehm, J. F. Saunders III, D. E. Snyder, and the Ecology of Fishes class of 2009.

Use and Features of the Book

This book can be used as a "stand-alone" first course in fishes, as enrichment to an ecology curriculum, as an aquatic option in an environmental science program, or as an ecology text in a fisheries school. It is intended for an upper division undergraduate course, but it also can be used at the graduate level, especially with additional assignments. The book will supply more than enough material for an entire semester. Outside the classroom, I hope that the book will be useful as a refresher for professionals in government and private agencies, consulting firms, and academic institutions. I also hope that environmental lawyers, politicians, and inquisitive hobbyists will find helpful information in it.

As a class, Ecology and Conservation of Fishes has attracted a mix of undergraduate, graduate, and extension students whose interests and concerns are diverse (ranging from journalism, economics, psychology, and law to diving, fishing, aquaria, environmental and fisheries sciences, biology, ecology, and ichthyology). If used as a first course in fishes, course prerequisites should be general, and the approach must be broad. In this case, the three chapters on fish diversity will be needed. On the other hand, if used in a fisheries science program, the book can be used as a companion course with ichthyology, and the fish diversity chapters can be assigned only as a review.

It is a daunting task to consider the factors that influence the distribution and abundance of fishes in this rapidly changing world. In making the effort to do precisely this, I have provided real case studies of fishes in all of the major aquatic systems. These can be updated as new information is obtained. Some of these case studies are "ecological disasters of varying types." However, I believe the examples chosen are truly representative of the current state of such systems, and all of them can provide valuable lessons. At the end of each chapter, there is a short summary of the major conclusions. The summary and case studies are provided to reinforce major concepts.

Three chapters are provided to acquaint students with the names and adaptations of major fish groups and some high-profile fishes. However, some students may not have a good grasp of the morphology of the fishes or characteristics of fish groups. To overcome these difficulties, a fish identification supplement is included in the appendix. The supplement can be assigned, covered in a lecture with demonstration specimens, or used as a special laboratory exercise. Identification of fishes and fish assemblages is aided by black and white pictures in the text and a 16-page insert that contains full color plates. Teaching videos such as "Eyewitness: Fish" and trips to an aquarium are very useful teaching aids, as are online assignments. Videos referenced at the end of selected chapters are enjoyable and extremely helpful in understanding the systems and fishes concerned.

Most students are unfamiliar with Greek and Latin terms. Thus, there is a lexicon of Greek and Latin words provided to help them. Also, a glossary will aid students with troublesome terms and concepts.

The student also will benefit from applying theoretical concepts to real-world situations. An "Ecological Concepts" section is given in appropriate chapters, and specific concepts are emphasized. References to pertinent reviews and specific papers should help students who wish to pursue topics more deeply.

Author Biography

Harold M. Tyus is Emeritus Research Scientist at the Center for Limnology, Cooperative Institute for Research in Environmental Sciences, University of Colorado at Boulder, where he taught Ecology of Fishes in the Department of Ecology and Evolutionary Biology. He is also adjunct professor of environmental policy and management at the University of Denver, where he teaches and serves as a faculty advisor. Dr. Tyus received his academic training in the Department of Zoology at North Carolina State University, with the aid of a National Science Foundation fellowship and a scholarship from the National Wildlife Federation. He was also affiliated with the North Carolina Cooperative Fishery Unit, earning an MS studying sunfish phylogenetics and a PhD studying population dynamics and migrations of river herring. His minor concentration was in water resources management. Dr. Tyus is a retired researcher and manager for the U.S. Government, serving 23 years with the Army Corps of Engineers and Fish and Wildlife Service. During that time, he was involved with environmental impact assessment and studied a wide variety of aquatic habitats, fishes, and human-induced changes in waters of the United States, from east coast oceans, estuaries, and wetlands to southwestern desert rivers. He has written and edited numerous scientific papers on fishes and government documents on fish ecology and conservation, including listing and recovery plans for endangered species. He was a member of the Colorado River Fishes Recovery Team for 12 years, and he has been a consultant and science advisor for industry and government. His professional affiliations include the Desert Fishes Council, the American Society of Ichthyologists and Herpetologists, and the Society for Conservation Biology. He is a Fellow of the American Institute of Fishery Research Biologists and a Life Member of the American Fisheries Society, which has certified him as a Fishery Scientist and Fisheries Professional.

A Lexicon of Greek and Latin Word Roots Used in this Text

(Terms are abridged from Borror 1960 and Moore and Moore 1997)

a, an = lacking, without
acanth = spine
arthr = a joint
benth = bottom, depths
bi = two
bio = life
brachi = arm
branchi = gill
carn = flesh
caud = tail
ceno = recent
cephal = head
cerc = tail
chil, cheil = lip
chondr = cartilage
chord = string
clistic = enclosed
cten = comb
cyclo = circle
de = down
dendro = branches
dent, dont = tooth
derm = skin
deutro = second
di = across, double
dors = the back
dromo = running
dys = bad
eco, oikos = house
ect = outside
elegan = splendid
end = within
eu = good
eury = broad
falc = sickle
genus = a race
geo = earth
gladi = sword
glob = ball

gloss = tongue
gnath = jaw
gul = throat
haem = blood
hemi = half
herb = grass
hetero = different
hist = tissue
hol = whole
homo = alike
hyper = above
hypo = under
ichthy = fish
in = not, without
infra = below
inter = between
iso = similar
lepis = scale
lepto = thin, weak
macr = large
mer = a part
mes = middle
meta = next to
micro = small
mon = single
morph = shape
myo = muscle
nect = swimming
neo = new
nephro = kidney
not = back
odont = tooth
ost, osteo = bone
otic = of the ear
ov = an egg
paleo = ancient
par, para = close to
parous = have young
peri = around

phag = to eat
phor = to bear
phot = light
plac = platelike, flat
platy = broad
pod = foot
poikil = many
post = after
potam = river
prim = first
pro = before
pseudo = false
pter = wing, fin
ptych = folded
quadra = squared
rhin = nose
rhynch = snout, beak
sarc = flesh
sclera = hard
som = body
spatula = spoon
squam = scale
sten = narrow
stom = mouth
super = above
sym = together
tele = perfect, entire
tetr = four
tri = three
trich = a hair
troph = food
ura = tail
vass = a vessel
velum = a cover
ventr = belly
viv = live
xiph = sword
zoo = animal

PART I
Introduction

CHAPTER 1

Ecology of Fishes: Content and Scope

HISTORY OF ECOLOGY

Ecology had its beginnings in natural history. Early humans sought information about organisms and their interactions with the environment because they wanted to be better hunters, fishers, and gatherers. The meaning of the word *ecology* has its origins from the word *oikos*, which means a home or a place to live. Assuming that some organism(s) live in this place, a literal translation of *ecology* (*oikos* + *ology*) would be the science or study of life in its home or place. The study of all life in all of its places would be an enormous field of study, so ecology is a broad field with many subdivisions. For example, "life" can include one organism or various taxa, including species or populations. The "home" or "place" can be terrestrial or aquatic, or even finer divisions such as ecosystems and habitats. Also, we might study the adaptations of organisms to their environment, as reflected in physiology, behavior, feeding, reproduction, and so on. Another subdivision might consider the role of environmental conditions in shaping adaptation as reflected in the fossil record. Finally, a branch of ecology is devoted to sorting out the role of humans on the physical (nonliving or abiotic) and living (biotic) components of the environment.

In a general or fundamental sense, ecology may be defined as "the study of structure and function of nature" (Odum 1971), but most biologists would probably define ecology as the study of "the interrelationships between organisms and their environment and each other" (Lawrence 1989). But there are many subdivisions of ecology, and other definitions have been provided to more closely represent them. Krebs (1972) identified the basic ecological problem as the determination of "the causes of the distribution and abundance of organisms." Obviously, the causes include abiotic and biotic environmental factors and the capacity or adaptations of the organism to respond to those factors.

Ecology can have a descriptive or a functional approach. A descriptive approach seeks to describe an ecosystem from producers through decomposers, investigating the interrelationships between its components. A functional approach is more focused, addressing specific relationships and general problems. Ecology also can be taught from a theoretical and an applied standpoint. Hypotheses, theories, and models have been developed to explain various phenomena observed in the science of ecology. Some theoretical concepts have never been fully accepted in a working sense, but almost all can serve in providing a basis for understanding. Such concepts also are useful in the field of applied ecology as a basis for resource management, conservation, and preservation.

As human populations increased, their effects on natural systems have had far-reaching impacts on ecosystems, and new fields have emerged in ecology to address these impacts. Classified collectively as *applied ecology*, these fields include environmental science, conservation biology, landscape ecology, restoration ecology, and ecosystem management. Thus, applied ecology considers the affect of humans on every aspect of the natural world (e.g., Hinckley 1976). Recent developments in rapidly emerging areas of applied ecology were effectively encapsulated by Smith and Smith (2001) and Krebs (2008).

FISH ECOLOGY EXPLORED

What then is the scope of ecology of fishes? I prefer to define it as the study of those factors that influence the distribution and abundance of fishes. In this sense, factors are abiotic and biotic components of the environment, those conditions that Darwin and Horan (1979) associated with the struggle for existence. Clearly, ecology of fishes is a study of aquatic ecosystems, fish adaptation, natural selection, and fitness.

Many papers dealing with various aspects of fish ecology in the United States were published after the turn of the twentieth century, but more complete treatments of the subject have been published only recently. Early on, regional, state, and local guides to fishes, fishing, and fisheries included much descriptive information on fish ecology, but texts more concerned with general aspects of fish ecology were marketed in natural history or fishery management fields. Descriptive and functional information on fish ecology also appeared in fish biology (Kyle 1926; Lagler 1952; Bond 1979) and ichthyology texts (e.g., Lagler et al. 1962; Moyle and Cech 1982). However, it was not until the latter part of the century that texts providing more experimental approaches appeared (e.g., Bone and Marshall 1982; Wootton 1990; Jobling 1995; Diana 1995).

Ecology texts are written with different purposes, goals, and perspectives. *The Ecology of Fishes* by Nikolsky (1963) provided my first exposure to fish ecology as a major field of study. The focus of that book was to explore concepts of ecology, with the goal of developing "the biological basis of a rational fishery." Its purpose was to aid human exploitation, ostensibly according to ecological principles. Other authors also have focused on the ecology of exploited fish populations (e.g., Pitcher and Hart 1982), but developing "rational" fisheries (i.e., sustainable fisheries constructed on a scientific basis) has been elusive, and recent texts have been oriented more toward understanding fundamental ecological problems (Jennings et al. 2001).

In simple terms, the basic scope of fish ecology considers use of space (i.e., habitat) by the fish and how the fish obtains and uses energy to maximize fitness. There are two components to this: what constraints are imposed on the fish by the environment and how the fish responds with adaptation.

There are at least two overlapping areas of study reflected in this text: ichthyology and ecology. The study of fish, ichthyology, is an academic approach that includes fish systematics, anatomy, physiology, evolution, and ecology. As taught in most institutions today, its focus is fish taxonomy and phylogenetic systematics. An old science, by the time Linnaeus began his taxonomic work, a fish classification system had already been developed.

As previously indicated, ecology is a broad field, combining many disciplines and approached with different perspectives. I have organized this text into an introduction and six additional parts that cover the ecology of fishes: evolutionary ecology, fish diversity, freshwater, estuarine and marine ecosystems, adaptation, and applied ecology.

Anthropogenic change has resulted in a loss of fish habitat and diversity across the planet; and a new synthetic field of study, *conservation biology*, has emerged in response. Conservation biology represents a specific discipline in the conservation movement that stresses scientific rigor, multidisciplinary approaches, and innovative measures to address the biodiversity crisis. Meffe et al. (1997) provided three guiding principles of conservation biology that are paraphrased here: (1) evolutionary change: evolution is the basic axiom that unites all of biology; (2) dynamic ecology: the ecological world is dynamic and characterized by nonequilibrium; and (3) the human presence: humans must be included in conservation planning. Adherence to these principles was a priority in writing *Ecology and Conservation of Fishes*.

WHAT IS A FISH?

The worldwide fish fauna includes a broad diversity of organisms with various combinations of shared and new characters. Popular notions of the fish have been conveniently summed up by a poem in *The Hobbit* by J. R. R. Tolkien (1965):

Alive without breath,
As cold as death;
Never thirsty, ever drinking,
All in mail, never clinking.

An interesting poem and the appropriate answer no doubt saved Bilbo Baggins's life; however, the riddle is an ichthyological nightmare! A story of half-truths when examined closely. Let us address each line separately:

- *Alive without breath*—This refers to breathing air directly for respiration. The belief is that fish take in oxygen from the water and not from air. However, some fishes have perfectly good lungs, and many fishes have an open swim bladder that can be used to extract oxygen from air. Other fishes have dispensed with lungs entirely and have more recently derived organs that allow oxygen to be taken directly from the atmosphere.
- *As cold as death*—It is true that most fishes are about the same temperature as the water in which they live, and their temperature would approximate that of a corpse. However, some fishes (such as sharks and tunas) regulate their internal body temperature by taking advantage of heat produced by sustained swimming: incoming blood from the gills is warmed by passive heat transfer from outgoing blood.
- *Never thirsty, ever drinking*—Marine fishes live in seawater, which is a medium of high salt content compared with their blood. They have to drink seawater to replace water that is passively lost to the salty medium, and then excrete concentrated salt back again using special organs. However, freshwater fish have just the opposite problem. The do not "drink" regularly.
- *All in mail, never clinking*—This one is easy. Mail is a type or armor made by interlocking small chains together like fish scales. The analogy would apply to fishes that have bony ridge scales. Sorry, but many fishes (e.g., catfish) do not have scales.

Let us not be too harsh on Tolkien. His intentions were to be entertaining, not scientific, but truly, what is a *fish*?

As pointed out by Nelson (2006), some authorities would restrict fish to only include jawed bony fishes, some authorities have added sharks and their close relatives, and some would also include jawless hagfishes and lampreys. However, if the term is restricted to a monophyletic group, then only the recent ray-finned fishes would be acceptable as fish.

In a general sense and in recognition of the great diversity of fishes, I endorse Nelson's (2006) simplified, if artificial, definition of fishes as "aquatic vertebrates that have gills throughout life and limbs, if any, shaped like fins." This is a simple but workable concept. The definition is artificial primarily due to a long evolutionary period and an incomplete fossil record. Thus, the term *fish* is used to designate an assemblage that is paraphyletic; it is not a monophyletic group as a whole (Nelson 2006). The term *fish* also is not used as a taxonomic group; there are plenty of names for fishes without it. Proper use of terms has "one fish" for one species or an individual of one species. Individuals of more than one species are given the term *fishes*. Thus, a bucket of "fish" is a generic term to mean a bucket with one or more species of fish. If there is a bucket of "fishes," then it most certainly should contain more than one fish species.

How do we recognize a fish? Draw two curved lines (one curved down and the other curved up) allow them to meet at one end but overlap on the other and then ask anyone, "What is it?" They will recognize a fish instantly. Almost all extant fishes are streamlined, with a torpedo-shaped appearance because of the need for locomotion in water, which is an extremely dense medium compared to air. Of course, we add a few fins and scales and then we have something like a recently evolved fish. However, there are subtle differences that we will learn to appreciate, such as those shown in Figure 1.1, which displays prominent features of fish morphology in a minnow and a largemouth bass.

Having made the point about streamlining, it must be immediately followed by the knowledge that many fish are not streamlined for fast swimming because they have no need for it. Benthic fishes that hide on the bottom and ambush their prey are dorsoventrally flattened like a pancake. Fishes that hunt and have need of escape in weedy ponds are likely to be short and laterally flattened for quick turns. Fishes that hunt in water with structures, such as kelp forest or sargassam weed, are likely to be camouflaged as part of the surroundings for protection and ambush.

Where do fish occur? An almost satisfactory answer is wherever there is liquid water and there is access. Fishes occur from the Arctic Circle to the equator, and from high mountains to the ocean abyss. They occur in freshwater, in salt water, in surface water, in ground water, and sometimes in no water at all!

Finally, there are the inevitable questions: What is the largest or the smallest fish? The largest fish is the whale shark (*Rhincodon typus*; Figure 1.2), with a length of at least 18.3 m (60 ft) (Robins and Ray 1986), with the basking shark (13.7 m) in second place. Choosing the smallest fish is a little more difficult because smaller fishes seem to be popping up in different taxa. I give the smallest fish award to Pietsch (2005), who reported that adult males of a deepsea anglerfish (*Photocorynus spiniceps*) are only 6.2 mm (Figure 1.3) and are parasitic on larger females (46 mm). In case of free-living fish, Watson and Walker (2004) reported a 7-mm gobi (stout infantfish, *Schindleria brevipinguis*) from the Great Barrier Reef in Australia. Kottelat et al. (2006) also found a 7.9-mm cypinid fish (*Paedocypris progenetica*) living in Sumatra.

Figure 1.1 Largemouth bass *Micropterus salmoides* (a), a higher teleost, and golden shiner *Notemigonus crysoleucas* (b), a lower teleost. (Courtesy of Raver, D. and the U.S. Fish and Wildlife Service (USFWS), National Conservation Training Center (NCTC).)

Figure 1.2 Largest extant fish, the whale shark *R. typus*, shown with diver for size comparison. (Courtesy of Cada, R. N., fishbase.org, www.fishbase.org. Image from Wikimedia Commons at http://commons.wikimedia.org/wiki/File:Rhtyp_u0_white_bg.gif.)

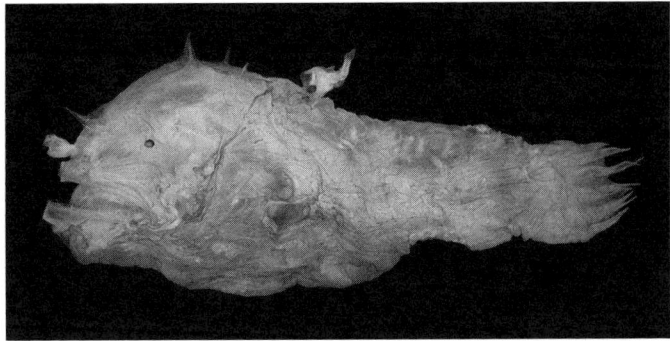

Figure 1.3 Preserved specimen of female ceratid anglerfish *P. spiniceps*, with attached parasitic male. (Courtesy of Pietsch, T. W.)

FISH: THE FIRST VERTEBRATE

Humans inherited most of their morphology from fishes, including almost all of their internal organs. The first fishes were chordates and shared some very important features of chordate ancestors. But the first fish, our first recognizable ancestor, was different. It was remarkable and it has been considered a radical, even a reckless experiment of nature (Curtis 1949). What made this organism so different? In a layman's sense, compared with the arthropods (the main thrust of animal evolution if we consider the number of species prevalent today), it was outside in, downside up, and backward! Nonscientists in general are apparently so unaware of differences between vertebrates and arthropods, that the gut of the shrimp exposed on the back of the tail when shell is removed is named its *vein*. We even purchase a "deveining tool" to remove it after we first remove its skeleton, which of course is on the outside, and apparently no-one seems to miss the backbone when using the tool.

Vertebrates first appeared with the origin of fishes. Thus, an understanding of fish evolution is fundamental for understanding origins of tetrapods, that is, amphibians, reptiles, birds, and mammals. However, the evolution of tetrapods was arguably a very small step for fishes at the time, considering the number of extant fish species (three of all five vertebrates) and enormous fish diversity represented by more than 400 million years of evolution. After all, only one species of tetrapod poses a significant threat to them. Also, it is important to note that extant fishes are just as modern as the most recently derived vertebrate although they represent an ancient lineage.

SUMMARY

Ecology had its beginnings in natural history as humans sought information to use in understanding and exploiting natural resources. Ecology is a broad field, with many subdivisions of study that focus on organisms, environments, adaptations, behavior, evolution, and anthropogenic effects (applied ecology). In this text, we study the ecology of fishes: the study of those factors that influence the distribution and abundance of fishes. By fish, we mean "aquatic vertebrates that have gills throughout life and limbs, if any, shaped like fins" (Nelson 2006). Fish are adapted for a life in water, and most are streamlined; however, they can vary greatly according to adaptations mainly for feeding and reproduction. Fish occur widely on the planet, from high mountains to ocean abysses. Fish were the first vertebrates, which appears to be a remarkable experiment of nature. An understanding of the evolution of fish is an aid in understanding the rise of tetrapods and the great diversity supporting the evolution of fish. The effects of humans on fish, fish populations, and their habitats have been so pervasive that anthropogenic changes are a main focus of this book.

Recommended teaching aid: Eyewitness video: *Fish!* BBC Lionheart (1994).
Further reading: Moyle 1993.

PART II

Evolutionary Ecology of Fishes

CHAPTER 2

Aquatic Evolution, Origins, and Affinities

AQUATIC EVOLUTION

Earth is unique among the planets in our solar system. An alien space traveler entering its system would quickly determine from a spectral and visual evaluation that most of its surface is covered by water and there is free oxygen in its mostly nitrogenous atmosphere. Presumably, the traveler would decide that life on this "blue planet" would evolve in water, and organisms would at least be tolerant of the effects of free oxygen.

In truth, life did have its inception in water and it took about 3 billion years for life to invade the continents. The aquatic environment was more suitable it seems for evolution, for at the beginning of this period there was no free oxygen in the atmosphere. It was by the action of chemosynthetic then later photosynthetic organisms living in water that oxygen was produced as a by-product of metabolism. After life appeared, it took about 1 billion years to oxidize all of the ferrous iron of the crust to ferric condition, and this allowed oxygen to accumulate in the atmosphere.

Water is the biological medium. It is present in all living organisms on Earth. Some organisms live in it, and the rest of us carry it around with us in tissues or bottles. It is a modifier of temperature and a major determiner of climate. Only the presence of water, the universal solvent, facilitated the chemical reactions that were necessary for life to develop and evolve. It is the reason that Earth is habitable by life as we know it. We will learn more in the next chapter about this amazing fluid that we generally take for granted. For now it is important to point out that evolution of the vertebrates was waterborne for a very long time.

EVOLUTIONARY ECOLOGY

Evolution is the development of new types of organisms from old ones by accumulating differences over time; in Darwin's terms, "descent with modification." Ecology is the study of interrelationships between organisms and their environment (or as used here, the study of factors that influence the distribution and abundance of fishes). By combining the two disciplines, evolutionary ecology is the study of interrelationships (e.g., what and how various factors have influenced the distribution and abundance of fishes) over time. To understand the evolution of fishes, it is necessary to identify taxa and be able to follow them through time, determine adaptations (principally through morphology and analysis of deposits in which they occurred), understand how environmental conditions would have affected the organism, and appreciate geological time scales. Central to the issue is adaptation for survival, for if fishes do not survive they cannot reproduce and there would be no evolution.

Survival of organisms over time is affected by both intrinsic and extrinsic factors. Intrinsic factors are within the organism; they are predictable and therefore *deterministic*—a term that means regulative within limits. The term *homeostasis* refers to physiological stability due to negative and positive feedback. This regulation also has been likened to steady-state or fluctuation within set-points. Our body temperature provides a good example. A person is considered abnormal (and probably sick) if their body temperature fluctuates over a degree above or below 98.6°F.

Extrinsic factors are evolutionary, environmental, random, unpredictable, and therefore *stochaistic*. The term *homeorhesis* refers to the more dynamic state of equilibrium appropriately given to ecosystems, in which there is a pulsing rather than a steady state. Such a condition is greatly influenced by positive feedback and lacks set-point controls (e.g., see Odum 1997). Because of stochaistic events, global conditions are far from constant. Earth experiences wide fluctuation in physical conditions such as temperature changes, and local fluctuations in climate such as tornadoes and hurricanes are common. In the extreme, the globe has experienced massive and violent occurrences such as volcanoes, massive lava flows, ice ages, asteroid/comet encounters, and so forth. Because of the long evolutionary history of fishes, they have been exposed to many such actions. Fishes also experience more subtle biotic forces due to intraspecific and interspecific competition and predation. An evolutionary ecologist is faced with evaluating how such abiotic and biotic forces affect species during their evolution.

ORIGINS AND AFFINITIES OF FISHES

Fishes are derived chordates, sharing and improving characters that first appeared in the phylum Chordata. Earliest chordates first appeared earlier than 540 million years ago (mya) from organisms that possessed a coelom, a fluid-filled body cavity. Early coelomates diverged into two distinct lines, depending on the fate of the blastopore in embryonic development. In one line (protostomes; i.e., first mouth) that developed into the arthropods the blastopore became the mouth, and a second opening emptied the gut. However, in the other line (deuterostomes; i.e., second mouth), which includes the chordates, the blastopore became the opening for the gut and the second opening became the mouth.

In addition, chordates have at least five other derived characteristics that are present in one or more life stages (Figure 2.1): notochord, a hollow dorsal nerve cord, pharyngeal slits, a muscular

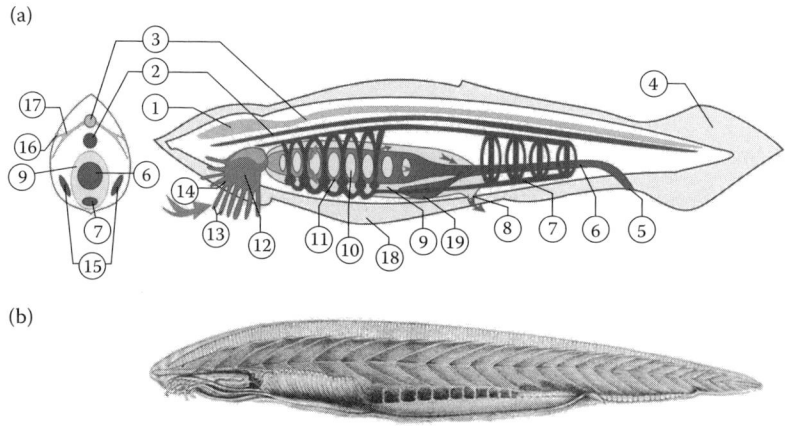

Figure 2.1 *Branchiostoma* (amphioxus): anatomy of a chordate. 1 = cephalization; 2 = notochord; 3 = dorsal nerve cord; 4 = caudal fin and postanal tail; 5 = anus; 6 = gut; 7 = circulatory system; 8 = abdomen; 9 = lacuna; 10 = gill; 11 = pharynx; 12 = mouth; 13 = cirri; 14 = space; 15 = gonads; 16 = photosensor; 17 = nerves; 18 = body wall; 19 = sack. ((a) Courtesy of Citron, Wikimedia Commons, http://commons.wikimedia.org/wiki/File:Branchiostoma_cultellus.jpg. (b) Courtesy of Jaworski, P. M., Wikimedia Commons, http://commons.wikimedia.org/wiki/File:BranchiostomaLanceolatum_PioM.svg.)

AQUATIC EVOLUTION, ORIGINS, AND AFFINITIES

postanal tail, and segmented muscles (myomeres), which occur along the body and throughout the tail. These characteristics were acquired by the first fishes, which greatly improved upon them and transforming them into vertebrate characters. Thus, they require further elaboration:

- The *notochord* is a fluid-filled rod encased by a fibrous sheath. It extends along the long axis of the body to provide attachments of muscles. Its function is to provide a muscular support and contraction from one side of the body to the other. The resultant motion sweeps the tail laterally and provides locomotion.
- In chordates, a hollow (or tubular) *nerve cord* lies *dorsal* to the gut. The cord is filled with fluid as opposed to the solid ventral nerve cords of other groups such as arthropods.
- The chordate *pharynx* also has a series of openings that appear as slits. These slits were used mainly for feeding. Cilia lined the slits and were used for moving food, and mucus was produced for retaining particles of food.
- Chordates have a *muscular postanal tail*. This flexible tail extends beyond the anus and greatly improves locomotion.
- Locomotion also is aided by a series of "V"-shaped *myomeres* or segmented muscles that lie along the notochord and occur on both sides of the body. These myomeres will become more complex structures as evolution progresses (Figure 2.2).

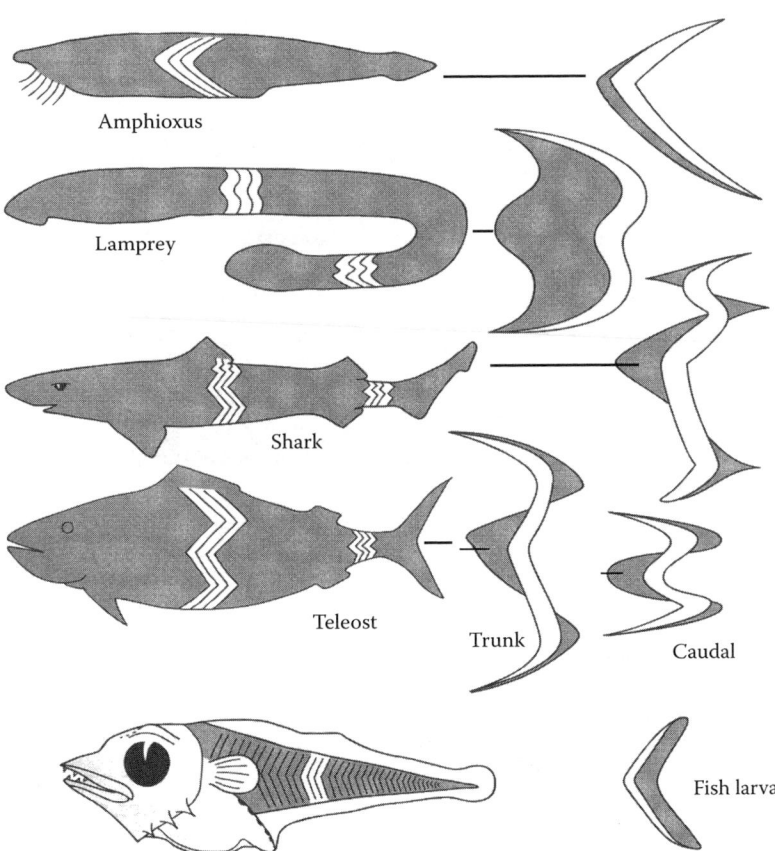

Figure 2.2 Muscle segments (myomeres) in fishes. (From Bone, Q., and Moore, R. H., *Biology of Fishes (3rd Ed.)*, Taylor & Francis Group, New York, 2008. With permission.)

The origin of chordates has been controversial, and at least three theories have been argued (Kardong 1998). It was long believed that that fishes evolved from prevertebrate chordates similar to *Amphioxus* (also known as *Branchiostoma*). It was believed that chordates evolved from ancestral tunicate larva by paedomorphosis (in which larval forms became sexually mature adults). But the discovery of *Pikaia,* a chordate from the middle Cambrian period (Figure 2.3) and hundreds of specimens of other free-swimming species from the early Cambrian, some of which are considered prechordates, casts doubt on the paedomorphosis theory (reviewed by Pough et al. 2005). Now it appears that free-swimming forms were ancestral to the invertebrate chordates and in turn to early fishes, while the sessile adult tunicate appears to be more recently derived.

Vertebrates first appeared in the fossil record in the form of jawless fishes, known as *agnathans* (a = without + *gnatha* = jaws). Early fossil agnathans (e.g., *Myllokunmingia* and *Haikouichthys*) (Figure 2.4) appeared about 540 mya in early (Lower) Cambrian deposits in China (Shu et al. 1999). Previously, the best evidence of the first fishes was in Early Ordovician deposits in several locations, which yielded early vertebrates known as *ostracoderms* (shell skin; Figure 2.5). The ostracoderms were early vertebrate agnathans that had armor of bony shields or plates. Prevertebrate chordates were feeble swimmers and suspension feeders, but these new agnathans developed into active deposit feeders. These ancient fishes had no jaws, but the ancestral pharynx was strengthened by muscles to become a more powerful suction device.

The development of bony armor in the ostracoderms provided good material for fossilization. With the appearance of external and internal bone in fishes, paleontologists have been able to construct a good record of their evolution and relate their distribution and abundance to indicators of environmental conditions.

Figure 2.3 *Pikaia*, a chordate from middle Cambrian deposits. (Courtesy of Tamura, N., Wikimedia Commons, http://commons.wikimedia.org/wiki/File:Pikaia_BW.jpg.)

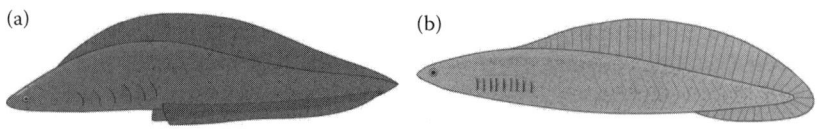

Figure 2.4 Two early fishes from lower Cambrian (ca. 450 mya): (a) *Myllokunmingia* and (b) *Haikouichthys*. (Courtesy of Anteater, G. B., Wikimedia Commons, (a) http://commons.wikimedia.org/wiki/File:Myllokunmingia.png and (b) http://commons.wikimedia.org/wiki/File:Haikouichthys4.png.)

Figure 2.5 Ostracoderms: heterostracans (a) *Arandaspis* and (b) *Pteraspis*. (Courtesy of Tamura, N., Wikimedia Commons, (a) http://commons.wikimedia.org/wiki/File:Arandaspis_NT.JPG and (b) http://commons.wikimedia.org/wiki/File:Pteraspis_NT.jpg.)

PALEOECOLOGY OF FISHES

What environmental factors influenced the distribution and abundance of early fishes? This is difficult to say because of the long passage of time, but some abiotic conditions stand out. As we shall discuss later, massive events on a global scale have been linked with major extinctions and loss of diversity. These events included ice ages, changes in sea level, volcanism, moving continents, changes in ocean circulation, asteroid strikes, and so on. But did biotic components of the early fish environments affect the first fishes on a competitive or predatory basis? This is difficult to interpret from fossils because of several factors. Passage of time in geological scales has obscured the fossil record, other soft-bodied organisms do not make very good fossils, and their presence would be difficult to detect. But there are other problems with potential reconstruction of aquatic life over millions of years.

Fishes evolved in a marine environment, where sedimentation is slow and the corrosive effects of salt water dissolve hard structures. But more importantly, most fossil evidence from ancient oceans has been destroyed by plate tectonics. Whereas parts of the continents have been around for about 3.8 billion years, sea floor spreading and subduction has resulted in a younger crust. Nowhere is the ocean floor (i.e., lithosphere) older than about 150 mya (Scotese 2004), and very little has been dated even as old as 125 mya. This limits the presence of marine fossils to the margins of continents that supported ancient shallow seas, and doubtless an incomplete picture has been provided (e.g., Maisey 1996). In some instances, we are so limited that we only infer causes of adaptations from the presence or absence of other species. But because of the long period of fish evolution, we *can* document evolutionary trends in fishes and relate them to planetary conditions.

For example, we know that ostracoderms, the first major radiation of true fishes, lived in warm shallow seas and invaded shallow freshwater systems. These fishes had a shell of bone—an outer skeleton to complement their inner one. The function of this bony case has been attributed to a need for a mineral supply (calcium phosphate), as a platform for sensory systems, to stabilize the body in swimming, and as armor to deter predators. Perhaps the answer to its function lies with all of these requirements. But we can infer that minerals were in short supply; there was a need for a sensory system. Additional sensory organs were presumably needed to detect food and/or predators, and also for body protection against enemies.

What were these potential enemies? Prime suspects are arthropods known as eurypterids, which were aquatic relatives of scorpions, because fossil eurypterids and ostracoderms co-occur. It is also possible that other soft-bodied predators or parasites plagued the ostracoderms, but none have been found to date. We believe that early agnathans had competitors and predators, but we speculate that they did not use speed to escape predators or to catch prey because they obviously were not fast swimmers. As muscular and powerful as some ostracoderms appear to be, they lacked proper fins to exert much control over finer points of locomotion. Thus, their roles emerge as roving mud filters whose sensory system gave them plenty of time to escape more sessile threats such as aquatic arthropods. In addition, they must be given credit for beginning a so-called arms race, and we give them credit for being the first "tank" of the fish world. No doubt their world changed with the appearance of jawed fishes. These more derived creatures likely preyed on them and participated in the arms race until a new evolutionary trend appeared: speed instead of armor.

SUMMARY

Evolution on Earth began in water; and survival of organisms, once evolved, was greatly dependent on extrinsic factors such as temperature changes and fluctuations in climate. Global conditions fluctuate widely in homeorhesis (pulsing + state), which is more dynamic (no set points) than homeostasis (same + state) and susceptible to positive feedback. Global conditions can be extreme because of massive volcanic eruptions, ice ages, asteroid strikes, and so on. In spite of all these global changes, the first fishes evolved from ancestral chordates over 450 mya. These fishes shared at least five features with the chordates and improved upon them. It is believed that free-swimming chordates gave rise to early fishes, but the appearance of the first fish is obscured with time—it may extend back over 500 mya. The first major radiation of fishes were agnathans known as ostracoderms, present in the Early Ordivician. These fishes had true bone, which aided fossilation and study of their radiation and diversity. The bony case also was presumably used as protection against predators, possibly the start of an evolutionary arms race that they lost.

Further reading: Classic: Lanham 1962.

CHAPTER 3

Aquatic Environment

AQUATIC ECOSYSTEMS

Water covers about 75% of Earth's surface and occurs as part of a hydrologic cycle. Due to the vast surface area, evaporation from the oceans exceeds that from continents. In addition, the higher elevation of the land results in a greater precipitation rate as water vapor rises and cools (Figure 3.1). Thus, precipitation exceeds evaporation for terrestrial ecosystems, resulting in a surplus of freshwater on the continents. Also, the present configuration of the continents is widespread relative to the supercontinent Pangaea, and this has resulted in a more uniform distribution of fresh surface water across the planet. Small rivulets form from rain, contribute to streams, large rivers, and lakes, and finally input into estuaries and the oceans. The greatest division in these aquatic environments is salinity. Falling on the continents from evaporative processes, the water is almost pure. In the oceans, however, the water is very salty, which prohibits its use for drinking by most terrestrial organisms, including humans (although some marine birds and other organisms have evolved in ways that allow them to use seawater directly). Accordingly, one of the major differences in aquatic environments, which can limit the distribution and abundance of fishes and divides the planet into two large aquatic environments, is the salinity of water. There are freshwater ecosystems and saltwater (marine) ecosystems. Most of this book is concerned with the abiotic and biotic attributes of these ecosystems and the adaptations of the fishes that live in them.

Freshwater environments include streams, rivers, ponds, lakes, reservoirs, and ground water, all of which can be considered as ecosystems occupied by fishes. However, distribution, abundance, and adaptations of fishes also depend on latitudinal distances from the equator toward the poles. On average, more fishes occur in warm climates than in colder ones. Saltwater environments also include various ecosystems, but these are classified in different ways. There are nearshore intertidal and subtidal systems, there are oceanic pelagic systems, and there are deep-sea systems that reach into the abyss. Fishes in these systems are influenced by the relative abundance of food and environmental extremes in light, water pressure, and other environmental components. Between the great freshwater and oceanic environments are estuaries, in which fishes contend with variable and mostly adverse physical conditions to take advantage of the abundant food there. A common factor in all of these systems, no matter where fishes live, is the adaptation to life in water as a medium. This is the basis for aquatic life.

PROPERTIES OF WATER

Evolution of fishes began in water—simple, common H_2O. Water is absolutely necessary for life, and about 80% of living cells are composed of water. From a human perspective, water is an alien environment, and it is difficult for us to appreciate living in an aquatic habitat. However, fishes are superbly adapted to water and live almost everywhere there is water, from mountain high to the

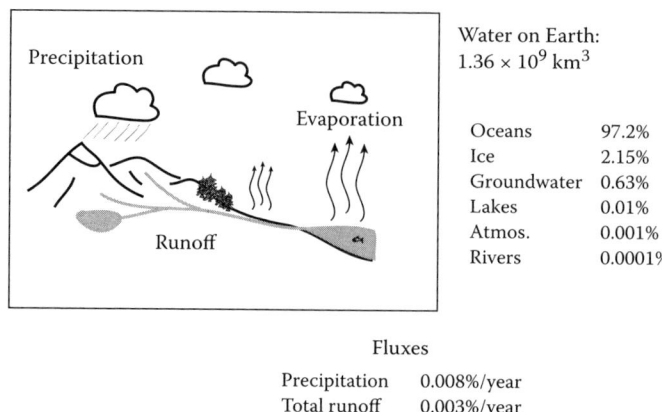

Figure 3.1 Hydrologic cycle. (Courtesy of J. H. McCutchan, Jr.)

ocean abyss. As a medium, water is much different than air, and the properties of water provide both advantages and disadvantages for the fishes that evolved in it as compared with terrestrial organisms.

Water is a remarkable substance. Compared with other compounds, it is different in many ways. Of extreme importance is the property of each molecule, because these molecules do not react as isolated molecules in their liquid or solid phases. Instead, water molecules tend to occur in a bonded condition. This condition results in emergent properties; that is, bonded molecules behave differently than individual ones.

A water molecule is the product of a combination of one oxygen and two hydrogen atoms in a covalent bond (Figure 3.2). Its configuration is well known but warrants review: Both hydrogen atoms occur on one end of the molecule, and it is called a polar molecule because the nature of this arrangement results in an asymmetrical electronic configuration: the molecule has a slightly positive charge on the oxygen end and a slightly negative charge on the hydrogen end. These charges are

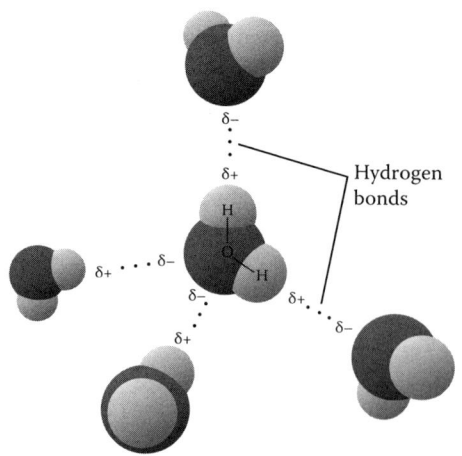

Figure 3.2 Water molecules showing covalent and hydrogen bonding. (From Manas, M. Wikimedia Commons, http://commons.wikimedia.org/wiki/File:3D_model_hydrogen_bonds_in_water.jpg.)

sufficient to attract other oxygen molecules in clusters of one central and four outside, a condition called hydrogen bonding. Water molecules attract each other and tend to stick together because of this bonding, which occurs constantly in solution, resulting in cohesion. This is an extremely dynamic phenomena, however, because the bonds shift about 10^{12} times per second for liquid water at 0°C, thus producing "quasi-stable polymers" (Wetzel 1975). At the water surface, the bonding is intense and results in an attempt for all of the molecules to bond with the ones below them, resulting in a film of surface tension.

Hydrogen bonding of water molecules results in a very high specific heat relationship (i.e., energy required to raise the temperature 1°C) and a high heat of vaporization, because great amounts of heat are required to disrupt the bonds before heat can be absorbed. Very few substances have a specific heat as high as that of water, and because of the ability of water to absorb and retain large amounts of heat, a relatively stable environment is thus provided for fishes.

Pure water has an interesting density relationship as well. Freshwater reaches its maximum density at about 4°C (Figure 3.3). Thus, water at the bottom of freshwater lakes will be liquid. Hydrogen bonds result in a lighter solid (ice) because the lattice caused by bonding spreads the molecules further apart compared with when in liquid phase. At 0°C (32°F), bonding in ice molecules would be only about one-half as dynamic as for the liquid phase. The lighter ice floats, and this property is extremely important in aquatic habitats; otherwise, lakes could freeze solid in winter.

Water as a medium provides some real advantages for fishes compared with living in air. Water has a high heat capacity due to hydrogen bonding; thus, temperature fluctuates less in water than in air. Water is cohesive, dense, and incompressible. Because of this, fishes can detect movement and sound perception is about three times faster. Water as the universal solvent carries many nutrients in solution for the fish to obtain directly, which also makes electroreception possible. Finally, the sense of smell is enhanced.

Water also poses difficulties that we hardly imagine. Water is about 800 times as dense as air, resulting in great pressure at depths. Its viscosity is about 100 times greater than air, making locomotion much more difficult. Viscosity also depends on the temperature, doubling as temperature falls from 25°C to 0°C.

In air, oxygen concentration (dissolved oxygen, DO) is about 21% (210 ml/l), whereas DO in water usually does not exceed about 10 ml/l, which is only 1/21 that of air. In shallow, stagnant

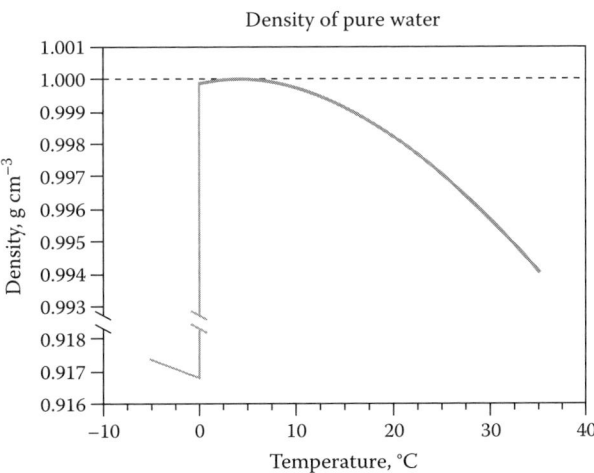

Figure 3.3 Density of water changes with temperature. (Courtesy of McCutchan, J. H., Jr.)

waters there may be little or no oxygen in water, yet we can find fishes such as gar and bowfin living there. These fish have the ability to gulp air into a vascularized swim bladder and survive conditions of very low oxygen concentrations. Fish usually live in water that has a temperature range of 14°C–19°C, and in those conditions, the DO ranges from 10 to 9 ml/l (Wetzel 2001).

Water varies in its concentration of salts, and this can cause osmoregulatory problems for freshwater and marine species. Freshwater fishes are hypertonic to the water they live in; thus, an osmotic pressure gradient is created, and water enters into the fish, primarily through the gill membranes. The reverse is true for marine fishes, which (with the exception of hagfishes) are hypotonic to the salt water and lose water to the environment. The good news is the solvent properties of water. Called the universal solvent, water has various salts and organic compounds from which fishes can benefit. Through passive or active intake, fishes can directly acquire many of the nutrients that they need. The affinity of water for high concentrations of chloride and sodium is obvious and used as the basis for dividing aquatic systems into marine and freshwater types. However, the presence of other compounds such as sulfates and carbonates of sodium, calcium, magnesium, and potassium, plus dissolved organic matter, is very important to fishes. Demand for major nutrients of nitrogen, phosphates, and silica can be so high that these compounds are depleted in surface water and regenerated at depths due to the availability and breakdown of organic matter.

Light is both reflected from the surface of water and absorbed. Depending upon the angle of incidence, 2%–15% of the sunlight is reflected, and over 53% of the light is absorbed in the first meter. Light is virtually extinct at depths of about 1000 m. However, water also distorts colors due to variable color penetration.

Fishes make a great use of color for several purposes, including communication and camouflage. However, all light waves are not absorbed at the same rate or penetrate to the same depth (Figure 3.4). Red light is absorbed entirely at only a few meters' depth, and only blue/violet penetrates to the greatest depths, giving water its blue color (e.g., Thurmond and Trujillo 2002).

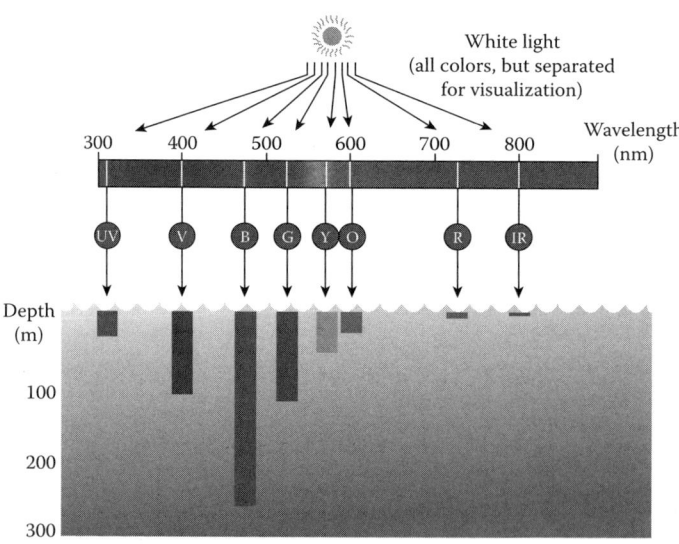

Figure 3.4 Spectral penetration of light in water. (From http://disc.sci.gsfc.nasa.gov/oceancolor/additional/science-focus/ocean-color.)

SEAWATER

The characteristics of seawater differ in some respects from freshwater. Seawater is influenced by the properties of water as a pure compound and also by dissolved substances (solutes). Solutes in seawater are dominated by sodium chloride (NaCl), which is present as about 86% of the total (chloride 55%, sodium 31%) (Castro and Hubert 2007). In addition to NaCl, seawater includes sulfate, magnesium, calcium, potassium, bicarbonate, bromide, borate, strontium, and fluoride (the top 10 constituents). The salt content of a volume of water is referred to as salinity and is expressed in parts per thousand (ppt) or practical salinity units (psu). The average salinity of seawater is about 35 ppt, and seawater is about 96.5% pure water and 3.5% salts. An interesting fact is the constant proportions of the major dissolved constituents in seawater, regardless of the salinity. This is the Principle of Constant Proportions. Stated in another way, the relative amounts of ions in seawater always remain the same even though the density changes (e.g., chloride = 55.04%) (Thurmond and Trujillo 2002). Using this principle, it is easy to convert the chloride concentration (electrical conductivity) to salinity (by multiplying with a constant 1.80655). Dissolved salts also increase the density of seawater compared to freshwater, which increases pressure with depth and makes fish more buoyant.

Seawater does not reach its maximum density at about 4°C, which occurs in freshwater. In fact, it simply gets heavier as it cools. Although sea ice does freeze (at about −1.8°C), it has interesting properties and does not maintain its integrity. As mentioned here, water molecules are held together by hydrogen bonding, and in ice, they form a hexagonal latticework arranged in layers. These layers are spaced further apart in ice than in liquid water, which results in less density and allows the ice to float. But there is another phenomenon that occurs as well. The structure tends to exclude solutes such as NaCl, whose crystals are cube shaped (Thurmond and Trujillo 2002).

As seawater freezes slowly at "higher" low temperature (e.g., −6°C), most of the salt is excluded from the ice in the form of a dense brine (i.e., brine rejection), which sinks into the water below. This process varies with temperature and the rate of freezing, but sea ice contains only about 30% of substances dissolved in seawater, which would be about 4 ppt salinity (Thurmond and Trujillo 2002). As a result, the ice that is left is very low in salinity and less dense than the water below it. If freezing occurs quickly at a very low temperature (e.g., at −40°C), more salt is trapped in the ice and the ice is more salty (about 10 ppt). However, as sea ice ages, more brine leaches out, and multiyear ice can attain very low salinities; the salinity of new sea ice is about 7–14 ppt, but multiyear ice ranges from about 0 ppt at the top to about 4 ppt toward the bottom (Gross and Gross 1995). This low level of salinity permits polar bears, birds, and other such animals to drink from melted ice pools at the surface.

The fate of the brine solution is perhaps more important. The very dense brine water mixes with surface waters, and since the process occurs at the surface of the poles, it is very cold. These are the coldest, saltiest, and thus the densest waters on the planet. This very dense solution slowly sinks and becomes part of the deep ocean circulation system, which we will discuss in more detail later.

FISH IN WATER: WHERE IS THE GRAVITY?

The previous discussion about the properties of water should make it evident that life in aquatic and terrestrial systems is fundamentally different. Major adaptations have to be made due to differences between water and air as media. Basic considerations involve gravity.

Leonardo da Vinci and others who followed him have pondered over the question: "Why is the fish in the water swifter than the bird in the air . . . it ought to be the contrary since water is heavier and thicker than the air . . ." (Marshall 1966). Perhaps fish are not quite as fast as our fastest avian relatives, especially when birds are locked into a dive, but tuna can attain a respectable speed of about 50 miles/h (81.4 km/h) (Bone et al. 1995). There is a related question: How can you catch a 20-pound fish on a 5-pound test line (Curtis 1949)? The answer to the last question should be "You don't, unless you have a landing net," which provides a good clue for both questions.

Although water as a medium is much denser than air, its buoyancy acts to greatly reduce the effect of gravity on the fish. Simply put, a 20-pound fish weighs about 1 pound in salt water, which is analogous to a 200-pound man having his weight reduced to only 10 pounds. Furthermore, most fishes have a gas-filled swim bladder that renders the fish neutrally buoyant. Thus, the fish can ignore gravity and spend its energy literally shoving its way through the dense and viscous liquid, while a bird must spend a large part of its energy to retard falling from the sky. The bird can sail more easily in the air once motion is started, but air is so light that there is not much to shove against, and extended wings also add to the drag component.

Water density acts to reduce the effect of gravity on the fish, but it also has a negative effect due to its high viscosity: fast or continuously swimming fishes have adapted their morphology to reduce drag and conserve energy. This is accomplished by general body shape. However, this shape is not static, and although rockets and submarines imitate the shape of a fish, they are rigid structures propelled with a screw. Swift fishes are pliable and undulate through the water, making their efforts to swim doubly effective. Friction and turbulence is reduced by various means, including recessed fins, but fish mostly swim with their skeleton and musculature. Powerful muscle segments are aligned as symmetrical units along the backbone of the fish. These segments contract in alternate succession. It is the body that swims, and with possible exception of a massive caudal fin in some advanced forms, fins are not absolutely essential for swimming.

But all fishes are not fast-swimming predators, and their morphology is in tune with their many professions. Some of the early fishes were eel- or snake-like bottom crawlers, sunfishes are laterally compressed for fast short turns to pluck invertebrates, and bottom-cruising and bottom-hiding predators are usually flattened in a pancake fashion (a dorsoventral compression) and "fly" in the water. Fast marine chase predators such as tuna and billfishes have huge and powerful caudal fins, and the armored (and poisonous) puffers move slowly on stubby fins, and so forth.

Free of most gravitational influence, life in water is truly life in a 3-D environment. It is difficult for humans to relate to this concept because gravity prevails for all terrestrial organisms and we tend to think of habitat in terms of surface area. But for fishes, living in deep water systems at least, habitat is more related to volume instead of area. In a terrestrial analogy, this means that life in the aquatic habitat would include niches that would be occupied by bats, birds, and insects in the terrestrial biome. This partially explains why fish diversity is so great.

SUMMARY

Planet Earth is predominately covered by water. As a result, precipitation exceeds evaporation over land, providing an excess of fresh surface water. Streams and rivers transport much of this water into the oceans, bringing with it dissolved and suspended inorganic and organic matter. Water constitutes about 80% of the composition of animals, and it is present in all living things. Water facilitated chemical reactions necessary for life to develop and evolve. It is the reason that Earth is habitable. A small molecule (H_2O), it has interesting properties. It usually occurs in clusters of five molecules, linked by hydrogen bonding. Pure water is most dense at about 4°C; thus, ice floats, and water at the bottom of a lake does not freeze. Seawater does not freeze per se, but as it gets cold, ice forces out the salt through a process called brine rejection. Consequently, the salinity of sea ice is about 10 ppt instead of 35 ppt. Other properties of water also include a high heat capacity, enabling it to resist changing temperature. Finally, it has the ability to dissolve many substances; hence, it is the universal solvent. Because of its properties and the substances that it contains, water presents some problems for aquatic organisms. Osmoregulatory problems occur with water passively entering freshwater fish and leaving saltwater fish. Water is 800 times as dense as air and 100 times more viscous; thus, it impedes motion. Also, oxygen saturation is about 1/21 less than in air. About 53% of the sun's rays that penetrate water are absorbed in the first meter, and light waves are absorbed differently. Red is lost quickly, and the most penetrating color is blue/violet. Fish are morphologically adapted to life in water, and most are streamlined to reduce friction. However, some are not streamlined; many are flattened, which is an adaptation for obtaining food. Fish are so well adapted to water that they can swim almost as fast as a bird can fly. This is due to gravity, and most fish are close to neutral buoyancy. On the other hand, most of the energy expended by a bird is to prevent falling out of the air!

Further reading: In depth: Bond 1996, Moyle and Cech 2004.

PART III
Fish Diversity

CHAPTER 4

Diversity 1: Chordates to Sharks

INTRODUCTION

A basic understanding of the diversity of fishes will be very useful in understanding their place in the structure and function of nature. The following three chapters explore the evolution of major fish groups of interest as an aid in understanding fish adaptation. The purpose of this section is to acquaint the reader on how nature has shaped the evolution of fishes and to know how major groups of fishes differ. As each major ecosystem is described later, these fishes will become more familiar. Much of the information about fish and the identification of extant fishes also is presented in an exercise given in the Appendix to this book.

FROM CHORDATE TO VERTEBRATE

The first fishes modified and enhanced ancestral chordate morphology to radiate into more active and diversified aquatic niches (Table 4.1) and became the first vertebrates (reviewed by Pough et al. 2005). The anterior end of the dorsal nerve cord was expanded (cephalized) to produce a brain, which was protected by a cranium of fibrous tissue, cartilage, and bone. To make room for the new cranium, the length of the notochord was reduced, and it did not extend past the cranium. The nerve cord also was protected by structural elements of cartilage or bone. Pharyngeal pores of collagen that were covered with cilia and used for feeding were elongated into slits, strengthened with cartilage and muscle and then provided with gill filaments to aid in respiration. The muscular postanal tail was retained, but the simple V-shaped myomeres were improved with more complex W-shaped myomeres to aid swimming (Figure 2.2).

Later, fishes also derived new characters, of which dermal bones, paired fins, and true jaws are the most obvious. Dermal bones first appear as bony plates, presumably used as armor. These plates would be replaced by a lightweight and flexible armor of scales. At first, nonpaired median fins were used for thrust, but these were later enhanced by the development of paired fins to allow more complex movements. The first jaws appeared from modified gill arches, and these also were modified to produce a very sophisticated structure that could be used for biting, grasping, grinding and as a suction pump.

Small prevertebrate chordates relied on cutaneous respiration, absorbing oxygen directly through the skin. However, as fishes radiated into new roles in the aquatic community, they developed a larger size. This required a better system for supplying oxygen, water, and dissolved substances to internal organs that also could remove carbon dioxide and other metabolic wastes. So fishes developed the heart as a functional pump and supplied the body fluid (blood) with a respiratory pigment (hemoglobin). Although this overview of vertebrate ontogeny is simplified, attaining these features now sets the stage for exploring the diversification of fishes, which is presented in Figure 4.1.

Table 4.1 Fish (Vertebrate) Origins

Fish were derived from an echinoderm line of evolution
Fish had a nonvertebrate chordate ancestor (phylum Chordata)
Early fishes shared and improved four chordate features:

Chordate	Vertebrate
Notochord to tip of head	Cranium anterior to notochord
Pharyngeal slits (collagen for feeding)	Slits supported by cartilage and muscle for respiration
Dorsal nerve cord	Protected by cartilage or bone
Postanal tail has V-shaped myomeres	Tail has W-shaped myomeres

Craniata: vertebrates in a general sense—they have only V-shaped myomeres, one semi-circular canal, accessory hearts, one nostril, fibrous cranium, no supporting vertebral elements, isotonic body fluids. Living representative—hagfishes: very successful, alive, and well.
First major vertebrate radiation: ostracoderms (shell skin). True vertebrates with cranium of cartilage or bone, two or more semicircular (SS) canals, and a well-developed heart.

AGNATHANS: HAGFISHES AND LAMPREYS

Vertebrate and vertebrate-like fossils are being retrieved from Cambrian deposits (e.g., *Myllokunmingia* from Early Cambrian) (Shu et al. 1999), but it is difficult to work with such very old material, and the phylogenetic status of these animals is still being evaluated. However, it is indisputable that the first major radiation of fishes was agnathans. These fishes had no true jaws (i.e., derived from gill arches), no paired fins, one to two semicircular canals in the ear, gills that open through pores, and other ancestral features. But there is much morphological diversity represented by them, and their external appearance ranges from naked, wormlike animals to armor-encased small monsters. The first great wave of these agnathans occurred during the Ordovician, about 500 million years ago (mya). Species diversity among agnathans is indicated by the assignment of seven superclasses (with one class each) of organisms, while only one superclass (with five classes) is given to all of the jawed fishes (Table 4.2). However, agnathans are poorly represented by only three extant families: one family of hagfishes and two families of lampreys (Nelson 2006).

An early offshoot from the main line of vertebrate evolution, hagfishes (class Myxini) are probably the sister group of vertebrates (Figure 4.1). They are considered to be craniates but represent vertebrates only in a very general sense. They have a cranium, but it is made of fibrous tissue and not of cartilage or bone. In fact, hagfishes lack bone entirely; there are no vertebral elements, only one semicircular canal in the ear (used for balance), and no dorsal fin (Figure 4.2). On the other hand, hagfishes have been very successful organisms, as reflected by their very long existence. Perhaps this is a case of finding and dominating a niche early. It seems that their niche has not been contested by teleosts, and at least part of their success is due to some interesting adaptations. Hagfishes are able to tolerate hypoxia, their wormlike shape and strange teeth assist in feeding, and production of slime provides defense against predation (Ilves and Randall 2007).

Hagfishes are marine scavengers. They also are called "slime eels" because of their ability to secrete very large quantities of mucus, which may be used as a defense against predation and for feeding. Hagfishes feed on dead animals by implanting their horny "teeth," forming a knot with their body, and then moving the knot forward on the slimy body to press on the animal, thus tearing off chunks of flesh. Fossil hagfishes have been dated to at least 300 mya, making them one of the most successful organisms. It is instructional to note that they have adjusted to their benthic marine niche through the secondary loss of some structures such as eyes and by solving osmoregulatory problems through avoidance: They have body fluids that are isotonic with salt water.

Hagfishes are abundant in some locations, and there are commercial fisheries off Korea and Japan valued at $100 million (Helfman et al. 2009). Efforts have been centered on the northwestern Pacific and elsewhere. This includes the inshore Atlantic Ocean off Gloucester, Massachusetts,

DIVERSITY 1: CHORDATES TO SHARKS

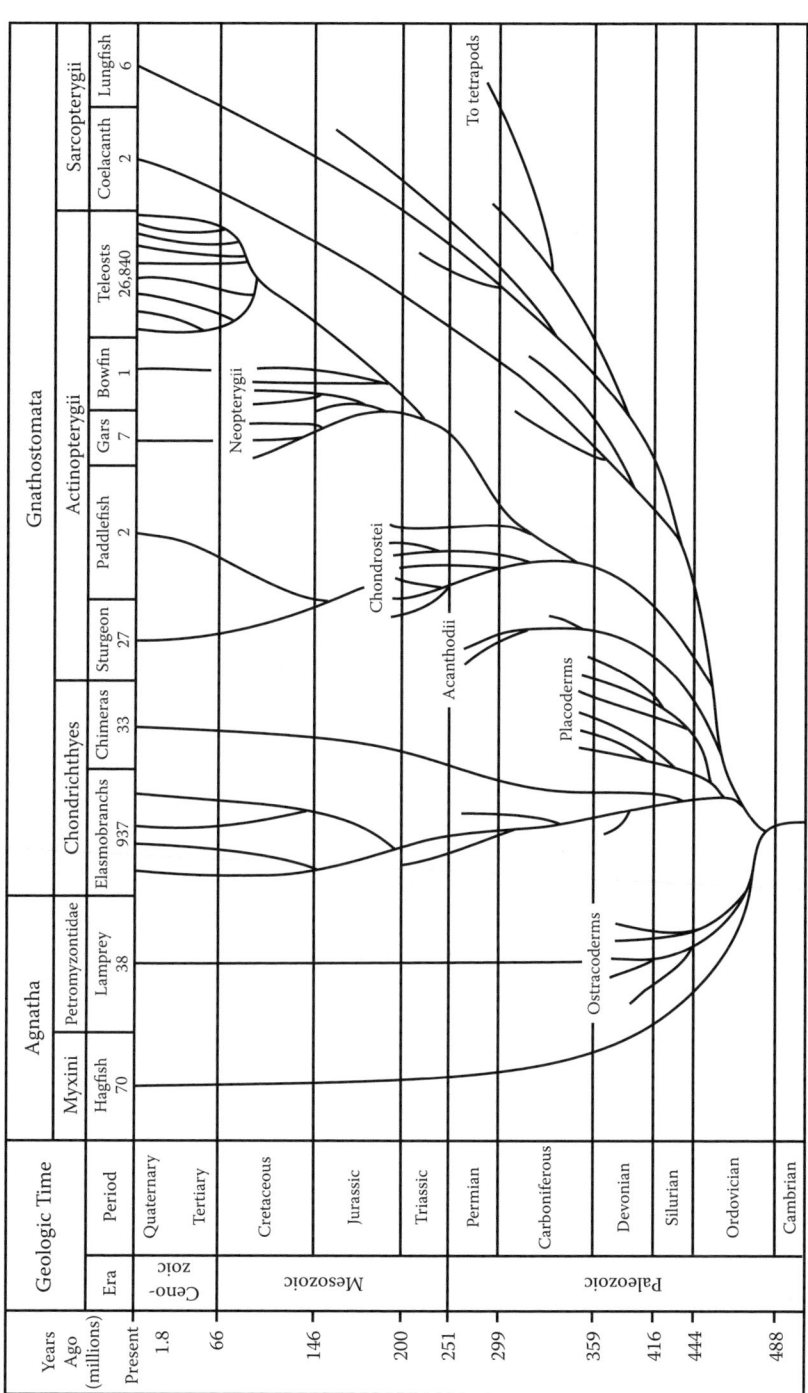

Figure 4.1 Evolutionary relationships among major groups of fishes. (Data and relationships from Long, J. A., *The Rise of Fishes: 500 Million Years of Evolution*, John Hopkins University Press, Baltimore, 1995; Nelson, J. S., *Fishes of the World* (4th edition), John Wiley & Sons, Inc., Hoboken, NJ, 2006. Format modified from Moyle, P. B., *Fish: An Enthusiast's Guide*, University of California Press, Berkeley and Los Angeles, 1993.)

Table 4.2 Affinities and Major Groups of Fishes

Phylum Chordata: includes acraniates, craniates, and vertebrates
Subphylum Urochordata—tunicates
Subphylum Cephalochordata—amphioxus, pikaia (extinct)
Subphylum Craniata (Vertebrata): includes seven superclasses of agnathans and one superclass of gnathostomes (five extinct superclasses of agnathans include conodonts to ostracoderms)
Superclass Myxinomorphi—hagfishes (craniates)
Superclass Petromyzontomorphi—lampreys (vertebrates)
Superclass Gnathostomata—vertebrates with jaws
Class Placodermi—placoderms (extinct)
Class Chondrichthyes—ratfishes, sharks, rays
Class Acanthodii—acanthodians (extinct)
Class Sarcopterygii—lungfish, coelacanth, and tetrapods
Class Actinopterygii—rayfin fishes
Relict bony fishes—sturgeons, paddlefish, gar, bowfin
Teleost fishes
Lower teleosts—eels, clupeids, cyprinids, catastomids, Ictalurids, salmonids, etc.
Higher teleosts—perciformids: sunfishes, perches, tuna, porgies, etc., also flatfishes and tetraodonts

Source: Nelson, J. S., *Fishes of the World* (4th edition), John Wiley & Sons, Inc., Hoboken, NJ, 2006.

where I was fishing for cod a few years ago. I noticed a large ship dumping 55-gallon drums overboard and thought that this was ocean dumping. When I called the mate over and asked them what was going on, he told me that they were commercial fishing with barrel traps for slime eels. I assumed that the drums had bait in them and the hagfishes would crowd into them to eat it. This must be a good way to obtain hagfishes; however, it requires a great many drums and a lot of effort setting and retrieving them. The commercial value of hagfishes is due to their fine leather; they are used for extremely expensive eelskin items (used in handbags, golf bags, shoes, etc.), and millions are caught annually.

The only extant agnathan that is representative of early vertebrates (perhaps anaspids) (Long 1995) is the lamprey (class Petromyzontida), the only living agnathan with a dorsal fin and the only living vertebrate with a single nasal opening high on top of the head in front of the eyes (Figure 4.3). These agnathans are true vertebrates, having a cranium of cartilage, cartilaginous vertebral elements, two semicircular canals, and other features. However, these fishes are considered specialized and degenerate parasites, although some species have become nonparasitic as a derived trait. Parasitic forms attach to their prey using suction created with an oral disk, use a rasping tongue to dig into the prey, and suck out body tissue and fluids. Nonparasitic forms do not feed as adults, and

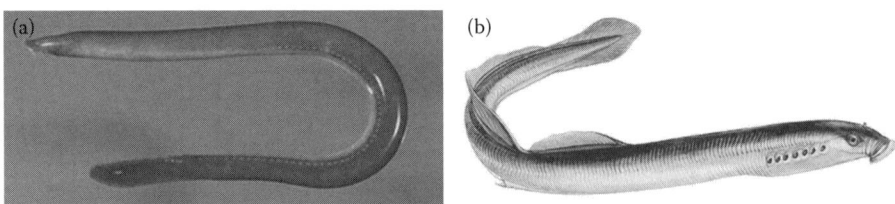

Figure 4.2 Comparison of hagfish *Myxine glutinosa* (a) and sea lamprey *Petromyzon marinus* (b). ((a) Courtesy of U.S. National Oceanic and Atmospheric Agency (NOAA); Keith, C. (hagfish), http://commons.wikimedia.org/wiki/File:Atlantic_Hagfish_(Myxine glutinosa).jpg. (b) U.S. National Oceanic and Atmospheric Agency (NOAA), Anon (lamprey), http://commons.wikimedia.org/wiki/File:Flussneunauge.jpg.)

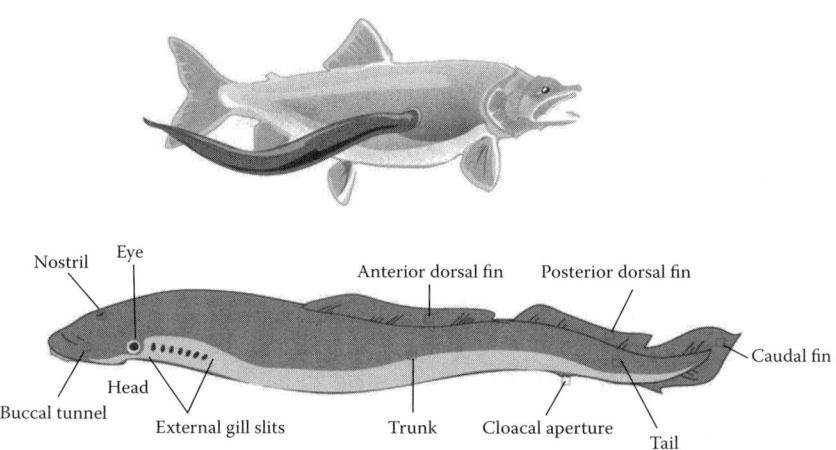

Figure 4.3 Morphology and parasitism: sea lamprey, *P. marinus*. (Courtesy of Hats, L., Wikimedia Commons, http://commons.wikimedia.org/wiki/File:Lamprey_illustration_side.png.)

all lampreys die after spawning. Lampreys have a complicated life cycle, and their young hatch, burrow into the substrate, and spend 3–4 years as free-living larvae (ammocoetes). The larvae then undergo a 2-year metamorphosis to become a parasitic adult. Nonparasitic forms may have a larval period of 6 years and require another 6 months for metamorphosis (Helfman et al. 2009). Lampreys also occupy an unusual life history (Ilves and Randall 2007). Although some are parasitic, all utilize a larval period during most of their life that is not. They are good swimmers and tolerate low oxygen levels.

Lampreys are not very popular in the United States because of their invasion into the Great Lakes via human-made canals. Once there, they proceeded to almost eradicate large fishes such as the lake trout and some whitefish. Special chemicals were developed to poison the larvae, reducing Great Lake populations by 90% from a 1961 peak. Control of this species continues to date.

CONODONTS AND OSTRACODERMS

Conodonts (cone + teeth) are tiny agnathans that existed from the mid-Cambrian to the late Triassic (Figure 4.4). These animals have only recently been accepted as true fishes based on the presence of a cranium, notochord, bone, and tooth structure. In this case, they appear to be more advanced than either hagfishes or lampreys. Their phylogenetic position is questionable, but they appear to be a class of very early vertebrates. They were numerous and diverse during the Paleozoic (Janvier 2007).

All of the remaining four classes of agnathans (i.e., pteraspids, anaspids, thelodonts, and cephalaspids) are small fishes that ranged from about 10 cm to 2 m in total length. They had a covering of dermal bone that occurred in shields, plates, and scales. These four classes have at one time or another been lumped under the common name of ostracoderms or "shell skins" (e.g., see Pough

Figure 4.4 Conodont. (Courtesy of Zica, M., Wikimedia Commons, http://commons.wikimedia.org/wiki/File:ConodontZICA.png.)

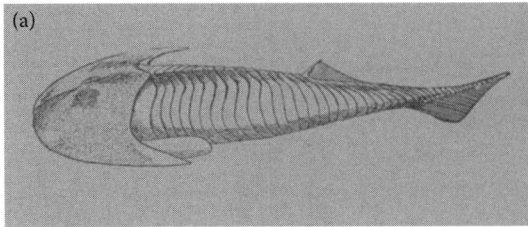 

Figure 4.5 Ostracoderms (upper Silurian and Devonian): *Cephalaspis* (a) and *Doryaspis* (b). ((a) Courtesy of Lucas, F. A., Wikimedia Commons, http://commons.wikimedia.org/wiki/File:Cephalaspis.jpg. (b) N. Tamura, Wikimedia Commons, http://commons.wikimedia.org/wiki/File:Doryaspis_NT.jpg.)

et al. 2005). Nelson (2006) has defined the term ostracoderm as a general one that means "fossil armored jawless fishes."

The ostracoderms dominated the great radiation of agnathans, which occurred during the early Ordovician (about 500 mya). The earliest of these fishes were the pteraspids (wing + shield), which had bony plates covering the head and gills (Figure 4.5). Albeit suspension feeders, they replaced the early ciliary pump with new encircling musculature and used cartilage for support. These fishes had a hypocercal tail and were obviously slow swimmers, perhaps plowing through sediments in search of food.

A larger and more successful group of ostracoderms was the cephalaspids (head + shield; also referred to as the osteostraci), which may be related to lampreys. These fishes had a solid head shield and more sophisticated "fins"—or paired pectoral flaps (Figure 4.5).

All of the cephalaspid ostracoderms are of the same general size, and the head shields lack sutures with no evidence of growth rings. This suggests that the cephalaspids had a larval form. This feature, plus the internal anatomy of the head shield, suggests that lampreys may have descended from them (reviewed by Helfman et al. 2009), but there is increasing evidence suggesting that ostracoderms are more closely related to gnathostomes than either hagfishes or lampreys (Janvier 2007). There may have been several uses of the ostracoderm shield, such as for mineral storage, a platform for sensory apparatus (perhaps electroreception), an aid to swimming on the bottom, and protection against predators.

It has been popular to speculate that the loss of ostracoderms during the late Devonian mass extinction was due to the radiation of gnathostomes (i.e., the placoderms), which would prey on them and compete with them. However, Pough et al. (2005) suggest that the reduction in the numbers of ostracoderms at the end of the early Devonian was likely due to the lowering of the global sea level, and their extinction during the late Devonian occurred with the mass extinction event that affected so many marine organisms. The extinction was likely caused by bad luck rather than by placoderms because the latter suffered extinction also at that time. Whatever the case was, the late Devonian mass extinction left only hagfishes and lampreys to carry on the agnathan lifestyle.

EARLY GNATHOSTOMES

The gnathostomes (jaw + opening) emerged from the fossil record with two major derived structures, jaws and paired fins. Jaws were useful for grasping, holding, and manipulating objects; building nests; and, with the development of teeth, cutting. Developed from anterior gill arches, simple jaws were not fused with the cranium (Figure 4.6), but in time, they were strengthened and then converted to perform complex tasks. Now, fishes could hold mates during reproduction, line up their prey before swallowing them, and perform many tasks that would be impossible for the

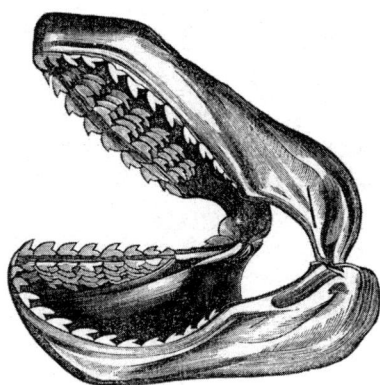

Figure 4.6 Shark jaws. (Courtesy of Wikimedia Commons, http://commons.wikimedia.org/wiki/File:PSM_V04_D076_Infant_shark_jaw.jpg.)

agnathans from which they evolved. Paired ventral and pectoral fins also were a great advantage in maneuvering, reducing the effects of yaw, pitch, and roll while swimming. In addition, a well-developed heterocercal tail provided the necessary lift and propulsion—some with an abrupt bend to facilitate burst swimming.

The fossil record is not enough to conclusively determine the earliest jawed fishes, and three classes, placoderms, chondrichthyans, and acanthodians (an early bony fish), appeared at about the same time (late Ordovician to early Silurian). The placoderms, a sister group of all of the other jawed vertebrates (Nelson 2006), were an early offshoot of the main line of vertebrate evolution and left no known living descendants.

Placoderms (plate + skin) grossly resemble the ostracoderms from which they evolved due to the presence of a bony shield, but the resemblance is only superficial (Figure 4.7). Placoderms are very different and caused great confusion among early taxonomists who tried to fit them into evolutionary trees. The great vertebrate paleontologist A. S. Romer (1966, 1970, 1977) called them "weird" and even remarked words to the effect that it would be better (I assume he meant that tracing the evolution of vertebrates would have been easier) "if they never existed!"

Early placoderms had massive dermal armor and appeared to be benthic feeders. They had unique jaw musculature and platelike teeth that resemble no other gnathostome. However, they also had three semicircular canals, paired fins, claspers, gills covered with ectoderm, and an operculum.

Collectively, the placoderms were an odd mixture of forms. One large predatory group of placoderms, the arthrodires (joint + neck) had a special craniovertebral joint and a space between the head and trunk shield (Figure 4.8) that allowed the head to be raised, thus producing a huge gape, possibly for feeding on benthic organisms. One of these fishes was the gigantic late Devonian arthrodire *Dunkleosteus,* which reached lengths of up to 10 m. However, smaller forms, including

Figure 4.7 The Devonian arthrodire *Dunkleosteus*. (Courtesy of Bogdanov, D., Wikimedia Commons, http://commons.wikimedia.org/wiki/File:Dunkleosteus_interm1DB.jpg.)

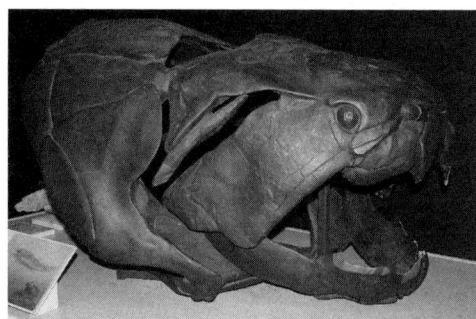

Figure 4.8 Constructed skull of *Dunkleosteus*, showing opening above intercranial joint and platelike tooth plates. (Courtesy of Liber, C., Wikimedia Commons, http://commons.wikimedia.org/wiki/File:Dunkleosteus_skull_QM_email.jpg.)

antiarchs, were more flattened and had stilt-like pectorals, while others still were chimera-like in appearance.

The large-sized and platelike crushing "teeth" of some placoderms (i.e., arthrodires) no doubt made them fearsome predators. As I was walking on the elevated boardwalk in Everglades National Park, I was impressed by the number of very large yellow slider turtles and witnessed a scene hard to forget. A large alligator swam close to the boardwalk, grabbed a slider in its jaws, and proceeded to crush it with a large sickening crunching sound—which was accompanied by a great squirting of liquid matter that ejected out both sides of the gator's jaws. It seems certain to me that a similar drama must have occurred millions of years ago, and that many ostracoderms must have furnished crunchy snacks for monster placoderms.

A remarkable find was announced as this manuscript was being written. A fossil placoderm (Ptyctodontodae) dating to 380 mya was discovered in the Gogo Formation of Western Australia, and it contained a fully formed fetus and umbilical cord. Subsequently, the investigators (Long et al. 2008) found another placoderm in collections from the same formation that contained three embryos. Without question, vivipary (live birth with young nourished by the mother) was a mode of reproduction all the way back to the Devonian, and these finds suggest that vivipary was first derived in the placoderms. The species was named *Materpisces* (mother + fish) *attenboroughi*. Placoderms experienced great losses of species in the late Devonian extinction and disappeared by the close of the period. Nothing exists today that resembles them.

CHONDRICHTHYES

The class Chondricthyes (cartilage + fishes), the sharks and their relatives, the rays and chimeras, appeared in the late Ordovician, and they are arguably the next in line in vertebrate evolution. The group consists of about 970 extant species (Nelson 2006), and it is divided into the elasmobranchs (sharks and rays) and the chimeras.

All chondrichthyans have an internal skeleton and a cranium of cartilage, no ossified bone, simple (true) jaws, paired fins, and a spiral valve in the intestine. Special features include placoid scales, internal fertilization, high urea concentration in the blood, an oily liver, and no swim bladder. The cartilaginous skeleton, oily liver, and huge pectoral fins are part of an integrated and specialized response to maintaining buoyancy in their marine environment, which we will discuss in more detail.

The chimeras (holocephali = whole + head; Figure 4.9) are an extant group of about 30 species. They have the upper jaw fused with the cranium and a gill cover over the four gill slits. They also have toothy plates that give them a ratlike appearance, thus the common name "ratfish." The group

DIVERSITY 1: CHORDATES TO SHARKS

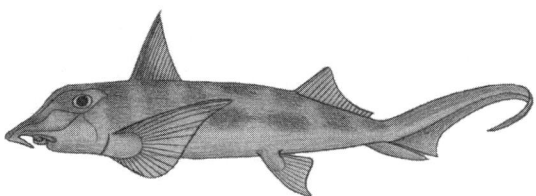

Figure 4.9 The chimera (ratfish) *Callorhinchus callorhynchus*. (Courtesy of Tambja, Wikimedia Commons, http://commons.wikimedia.org/wiki/File:Callorhinchus_callorhynchus.jpg.)

occurs in ocean depths worldwide, where they mainly feed on invertebrates. The chimera was a monster in Greek mythology that had the head of a lion, the body of a goat, and the tail of a snake. It also can mean a "foolish fantasy." If anything, the ratfish looks like a foolish fantasy: the head has a rodent-like appearance with protruding ratlike teeth; the dorsal fin is large and sharp, with a poison gland; there is a cephalic clasper in the male; and it has a naked body, strange operculum, and a long thin ratlike tail with no apparent caudal fin (Figure 4.6). Its phylogeny is somewhat of a mystery, but chimeras bear some resemblance to placoderms, and they may share an agnathan ancestor.

The elasmobranchs (platelike + gills) include 937 extant species, about equally divided between the sharks (403; Figure 4.10) and their benthic-adapted descendants, the rays (534; Figure 4.11). Most of these are marine predators, but 28 species are known to occur in freshwater (Nelson 2006). All have five to seven separate gill openings which are located on each side of the body in sharks and on the ventral side of rays. Almost all of these fish are large predators (>0.5 m), and the whale shark (*Rhincodon typus*) is the largest fish, attaining sizes of 12 m or more.

The sharks and rays have radiated to fill a broad range of predatory niches in the marine environment. They are widely diversified in habitat utilization, tirelessly searching for a variety of food items from coastal to oceanic realms, from the neritic zone to the abyss. Chimeras seem to be more limited to deep marine habitats, where they use their crushing teeth to prey on marine invertebrates.

The success of sharks is attested to by their long persistence of more than 400 million years. It is pertinent to note that sharks from the Devonian, including the well-known *Cladoselache* (a branch + shark), have some features that appear primitive compared with modern forms, but in most aspects, it bears close resemblance to recently derived species, especially in general body form (Figure 4.12). *Cladoselache* had very large eyes, a large terminal mouth, broad paired fins, and a powerful caudal fin. It could do with a more streamlined snout, a better jaw, and more movable dorsal and pectoral fins, and it lacked scales over most of its body, but it was no doubt a highly effective and very fast sight-feeding predator (Springer and Gold 1989). *Cladoselache* was about

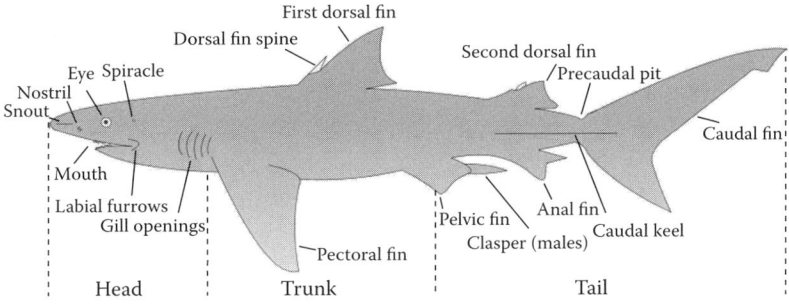

Figure 4.10 Morphology of a representative shark. (Courtesy of Huh, C., Wikimedia Commons, http://commons.wikimedia.org/wiki/File:Parts_of_a_shark.svg.)

Figure 4.11 Little skate, *Loeucoraja*. (Courtesy of U.S. NOAA, http://www.noaa.gov/, Martinez, A., http://stellwagen.noaa.gov/education/adulted/fished-sharks.html.)

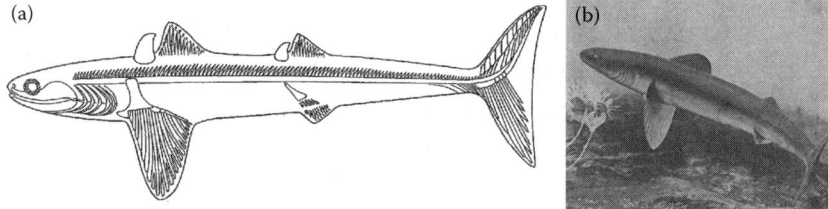

Figure 4.12 The Devonian shark *Cladoselache*. ((a) From Springer, V. G., and Gold, J. P., *Sharks in Question: The Smithsonian Answer Book*, Smithsonian Institution Press, Washington, DC, 1989. With permission. (b) Courtesy of www.50birds.com/extan/gextanimals8.htm.)

4 feet in length and perhaps also used its speed to escape the more ponderous placoderms. One thing is for sure: this fish demonstrates that important marine adaptations were already in place by the late Devonian.

Sharks and rays are not as plastic in trophic adaptations as bony fishes, and most species are predators. This limits their niche diversity, but as predators, their success is undisputed. This success is due in part to a combination of adaptive characteristics that include solutions to various problems of marine life, especially their ability to conserve energy (about 1/3 that of teleosts at rest; Helfman et al. 2009) and to protect their large precocial young. It will be useful to contrast some of these adaptations with those of the bony fishes (presented in the next chapter), so here are a few important ones:

Flotation: There is no evidence of a swim bladder in chondrichthyan fossils, and apparently, solutions to the buoyancy problems were developed long before those used by bony fishes. Of course, one solution was to live on the bottom, and the rays have been very successful at doing this. A few sharks have learned to gulp air at the surface for buoyancy, but other solutions emerged. A partial solution was to lighten the skeleton with cartilage, which has only 55% the specific gravity of bone. In addition, the use of large winglike pectoral fins provided hydrodynamic lift while swimming, aided by the pointed snout and a highly efficient, adaptable heterocercal tail. Finally, storage of a relatively large amount of fat in the liver appeared to solve the problem. For a predator, the oil-based solution seems to be especially satisfactory because a constant size is insured by incompressible oil regardless of depth that might be encountered in chasing prey.

Osmoregulation: Chondrichthyans have the ability to concentrate solutes in the blood at levels close to or even higher than seawater due to the retention of nitrogenous wastes such as urea and trimethylamine oxide to which the gills are almost impermeable. In addition, a rectal gland is used for the excretion of sodium and chloride ions.

Reproduction: There is no free living larval stage. Instead, much energy is provided to a small number of large and active young, which are protected by heavy fibrous egg cases or retained internally until they have attained adult features. Internal fertilization is possible because of the development of claspers and saves energy due to a reduction of egg loss.

Feeding: All are carnivores, but there is a wide adaptation in dentition and trophic adaptability. In addition, sensory systems are highly developed, including large sensitive eyes, exceptional olfaction, well-developed inner ear and lateralis (the lateral line system), and an extensive system for electroreception.

Respiration: Unlike their relatives, the ratfishes, elasmobranchs do not have their gill openings confined by an operculum. This facilitates ram ventilation if the shark swims with its mouth open, and gives a fearsome appearance to those with large teeth (such as tiger and great white sharks). Ram ventilation would cost little extra energy for a shark that needs to swim to maintain neutral buoyancy. In bottom forms, respiration was greatly aided by retention of a portion of the first gill opening known as a spiracle. This remnant gill opening exists between the upper jaw and first remaining gill arch in many sharks and in the rays, which provides the ability to suck in water through the spiracle and out the ventrally placed gill slits—especially useful when hiding in the substrate.

The continued success of the chondrichthyans even after the appearance of the advanced teleost fishes gives testimony to their effectiveness as predators. Perhaps sharks were perfectly suited for their arrival: most sharks consume 70%–80% bony fishes in their diets (Helfman et al. 2009).

SUMMARY

The first fishes were agnathans, lacking true jaws. They improved on chordate features and developed new ones such as a brain, gills with filaments to aid in respiration, and improved myomeres to aid swimming. Agnathans were diverse; at least seven superclasses have been described. However, agnathans are represented today by only three living families, all highly specialized: hagfishes and two families of lampreys. Hagfishes (slime eels) are primitive marine scavengers that lack a dorsal fin, a larval stage, bone, and a cranium. They have only one semicircular canal and blood that is isotonic with seawater. Lampreys resemble hagfishes, but they are true vertebrates with a dorsal fin, bone, a cranium, and two semicircular canals. They are specialized and degenerate parasites, although the larvae and adults of some are free-living. Of the extinct agnathans, conodonts have a questionable phylogenetic position. The rest (four classes) are lumped as "ostracoderms." Extinction of the ostracoderms has been linked to the rise of jawed fishes (gnathostomes), the first of which were placoderms. New characters of dermal bone, jaws, paired fins, and other changes resulted in a very different animal that radiated into new roles. The placoderms were an odd mix, but all had three semicircular canals, true jaws, and paired fins. Recent fossils document them as one of the first livebearers. Many placoderms were small; however, later forms included the gigantic Devonian *Dunkleosteus*. Placoderm extinction may be linked with the radiation of sharks and their relatives (class Chondrichthyes). An ancient group, the chimeras or ratfishes are strange chondrichthyans that may have been related to the placoderms. About 30 species cruise the ocean depths today. Sharks and their close relatives, the rays, are elasmobranchs, differing from ratfishes by having platelike gills and other features. Most sharks are wide-roving predators, while the rays are benthic. The success of elasmobranchs is attested to by their persistence for 400 million years, due to adaptations that provide different solutions to common problems shared by the more recent fishes: flotation, osmoregulation, reproduction, feeding, and respiration.

Further reading: Maisey 1996.

CHAPTER 5

Diversity 2: Teleostomes to Bony Fishes

RADIATION OF TELEOSTOMES

The rise of jawed fishes (gnathostomes) began about 440 million years ago (mya) with the radiation of placoderms and chondrichthyans (chimeras, sharks, and rays). However, those fishes were soon joined by a more advanced group collectively known as *teleostomes* (whole + mouth). By the Silurian period, three new and closely related classes of teleostome fishes had joined the existing fish fauna of agnathans, placoderms, and chondrichthyans (Figure 4.1). The fates of these new fishes were very different. The Acanthodii (spiny + ones) suffered total extinction; the Sarcopterygii (flesh + fin), most commonly called lobe-fin fishes, is represented today by a few relict species (one line suffered pseudoextinction in giving rise to the tetrapods); and the Actinopterygii (ray + fin) expanded to develop the main line of fish evolution.

These three classes of new gnathostomes are thought to represent a monophyletic group, and have been placed in the grade Teleostomi (Nelson 2006). In a general sense, the term refers to the advanced jaw, in which supporting elements and dermal bone transformed the simple jaw into a highly sophisticated tool. Phylogenetically, the teleostomes have in common several other anatomical features, including endochondral bone (formed from cartilage), branchiostegal rays (supports for the gill membrane), and new bones in the ear, called otoliths (ear + a stone; used for sensing gravity). The acanthodians and actinopterygians have three otoliths (sagitta, lapillus, and asteriscus), and one otolith is present in each of the three membranous sacs of the inner ear (i.e., the sacculus, utriculus, and lagena, respectively). Sarcopterygians only have one or two otoliths.

I have avoided use of the term "osteichthyes" (bony + fishes) here. Although the term has been used for a long time and it is well known, it has been used in different ways over the years, and definitions vary. Osteichthyes is sometimes used only for the ray-fin fishes, and also, the term is used to include the lobe-fins (and tetrapods). Obviously, the term "bony fishes" can cause confusion when the tetrapods are included with them, but the use of *sarcopterygian* seems to be clear. Here, we use the three teleostome classes whose meanings are open to little speculation.

Teleostomes were diverse and well represented throughout aquatic habitats by the Devonian period. It is doubtless that there were competition and predation between teleostomes and other extant fishes during this time. Almost all of the agnathans and all of the placoderms were extinct at the end of the Devonian, but all three classes of the teleostomes (and the chondrichthyans) continued their radiation well into the future. This suggests that the more recently derived jawed fishes may have been a factor in the loss of these major groups, perhaps in synergy with the adverse affects of global climate change.

ACANTHODIANS: SPINY ONES

Known as "spiny sharks" by some early workers, the Acanthodii is a controversial class of extinct fishes that has been placed by earlier taxonomists with different groups: placoderms, chondrichthyans, and as an intermediate between the cartilaginous and bony fish. The group is now recognized as an ancestral form of bony fishes, and it may be the earliest known fish to include dermal bone as part of the jaw. Three groups of these fishes have been reported (Long 1995; Nelson 2006). Some were predators that, except for stout spines, resembled some of the placoderms (Figure 5.1). However, the groups that persisted the longest were filter feeders. The acanthodians were apparently victims of the End Permian mass extinction.

Acanthodians had large spines located at the origin (leading edge) of dorsal, anal, and paired fins, and there were additional spines between the paired fins in some forms (Figure 5.1). The spines located between paired fins were connected to the body with a web of skin. This feature has been used to support the development of the Fin Fold theory of the origin of paired fins (reviewed by Kardong 1998).

SARCOPTERYGIANS: LOBE-FIN FISHES

The class Sarcopterygii contains two subclasses of fishes: the coelacanths (hollow + spine or thorn), with two living species, and the lungfishes, with six living species (Figure 5.2). Only two of these lobe-finned fish groups exist today, and they live in such extremely different places as deep-water marine and shallow freshwater systems.

In the lobe-finned (lobe = a fleshy anatomical projection) fishes, the fins have fleshy bases that extend out onto the paired (and some unpaired) fins. The paired fins are attached to the body with a single bone (e.g., humerus or femur) (Maisey 1996). Supporting skeletal elements in lungfishes include a series of ball and socket joints, and there is a recognizable femur and humerus in coelecanths (Kardong 1998). It is thought that these fins are evidence of ancestry to the first terrestrial

Figure 5.1 Acanthodians of the early Devonian included *Mesoacanthus pusillus* (top), *Parexus falcatus* (middle), and *Ishnacanthus gracillis* (bottom). (Courtesy of Fink, S. F., Wikimedia Commons, http://commons.wikimedia.org/wiki/File:Mesacanthus_Parexus_Ischnacanthus.JPG.)

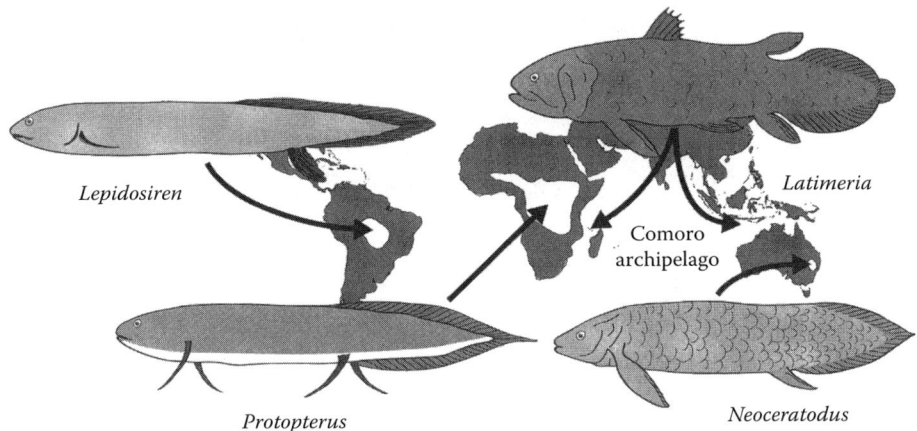

Figure 5.2 Geographical separation of coelacanths and lungfishes. (From Bone, Q., and Moore, R. H., *Biology of Fishes. 3rd Edition*, Taylor & Francis Group, New York, 2008. With permission.)

vertebrates, the tetrapods, which originated >350 mya as an ancient offshoot of the main line of fish evolution. The origin of these lobe-fins has been attributed to a switch in gene expression (Ahlberg 1972).

The relative degree of success attributed to the lobe-finned fishes must be evaluated from two perspectives. From the human point of view, this was an overwhelming triumph—clearly the way to terrestrial exploitation. However, what was a giant step for tetrapods would have been a tiny one affecting fishes (e.g., see Long 1995) and their aquatic habitats (covering about three-fourths of the planet) had it not been for the evolution of humans. To give credit where due, fishes invaded land and air millions of years before tetrapods flourished. The overwhelming numbers, distribution, and diversity of fishes attest to their triumph, and further radiation of ray-fins suggests that the genetic horsepower of fishdom was not overtaxed by the effort. From the fish perspective, it has to be judged a failure or, at best, a minor event, based on their few surviving lobe-fin relatives: The tetrapod radiation was simply a divergent sideline in fish radiation to land and air. After all, the fish empire developed other adaptations for dealing with those terrestrial habitats, as will be discussed later. For our purposes, we shall consider the radiation of tetrapods, which occurred during the Devonian, as more fish diversification during the Age of Fishes.

Tetrapods may have evolved from a long extinct lobe-fin fish, such as an osteolepidiform-like ancestor (possibly *Panderichthys*) (Long 1995); however, Zhu and Yu (2002) reported another fossil that also could be a common ancestor of lungfishes and tetrapods. Among the extant lobe-fins, lungfishes may be more closely related to tetrapods than are the coelacanths (Gorr and Kleinschmidt 1993).

Coelacanths

The coelacanth story is well known (e.g., see Balon et al. 1988; Thomson 1991; Erdmann et al. 1998). Considered a "living fossil," the first live specimen recorded in history was captured in waters off east Africa in 1938. In 1998, another coelacanth, now recognized as a separate species, was captured in Indonesian waters about 10,000 km from where the first one was found (Erdmann et al. 1998).

Coelacanths are lobe-finned, live-bearing marine species and the only living chordates with an intracranial joint. It has a fat-filled vestigial lung, cosmoid scales, well-developed elastic and

Figure 5.3 Preserved specimen of the coelacanth *Latimeria chalumnae*. (Courtesy of Fernandez, A., Wikimedia Commons, PD, http://commons.wikimedia.org/wiki/File:Latimeria_Chalumnae_-_Coelacanthe_-_NHMW.jpg.)

unsegmented notochord, external nostrils, and a diphycercal caudal fin that has three lobes (Figure 5.3) (Nelson 2006). The coelacanth is thought to be extant for at least 60 million years. Living specimens have a beautiful steel blue color with white spots.

I saw my first coelacanth at the Smithsonian Institute, courtesy of my friend Dr. Wayne Starnes. I distinctly remember when Wayne opened the lid to the large tank full of preservative, and for the first time, I viewed the creature. Immediately I thought of a passage from *The Run* (Hay 1959, p. 21): "I had the feeling too that I was looking at a professional from an old water world, a new agent of old assurance, deserving profound respect." I had not read that passage for many years, but I remembered every word under those circumstances.

The coelacanth has very interesting adaptations that appear intermediate between those of the chondrichthyans and some bony fishes. But by far the most interesting is its use of the lobe fins, which include the paired fins and median unpaired fins except the dorsal fin. Hans Fricke (Max Planke Institute) was able to capture the feeding behavior of the coelacanth on video, which I was able to view. The coelacanth does not crawl around on the bottom in search of food. Instead, it uses all of its fins to drift suspended (at depths of about 140–200 m) just above a lava-encrusted/boulder-strewn floor with cavities, crevices, and caves. Apparently, there are strong currents in this habitat, and the coelacanth uses the powerful fins to twirl around with its head down, searching for food in a very uncharacteristic (for fishes) twirling, almost dancing motion.

Lungfishes

The lungfishes have functional lungs as well as gills, and they are represented today by only three families, three genera, and six species. However the lungfishes have persisted for about 400 million years. Lungfish casts have been found as early as the late Paleozoic, and lungfishes have changed little since the Mesozoic (Long 1995). Their recent distribution on three widely separated continents, Australia, South America, and Africa (Figure 5.2), has been linked to continental drift after the breakup of Gondwanaland, because the fish cannot tolerate salt water.

Lungfishes have two otoliths, functional lungs, cycloid scales, powerful jaws, and a diphycercal caudal fin that is continuous with the anal and dorsal fins. Diet varies among the species, but the diet of the African lungfishes is reported to be more carnivorous than that of other extant species (Moyle and Cech 2004).

The Australian lungfish has only a single lung and uses its gills almost exclusively. It is the most ancestral lungfish and has changed little in over 100 million years. This form does not form burrows or estivate; instead, it has large and powerful lobe fins that it uses for maneuvering through heavy vegetation, mud, and debris in shallow water (Figure 5.4). The remaining, more derived lungfishes have two lungs, and their larvae have external gills. The South American form is adapted more for surviving episodes of very dry conditions in muddy burrows until rainy periods resume. The African lungfish, the most derived species, has smaller gills and relies mainly on lungs for oxygen.

DIVERSITY 2: TELEOSTOMES TO BONY FISHES

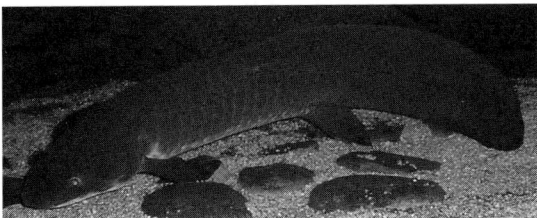

Figure 5.4 Australian lungfish *Neoceratodus forsteri*. (Courtesy of Tannin, Wikimedia Commons, http://commons.wikimedia.org/wiki/File:Australian-Lungfish.jpg.)

This form has filamentous paired fins, and it encysts in mud where it can estivate for as long as 4 years.

ACTINOPTERYGIANS: RAY-FINS

The class Actinopterygii (ray + fin) includes the largest group of living vertebrates, and it is the main line of fish evolution. Ray-fin fishes have fins composed of membranes separated by bony rays. The muscles that control these fins are retained within the body. Thus, the fish is more streamlined and offers less drag (swimming resistance). In most of the recent fishes, the fins are used mainly for fine maneuvering, not propulsion. There are exceptions to this, especially in very recently evolved tetraodonts (box fishes). Lobe fins are much different in appearance and use: Their supporting muscles and bony elements are on the outside of the body, and the fins are used for locomotion. However, in addition to the teleostome characters mentioned above, the ray-fins established new directions in evolution that culminated in the highly successful teleosts (whole + bone).

The ray-finned fishes derived new structures as improvements, differing from their ancestral forms in numerous ways (Table 5.1): They increased their swimming ability to perform rapid and complex movements, developed efficient respiratory systems, and were able to utilize a wide range of foods. All ray-fins have three otoliths, bony ridge scales, and branchiostegal rays. Their evolutionary adaptations are summarized below, highlighting the changes in various structures:

- Scales: Bony plates and platelike scales were replaced with cycloid, or ctenoid scales (Figure 5.5). These were thin overlapping discs that reduced heaviness and increased flexibility. Some forms later embedded, reduced, or lost the scales.
- Branchiostegal rays: The branchial cavity was made more flexible, and bony rays developed to strengthen the opercular membrane. This allowed a seal, which enabled suction feeding and efficient breathing under a wide range of conditions.
- Swim bladder: A pouch off the esophagus was used as a lung, but later developed into a swim bladder. Open at first and filled or discharged through the mouth, it became a highly efficient hydrostatic organ, enabling the fish to remain at a certain depth with little energy expenditure. In some forms, the swim bladder was used as a hearing aid.
- Jaws: Rigid toothed structures became more flexible with the addition of dermal bone and rearrangement of elements to allow a wide range of trophic adaptations. Some fish lost teeth, whereas others increased tooth diversity by using pharyngeal teeth (located on the last gill arch) to hold or crush prey.
- Caudal fin: Heterocercal tails were replaced by an abbreviated fin and then by a new homocercal caudal fin that provided more precise control of fast swimming and stopping.
- Fins: Forward migration of supporting elements of the pelvic fin and lateral placement of the pectoral fin provided greater maneuvering for feeding, courtship, nest building, and other purposes. In addition, erectile spines were part of the paired and unpaired fins as antipredator devices. These spines differed from earlier forms in that they did not interfere with other functions.

Table 5.1 Ray-Fin Fishes

Evolution of ray-fin fishes (class Actinopterygii) was accompanied by the emergence of new characters that enabled the emergence of teleost fishes. The ray-fin line of evolution resulted in rapid and complex movements, more efficient respiration, and use of more foods. The following highlights the transition:

- Bony plates and denticles were replaced by overlapping disc-like scales of thin bone. This provided protection with reduced weight and flexibility in movement.
- Development of bony supports in the opercles (i.e., branchiostegal rays) enabled suction feeding and improved respiration.
- The early lung became a swim bladder, useful in maintaining buoyancy, and having lost its connection with the gut, it was regulated by the blood. The resulting organ also was useful under higher pressure and in amplifying sound vibrations.
- Heavy toothed jaws were replaced by a more flexible system that was useful in many new feeding adaptations. The jaws became protractile, jaw teeth were lost in some forms, and teeth also appeared in the pharynx (last gill arch).
- The heterocercal caudal fin was abbreviated and then replaced by a homocercal tail with new supporting bones. The resulting tail facilitated fast bursts of speed.
- The position of the paired fins began to change to allow more maneuverability. The pectoral fin shifted higher on the side of the body, and the pelvic fins moved more toward the front.
- A final touch for the new body plan was the development of spines in the paired and median fins for protection against predators. The spines could be folded down when not needed, and did not interfere with other body functions.

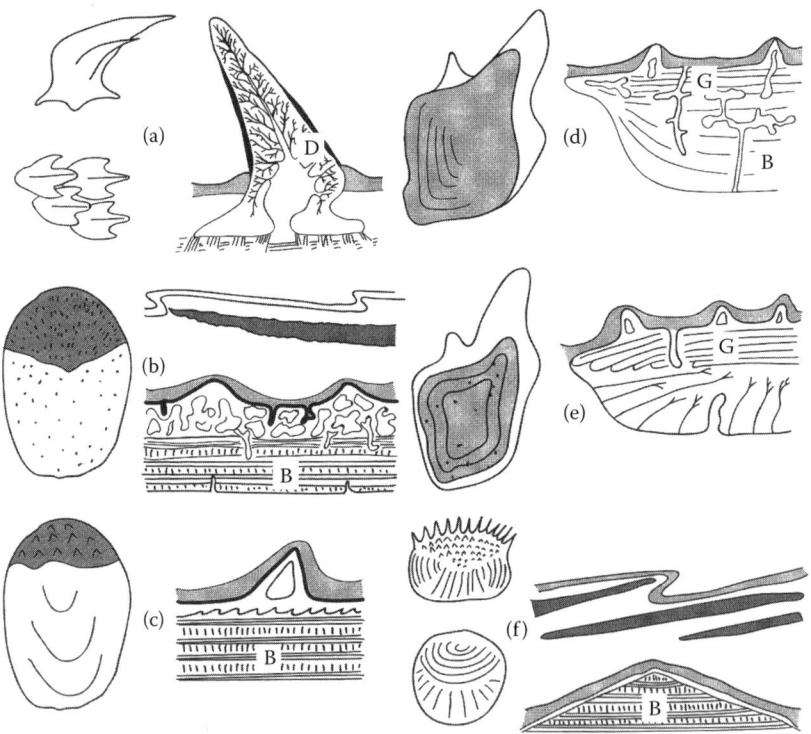

Figure 5.5 Fish scales of different taxa. Scales and sections are (a) shark, (b) lungfish, (c) coelacanth, (d) bichir (e) gar, (f) teleost (ctenoid and cycloid). B = bone, D = dentine, G = ganoin. (From Bone, Q., and Moore, R. H., *Biology of Fishes. 3rd Edition*, Taylor & Francis Group, New York, 2008. With permission.)

All of the above improvements are present in the most derived forms, the teleosts, but ray-fin improvements did not occur all at once, and some relict bony fishes persist. These nonteleostean ray-fin fishes (they lack the homocercal caudal fin) allow us to visualize how some of these changes occurred.

Relict Bony Fishes

Two groups of relict fishes have living representatives in North America, the Chondrostei (cartilage + bone) is represented by sturgeons and paddlefishes, and the Neopterygii (new + fins) is represented by the gars and bowfins. The sturgeons and paddlefishes (Figure 5.6) are the oldest living ray-fin fishes, representing two distinct lines. Paddlefishes presumably diverged from sturgeons in the Jurassic, at least 200 mya.

Sturgeons appear quite primitive, with long heterocercal tail and five rows of scutes (bony plates) down the body. The scutes have sharp edges and denticles in some species. The fish is well adapted for benthic feeding, and four sensory-laden barbels occur in front of an inferior and protrusive mouth. In the species I have seen, the gill cover is incomplete, and the gills are exposed at the upper end of the opercle, showing that suction feeding would be impossible. Sturgeons also have mostly cartilaginous skeletons. There are two families, six genera, and 27 species. Anadromous and potamodromous migrations occur.

Sturgeons are especially important as a source of caviar and boneless meat. All sturgeons have been subjected to intense fishing pressure, and the status of some of them is precarious (e.g., lake sturgeon, pallid sturgeon). The largest of freshwater fishes, *Huso huso* (beluga), attained ages of more than 100 years and sizes of 8.5 m and 1300 kg. The persistence of sturgeons may be related to their benthic feeding, in which the teleosts have no advantage. Heavy armor provides extreme protection and is not a problem with life on the bottom (Ilves and Randall 2007).

Paddlefishes are closely related to sturgeons, but they have a paddle-like snout and do not have the bony scutes of the sturgeons. Instead, the skin is mostly naked, and there are only a few small ganoid scales on the upper caudal fin and perhaps elsewhere. Only two species are extant, from the Mississippi and Ganges (China) river systems. The North American form is a filter feeder that strains the water with its minute gill filaments, presumably using sensory organs in the snout to locate denser volumes of prey. The snout also may aid the fish in balancing in the water when it is feeding with its gape-like mouth open. It continues to be sought in the Mississippi and Missouri river mainstreams for food. Not classified as endangered, it has been subjected to intense overfishing, and its potamodromous migrations have been blocked by dams.

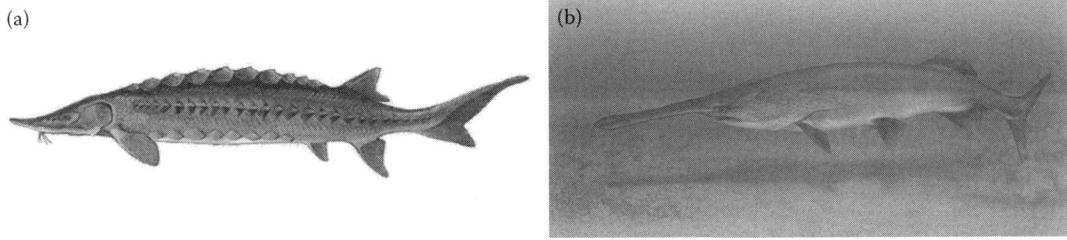

Figure 5.6 Relict ray-fin fishes (Chondrostei): (a) Atlantic sturgeon *Acipenser oxyrhynchus* and (b) paddlefish *Polyodon spathula*. ((a) Courtesy of Raver, D., U.S. Fish and Wildlife Service (USFWS). Accessed by usa.gov. (b) Courtesy of Knepp, T., U.S. Fish and Wildlife Service (USFWS), http://commons.wikimedia.org/wiki/File:Paddlefish_Polydon_spathula.jpg.)

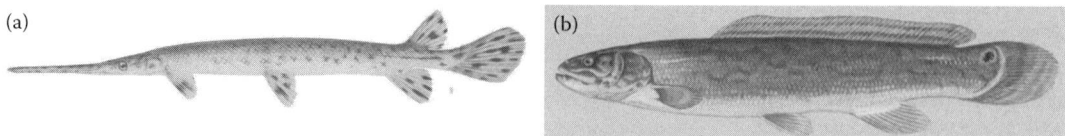

Figure 5.7 Relict ray-fin fishes (Neopterygii): (a) longnose gar *Lepisosteus osseus* and (b) bowfin *Amia calva*. ((a) From Edmonson, E., and Chrisp, H., New York State Department of Environmental Conservation (NYSDEC) as part of the Freshwater Fish Gallery, http://www.dec.ny.gov/animals/52634.html. With permission. (b) Courtesy of Raver, D., U.S. Fish and Wildlife Service (USFWS). Accessed by www.usa.gov.)

Gars and bowfins (Figure 5.7) have fin rays equal to their supporting elements (i.e., pterygiophores). They have an abbreviated heterocercal tail, and the swim bladder is heavily vascularized for use as a lung. Both species are common in warm coastal areas of the southern United States.

Gars are lie-in-wait (ambush) predators that have an elongate mouth with needlelike teeth and heavy platelike ganoid scales. Three branchiostegal rays are present. Some U.S. species are large, such as the alligator gar, which can attain sizes of 10 ft and 300 lb. These fishes are unpopular with net fishermen because of the severe damage they can do when entangled.

Bowfins are predaceous fish of sluggish, warm waters. This fish has 10–13 branchiostegal rays, cycloid scales, a heavy reptilelike skull, and a large dark spot on the upper part of the caudal peduncle. I can attest to the effectiveness of the vascularized swim bladder because I observed a large bowfin swimming blithely through a shoreline area (cove) treated with a fish toxicant (i.e., rotenone, which affects the gills) and out into a lake. The fish was clearly seen swimming near the surface, and it seemed perfectly unaware of any problem.

SUMMARY

Placoderms and chondrichthyans were early gnathostomes. By about 440 mya, three new classes of fish, collectively called teleostomes, had emerged to swim with them. Teleostomes had advanced jaws and shared several new characters, including otoliths and branchiostegal rays. The fate of the three classes of teleostomes was entirely different: Acanthodians survived longer than placoderms but went extinct at End Permian. Sarcopterygians (lobe-fins) are represented by the coelacanths and lungfishes, the more advanced fishes lost to pseudoextinction, and became tetrapods. The remaining group, the actinopterygians (ray-fins), is the main line of fish evolution. The coelacanth is considered a "living fossil," extant for about 60 million years. Two species are large live-bearing marine fishes with ancient features such as an intracranial joint, fat-filled lung, and diphycercal tail. Lungfishes have functional lungs, two otoliths, and a continuous diphycercal caudal fin. There are six extant lungfishes in three families, and they have a Gondwanaland distribution. Ray-finned fishes (Actinopterygii) have predominantly streamlined bodies and membranous fins with bony rays. Their adaptations include a protractile mouth, flexible branchial cavity, closed swim bladder, flexible jaws, heterocercal tail, location of paired fins, development of erectile spines, and other characteristics. These changes did not happen at once, and there are two groups of non-teleostean ray-fin fishes. One group (Chondrostei = cartilage + bone) of sturgeons and paddlefishes diverged more than 200 mya. Sturgeons are important for flesh and caviar. They are benthic feeders, with heterocercal tails, cartilaginous skeletons, and five rows of bony scutes. Paddlefish are filter feeders, have mostly naked skins and heterocercal tails, and also are sought for food and caviar. The other group (Neopterygii = new + fins) consists of gars and bowfins. These fish have new fins with supporting elements for each fin ray and abbreviated heterocercal tails. Gars are ambush predators with needlelike teeth and an armor of ganoid scales. Bowfins have heavy reptilelike skulls and large cycloid scales. Both gars and bowfins have vascularized, open swim bladders and can supplement oxygen by gulping air.

Further reading: Helfman et al. 2009, McKenzie et al. 2007, Paukert and Scholten 2009a, and Van Winkle et al. 2002.

CHAPTER 6

Diversity 3: Teleosts

DIVERSITY AND ADAPTATION

The teleost (whole + bone) fishes are not as menacing as sharks, but their numbers attest to their success. They are fierce competitors if not fierce predators, which many of them are. Lanham (1962, p. 77) pointed out that teleosts have radiated into niches formerly used by more primitive fishes and have expanded their ranges into habitats that may not have been occupied previously:

> The catfish grovel on the bottom like the ancient ostracoderms; the tuna disputes the high seas with the great pelagic sharks; the mudskippers may well be better air-breathers than the ancient lungfishes. The teleosts have invaded highly varied environments: Arctic ponds that thaw for only a few weeks of the year; hot springs of the desert; the deep sea floor. Some have even learned to glide in the air to escape their enemies.

Teleosts were present by the late Triassic (ca. 220 mya), and they are the most numerous and diversified vertebrates. There are at least 26,840 teleost species, and they constitute 96% of all living fishes (Nelson 2006). Common to all teleosts is a newly derived homocercal tail that appears more symmetrical on the outside than the heterocercal tail of their ancestors (Figure 6.1), but it is actually asymmetrical on the inside due to the presence of new skeletal bones (i.e., uroneurals) that act as supports to stiffen the caudal fin. Other features include further development of the gas bladder into a highly efficient hydrostatic organ changes in bones of the mouth, and development of a branchial pump for feeding and other purposes.

As will be shown in this chapter, adaptations developed by teleost fishes stagger the imagination. They occur in all sizes and can utilize food sources from all trophic levels (i.e., detritivores to top carnivores). They have different reproductive strategies: some produce a large number of eggs that are not cared for, some produce a smaller number that are guarded, and others are live bearers. They also have different ways of coping with abiotic and biotic environmental factors, through morphological, physiological, and behavioral adaptations. But according to Marshall (1966), one important feature is a small average size of less than 15 cm, which gives them a better chance at "finding a suitable niche." Small fish can survive by eating small food items; they can live in restricted spaces, such as in coral reefs; and they also can live in places with little food, such as the deep sea.

There are 448 families in the 40 orders of teleosts identified by Nelson (2006), and coverage of all of them is impractical and unnecessary here. I have selected a few orders for study, based on their ecological significance, exploitation by humans, or their sheer abundance. In most cases, species selected are native to North American waters.

Early teleosts such as herrings, minnows, and trout have a greater resemblance to one another than they do to the more derived groups, such as perches. The earlier teleosts, which have historically been called "lower teleosts" due to their origins, have in general the following structures: (1) There is only one dorsal fin, and it is of soft rays. (2) There are no spines in the dorsal, pelvic,

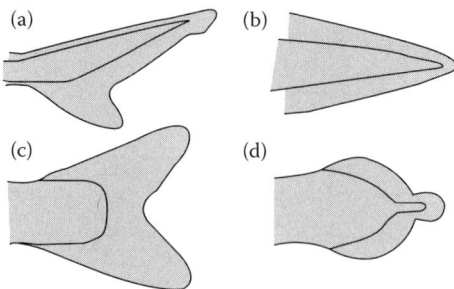

Figure 6.1 Caudal fins of fishes. (a) Heterocercal (sharks), (b) protocercal (lungfishes), (c) homocercal (teleosts), and (d) diphycercal (coelacanth). (Courtesy of Wikimedia Commons, http://commons.wikimedia.org/wiki/File:PletwyRyb.svg.)

Table 6.1 Lower versus Higher Teleosts

Structure	Lower Teleost	Higher Teleost
Dorsal fin	One, no spines	Two, with spines
Pelvic fin	Near the anal fin	Near the pectoral fin
Adipose fin	Sometimes	Never
Pectoral fin	Ventral	Lateral
Scales	Mostly cycloid	Mostly ctenoid
Gas bladder	Open to mouth	Closed (open to blood)

Source: Nelson, J. S., *Fishes of the World. 4th Edition*, John Wiley & Sons, Hoboken, NJ, 2006.

Note: Emergence of teleosts in the evolution of the ray-fin fishes was based on the development of new bones in the tail, a mobile premaxilla to permit protrusion of the mouth, and other features. This table contrasts selected features of ancestral teleosts, with representatives of the definitive teleost typical of most percoid fishes.

or pectoral fins. (3) The pelvic fin is abdominal in body position. (4) The pectoral fin is located on the ventral part of the body. (5) The scales are of the smooth, cycloid type. (6) The swim bladder is of the open type (physostomus). In contrast, the more recently evolved perches or "higher teleosts" have the following structures in common: (1) There are two parts in the dorsal fin or two distinct dorsal fins, one spiny and the other with soft rays. (2) Dorsal, anal, and pelvic fins have sharp spines. (3) The pelvic fin is thoracic or jugular in location. (4) The pectoral fin is located on the side of the body. (5) The scales are usually of the rough, ctenoid type. (6) The swim bladder, if present, is of the closed type (physoclistus). External differences in these two groups are apparent in Figure 1.1.

The terms "lower teleost" and "higher teleost" have been widespread in use for various phylogenetic comparisons (e.g., see Gosline 1965; Matthews 1998). The terms are used here only in a general comparative sense (Table 6.1). Nelson (2006) provided a more in-depth comparison of the two groups, using orders of fish that contain the salmonids and minnows as typical lower teleosts, and contrasting those groups with the perches.

LOWER TELEOSTS

Bonytongues

There is only one family of extant bonytongues, and it has only eight species. Bonytongues comprise an ancient freshwater group (order Osteoglossiformes), and it has a Gondwanaland distribution

DIVERSITY 3: TELEOSTS

Figure 6.2 Bonytongues: (a) Red and silver arowana (*Sclerophages* and *Osteoglossum*) and (b) pirarucú (*Arapaima*). ((a) Courtesy of McCutchan, J. H., Jr. (b) Courtesy of Bjoertvedt. Wikimedia Commons, http://commons.wikimedia.org/wiki/File:Arapaima_gigas_01.JPG.)

(found only in Africa, South America, and from Southeast Asia to Australia). Two species, the arapaima and the arowana, are well-known predatory South American species that are commonly kept in large aquaria (Figure 6.2). These fish have dorsal and anal fins located posteriorly for a heavy thrust, and the arowana has a surface-oriented jaw. The silver arowana (*Osteoglossum bicirrhosum*) also is known as "monkey fish" because it can jump out of the water to catch prey.

Arapaima gigas of South America (also known as pirarucú) resembles the arowana, except for its smaller, depressed head. It is one of the largest scaled freshwater fishes, growing to lengths exceeding 4 m. This fish has large cycloid scales that can grow to 6 cm in diameter. The scales are sold for use as a sort of nail file, and I use some of them to show my classes that bony scales really are bony! The meat of the fish is reported to be boneless, and it is a popular food and sport fish.

Eels

Eels do not look much like teleost fishes (Figure 6.3): They are elongated, snakelike fishes, with a continuous dorsal, caudal, and anal fin. They do not have pelvic fins, and even the pectoral fins are absent in some taxa. Moray, conger, and American and European eels are well known. These fishes are very adept at moving through narrow openings and may live in rocky cavities from which they venture out to feed. Other members of the group (Anguilliformes) move through muddy substrates, and some are pelagic.

Eels and their relatives (e.g., tarpons, tenpounders, bonefishes) have a special ribbonlike larva (Figure 6.4) that was historically mistaken for another organism. The leptocephalus (small + head) larvae were first linked to the adult eel in 1896, and the spawning area for the American and European eels were found by Schmidt in 1922. Spawning areas are limited and still being

Figure 6.3 American eel *Anguilla rostrata*. (Courtesy of Edmonson, E., and Chrisp, H., New York Biological Survey, http://www.dec.ny.gov/animals/52634.html.)

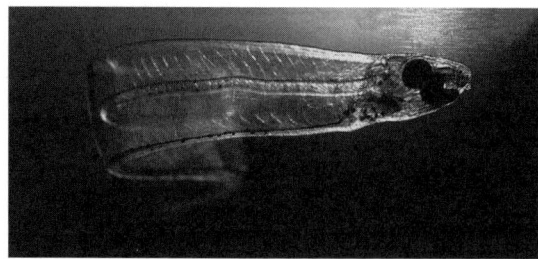

Figure 6.4 *Leptocephalus* larvae of conger eel. (Courtesy of Kils, en.wikipedia.org, Wikimedia, http://commons.wikimedia.org/wiki/File:LeptocephalusConger.jpg.)

discovered (e.g., Tsukamoto et al. 2004). Spawning habits of the American eel is used as one of the best examples of catadromy (migrating from freshwater to marine systems to spawn). Eels are used for food worldwide, eaten raw or cooked and considered a delicacy by many.

Herrings, Sardines, Menhaden, and Anchovies

Technically speaking, this order of fishes (Clupeiformes) is characterized by sensory canals that join the ear to the swim bladder to amplify sound. Herrings and other members of the family Clupeidae are easily recognized by their sleek appearance, light countershading, and rough keeled belly (Figure 6.5). Their sleekness comes from the smooth overlapping cycloid scales and lack of spines. The rough or "saw-toothed" belly comes from midline scutes. Everyone who has eaten a can of sardines or an anchovy pizza has come face to face with the herring (Clupeidae) and anchovy (Engraulidae) families. The anchovies can be separated from the herrings because of their huge (relatively speaking) mouth (Figure 6.5).

Most of these fishes are small (discounting the giant herring in the fossil record) planktivores or pick out and eat slightly larger prey, and they live in the epipelagic zone of the ocean or in lakes. Some marine clupeids such as the Atlantic herring, sardines, and menhaden spawn in open water, where they release buoyant eggs, but other species such as alewives, shads, and blueback herring are anadromous, ascending coastal streams to release their adhesive eggs in flooded freshwater bottomlands.

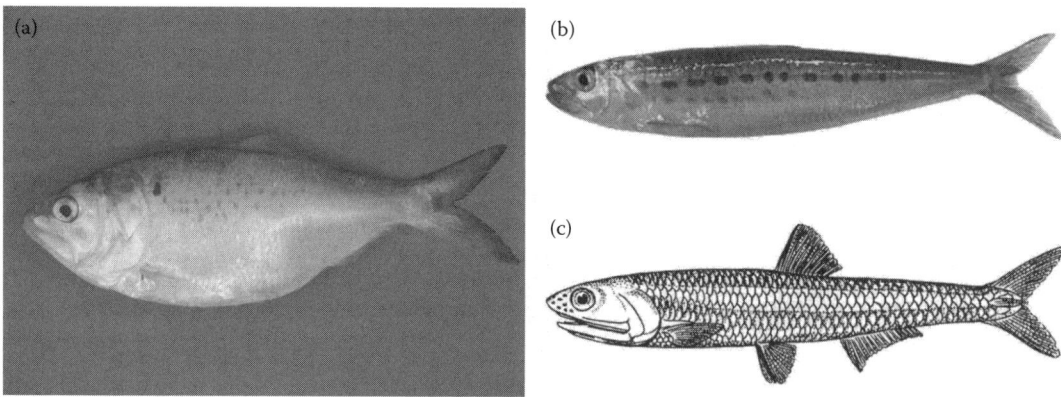

Figure 6.5 Clupeiforms—commercially important: (a) Gulf menhaden *Brevoortia patronus*, (b) Pacific sardine *Sardinops sagax*, and (c) northern anchovy *Engraulis mordax*. ((a) Courtesy of Boumje, J., Wikimedia Commons, U.S. National Oceanic and Atmospheric Agency (NOAA), http://commons.wikimedia.org/wiki/File:B.patronus.jpg. (b) Courtesy of Foresman, P. S., from http://commons.wikimedia/wiki/File:Anchovy(PSF).png. (c) NOAA, http://www.nmfs.noaa.gov/fishwatch/species.)

All of these species mentioned here are extremely important in commercial fisheries in the waters of the United States (and elsewhere) and will be discussed in more detail later. Clupeids comprise over one-fourth of the world's catch of commercial fishes. The catch is used directly for human food (sardines, kippered snacks, etc.), and it also is used to make fish meal and fish protein concentrate for a variety of uses (food additives, fish feed, and pet food).

Minnows, Suckers, Characins, and Catfishes

These families belong to the superorder Ostariophysi (little bone + bladder). They have an alarm substance (Schreckstoff) first noted by Karl von Frisch in 1938. Skin injury releases the substance, which causes fright reaction in other fish. Other features of the group are breeding tubercules, an open swim bladder, abdominal pelvic fins, and a protractile upper jaw in most species. Ostariophysi is the most successful freshwater fish group, with five orders and about 8,000 species. Many families of this superorder are placed in the series Otophysi (ear + bone) because small bones called Weberian ossicles connect the ear with the swim bladder to enhance sound reception. The minnow family (Cyrinidae) is the largest of the freshwater fishes (i.e., 220 genera and 2,420 species). Most freshwater fish species in the United States are otophysians.

Minnows and suckers are closely related freshwater fishes that lack jaw teeth (Figure 6.6). They only have pharyngeal teeth, which are located on the last gill arch. The largest North American minnow is the piscivorous (but endangered) Colorado pikeminnow, which attained historic sizes of 1.5 m (5 ft) and 30 kg (80 lb) (Jordan and Evermann 1923). The introduced common carp from Europe is a well-established exotic minnow in North America and is considered a pest. Suckers also can be large fishes (Figure 6.7), and sizes of about 600 mm (2 ft) are not uncommon. Suckers have inherited an unjust reputation as "trash fish." However, their role in the ecosystem can be pivotal. Most serve as decomposers, recycling detritus (pieces of dead plants and animals) in their body wastes and converting it into fish flesh, which can be consumed by carnivores. Freshwater spawning migrations (potamodromy) are common in suckers. Suckers and minnows are very successful fishes, as evidenced in some river systems, such as the upper Colorado River basin, where the post-Pleistocene (extant) native fish fauna includes only minnows and suckers. However, fossilized remains of some other fish groups also have been found.

Characins and catfishes comprise two large orders of freshwater fishes, and many of the fishes in these two groups have an adipose fin, which is a small fatty fin with no rays that is located between the dorsal and caudal fins. This fin appears to be a relict structure. Its function is unknown,

Figure 6.6 Cyprinids—minnows and carp: (a) Fathead minnow *Pimephales promelas* and (b) common carp. *Cyprinus carpio*. (Courtesy of Raver, D., U.S. Fish and Wildlife Service (USFWS).)

Figure 6.7 Catostomids—suckers: Mountain sucker *Catostomus platyrhynchus*. (Courtesy of Krebit, H., Yellowstone National Park, U.S. National Park Service, www.nps.gov/features/yell/slidefile/fishherps/fish/images.)

Figure 6.8 Ictalurids: Black bullhead *Ameiurus melas*. (Courtesy of Raver, D. and USFWS, NCTC)

but Moyle and Cech (2004) suggest that it may be useful for larval swimming. Characins include piranhas, pacus, and other fishes popular to the aquarium trade. They are found mainly in Africa and South America. Catfishes are very familiar to most people because of their unique morphology (Figure 6.8). They have catlike barbels around the mouth and interlocking spinelike antipredator structures in the dorsal and pectoral fins. North American freshwater (i.e., bullheads, channel and flathead catfish, madtoms, and cavefish) and marine catfishes (e.g., hardhead and gafftopsail catfish) appear similar and have no scales. Catfish families elsewhere can be covered with bony plates.

Whitefish, Arctic Grayling, Trout, Salmon, and Pike

Arguably, the most well-known and sought-after coldwater fishes are the salmonids (Salmoniformes). This group was restricted to the northern hemisphere, where many forms were circumpolar. The main line of salmonid evolution has been traced to *Eosalmo* (about 40 mya), with subfamilies of whitefishes and ciscos and with the graylings as early offshoots. Chars emerged as an early salmonid offshoot, followed by the Atlantic and Pacific salmon (Figure 6.10), the latter appearing more recently (ca. 6 mya) (Nelson 2006) and giving rise to the cutthroat trout. Chars are extremely cold hardy forms, and the arctic char (*Salvelinus alpinus*) has been given the distinction of the most northern fish (Helfman et al. 2009). In general, salmon and trout can be recognized by their general body shape, coloration, scales, and dorsal fin. Salmon and trout tend to be more brightly colored and have very small scales and a shortened dorsal fin; whitefishes (Figure 6.9) are more uniform in shading, have a silvery color, look less sleek, and have large scales and a shortened dorsal fin; and graylings have medium-size scales in contrast with the others and a longer, sail-like dorsal fin. Contrary to popular opinion, the salmonids are not advanced forms; they are lower teleosts and are

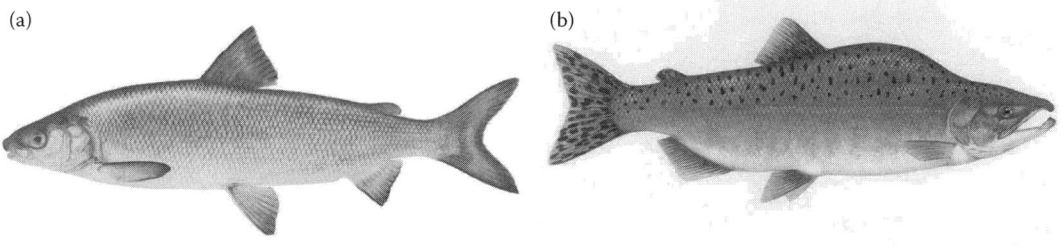

Figure 6.9 Salmonids—high-profile sport fishes: (a) Lake whitefish *Coregonus clupeaformis* and (b) pink salmon *Oncorhynchus gorbuscha*. ((a) By Edmonson, E., and Chrisp, H., courtesy of New York State Department of Environmental Conservation. (b) By Knepp, T., courtesy of USFWS, http://commons.wikimedia.org/wiki/File:Pink_salmon_FWS.jpg.)

Figure 6.10 Pikes and pickerels: Northern pike *Esox lucius*. (By Knepp, T., USFWS, http://images.fws.gov/defalt.cfm?)

placed under the superorder Protacanthopterygii (first/primary + spiny fin) and have an open gas bladder, cycloid scales, and nonprotractile mouths.

Trout (principally rainbow and brown trout) have been subjected to fish culture and transplanted around the world primarily for sport fishing. Atlantic salmon is an important food fish, and aquacultural interests are raising them in several marine locations for global marketing. Pacific salmon also have been cultured, primarily in an effort to slow the decline of native stocks in the American northwest.

The pikelike fishes (Esociformes), including pikes, muskellunges, and pickerels (Figure 6.10), have a very distinctive elongate and flattened (i.e., duckbill-like) rostriform head and a body shape typical of the lie-in-wait or ambush predator (the dorsal and anal fins are posterior for thrust). The Northern pike is a cool water fish, with a circumpolar distribution. Pickerels are warmwater fishes that are typical of the Atlantic coastal plain. The fast attack and large size of most of these fishes make them a popular sport fish.

CODS AND ANGLERFISHES

Cods (Figure 6.11) and their relatives, the haddock, hake, and cusks (Gadiformes), comprise about one-fourth of the commercial marine harvest worldwide. However, these benthic fishes have been severely overharvested in most, if not all, locations. These fishes have multiple or long and continuous dorsal and anal fins with no spines, but the mouth is protractile in some species; pelvic fins are located under or anterior to the pectorals (thoracic or jugular), and the gas bladder is closed. These are more recently derived characteristics, and these fishes are under a group known as Paracanthopterygii (near, like + spiny fin). Deep sea benthic fishes of this group include the grenadiers.

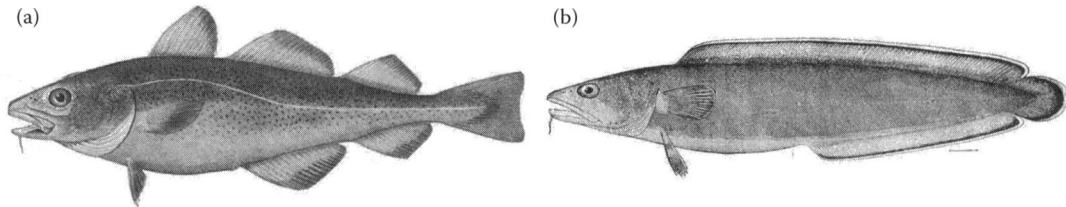

Figure 6.11 Cods *(Gadids)*: (a) Atlantic cod *Gadus morhua* and (b) cusk *Brosme brosme*. ((a) Courtesy of NOAA, http://www.photolab.noaa.gov/historic/nmfs/figbp314.htm. (b) From Goode and Bean (1896), http://commons.wikimedia.org/wiki/File:Brosme_brosme.jpg.)

Figure 6.12 Lophiiformes: This anglerfish "stands" on a lava rock, hunting for prey. (Courtesy of NOAA, https://oceanexplorer.noaa.gov/explorations/05fire/logs/april22/media/anglerfish.html.)

Anglerfishes (Lophiiformes) may have lost some characteristics such as a gas bladder, but they have additional features (Figure 6.12): the first dorsal ray is transformed into an illicium with esca (i.e., a line and bait), and they are bioluminescent. In general, there are shallow benthic forms (e.g., batfish and goosefish) and deepwater pelagic forms (e.g., deep sea anglerfish). The deep sea form is of interest here because of its interesting life history patterns and its ability to survive in the bathypelagic zone. Cods and anglerfishes will be the subjects of case studies.

HIGHER TELEOSTS

As previously indicated, *higher teleost* as used here refers to the more derived teleosts and does not imply any particular taxonomic group. Some may include superorder Acanthoptergyii (spiny + fins) or series Percomorpha as higher teleosts, but all would probably agree that fishes included in the suborder Percoidei (Nelson 2006) would qualify even if they do not like the gross lumping.

In this brief survey of the diversity of fishes, I have omitted several interesting spiny ray fish groups, such as the mullets, silversides, killifishes, topminnows, pupfishes, and livebearers. However, the ecological significance of these fishes will be discussed in various later sections.

Perciformes

Perciformes is the largest vertebrate order, and it includes the most numerous fishes in the oceans of the world, with 160 families and 10,033 species (Nelson 2006). The most important subdivision of this order is the Percoidei, a suborder that contains some freshwater fishes and most of the well-known marine and estuarine fishes, including sea basses and closely related sunfishes, perches,

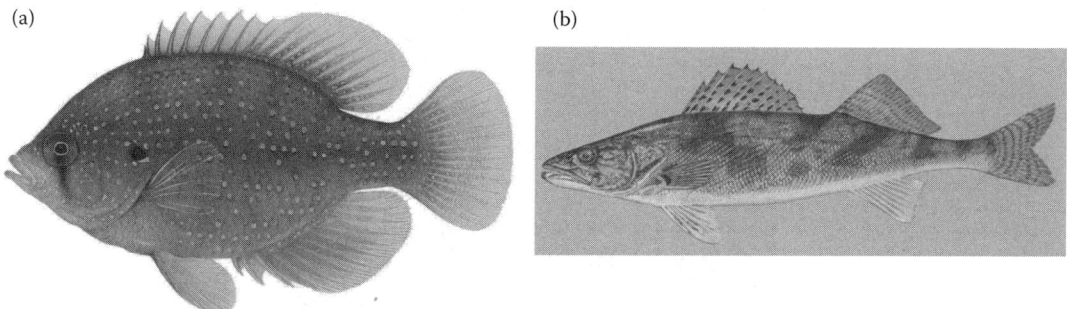

Figure 6.13 Perciformes—sunfish and perch families: (a) bluespotted sunfish *Enneacanhus gloriosus* and (b) sauger *Sander Canadensis*. ((a) By Edmonson, E., and Chrisp, H., courtesy of New York Biological Survey, New York State Department of Environmental Conservation. (b) By Raver, D., courtesy of USFWS. Accessed by usa.gov.)

bluefishes, dolphins, jacks, snappers, grunts, porgies, croakers, butterflyfishes, and angelfishes, to include just a few of the 79 families. Characteristics of these fishes include spines in dorsal, anal, and pelvic fins; two dorsal fins; ctenoid scales; thoracic pelvic fins; premaxilla only present in the upper jaw; closed swim bladder; and so on (Figure 6.13) (Nelson 2006).

Other more derived groups of perciforms that are of interest here include (suborders) cichlids; mullets; and mackerels, tunas, and billfishes. All will be discussed in their respective aquatic environments.

Flatfishes and Tetraodonts

The two remaining orders of fishes include a recent offshoot, the flatfishes, Pleuronectiformes (pleuronectid = side + swimming), and presumably the most recently derived teleosts, the Tetraodontiformes (tetraodonts = four + teeth).

Figure 6.14 California halibut *Paralichthys californicus*. (Courtesy of Hsiano, R., San Francisco Bay. Wikimedia Commons, http://commons.wikimedia.org/wiki/File:3_Foot_California_halibut.jpg.)

Figure 6.15 Tetraodont whitespotted puffer *Arothron hispidus*. (Courtesy of Smith, M., Churaumi Aquarium, Wikimedia Commons, (http://commons.wikimedia.org/wiki/File:Arothron_hispidus_-_Churaumi_Aquarium.jpg.)

Flatfishes include the flounders, sole, and tonguefishes. This group has extended dorsal and anal fins, and it has the distinction of having adults that are not bilaterally symmetrical. The young flatfish looks like a normal fish, but in developing into an adult, one eye migrates (left eye in some forms, right eye in others) across the top of the skull to the other side of the head. At that point, the fish abandons upright swimming and takes on a benthic lifestyle by laying on the side without an eye. In this context, they are an example of convergence with the rays as a lie-and-wait benthic predator (Figure 6.14). With delicate white flesh, they are a popular food fish.

Tetraodonts include triggerfishes, filefishes, puffers, box fishes, and molas (Figure 6.15). These are slow-swimming fishes that tend to rely on spines, bony armor, or poisonous tissues or discharges to deter (or kill) predators. The small mouths and fused "teeth" are for picking and crushing. As an aside, humans have learned how to prepare the delicate flesh of some poisonous puffers (*Takifugu*) for consumption; however, many deaths resulting from the practice have been reported in Japan (Bond 1996). Evidently, the meat of this fish produces a narcotic "high" (Helfman et al. 2009).

The most recent fishes and thus, for now, the endpoint of teleost evolution are four species of the mola or "ocean sunfish." These are very large and strange fishes (Figure 6.16) whose small altricial

Figure 6.16 Ocean sunfish *Mola mola*. (Courtesy of Norman, P.-O., Nordsøen Oceanarium, Hirtshals, Denmark, Wikimedia Commons, http://commons.wikimedia.org/wiki/File:Sunfish2.jpg.)

larvae can reach adult sizes of 1000–1500 kg (Moyle and Cech 2004; Nelson 2006). Molas have lost the caudal fin and have developed huge dorsal and anal fins of about equal sizes, which they use to scull through the water. These fishes occur throughout the epipelagic zone of oceans and include mostly jellyfish in their diet. Molas are known to be extremely fecund and reportedly can produce as many as 300 million eggs (Moyle and Cech 2004), which makes it the most fecund vertebrate.

SUMMARY

Teleost fishes are numerous and diversified, comprising 96% of all living fishes. Ray-fin fishes provided the platform upon which the teleosts evolved through a number of stages: a newly derived homocercal caudal fin, change from an open to a closed air bladder, a branchial pump, improved scales, relocation of fins, and added spines. Teleosts occur in all sizes, utilize foods from plankton to detritus, feed at all consumer trophic levels, and live from the tropics to polar regions. First to appear were the lower teleosts, familiar herrings, minnows, and salmon, with their smooth cycloid scales and single dorsal fins. Early teleosts include bonytongues, such as the South American arowana and arapaima, which are often kept in aquaria. These large ancient fish have median fins positioned around the tail for thrust: arowanas jump in the air to pluck food from vegetation. Another group of lower teleosts is the catadromous eels and their relatives, such as the tarpon, with very different adults but similar (leptocephalus) larvae. Fishes more typical of lower teleosts are the commercially important clupeids: herrings, menhaden, and sardines. These sleek fishes are mostly plankton feeders and can occur in schools of millions. Dominating freshwater are the ostariophysians, mostly freshwater families of suckers, minnows, characins, and catfishes. The group is known as the "hearing aid" fishes because the ear is connected to the swim bladder for sound amplification. Also, they possess pheromones called "shreckstoff," or fright stuff, which is used to warn others of danger. Trout, salmon, and relatives are so well known that they need little description. Hardy coldwater fishes, many are brightly colored and of interest to anglers. They also are considered to be a good prototype for higher teleosts. The salmonids have lower features of an open swim bladder, cycloid scales, a nonprotrusible mouth, and abdominal pelvic fins. Cods and anglerfishes are "almost" higher teleosts and are referred to as para (almost)-spiny fins. Important food fishes, cods and their benthic relatives have pelvic fins that have moved forward to the thoracic or jugal position, and their swim bladder is of the closed type. Anglerfishes are related, but much has changed in their morphology to adapt to life in the deep benthic or pelagic realm, where they catch prey by use of a lure. Deep pelagic forms also may have parasitic males. Finally coming to the higher teleosts, we find the largest order of vertebrates, with 160 families and 10,033 species of fishes. From sunfishes and perches in freshwater to bluefish, dolphins, billfishes, jacks, snappers, porgies, and so on, almost all have the closed swim bladder, ctenoid scales, protrusible mouth, branchial pump, two dorsal fins (with one spinous and one that is not), forward and lateral movement of paired fins, and a homocercal tail. However, some fishes have lost or changed one or more of these characteristics, such as loss of swim bladder in the tuna and readaptation of the dorsal fin in the remora, and some have changed so much that they look bizarre or ancient. These include the last two groups to evolve, the apex of fish evolution: the flatfishes, whose eye moves from one side to the other, and the tetraodonts with four platelike teeth. Tetraodonts include boxfishes, which resemble ostracoderms because of their rigid case of bone, and the huge ocean mola (ocean sunfish), whose sizes can reach 100–1500 kg and is the most fecund vertebrate, laying as many as 300 million eggs. This fish has lost its caudal fin, sculling along in the open ocean with huge pelvic and anal fins and eating jellyfish and other items that it can catch.

CHAPTER 7

Radiations, Extinctions, and Biodiversity

LIFE ON EARTH HAS NOT BEEN EASY

Evolution of life on Planet Earth has not been constant; rather, there have been pulses of new organisms and losses of old ones. This was recognized by early nineteenth century geologists, including Adam Sedgewick, a Cambridge geology professor. He found deposits of fossils in strata representing an early period that he named the Cambrian (i.e., the so-called Cambrian explosion), and he also named the Paleozoic Era in 1838 (Benton 2003). Phillips and Gould (1860) also noted that fossil species were very different during a certain period of the Phanerozoic (abundant + life; the last 540 million years (my) when larger fossils were present) and divided it into three periods accordingly: Paleozoic (old life), Mesozoic (middle life), and Cenozoic (originally Caenozoic = recent life). The presence of these different fossil layers was later attributed to episodes of mass extinction and replacement by new forms. The extinctions were presumably due to global catastrophic events that resulted in ecosystem-level extinctions.

We now understand that many periods of increased extinction rates have occurred in the fossil record and at least 17 minor (20%–30% species loss), four intermediate (about 50% species loss), and one major (about 90% species loss) mass extinctions have been identified (Figure 7.1). The five largest extinctions, called the Big Five (in which the extinction rate was double or more than minor ones), occurred at various intervals: end of the Ordovician (440–450 million years ago (mya), Devonian (360–370 mya), Permian (245–250 mya), Triassic (210 mya), and Cretaceous (65 mya) periods. Three events are particularly interesting for fish paleoecology because major fish groups disappeared during those events: the Ordovician, Devonian, and Permian extinctions (Table 7.1).

Aquatic life flourished during the Ordovician, mostly benthic invertebrates, but jawless fishes also appeared in several forms. However, the end of that period went through a global climate change, with a temperature drop, glaciations, and a falling sea level. At that time, faunas typical of polar regions were found near the equator, and warmwater faunas were lost (Benton 2003). During the Ordovician extinction (about 440–450 mya), the second largest mass extinction of marine life occurred, when at least 100 families of marine invertebrates disappeared. At that time, there was a great loss in reef animals and the loss of most Ordovician ostracoderms (except thelodonts) (Maisey 1996).

The Devonian period was marked by diversification in fishes, but it also was marked by three extinctions. The Late Devonian mass extinction (360–370 mya) was the worst, and about three-fourths of the fishes were lost (35 of 46 families). The ostracoderms vanished, followed by 15 of 22 placoderm families, 10 families of lobe-fin fishes, and some sharks and ray-fin fishes (Maisey 1996). At the end of that period, the global sea level abruptly dropped after it reached its maximum height and anoxic conditions occurred on the seabed. Many marine invertebrates were affected, and there could have been two asteroid strikes (Masaitis 2002; Schieber and Over 2005). The close of the Devonian saw the loss of all of the ostracoderms and virtually all placoderms but an increased

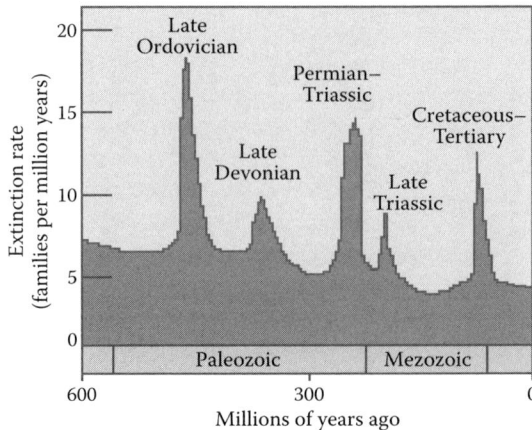

Figure 7.1 Extinction rate for families of organisms, showing the "Big Five" events. (Courtesy of National Aeronautics and Space Administration, http://search.usa.gov/search/images?query=extinctions.)

radiation of jawed fishes (sharks and teleostomes). Some invasion of freshwater systems by placoderms occurred during the close of the Devonian, suggesting poor marine conditions or heavy competition. The Late Devonian mass extinction had the greatest effect on the marine communities, and 22% of all marine invertebrate families were lost. This extinction may have produced extinction pulses among various groups that lasted for 3 my.

The great Permian extinction resulted in a loss of about 50% of families and 90% of marine invertebrate species. The mass extinction has been related to an asteroid strike (Becker et al. 2004), a long period of volcanism, atmospheric changes, and climate alteration produced by a large connected sea. At that time, the sea level recorded the greatest drop ever detected, and there were anoxic conditions on the seafloor. Both shallow and deep waters had low levels of oxygen. There was a drop in marine productivity and a temperature rise in the oceans. During that period, the few remaining acanthodians vanished, as did almost all of the lobe-finned fishes. At the end of the Permian period, there was a single global continent (Pangaea) that produced extreme conditions on land.

How did the three mass extinctions just described affect the evolutionary ecology of fishes? We may never know for sure, but we know some facts, we can relate some phenomena with theories, and we can speculate.

Table 7.1 Mass Extinctions Affecting Fishes

Period	Date (mya)	Cause	Loss
Permian	250	Asteroid (?), low sea level, anoxic seafloor, climate change	Acanthodians, 90% of marine invertebrates, some lobe-fins
Devonian	365	Asteroid (?), low sea level, anoxic seafloor, climate change	Ostracoderms, 35 of 46 fish families, most placoderms
Ordovician	440	Climate change, low temperatures, glaciations, falling sea levels	Most ostracoderms, many marine invertebrates

Source: Maisey, J. G., *Discovering Fossil Fishes*, Westview Press, Boulder, CO, 1996. With permission. Benton, M. J., *When Life Nearly Died: The Greatest Mass Extinction of All Time*, Thames & Hudson, Ltd., London, 2003. With permission.
Note: Dates are approximated.

FISH EXTINCTIONS AND A FEW QUESTIONS

Nelson (2006) identified 27,977 living fish species and estimated that the total number of extant fishes would exceed 32,500. This means that fish diversity exceeds the tetrapods (about 28,000 amphibians, reptiles, birds, and mammals). However, extant fishes do not represent all of the known diversity in fishes, because large groups of fishes are extinct. What kind of fishes exists today? As shown in Figure 4.1, ray-fin fishes (class Actinopterygii) clearly dominate the present fauna, with 26,891 species, and recently derived teleost fishes constitute 96% of all extant fish species (26,840 species). Of the remaining living fishes, there are only about 970 sharks and their relatives (class Chondrichthyes), eight lobe-fin (class Sarcopterygii) fishes, and arguably only one group of vertebrate agnathans (38 lampreys). Thus, five of seven known classes of vertebrate (and craniate) agnathans are extinct, and two of the five classes of gnathostomes are extinct (and the remaining lobe-fin fishes are represented by only two orders). In summary, of 12 classes of fishes, only five are extant, and two of these are underrepresented.

Extinctions of species have been so common in the past that they have been referred to as "a way of life" by Raup (1991), who has calculated from Phanerozoic fossils (last 540 my) that only about one in 1000 species survives today, or an extinction rate of about 99.9% (survival rate of 0.001). Making a gross speculation that fish extinction would be similar, the present estimated number of fish species (32,500) would be remnants of the staggering number of 32.5 million fish species. What can these radiations, extinctions, and survivors tell us about evolutionary ecology?

Fishes have persisted for about 500 my and provide a basis for judging how evolutionary events have influenced the diversity of complex organisms. How did global mass extinctions affect the persistence of ostracoderms, placoderms, acanthodians, and lobe-fins? There are several questions we would like to seek answers for: (1) Did abiotic changes in the planet cause the extinctions? (2) Were biotic environmental pressures also a factor? (3) Were fishes unaffected by the extinctions and instead were lost only to pseudoextinction? (4) Were species that survived for a long time more likely to persist than species of more recent origin? (5) Did extinctions allow greater diversity to occur in a less crowded world? (6) Are there lessons to be learned from long term survivors? The following discussion will attempt to answer these questions.

Abiotic Change

Shallow water forms would have been greatly affected by falling sea levels (possibly falling hundreds of meters) (Maisey 1996) and extreme cold conditions (the most extreme global glaciations known) at the close of the Ordovician. This could have caused the loss of the original ostracoderms, as shallow seas and continental margins were drained or frozen.

Sea level was high during most of the Devonian; however, it abruptly declined at the end of the period. Extreme anoxic conditions occurred on the continental shelf and extended even to the deep seabed. This could have been the death of the benthic feeding fishes such as the ostracoderms, and if the placoderms were feeding on them or on other benthic organisms, they also would have suffered. Presumably, this explains why pelagic sharks and ray-fin fishes did not suffer as greatly.

A large connected sea and loss of shallow inland seas occurred during the End Permian extinction due to Pangaea's formation. This could have adversely affected the lobe-fin fishes. Anoxia on the seafloor eliminated most benthic invertebrates and could have disrupted aquatic food chains for the acanthodians. Low oxygen levels could have eliminated intolerant species.

Did abiotic change cause extinctions? All of the physical changes above could certainly have affected the distribution and abundance of ancient fishes during these three extinction periods, but we may never know if abiotic factors alone caused the extinctions. Also, it is difficult to pin extinctions on any one physical factor, and extinctions likely were caused by more than one factor during each or among all of the extinctions. It is worth noting that this has been dubbed the "murder on the Orient Express hypothesis" because all of the suspects are guilty (Benton 2003).

Biotic Factors

Loss of the original Ordovician ostracoderms may have been related to abiotic conditions, but competition by benthic placoderms and predation by large pelagic placoderms cannot be ruled out as contributing factors. Also, radiation of the more agile thelodonts (feeble + teeth) during the latter part of the Ordovician may have posed a competitive threat.

Arthrodires (jointed + neck) included about two-thirds of the placoderms. These fearsome predators, of large sizes, with a huge gape and slicing tooth plates, were a menace to anything smaller and slower, although the tooth plates were likely useful against large arthropods and their own kind. The aggressive nature of these beasts was discussed by Maisey (1996) with reference to "large wounds and deep gouges" in their bony armor. The deep puncture marks (unhealed) came from placoderms attacking placoderms; these give evidence to their aggressive predatory behavior, even toward their own species. But these fearsome beasts appear to be highly specialized in their communities, and specialization, once attained, seems to be sustained in evolution. This principle, Dollo's law, is attributed to paleontologist Louis Dollo in 1890. In general, it states that evolution is not reversible and that specialized characters, once derived, will not revert to general ones (reviewed by Gould 1993; Martin 2004). Placoderm extinction was no doubt related to geological events, but competition by chondrichthyans, such as bradyodonts (related to chimeras), could also have been a factor (Janvier 2007).

Did biotic factors result in or hasten extinctions? Evolution of jawless fishes has been considered an "arms race" (perhaps in defense of placoderms or eurypterids) by many scientists over the years. It is certainly possible that agnathans were running second to placoderms and that both groups lost the race to the new ray-fin fishes, which followed a different evolutionary path. Instead of being more and more specialized for body protection, the ray-fins developed the speed and maneuverability to outrun their enemies. Also, the ray-fins developed a high degree of trophic adaptability, from being detritivores to being top carnivores. Thus, they likely were not as restricted in their use of food items or habitats. Under the added stresses of poor environmental conditions, the ray-fins could be expected to win in some situations. This scenario would apply across all time periods regardless of mass extinctions, and the present teleost diversity supports this hypothesis.

Are some species vulnerable to extinction because they have certain biological traits? In an excellent review, McKinney (1997) found that species that have been identified as most vulnerable to extinction have a large body size and are highly specialized. The large body size trait did not hold up to scrutiny, but the concept that specialized species with narrow niches are extinction prone is "arguably the most fundamental concept in the history of thought on extinction risk" (McKinney 1997, p. 500). This, coupled with other traits such as feeding at high trophic levels, suggests that the loss of the placoderms was predictable.

Pseudoextinction

Gradual transformation of one species into another by the process of phyletic transformation could result in pseudoextinction. In this case, a species is lost in the fossil record as it undergoes enough change to be classified as a new organism (Raup 1991). It is important to note, however, that the genetic material is passed on.

Did fishes disappear because of pseudoextinction? Except for the possibility that the lampreys descended from one group of ostracoderms, all the rest of the agnathan fishes, placoderms, and the acanthodians disappeared from the fossil record completely, leaving no similar forms that have ever been detected. In this case, pseudoextinction can be ruled out.

Persistence of Survivors

It seems logical that if a species can survive an extinction episode, it would be more capable of persisting over time. Raup (1991) addressed this idea relative to "bad genes or bad luck" and decided that most extinctions are due to bad luck: species die out because they are exposed to stressors that they have not experienced before in their evolution and are thus not available for natural selection to develop a genetic defense.

The "law of constant extinction" (i.e., Van Valen's law) also addresses this issue. As discussed in Martin (2004), Leigh Van Valen in 1973 reported from a study of fossils that the probability of a taxon going extinct remains constant regardless of its age. This is presumably due to a continuing evolutionary arms race between species due to competition. Called the Red Queen hypothesis (from Lewis Carroll's book *Through the Looking Glass*), in which the Red Queen has to continue running just to stay in the same place, this hypothesis posits that a species needs to spend energy in order to maintain its fitness in a coevolving system. In other words, a species' niche will be exposed to abiotic and biotic pressures as the environment changes over time, and it must expend energy to adapt to these pressures. In some cases, this development will result in specialization that is likely to place species at a disadvantage during catastrophic events. This notion is similar to an earlier concept (i.e., Romer's Rule; see Glossary).

Are species that survive mass extinctions more likely to persist? Although we have yet to decide the precise causes of mass extinctions (with the possible exception of the K-T event 65 mya), extinctions are ostensibly due to several causes (the "Murder on the Orient Express hypothesis," in which all known suspects are presumed guilty) (Erwin 2006). In this case, species surviving one mass extinction might not experience the same impacts as those of another one; thus, it would be difficult, if not impossible, to avoid extinction due to catastrophes, especially in localized species. This is especially true since the frequency of occurrence of the events is on the order of tens to hundreds of million years.

Enhanced Radiations

The recovery time for species diversity to return to levels that existed before a mass extinction has been estimated at about 10 my for an intermediate-level extinction and about 100 my for a major extinction (Benton 2003). There seems to be some expectations that recovery would be enhanced and occur more quickly due to the presence of "vacant niches." However, this concept has found disfavor.

Erwin (2006) provided an insightful discussion about the idea of vacant niches by constructing and then rejecting a "chessboard" model. On the chessboard, each space is occupied by a species and represents its "niche." Extinction is seen as a process of removing players, which then frees up "niches." The idea of recovery is to replace each species with a new one by filling up the spaces. However, he points out that during a mass extinction, the board collapses and a new game with different rules begins. He also reminds us that organisms modify the environment and that niches can be created by organisms, such as in mutualism.

Were further fish radiations enhanced after mass extinctions? I like to use the example of a coral reef that provides for a multitude of overlapping niches, or a forest that creates microclimates beneath the canopy: no forest trees, no canopy, no microhabitat, and a loss of many niches. Niches can disappear with the loss of organisms, and the recovery process will then have to rebuild them. For truly catastrophic events, it seems that the answer to the question is "no."

Lessons from Long-Term Survivors

Very few relict freshwater fishes have persisted to the present; these include lampreys, lungfishes, sturgeons, paddlefish, gars, and bowfins. Adaptations of these fishes are highly varied, but most of these fishes are large, have vascularized gas bladders, have protective scutes or scales, are migratory, and live in large warmwater systems with dynamic but predictable conditions.

Ilves and Randall (2007) noted that the success of primitive fishes was related to their niches. Teleost adaptations for feeding and movement are not very important in the niches that primitive fishes occupy. Also, sensory skills, energy conservation, reproductive mode, and the capability to deal with hypoxia in various ways all seem to be important attributes.

Relict marine fishes include the hagfishes, coelacanths, ratfishes, and primitive sharks. These fishes seem to be highly specialized in some aspects, including feeding, and very primitive in comparison to modern forms, but they have been well adapted so far. All of the examples are relatively large in size, have good morphological defenses against predation, and most have precocial young. In addition, perhaps David Raup (1991) would say that these survivors had good luck!

ECOLOGICAL CONCEPTS

- *Dollo's law*—Attributed to Louis Dollo in 1890, it states that specialization (e.g., in feeding and habitat requirements), once attained in evolution, is irreversible (i.e., evolution is not reversible).
- *Highly specialized species are extinction prone*—Specialization is implicated in extinction vulnerability and selectivity (e.g., McKinney 1997).
- *Murder on the Orient Express hypothesis*—This was attributed to D. H. Erwin (2006) from an Agatha Christie novel in which all suspects were guilty.
- *Red Queen hypothesis*—This was attributed to Van Valen (1973), who got the idea from a novel by Lewis Carroll. The Red Queen has to continue running just to stay in the same place. In this analogy, a species has to expend energy to compete with others and to keep up with abiotic and biotic environmental change.
- *Romer's Rule*—Attributed to A. S. Romer by Hockett and Ascher (1964), this states that changes in organisms are generally conservative and aid in maintaining a traditional way of life in an altered environment.
- *Law of constant extinction*—Also known as Van Valen's law, this states that the probability of extinction is constant regardless of the age (persistence) of a species.
- *The Chessboard model*—Attributed to Erwin (2006), this concept of recovery suggests that there are vacant niches after a mass extinction. However, the catastrophic force destroys the board as well as the pieces on it. Although there may be survivors, they face a new game with different rules.

CASE STUDY: FISHES OF FOSSIL LAKE

Fossil Butte National Monument is located in southwestern Wyoming, about 16 km northeast of the city of Kemmerer. A large inland sea was present in this vicinity during the Cretaceous, and it extended from Alaska to the Gulf of Mexico. This sea disappeared about 55 mya, and the strata buckled, folded, and uplifted with the formation of the Rocky Mountains. In the process, inland freshwater lakes appeared in basins between mountain ranges. These lakes disappeared during the Eocene, leaving accumulated lake sediments (shale) that covered 88,000 km^2 and extending to over 600 m in thickness (Jackson 1980). This accumulation is known as the Green River Formation. Millions of fishes have been perfectly preserved in limestone-bearing sediments of Fossil Lake, and many of these have been linked with mass mortalities (occurring under catastrophic conditions, presumably due to anoxic conditions and/or blooms of toxic bluegreen algae). Species include herrings (Figure 7.2), bonytongues (Figure 7.3), perches (Figure 7.4), gars, bowfins, paddlefish, stingrays, and others, and identifies 50-mya piscivores (Figure 7.5). Fossils may be viewed at the National Park Service visitor facility or at a private quarry. The Ulrich Fossil Quarry there provides a guided service for digging your own fossil fishes (for a small price), which you may keep.

Figure 7.2 Mass kill of the clupeid fish *Knightia*. (Courtesy of U.S. National Park Service, Fossil Butte, www.nps.gov/fobu/photosmultimedia/GreenRiver-Formation-Fossils.htm.)

Figure 7.3 Eocene bonytongue *Phareodus encaustus*. (Courtesy of U.S. National Park Service, Fossil Butte, www.nps.gov/fobu/photosmultimedia/GreenRiver-Formation-Fossils.htm.)

Figure 7.4 *Priscacara serrata*, a perciform fish. (Courtesy of U.S. National Park Service, Fossil Butte, www.nps.gov/fobu/photosmultimedia/GreenRiver-Formation-Fossils.htm.)

Figure 7.5 Predator–prey relationships from the Eocene: *Diplomystus dentatus* with a *Knightia* lodged in its mouth. (Courtesy of U.S. National Park Service, Fossil Butte, www.nps.gov/fobu/photosmultimedia/GreenRiver-Formation-Fossils.htm.)

SUMMARY

Life on Earth has changed over time, with the loss of species and the appearance of new ones. Three major radiations of species were noticed in the fossil record as early as the 1860s. As a result of this, geologic time has been divided into three periods: the Paleozoic, Mesozoic, and Cenozoic periods. These declines and radiations have been attributed to mass extinctions, of which two were responsible for deliniating the three periods above: the Permian extinction (245 mya) and the K–T extinction (65 mya). However, many more extinctions have been documented, including about 17 minor and five major ones that resulted in a loss of 50% or more of species. Three of these were important for fishes during the Ordovician extinction in which most of the ostracoderms were lost; the Devonian extinction in which 35 of 46 fish families disappeared, which were most of the placoderms and the rest of the ostracoderms; and the Permian extinction, in which the acanthodians and most of the lobe-finned fishes vanished. In total, of the 12 classes of fish present, seven went extinct, leaving only five classes, of which two are underrepresented. From analysis of the Phanerozoic (good fossils) period, there has been an extinction rate of 99.9%, which means that the present fish faunas are survivors from perhaps 32 million fish species. What happened to all these fish? We can evaluate causes and outcomes based on the meager data available: (1) Abiotic changes could have been responsible for extinctions, but most likely, the causes were due to more than one factor. (2) Biotic changes also are likely, and an example is the "arms race" and specializations of ostracoderms and placoderms that faced a new concept with the appearance of the swift ray-finned fishes. (3) Pseudoextinction seems unlikely as a major factor. (4) Survivors of one extinction are not any better prepared for another extinction (Van Valen's law). (5) Recovery after a mass extinction is not enhanced by "vacant niches" due to the probable loss of ecosystem structure and function. (6) Lessons from long-term fish survivors: Find a way to make a living that does not compete with the teleosts. In freshwater, some large fishes that can withstand low oxygen, have a large range, can disperse long distances, feed at lower trophic levels, and have body protection have persisted. In the marine environment, it seems that a premium has been placed on precocial young, while most or all of the same features mentioned for the freshwater fishes also apply.

Further reading: Long 1995.

PART IV

Freshwater Ecosystems

CHAPTER **8**

Zoogeography of Fishes

PATTERNS AND SPECIES DIVERSITY

Georges-Louis Leclere, Comte de Buffon, is given credit for developing the first principle of biogeography (also known as Buffon's law). He observed that different plants and animals occurred in different geographical regions even though the climates in those regions were the same. The principle of evolution, which lineages change in time and space, is attributed to Charles Darwin and Alfred Russell Wallace. With just these two principles, it was possible to form the Science of Biogeography, and we are still trying to figure it all out. This chapter will discuss probable causes for the present worldwide distribution and diversity of freshwater fishes. Subsequent chapters will evaluate factors associated with the distribution and abundance of freshwater and marine fishes in specific systems and habitats.

We have been discussing the evolution of fish diversity and major factors associated with radiations and extinctions. However, we have not focused on fish diversity in regional or local habitats, which can be expressed as species richness, the number of species present in an area or community. This also has been referred to as "biodiversity," a relatively new term (dating back to 1988) (Benton 2003) and was first used for the "variety of species in an ecological community" (Krebs 2008). Because of the recent explosion of conservation biology, use of *biodiversity* has been greatly expanded (to be covered in chapter 30). Thus, we will simply use the term "diversity" or "species richness" to indicate the number of fish species present as we compare faunas of different geographic areas.

Zoogeography is concerned with the geographic distribution of animals, which occurs in time as well as space. This can be visualized as global patterns of species richness as reflected by the numbers of species or taxa. These patterns are greatly affected by environmental conditions that vary considerably on the planet with respect to time and location. Location is especially important as one travels from the equator to the poles. In this case, species diversity in general has been shown to decrease with increasing distances away from the tropics.

FACTORS AFFECTING DISTRIBUTION

Freshwater fishes are widely distributed, occurring from low to high altitudes, in cold, hot, wet, and dry conditions, and from the equator to the Arctic Circle. However, they are not uniformly distributed: All species do not occur on all continents, and species on continents do not occur in all suitable continental locations. In a general sense, one could say that they occur in certain areas because of their adaptation to the conditions at that location. But that could only be partly true. Invasions of nonnative fishes into new locations and continents (e.g., common carp in North America) have occurred across the globe, demonstrating that many fishes have a potential for a greater distribution than their native habitat. The question is "Why are fishes distributed the way they are?"

Fishes occur in various locations because of at least three reasons: (1) They were born (evolved) there. (2) They moved there by dispersal. (3) They were moved there by some outside force (i.e., vicariance).

Of course, fishes also may be missing from an area where they previously occurred due to catastrophic events or they may be prevented from occurring in a location due to barriers. This prompts questions: What conditions are associated with dispersals or extirpations? Also, are there features of different aquatic systems that facilitate or discourage the establishment of new fish populations? These questions can be addressed with respect to present global conditions, but due to the long period of fish inhabitance on the planet, answers must be developed from a perspective based on geological time scales.

With the possible exception of very large, deep lakes, most freshwater systems should be viewed as "aquatic islands in a terrestrial sea" due to their great exposure to fluctuating climatic conditions on land. Thus, global cooling or heating can have a profound effect on freshwater systems, especially in periods of continental glaciations. Tectonic events that raise or lower land elevation, such as the upheaval of the Rocky Mountains, also can change climatic condition, as can continental drift if land areas are transported to different climates. Finally, pluvial episodes can cause great changes in freshwater habitats, converting streams to lakes, uniting drainage basins during wet periods, and causing deserts in interpluvial periods.

Dispersal of fishes also can be enhanced or blocked by barriers such as rise or fall in sea level. For example, dispersal of freshwater fishes and other organisms (including humans) between Asia and North America via Berengia was aided by low sea levels. Fish dispersal also is affected by tectonic forces, such as the upheaval of mountain ranges or the joining or separation of continental plates.

Not only distribution patterns but also more severe changes such as extirpations or extinctions of fishes have been related to global climate change. Periods of continental glaciations, volcanisms, and asteroid strikes are adverse, catastrophic events that have been implicated in extinctions.

A word about the marine environment: changes in global climates take longer to have an effect on marine environments; there are fewer barriers to dispersal than in freshwater, and effects of plate tectonics on them are mostly unknown. However, fish dispersal in marine systems is affected by ocean currents and barriers imposed by continents, reefs, and islands. Certain habitats such as those in reef systems also are semi-isolated in oceans by large pelagic systems. Oceanic realms also pose other constraints in the physicochemical (e.g., temperature, oxygen, pressure, sea level) and biological components (e.g., food supply, etc.) of that environment. Thus, the overwhelming number of fishes occurs in areas of greatest productivity, such as continental shelf systems, other shallow areas such as coral reefs, and areas of upwelling. In addition, the discovery of hydrothermal vents and associated communities has resulted in more interest in studying the ocean floor. Six seafloor regions are now known, based on the assemblages of organisms. For more information about the field of marine biogeography, see a review by Briggs et al. (2004) and results from the Ocean Explorer program at OKEANOS explorer (http://oceanexplorer.noaa.gov/okeanos). We will evaluate factors associated with marine fish distribution and abundance in later chapters, with emphasis on extant fishes in definable habitats.

Although we have addressed factors associated with the distribution of fishes, there is no simple answer to what controls species diversity across spatial and temporal scales. As shown by Krebs (2008), there are several likely explanations for each region, which we will consider later.

ADAPTATION

Freshwater organisms are confined by the terrestrial environment to their native drainage basins, and they can only spread to other locations as the land or aquatic system changes. But there are inherent physiological attributes of fishes that also constrain movement between aquatic systems. Arguably, the most important of these is the degree of salt tolerance, perhaps followed by the ability to adapt to different water temperatures, oxygen levels, and so forth. Salinity and temperature extremes are overriding factors that affect worldwide distribution and will be discussed here.

It was George Myers (1951) who provided a very useful tool to zoogeographers by identifying those fishes that were confined only to freshwater due to a low tolerance for salt concentrations, and those that were not. Myers related the tool in papers published from 1928–1951. I rely on a review and discussion of Myers' work by Darlington (1957) in interpreting and relating the concept.

Myers divided fishes from around the world into three groups, called divisions, based on their occurrence and tolerance for salt. (Note: I will only provide the names of the fishes from North America here.) He placed all fishes that were confined to freshwater in a *primary division*, which only contained bony fishes. This division included the North American paddlefish, bowfins, mooneyes, many ostariophysian fishes, pikes, sunfishes, perches, darters, and a few others. Darlington (1957) added a few pertinent points. He indicated that seven-eighths of the primary division species are ostariophysians, likely due to an array of adaptations for freshwater life, including the famous Weberian apparatus. Presumably, the marine catfishes (the only group of ostariophysians that have salt tolerance) had freshwater ancestors. Continuing the story, fishes that are usually confined to freshwater but occasionally enter the sea were placed in a *secondary division*. In North America, this includes gars, cyprinodonts, and cichlids. These fishes would have the ability to survive short-distance travels in the ocean.

Finally, fishes that were considered very tolerant to salt were placed in the *peripheral division*. Many fishes were placed in this group, and most of these were teleosts. Many of the peripherals are marine fishes that enter freshwaters occasionally; some are migratory, such as lampreys, eels, sturgeons, and salmonids, and others are freshwater fishes that are recently derived from marine ancestors, such as cottids (e.g., sculpins).

Myers' concepts have proven most useful in determining which factors may be implicated in the global distribution of fishes and, in turn, in assessing the movements of continents as we shall see below. The original concept remains a good one, and the assignment of fishes to the various groups also has been expanded, refined, and updated. An excellent example is Berra (2007), who identified the primary, secondary, and peripheral characteristics of 138 fish families, as well as the geographic areas inhabited by each. Moyle and Cech (2004) used the concept but refined and qualified the terms. Their "freshwater dispersants" basically combine the primary and secondary division of Myers, and their "saltwater dispersants" include the peripheral division. Cohen (1970) classified 33% of all fishes in the primary division, about 8% in the secondary division, about 58% as marine fishes, and 0.6% as diadromous.

Worldwide distribution of freshwater and marine fishes also has been profoundly influenced by temperature. In freshwater temperate systems, fish communities have been divided into three groups: coldwater, coolwater, and warmwater communities. Perhaps the most difficult to define are the coolwater fishes, and in North America, this group is bounded to the north by the coldwater salmonids and to the south by the warmwater centrarchids. Coolwater fishes as a group were first recognized in the 1960s, and major recreation and commercial fisheries included at least five species: "walleye, sauger, yellow perch, northern pike, and muskellunge," as recognized by Kendall (1978). In marine fishes, coolwater fishes are not generally recognized, but the polar, temperate, and tropical ones are.

CONTINENTAL MOVEMENT

It was noted as early as 1596 by map makers (e.g., Abraham Ortelius) that the physical configurations of Africa and South America suggest that they had once been joined. After much speculation about the matter, German scientist Alfred Wegener published his controversial treatise on *The Origin of Continents and Oceans* in 1915 (e.g., see Wegener 1966). He reported on disjunct continental distributions of fossils and postulated that all of the continents had once been joined in a giant supercontinent that he named Pangaea (all + Earth). It was not until the 1960s that the controversy was settled, mostly due to sonar technology and investigation of the seafloor.

Plate tectonics (continental drift) is now a fundamental concept of geology, evolution, and ecology. Continental movements over the past 150 million years (my) have had a great influence on global patterns of fish diversity and abundance. I provide a summary and discussion of major events from an ecological perspective.

In a greatly simplified nutshell, the outside layer of the Earth contains continents and the ocean floor, which are divided into plates. The plates "float" over a more dense and plastic part called the asthenosphere. Heat from the Earth's core warms the asthenosphere and causes magma to rise to the surface at the junction of plates at oceanic ridges, spreading the plates out from these locations (sea-floor spreading). When the spreading plate encounters another plate, one will plunge under the other (subduction). As these plates move, they also transport the continents that rest on them. Thus, while the floating continents have remained for 3.8 billion years (by) and changed little in the past 2.6 by, the ocean floor is subducted every 150 my (Scotese 2004), effectively destroying older fossils.

The supercontinent of Pangaea presumably was united until about 180 mya, when it began to separate into two portions named Laurasia (after a prominent North American landmark, the St. Laurence River) and Gondwanaland (a province in India). As the continent divided, it also divided its endemic freshwater fish fauna, which at that time included paddlefish and sturgeons, early neopterygians (ancestors to gars and bowfins), and presumably ancestral teleosts. Subsequently, the land masses slowly moved to their present positions over 200 my, taking the fish with them (Figure 8.1). However, during times of lowered sea levels, land bridges formed between many continental areas,

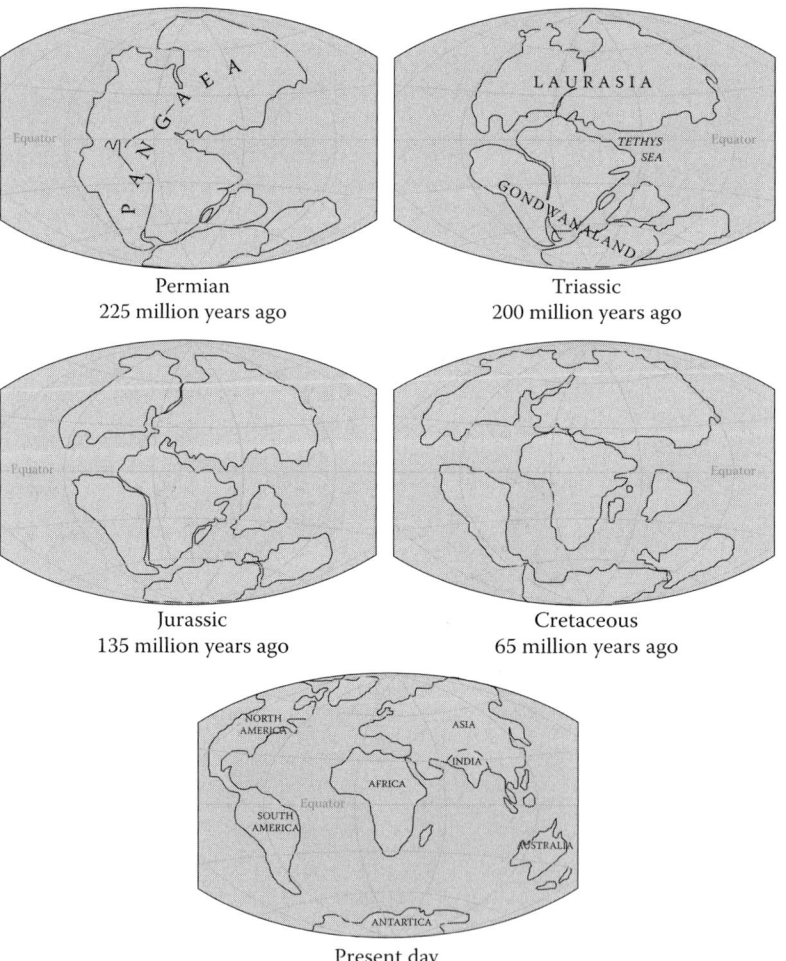

Figure 8.1 Continental movements from Pangaea (225 mya) to present. (Courtesy of U.S. Geological Survey, http://search.usa.gov.images?query=gondwanaland.)

and all of those connections would be impossible to detect. Approximate dates of continental movements are provided by Scotese (2004):

- *Late Jurassic (ca. 160 mya)*—The northern continent of Laurasia began to separate into NA and Asia (i.e., North America + Europe and Northern Asia). The southern continent of Gondwanaland was composed of SA, AF, MAD, IND, AUS, ANT, and NZ (i.e., South America, Africa, Madagascar, India, Australia, Antarctica, and New Zealand). Both of the continents thus formed could have received some or all of the Pangaean fish fauna.
- *Early Cretaceous (ca. 130 mya)*—Laurasia was divided as above until ca. 90 mya, when NA was effectively separated into two by the midcontinental seaway. Global climate was warm during the Cretaceous, and there could have been extensive fish migrations in polar regions. By this time, IND (perhaps with MAD) separated from Gondwanaland, and by 100 mya, AF and NZ had separated from SA, ANT, and AUS. At this time, AUS was effectively isolated from SA by ANT (they finally separated in the Eocene). The otophysian fishes radiated in SA + AF during the early Cretaceous and spread to Eurasia during the end of the period. Esocids, salmonids, and other groups also appeared.
- *Late Cretaceous (ca. 70 mya)*—Laurasia comprised of Asia, western NA, Eastern NA, and EU (i.e., Europe). North temperate minnow, sucker, and perch families diversified. Gondwanaland was completely severed into modern continents, NZ and MAD. At this time, AF came close to or united with EU.
- *Eocene (ca. 50 mya)*—NA was separated, and endemic sunfishes, ictalurid catfishes, and darters evolved. IND came in contact with Asia.
- *Miocene (ca. 18 mya)*—Laurasia increased in size by uniting with IND. Evidently, IND collided with great force, forming the Himalayas; the ancient IND fishes were apparently overwhelmed and replaced by invading ostariophysians. AUS moved northward.
- *Pliocene (ca. 3.5 mya)*—The Isthmus of Panama functioned as a land bridge connecting North and South America for the first time since Pangaea. Western NA was reconnected with Asia.
- *Holocene (ca. 8000 years ago (ya))*—This marked the close of the Pleistocene. Sea levels rose, and Asia was separated from western NA.

FISHES OF ZOOGEOGRAPHIC REGIONS

Animal distribution patterns have been studied historically by various zoogeographers, including the famous Alfred Russell Wallace. Wallace (1876) published *The Geographical Distribution of Animals*, in which he divided the world into six zoogeographic regions and established the famous "Wallace's Line" separating two regions (Asian and Australian) (Figure 8.2). Fish distribution in the regions has been the subject of much study and revision (such as Weber's Line), and the regions have been combined in various ways. Early ichthyologists were quick to reduce these regions to reflect climatic conditions, recognizing that fish faunas could be separated into northern, equatorial, and southern zones. Wallace's zoogeographic regions have been variously combined into these three zones (reviewed by Bond 1996). Wallace's zones remain useful for evaluating the evolutionary ecology of many groups, including fishes. Species distribution among these regions was summarized by Helfman et al. (2009) and Moyle and Cech (2004). It is noted that fish distributional data have been recently updated due to the discovery of additional species and are expected to change even more in the future. I have prepared the following brief synopses:

- *Ethiopian (African) Region*—Fish fauna is very diverse. There are about 3000 freshwater fish species and many ancient forms, including lungfishes and bonytongues. Freshwater dispersants (primary and secondary divisions of Myers) comprise 95% of the species, and more than one-third are otophysians. The most diverse family is the Cichlidae. The region has about 76 fish families (Moyle and Cech 2004), with a high level of endemism (perhaps 50%) (Berra 2007). The island of Madagascar presents an interesting fauna. Presumably isolated for more than 100 my, it separated

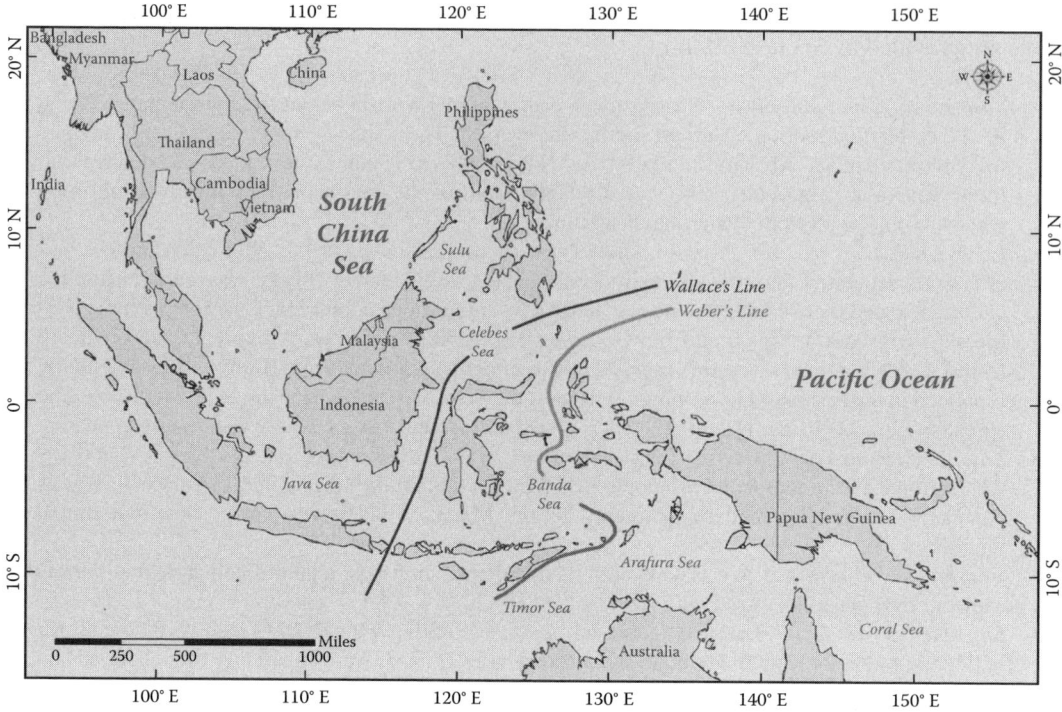

Figure 8.2 Wallace's Line as the division between the Asian and Australian fauna. (Courtesy of U.S. National Oceanic and Atmospheric Administration, http://oceanexplorer.noaa.gov/okeanos/explorations/index2010/Indonesia-USA/.)

from India at that time (after both had separated from Gondwanaland) and may provide a clue to the ancient Indian fauna that was lost after India collided with Asia. The fish fauna in Madagascar is more closely related to that in India and Southeast Asia than that in Africa (Moyle and Cech 2004).
- *Neotropical Region*—South America was an isolated continent for millions of years and the best example of a Gondwanaland remnant. There are over 3600 obligate freshwater fishes, dominated by characins (1800) and catfishes (1400); however, there are undoubtedly many undescribed species that remain (Moyle and Cech 2004). There are no cyprinids. Central America is included in this region, and the fish fauna there is transitional in nature. The region has a high level of endemism (perhaps 69%) (Berra 2007).
- *Oriental Region*—Peninsular India and Southeast Asia may have as many as 3000 species of fish, dominated by over 1000 cyprinids (Moyle and Cech 2004). India was an island for millions of years before uniting with Asia, but none of its ancient fauna survived after the linkage. Presumably, climate change and invasion by more advanced (competitive) Asian cyprinids overwhelmed the ancient fishes.
- *Palaearctic Region*—This region covers the remainder of Eurasia outside of the Oriental region. There are perhaps 500 fish species. Arctic portions of the region are dominated by anadromous fishes, but about 80% of the European fishes are in the primary group (Moyle and Cech 2004). The region has about 30 families, with 10% endemism (Berra 2007).
- *Nearctic Region*—This covers North America to the Mexican plateau, including three subregions: the Arctic Atlantic, Pacific, and Mexican transition. The best documented of all regions, there are at least 1061 fishes in 56 families (Moyle and Cech 2004). Hocutt and Wiley (1986) provided a detailed account of North American zoogeography. Ancient fishes are well represented by paddlefish, sturgeons, gars, and bowfins, presumably extant due to the Mississippi River system, which served as

a refuge during times of glaciation. However, in western North America, small unstable basins and isolation resulted in high levels of fish endemism (80% or more) in some areas.
- *Australian Region*—Long isolated from other continents, there are only four species of ancient freshwater fishes from three families. All the rest of the fishes in the entire region are derived from diadromous or marine families. Included in this region is New Zealand, where there are no primary fish species represented in its fauna. Australia has more than 200 native fishes that spend portions or all of their life cycle in freshwater and an additional 150 species of marine fishes that spend some part of their lives in estuarine or freshwater (Moyle and Cech 2004).

The greatest fish diversity and overwhelming abundance occur in the tropics (Turner and Hawkins 2004). In addressing this, the Ethiopian and Neotropical regions have been studied together. The continents of Africa and South America (and also in consideration of other parts of Gondwanaland such as Australia and Madagascar) give the best representation of the ancient Gondwanaland fauna: lungfishes, cichlids, and bonytongues. These two regions also include the vast majority of tropical systems and support overwhelming numbers of freshwater fishes (at least 7000 species) compared with the remainder of the planet. Reasons for this are presumably related to long growing seasons and even climate and abundant resources (Krebs 2008; Diana 2004). It seems apparent that abundant resources produced in the even tropical climate over a long period of time would result in a proliferation of species, leading to species interactions and adaptations of species to each other (reviewed by Turner and Hawkins 2004). Accordingly, tropical species are expected to be mostly specialized and adapted to biotic factors, and many have mutualistic interactions.

The Nearctic and Palaearctic regions have been united into the Holarctic region by some authors in order to more adequately describe fish distribution in the northern parts of these continents, which have been joined in various ways over millions of years. Five families of freshwater fishes occur only in the Holarctic region: paddlefish, pikes, mudminnows, suckers, and perches. These fishes are said to have a Laurasian distribution.

The Holarctic represents a cold temperate region occupied by about 2000 freshwater species. It is instructional to compare the diversity of this area with that of a similar-size land mass in the tropics. If we use the combined area of the South America and Africa continents to do this, we see that fish diversity of the Holarctic is less than half that of the more tropical continents. In view of short growing seasons, lower productivity, and cold climate, we could hypothesize that the lower fish diversity would be due to resource limitation and abiotic stressors. If so, we might expect adaptation to abiotic factors. This hypothesis will be explored further as we pursue our course of study.

Islands also can provide interesting insights into ecological conditions. New Zealand was separated from Pangaea very early, subjected to cold temperatures, and remained effectively isolated for 150 my. During this time, it lost all of its freshwater fishes, and its freshwater fauna includes only 27 species of marine-related fishes. Australia, larger than New Zealand and perhaps less subjected to cold conditions, has only four obligate freshwater species representing three ancient families. India preserved its ancient fauna apparently until it collided with Asia, when, it is assumed, a combination of climatic changes due to the Himalayas and an invasion of otophysian fishes eradicated them. On the other hand, Madagascar presumably separated with India, but it has remained isolated and maintained its ancient fauna of 127 fish species, of which about 60% are endemic.

VICARIANCE BIOGEOGRAPHY

There are two approaches to the study of biogeography, *descriptive* and *interpretive*. The preceding documentation and discussion about regional distribution are mostly a descriptive exercise. In this section, we discuss more interpretive approaches to the global distribution of fishes.

Until Wegener proposed his theory of continental drift (i.e., in 1912), the distribution of terrestrial animals was explained by postulating the existence of cross-oceanic land bridges and sunken continents: The continents were assumed to be stationary, and animals dispersed with these implausible connections (Briggs et al. 2004). This "dispersal hypothesis" was firmly entrenched when Wegener (1966) made his argument (i.e., the "vicariance hypothesis"). For about 50 years, the vicariance hypothesis was rejected and even ridiculed, with the more "simplistic" argument of dispersal prevailing. Although the validity of plate tectonics is now firmly established, the two interpretive approaches still remain; however, the vicariance school has greatly expanded and further subdivided.

Presently, the first interpretive approach, now known as ecological biogeography, concentrates on the study of the dispersal of organisms and the mechanisms involved. As an example, freshwater fishes were able to disperse between North and South America and between Asia and western North America using land bridges. Ecological biogeography would study the fish movements, adaptations, and the conditions that facilitated successful invasion. Results of this study would address the dispersal hypothesis; that is, a disjunct taxon would result from dispersal of fishes from one area to a separate one. In this case, both areas are separated to some degree when occupied.

Historical biogeography is the other interpretive approach to understanding the distribution of organisms by exploring past evolutionary and geological relationships associated with disjunct populations of a species, presumably caused by geological events. For example, freshwater fishes cannot tolerate salt water but were dispersed around the globe by the movement of continents, thus separating ancestral fish populations. Such fragmentation of an ancestral stock by tectonic movements occurs due to *vicariance*. The vicariance hypothesis states that disjunctly distributed taxa are produced by separation of formerly contiguous areas occupied by the ancestors of the disjunct group; that is, areas were not separate when occupied.

There are at least two approaches to the study of historical (vicariance) biogeography, phylogenetic biogeography (phylogeography; reviewed by Riddle and Hafner 2004), and cladistic biogeography (reviewed by Humphries and Ebach 2004). Both approaches consider changes in geography and phylogenetic systematics in relation to disjunct stocks. Phylogenetic biogeography uses the geographic distribution and phylogenies of organisms to explore allopatric speciation, and cladistic biogeography looks for relationships among areas by interpreting phylogenetic relationships of species in the areas under consideration. Both of these approaches are discussed by Brooks (2004).

Matthews (1998) provided an excellent discussion of the history and significance of historical biogeography in fishes. In addition, Helfman et al. (2009) provide marine examples, including study of Spanish mackerel vicariance across the Isthmus of Panama, and Echelle et al. (2005) explored phylogenetic relationships of pupfish in an arid region of southwestern North America.

PLEISTOCENE GLACIATION

Pleistocene glaciation was the last global climatic catastrophe that greatly affected the distribution of organisms. It was a worldwide event, but I will restrict comments about its effects to North American freshwater fishes, which have been more intensively studied by others (e.g., Hocutt and Wiley 1986 and references therein; Matthews 1998).

The Pleistocene began about 2 mya following about 1 million years of dropping global temperature, which could be linked to a comet impact. Several major ice sheets advanced during the epoch, including many smaller ones as well. Continental glaciers covered about one-third of North America, and higher mountainous areas experienced snow and ice with depths of 2 km or more. The last ice sheet reached its maximum size about 18,000 ya and retreated by at least 8000 ya. Glacial episodes stored high amounts of frozen water, and the sea level dropped up to 100 m during those times. Falling sea levels allowed for the passage of freshwater fishes (e.g., genera *Catostomous*, *Dallia*, *Lota*) across Berengia and affected marine fishes by widened coastlines and connected islands.

As the ice sheets came and went, fishes retreated and advanced with them. Some coastal areas that were not covered with ice served as refuge for coldwater fishes, while warmwater fishes survived by moving south when possible. During pluvial events and glacial melting, some areas experienced heavy rainfall, and huge lakes formed in the southwestern United States.

Glacial impacts had a profound effect on freshwater fish distributions. Ice sheets covered many rivers, lakes, and ponds, killing all of the fish in them. In areas not directly affected by the ice, colder temperatures prevailed. Coldwater fishes that moved south became isolated in high mountainous areas as they retreated. Warmwater fishes also became isolated to ponds and lakes scoured out by glaciers that filled with water for a time and then froze again. Fishes in the western states were greatly reduced due to isolated and unstable basins and lack of refuge from harsh conditions (Briggs 1986). Fishes that survived were adapted to life in huge rivers and lakes. Fishes in the Mississippi River basin fared better (Smith 1981a; Robison 1986), and fish populations retreated and advanced in northern areas with the glaciers. Even representatives of the ancient Laurasian fauna survived there. However, the landscape and freshwater habitats were profoundly changed, as Pleistocene basins that were filled then emptied (e.g., leaving the Great Basin isolated and also forming the Laurentian Great Lakes).

Pleistocene events had a great effect on North American freshwater fishes, which can be observed by looking at fish faunas in various regions. One study, done by Moyle and Herbold (1987), noted that streams in the eastern United States are dominated by fishes whose adults are small to medium in size (<300 mm standard length), but adult fishes in the western United States are mostly large species. The authors proposed that the Missouri/Mississippi system refugia allowed the persistence and continued speciation of all sizes of fish and allowed reinvasion as glaciers came and went. As a result, eastern streams are dominated by large bass (Centrarchidae), pike, catfish, and suckers. Large cyprinids there are rare, comprising only about 2% of the fish population. On the other hand, harsh conditions, isolated basins, and a lack of refuge from big river and lake conditions during the Pleistocene resulted in selection for a different life history strategy in the western United States. Adverse conditions were selected for large size and longevity. The western fauna is dominated by large salmonids in the northwest and large cyprinids in the southwest. Large cyprinids in the western United States comprise about 41% of the post-Pleistocene fish fauna. I also note that migration is an important component of the large western fishes.

THE FUTURE

Planet Earth has experienced great climatic fluctuations in the past, and it is sure to have more in the future. The last stable period, from the melting of glaciers at the end of the Pleistocene to the present, has allowed fish and human populations to flourish. However, humans are speeding up warming conditions globally and changing the planet in other ways as well. One of these changes is the introduction of nonnative fishes that compete with and prey on native fish communities. It has been predicted that the adverse effect of these invasive species will increase with global warming (Rahel and Olden 2008). As we continue on to chapter 32, we will see how humans have affected fish habitats and populations and speculate on how changing climate may affect fish populations in the future.

ECOLOGICAL CONCEPTS

The first six concepts were presented by Krebs (2008) and are related to global species distribution and richness (diversity), which I have related in turn to zoogeography with a brief explanation:

- *Evolutionary speed hypothesis*—Diversity appears to be favored by a long evolution in favorable conditions. This seems to be the case in the tropics. Note that this hypothesis is the same as the stability-time hypothesis, which we will refer to later.
- *Geographical area hypothesis*—Larger areas likely contain more species in habitats and microclimates; hence, greater diversity would seem likely. This has been discussed by others with respect to island biogeography.
- *Interspecific interactions hypothesis*—Distribution and abundance of a species is affected by interspecific competition. However, predation can reduce intense competition and result in a higher diversity of prey species in a given area. This is difficult to prove in fish.
- *Ambient energy hypothesis*—Favorable climatic conditions of solar radiation, temperature, and water can promote speciation and increased species diversity. The antithesis is that diversity, distribution, and abundance can be reduced in areas with stressful conditions.
- *Productivity hypothesis*—Higher productivity in large enough areas can promote speciation and increased biodiversity. This presumes natural areas rather than monocultures and that abundant resources will reduce competition even though there are overlapping fundamental niches.
- *Disturbance hypothesis*—A moderate degree of habitat disturbance can reduce competition and encourage speciation. This is the "edge effect" and could result from human disturbance or the maintenance of some seral stages (e.g., due to fire in forest systems). In this case, nonequilibrium stages of a system are expected to be more diverse.
- *Dispersal hypothesis of biogeography*—Global distribution of organisms occurred by the dispersal from one area to another, somewhat separated area over land bridges affected by sea levels.
- *Vicariance hypothesis of biogeography*—Global distribution of disjunctly distributed taxa occurs because contiguous areas once occupied by their ancestors were separated by tectonic forces, such as continental drift.
- *First principle of biogeography*—Also known as Buffon's law and attributed to the Compte de Buffon, this recognizes that different species were found in different regions even though the climate was the same.

SUMMARY

Zoogeography of fishes is concerned about global patterns of fish distribution and species richness. Fishes are widely distributed from hot to cold climates, from low to high altitudes, and from surfaces to depths. Fish tolerance to salinity has been used to place fish in at least three groups: the primary division, which includes all salt-intolerant (obligate freshwater) fishes; the secondary division, which includes somewhat salt-tolerant fishes; and the peripheral division, which includes very salt-tolerant fishes (marine). Freshwater fishes also can be cold, cool, or warmwater species, and marine fishes may be polar, temperate, or tropical species. Plate tectonics, specifically continental drift, is responsible for the distribution of many different fishes. Sturgeons, bonytongues, and lungfishes were extant on the supercontinent of Pangaea and were then transported to different places as Pangaea broke up. Later, fishes that originated from Gondwanaland and Laurasia were transported within the southern and northern hemispheres, respectively. Fish distribution also was influenced by land bridges that formed during periods of continental glaciation. As a result, continents have both a mix of early-evolved fishes shared with other continents and fishes that emerged after continents separated, which are endemic to only one continent. Alfred Russell Wallace identified six faunal regions: the Ethiopian (African), Neotropical (South American), Oriental, Palaearctic, Nearctic (North American), and Australian. A comparison of the colder northern continents (Nearctic and Palaearctic regions) with the tropical ones (Neotropical and Ethiopian) demonstrates that cold northern regions have a less diverse fish fauna of only about 2000 species, while the warm tropical regions have a much more diverse fish fauna of about 7000 species. Also, cold, isolated islands of Australia and New Zealand are very poor in species richness, with no freshwater fish remaining in

New Zealand and only four obligate freshwater fishes remaining in Australia. Two different schools of thought emerged for explaining fish distribution: One (ecological biogeography) supported the dispersal hypothesis, which states that fish moved from one location to another via land bridges, and the other (historical biogeography) supported the vicariance hypothesis, which states that fish were transported by the breakup and drift of continents. Both schools are pertinent, and the latter has expanded and subdivided. Another force that has affected the global distribution of fishes is Pleistocene glaciation. The last ice sheet reached a maximum thickness about 18,000 ya and had retreated by about 8000 ya. Falling sea levels allowed freshwater flows over the shallow Berinig straits (Berengia), allowing passage of freshwater fishes. Fishes in North America fared best in the Mississippi River drainage, which did not freeze in southern states, thus providing refuge for reinvasion. On the western coast of North America, cold hardy Pacific salmonids established many coastal populations of anadromous fishes. The Earth has experienced some great fluctuations in climate in the past and will do so in the future. Hopefully, our understanding of past changes in fish distributions will be an aid in predicting the future effects of global climate change.

CHAPTER 9

Lotic Systems: Flowing Water and the Terrestrial Environment

A DROP OF RAIN

Freshwater systems occur in terrestrial environments as a result of precipitation, water storage, and gravity. We can understand how freshwater systems form by following a drop of rain. The drop falls due to gravity, and that force also moves water downhill. Assuming our drop falls on a high mountain, it is most likely to strike the earth. If the soil is porous or dry, moisture may soak into the ground to be stored or evaporated. As water accumulates in porous soil, gravity will pull it deeper until it encounters a nonporous stratum in which it will accumulate, or if slanted, water may emerge as a spring or seep where the stratum is exposed. If the drop falls on a pond or a lake, it will be stored there for a time and will evaporate, seep into the ground to become groundwater, or be transported out of the lake with seasonal flooding.

If there is much rain and low evaporation, our drop would flow downhill in tiny rivulets, coalesce into bigger ones, and so forth, until a small headwater stream is formed. From that stream, it is possible for a raindrop, thus entrained, to find its way from the mountain top to the sea. This encapsulates the "lotic" or flowing water ecosystem. The net water flow is unidirectional (downhill) from source to end. It also is our framework for organizing and discussing aquatic systems.

FLOWING WATER

Water flow is the dominant feature of all lotic ecosystems. Water flows to the lowest point available and takes the path with the least resistance. The flow velocity (e.g., in centimeters per second) varies according to the amount of water being transported, the gradient (i.e., the drop in elevation per distance of stream travel), the cross-sectional size and configuration of the stream channel, and the type of substrate (fine particles (smooth) vs. large particles (rough)). Needless to say, water is a dense medium, and it can acquire a great amount of energy as it is transported down a gradient. Higher gradient produces faster flow and increases the ability of water to entrain and transport sediments.

At first, a mountain stream will transport inorganic particles (sediments) of various sizes depending on the flow velocity. A flow of about 10 cm/s will transport silt, 25 cm/s will entrain sand, and 50 cm/s will carry gravel along the bottom (Nielson 1950). As the gradient becomes less, or the volume of water decreases, the flow rate drops and the stream can no longer transport larger-size particles. As a result of varying flow rates, particles are not distributed in the same place, resulting in a differential distribution of substrate and heterogeneous stream bottoms. Also, the evolution of streams results in the channel lengthening by meander as it slows, and more material is transported. Thus, streams increase in length as they age.

Streams may entrain and move particles at one flow rate and deposit particles at another. With this knowledge, stream systems may be divided longitudinally into three zones relative to sediment transport: An upper drainage zone that is erosional (i.e., sediment loss), a middle zone of sediment transfer, and a lower zone of sediment deposition. Distribution of the stream fauna also is related to these three zones.

A stream will either join other streams or receive tributary input as it travels. In the process, streams grow larger and have different characteristics. Thus, it is convenient to rank them (Strahler 1957): very small tributaries with no branches are first-order streams, a junction of two first-order streams produces a second-order stream, the junction of two second-order streams produces a third-order stream, and so forth—a method developed by R. E. Horton (Leopold 1994). However, with stream order, streams increase in size exponentially (Figure 9.1). For example, the largest stream in the United States is the Mississippi River, with an order of 10. Stream order gives a rough approximation of discharge but may be misleading when comparing streams from different regions (e.g., arid vs. mesic). Hughes and Omernik (1981) compared six fifth-order streams and pointed out the great variation in size of basin and mean annual discharge. These authors suggested other alternatives to the use of stream order, especially mean annual discharge per unit of drainage basin area.

In my experience, the concept of stream order is most useful for characterizing and comparing smaller streams within a basin. Also, stream order may be used in a general way as an indicator of the size of larger streams. However, for larger streams and rivers, there are better ways to characterize them: their average annual discharge, size of the drainage basin, annual discharge per size of the basin, and the length of the stream. Three basic attributes of streams are presented in Table 9.1, which shows that in the largest rivers of the world, stream discharge is closely ranked with size of the drainage basin.

As a stream develops, litter from vegetation will be entrained, providing an input of detritus (particulate organic matter produced from once living organisms), and the solvent properties of water will dissolve compounds from soil and rocks. Dissolved compounds are transported downstream and aid in developing populations of instream producers, such as various species of algae. A stream community of invertebrate organisms will use the plentiful supply of oxygen in the shallow waters to decompose the detritus into nutrients, which include dissolved organic compounds (carbohydrates, urea, amino acids, fatty acids, etc.) and inorganic compounds needed to support life, such

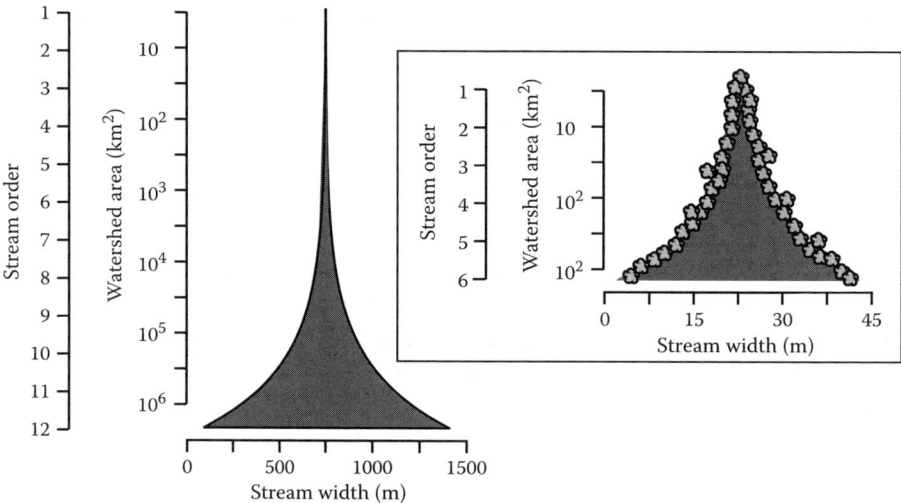

Figure 9.1 Stream order and size of basin. (Courtesy of McCutchan, J. H., Jr.)

Table 9.1 Ranking of Selected Large Rivers of the World

River	Basin[a]	Length	Discharge
Amazon	1	3	1
Congo	2	8	2
Mississippi	3	2	9
Nile	5	1	36
Yangtze	9	5	3
St. Lawrence	15	15	17
Columbia	23	28	18

Source: Leopold, L. B., *A View of the River*, Harvard University Press, Cambridge, MA, 1994; Sheehan, R. J., and Rasmussen, J. L., *Inland Fisheries Management in North America*. 2nd Edition, 529–559, American Fisheries Society, Bethesda, MD, 1999. With permission.

[a] Basin is total area of drainage.

as nitrates and phosphates. Finally, larger organisms such as fish will appear to consume the algae and stream invertebrates. This outlines the basic structure of a small lotic system.

Streams change as they flow from high elevations to the sea, and there have been several ideas about how to characterize longitudinal zones of river systems, including faunal changes associated with differences in elevation. At high elevations and in the arctic, headwaters (i.e., upper reaches) may originate from melting glaciers (kryal zone) or springs (krenal zone). In these systems, water temperatures are at or near the freezing point (2°C–4°C), and there are no fish or aquatic plants (Ward 1994). Proceeding downstream, flows from the rhithral reach originate mostly from precipitation and melting snow. This reach (the rhithron) extends downstream to the end of the trout zone. It is delineated by mean monthly water temperatures that range from >4°C to 20°C, and sometimes it is further divided into an additional three zones. Finally, the remaining downstream portion of the stream is called the potamnal reach (or the "potamon," which also can be subdivided into three zones). This reach is characterized by mean monthly water temperatures that rise above 20°C. In most locations, a river can be adequately defined by the rhithron and potamon alone. The rhithron is cold, fast, and coarse in substrate; its energy comes from outside the stream, and its food webs are mostly benthic. On the other hand, the potamon is warmer, slower, and finer in substrate; its energy is mostly produced in the stream, and its food webs may be mostly planktonic.

CHARACTERISTICS OF STREAMS

Water Is (Almost) Always Moving

As we have discussed, water in a stream moves downhill with a speed related to the stream gradient. However, flow also is affected by the configuration of the channel, and seasonal or daily flow increases can exceed channel capacity. When that occurs, the stream moves laterally into the floodplain as well as longitudinally down the channel. In larger and slow-moving rivers, the water surface area can be large enough to be affected by wind, which in some cases can move surface water upstream. Also, higher flows in one tributary than in another (perhaps due to dam releases) can also move river water upstream in tributaries—an anomaly that I experienced while rafting in the Yampa River of Colorado many years ago!

Streams vary in flow characteristics due to their water supply sources, which may only maintain surface flow seasonally. In this case, it is customary to identify the type of stream by its ability to maintain flow during the year. *Permanent* streams flow all year because they receive water mostly from subsurface reservoirs. *Intermittent* streams flow seasonally because they receive mainly surface water inflows that depend on precipitation—during wet years, these can be permanent. *Interrupted* streams may flow all year, but sections may flow underground to appear later as surface flow.

Channel Complexity

Natural stream channels are not straight because of differences in flow resistance offered by the substrate and within channel features, which can result in differential erosion and deposition. Silty shorelines are easily eroded into a wide channel. As hard strata or large substrates such as boulder or bedrock are encountered, eddies and rapids can occur. Large obstructions such as boulders cause an interruption in laminar flow, forming eddy conditions and providing local conditions of little or no flow. All of these influence the degree of sinuosity (Wesche and Isaak 1999): fast-flowing streams are relatively straight, but as the gradient decreases, the degree of sinuosity increases so that the stream increasingly meanders across the landscape (Figure 9.2). With increasing sinuosity comes an increase in habitat complexity, especially with braided channels, which are very interactive with the floodplain (Figure 9.3).

A stream can be viewed as a complex of littoral habitats, which are very productive because light can penetrate through the water column all the way to the bottom. These within-channel features include: *runs* where the substrate is some distance from the surface, *pools* that are deep and have wide runs and occur below constrictions in the channel, *riffles* that are shallow runs constricted by depth and bottom substrates that ripple the water surface, *eddies* that are areas of recirculation around an obstruction on point bars, and *rapids* that occur in very fast turbulent water as the current sweeps over very large boulders or through narrow channels. However, many other channel types are recognized (Armantrout 1998). Associated with channels are side channels that can be quite different. Some side channels may be permanent, but streams with a natural hydrograph and spring flooding usually have a dynamic condition in which side channels during low-flow periods may be partially dewatered and serve as backwaters.

Riparian and Floodplain Features

Alongshore (riparian) vegetation is typical of streams and may be thought of as a buffer between the aquatic and terrestrial systems. Depending on the river system and degree of flooding, the

Figure 9.2 Sinuosity shown by a meandering stream. (Courtesy of U.S. Geological Survey (USGS).)

LOTIC SYSTEMS: FLOWING WATER AND THE TERRESTRIAL ENVIRONMENT

Figure 9.3 Sinuosity resulting in a braided stream. (Courtesy of USGS.)

vegetation may be divided into various zones. Riparian vegetation aids the stream system by providing cover, bank stabilization, food sources (organic matter and terrestrial organisms), and insulation from solar radiation.

A stream ecosystem also includes its floodplain. Backwaters, alongshore embayments, flooded bottoms, and associated seepage wetlands may be connected with a stream, but they are considered ephemeral because they are not usually flooded each year. Some permanently flooded areas may include oxbows, sloughs, and swamps.

Hydrology

Water dominates the surface of the Earth, and it occurs in a global water cycle. As water evaporates from the surface of oceans through solar energy, it is transported with prevailing wind patterns. When moisture-laden air encounters continents, it is forced to rise in elevation, which cools the air and reduces its capacity to hold water. This results in an excess of precipitation on the continents: about 23% of world precipitation falls on continents, but only 16% is lost due to evapotranspiration (Ricklefs 1990). Continental runoff via streams and groundwater returns enough water back to oceans to make up its water deficit. We are lucky to have the present configuration of the continents because the distribution of freshwater could be much more limited if continents were united, as in Pangaea.

Lotic systems are the primary agent for transfer of surface water from one place to another, within continents and from continents to oceans. In this case, stream systems connect reservoirs of water that may be temporarily stored as snow or ice, in the soil, in underground aquifers, and in ponds, lakes, and streams. Water also is stored in the atmosphere and in organisms. Obviously, this water is made available to streams in different ways and at different times during the day or year. Thus, a hydrologic water cycle (i.e., the amount of water that flows through a system over time) will vary depending on the size and soil characteristics of the drainage basin, amount of precipitation,

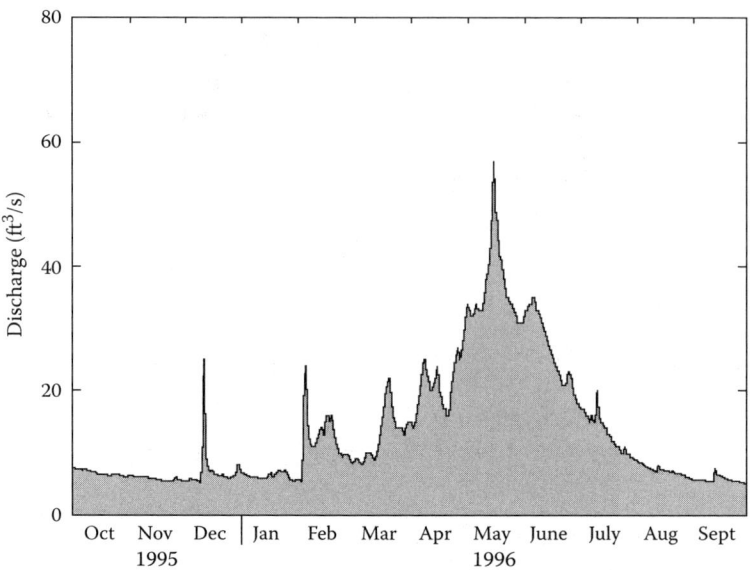

Figure 9.4 An annual water distribution hydrograph for a snowfed river, showing daily mean discharge for Incline Creek during the 1996 water year, representative of streams in the Lake Tahoe Basin. (Courtesy of USGS.)

and so on. It is the seasonal variation of water input into streams that is of great consequence in most regions of the world, and this can be illustrated in an annual stream flow hydrograph. As an example, consider the hydrograph of a north temperate (coldwater) stream. An annual hydrograph would have a definite pattern produced by water storage as snow in winter to be released as snowmelt during warmer months. When the surface supply of snow is exhausted, the stream is fed by soil moisture. An example of snowmelt hydrograph is provided in Figure 9.4, which depicts a typical water year from October 1 to September 30. A tropical stream might look very different from this hydrograph due to more or less constant rainfall.

Trophic Status and Energy Transport

Streams are predominately heterotrophic (other + food) and receive most organic matter from the terrestrial system in the form of litter (leaves and woody debris) and detritus (particulate matter and organisms). However, some streams have a considerable amount of autotrophic (self + food) production within the channel from algae. This production is especially important in mountain streams that flow through forests of conifers (McCutchan and Lewis 2002). The conifer needles are very difficult to decompose, and the stream biota may depend on algae as food. In either case, there is a unidirectional and biologically mediated cycling of nutrients that occurs along the length of the stream. In this process, litter and detritus are fragmented, leached, and finally decomposed to nutrients, which are then used by stream biota for growth. This cycling process is affected by river flow, and litter, detritus, and nutrients are transported downstream. Movement of a cycle in distance would then become a spiral, and the stream process of carbon and nutrient cycling is referred to as *resource spiraling*.

Spiraling of nutrients and carbon has at least two effects on stream ecosystems, as discussed by Elwood et al. (1983): how energy is supplied in time and space, and the way in which stream organisms use energy supplies. It was once believed that streams were not nutrient limited due to the transport of resources through the system. However, this is not true, and more recent studies have investigated the role of nutrient spiraling in streams.

Because decomposition occurs almost entirely upon or in the substrate, the trophic structure of the stream depends heavily on the efficiency of the benthos. One area of fairly recent study is saturated sediments associated with the flowing stream and on its bottom and sides known as the hyporheic (below + flow) zone. This zone is an important part of the stream and connects the stream with groundwater. Riparian vegetation is rooted in the hyporheic zone on its sides, and benthic organisms live in and upon it. This zone can be of great importance for supporting floodplain habitats needed for fish-rearing areas. It also can support an abundant and diverse fauna of microbes and invertebrates known as the *hyporheos*. The hyporheos is an important part of the trophic structure of streams and of the biogeochemical processes (e.g., see Edwards 1998).

ECOLOGICAL CONCEPTS

- *River continuum concept (RCC) (Vannote et al. 1980)*—Although different parts of a stream have different attributes, stream systems are connected along their entire length, where physical and biological processes occur in a predictable continuum of stream order and size (Table 9.2). In this concept, the upper regions of river systems originate in small coldwater streams (order 1–3) in forested regions. Tree cover shades the stream and also provides inputs of litter and woody debris. This organic matter is shredded and decomposed by benthic invertebrates, microbes, and fungi to provide most of the energy available to support the stream system. As the stream progresses down to a medium size (order 4–6), there is an increase in light and temperature and an increase in autotrophy due to the growth of algae. Here, aquatic species are adapted to warmer water and are composed mostly of collectors and grazers. Further downstream, as the stream increases to a large size (order 6–12), the system is supplied mostly with fine organic matter originating from headwater areas, and it reverts to heterotrophy. Consumers are mostly benthic organisms (Figure 9.5). Although lotic sys-

Table 9.2 Upper, Middle, and Lower Sections of Rivers Related to River Continuum Concept

Features	River Sections		
	Upper	Middle	Lower
P/R	<1	>1	<1
Food source	Imported	*In situ*	Transport
Nutrients	Low	High	Low
Attached algae	Sparse	Abundant	Sparse
Plankton	Absent	Absent	Present
Litter	Much	Some	Little
Inv. Shredders	Many	Rare	Absent
Inv. Collectors	Many	Many	Dominant
Inv. Grazers	Few	Many	Absent
Inv. Predators	Few	Few	Few
Fish food	Inverts.	Fish/inverts.	Plank./benth.
Biodiversity	Low	High	Low

Source: Adapted from Vannote, R. L. et al., *Canadian Journal of Fisheries and Aquatic Sciences*, 37, 130–137, 1980; Ward, J. V., *The Rivers Handbook: Hydrological and Ecological Principles. Volume One*, 493–510, Blackwell Scientific Publications, London, 1992.

Note: P/R, photosynthesis/respiration ratio; Inv., invertibrate; Inverts., invertebrates; plank., plankton; benth., benthos.

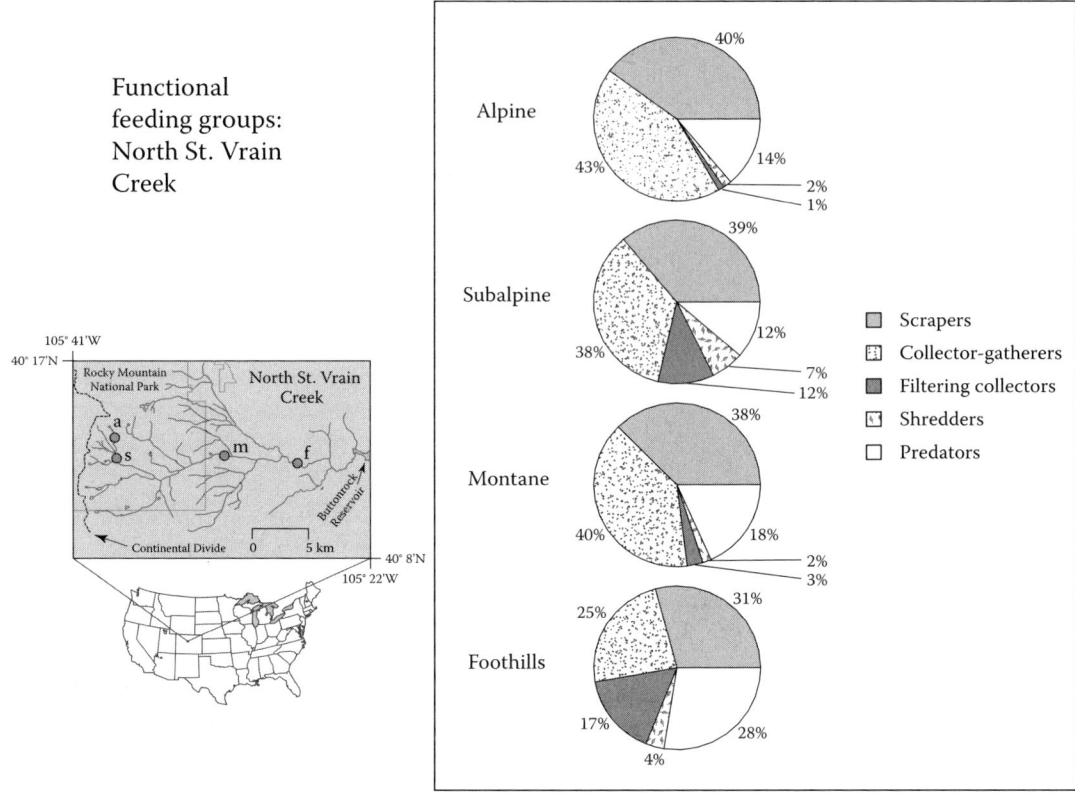

Figure 9.5 RCC in the St. Vrain River, Colorado. a, alpine; s, subalpine; m, montane; f, foothills sampling locations. (Courtesy of McCutchan, J. H., Jr.)

tems are extremely diverse and it is difficult to make generalizations about the trophic structure, the RCC serves as an effective paradigm to evaluate energy pathways affecting autotrophy/heterotrophy and how organic inputs affect stream biota (Allan 1995).

- *Flood pulse concept (Junk et al. 1989)*—Streams have a lateral component that waxes and wanes with a seasonal flow regime. This concept (Figure 9.6) considers seasonal flooding to be a predictable event for stream organisms, which are adapted to it. In this case, the floodplain is considered the most important ecological feature of the river system. The floodplain provides nutrients, sediments, nursery areas, and forest products into the channel while supporting riparian forests and a variety of wetland habitats. Fish ecologists have addressed these issues using landscape perspectives (e.g., Schlosser 1991). A drawback of most attempts to conceptualize stream systems is difficulty in application to all streams (Johnson et al. 1995). For example, the RCC was developed from pristine (undisturbed) deciduous forest streams; thus, adjustments have to be made for systems that depart from this model, such as exposed coldwater streams in coniferous forests, or for streams that have no cold headwaters (e.g., see Ward 1992). Although few pristine rivers survive in the world, these concepts have utility in understanding how natural lotic systems function and for evaluating conditions and formulating mitigating measures.
- *Resource spiraling concept (Webster 1975)*—This concept concerns the use, cycling, and transport of carbon and nutrients in a stream ecosystem. The role of cycling with stream transport was studied first, perhaps due to the belief that streams were not nutrient limited. Of course, this may not be the case, and more recently, *nutrient spiraling* emerged as an important field of study. Stream organisms benefit from being constantly washed with food, nutrients, and oxygen, while feces, ammonia, and other desirable by-products of metabolism are washed downstream.

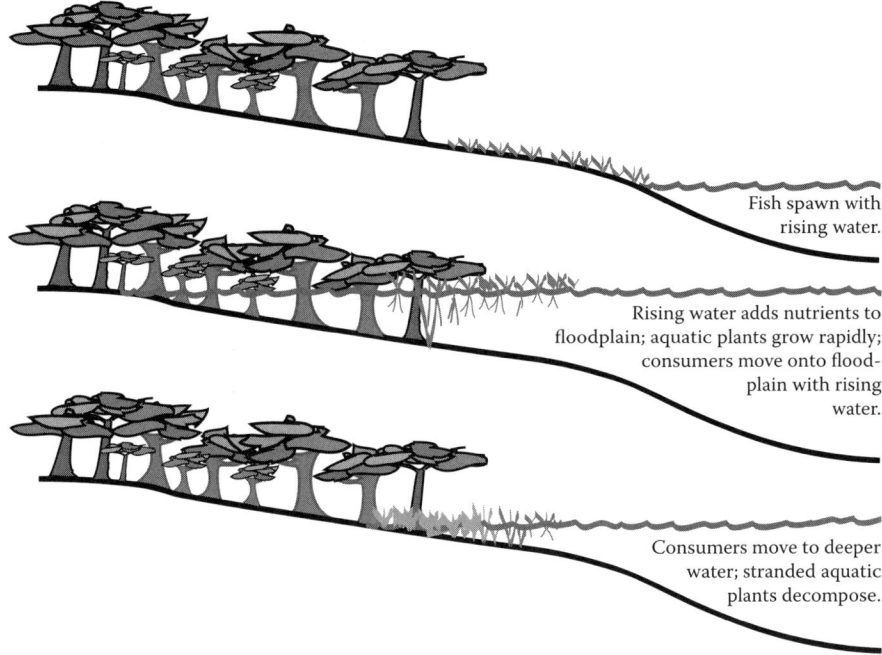

Figure 9.6 Flood pulse concept. (Courtesy of McCutchan J. H., Jr.)

- *Serial discontinuity concept (Ward and Stanford 1983)*—Most of the river systems in the United States have been subjected to flow regulation of one type or another, usually by construction of dams and the impounding of reservoirs. This concept provides a better understanding of these regulated lotic systems, which are treated as large experiments where the continuum and spiraling processes have been affected in some way.
- *Ecosystem simplification (Doppelt et al. 1993)*—This concept refers to loss of ecosystem diversity and complexity due to human effects that reduce the ability of a stream to repair itself. Examples of ecosystem simplification includes alteration of flow regimens, habitat destruction by damming, flooding, or draining, degrading water quality, overharvesting of native species, and introduction of nonnative species.

SUMMARY

Lotic systems are streams of all sizes and are characterized by flowing water at least some of the time. Streams can be permanent (flow all year), intermittent (flow part of the year), or interrupted (sections flow underground). The rate of flow and the amount of water transported depend on stream gradient and basin hydrology. As water flows, it also transports dissolved and suspended material with it. Typically, streams can be divided into an upper drainage zone that erodes sediments, a middle zone that mostly transfers sediments, and a lower zone that deposits sediments. Streams also transport organic matter in the form of leaves and detritus (broken up pieces of leaves, etc.). It also is convenient to rank streams to approximate size. In this case, very small tributaries with no branches are first-order streams; the junction of two of the first-order streams then produces a second-order stream, and so forth. A stream also can be viewed as a complex of littoral habitats, runs, pools, riffles, eddies, and rapids. A stream includes all or most of the following: its floodplain, alongshore embayments, flooded bottoms, seepage wetlands, oxbows, sloughs, and swamps. Streams are heterotrophic systems and thus receive organic matter from the drainage basin in the form of litter and detritus. However, streams also have autotrophic production usually from algae within its banks. As the organic matter is cycled through stream organisms, it also is being transported downstream, resulting in a nutrient spiral. Associated with the stream bottom and involved with decomposition is the hyporheic zone, which connects the flowing stream with groundwater. Some concepts about streams include the RCC, in which streams are viewed as a predictable continuum of physical and biological processes; the flood pulse concept, which considers the floodplain and riparian zone to be important features; the serial discontinuity concept, which incorporates dams and reservoirs into the river continuum; and the ecosystem simplification concept, in which stream diversity and complexity are degraded by human activity, such as channelization.

Further reading: Hynes 1970.

CHAPTER **10**

Coldwater Streams

STRUCTURE AND FUNCTION

A headwater stream is the origin of most rivers. These streams are small, first-order streams that have few or no tributary, have incised channels that are eroding, and are found to be associated with seeps or springs (Armantrout 1998). In higher elevations, headwaters are very cold mountain streams with fast flows and relatively straight channels. The channels are incised in mostly coarse inorganic soils, and the flowing water is clear. Coldwater streams are usually small ones of stream order 1–3 and greatly affected by conditions of the terrestrial environment, including daily fluctuations in ambient temperature. Due to higher elevation, the water temperature is cold and may not exceed an average annual high of 15°C, with a maximum high temperature that does not exceed 20°C. This stream reach also is known as the *rhithron*: a reach of low temperature, fast and turbulent flow, and substrates of gravel, cobble, or boulder (Armantrout 1998).

Winters are long in headwater streams, and conditions can be especially harsh due to ice. The more stable streams and deeper reaches can be temperature-stratified and covered with ice for long periods. However, in streams of highly variable flow, ice may form throughout the water column (known as frazil ice), and surface ice may be broken up and reformed many times during the season. As a result, broken pieces of surface ice and frazil can partially or even completely block steam flow, causing lateral flooding, erosion, and filling, and fast flowing, turbulent runs. In addition, riffles can slow and catch frazil, resulting in the formation of anchor ice and obstruction. All of these changes can directly or indirectly affect stream biota, including fishes.

However, it is important to understand that fish populations will move in response to undesirable conditions induced by ice or temperature, and as the seasonal high temperatures develop, trout will be concentrated in upper (cooler) portions of a drainage. Coolwater and perhaps even warmwater fish populations also will expand upstream into the former trout habitat. Temperature effects on fishes were discussed by Winger (1981), and temperature may be a lethal, controlling, and limiting factor for fishes (as defined in Fry's Paradigm—see chapter 21).

At extremely high elevations, coldwater streams will form above the tree line (above the 300-m mean sea level or 10,000 ft above mean sea level), but most streams will soon descend within a forest. Tree cover is an important component of a headwater stream ecosystem because the cover reduces the duration and extent of temperature extremes and furnishes the stream with a source of allochthonous (elsewhere + earth + full of) organic matter in the form of leaf litter and woody debris. Also, with increasing vegetation comes an input of forest invertebrates that can serve as food for fishes.

In high elevations or exposed situations, the coldwater stream may be autotrophic, obtaining autochthonous (self + earth + full of) organic matter from algae attached to within-channel cobbles and boulders. However, this soon changes as the stream enters a wooded area and receives inputs of litter from runoff and wind, primarily large-sized material (>1 mm) that is known as coarse particulate organic matter (CPOM). It is this input of allochthonous CPOM that fuels the stream biota and enables the stream to becomes heterotrophic (photosynthesis/respiration ratio [P/R] <1.0) as the stream invertebrates feed on it.

Litter and woody debris enter into a stream from the drainage basin, and it is incorporated into a benthic food web (Figure 10.1). This material is caught up in the substrate or entrained in eddies and become water soaked. Some of this litter, such as conifer needles and woody debris, are difficult to decompose. However, decomposition of deciduous leaves occurs much faster. Soluble organic matter (dissolved organic matter) is leached out in a few days, starting the process of decomposition and leaving a softened leaf. The leaf is then colonized by bacteria and fungi that begin to decompose it further. Soon, these small organisms form a rich layer of protoplasm and associated detritus (a biofilm) on the surface of the leaf. In time, larger aquatic invertebrates attack the leaf by biting, gouging, boring, or otherwise shredding it into smaller particles (<1 mm) known as fine particulate organic matter (FPOM). The invertebrates benefit from the organic matter in the leaf, but especially from the rich layer of microbes and FPOM that coat the leaf. As these shredders, such as caddis flies (Order Tricoptera) and stone fly (Plecoptera) nymphs, do their work, some of the fragments along with the feces they produce enter the stream flow as drift. Further downstream, the FPOM is sought after by collectors. These include filterers such as blackfly (Diptera) larvae and net spinning caddisflies, and gatherers such as midge (Diptera: Chironomidae) larvae.

In addition to the allochthonous food source just mentioned, there also is another trophic web in the stream that utilizes the autochthonous organic matter. Scrapers, including beetle larvae, and mobile caddis flies feed on the bacteria and algae that occur on cobbles and boulders. In lower portions of streams, there may be native suckers that graze on attached algae and detritus.

The stream fauna includes predators as well as herbivores and detritivores. Stream predators eat a variety of prey items, mostly aquatic and terrestrial invertebrates. Predatory stream invertebrates include larger stone flies, and the large, fearsome dobsonfly (hellgrammites Megaloptera) larvae. Larger aquatic predators include sculpins and trout, which in turn are fed upon by birds and mammals.

Aquatic insects are not effective swimmers in fast streams, they have attachment organs, such as hooks, or suckers, and they avoid the current by seeking cracks and eddies or hiding under rocks. To avoid predation, many are disguised, flattened, or wear protective armor. Some species such as

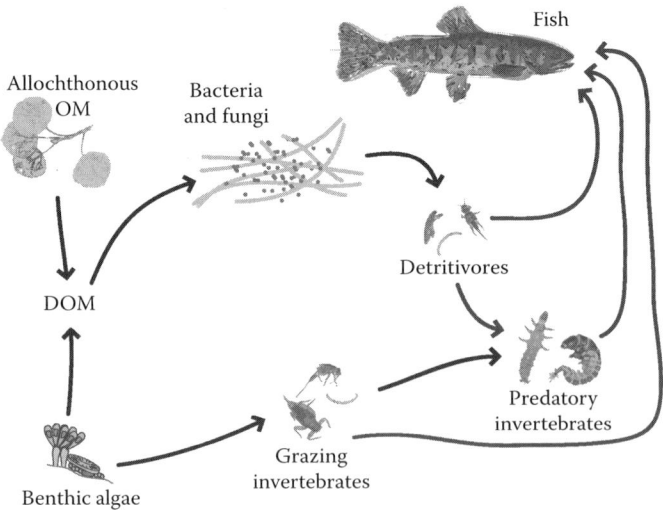

Figure 10.1 A benthic food web. (Courtesy of McCutchan, J. H., Jr.)

blackflies glue themselves to rocks. When insects move to find a better patch to feed upon, they use the current to drift to new areas. They are adapted to move at night when fish cannot see them.

COLDWATER FISHES

North American trouts are familiar residents of headwater streams that are derived from cold-hardy marine ancestors. Pacific and Atlantic salmon separated from a common ancestor in the Miocene (perhaps 15–20 million years ago (mya)) (Behnke 2002), producing *Oncorhynchus* in the Pacific and *Salmo* in the North Atlantic region. Later, near end Pliocene (about 2 mya), rainbow and cutthroat trouts emerged. As coldwater fish, trout prefer optimum water temperatures of about 4°C–15°C (Armantrout 1998). Also, some trout and all salmon are migratory (anadromous) between marine and freshwater environments. The following points are pertinent: Salmon invaded and exploited unoccupied habitats that were left after glacial retreat at the close of the Pleistocene. By ascending to the headwaters of coldwater streams, they reproduced in habitats that could not support adults all year. Also, small coldwater streams are usually devoid of other predaceous fishes and are good habitats for spawning and rearing small salmon or trout. In streams that are connected to the ocean, salmonids are mostly anadromous. In streams now isolated from oceans, they will adapt to cooler regions and become resident trouts.

High elevation (first-order) coldwater streams in North America may not support fish because of the small size of the stream and insufficient habitat and food. However, when they do support fish, they are always occupied by trout and are known as "trout streams." Trout eggs have ample quantities of yolk, and the young trout are large enough to consume aquatic insects upon hatching, which are the main food source of trout during most of the year. However, the trout are adaptable, and as they feed on drifting organisms, they may consume as much as half of their diet in the summer by eating terrestrial insects (reviewed by Hynes 1970).

As the headwater stream becomes larger and warmer, it also may support other fish species, depending on the location. Sculpins and mountain suckers occur in coolwater streams, while dace, redside shiners, creek cub, and darters appear in warmer, higher-order streams. Two of the largest nonsalmonid fish that occur in trout streams are the sculpins and mountain suckers. Sculpins (cottids: fish of marine ancestors [Figure 10.2]) are cold tolerant and most likely to co-occur with trout in cooler stream sections. Unlike trout, which are sight-feeding predators, sculpins are benthic oriented and feed by seeking out prey among bottom substrates. Mountain suckers (Ictalurids: primary few fishes) have cartilaginous scrapers as mouth parts and feed by removing attached algae and invertebrates off cobbles and boulders.

The word "trout" should be used only as a generic term to indicate a freshwater member of the salmonid family that occurs in freshwater and looks like ... a trout! There are three different groups (genera) of "trouts" native to the United States: char (*Salvelinus*; Arctic char and brook, lake, and Dolly Varden trouts), Atlantic salmon (*Salmo*; closely related to the European brown trout), and

Figure 10.2 A coldwater nonsalmonid: slimy sculpin *Cottus cognatus*. (Courtesy of Edmonson, E., and Chrisp, H., New York State Department of Environmental Conservation.)

Pacific salmon (*Oncorhyncus*; chinook, coho, pink, chum, and sockeye salmon, and also rainbow, redband, golden, and 14 subspecies of cutthroat trouts) (Behnke 2002). I will identify the four most abundant and well known of the trouts present in the United States:

- Brook trout (Figure 10.3), native to eastern United States, are cold-hardy char, and it would be best to call them brook "char" (Behnke 2002). It is a dark fish with very small scales, light-colored wormlike marks on its body, and white edges to the ventral fins. This fish is closely related to the Arctic char, which has the most northern distribution than any freshwater fish (80°N; Ellesmere Island, Canada). Char are fall spawners, and their young emerge in early spring to take full advantage of short cold summers.
- Brown trout (Figure 10.4), native to Europe, have been transplanted throughout the United States. Its close relative, the Atlantic salmon, is native to the northeastern United States and western Europe. Brown trout have a light body and a head with dark spots, some of which may have halos. The spots are not present on the caudal fin, and there are red spots on the side. This hardy fish spawns in the fall and tolerates warmer conditions than most trout.
- Rainbow (Figure 10.5) and cutthroat trouts are freshwater offshoots from the Pacific salmon and can be hard to tell apart. Both have dark spots on a light body, including the caudal fin. Larger adult rainbow trout have a wide red band down the side, and a yellow or orange cutthroat mark beneath the jaws. The cutthroat has an intense bright red cutthroat mark, a less distinct side band, and larger spots that may be mostly on the posterior of the body. Rainbows and cutthroats are spring spawners.

Figure 10.3 Brook trout (char) *Salvelinus fontinalis*. (Courtesy of Raver, D. and USFWS, National Conservation Training Center (NCTC).)

Figure 10.4 Brown trout (an exotic) *Salmo trutta*. (Courtesy of Raver, D., and USFWS, NCTC.)

Figure 10.5 Rainbow trout *Oncorhyncus mykiss*. (Courtesy of Raver, D., and USFWS, NCTC.)

CONSTRAINTS ON TROUT

Trout are cold-tolerant fishes adapted for life in the harsh environment of headwater streams (Table 10.1). It is probable that food resources may be limited during some periods. However, Behnke (1992) has long contended that trout abundance in high-gradient streams may be limited by their habitat requirements more than food. He pointed out that trout require four kinds of habitat according to their life history needs:

- Egg survival requires suitable spawning gravels. With accelerated erosion, spawning substrates can be smothered with fines, leaving only boulders and rubble, which are not acceptable.
- Young trout require protective cover and low-velocity water during the first few months after hatching; otherwise, they may be swept downstream during high-flow events.
- Adult habitats may be the biggest problem with habitat limitation. Adults require deep water of slow velocity, with protective cover for resting, but they also have to be near fast water for feeding.
- Overwintering habitats historically were provided by beaver ponds and similar deep areas with protective cover such as large boulders in deep pools. In streams with no overwintering capacity, trout may migrate to larger streams of lower elevation.

One of the most difficult periods for coldwater stream fishes is the long winter and associated adverse effects of ice. Brown et al. (2011) presented a compelling review of the effects of ice on stream fishes. Essentially, the formation, buildup, and breakdown of surface ice, frazil ice, and anchor ice can change fish habitat, resulting in increased fish movements. This expenditure of energy during the least productive time of year, compounded by the stress of dealing with the ice and finding suitable habitat, is a primary source of overwintering mortality. Anthropogenic changes in streams have exacerbated this problem, especially by reducing the amount of deeper water.

Table 10.1 Characteristics of Coldwater Streams

General: Small aquatic systems greatly influenced by the surrounding terrestrial environment (i.e., drainage basin/watershed). Classified as headwater streams (1°–3°) but also may include larger streams of high elevation (<6°).

Structure: Shaped by amount, duration, timing, extent of runoff, stream gradient, and substrate types.

Conditions: Stressful due to the following: cold water temperatures (<20°C), erosional substrate, fluctuating water levels, and little buffering due to mostly inorganic soils. Shallow streams exacerbate temperature extremes. Overwintering mortality of stream organisms can exceed 50%.

Habitats: Narrow streams with short pool–riffle–pool sequences. Riffles: sites of primary production. Pools: decomposition sites. Eddies: recirculation areas that trap food and provide habitat for larger fishes. Trout need four habitats: spawning, rearing, adult, and overwintering. Deeper habitats may be limiting.

Function: A mostly heterotrophic system, with some autotrophy. Major energy source: allochthonous organic matter. Energy flow: spirals. Food webs short: linked to processing organic matter, grazing on algae. Autotrophs: vascular plants (litterfall, blowdown) and benthic algae. Consumers: 1° = herbivorous insects and trout. 2° = predaceous insects and fish (sculpins and trout). Detritivores = microorganisms, insects, and suckers.

Fauna: Fishes: mostly salmonids—freshwater and anadromous forms, and sculpins. Suckers, dace, and minnows in summer from coolwater zones. Historically native trouts are top predators. Fish diversity is very low, which may be as few as one or two species in headwaters. Stream invertebrates: aquatic insects. Flora: mostly benthic algae and riparian vegetation. As stream intergrades into warmer downstream areas, fish diversity increases.

Problems: Fragile systems easily degraded by human activities. Water quality may be degraded by elevating temperature, lowering dissolved oxygen (DO), siltation, contaminants, and nutrient loading (sewage). Channelization, impoundment, and diversions affect water flow and runoff characteristics. Removal of beaver dams and clearing snags increases downcutting of bed and can dry up stream channels. Introductions of nonnative trouts and other coolwater fishes can adversely affect native cutthroat trouts and suckers through predation, competition, and hybridization.

Coldwater streams are remote, small, fragile systems that are vulnerable to habitat alteration and other effects caused by humans. Many trout streams have been damaged, and trout populations have been lost or seriously depleted by dewatering, damming, erosion and cutting down of stream channels, destruction of beaver dams, siltation due to road building and forestry practices, exposure and overheating due to removal of the tree canopy, deterioration due to livestock grazing, and so on. Fortunately, habitat renovation programs as simple as fencing livestock from riparian zones can be effective mitigating measures. Another serious problem has been indiscriminate overfishing, which very early on leads to population depletion. Anglers then demand stocking efforts, which have resulted in long-term problems due to biological factors as well, mainly because the native brook and cutthroat trouts are more vulnerable to angling and are soon fished out, leaving the nonnative hatchery fish (Table 10.2). Behnke (2002) ranked the fishing vulnerability of trout from highest to lowest as cutthroat, brook, rainbow, and brown trout, and presented catch data from a catch/release study in Wyoming. In that study, eight rainbow trout were captured for every 1.2 brown trout, indicating that brown trout were six times more difficult to capture.

Competition and predation from primary freshwater fishes are threats to trout populations, especially in transition zones between cold- and warmwater stream reaches, which may be called coolwater zones (water temperatures 10°C–21°C) (Armantrout 1998). But in coldwater situations, trout have an advantage. For example, Reeves et al. (1987) found that redside shiners declined in a controlled stream channel occupied by juvenile steelhead at water temperatures of 12°C–15°C. The situation was much different at warmer temperatures; at 18°C–22°C, steelhead production declined by 54% in the presence of redside shiners. Small species of primary fishes (e.g., creek chub, shiners) could pose a threat to trout eggs and fry, but the trout reproduce at even colder temperatures. Rainbow and cutthroat trouts have been observed spawning at 6°C–9°C, and the cold-adapted greenback cutthroat eggs incubated at 8°C can hatch in only 32 days (Behnke 1992). These temperatures would certainly preclude egg or larva predation by warmwater fishes. Warmwater predators are more likely to affect migrating salmonids (such as salmon smolts) in large and usually altered river systems.

Trouts are adversely affected when nonnative trouts are introduced into their stream. Pearsons (2008) provides a good discussion of the problems associated with the introduction of hatchery fish, which include competition, predation, hybridization, and introduction of diseases. Whirling disease, *Myxobolus cerebralis*, is a devastating disease that affects young trout, especially those derived from Pacific salmon, resulting in disfigurement and usually death. It is a European disease, so brown trout are tolerant of it, while the native North American cutthroat and rainbow trouts are not.

Table 10.2 Vulnerability of Salmonids to Angling

Species	Fishing Effort Expended to Catch Half of the Population (h)	Relative Ease of Capture
Lake trout	500	Very difficult
Brown trout	400	Difficult
Rainbow trout	280	Moderate
Arctic grayling	250	Moderate
Cutthroat trout	20	Easy
Brook trout	10	Very easy

Source: Griffith, J. S., *Fisheries Management in North America. 2nd Edition*, 481–504, American Fisheries Society, Bethesda, MD, 1999.

One highly visible case of unauthorized (and illegal) stocking of a nonnative trout has occurred in Yellowstone Lake (Yellowstone National Park). Lake trout (Figure 10.6) were discovered in the lake in 1994 and considered a threat to the natural ecosystem dominated by native Yellowstone cutthroat trout (Figure 10.7). The life histories of these fish are quite different. The native cutthroat trout migrate out of the lake on upstream spawning runs. At that time, adults are vulnerable to capture by such animals as osprey and black and grizzly bears, and the death of spawners provides nutrients for upstream areas used by young trout.

Lake trout reproduce in lakes, remain in deeper waters, can reach a much larger size, and are very predaceous on other trout. Lake trout have been responsible for the decline of cutthroat trout populations in other western lakes, so efforts to control them were started by the park. In 1996, the park found that each lake trout consumed about 41 cutthroat trout annually, amounting to an estimated 15 metric tons of cutthroat trout (about 14% of the population). It was determined that without a control program, the number of lake trout and their predation would have resulted in a loss of 23 metric tons of cutthroat trout in 1999 alone (Ruzycki et al. 2003). More than 100,000 lake trout were removed from the lake from 1994 to 2004, and intensive effort in 2004 produced 26,000 lake trout (Koel et al. 2005). Unfortunately, the number of Yellowstone cutthroat trout declined by more than 90% in a 5-year period due to lake trout predation and the whirling disease that was introduced in 1998. At present, there is no certain way of ridding the system of lake trout. Without the removal program for lake trout, it is likely that the native trout would be functionally extirpated from the lake (Behnke 2002).

Figure 10.6 Lake trout *Salvelinus namaycush*. (Courtesy of USFWS, http://commons.wikimedia.org/wiki/File:Fisher-holding_Lake_trout.jpg.)

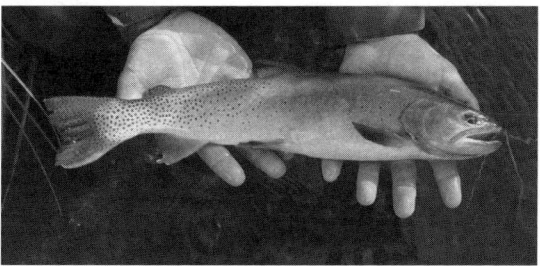

Figure 10.7 Yellowstone cutthroat trout, native to the park. (Courtesy of U.S. Department of Agriculture Forest Service Region 2 http://commons.wikimedia.org/wiki/File:Yellowstone_Cutthroat_Trout.jpg.)

CASE STUDY: GREENBACK CUTTHROAT TROUT

Ancestors of Colorado trouts were undoubtedly Yellowstone cutthroat trout that crossed the continental divide to the west and entered the upper Green River drainage of the Colorado River. This fish became the Colorado River cutthroat trout, which expanded its range in all directions. Some of these fish crossed the continental divide and gave rise to the east-slope trout of the South Platte and Arkansas River drainages in Colorado. In all, three cutthroat trout subspecies developed on the east slope of the Rocky Mountains in Colorado: the greenback, yellowfin, and Rio Grande cutthroat trouts (Figure 10.8). Unfortunately, the yellowfin is extinct, and the remaining two are restricted to a tiny portion of their native ranges.

Native cutthroat trouts are opportunistic feeders and consume a wide variety of food items. In-stream foods include the expected aquatic insects and fish. But in low-ration headwater streams, they also may consume terrestrial prey such as insects that fall into the stream, mice, frogs, and snakes (e.g., Varley and Schullery 1998). Such voracious feeding habits may be necessary to ensure survival in an unpredictable and harsh aquatic system. However, their aggressiveness would get them in trouble. By all accounts, the native trouts were extremely abundant and easy to catch all along the front range of the Rocky Mountains. This was examined in Yellowstone National Park, where Varley and Schullery (1998) reported that the cutthroats were twice as easy to catch as wild brook trout (an easily caught fish), and about 18 times as easy to catch as brown trout. As a result, the native cutthroat trout, so successful at invading across the Rocky Mountains and subsisting in harsh conditions of cold mountain streams, were no match for human predators and the habitat changes they made in nineteenth-century Colorado. By the end of the nineteenth century, the number of native trout had declined.

Attempts were made to cultivate the greenback trout in hatcheries, but the effort failed (U.S. Fish and Wildlife Service [USFWS] 1998) due in part to their wild behavior. To meet the needs of the angling public, Colorado took advantage of the availability of eggs and fry of other trout species that were made available by the new transcontinental railroad. In 1890, rainbow trout, brook trout, lake trout, and Atlantic salmon had been introduced into Twin Lakes, home of the large and highly desirable yellowfin cutthroat. By that time, the yellowfin was extinct, and the greenback had been extirpated from the Arkansas basin. Fish culture flourished in the early twentieth century, and as an example, in 1912, over 11,280,000 young nonnative trout were stocked from Colorado hatcheries (Wiltzius 1985), including the cold-hardy eastern brook trout. By 1937, the greenback cutthroat trout was thought to be extinct.

Fortunately, graduate students at the University of Colorado's Mountain Research station caught a strange trout 16 years later (1953) in Como Creek that fit the description of the "extinct" trout. At that time, student Bill Rickard sent it to the U.S. National Museum, where the identity of the fish was

Figure 10.8 Greenback cutthroat with parr marks. (Courtesy of USFWS, http://search/usa.gov.search/images?query=greenback_cutthroat_trout.)

confirmed as the greenback cutthroat trout (Mitchell 2000). In 1973, the fish was among the first to be protected under the new Endangered Species Act, bringing the fish under federal protection.

An intensive effort was made to protect the Como Creek population of greenbacks, which was found to be a pure strain. More searches also were made for the fish throughout the drainage, and some populations were found isolated above impassable barriers. However, when suspected greenbacks were found coexisting with rainbow trout, many were found to be hybrids.

The USFWS (1998) identified the major factor in the decline of the greenback cutthroat trout to be the introduction of nonnative salmonid species. Rainbow trout (in the same genus) freely hybridize with them. The very aggressive and larger brown trout (native to Europe) displace them from habitat and they also eat them. The cold-hardy brook trout (char) have a competitive advantage, and contrary to most situations of competition, the mechanism outlined below seems evident.

Brook trout spawn in the fall of the year, and the fry hatch about the same time that the greenback trout begin spawning. This gives them a size advantage and also provides a longer period of growth in an environment that has a short growth period. The yearling brook trout may attain lengths of 30 mm or longer compared with the size of young greenbacks in October. Moreover, size is linked with behavior. As in other trouts, there is a dominance hierarchy effect in which smaller fish will yield to larger fish for stream position. This dominance is apparent during minimum flows in winter, when the brook trout are found in more favorable stream habitats (Fausch and Cummings 1986).

A program was initiated by the USFWS and the U.S. National Park Service (USNPS) in Rocky Mountain National Park to replace nonnative trout with the greenback. This program was highly successful in isolated lakes in which nonnative trouts were removed by a toxicant (antimycin; Bruce Rosenlund, USFWS, Golden, CO) and replaced with greenbacks. However, it was not successful in streams that were stocked with brook trout because a complete kill was difficult. If only a few brook trout survived, they would soon overcome the greenback population (Behnke 1992).

The greenback recovery effort has been hailed as a success, and the fish has been downlisted from an endangered to a threatened status. In breeding coloration, the greenback trout is very striking (Figure 10.9), so it came as no surprise that it was officially selected in 1994 to be the state fish of Colorado (Johnson 2007). But with the continuing presence of other trouts, there is only a limited amount of isolated habitat left for the fish. Other developments highlight problems inherent in recovery programs. It has been discovered that some hatchery fish that were stocked as greenbacks for recovery purposes were misidentified and actually were Colorado River cutthroat trout instead of greenback cutthroat trout (i.e., five of nine populations studied) (Metcalf et al. 2007). This has

Figure 10.9 Greenback cutthroat trout in breeding coloration. (Courtesy of USFWS, Golden Colorado Field Office.)

reduced the number of populations of the greenback cutthroat and presumably lowered prospects for future recovery of the fish. Such disturbing events may have delayed the preparation of a revised recovery plan to replace the existing one (USFWS 1998), which is badly outdated. In the absence of a revised plan, there is a more recent update of the status of the fish and recovery actions, prepared by Myklebust (2006). This update is a draft status review submitted to the USFWS for consideration in preparation of a new official plan.

SUMMARY

Coldwater streams are small streams of order 1–3, with a water temperature that usually does not exceed an average high of 15°C or a maximum temperature of about 20°C. This reach is characterized by low temperature, fast and turbulent flows, and gravel to boulder substrates. If above the tree line or exposed by timbering, the coldwater stream can be mostly autotrophic (P/R > 1.0) due to in-stream productivity. However, as streams enter forests and receive CPOM in the form of leaf litter and woody debris, they become heterotrophic (P/R < 1/0) via a benthic food web. Stream invertebrates shred the CPOM into FPOM, consuming the plant material along with a biofilm of bacteria and fungi. Waste particles are washed downstream as FPOM and consumed by collector invertebrates. Algae and invertebrates also are grazed by invertebrates and suckers. Aquatic insects are flattened to aid in hiding, but when food has been consumed, they enter the drift—to be transported to a new area. It is during this drift period that they become most vulnerable to predation from trout. The most abundant and well-known trout in the United States are a strange mix of species maintained by stocking of the following: brook trout; char (*Salvelinus*), native to the eastern United States and Canada; cutthroat trouts (*Oncorhyncus*); Pacific salmon, native to the western United States; and brown trout (*Salmo*), native to the North Atlantic and Europe. Trouts may be limited by habitat availability, because they require four kinds of habitat: spawning, rearing, adult, and overwintering habitats. Coldwater streams are small and vulnerable to habitat alteration by humans due to draining, siltation, overheating due to logging, erosion, pollution, and bank destruction by livestock. Fish introductions for sport fishing also have caused problems for native trout, by predation, hybridization, and competition with nonnative fishes. Illegal stocking also has been a problem, such as in Yellowstone Lake, where native Yellowstone cutthroats are being displaced by stocking of lake trout. A case study of the endangered greenback cutthroat trout of Colorado presents a restoration program for the fish, using isolation portions of streams. Nonnative fishes are removed from these stream sections with toxicants and replaced with the native greenback cutthroats.

Recommended video: (DVD) *The Incredible Journey of the Greenbacks*, by Hugh Gardiner and Paul Herlinger. 2006. Borderlines Products. (Available from Amazon.com.)

CHAPTER 11

Fishes of Warmwater Streams and Rivers

A WARMWATER FISH VIEWPOINT

Historic fish studies and management actions in lotic systems of the United States were concentrated on coldwater streams in states where they existed. This focus reflects the ease of working with small streams, but also, it was due to an emphasis placed on trout fishing. That is not to say that warmwater streams were not fishing streams. On the contrary, warmwater streams were much more likely to be fished because they were near human centers of civilization. In 1970, fishing opportunities in the United States included 489,000 km (305,625 miles) of warmwater stream fishing in 49 states, and it was the only form of stream fishing in 29 states (Funk 1970). Warmwater streams were abundant, and everyone fished in them. But fishing for the brightly colored trout had an aura of the mystic: One only has to read the writings of early proponents or read *A River Runs through It* (Maclean 1989) to get the idea. The trout (lure) fisherman was the gentleman while the bream (worm) fisherman was a few classes lower. When trout streams declined due to overfishing and habitat loss, state and federal agencies became involved in studies to learn more about these streams and to restore fishing opportunity. It also is pertinent to note that funds for study or restoration of streams were available almost exclusively from sales of fishing licenses; thus, "nongame" fishes, unfortunately called "trash fish," were not deemed very important either.

When I first moved to the intermountain west and attended local chapter meetings of the American Fisheries Society, I found that almost all of the members were trout biologists, who (it seemed to me) thought that all fishes (warmwater included) should behave like trout and that nongame fishes were an invention of the devil. Even limnologists placed little emphasis to streams, probably because they were open systems and very difficult to study without considerable economic support. Warmwater stream ecology really came of age in the United States after construction of major dams and reservoirs were implicated in a loss of native fishes, including species that were protected by provisions of the Endangered Species Act of 1973. For the first time, substantial sums of money were made available to study nongame fishes in warmwater streams and large rivers.

THE STREAM CONNECTED

The river continuum concept tells us that the streams and rivers are connected as they flow down gradient, that is, from cold to warm climates. Thus, coldwater streams, where they exist, are the upstream part of the lotic ecosystem. In review of the past chapter, these are high-elevation, high-gradient, small streams of the lowest order (1–3). They have narrow channels that are shaded in forested areas, which helps to keep them cold (i.e., 4°C–15°C is optimum for salmonids), even in the summer, when they rarely exceed an average maximum daily temperature of 20°C. Soils in higher basins are mostly inorganic and poorly buffered. The trophic structure is dominated by inputs of allochthonous organic matter in the form of leaf litter and woody debris (i.e., coarse particulate organic matter [CPOM]). The fish populations in the upper reaches of these streams are cold-hardy fishes of marine origin (e.g., salmon and sculpins) whose numbers are increased by smaller coolwater fishes as stream order (and temperature) increases downstream. Although physical conditions in these small streams are harsh, water flows year round from snowmelt and groundwater, and flow peaks with spring snowmelt.

In contrast, warmwater streams may begin as medium-size streams (order 4–6) of lower elevation and warmer (15°C–27°C) temperatures, which are optimum for most warmwater fishes (Armantrout 1998). These streams are wide enough to be exposed to the sun, and daily temperatures are markedly increased, so they exceed an average monthly temperature of 20°C each year. Some streams in arid environments may become intermittent during low-base-flow periods. Winger (1981) identified and compared 22 characteristics of coldwater and warmwater streams, and for most characteristics, the warmwater stream is just a larger version of the tributary coldwater stream. However, there are a few characteristics that stand out (Table 11.1). In physical characteristics, warmwater streams have higher water temperature, lower gradient, lower elevation, higher turbidity, have more shade and bank cover, and have a larger percentage of finer organic material present [fine particulate organic matter (FPOM)]. In general, soils are more complex, and streams have a greater buffering capacity. In cooler regions, the warmwater stream may have ice cover, but it is much more stable than in the smaller, coldwater stream due to some amount of stratification. However, even in large rivers, the effects of dam regulation, especially for peaking power, can cause ice breakup and result in extra fish movement, which is presumably a source of increased winter stress.

Fish diversity is high in the warmwater streams of the United States, and various species of darters (percids), bass and sunfishes (centrarchids), minnows (cyprinids), suckers (catastomids), and catfishes (ictalurids) are usually present. In biological characteristics, warmwater fishes are more affected by interspecific competition (instead of intraspecific competition), and there are more predatory fish present. In addition, there are a score of avian (e.g., herons, egrets, and ospreys), mammalian (raccoons and otters), and reptilian (snakes) predators that prey on stream fishes.

Table 11.1 Coldwater versus Warmwater Streams

Factors	Coldwater	Warmwater
Physical		
Gradient/elevation	High	Low
Pools/riffles	Short/many	Long/few
Stream order	Low (<3)	Higher (>3)
Temperature	Cold (<20°C)	Warm (>20°C)
Discharge	Low	Higher
Seasonal change	Can be abrupt	Gradual
Substrate	Gravel–rubble	Mud–rubble
Turbidity	Low	Greater
Organic matter	Coarse (CPOM)	Fine (FPOM)
Ecological		
Fish diversity	Low	High
Fish competition	Intraspecific	Interspecific
Shade and cover	Much	Less
Fish food	Macroinvertebrates	Plankton to fish

Source: Winger, P. V., *Warmwater Streams Symposium: A National Symposium on Fisheries Aspects of Warmwater Streams*, 32–44, Southern Division of the American Fisheries Society, Bethesda, MD, 1981. With permission.

In this account, we focus on temperate streams, but it is important to note that riverine fish diversity is greatest in tropical floodplain rivers, such as the Amazon and Orinoco rivers of nearby South America. Briefly contrasting temperate and tropical streams, tropical streams have more abundant and diverse fish communities. This is due to warm stable climates and abundant precipitation, which produce year-round growing seasons (as previously discussed under biogeography). Thus, it is easy to imagine that adaptable fishes would occur and feed at every trophic level. With abundant plant production and herbivory, more detritus would be produced, and consequently, there are more detritivores of all sizes. Thus, in contrast to temperate zone fish communities, which are mostly second- and third-level carnivores, the tropical fishes would include herbivores, carnivores, detritivores, and fishes feeding on detritivores that occur in the biofilm of plants and substrates. In such a rich stream environment, we also expect and find great numbers of omnivores. In this case, we can predict that fishes would consume algae, microinvertebrates, protozoa, and bacteria all at the same time!

Ecologically, the structure and function of a temperate warmwater stream also differ from those of a coldwater tributary. The coldwater stream is heterotrophic (photosynthesis/respiration ratio (P/R) <1), and its trophic structure is dominated by large allochthonous inputs of CPOM. Much of this input will be converted to FPOM and dissolved organic matter (DOM) by the benthic shredders and microbial decomposers, and some of it will be transported downstream. As the stream increases by tributary input to a small river, the stream widens, water velocity declines, and the streambed accumulates waterborne sediments (i.e., it becomes alluvial). Sediment deposition is coarse at first and then progressively finer, especially in pools, where quantities of FPOM begin to accumulate. At this point, the benthos is dominated not by invertebrate shredders but by collectors that work on the fine sediments. In quiet reaches and recirculating eddies, there are likely to be blooms of plankton utilizing the DOM, and as the sediment is deposited, the water becomes less turbid, promoting algae and aquatic plant growth. The stream has entered into an autotrophic phase with P/R >1.

Large rivers (order >6; the Mississippi River is order 10) are larger versions of the medium-size warmwater stream in the same region, but habitats are more heterogeneous and there is a more extensive floodplain system. The largest rivers, or great rivers, have drainage basins of >3200 km^2 (Simon and Emery 1995). Drainage area, discharge, and length of 14 of the largest rivers in North America are provided by Sheehan and Rasmussen (1999), and the same information for 211 of the largest rivers was given by Leopold (1994). Table 9.1 contrasts these attributes for some selected rivers.

The natural flow regime of rivers is important to ecosystem function. This flow regime depends upon the climate and nature of the drainage basin, which typically has a high spring flow for most temperate locations. During high-flow events, flooded terrestrial areas can provide substantial cover and food for stream fishes, and as flow subsides, there are likely to be abundant backwaters or alongshore embayments caused as the receding waters partially expose cutoff side channels and eddies. Fishes in general incorporate the floodplain into their life history patterns, and there is likely an ontogenetic (life stage) separation in use of space and food in most fishes.

One aspect of warmwater streams and larger rivers that needs to be emphasized is the integration of the floodplain as a resource base for fishes. The riparian habitat is occupied by a host of other animals that provide food for stream fishes. This food is made available by a variety of different mechanisms, including falling into the water (such is the case with flying insects in particular) and being washed into the water by lateral flooding or surface runoff. In the case of major floods, the whole floodplain can be available for fishes, and some move considerable distances away from the main channel and search for food (such as the Colorado pikeminnow, which presumably captured and consumed prairie dogs, rabbits, chickens, and other large prey) (Beckman 1952). Another interesting mechanism is direct migration into streams by terrestrial animals. Some terrestrial animals, including the flightless Mormon cricket *Anabrus simplex* (Tettigoniidae) of the western United States, will enter the water and swim across it. My friend W. L. Minckley, I, and other U.S. Fish and

Wildlife Service employees documented several bands of 1 km² or more entering the Green River during migration. At our estimated 10 to 20 crickets/m², one of these bands could have constituted 30 to 60 metric tons (Tyus and Minckley 1988). The large crickets are a high-quality food item and average about 58% protein and 16.5% fat. It was no wonder that all 11 fish species we captured were gorged on crickets. Periodic contribution of such a food source could be significant to the nutrition of the long-lived (i.e., 30 years and more) endemic fishes.

STREAM FISHES

It is not possible to identify all of the combinations of fishes that exist in warmwater streams and rivers across the United States. However, there are patterns that largely depend on regional climate and distribution of the fish fauna that remained in various geographic locations after the Pleistocene. An accounting of the fish distribution also is compounded by the many introductions of fishes across the nation, most of which occurred during and after the late nineteenth century.

Funk (1970, p. 142) provided an excellent map (Figure 11.1) of regions in the United States and southern Canada that support fishes and those that do not. His map showed that the eastern part of the United States, from the western third of Kansas to the eastern seaboard, supports warmwater fish, excluding a narrow band of the Appalachian Mountains, which supports coldwater fish. Higher elevations of the Rocky Mountains west and northwest to the Cascades and Sierra Nevada ranges support coldwater fishes, and the eastern plains and much of the lower elevations of the western United States have only intermittent streams and do not support fishes year round.

It is clear from this map that all freshwater streams do not originate in snow-covered mountains. In the north temperate regions of the United States flat, glaciated regions have marshy lakes, ponds, and sluggish rivers that originate in "lowlands." Such areas are cold in winter, and the lakes and streams are frozen over with layers of ice and snow. In the summer, however, the temperatures rise to well over 20°C (10°C–21°C optimal, may exceed 26°C) and are thus unsuitable for coldwater fishes. Instead they are inhabited by coolwater fishes such as esocids (northern pike), percids (yellow perch, walleye, and sauger), and centrarchids (smallmouth and largemouth basses). In the southern United States, streams may originate as warmwater streams flowing out of swamps, marshes, or clear springs. These headwaters support populations of centrarchids (smallmouth, largemouth, rock bass, and other sunfishes) and ictalurids (channel catfish).

Freshwater fishes are most diverse in the Mississippi River basin and its tributaries, which drain about 50% of the United States. Fish populations have the greatest diversity in the mid- to upper portions of that basin, and the state of Tennessee supports the richest fauna, with 302–319 species, of which 277–297 are presumed native (Etnier and Starnes 1993). The fishes there comprise 29 families, 18 orders, and are dominated by percids (93 species, including 90 darters), minnows (83), catfishes (24), suckers (21), and sunfishes (21). In that region, coldwater streams support native brook trout, upper warmwater zones are classified as smallmouth/rock bass streams, and lower warmwater reaches are largemouth bass/sunfish streams. There are no obligate lacustrine species because the lakes there are either oxbow lakes or shallow and swampy (e.g., Reelfoot Lake). In the more southern part of the Mississippi River basin, the state of Arkansas has only warmwater streams (and oxbow lakes) that support 214 species, including 197 natives. As in Tennessee, the fish community is dominated by five families that comprise 76% of the fauna: minnows (66 species), percids (41, including 39 darters), sunfishes (20), catfishes (19), and suckers (18).

Typical warmwater fish communities of the eastern United States coastal region can be classified into at least three groups of index species: headwater streams: smallmouth bass/rock bass/redhorse/stonerollers; lower-gradient streams: sunfishes/catfish/common carp/carpsucker; and coastal streams: largemouth bass/sunfishes/pickerals. In the coastal region, streams enter into various estuaries, where the number of fishes increases due to the presence of euryhaline marine and anadromous

FISHES OF WARMWATER STREAMS AND RIVERS

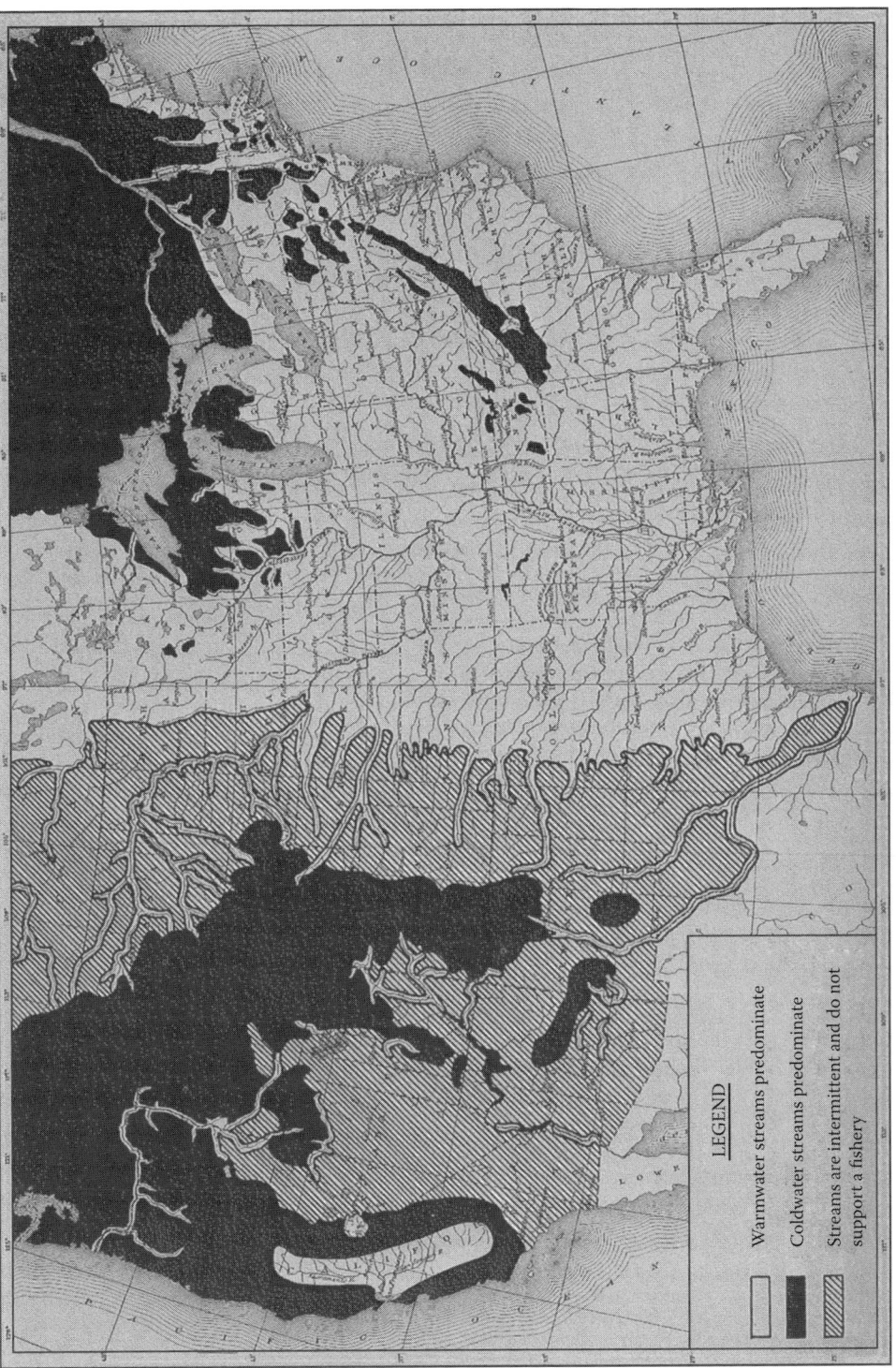

Figure 11.1 Stream fishing in the continental United States. Image shows locations of streams which support warmwater fish, those that support coldwater fish, and streams that do not support a fishery. (From Funk, J. L., *A Century of Fisheries in North America*, 141–152, American Fisheries Society, Washington, DC, 1970. With permission.)

fishes. Historically, lower parts of rivers and smaller streams were inundated in spring with migrating herrings, shad, striped bass, and Atlantic salmon.

The intermountain western United States was isolated from the center of fish distribution (Mississippi/Missouri River system) during the Pleistocene and has a remnant warmwater native fish population composed mainly of minnows and suckers. Whereas fish populations in the central and eastern portions of the United States have not been greatly affected by nonnative fish introduction, fish populations of the west are greatly changed by introductions of nonnative warmwater fishes, and native fish populations there are mostly unstable. The same is true for southwestern desert fishes whose native streams have been overwhelmed by nonnative fishes.

To the far west, the Sacramento San Joaquin drainage, the center of fish evolution in California, supports 28 native fishes; however, at least half of these are extinct or vulnerable to extinction (Brown and Moyle 2005). The formerly abundant Pacific salmonid populations of the west coast, which migrated to and spawned in most of the north temperate streams and rivers, have experienced severe declines, and more than 200 anadromous stocks are declining or vulnerable to extinction (Nehlsen et al. 1991). Finally, Washington State is a good example of a fauna severely depressed due to Pleistocene glaciation and isolation. It has an inland fauna of 91 fishes, of which 41 are introduced (Wydoski and Whitney 2003). Primary fishes there include only 11 minnows and five suckers; by my count, the rest are recently derived from marine ancestors.

LARGE RIVER FISH FAUNAS

Unfortunately, there is an easy way to identify a large river system, and that is to just look for the dams and reservoirs! There are not very many large river systems left in the United States that are still free flowing. Dams block spawning and feeding migrations of fishes, and flow regulation changes the natural flow regimen by cutting off the peaks and raising low-flow periods. With hydroelectric power generation, there are artificial peaks on a daily basis, changing the structure and function of a stream. As more dams and reservoirs are built on consecutive reaches of the same stream, there is not enough time for recovery, and the river as a natural system is lost. In addition to dams and reservoirs, there are many other human-induced changes in river systems, and their adverse effects have been substantial.

The American Fisheries Society hosted a symposium on *Historical Changes in Large River Fish Assemblages of the Americas* (Rinne et al. 2005) and documented changes in selected river systems. The decline of native large river fishes in the United States due to human-induced impacts is well documented in this volume, and the editors introduced the symposium by presenting some statistics. In general, it was estimated that 81% of streams have been adversely affected by humans, and my assessment is that perhaps as much as 50% of wadable streams across the nation have suffered from ecosystem scale alteration. For example, native fishes have been affected as follows: 80% of southwestern fishes are threatened or endangered; 63% of California fishes are extinct, threatened, or endangered or vulnerable; 52% of Missouri River fishes have declined in abundance in Wyoming; 34% of stream fishes in Illinois have been decimated or extirpated; 28% of stream fishes in the southeast are vulnerable, threatened, or endangered; and 23% of anadromous Pacific salmonids that once supported commercial fisheries are vulnerable or at risk of extinction despite management efforts (Hughes et al. 2005; Rinne et al. 2005 and authors therein).

The problems caused by water resources development and industrial pollution in large rivers are now well understood, and millions have been spent on some rivers to clean up water pollution (e.g., Hudson and Connecticut rivers). However, hundreds of dams and reservoirs are now considered to be permanent features of most rivers in the United States, and prospects for dam removal as a way of river renovation are not promising.

Of all the major rivers in the United States, perhaps the saddest tale is the fate of the lower Colorado River system. Arguably, the Colorado River is the most regulated river in the world. It no longer flows through its once productive estuary to enter the Gulf of California. Instead, the last of its water is taken at Morales Dam, and there is no surface flow to the sea. The river now functions like a great basin, in which all of the water in it is stored or evaporated.

In an excellent publication by Mueller and Marsh (2002), historical circumstances leading to the collapse of the Colorado River ecosystem and its replacement by a human-dominated system of locks, dams, and diversions are carefully considered. The fish fauna of the lower Colorado River was unique: 75% of its fishes were endemic. However, these authors consider the river and its native fish populations to be lost: perhaps as many as 95% of the fish density in the lower Colorado River today is nonnative. One ray of sunshine is a disjunct population of endangered humpback chub (*Gila cypha*) in the Grand Canyon. At present, it is isolated in the Little Colorado River from warmwater predators by very cold water released into the mainstream river from Glen Canyon Dam. However, persistence of that population is in no way guaranteed, especially with warming trends in the basin.

In the upper Colorado River basin (above Lake Powell), remnants of the native Big River Fish Community still exist; however, four of seven fishes of this community are federally listed as endangered, and one of those should be considered extirpated. Recovery programs there are in a desperate struggle to save the remaining fish community in the face of continuing water resource development and proliferation of aggressive nonnative fishes such as northern pike, smallmouth bass, channel catfish, and several predaceous smaller fish (e.g., see Valdez and Muth 2005).

CASE STUDY: THE NORTH AMERICAN PADDLEFISH

As stated previously in chapter 4, paddlefishes are one of our oldest ray-fin fishes. The paddlefish line of evolution presumably diverged from the sturgeons in the Early Jurassic, about 200 million years ago (Maisey 1996). A fossilized paddlefish (*Crossopholis*) discovered in the Green River shale deposits in Wyoming (Long 1995) shows that the extant fish has changed very little in 40 million years. There are only two living paddlefishes in the family (Polyodontidae), the North American paddlefish (*Polyodon spathula*) (Figure 11.2) and the Chinese paddle fish (*Psephurus gladius*). Both species are fishes of big rivers, the Mississippi River basin of the United States and the Yangtze River of China. It is unknown how long the fishes have been separated, but the extinct *Crossopholis* bears closer resemblance to the Chinese form, and a primitive paddlefish (*Protopsephurus liui*) has recently been discovered from Lower Cretaceous deposits in China (Berra 2007), suggesting that the North American fish may be the more recently derived one. Regarding status, the Chinese fish has been greatly reduced in distribution and abundance and is considered hopelessly endangered. The North American form will be discussed here in detail.

The North American paddlefish has been a very successful fish based on its 200-million-year persistence. But its success is not only measured in persistence. Less than 100 years ago, it was widespread and abundant in the mainstream and all large tributaries of the Mississippi River basin (e.g., Jordan and Evermann 1900). Due solely to human impacts, its numbers have declined in most

Figure 11.2 The North American paddlefish, *Polyodon spathula*. (Courtesy Raver, D. and U.S. Fish and Wildlife Service, NCTC.)

areas. However, this is a resilient fish: the paddlefish still occurs over most of its historic range, and it is still found in 22 of the 26 states in which it has been documented. Populations are currently increasing or stable in 17 states, declining or status unknown in five states, and extirpated from four states. Sport harvesting has recently occurred in 14 states, and commercial fisheries have been reduced from 12 states in 1983 to seven states in 2006 (Bettoli et al. 2009). In addition, 10 states have stocked hatchery-reared paddlefish to bolster populations or recover paddlefish populations (Grady and Elkington 2009). Overfishing has decimated paddlefish populations in the past. All states that have paddlefish now regulate harvest, if it is allowed at all. However, several authors have pointed out that the effectiveness of regulations was subject to question (e.g., Hansen and Paukert 2009).

Jordan and Evermann (1923) included the North American paddlefish in their book *American Food and Game Fishes*, indicating the importance of the flesh and especially of the roe as food. In describing the meat of the paddlefish, Etnier and Starnes (1993) described it as excellent, stating that when baked, it is similar in mild flavor and consistency to chicken. However, the real commercial value for paddlefish is for the roe, which makes very good caviar. During the 1980s an embargo affected the supply of beluga caviar into the United States, and paddlefish caviar was substituted, selling at $300 or more for a 14-oz. tin (Willis 1993). Overfishing and illegal taking (poaching) have been a problem for paddlefish populations. For example, Pflieger (1997) points out that a decline in sturgeon fishing in the Mississippi valley resulted in the fishing industry turning to paddlefish. By 1899, the paddlefish harvest peaked at almost 2.5 million pounds, and the largest occurred in the Mississippi River (Gengerke 1986). Thereafter the harvest declined for many years due to overfishing. Carlson and Bonislawsky (1981) provided commercial harvest data from eight locations in 1874–1975. Quinn (2009) reported on recent harvests (i.e., 2000–2006) and reported that the largest commercial catches came from Arkansas, Kentucky, and Tennessee.

Paddlefish are vulnerable to overfishing for several reasons. They are easy to catch. These are large fish, and they tend to aggregate in spawning runs and overwintering areas. With the construction of dams and reservoirs, the fish tend to accumulate in tailwaters and other areas that make them even more vulnerable (e.g., Jennings and Zigler 2009; Quinn 2009). They have low recruitment due to the long time required to reach sexual maturity, which varies according to latitude (6–14 years in females) and the lack of annual spawning (females may spawn only once in 4–7 years) (Russell 1986; Jennings and Zigler 2009).

Another factor in the present environment is habitat degradation (Sparrowe 1986), a lack of spawning and nursery areas (Carlson and Bonislawsky 1981; Gerken and Paukert 2009), and poor environmental conditions affecting the reproductive effort. There are at least three specific requirements for spawning, and the larvae have specific needs as well. For spawning to occur, there must be proper temperature (about 50°F) at the right time in the spring (photoperiod), and there must be a significant rise in water level to flood the gravel bars used for spawning (Jennings and Zigler 2009). If all these events do not occur simultaneously, the females will reabsorb their eggs (Russell 1986). Finally, the water level rise must be sustained for more than a week to prevent stranding and death of the larvae: the larvae need time to hatch, absorb the yolk sac (so they can develop mouth parts for exogenous feeding), and be swept downstream into more favorable nursery areas.

Presently, habitat destruction and river modification are the most obvious changes affecting distribution and abundance of paddlefish. Adult paddlefish require quiet, slow water rich in zooplankton for feeding and extensive gravel bars for spawning. The historic large river environment with its many oxbows and backwaters provided an excellent habitat for paddlefish foraging, and the areas of the main channel with gravel bars provided a good spawning habitat (e.g., Southall and Hubert 1984). Unfortunately, in the rush to develop the Mississippi, Missouri, and other rivers, there was no consideration given to protecting the great populations of sturgeons and paddlefish. As a result, populations of these fishes have suffered enormous declines. Stream channelization, construction of dikes and levees, and drainage of floodplain lakes have destroyed much of the feeding habitat. In addition, construction of reservoirs has inundated most of their spawning habitat. For example, a large population of paddlefish inhabited the Osage River and Lake of the Ozarks complex. However, the construction and closing of Truman Dam in 1977 flooded the spawning habitat of the population, and paddlefish there are being maintained only by hatchery stocking (Pflieger 1997).

Large fishes can cope with the energetic costs associated with life in a large river environment, which includes dispersal and migration to different locations during the year. Paddlefish are large as adults, reaching sizes of at least 72 kg (160 lb) and 2.2 m (7 ft) (Jennings and Zigler 2009), and are fully capable of making the best use of a large environment. Adults have been documented moving distances of up to 2000 km (1250 miles) downstream (Unkenholz 1986), and distances of 160 km (100 miles) or more are not uncommon. The fish select areas that are at least 1.2 m (4 ft) deep and prefer depths of 3 m (10 ft) or more in winter (Russell 1986). Moreover, spawning fish need gravel bars that are deep enough for them to deposit the adhesive eggs into, and the bars need to remain flooded until the young have had time to hatch and emerge. After emergence, the young are swept downstream into a quieter habitat, where they feed on individual prey, developing into filter feeders later. Downstream dispersal of young paddlefish can be extensive, and travel of 100 miles in 1 month has been documented (Russell 1986). However, in the present environment, the young may be displaced into the upper ends of reservoirs (e.g., Lake Sakakawea), where they can fall prey to introduced gamefish, such as the juvenile walleye and the northern pike. Although adult paddlefish are not vulnerable to predation by other fish, the young fish are. The newly hatched fry have no rostrum or paired fins. They remain attached to a large yolk sac for several days and are feeble erratic swimmers.

How have paddlefish been so successful and survived for so long? Paddlefish adaptations address both biotic and abiotic threats. They are large and long-lived species (30 years or more), which enables them to reproduce in the most favorable years. They are virtually immune to predation due to their size and are capable of long-distance migration to avoid climatic extremes and poor local conditions. They spawn in areas that provide downstream transport of the young to more suitable conditions, and presumably, natural selection has resulted in a homing behavior for returning to natal areas.

Although the adults are clearly specialized in morphology for life in the large river system, they appear to exhibit an unusual feeding strategy (Rosen and Hales 1981). As filter feeders, they no doubt sought after and preyed upon zooplankton that were very abundant in the historic system. However, it appears that they consume anything that is small enough to catch and large enough to be filtered out. For example, stomach contents taken from many fishes over the years show that they will consume a variety of organisms, including terrestrial and aquatic insects (mostly mayfly naiads), fish (darters and threadfin shad), phytoplankton, and zooplankton (Ruelle and Hudson 1977). The fish appear to be indiscriminate (Robison and Buchanan 1988), at least sometimes consuming anything larger than about 0.25 mm, including algae, detritus, and sand. In this case, the fish might digest particles of once-living organisms and also strip the biofilm coating off debris. Also, it is important to note that paddlefish are attracted to and consume commercial fish food, especially if such artificial diets remain suspended in the water (Graham et al. 1986). Thus, from a trophic perspective at least, I would classify the paddlefish as a generalist, feeding on anything the right size (small).

As noted above, the paddlefish was recently widespread and abundant throughout the mainstream rivers of the Mississippi River basin, and as pointed out by Raup (1991), widespread and abundant animals are very persistent (even when faced with catastrophes) due to a large interconnected population size. All of the above life history attributes of this fish point to continued success in the natural river system where it evolved.

But what is the prognosis for their continued survival in the present system? The Mississippi River and its tributary rivers were subjected to massive, system-wide changes with little or no concern for what might happen to the natural resources there. In truth, many residents of the area looked forward to having recreational fishes in the many large reservoirs that were created. And one could argue that we did not know enough about the life history or habitat needs of the paddlefish to even make a compelling argument for mitigation. Thus, in spite of all their superb adaptations to big river existence, the paddlefish are at risk. The very adaptations that have served them so well for many millions of years are now causing their decline in the altered river environment.

Their large size makes them more vulnerable to capture by humans. Spawning migrations have been blocked by dams and spawning areas obliterated by flooding. Food supplies have been decreased by drainage of the floodplain and construction of levees. Their young, when there are any, are delivered into the upstream portions of clear water reservoirs that are inhabited by aggressive sight-feeding predators (stocked by the very agencies that are pledged to protect the paddlefish). In the absence of a definitive study that shows there are no impacts of introduced gamefish on the young paddlefish, predation seems an obvious problem. Finally, the use of hatcheries to maintain paddlefish in some locations has provided a temporary solution, but the process is fraught with many uncertainties.

Are paddlefish threatened or endangered? Two major symposia on paddlefish sponsored by the American Fisheries Society have met to consider management of the fish. According to conclusions of the 1983 paddlefish symposium (i.e., "Paddlefish—A Threatened Resource?") (Anderson 1986), the answer is not yes or no, but "not yet." In this case, the symposium leaders made the following recommendations, which I paraphrase: (1) There is a need for a regional approach to manage paddlefish populations instead of each state by itself, and more attention needs to be placed on identifying and protecting spawning sites. (2) Better data needs to be collected on population sizes and locations for each state, especially where commercial taking continues. (3) More reasonable management objectives need to be developed; otherwise, management is haphazard and ineffective. (4) Efforts should be focused on preventing further habitat deterioration and controlling harvest so that stocking is not needed. Artificial propagation and stocking is not a cure all.

The 2006 symposium ended by pointing out areas where progress was made to address the above recommendations, such as the participation of all states with paddlefish in the Mississippi Interstate Cooperative Resource Association, but acknowledged that the same threats mentioned in 1983 still exist for the paddlefish, and new ones have emerged, such as invasive species, climate change, and river management (Paukert and Scholten 2009b). There have been some positive steps taken to protect paddlefish, including protection (in 1992) under Appendix II of the Convention on International Trade in Endangered Species of Wild Fauna and Flora (CITES) and its reclassification as a "species of concern" under the Endangered Species Act, mainly because data on population sizes of the fish are lacking (Jennings and Zigler 2009). All states that have paddlefish now regulate harvest, if it is allowed at all. However, several authors have pointed out that the effectiveness of regulations was subject to question (e.g., Hansen and Paukert 2009).

I believe that the recommendations of the 1983 symposium are just as pertinent today as they were 27 years ago. If the paddlefish is a model for the fate of large river fishes in the Mississippi drainage, then we need a better model. We should not allow this amazing fish to vanish as the system continues to be managed piecemeal. Hard choices need to be made in protecting and maintaining paddlefish in areas where there is some semblance of a natural system and to identify and address all limiting factors that could reduce populations or further limit range.

SUMMARY

Warmwater streams are medium in size (stream order 4–6) and, as part of the stream continuum, connect the coldwater stream to large rivers. These streams occur at lower elevation, and the stream channel is wide enough that solar radiation warms the water to exceed an average monthly temperature of 20°C. Although similar to lower parts of the coldwater stream, there is a lower stream gradient, more complex soils, and other features in addition to warmer temperatures that differentiate the warmwater stream. One of these is a large input of organic matter in the form of FPOM (instead of CPOM), and the predominance of collector instead of shredder invertebrates. As the stream widens and slows, quiet areas can support plankton, algae, and aquatic plants, and it enters into an autotrophic phase (P/R > 1). However, all streams do not originate from coldwater ones. In the northern Midwest, glaciated areas of marshy lakes, ponds, and sluggish headwaters support coolwater fishes such as northern pike, yellow perch, and smallmouth bass. Although cold in winter, they are too warm in the summer to support trout (10°C–21°C optimal, can exceed 26°C). In the south, headwater streams can flow from swamps, marshes, or springs. These streams support sunfishes, including largemouth bass. In general, fish diversity in warmwater streams is high in comparison with coldwater ones. Depending on region, latitude, and elevation, warmwater stream fishes include resident darters, sunfishes, minnows, suckers, and catfishes. In addition, upper reaches of warmwater streams can support trout from upstream and whitefish. In coastal areas, warmwater streams and rivers can support transient diadromous fishes as well. Fishes of warmwater streams are more affected by interspecific competition and predation with other fish than those of coldwater streams, and predation by avian, mammalian, and reptilian species can be high. Large rivers have a stream order >6, they can have extensive floodplains, and seasonal flows can vary greatly. Riverine fishes will move laterally to take advantage of the increased resource base that is accessible due to flooding. Over 80% of rivers and streams in the United States have been adversely affected by humans, and perhaps 50% of smaller streams have suffered ecosystem-scale damage. Largest rivers have been altered by construction and operation of dams and reservoirs. The usual effects of a dam are blocked fish migration, fragmented ranges, and changes in the seasonal flow regimen, while the reservoir basin and dam tailrace converts the river into an artificial system. The native riverine fish faunas in reservoirs are replaced by introduced species that are adapted to living in the new systems (i.e., "preadapted"). In a case study, the present status of the North American paddlefish, a well-adapted riverine fish, is presented as an example of how system changes are affecting native riverine fish. Extant for perhaps 200 million years, this formerly widespread and abundant fish is slowly dwindling in number throughout its range due to overfishing and human-induced changes in its environment. Management and conservation of this fish are almost entirely controlled by 22 separate states. Although several authors consider population data to be inadequate for regulatory purposes, 14 states allow sport fishing, and seven states permit commercial exploitation. Hard choices need to be made in the near future to reduce threats and protect habitats required by this incredible animal.

Further reading: Mueller and March 2002.

CHAPTER 12

Lentic Systems: Standing Water

THE DROP IS STORED (TEMPORARILY)

As we continue to follow our drop of water from the mountain to the sea, it may be stored for a time, either in a body of surface water or in groundwater. We postpone a look at groundwater storage to first investigate storage in a standing body of freshwater, thus leaving the lotic system and entering a lentic (slow) system. Lentic systems are also called "standing water" systems because the water flow may be interrupted by a barrier of some kind. However, there is almost always some flow out of the system, whether over a natural dam or into subsurface flow.

Lentic systems are bodies of water in which incoming water from precipitation, streams, surface runoff, or subsurface sources (groundwater) is stored in open or closed basins. In closed basins (in North America, these are likely due to glacial events), present environmental conditions do not allow stored surface water to flow out of the basin. Instead, evaporation and groundwater recharge is greater than precipitation and surface or subsurface inflows—thus, these systems are drying up. The largest closed basin in North America is the Great Basin, which covers about 520,000 km^2 in the states of Nevada, Utah, California, Idaho, Oregon, and Wyoming. This basin was actually a system of thousands of Pleistocene lakes that have mostly dried up. The best known of these lakes is Pleistocene Lake Bonneville, which covered about 50,000 km^2. As this lake dried, it left behind the Bonneville Salt Flats and the Great Salt Lake. Great Salt Lake once supported Utah cutthroat trout, but its waters have become so salty (23%–27%) that fish cannot live in the open lake (Sigler and Sigler 1987).

Open basins have an outlet at the lowest point, and water can flow downstream either seasonally with wet conditions or, in some cases, all year. In most cases, the open water body will receive one or more streams and will have one outlet that overflows the lake bed at the surface. Also, there usually is some water loss through subsurface seepage. Fish access to both open and closed systems can occur from upstream, but only from downstream for the open system, usually during high-flow periods. However, closed basins are usually accumulating salts due to evaporation, and stored water may not be acceptable to some fish, aquatic plants, or other aquatic life.

Water residence time varies between lentic systems, but in general, natural lakes have a relatively long period of storage before the lake volume is replaced by inflow. Smaller systems such as ponds can have substantial flow-through during seasonal flooding and then have very little flow during the remainder of the year, when it is stored. The storage period is very important for lake productivity, especially for artificial lakes, which will be discussed in a later chapter. The Great Lakes of the United States and Canada provide some insight into water residence time of natural lakes: Erie (2.6 years), Ontario (6 years), Huron (22 years), Michigan (99 years), and Superior (191 years), with the average resident time among lakes being about 64 years (U.S. Environmental Protection Agency 2008). A more complete accounting of water residence time is provided by Kalff (2002).

Regarding the storage of water, freshwater lakes store an exceedingly small amount of water, about 0.3% of the total water on the planet (Kalff 2002). Of great interest is the location of that water, because 95% is contained in only 145 freshwater lakes, and most of it is within 20 of the largest lakes. We will study the fishes of three global Great Lake systems due to their great significance for fishes: Lake Baikal = 23% (23,000 km^3) of water contained in lakes worldwide; Laurentian Great Lakes Superior, Huron, Michigan, Erie, and Ontario = 22.5% of water; and the East African Great Lakes Victoria, Tanganyika, and Malawi = 26.5% of the water. Just two of these Great Lakes, Baikal and Tanganyika, contain almost half (i.e., 44%) of the total amount of the surface supplies of freshwater on the planet (data obtained from Kalff 2002).

STANDING WATER ECOSYSTEMS

Lentic systems can be divided into four main types of standing water bodies: lakes, ponds, swamps, and marshes. Of these, the latter two are shallow water systems and easy to define. Swamps are lower-elevation wetlands that support the growth of trees and shrubs, and marshes are large wet areas with grasses and sedges. Lakes and ponds, on the other hand, exist in a continuum from large and deep open basins to very small shallow bodies that are dominated (closed in) by plant growth. Because of this, it is so difficult to determine when a lake becomes a pond that Reid (1961) noted that "lakes and ponds defy precise definition." The problem is due in part to the many ways that lakes and ponds are formed, the changes as they age, and the differences in the use of terms across disciplines and geographic areas.

Even though lakes and ponds are so similar, some distinctions can be made. Lakes and ponds are semi-isolated bodies of water not connected directly to the marine environment, but early definitions (reviewed by Welch 1952) restricted lakes to (1) bodies of water that are deep enough to be thermally stratified, (2) deep bodies of water in which the littoral zone does not extend to the middle, (3) bodies of water so large that they have a barren wind-swept shoreline, and (4) large and open bodies of water in contrast with very small ponds that have extensive growth of aquatic plants. Unfortunately, these well-intentioned definitions have not been accepted, and the common practice today has been to include both natural and artificial bodies of water as lakes and to use the term "lake" when referring to a large body of standing water with a surface area exceeding some value.

In my opinion, the best classification of lakes and ponds according to surface area was presented by Kalff (2002), in which he classified lakes into great, large, medium, and small, and ponds as large or other. He reported 19 great lakes (surface area exceeding 10,000 km^2), 1504 large lakes (100–10,000 km^2), 139,000 medium lakes (1–100 km^2), more than 1 million small lakes (0.1–1 km^2), and more than 7 million large ponds (0.01–0.1 km^2). However, these size standards have not been used universally. As an example, all bodies of standing water that are larger than 4 ha have been classified as "lakes" (Hayes et al. 1999). However, large artificial lakes seem to be different and are classified as 200 ha or more, with artificial ponds size at less than 10 ha (Summerfelt 1999). To confuse matters further, natural lakes and artificial impoundments also have been lumped together as "lakes" (Cooke et al. 2005).

Hutchinson (1957) provided a detailed summary of 76 different types of lakes based on their formation. We hardly need to go into that detail here, but it is instructional for us to consider the major ways in which lake basins form. I find that there are 10 major ways in which lakes are formed, most of which are due to geological disturbances. In North America, lakes are formed primarily due to glacial action (reviewed by Welch 1952; Reid 1961; Wetzel 2001). The following summarizes nine of the most common actions forming natural lakes, plus an added one for artificial lakes constructed by humans:

LENTIC SYSTEMS: STANDING WATER

- Glacier: from scouring, damming, melting of buried ice, and so on—cirque lakes, kettle lakes, pluvial lakes
- Tectonic: due to movement of the earth's crust—rift or graben lakes
- Volcanic: by lava dams, flooded calderas—crater lakes
- Landslide: that of dam valleys—usually transitory due to erosion
- Solution: when water dissolves limestone deposits—sinkhole lakes
- Wind: makes eroded depressions—playas and dune lakes
- Stream: due to cutoff, erosion pool, lateral isolation—oxbow, plunge pool, and lateral lakes
- Organic: by dams of plants and woody debris—erosional log jams or beaver dam lakes
- Meteorite: in depressions from collisions—impact crater lakes
- Human: build artificial dams, unnatural structures—reservoirs and farm ponds with bottom discharge

The preceding data show that lakes are formed in diverse ways, and the concept of a lake must be broad. Welch (1952, p. 28) very effectively closes this subject with the following:

... lakes are large, medium or small; deep or shallow; protected or unprotected; with or without tributaries and outlets; fresh, brackish, or salt; acid, neutral, or alkaline; hard, medium, or soft; turbid or clear; surrounded by bog, swamp, forest, or open shores; high or low in dissolved content; with or without stagnation zones; with marl, muck, sand, or false bottoms; with or without vegetation beds; with high, medium, and low biological productivity; young, mature, or senescent; and so on.

CHARACTERISTICS OF LAKES

Perhaps the best way to characterize lake systems is to compare them with a system that we already know: in this case, the stream system. Table 12.1 contrasts fundamental differences in lotic versus lentic systems. In physical appearance, natural lakes are deeper and wider, with little or no current or gradient. In nutrient balance, natural lakes tend to contain (cycle) nutrients within the system rather than flush them downstream. In trophic structure, natural lakes are mostly autotrophic, with autochthonous organic matter predominating. In chemistry, natural lakes may be oxygen and light limited at depth, but temperature is more stable. In the evolution of systems, lentic systems progress toward ponds, swamps, and, finally, dry land as they accumulate sediment and organic matter from various sources.

Table 12.1 Lotic versus Lentic Systems

	Rivers	Lakes
Physicochemical Factors		
Water movement	Due to gravity	Due to wind
Temperature regimen	Uniform top to bottom	May be stratified
Temperature changes	Wide range	Moderate
Dissolved oxygen	Rarely limiting	May be limiting
Water fluctuations	Can be abrupt	Gradual
Chemical balance	Usually buffered	Can be a problem
Turbidity	May be great	Seasonal
Fish Adaptations		
Morphology	Streamlined	Variable
Movements	Mostly horizontal	Mostly vertical
Adult habitat	Mostly benthic	Mostly pelagic
Larval food	Macroinvertebrates	Plankton

Temperate lakes can undergo stratification into three definable areas or zones depending upon the density of water, which varies according to temperature. This stratification affects other water parameters such as oxygen and nutrient content, which in turn can greatly affect the lake fauna, including fishes.

Deep lakes experience differential heating due to solar radiation and ambient air temperature. For this reason, it is instructional to divide the water mass into three parts: the epilimnion (upper lake), the metalimnion (middle lake), and the hypolimnion (lower lake; Figure 12.1). The reason for this is the difference in the distribution of temperature in the lake and the density properties of water. Most large temperate lakes mix twice each year. In these dimictic (twice mixing) lakes in the summer, solar energy warms the surface of the lake above 4°C (the temperature of maximum density for freshwater); the water becomes progressively lighter than the water below it, and it floats on the top (Figure 12.2). The epilimnion continues to warm as the summer progresses. However, with depth, the solar heating drops, and the water becomes progressively colder until it reaches 4°C, producing a thermocline. In most locations, wind is light, and only the surface waters are mixed. With increasing depth, there comes a point at which mixing from the surface becomes negligible, and this point is further defined by an increasing density of water as it cools to 4°C. The thermocline

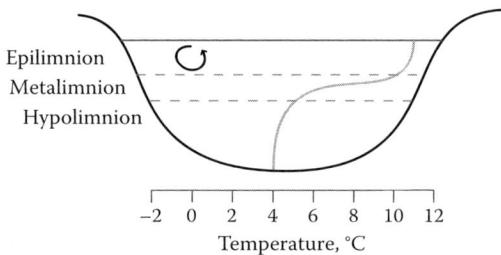

Figure 12.1 Lake temperature stratification. (Courtesy of McCutchan, J. H., Jr.)

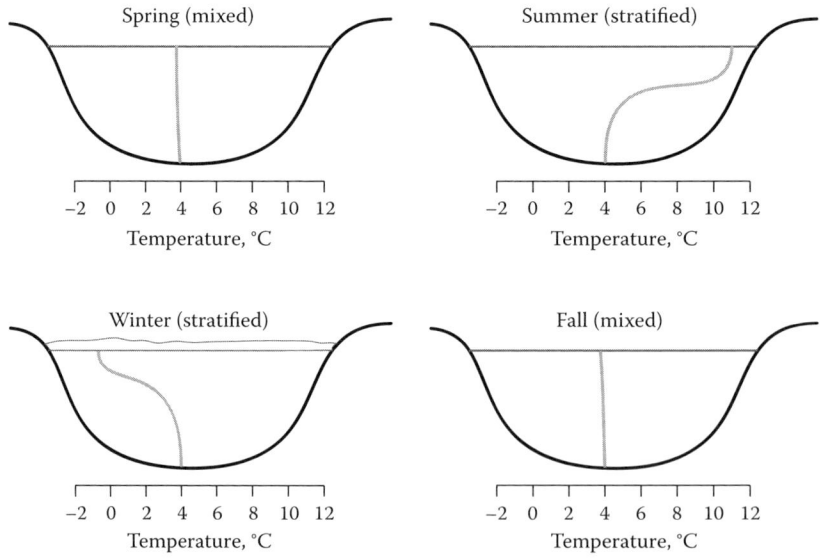

Figure 12.2 Annual cycle in a dimictic lake. (Courtesy of McCutchan, J. H., Jr.)

defines the location of the metalimnion, and when the water mass below reaches a uniform 4°C, this area is defined as the hypolimnion, which is a deep cold area that is effectively isolated from the warm surface. At this point, the lake is thermally stratified. In autumn, conditions change. As the air cools and the rays of the sun become weaker, the surface loses heat, water becomes more dense, and convection currents progressively warms the deeper water until a uniform temperature of 4°C is reached and the lake can freely mix throughout. As winter comes, the surface waters are cooled below 4°C and float on top. As the cooling progresses, an inverse temperature situation can occur with the warmer and thus denser water below, especially with ice cover. In spring, warming conditions melt the ice, surface water is again warmed to 4°C, and it tends to sink downward, aided by wind action, until the lake is once again uniform and mixing is no longer impeded by differing water densities.

In spring and autumn, seasonal circulation of water in the entire lake is referred to as overturn (or turnover). Nutrients produced by the process of decomposition and accumulated organic matter can be recycled to upper waters at this time (Figure 12.3). During winter and summer, however, another condition may occur, which can be particularly important to fishes. Winter stagnation at depths can result in the depletion of oxygen due to decomposition occurring on the bottom and winterkill of fishes. A similar condition can occur in warm climates in the summer due to summer stagnation and high rates of metabolism.

A final and extremely important characteristic of lakes is their relative state of impermanence. The oldest lake is Lake Baikal (25 million years (my)), followed in turn by Lake Tanganyika (11 my), but these are exceptional. Most lakes are ephemeral, and they are of very recent origin. Measured on geological time scales, only a few lakes date back to the Tertiary period, and most lakes were formed during the Pleistocene (Ruttner 1963). Due to sedimentation and tectonic forces, lakes are present for only a relatively short time, and as they disappear, the fauna that are endemic to them may also disappear.

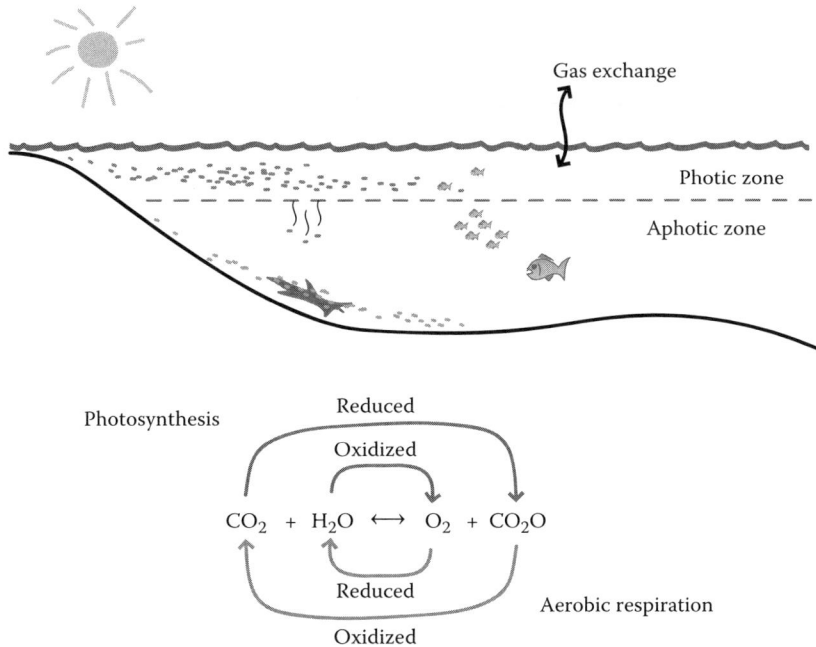

Figure 12.3 Hypoxia in a stratified lake. (Courtesy of McCutchan, J. H., Jr.)

STRUCTURE

Lentic systems have relatively distinct zones based on the density stratification and rate of photosynthesis. In shallow shorelines, the light will penetrate all the way to the bottom; this is the littoral zone. In deeper lakes, the limnetic (or euphotic) zone supports photosynthesis in open water, where the lowest part of that zone will correspond to the deepest level of effective light penetration. At that point, the amount of energy expended by photosynthetic organisms is equal to the amount of energy produced by photosynthesis, that is, the compensation level. Below that level is the profundal zone. The littoral zone is also called the zone of rooted, emergent, and floating plants. This zone has cover and produces much food that can be utilized by small fishes and other organisms. The limnetic zone is used by planktonic organisms and the animals that prey on them, including planktivorous fishes. The profundal zone may be extremely important to coldwater fishes in the summer, and it provides access to the bottom for fishes that feed on benthic organisms or organic matter.

FUNCTION

Lentic systems are relatively complex. Due to the great diversity of lakes, there are combinations of autotrophy and heterotropy within them. Autotrophic components included the aquatic vascular plants, filamentous algae, and phytoplankton. Depending on lake size and other considerations, many receive large inputs of nutrients and organic matter from the streams and groundwater within the drainage basin, and they also receive dissolved and particulate matter from wind, rain, and snow. As we shall see in the case studies, the detrital food chain can be of supreme importance in some lakes.

Because of the great diversity among lakes in their degree of productivity, it is helpful to classify them by their nutrient status: Oligotrophic (low food) lakes are nutrient-poor systems characterized by a low surface-area-to-volume ratio. In these systems, water is clear, sediments are mostly inorganic, and they are well oxygenated. On the other end of the nutrient spectrum are the eutrophic (rich food) lakes, which generally have a high level of nutrients, large surface-to-volume ratio, colored and/or turbid water, organic matter in the sediments, and possibly insufficient oxygen during the year to meet the needs of within-lake decomposition. Eutrophic lakes occur as lakes age, but this condition can be accelerated by humans due to agricultural runoff of fertilizers, discharge of sewage, and so on. I am very familiar with another type called dystrophic (badly nourished) lakes, from my research work in coastal North Carolina. These brown-water lakes are so high in humus that there is a lack of nutrients and plankton, but most are shallow and they can be high in productivity due to growth of rooted aquatic plants (macrophytes).

It is tempting to think of freshwater lakes as small oceans. This is a bad idea (e.g., see Magnuson 1988). Although there are similarities, lakes should be thought of as islands, because from an ecological perspective, lakes are usually not well connected and fish movement between them is restricted. Also, environmental conditions in lakes are more extreme. Fish recruitment in lakes are thus affected by extinction forces, such as low dissolved oxygen (DO), little connectiveness, and slow reinvasion. In contrast, recruitment in oceans is encouraged by colonization of connected areas.

FISH IN LAKES

Fish are of enormous importance to humans for subsistence, recreation, and economic value, and much attention has been devoted to their study. Unfortunately, fish have been treated almost entirely in a void, relying on population dynamics and exploitation models to regulate fisheries in lakes and oceans and with little knowledge or understanding about limiting factors or regard for the ecosystem upon which they depend. Two schisms developed: (1) fisheries without limnology and (2) limnology without fish! Neither is very helpful. The first instance will be made clear as we discuss commercial exploitation of ocean fisheries, which will have to wait for now; the second can be related to the amount of material about fish that is presented in limnology textbooks, which historically have averaged less than 1.5% of the pages (average of 3 pages).

Limnologists and fishery scientists understand that fishes are integral to aquatic systems, especially to freshwater ones. Although the amount of energy flowing through the fish community is small with respect to the entire system, fishes do high-quality work ordering the system, and their presence or absence can have a great deal of effect. It is pertinent to point out that fishes can be herbivores, carnivores, or detritivores and that they have a great deal of trophic adaptability (i.e., capable of feeding at more than one trophic level).

Young lakes are occupied by the same fishes that occupy streams, and in the very old lakes, there are likely to be endemic fishes that have evolved there from stream ancestors. Stream linkages may be easily detected in most instances, for there will be movement between them by the lake fish, especially during spawning. However, there are major differences in fish adaptations between temperate and tropical lakes, which we will investigate.

In the northern portions of the United States, deep lakes were occupied by lake trout, ciscoes, and whitefishes, while shallow lakes are occupied by smallmouth bass, pike, walleye, and perch. In the southern portions of the nation, shallow lakes are mostly inhabited by largemouth bass, a variety of sunfishes, bullheads, gar, and bowfin. Fishes of North American lakes were clearly divided into two groups by Bennett (1971): (1) fishes characteristic of oligotrophic (trout) lakes (production <20 lb fish per surface acre) included trout and salmon, whitefish, coldwater herrings, and perhaps grayling, and (2) fishes characteristic of eutrophic (bass) lakes (production 100–150 lb) included largemouth and white basses, crappies, bluegills and other sunfishes, buffalo, channel catfish, bullheads, carp, and suckers. In two-story lakes, it is now common to find bluegills and largemouth bass in the littoral areas and rainbow trout in the hypolimnion.

Taking a regional approach, the northeastern U.S. lakes are mostly glacial and include deep trout lakes and shallow bass lakes. In the southeast, there is an estimated 955,000 acres of shallow lakes occupied mostly by bass, sunfishes, and bullheads. In the north central region, there are over 40,000 natural lakes larger than 10 acres (Bennett 1971), many parts of which are yellow perch and walleye dominated, although largemouth bass and bluegills have been widely stocked in this region. Natural Rocky Mountain lakes are high in elevation (above 4000 ft) and are now stocked mostly with native and nonnative trouts and salmon. Natural lakes in the southwest are rare and originally had native killifish, pupfish, and other hardy species.

Theoretically, fish production in lakes has been linked with two conceptual models, the *trophic state* model, which holds that nutrient levels are a limiting factor for fish production, and the *food web* model, which holds that species interactions and the structure of the lake community itself function as limiting factors (Hayes et al. 1999). As lakes and reservoirs age, their nutrient levels also change due to natural processes associated with productivity and, in many cases, nutrient loading by humans, leading to a state ranging from oligotrophy to eutrophy, as discussed earlier. Increasing productivity can lead to expansion of the fish biomass as a part of the food chain. However, food webs also will be affected during this process, and one could expect that the amount of fish biomass at the various trophic levels would depend on the productivity, while the number and type of predators would shape fish standing crops in the lower trophic levels. However, the availability of the physical component of habitat also can be limiting (reviewed by Hayes et al. 1999). These models are simplistic but helpful in conceptualizing fish production. Both models are related and thus interactive and would have to be considered together. Also, just observing an isolated lake is not a good idea. Lakes are complicated, and the nature of the basin and the inflows and outflows have to be considered in real systems.

ECOLOGICAL CONCEPTS

- *Lake as a Microcosm*—The term "ecosystem" was proposed by A. G. Tansley (1935), but Stephen A. Forbes had essentially developed the concept in his famous paper, "The Lake as a Microcosm," many years before then (Forbes 1887). Forbes characterized lakes as understandable units of nature that have remained isolated and relatively unchanged in their "organic interactions" for geologic time scales. He also envisioned that aquatic animals in lakes "are, as a whole, remarkably isolated" and "independent of the land about them" (p. 77). However, this part of Forbes' work should not be taken out of context. It is clear later in the paper that he considered the floodplain of "fluviatile lakes" to be an integral part of the microcosm and discussed the role of fish migration in connecting marshes, shallow, and deep lakes. Forbes also discussed the interconnectivity of lake organisms, which includes a discussion of trophic adaptability in largemouth bass (black bass in his paper), and identifies detritivory in fishes. Thus, natural lakes have been studied intensively as ecosystems. Although we now appreciate more fully the contributions of watershed and wind effects on lakes, Forbes (1887) provided a basis for the development of the field of ecology, introducing the ecosystem concept, food webs, population and community ecology, flood pulse, and the role of natural selection on lake organisms. Recently, the microcosm concept has been criticized for its narrow focus and lack of relevance to ecosystems (e.g., see Carpenter 1996), but the concept persists today as a paradigm for the study of various aspects of lake ecology (see response to Carpenter 1996 by Drenner and Mazumder 1999).
- *Ecological Succession and Evolution of Lake Ecosystem*—Lakes undergo changes in physical and community structure as they age. New lakes are typically oligotrophic, clear, and deep, while older lakes become more eutrophic, turbid, and shallow. These changes are due to accumulations of inorganic and organic matter, which results in the gradual filling of the basin. This filling continues from lake to pond to swamp and, finally, to dry land. Lake filling is due to sediment accumulation from various sources, including wind-borne materials, sediment transport by runoff and stream input, wave action on shorelines, and the accumulation of organic matter and animal remains. The process has been well studied in U.S. bog lakes (e.g., Gates 1942, reviewed by Wetzel 2001).
- *Lake Stratification*—Thermal stratification of lakes due to the density properties of water was discovered by Simony in around 1850 (Welch 1952). In the deeper lakes of the world, this phenomenon greatly affects the structure and function of the lentic ecosystem, especially in temperate climates. Thus, lakes have been classified based on thermal condition, as succinctly addressed by Ruttner (1963): Forel attempted the first lake classification in 1901 and grouped lakes based on surface temperature: polar Lakes (<4°C), temperate lakes (<4°C to >4°C on a seasonal basis), and tropical lakes (>4°C). This classification was found lacking, so Hutchinson and Löffler (1956) identified six types of lakes based on lake mixing. Lewis (1983) developed the system presently in use based on eight types of mixing: amictic (= without mixing), cold monomictic, cold polymictic, discontinuous cold polymictic, warm polymictic, discontinuous warm polymictic, dimictic, and warm monomictic lakes.
- *Trophic State Model*—Fish production in lakes has been related to the changes in productivity that occur with aging, from oligotrophic, to mesotrophic, and then to eutrophic states. The trophic states represent the amount of nutrients. This is a very broad and general concept that can be difficult to apply depending on how "nutrients" are measured and which nutrients are limiting. Also, lakes are not isolated bodies of water.

- *Food Web Model*—The role of fish predation is considered to be the main force structuring prey communities. This model resulted from studies of fish introduced into previously fishless ponds and from seeing the effect on zooplankton.

CASE STUDY: LAKE BAIKAL

Also known as the "Pearl of Siberia," Lake Baikal is the oldest and deepest freshwater lake on Earth. It also has the largest volume of water: Lake Baikal contains 23,000 km^3 of water, which is equivalent to the contents of all five of the Laurentian Great Lakes combined! This oligotrophic lake is considered to be a national treasure to the Russians, who compare its beauty with that of the Grand Canyon. The lake was showcased in a very interesting and informative article in *National Geographic* (Belt 1992), and information about the lake can be obtained from many Web sites.

With a shoreline of over 2000 km and a length of 635 km, Lake Baikal is a deep blue lake set in granite outcrops and green coniferous forest. It is truly a magnificent lake (Belt 1992). However, its true depth is even more amazing. Although presently filled with sediment to depths exceeding 1600 m, the depth to the floor of the rift that created the lake 25 million years ago (mya) is 9000 m (9 km). Even in its present condition, the lake is so vast that if its 336 tributaries were cut off today, it would take another 400 years of its present discharge (the Angara River) to empty it. Even with such depths, its cold water and oligotrophic history have resulted in enough oxygen at depths to support an endemic benthic fish community.

Lake Baikal has produced about 1500 animals in 25 my of which 80% are endemic. There were 56 native fishes in the lake, but one of these, the Davatchan (a char), has been extirpated (Sideleva 2003). Of great interest is the presence of the oldest known species flock of fishes, composed of cottoids (sculpins) that have persisted from the Pliocene or early Pleistocene. It is thought to be composed of three families, 12 genera, and 29 species (Hunt et al. 1997), and Kontula et al. (2003) recently reported an endemic flock of 33 species, with strong support for monophyly of the group.

The best known fish of Lake Baikal is the omul (Figure 12.4), a large whitefish that supports most of the commercial fishing catch. Unfortunately, numbers of the fish have recently declined, along with those of a tiny crayfish (epishura) that filters the upper level of the lake. The decline of these species has been related to pollution from a huge pulp-and-paper plant constructed on the lake, poaching, and unwise development of the lake's shoreline. According to an opinion and analysis news release (Sinitsyna, 2008), the Russian government is very concerned that the plant is degrading the lake ecosystem due to huge waste discharge (41 million m^3) and the leaching of almost 400,000 tons of ash stored on the lake shore. Local, regional, and international environmental agencies and private groups also are very concerned and a closed-loop water system is being

Figure 12.4 Omul, the Baikal whitefish. (Courtesy of Octagon, http://commons.wikimedia/wiki:File:Thymallus_ baikalensisBaikal-Museum.)

developed for pulp and paper mill wastes, and sewage from the town of Baikalsk. Hopefully, effective management efforts can be developed to maintain the integrity of this unique and magnificent ecosystem while there is still time.

SUMMARY

Lentic ecosystems include swamps, marshes, ponds, and lakes. Swamps are shallow, lower-elevation wetlands that support trees and shrubs. Marshes are very shallow wetlands with grasses and sedges. Ponds and lakes exist more in a continuum; however, in general, ponds are small lentic systems in which the littoral zone occurs all the way to the middle, whereas lakes are deep enough to be thermally stratified. Natural lakes are the largest type of lentic system, and in North America, most natural lakes are formed by glacial action. Differences in water residence time can vary due to lake volume and gradient. In the interconnected Laurentian Great Lakes, Lake Erie has a water residence time of only 2.6 years, while Lake Superior stores water for 191 years. Such lakes are considered open basins. Closed basins have no surficial outflow, such as in the Great Basin in the western United States. Perhaps the best way to characterize lentic systems is to compare them with lotic ones: Natural lakes are deeper and wider, have little or no water flow, tend to cycle nutrients instead of flushing them downstream, are mostly autotrophic, and are more stable in temperature regimen. Lakes also retain sediment, which eventually begins to fill them as part of an ecological succession (or "evolution") from lake to pond, swamp, marsh, and, finally, dry land. Deeper lakes in the temperate zone can undergo temperature stratification into three zones: the epilimnion, metalimnion, and hypolimnion. Due to the density properties of water, wind strength, and duration, they may be thermally stratified, usually mixing only twice each year (dimictic). This mixing is called overturn and typically occurs in autumn and spring. Lakes also have zones based on depth, which include the littoral, limnetic, and profundal zones. Lakes may be nutrient poor (oligotrophic), moderately rich (mesotrophic), nutrient rich (eutrophic), or badly nourished (dystrophic, brown water). Fishes of recent lakes are local riverine species; however, systems such as bottomlands and coastal lakes can be occupied seasonally by diadromous species. Large, very old lakes usually have some endemic fishes. In the United States, deep northern lakes are occupied by lake trout, shads, ciscoes, and whitefishes, while shallow northern lakes are occupied by smallmouth bass, pike, walleye, and yellow perch. Southern lakes are usually shallow and occupied by largemouth bass, sunfishes, bullheads, gar, and bowfin. Fish communities of oligotrophic lakes are mostly salmonids, and those of eutrophic lakes are mostly sunfishes, catfish, and suckers. Fish production has been linked with a trophic state model, in which nutrients limit production, and a food web model, in which species interactions in the community limit production. Lakes have been considered as "microcosms." The concept still survives as a paradigm for lake study. Lake Baikal, the "Pearl of Siberia" and the largest lake in the world by volume, is used in a case study. Formed about 25 mya, this oligotrophic lake has produced a species flock of about 30 endemic species of cottoids (sculpins). Location and operation of a large pulp-and-paper plant on the lake have been a concern.

CHAPTER 13

Fishes of Temperate and Tropical Great Lakes

GENERAL

Fish communities of lakes are influenced by many factors. Some of the most important ones are the age of the lake; its size relative to depth, volume and surface area; elevation; latitude; and trophic status. It is also important to consider the capability for rapid speciation of fishes in older lakes, especially in some fish families—a topic to be discussed when we consider reproduction of fishes. For now, major influences on fish communities in even the largest of lakes are occurring due to the actions of humans. For comparison purposes, two major types of lakes that are important to fishes stand out: temperate and tropical lakes (Table 13.1).

FISHES OF TEMPERATE LAKES

Natural temperate lakes are predominately of glacial origin. They are comparatively recent (i.e., Pleistocene), shallow, and thus ephemeral due to the normal process of lake filling with sediments from inflowing streams, soil from land runoff, and wind-borne particles. Fishes of temperate lakes are exposed to changing conditions (e.g., temperature, turbidity, nutrients) on an annual, seasonal, and, for shallow water species, perhaps daily basis. Thus, native fishes are usually well adapted to the demanding climatic conditions of the temperate zone. Fish species of temperate lakes tend to be the same as or recently derived from ancestors in the rivers and streams that flow into them. With few exceptions, the fishes are more of generalists and not uniquely specialized for lentic life. This is not a bad strategy, for as lakes fill and change, the fishes that live in them can disperse elsewhere.

In addition to the abiotic components of the environment, fish communities of smaller lakes in the temperate zone have been greatly altered by introductions of many nonnative fishes over the past 150 years. As a result, historic fish communities of lakes may be hardly recognizable. For example, all across the United States, coldwater lakes of higher elevation (some historically having no fish in them) have been stocked with rainbow, brook, and brown trouts (and deep lakes with lake trout, e.g., Yellowstone Lake). Coolwater lakes have been stocked with northern pike, walleye, yellow perch, and smallmouth bass. Warmwater lakes have been stocked with largemouth bass, other sunfishes such as bluegills and crappies, catfishes, and carp. In addition, even smaller nongame fishes, including suckers, minnows, darters, sticklebacks, and so on, also have been stocked, usually from bait buckets or from warmwater hatchery ponds. In some cases, a remnant of the native fish fauna persists; in others, it does not. Therefore, it can be easier to predict which nonnative fishes you might find in a lake than to predict which native fishes might remain.

Fish communities of the larger temperate lakes may have persisted long enough to have more lake-adapted fishes. In general, these lakes (and especially oligotrophic ones) have been less affected by introductions. As an example, Lake Baikal, as previously discussed, is the largest

Table 13.1 Temperate versus Tropical Lakes

Temperate Lakes
Fish distribution and abundance mostly influenced by physical conditions
Temperature may be cold, varying greatly between seasons
Precipitation results in dependable runoff in most locations
Lakes tend to be recent and ephemeral in geologic time scales
Lakes formed or affected by continental glaciation
Fishes tend to be generalists, few endemics; life histories may include stream, many migratory forms; fish diversity usually low
Fish food webs simple; most are carnivores
Fish life histories include both r- and K-selected species; rapid growth in some situations, but delayed reproduction and large size in others
Tropical Lakes
Biological factors can be more important in controlling fish distribution and abundance than physical factors
Temperature is mostly constant, with diurnal fluctuation
Precipitation can be limiting due to wet/dry cycles
Lakes tend to be permanent, geologically stable, and large lakes tend to be very old; no effects due to glaciations
Fishes tend to be specialized; many endemics; fish mostly nonmigratory; fish diversity very high
Fish food webs complex, includes fish herbivores and detritivores
Fish life history tends to be r-selected, favoring early reproduction and rapid growth

freshwater Great Lake. Although it has been affected by humans, its fish community is largely intact. Equivalent in volume (but much larger in combined surface area) to Lake Baikal are the five Laurentian Great Lakes of North America. These lakes have been variously affected by introductions and other human impacts. Due to ecological, economic, and sociopolitical factors, these lakes are so important that all students of fish ecology should be acquainted with them and the changes that have occurred.

The Laurentian Great Lakes: A History of Change

The Great Lakes region (Figure 13.1) of the United States and Canada consists of a vast area of five Great Lakes, smaller lakes, rivers, and waterways that are usually discussed in the concept of a Great Lakes–St. Lawrence zoogeographic province. The Great Lakes themselves cover an area of about 94,700 square miles, with a shoreline of more than 10,000 miles. The Great Lakes region was covered with ice during the Wisconsin Age until about 14,000 years ago (ya). Its waters are vast; however, they are a remnant of past pluvial conditions and should not be considered as a renewable resource. The huge surface area of the lakes experiences a great amount of evaporation, and consequently, water inflow in the system barely exceeds outflow. Evolution of the Great Lakes during and after the Pleistocene is presented in detail by Underhill (1986) and in a recent revision of Hubbs and Lagler (1958) completed by Gerry Smith (Hubbs et al. 2004).

Hubbs et al. (2004) identified 215 fishes native to the Great Lakes, which comprise 161 species and 70 genera. However, there have been numerous additions to the fish fauna by stocking and invasion of nonnative fishes of all kinds, increasing the number of fish species in the lakes to over 200 species.

The Great Lakes region has been connected in various ways to the Mississippi River drainage since the Pleistocene, and its native fish fauna is mostly composed of postglacial invaders from that river. Fishes also have entered the Great Lakes from the Atlantic Coastal Plain and the Beringian Refugium from the unglaciated part of the Yukon Valley, Alaska (Underhill 1986). Hubbs et al. (2004) listed 6 alternative routes for immigration of cold, cool, and warmwater fishes into the Great Lakes basin. The Great Lakes also have several fishes that presumably evolved there from close

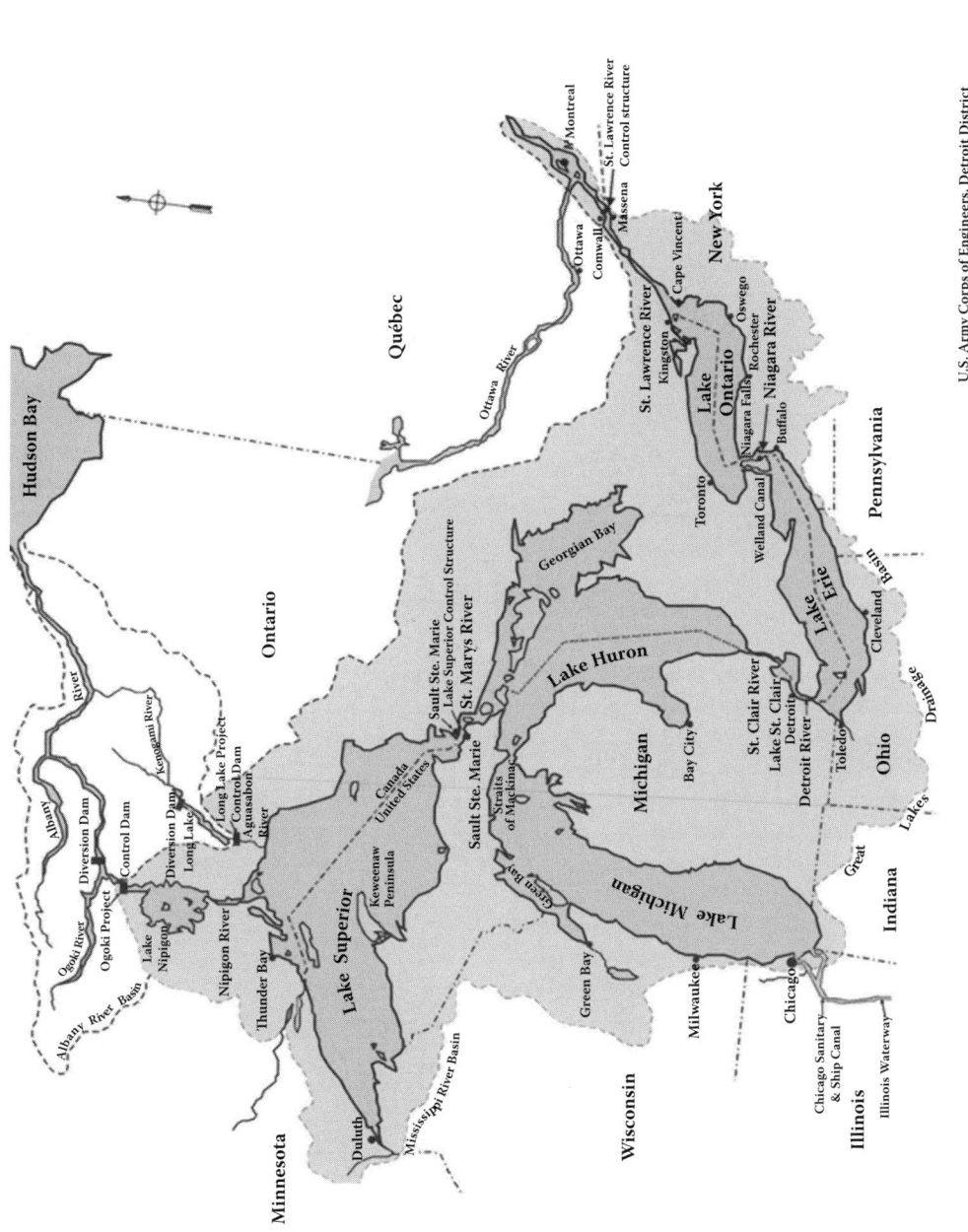

Figure 13.1 Great Lakes and drainage basin. (Courtesy of United States Army Corps of Engineers, http://commons.wikimedia.wiki/File:Great_Lakes_1.png.)

relatives that originally colonized the lakes, including blue pike, siscowet, and four endemic ciscoes (blackfin, shornose, kiyi, and deepwater). Also, many subspecies of ciscoes have evolved in the stable environments of the lakes.

The Great Lakes region, its lakes, and corresponding fish communities have gone through 200 years of anthropogenic changes, which have been well documented (e.g., entire chapters by Jude and Leach 1999; Diana 2004). Early settlers were impressed by the numbers and kinds of fishes in the various lakes, which included five interconnected Great Lakes (Figure 13.1) and other lakes and streams. Although the lakes are in the same basin and are connected by their trophic states, depths and water residence time vary (Figure 13.2). Of the three upper lakes, the oligotrophic Lake Superior is the deepest and has a long water residence time (191 years; see chapter 9). Lake Huron and Lake Michigan also are relatively deep and oligotrophic, and all three of the upper lakes supported mostly salmonids. Lake sturgeon, blue pike, and lake trout were found throughout the upper lakes. Lake trout and burbot were top predators, with whitefishes, ciscoes, and sculpins as prey. Walleye and yellow perch also were found in some areas. Lake Erie is a shallow, eutrophic lake with water residence of only 2.6 years, and walleye and blue pike were the top predators. Whitefish,

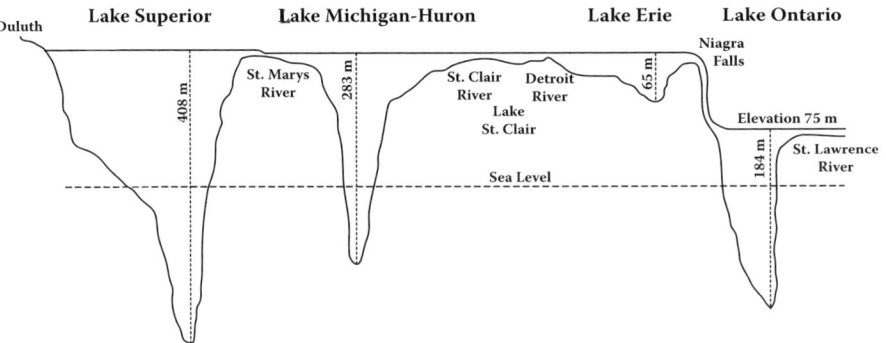

Figure 13.2 Depths and connections of the large lakes. (Redrawn from Jude, D. J., and Leach, J., 1999, *Fisheries Management in North America. 2nd Ed.*, 623–664, American Fisheries Society, Bethesda, MD. With permission.)

lake herring, lake sturgeon, blue pike, sauger, walleye, yellow perch, freshwater drum, and channel catfish were abundant. Finally, Lake Ontario is a relatively deep lake and is isolated downstream of the other lakes by Niagara Falls. The lake had an indirect opening to the Atlantic Ocean via the St. Lawrence River, which admitted Atlantic salmon into the lake as its top predator, and it lived in association with freshwater lake trout (Jude and Leach 1999). Invasion of advanced humans into the area has resulted in profound changes to these lakes in the past 200 years, and such changes have been reviewed by Smith (1970), Jude and Leach (1999), Diana (2004), Dempsey (2004), and others.

The first Europeans to settle in the Great Lakes region were fur traders, who traveled into the area in the early 1700s. They established trading posts and did little to interfere with the lake ecosystem. However, they were followed by farmers and tradesmen. By the early 1800s, they began draining the marshes, lumbering, commercial fishing, and constructing dams for mill ponds (a main source of industrial power in those days) on smaller tributaries. Human populations grew, and trade and industry flourished—unfortunately, all these activities resulted in discharges of solid and liquid wastes (including nutrients and chemical pollutants) into the lakes. These activities set in motion irreversible changes that resulted in a calamitous disruption of the Great Lakes ecosystem, because changes continued to build as human settlement and exploitation moved further inland from lake to lake.

Marsh drainage destroyed spawning and nursery habitats for some of the fishes. Lumbering resulted in deforestation that choked the streams and shorelines with sawdust, increased water temperatures, and siltation over stream spawning areas. Dams for mill power placed on tributary streams blocked spawning migrations, and the pollution from agricultural and cultural discharge added nutrients and organic matter into the lakes. All of these impacts began disrupting lake systems, especially those of the shallower warmwater lakes. For example, the natural aging process in Lake Erie has been advanced about 15,000 years in the last half of the nineteenth century due to these combined effects.

Continuous overfishing began to take its toll on native fish communities. Lake trout, lake whitefish, walleye, and blue pike (a walleye subspecies) were highly desirable for human food and were greatly affected by the late 1800s when commercial catches of these species peaked. Depression of these key predators undoubtedly caused disruptions in fish communities and resulted in changing fish assemblages. Fishing was intense early in the lower lakes, and by 1900, lake trout and Atlantic salmon were eradicated from Lake Ontario. Lake sturgeons (Figure 13.3) were almost eradicated throughout the lakes by 1900, which was due in part to their large size and long generation time. Initially regarded as "trash fish," they were discarded but were later avidly sought after for flesh and caviar; finally, however, incidental captures were killed because they damaged nets.

As national and local Canadian and U.S. governments became concerned about the declining harvest, some regulations were placed on the fishing industry. Perhaps these regulations could have rescued the fish communities; however, the largest species were reduced by recruitment overfishing (produced by using successively smaller nets as catches of larger-size fishes declined), and the opening of the Erie and Welland canals (1825 and 1829) permitted access of marine fishes into the system from the Atlantic Ocean via the Erie canal from Lake Ontario and via Welland canal from the Hudson River. These marine fishes would produce irreversible changes in the Great Lakes fish communities.

Figure 13.3 Lake sturgeon *Acipenser fulvescens*. (Courtesy of the National Oceanic and Atmospheric Administration, http://www.glerl.noaa.gov/pubs/photogallery/Fish/pages/1050.html.)

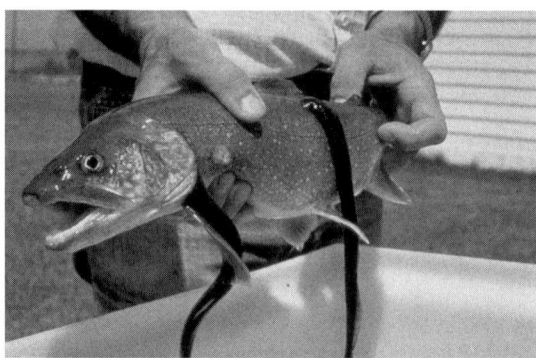

Figure 13.4 Parasitic sea lamprey attached to a lake trout (Courtesy of USGS, http://search.usa.gov/search/images/query?=sea_lamprey.)

The parasitic sea lamprey (Figure 13.4) is native to Lake Ontario or entered it via the Erie Canal (open 1825). The lamprey spread to other lakes after modification to the Welland Canal in 1931 (Courtenay, W. R., Jr., Personal Commnication, 2011). There was not much concern expressed about Lake Ontario, because the lake trout and Atlantic salmon were already extirpated by the time the lamprey was reported. Also, numbers of the parasite were obviously low, presumably due to a lack of good rearing areas for the larvae (called ammocetes, which require silty conditions and remain in tributary streams for up to 7 years). This changed later with deforestation and stream siltation. Spawning areas in Lake Erie also were not very satisfactory. However, by 1946, the lamprey had penetrated all the way to Lake Superior, finding good conditions to complete its life cycle in lakes Huron and Michigan. In those lakes, in combination with overfishing, it devastated the lake trout population, driving it almost to extinction, and also attacked other species, including burbot, walleye, suckers, larger ciscoes, and especially the largest whitefish (Smith 1970). Overfishing removed the larger fish, which could usually survive a lamprey attack. However, smaller fish usually do not survive an attack, so this was a very dangerous combination of impacts. By 1950, the premier fish, the lake trout, had been reduced by 99% in lakes Michigan and Huron; this loss was mostly attributed to the sea lamprey.

Smith (1973) points out that it was more than 50 years later that problems with the sea lamprey were recognized, and by then, it had already spread to all of the Great Lakes. In 1946, an investigation was initiated to develop a control method for it. A selective larvicide called TFM was developed and tested, and by 1961, the adult lamprey population was reduced to about 80% of its former abundance. Unfortunately, the lake trout population was virtually extinct in Lake Michigan where the catch of lake trout declined from 5.5 million lbs in 1946 to only 400 lbs in 1953 (Springer 2011) and was reduced by about 10% even in the oligotrophic Lake Superior. However, by 1973, there was evidence that the sea lamprey populations were at a sufficient level to inflict severe damage on the fish populations. Presently, TFM remains in use, and its effect combined with construction of barriers and sterilization of breeding males has reduced sea lamprey populations by 90% compared to historic values (Springer 2011).

The absence of larger predators in the lakes due to overfishing and lamprey parasitism caused a major disruption that allowed other invading fishes to thrive, including rainbow smelt and alewives. Rainbow smelt invaded the Great Lakes in 1910 after escaping from a lake where they were stocked. Now found in all of the lakes, these insectivorous fish compete with native fishes in the lakes for food (especially zooplankton), and they also consume fish eggs and larvae.

The alewife (*Alosa pseudoharengus*, Figure 13.5a) is an anadromous herring of the inshore Atlantic Ocean that can adapt to living in warmwater lakes. However, it is not a cold-adapted

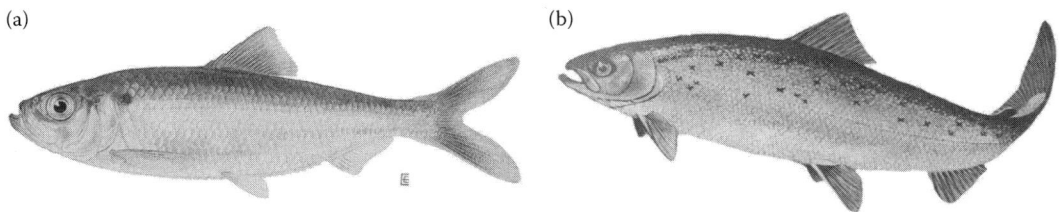

Figure 13.5 Alewife (a) and Atlantic salmon (b). (By Edmonson, E., and Chrisp, H., Courtesy of New York State Department of Environmental Conservation.)

species due to a marine-dominated life cycle. Alewives were present in Lake Ontario by the early 1870s, presumably stocked by accident in shipments of young Atlantic shad (Smith 1970), but also presumably, the alewife populations were kept in check by a combination of cold temperatures and predation by Atlantic salmon (Figure 13.5b) and lake trout (Jude and Leach 1999). With deforestation (increasing water temperatures) and extirpation of the large predators, conditions improved for this species, and they became the most abundant fish in the lake. Alewives invaded the upper lakes via the Erie and Welland canals but did not do well in Lake Erie due to the shallow water and very cold winter water temperatures. The fish was found in Lake Huron in 1932, but their numbers did not greatly expand until the 1940s, when the lamprey had reduced the numbers of large predators. Alewives were first found in Michigan in 1949, when they rapidly increased in numbers (Smith 1970).

Alewife populations exploded in lakes Michigan and Huron, where they were considered "superabundant" (Smith 1970). Adverse effects of alewives include competition with other species for food, displacement of natives due to sheer force of numbers, causing early mortality syndrome in salmonids due to thiamine deficiency, and predation on early life stages of other fishes (reviewed by Madenjian et al. 2008). The fish consume mostly larger plankton but also will eat larger prey, including fish eggs, larvae, and small fish. Due to their high abundance in the lakes, they caused a great reduction in the numbers of other small fish such as the lake "herring" (a cisco) (Figure 13.6), emerald shiner, and yellow perch, and they have inhibited successful reproduction in larger species (reviewed by Jude and Leach 1999). As alewife populations exploded to about 80% of the catch, the commercial fishing industry turned to its harvest. In the absence of other large predators, the alewives can sustain commercial onslaught; however, they are not adapted to cold temperatures and can experience huge winter kill, especially in shallow lakes where there is no deeper water to sustain them.

Fish management agencies in various states began a new lake experiment in the late 1960s in an attempt to increase recreational fishing. In 1966, the State of Michigan began releasing Pacific salmon, and over one-half million coho were introduced into Lake Michigan. Since that time,

Figure 13.6 Cisco *Coregonus artedi* (aka "lake herring"), a major forage fish for the upper lakes. (By Edmonson, E., and Chrisp, H., Courtesy of New York State Department of Environmental Conservation.)

many large predators have been stocked, amounting to about 30 million fish annually. Fishes stocked include coho, chinook, sockeye, and pink salmon. Pink salmon have reproduced and established sustaining populations in all of the Great Lakes (Jude and Leach 1999). Also stocked were lake trout, rainbow trout, and other species. Where there are abundant alewives to feed on, the salmon have done well. However, alewives can experience massive die-offs of hundreds of thousands in cold winters, and they have not developed large populations in lakes Superior and Erie.

System-wide problems have been rampant in the Great Lakes, including widespread pollution due to municipal and industrial waste discharge and problems with the fish populations. Efforts to correct these problems intensified, and different ways were sought to rehabilitate the system. Problems with obtaining results were linked with the difficulty of managing so large a resource with some 30–40 state, provincial, and federal agencies that need to coordinate their efforts (Smith 1973). There also have been constraints due to socioeconomic pressures (Dempsey 2004). However, progress was made, international and other agreements were signed, and water quality, in particular, improved. These positive signs have been partially obscured with further fish invasions and introductions that occurred in the late 1980s.

In 1988, the exotic zebra mussel (*Dreissena polymorpha*) (Figure 13.7) was discovered in Lake Erie, and its presence has been traced to ballast water discharged from a European ship into Lake St. Clair 2 years before. The zebra mussel spread into 11 states and into Canada in the next 4 years. Aggregations of 5000 to 30,000 mussels per m^2 are not uncommon in Lake Erie, where average-sized adults filter about a liter of water per day—and remove virtually all of the plankton from it. Water quality in the lake has improved dramatically, and water clarity has increased remarkably in some areas. But not only do the mussels remove plankton, they also deposit feces and pseudofeces on the bottom, and this accumulation of organic matter has changed benthic communities in the process. Unfortunately, zebra mussels are fouling organisms with attachment devices (byssal threads). The mussels attached to other mollusks, and they have eradicated the native mussels in lakes Erie and St. Clair by blocking shells and smothering siphon tubes.

Other changes in the lakes also have resulted from zebra mussels. Increased light penetration from their filtering has encouraged the growth of aquatic plants in littoral zones. Sunfish populations have increased in some of these areas. Invasive mussels also are providing some food for birds, but bird die-offs have been occurring due to consumption of invasive mussels and fish that are contaminated with botulism. Such an outbreak occurred in the fall of 2007, resulting in the deaths of over 2000 water birds in Lake Michigan, including 520 loons (Breederland and Daniels 2008). In some locations, increased mayfly hatches have been attributed to the mussels, and this translates to increased food for fish. However, the total impact of zebra mussels on Great Lakes fishes has yet to

Figure 13.7 Exotic zebra mussel. (Courtesy of the U.S. Geological Survey (USGS).)

be determined. At present, there is no known way of controlling their numbers, and final outcomes are years in the future.

Ballast water also has resulted in more introductions, such as the quagga mussel (a close relative of the zebra mussel but more tolerant to great lake conditions), spiny water flea (a large sticky copepod), and the ruffe (*Gymnocepalus cernuus*), a percid fish that is spreading to rivers and bays in Lake Superior. The ruffe has displaced perch and whitefish populations in Europe due to a high reproductive rate, feeding adaptability, and predator avoidance skills. As ruffe populations increase, yellow perch, shiners, and other forage fishes can be expected to decline, and states have taken management actions to deter their spread. Finally, recent fish introductions involve two species of gobies (round and tubenose gobies), first discovered in 1990. By 1995, these fishes had invaded all of the Great Lakes. These are hardy benthic fishes expected to displace sculpins and small percids and are likely to invade rivers outside of the lakes.

Finally, invasions of other fishes from the Mississippi River connection pose another potential threat. In particular, three species of Asian carp (silver, bighead, and black carp) seem poised to invade the Great Lakes through the Illinois River–Lake Michigan canal. These fishes are thought to be of such danger to the lakes that an electric barrier fence has been installed (at great cost) in the canal in an effort to stop them (Finney 2011). We will discuss Asian carp invasion more fully in chapter 31.

There is some good news in the Great Lakes due to the reestablishment of the lake sturgeon with fishes stocked from hatcheries, ongoing research to determine and improve habitat for the fish (Bryan 2011), and continued success of keeping the sea lamprey population down (Springer 2011). Also, the rebound of ciscos (previously known as lake herring) in Lake Superior is good news but unlikely in other lakes where alewives are abundant. Stockwell et al. (2009) considered alewives to be a major detriment for the rehabilitation of native fishes, while pointing out that management agencies have protected alewife stocks (i.e., known as the "alewife paradox"). The present management approach is to view alewife populations as important commercial fisheries and a means for sustaining the recreational sport fishery for Pacific salmon. Stockwell et al. (2009) postulated that conflicting management goals would "continue to produce poor results for native fish rehabilitation."

Problems presently affecting the Laurentian Great Lakes are much more extreme than presented here if contaminants are addressed in association with all of the other fish-related issues. Also, the issue of increased lake evaporation and heating associated with present global warming could pose additional constraints on the system due to its slim advantage in inflow versus outflow. There is one thing that is very clear, however: the historic Great Lakes ecosystem cannot be restored to its former condition. Not only are fish populations distorted, reduced, and overwhelmed with introduced species of predators, competitors, and parasites, but there already are five fishes that are extinct and five more that are included in the federal endangered species list. The present Laurentian Great Lake system is one of change and continuing instability for the foreseeable future. It is also clear that human changes in the system have improved conditions for the invasion of nonnative fishes and will likely continue to do so. Very recent attempts to focus management on native fishes include taking an ecosystem approach, such as proposed by Zimmerman and Krueger (2009). They used a conceptual model to develop hypotheses and research questions associated with reestablishing native deepwater species.

FISHES OF TROPICAL LAKES

As discussed in the zoogeography chapter, fishes are much more diverse in the tropics compared with the temperate zone. This occurs for several reasons, including year-round growing seasons and a relatively stable climate. Tropical lakes are not formed due to glaciation; the large tropical lakes were formed by tectonic forces or riverine processes. Fishes of the tropical great lakes have had more time to evolve; they generally benefit from a more stable environment and have great plasticity in feeding adaptations. As a result, tropical fish communities tend to be more diverse, the fishes are more specialized, and their populations are influenced more by biotic factors (other species) than abiotic factors (climatic conditions).

CASE STUDY: CICHLIDS OF EAST AFRICAN GREAT LAKES

The Lakes

The Great Lakes of Africa—Lake Victoria, Lake Tanganyika, and Lake Malawi—are the largest and most persistent lakes in the tropics (Barlow 2000). The lakes and other smaller ones were formed by tectonic forces in the Great Rift region of eastern Africa. The Great Rift is composed of eastern and western parts, and Lake Victoria was formed from a crustal uplift between them that tilted the lower end of its basin upward, trapping a large shallow basin. The other two lakes are deep. They are direct products of the rift, created as the Earth's crust was torn into a chasm of almost 3000 km in length. There are great differences, however, among the lakes (Table 13.2). Lake Victoria is relatively young; it was formed only about 14,000 ya. However, it has a huge surface area for a freshwater lake, second only to Lake Superior. Its depth is only about 100 m, and historically, it had oxygen all the way to the bottom. Lakes Tanganyika and Malawi are more similar to each other and unlike Victoria in several ways. Lake Tanganyika is very old, perhaps formed 9–12 million years ago, and it is very deep (almost 1500 m). Lake Malawi is younger (1–2 million years old) and about 700 m in depth. Because both of these rift lakes have great depth and long narrow profiles, they are permanently stratified and anoxic below 150 to 200 m. These three lakes are important for many reasons, one of which is their great diversity of fishes.

The Laurentian Great Lakes of North America, which have just been discussed, have about 215 native fishes (Hubbs et al. 2004). These lakes are physically comparable in surface area to the African lakes, but the diversity of fishes in the African lakes is astounding. No one really knows how many fish species there are in these lakes, but in the cichlid family alone, the three lakes contain at least 1000 species of cichlids (perhaps there are 1350 species in all) (Nelson 2006), and there are another 50 or so species of noncichlid fishes. Even more strange is the high level of endemism. Almost all of the cichlid species in each of the lakes are endemic there (Helfman et al. 2009), and the wide range of adaptations that have developed is truly incredible (Barlow 2000). The appearance of so many similar species in these lakes is remarkable, but even more so, it is thought that they appeared within only a few thousand years. The process has been dubbed "explosive speciation" due to the resulting production of flocks of species (or "species flocks"). The process was once thought to be the result of sympatric speciation or of allopatric speciation due to drying up separation and reinvasion of some smaller lakes. Now it is generally believed that it is a special case of allopatric speciation that can occur within a lake. (Note: The video *Lake Tanganyika: Jewel of the Rift* shows the likely mechanism.) The process occurs when cichlids are isolated by small events such as

Table 13.2 East African Great Lakes

Name	Rank	Depth	Age	Cichlids
Victoria	3	<100 m	<14,000 y	400+
Tanganyika	7	1470 m	9–12 my	300+
Malawi	9	700 m	<2 my	500+

localized flooding by tributaries or siltation that covers rocky bottoms inhabited by rock-dwelling cichlids. The fish do not want to leave the protective cover of rocks to swim over sandy flats, so the population becomes "isolated." Because of the scale, this is called *microallopatric speciation*. The speed of the process is enhanced by color selection by breeding females and the presence of both jaw teeth and pharyngeal teeth. There are other examples of species flocks (e.g., Lake Baikal), but none of them is as remarkable as the cichlid flocks of the three African lakes.

Cichlids

What is a cichlid? It is a fish in the family Cichlidae (Perciformes), which is related to and superficially resembles fishes in the surfperch family (Figure 13.8). Also, a generalized cichlid appears at first glance to be very similar to a green sunfish (Centrarchidae), a fish widely known across the United States. However, there are major differences to an ichthyologist: the cichlid has a single nostril on each side of the head instead of the sunfish's double nostrils, the lateral line is not continuous, and the cichlid generally will have more spines in the unpaired fins. Of course, there is much diversity in cichlid morphology. Another way to relate to the cichlids is to dine on a tilapia, a closely related genus, which has been widely cultivated and quite common in restaurants. Nelson (2006) tells us about classification problems for all those cichlids but reports 112 genera, including the genus *Tilapia* and the important African genus *Haplochromis*.

Of great interest here is the presence of the teeth on the jaws and also on the last gill arch, which are sometimes called "pharyngeal jaws." Unlike other perches, the cichlid pharyngeal jaw is constructed differently, and it is capable of handling a wide variety of foods. The presence of two types of jaws with a diversity of teeth types that differ between them is a key to understanding how such a diversity of feeding adaptations evolved (e.g., see Barlow 2000).

Dutch researchers have long been interested in fish evolution in Lake Victoria, because it developed its diversity of cichlids in less than 14,000 years (the lake was dry before then). Goldschmidt (1996) has written an interesting and informative text of his experiences and research. In the 1970s, the Dutch government provided aid to Tanzania in the form of a research trawler, which was shipped to the lake by railroad. They also set up a fish research station on the Mwanza Gulf in 1977.

In Lake Victoria, Goldschmidt (1996) and others studied a fish community dominated by perhaps as many as 500 species of haplochromine cichlids (locally identified as "furu"). The food web of these fish included just about every conceivable food adaptation, including some that are almost inconceivable. Their adaptations are greatly aided by the ability to use both jaw teeth and pharyngeal teeth to process food, resulting in the expression "trophic jack of all trades" (Barlow 2000, p. 227). Those furu species that Goldschmidt (1996) classified according to various food

Figure 13.8 Cichlid. (Courtesy of Drake, J., and Midgley, D., http://commons.wikimedia/wiki/File:Astatotilapia_latifasciata.png.)

eaten show feeding at all consumer trophic levels, and his classification is presented below (for more detail, see Kaufman 1992):

Mud biters	Feed on diatoms and bluegreen algae filtered from mud
Algae scrapers	Scrape algae off rocks—rock grazers on *aufwuchs*
Algae eaters	Scrape plants—periphyton grazers
Leaf choppers	Chop off pieces of plants
Snail crushers	Crush whole snails with pharyngeal jaws
Snail shellers	Seize soft parts of snails and pry them out of shells
Zooplankton eaters	Suck in crustaceans directly with protrudable mouth
Insect eaters	Sift insect larvae from mud
Shrimp eaters	Eat shrimp directly in deep waters
Fish eaters	Ambush or pursue predators of other fish
Pedophages	Ram mouth brooders to release young and consume them
	Snout-suckers—attach to mouthbrooders and suck out fry
	Egg snatchers—raid nest of whole nesters
Scale eaters	Rip off scales (left and right handed fish)
Cleaners	Eat parasites on other fishes

The food habits and implied adaptations mentioned above arose in isolation in Lake Victoria, the youngest lake of the rift valley. It is incredible that these fishes could have evolved so many adaptations in such a short time there. Unfortunately, these fish have been greatly affected by the actions of humans, and the entire community is being lost.

The Nile Perch Arrives

In the early 1950s, British officials were trying to assist Uganda with their fisheries and proposed the introduction of a predaceous fish into Lake Tanganyika. It is apparent that even then, overfishing was already happening or imminent. In this case, such a fish would hopefully consume the smaller and less desirable bony fishes, which were unused by fishermen, and convert this biomass into tasty fish fillets. On the surface, this seems like a great idea. A giant piscivore, the Nile perch (*Lates niloticus*) (Figure 13.9) was selected as a candidate for African great lakes, but this was fiercely opposed by Geoffry Fryer, an ecologist who presented some disastrous possible consequences to such action. Unfortunately, these logical warnings were ignored, and in 1954, a Kenyan fisheries officer stocked Nile perch into Lake Victoria (Goldschmidt 1996). The rest is history. The Nile perch began to multiply, and soon the older fish were large enough to feed on even the largest cichlid and every other fish they could catch. Nile perch in Lake Victoria attain sizes of 6 ft and 135 lb (Barlow 2000).

Researchers worked on while the Nile perch population grew and expanded its range. However, when they found more of their research fish in the stomachs of the Nile perch than in their nets, they began to be concerned. In samples from the Mwanza Gulf, Goldschmidt (1996) noted that by 1988, open water samples (at his station G) that formerly exceeded 9000 furu were severely depleted, and there was a 93% reduction in the number of species present. The fish populations also seemed to be affected differently among habitats. Fish populations in shallow waters were reduced by 70%, but rocky habitats occupied by rock-hiding cichlids were hardly affected. By 1990, 80 of 123 species were gone from the sampling location, and about 200 of 300 species were missing from other locations representing the whole lake.

Introduction of the Nile perch has been heralded as a great success in fish management by providing food. In Tanzania, the name for the fish (Sangara) translates as "savior" to the local people. In fact, Lake Victoria is the most productive lake on Earth, with 200,000 to 300,000 tons of fish

Figure 13.9 Nile perch. (Courtesy of smudger888, http://commons.wikimedia/wiki/File:Lates_niloticus_2.jpg.)

taken annually (Barlow 2000). However, a tragedy also has unfolded for the local people, as vividly portrayed on the video *Darwin's Nightmare* (Mille et une Productions 2004), in which the poverty-stricken, starving, and disease-ridden local people are forced to subsist on discarded fish scraps. How could have this happened in so short a time?

Historically, fishes caught in Lake Victoria were about 80% cichlids. This catch was preserved by air drying and was sold locally. By the 1980s, this began to change in various locations, and clearly by the early 1990s, throughout the lake, the catch was about 80% Nile perch and only 1% cichlids. Now there are indications that the harvest of Nile perch is dropping. Some contributing factors follow:

(1) Nile perch flesh is oily. It cannot be preserved by air drying alone, and locals had to burn wood to smoke the meat. The lake already was increasing in turbidity due to increasing human activity in the basin, but the deforestation has greatly increased lake turbidity and has increased nutrients. However, turbidity has reduced phytoplankton productivity, and there were fewer phytoplanktons producing oxygen and more phytoplanktons dying and rotting. As a result, there were more cyanobacteria and an invasion of water hyacinths, which have resulted in a lowering of oxygen in lake waters.

(2) Nile perch extirpated the easily captured cichlids—including deeper water planktivores and detritivores (it appears that some of the rock-hiding cichlids persist). This resulted in the explosion of freshwater shrimp (prawn, *Cardina nilotica*) populations due to less predation by cichlids and the availability of enormous quantities of organic matter as food. Stomach content analyses showed that smaller Nile perch switched to feeding on shrimp, and they in turn were consumed by larger Nile perch (Goldschmidt 1996).

(3) The vast quantity of accumulating detritus could not be consumed and recycled by the shrimp alone. Previously, the cichlids were able to do this and also feed on planktonic algae. Their removal by fishing aided in keeping oxygen levels high all the way to the bottom (<100 m). Now the accumulating detritus (including sinking plankton from the littoral zone), cyanobacteria, and decreasing phytoplankton productivity have resulted in significantly reduced lake oxygen levels.

(4) Before 1986, oxygen did not fall below 1 ppm in the lake, but after 1987, this level occurred at a depth of only 40 m. A remote submarine employed in the early 1990s produced videos showing only dead organisms (fish and invertebrates) below 40 m (Kaufman 1992). At this time, it was evident that most of the lake could no longer sustain aerobic life. Even though the shrimp are capable of tolerating relatively low oxygen levels, they obviously cannot live without it, and there are indications from die-offs that the anoxic level fluctuates, perhaps catching organisms unaware.

To review changes in the lake, about two-thirds of the native cichlids and other fish disappeared after the introduction of Nile perch, including most detritivores and planktivores. At that time, the stable food web, based on perhaps 350 fish species interacting at all trophic levels, was replaced by a much smaller one (Goldschmidt 1996). After shrimp-eating cichlids were reduced, shrimp increased by about 1000% and became the main food of young Nile perch, which then were cannibalized by larger Nile perch. The few other fishes remaining in any numbers include the introduced Nile tilapia (mbuta, *Orechromis niloticus*), which feeds on algae and detritus. and the dagaa (or omena, *Rastrineobola argentea*, a sardine-like cyprinid), which feeds on plankton higher in the water column. It appears that the shrimp can tolerate low dissolved oxygen and thus may have some refuge from Nile perch. What has resulted is a loss of the lake recycling process (cichlids) in Lake Victoria by Nile perch predation. How is this so? About 80% of the cichlid biomass, constituting a whopping 60% of the lake total, fed on detritus and mixed this biomass through the lake. This facilitated export out of the lake via terrestrial food chains, including humans (Kaufman 1992).

How can so many fish species suffer extinction in so short a time? Such major extinctions usually have been related to catastrophic impacts that affected organisms over ecosystem scales and geologic time. For large and widespread species, extinction has generally been linked to impacts that have never been experienced in the evolutionary history of species (e.g., Raup 1991). Clearly in this case, a combination of overfishing and the introduction of a giant predator provided the knockout punch ("a hard first strike," Raup 1991). As the conditions contributing to eutrophication continue, the murky water also seems to be causing problems that also could lead to extinction. It seems that the female cichlids that persist have a difficult time perceiving males of their own from other species, and hybridization is increasing (Goldschmidt 1996).

What about the human populations? Lake Victoria is suffering from galloping eutrophication. With its unstable condition and strained food web, there is no way to predict what the future may bring. Once looking upon it as their economic savior, the local people have no means of preserving the Nile perch, and they also cannot compete with the motorized commercial fishing industry. The prices of Nile perch fillets are not within reach of most of the people who live around the lake. As a result, almost all of the catch is refrigerated or frozen and exported by truck or airplane. Local people are forced to subsist, in part, on scraps obtained from the processing plants. The once plentiful cichlids are virtually nonexistent, as well as the other native fishes. Poverty and disease have been rampant.

Can this happen to lakes Tanganyika and Malawi? These lakes are colder than Victoria; their waters are much clearer, and the cichlids are more colorful. The lakes have long been valued for their diversity in aquarium fishes; the fishes are valuable, and their trade has persisted for many years (e.g., see Axelrod and Burgess 1986). However, both of these lakes also have a commercial fishery that has been increasing with an expanding human population. For example, Lake Malawi provides about 75% of the protein consumed in the country of Malawi. The human population size in that country was about 4 million in 1964 and increased to 9.5 million by 2008. Presently, the lake fishery supports about 35,000 fishermen and perhaps as many as 2 million individuals in the whole industry. Unfortunately, the fishery has declined in recent years, and there is worry that it is not sustainable. This problem has been compounded with the import and use of trawls and motorboats (Barlow 2000). Regulation of fishing in the lakes is a problem because there are multiple countries included in the shorelines.

SUMMARY

Fishes of temperate lakes are exposed to changing conditions (e.g., temperature, changing lake level, turbidity, nutrients) on annual, seasonal and, for shallow water species, daily bases. These fishes also are adapted to riverine conditions in the streams that feed into the lakes. Consequently, they tend to be generalists species. The largest complex of temperate lakes in the world is the Laurentian Great Lakes of the United States and Canada. These lakes have been profoundly affected by species introductions and habitat changes induced by humans. The first Europeans settled in the Great Lakes in the early 1700s. They found pristine conditions with abundant fishes: there were blue pike, lake sturgeon, lake trout, whitefishes, ciscoes, and sculpins in the upper lakes while the lower lakes had Atlantic salmon in Lake Ontario and walleye, blue pike, lake trout, sauger, yellow perch, freshwater drum, and so on in the others. By the early 1800s, the human population was draining marshes, lumbering, commercial fishing, and constructing dams and mill ponds in smaller tributaries for power. They also discharged raw sewage, solid, and liquid wastes into the lakes, setting in motion irreversible changes. Eutrophication aged Lake Erie by about 15,000 years. Overfishing eradicated Atlantic salmon and lake trout from Lake Ontario and lake sturgeon throughout the lakes. Opening of canals that connected the lakes directly with the Atlantic (via the Hudson River) and with the St. Lawrence seaway allowed access into the lakes for the parasitic sea lamprey that decimated the lake trout population, which had declined by 99% before an effective control program was developed. Also, anadromous alewives invaded the lakes and became superabundant in lakes Michigan and Huron in the absence of large predators. Pacific salmon were then introduced by state agencies to control alewives, which was met with some success. However, alewives are not adapted to such cold temperatures and are prone to die-offs. By the mid-1980s, several more introductions occurred in ballast water, and populations of the exotic zebra and quagga mussels, spiny water flea, ruffe (a percid), gobies, and other organisms have become established. Now there is the threat posed by three species of large Asian carp. As a result of all these changes, the historic ecosystem cannot be restored. Fish populations are distorted, reduced, and overwhelmed with all of the problems, which include habitat alteration and loss and predation, competition, and parasitism from nonnatives. Five fishes are extinct, and five more are on the endangered species list.

Tropical lakes, in contrast, are not of glacial origin and tend to be much older and more stable than temperate lakes. Tropical fish communities tend to be more diverse, fishes more specialized, and populations more susceptible to biological controls than to abiotic factors. A case study of African great lakes is presented to aid in understanding these differences and to explore some new problems. One of the largest lakes in the world, Lake Victoria, is of special interest. This lake has a huge surface area; it is relatively shallow, only 100 m in depth, and historically had oxygen to the bottom. Formed only 14,000 ya, the lake supported a species flock of perhaps 500 cichlid fishes, whose adaptations to every conceivable food item are enhanced by pharyngeal and jaw teeth. Unfortunately, introduction of a giant piscivore, the Nile perch, which can prey on even the largest cichlid, has caused the probable extinction of perhaps 200 species. This predation also has resulted in the loss of planktivores and detritivores that recycled the organic matter. Now the rotting detritus is accumulating with a resulting loss of oxygen. By the 1990s, the bottom 40 m of the lake had lower than 1 ppm oxygen and could not sustain aerobic life. Complex food webs involving hundreds of species have been reduced to large Nile perch eating smaller Nile perch that feed on shrimp. In addition, the incidence of fish die-offs increases due to circulation of hypoxic water. Lake Victoria is now suffering from "galloping eutrophication." With its strained food web and unstable condition, it is difficult to imagine anything but a continuing tragedy for both the cichlids and the local people.

Video: Nature/ecology—*Lake Tanganyika: Jewel of the Rift.* National Geographic (socioeconomic orientation); Millet et une Productions. 2004. *Darwin's Nightmare* (www.hubertsauper.com or at Amazon.com)

CHAPTER 14

Artificial Lakes and Groundwater Reservoirs

ARTIFICIAL LAKES: RESERVOIRS

Before we continue our aquatic journey down-gradient toward the sea, we must consider two other systems in addition to natural lakes that might store our drop of water. These systems are hard to characterize as fish habitat because they are so diverse in size and attributes. The most conspicuous and, thus, familiar of these are artificial lakes.

Humans have dammed, diverted, and stored water for their use since antiquity. Another mammal, the beaver (*Castor canadiensis*), also is a prodigious builder of reservoirs; however, beaver ponds are similar to natural "organic lakes" (e.g., Wetzel 2001). Early impoundments by humans also resembled beaver dams, but modern humans have perfected an entirely different process for constructing reservoirs of all sizes, including huge impoundments. Human skills at building large impoundments began to emerge around the turn of the twentieth century in the United States and reached perfection with the completion of the Hoover Dam in 1935 (at the time, it was viewed as a modern miracle). Humans have gotten so good at reservoir construction that huge reservoirs similar in size to the Laurentian Great Lakes have been proposed (Wetzel 2001). Once the reservoir building infrastructure was developed in the government and the industry, the rate of reservoir building increased exponentially. On a global scale, reservoir building has resulted in over 50,000 reservoirs with a dam height of 15 m or more (Kalff 2002), and the combined area of present impoundments and reservoirs exceeds the surface area of natural lakes (St. Louis et al. 2000). The combined storage capacity of reservoirs in the United States is astounding (Figure 14.1).

"Reservoirs" or artificial lakes are considered ecologically distinct systems (e.g., Wetzel 1990) brought about by a little human tinkering that sets them apart from natural lakes (e.g., a bottom [Figure 14.2] rather than a surface [Figure 14.3] discharge of water). There is a real need to increase awareness about how this new system functions because there seems to be little recognition of the differences and a lack of awareness of problems induced by hundreds of thousands of artificial lakes in the world. Reservoirs have had a tremendous effect on native fish systems globally, ranging from mildly negative to catastrophic. However, objective evaluations of the effects of such water development projects in the United States were not studied (or made public) until after passage of the National Environmental Policy Act of 1969. This act emerged at the end of the "golden age" of reservoir building in the United States, which occurred from the 1930s to the 1970s.

The National Research Council (1996), in working to revitalize educational programs, pointed out that reservoir construction was first accepted as a much needed way to provide hydroelectric power, flood control, and water storage. Adverse environmental effects of reservoirs were not fully appreciated until they were built and their impacts evaluated over time. Functions of reservoirs as systems and their effects on natural systems have only been determined as more and more of them are built and effects are known.

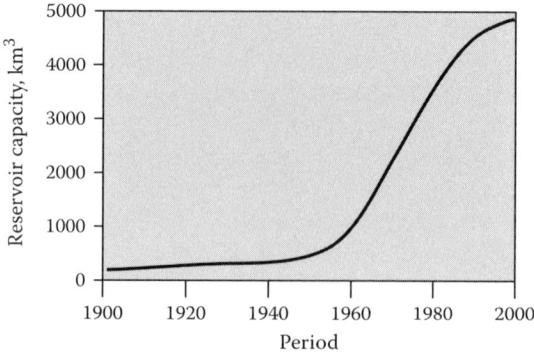

Figure 14.1 Reservoir storage capacity. (Courtesy of McCutchan, J. H., Jr.)

Figure 14.2 Surface discharge from a run-of-river reservoir. (Courtesy of McCutchan, J. H., Jr.)

STRUCTURE AND FUNCTION

Large reservoirs that are created from impounding streams and rivers appear to the uninitiated to be natural lakes, and this problem is compounded by naming them lakes (e.g., Lake Powell). However, an observant person will recognize that a concrete dam is not a natural structure or will fail to notice that it can be a long way from the top of that dam to the water discharge far below. In addition, a fisherman will also be aware that a natural lake tends to be roundish in shape, while a

Figure 14.3 Bottom discharge from a storage reservoir. (Courtesy of McCutchan, J. H., Jr.)

large riverine reservoir will tend to be more elongate, following the river channel that it impounds. These main features will go a long way in facilitating an understanding of the differences between these artificial systems and natural ones. Many authors (e.g., Cooke et al. 2005) have contrasted the structure and function of lakes and reservoirs. Table 14.1 contrasts attributes of natural lakes and reservoirs with respect to some of the more important differences that affect fish populations.

Table 14.1 Natural Lakes versus Reservoirs

Factors	Natural Lake	Reservoir
Physical		
Appearance	Lake (roundish)	Riverine (long)
Water level	Stable	Can be erratic
Pool	Seasonal minimum	Irregular cycle
Discharge	Warm surface outflow	Cold subsurface
Water residence time	Long, small inflow	Short, large inflow
Ecological		
System	Natural	Artificial
Riparian vegetation	Established	Little
Shade and cover	Much	Less
Nutrients	Cycle (microcosm)	Downspiral out
Productivity	Stable	Changing
Fish species	Native	Mostly introduced
Fish populations	Stable	Unstable
Fishery	Low management	Intensive management

Dam construction converts a portion of a stream or river into an artificial lake (reservoir). Considering the origin and result, many authors have commented on the intermediate ("hybrid") nature of reservoirs between lotic and lentic systems with respect to morphology, hydrology, nutrient inputs and cycling organic material, and phytoplankton production (reviewed by Kimmel et al. 1990). The hybrid analogy is useful when considering the origin of reservoirs, but it is most appropriately applied to mainstream reservoirs that have a high rate of flow through them. These "run-of-river" reservoirs are riverine and have a short water residence time (<1 year). Reservoirs with a longer water residence time are more like natural lakes.

Large riverine reservoirs can be divided into three segments: an upstream, or *riverine zone*; a midreach, or *transitional zone*; and a lower, *lacustrine zone* (Figure 14.4) (Cooke et al. 2005). The riverine zone tends to be narrow, with a high flushing rate and high suspended sediment load. In some locations, this zone is characterized by a very high rate of sediment deposit. Accordingly, the transitional zone is broader, deeper, and less turbid. The lacustrine zone is characterized as wide and deep, with low flow and highest light penetration at the surface (Kalff 2002). Water discharge from the reservoir occurs at the lower levels of this zone, which are usually thermally stratified and can be very low in oxygen. These discharges are usually much colder than the surface, low in sediment, and can be lacking in oxygen. When delivered below in the "tailwaters" of the dam, the riverine habitat tends to be scoured as sediment is picked up, and the water is cooled and infused with nitrogen gas due to the low dissolved gas content. Because of these features, the tailwaters of dams are an unsuitable habitat for warmwater fishes, and in some cases, the water is even too cold for trout.

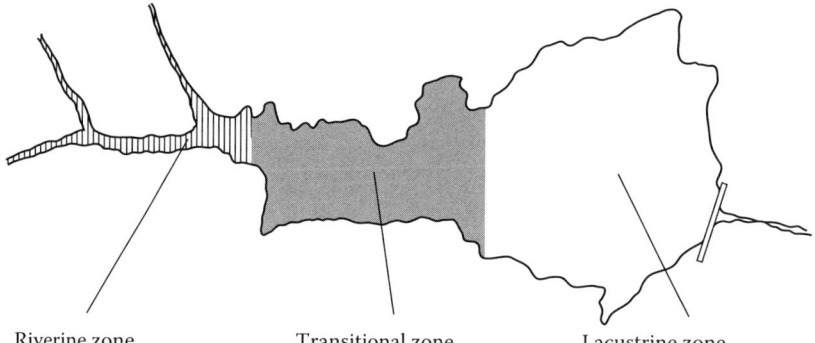

Riverine zone
- Narrow basin
- High flow
- High suspended solids, low light
- High nutrients, advective supply
- Light-limited photosynthesis
- Algal cell loss by sedimentation
- Organic matter supply allochthonous
- More "eutrophic"

Transitional zone
- Broader, deeper basin
- Reduced flow
- Lower suspended solids, more light
- Advective nutrient supply reduced
- High photosynthesis
- Algal cell loss by sedimentation, grazing
- Intermediate

Lacustrine zone
- Broad, deep, lake-like
- Little flow
- Clearer
- Internal nutrient recycling, low nutrients
- Nutrient-limited photosynthesis
- Algal cell loss by grazing
- Organic matter supply autochthonous
- More "oligotrophic"

Figure 14.4 Three longitudinal zones in a riverine-type reservoir. (From Cooke, G. D., et al., *Restoration and Management of Lakes and Reservoirs. 3rd Ed.*, Taylor & Francis Group, CRC Press, Boca Raton, FL, 2005. With permission.)

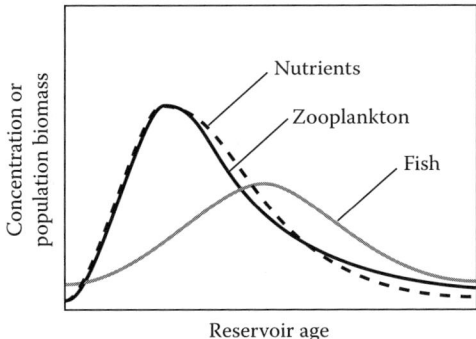

Figure 14.5 Trophic downsurge with reservoir age. (From Hayes, D. B., et al., 1999. *Inland Fisheries Management in North America, 2nd Ed.*, American Fisheries Society, Bethesda, MD, 1999. With permission.)

Because of the nature of reservoir impoundment, terrestrial areas are permanently covered with water, which slowly leaches out nutrients from the soil and flooded vegetation into the water. These nutrients cause a condition known as *trophic upsurge* within the first 5 to 10 years because aquatic populations, including fishes, expand in biomass, and as a result, sport fishing is very successful (Summerfelt 1999). Nutrients are cycled within a reservoir during its filling period as in a natural lake. However, as the reservoir ages, it goes through a protracted period of *trophic downsurge* throughout the remaining life of the reservoir, which is caused by the release of nutrients and organic matter downstream in a nutrient spiraling effect, similar to that of streams. This effect has been discussed by many authors (e.g., Hayes et al. 1999; Kalff 2002). A hypothetical rendition is provided in Figure 14.5.

In brief, the primary function of reservoirs is to capture and store water seasonally for year-round use. In the temperate zone, reservoirs store water during spring floods and release water in response to a flow regime that produces hydroelectric power, provides irrigation for agriculture, provides water for municipal use, and, of course, allows lacustrine recreation. Spring water storage provides some flood reduction downstream. Unfortunately, the effects of flow regulation on natural riverine systems change the physical habitat by reducing heterogeneity and change a dynamic system into a more static one. Part of the problem is related to a loss of flood regime, and the other is due to a loss of sediment, which is stored in the reservoir. The American Fisheries Society (AFS) has long been concerned about the conversion of free-flowing rivers to flow-regulated systems because of the continuing loss of native riverine fishes. Fish losses due to altered flow regimes have been so pervasive that AFS's policy is to adopt a "no net loss" policy in the remaining free-flowing stream ecosystems left in North America (Tyus 1990a; Rasmussen 1997).

ECOLOGICAL CONCEPTS

- *Reservoirs as Hybrid (Lotic + Lentic) Systems*—Artificial lakes are not natural lakes, but result from converting part of a natural river into an artificial system. This concept best fits large "run of river" reservoirs.
- *Serial Discontinuity Concept*—Construction and operation of dams interrupt the river continuum, but in predictable ways: Regulated rivers can be envisioned as an aquatic environment of alternating lotic and lentic systems. In order to address the changes caused by dams and reservoirs in regulated rivers, Ward and Stanford (1983) proposed this concept to deal with dam-related departures from the river continuum concept.

- *Winterkill and Summerkill*—Some environmental conditions can result in fish mortalities in natural ponds and lakes, but such mortalities appear frequently in artificial farm ponds. Conditions for both kills occur during periods of little or no water flushing, when nutrients are allowed to accumulate. In winter, ice cover may allow continued photosynthesis but restrict oxygen exchange at the surface. In this case, photosynthesis can use up oxygen and result in total or partial kills. During summer stagnation, especially in shallow eutrophic ponds, periods of hot overcast days can limit oxygen production, lower saturation, and result in fish kills, especially at night.
- *Artificial Fertilization*—An important management tool for fish production in impoundments, this practice has been used with organic fertilizer since antiquity. However, the use of inorganic fertilizers to increase fish production can be traced to the work of Swingle (1970) in the rather infertile ponds of southeastern United States. In essence, using inorganic fertilizer, it is possible to increase fish production four or five times while limiting growth of nuisance algae, but the effects of this practice can be highly variable (reviewed by Bennett 1971).

CUMULATIVE EFFECTS

Worldwide, about 200,000 square miles of land are inundated by artificial lakes. In the United States, there are more than 75,000 dams constructed with walls higher than 6 ft, and these dams store more than 60% of the entire river flow (Collier et al. 2000). The effects of all these reservoirs on natural ecosystems have been disastrous (e.g., see Helfman 2007), and these will be discussed more fully in the following chapters, especially with respect to native fishes of the Colorado, Missouri, Columbia, and Snake rivers. Adverse effects of single reservoirs can be mitigated to some extent. However, damming rivers destroys and fragments riverine habitat by blocking the movements and migrations of fishes, converting riverine habitat into an artificial lake, disrupting life history patterns and food webs of native organisms, and changing the downstream flow regimen (according to human needs for hydropower, irrigation, and flood control). It also results in the replacement of native fishes with nonnative competitors and predators that are stocked for recreational fishing. Finally, effects of multiple reservoirs can disrupt even the largest of natural systems, causing irreversible damage. Two examples of cumulative effects bear mentioning.

Cumulative effects of dam building have not been universally appreciated. This is changing thanks to authors such as Kalff (2002) and Helfman (2007), who have reported on the environmental damage caused by unconstrained reservoir building. The Caspian Sea, the largest inland body of water in the world, has been extremely altered by a cascade of reservoirs (11 large and hundreds of smaller ones) that were built on the tributary Volga River system by the former Soviet Union. Water storage in these reservoirs has reduced spring flow peak by about 40% in the Caspian Sea, reduced freshwater habitat by lowering water levels and increasing salinity from about 8 to 11 ppt, blocked spawning runs of anadromous fishes, and disrupted trophic relationships by upstream storage of nutrients, organic matter, and silica (needed by invertebrates for their exoskeletons). Among other things, this has led to a collapse of salmon and sturgeon populations due to stream blockage and loss of nursery habitat, and declines of freshwater fishes due to lower stream flow and increasing salinity.

The Aral Sea has suffered from a different problem. Upstream water diversion by dams has caused over 90% reduction in inflow (Helfman 2007). This reduction has decreased its surface area by 64% and its volume by 84% (Kalff 2002). In order to visualize such a loss in size, Helfman (2007) provides a map that shows the fishing village of Muynak, which has been left 100 km away from the water. Environmental effects due to this dewatering have been disastrous. Salinity has increased from about 10 to 50 ppt, causing declines in all trophic levels from phytoplankton to benthos. Commercially fished species ceased reproduction in the 1970s due to high salt content, and by the 1980s, the fisheries had collapsed. A smaller lake size also affects the climate of the region and results in a shorter growing season for agriculture. Problems in both of these large lakes also include a great increase in contaminants, lack of clean drinking water, and a resultant increase in human diseases and deaths.

FISH AND RESERVOIRS

The conversion of a natural river into a new artificial lake is generally unfavorable for native fishes. As discussed in our section on natural lakes, fishes in lakes tend to be the same species as found in the streams and rivers that feed them. Only in very old lakes have endemic lacustrine species evolved that spend their entire life cycles in a lake environment. A lack of fishes adapted to life in reservoirs after impoundment has resulted in fishery management programs that promote the stocking of nonnative, lacustrine-adapted forms to provide for more recreational fishery opportunities. Overwhelmingly, reservoirs extant today are dominated by nonnative fishes from the bottom to the top of the food web; thus, artificial lakes have artificial fish communities. As a result, fish ecology in reservoirs is extremely dynamic and human dominated. It is really a subject for fisheries management. However, we would be remiss here not to cover basic aspects of fish in impoundments and large reservoirs.

Impoundments

There are hundreds of thousands of small impoundments constructed in the United States. These are most abundant in the southeastern and central portions of the country, ranging in number from over 250,000 in Texas and Oklahoma to less than 25,000 in the more arid western states (Modde 1980; Flickinger et al. 1999). These impoundments are actually small ponds. According to the AFS, ponds can have a surface area of up to 8 ha (20 acres) (Armantrout 1998), but in my estimate, farm ponds used to produce fish generally range in surface area from about 0.5 to 2.5 ha. Farm ponds are constructed by damming small streams or springs, or from digging depressions such as gravel pits. Almost all smaller impoundments are used as farm ponds (Flickinger et al. 1999) that are meant primarily for livestock watering or irrigation. However, they also are universally built with fish production in mind. In this case, they are built with a drain (pipe) through the lower part of the dam, which enables the water level to be controlled or for the pond to be drained (via opening or closing a valve).

The history and development of fishery management of farm ponds is provided by Homer S. Swingle (1970), who was a pioneer in the field and developed the science of pond fish management in the southeastern United States. At first, streams were dammed to produce mill ponds for production of water power. These ponds provided fishing due to access and use by stream fishes, and they were drained and filled at various intervals. However, in the 1930s and 1940s, technology was available to construct dams on smaller streams, springs, or in areas of high surface runoff, and these areas did not have a dependable, adequate, or, in some cases, any number of fishes to populate them. In fact, there was little information on how to construct ponds to encourage fish production, what fishes to stock them with, and how to manage them. Without going into great detail here, an effective method was developed by H. S. Swingle, together with his colleagues and students at Auburn University. Their method, principally to establish largemouth bass/bluegill (Figure 14.6) communities, required proper pond construction, stocking schedules to initiate reproduction of the bass and then the bluegills, and fertilization techniques to keep production high (Swingle 1970). Other workers have applied these techniques to other locations in the country and

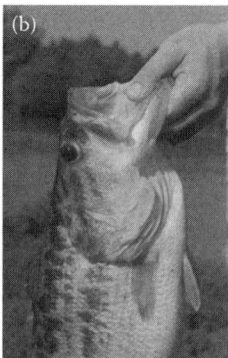

Figure 14.6 Bluegill (a) and largemouth bass (b) from farm ponds.

made improvements, especially in northern areas, by using other techniques such as incorporating other species and using different fertilization techniques (reviewed by Bennett 1971). Pond culture of fishes also has resulted in the development of commercial aquaculture in ponds, especially for raising channel catfish and other food fish (e.g., tilapia) and also for baitfish production.

Large Reservoirs

By 1970, almost all of the present large reservoirs had been constructed in the conterminous United States, and Jenkins (1970) provided a map of the location of reservoirs with a size of 500 acres or more. Presently, there are more than 75,000 dams built 6 ft or higher in the United States, and these cover more than 3% of the land surface and store 60% of the entire river flow (Collier et al. 2000). Technical ability to build large dams expanded greatly after the construction of the Hoover Dam, and thereafter, dam building occurred in a frenzy. In about three decades, virtually every river in the conterminous United States was dammed or its flow regulated by dams and diversions, and there are very few dam sites left that have a potential for development. From a socioeconomic perspective, these dams were generally viewed as a godsend. But some reservoirs declined in fish productivity in only a few years, and ecologists feared that after decomposition of organic matter in the catchment basin, large reservoirs might become biological deserts. Accordingly, fishery biologists began to develop strategies for reservoir management to enhance fisheries production, and by the 1950s, five levels of management were selected for implementation (Bennett 1971): (1) water level manipulation to favor or disfavor various species, such as lowering the reservoir to destroy carp spawn or to force young bluegill out of littoral vegetation, thus exposing them to bass (Summerfelt 1999); (2) addition of nonnative species to "fill in" perceived vacant portions of the food webs, such as threadfin shad to eat plankton and white or striped basses to eat the shad; (3) control of undesirable (rough fish) native or nonnative fishes to reduce their competition with game fish, usually with a fish toxicant; (4) increasing game fish populations and the harvest of sport fish by anglers; and (5) publicity to promote reservoir fishing and to attract anglers providing license dollars. These management objectives have, in general, persisted to the present day. However, other problems also plague fisheries management in reservoirs, and these have been discussed in detail by Hayes et al. (1999).

What fishes are placed in reservoirs? That depends upon the conditions, but various species have been stocked historically, depending on climate (Jenkins 1970), including trout, white bass, striped bass, walleye, northern pike, muskellunge, sunfishes (including large- and smallmouth bass, crappies, bluegills, etc.), and channel catfish. More recently, pacific salmon and various hybrids of those mentioned above, such as wipers (white × striped bass), also are stocked as sport fish. To this list, we also add some forage fishes, such as threadfin shad or invertebrates such as freshwater shrimp. Other fishes, of course, occur in reservoirs, including many native species that live in the natural riverine portions that remain upstream. These are principally native riverine fishes such as minnows and suckers, but there are also some introduced species such as common carp and small nonnative fish that have escaped the bait bucket. Judgment on what fishes to introduce has admittedly been flawed: Common carp were once widely introduced throughout the United States by Federal and state hatcheries (beginning in the 1870s) to satisfy the demands of European settlers, but they are now considered to be very undesirable—and the target of fish control measures by fishery agencies.

In the upstream riverine zone of large reservoirs and along the shorelines, one may find populations of warmwater and coolwater fishes that move between the reservoir basin and upstream. These include minnows (including shiners), freshwater drum, sunfishes (basses, bluegill, black crappie, etc.), perch, pike, walleye, catfishes, and so on. In the downstream lacustrine zone of reservoirs, the upper levels are inhabited by planktivorous fishes, such as shad, white crappie, white bass, striped bass, trouts, and salmon (especially kokanie). In lower levels of this zone, there may be inshore catfishes, bullheads, and carp, especially in warmer months or when oxygen is sufficient. In colder climates and deeper habitats, one may encounter trouts and salmon (especially the well-adapted lake trout). In cooler portions of the country where the hypolimnion may contain sufficient oxygen, there may be intensive efforts to manage a "two-story" fishery of coldwater fishes in the cold lower layers in summer, while warmwater fish are present in the epilimnion.

Fishery management in reservoirs has increased in sophistication, and various techniques are now being used for managing fish habitat (reviewed by Summerfelt 1999). Also, better construction practice has aided fish management, especially in smaller impoundments. However, poor reproductive success may limit the abundance of some fishes in reservoirs, including stream spawning salmonids (also, kokanie and some hybrids cannot reproduce), due to a lack of spawning habitat. Fishes that do spawn in the reservoir also may be affected by a short water residence (flushing rate) that produces downstream transport of eggs and larvae (Summerfelt 1999). As an example, in Lewis and Clark Lake (reservoir) on the Missouri River, Walburg (1971) found a staggering amount of daily (24 h) loss of small fishes (about 25 mm), which peaked at 10 million freshwater drum, 800,000 emerald shiner, 700,000 sauger and walleye, and 170,000 channel catfish per day. The input of reservoir-spawned fishes into downstream reaches of rivers and streams has not been fully investigated. However, in some locations such as the Yampa and Green rivers in Colorado and Utah, escape of aggressive nonnative fishes (i.e., smallmouth bass and northern pike) from reservoirs has resulted in competition and predation on endangered fishes.

GROUNDWATER

About 22.8% of global freshwater resources occur as groundwater, with a residence time or renewal rate averaging about 1400 years (Kalff 2002). About half of that amount is stored as actively exchanged groundwater (i.e., close to the surface), with a water renewal rate of about 300 years. This quantity seems small, but it is almost 40 times the amount stored in all of the freshwater lakes of the world (Wetzel 2001), and it is a valuable resource for humans.

There are many species of fish that live in groundwater of caves and at the surface where groundwater appears as springs. It sometimes seems incredible that fishes can be found in such places; however, we must remember that water levels have been much higher in recent geologic time, especially during and after the Pleistocene, when melting glaciers provided ample water for huge lakes and river systems. Although cave fishes are fascinating creatures (e.g., see Helfman et al. 2009), we will focus here on groundwater habitats produced in the desert environment of southwestern United States.

CASE STUDY: DEATH VALLEY AND DEVILS HOLE

Deserts are generally defined as areas in which evaporation exceeds precipitation, and deserts occur in all continents except for Antarctica. The Death Valley desert is part of the Great Basin, which is a huge area that includes most of Nevada and Utah, and portions of four other states. This large inland area has no drainage to the ocean, and vast amounts of water were entrapped there after the last ice age, the Pleistocene. Two vast lakes that existed in the basin are the well-known lakes Bonneville and Lahonta, which covered thousands of square miles and were quite deep (Sigler and Sigler 1987). The basin has been divided into several smaller areas, including a southern Mojave–Death Valley region. Death Valley, California, occurs as a trough along the Nevada border and once supported Pleistocene Lake Manly. The Mojave Desert lies further to the south.

As glaciers melted and climate changed to desert, huge lakes and vast rivers began to dry up. Death Valley, the hottest and driest place in the United States, lost almost all of its surface water but retained vast quantities of groundwater that accumulated in the fractured rock formations that underlie the area. This groundwater appears at the surface in some locations such as Ash Meadows, where spring oases occur as if by magic (Figure 14.7). These springs were historically occupied by 10 species of fishes that evolved during the last 10,000 years or so, including the "toothed carps" (Cyrinodontidae), represented by five pupfish, two topminnows (killifish; one now extinct), two minnows, and one sucker. As the Pleistocene waters dried up, one of these pupfish was left behind in an exposed groundwater pool that is now located high and dry on a side hill. It is presumed that its habitat was much larger, and over the past thousands of years, it has retreated further and further into a wide deep "hole" in the earth. This hole was formerly called "the miners bathtub," but today, it is called Devils Hole (Figure 14.8).

Devils Hole was made a separate part of Death Valley National Monument by proclamation of President Harry S Truman in 1952. However, this did not entirely protect the endemic pupfish that lived there. The history of Devils Hole and the fight to save the Devils Hole pupfish are presented in great detail by Deacon and Williams (1991) and will only be summarized here.

Figure 14.7 The author at a desert spring in Ash Meadows near Devils Hole National Monument.

Figure 14.8 Devils Hole National Monument. (Courtesy of the U.S. National Park Service, http://www.nps.gov/deva/naturescience/devils-hole.htm.)

Devils Hole Pupfish

The pupfishes comprise a monophyletic group of about 30 hardy little brackish and marine fishes of the genus *Cyprinodon* (Cyprinodontidae). In general, these fishes occupy fluctuating habitats that are too small, too warm, too salty, or too low in productivity to support other fish groups. The cyprinodonts also include about 20 species that occur only in deserts or semideserts (Miller 1981; Nelson 2006). As a member of this group, the Devils Hole pupfish (*Cyprinodon diabolis*) (Figure 14.9) coevolved with increasing conditions of desertification in the Death Valley–Mojave Desert region. Its evolution as a "stress tolerant" species was shaped by extreme thermal and osmotic factors, the unpredictability of the system, and low abundance of food and habitat, resulting in unusual combinations of physiological and behavioral traits (Soltz and Naiman 1978; Miller 1981) and a high degree of phenotypic plasticity (environmentally induced phenotypic variation: Lema 2008).

Figure 14.9 Devils Hole pupfish *C. diabolis*. (Courtesy of the U.S. Fish and Wildlife Service, http://commons.wikimedia/wiki/File:Cyprinodon_diabolis.jpg.)

Devils Hole is the highest and most isolated of all springs in Ash Meadows. It has been separated from other water bodies for perhaps at least 10,000 years. During this time, surface water around Devils Hole dried up, and its habitat shrank to a small exposed pool sunken about 15 m from the surface and surrounded by rock (Figure 14.4). However, this pool is associated with an underground cavern that is connected to a vast underground aquifer that filled about 12,000 years ago. Although water seems to be abundant at present, productivity has been severely limited and now consists only of a narrow window exposed to the direct rays of the sun, and food is provided solely by algae growth (and associated invertebrate species) on a narrow shelf (<20 m^2) in the water. The Devils Hole pupfish also uses this shelf for reproduction. The population size of the fish, which may live for only a year or two, historically expanded to a maximum size of about 800, and as low as 200 in winter, with an average size of about 400 (Soltz and Naiman 1978). This small (about 2 cm long) fish has the distinction of existing in the smallest habitat of any other vertebrate.

How has the pupfish adapted to conditions of life imposed by Devils Hole? The pupfish has no pelvic fins, its lateral line has been reduced, it has only cycloid scales, it has no territorial behavior, and it is considered a dwarf—it is small even for pupfish. All of these attributes can be related to living in a small and relatively low-productivity environment. One benefit of Devils Hole, however, is the ability of the fish to move lower into cooler water as the surface heats up—as opposed to fishes living in more shallow surface water, such as ponds or springs.

The Fight to Save the Fish

Protected by President Truman by proclamation and fenced by the National Park Service, the pupfish was considered safe by any standard. However, this was not to be so. Public land was sold to a corporate farm in Ash Meadows in 1967, which acquired state water rights and began to pump water into the desert for growing cattle feed. Accordingly, the level of water in Devils Hole began to decline, resulting in a public outcry over a potential loss of the pupfish. The outcry intensified, especially in 1970, when an Emmy Award–winning television special hosted by movie star Jack Lemmon was aired, in which the plight of the Devils Hole pupfish was discussed. This was followed by more media coverage, threats to the government, formation of the vocal Desert Fishes Council (a scientific body dedicated to saving desert fishes), action by a U.S. Department of Interior task force, and a lawsuit filed in 1971 by the U.S. Justice Department (see Deacon and Williams 1991 for much greater detail). The lawsuit was filed with the district court, which stopped further groundwater pumping in the immediate area. Although the water level somehow recovered in Devils Hole, pumping continued in more distant wells, so the water level began to drop again. After a continuing court battle and various appeals, the case was finally heard by the U.S. Supreme Court, who established an absolute minimum water level in Devils Hole to protect the fish, which also made further major pumping activities infeasible. However, there were many more battles before the Ash Meadows area was finally purchased and was incorporated into the Ash Meadows National Wildlife Refuge (Deacon and Williams 1991).

The U.S. Fish and Wildlife Service, National Park Service, and state agencies have been monitoring the Devils Hole pupfish since 1970 (Barrett et al. 2008). Unfortunately, there have been problems in maintaining the population size. During the period 1970–1995, the summer population size averaged about 324 pupfish. The population then declined in 1997–2004, falling from 275 to 171 fish. In 2004, two storms resulted in a runoff that flowed into the hole and buried about half of the spawning shelf in sediment. Spring surveys showed a decline of pupfish from 84 to 38 fish. Then the numbers slowly increased to 70 in 2009. During this time, fall surveys also showed an increase reaching 104 fish (www.fws.gov/nevada/protected_species/fish/dhp). It seems though that the precarious situation of these fish will continue even with conservation efforts.

Some fishes may exhibit a change in phenotype due to changes in environmental factors such as temperature and food. Recent conservation efforts have provided some explanation about the ability of the Devils Hole pupfish to survive in such restricted and barren environments. In this case, fish stocked in refuge ponds changed in morphology in response to conditions in their new environment (phenotypic plasticity), becoming longer and deeper in body, with smaller heads and eyes (Lema 2008). Also it appears that pupfish can adjust the amount of yolk given to the eggs in response to temperature, which affects the number of eggs and their size (review by Kaplan and Cooper 1984).

SUMMARY

Reservoirs are artificial lakes that are constructed to provide hydroelectric power, flood control, water storage, and recreation. They have attributes of both lotic and lentic systems, especially in the large "run of river" reservoirs that have a short water residence time. Reservoirs have resulted in problems for native fishes because only a few have been designed to favor them. Instead, river reaches are converted into an artificial lake, a dam blocks fish movements and migrations, downstream flows are altered and usually cold, and "preadapted" lacustrine fishes are introduced from elsewhere. Introduced fishes have supported large recreational fisheries for salmon and trout, white bass, and perch, largemouth and smallmouth bass, and sunfishes, but to the detriment of natural biodiversity. Reservoir building has become a major trend in the United States, and more than 75,000 dams over 6 ft high now store more water than all of the rivers can contain. Some rivers have been altered by the construction of multiple dams, and cumulative effects of damming are dramatically presented by the Aral and Caspian seas. Those systems have been seriously impaired by water storage and diversions that block fish passage in the river, while in the basins, salinity has increased and surface area and volume have decreased. In addition to large reservoirs, there are hundreds of thousands of small impoundments generally known as farm ponds. Contrary to reservoirs, most of these impoundments are constructed with fish production in mind. In warmer latitudes, they are either stocked with largemouth bass and bluegills, channel catfish, and tilapia, to name a few, or used for raising bait fishes. In addition to storage of water in reservoirs, a substantial amount is stored in groundwater. In total, 22.8% of global freshwater occurs in groundwater, and about half of that amount has a water renewal rate of about 300 years. Many fishes live in groundwater, either in caves or in springs, including desert fishes of southwestern United States. One is the Devils Hole pupfish, *C. diabolis*, a small cyprinodont that lives in Ash Meadows, an area of upwelling springs in the Death Valley–Mojave Desert region. The fish lives in a cavern that connects with a vast underground aquifer that was filled after the Pleistocene. Now isolated to a deep hole on a hillside, the pupfish survive only because the groundwater covers an exposed shelf that produces food and allows for the survival of their young. However, due to water pumping, the level of water in Devils Hole began to decline, until the U.S. Supreme Court established an absolute minimum water level in Devils Hole to protect the fish. Since then, the fish population has fluctuated, but the fish still survives.

PART V

Estuarine and Marine Ecosystems

CHAPTER **15**

Estuaries and Coastal Zone

COASTAL ZONE

The marine coastal zone marks the transition between terrestrial and marine environments. This is a dynamic area that is constantly changing due to an interface of natural forces. While upland portions of the continents have changed slowly over geologic time (hundreds of millions of years), marine shorelines have been substantially influenced by the movement of tectonic plates (i.e., continental drift) and Pleistocene glaciation. Drift has resulted in geological upheaval on the active continental margin due in part to seafloor subduction and produced mostly erosional processes in the passive margin or trailing edge of the continents over a few million years (Castro and Hubert 2007). More recently, sea level changes have had a great effect on low-lying shores (Figure 15.1). The shapes of coastlines we recognize today are the result of a 120-m rise in sea level that occurred less than 18,000 years before (Scotese 2004). In the past few decades, major storms have resulted in severe coastline alteration. Such storms can cause major physical changes, especially on beaches and ocean inlets. But there are natural processes that shape coastline habitats on a daily basis. What are these processes and how do organisms cope with them?

Marine coastlines are constantly affected by forces of nature. Three of these forces are pervasive: freshwater runoff from continents, tidal oscillations and ocean wave action, and wind. Included in the coastal zone are two major aquatic systems: (1) estuaries, diverse ecosystems where freshwater of the terrestrial environment and salt water of the marine environment join, and (2) inshore marine systems that occur as the sea meets the land directly. Because of their great importance to fishes, we will consider estuaries separately.

WHAT ARE ESTUARIES?

Assuming the drop of water we are following did not enter a closed basin, evaporate back to the atmosphere, or enter into groundwater storage, it eventually could proceed from mountains, across plains, piedmonts, and coastal zones, and enter into the upper reaches of an estuary. Because freshwater merges with salt water there, one might ponder whether such a place should simply be classified as a transition zone between freshwater and marine systems. However, estuaries are complex and differ in their shape, the degree of isolation from the ocean, the amount of freshwater they receive, the kinds and amounts of associated coastal ecosystems, and so forth. Some estuaries are clearly semi-isolated portions of the ocean and consist mostly of slightly diluted seawater. Others are dominated more by rivers and contain vast areas of brackish freshwater with low salt concentration (i.e., they are oligohaline). Also, estuaries can vary in appearance: some may be large "sounds" or smaller bays that are dominated by plankton-based systems; some have vast areas of emergent

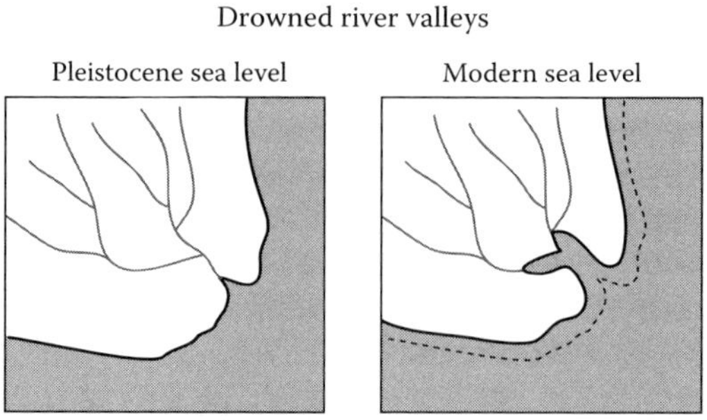

Figure 15.1 Formation of drowned river estuaries. (Courtesy of McCutchan, J. H., Jr.)

marsh vegetation; others may be dominated by large areas of inundated mudflats, grass flats, and shellfish reefs; and some are a mix of the above.

Ecologists consider estuaries to be special systems with unique and sometimes ephemeral properties. Estuaries are places of daily, weekly, seasonal, and annual changes in major environmental factors. They also are one of the most productive ecosystems on Earth; in aquatic systems, they are perhaps second in productivity only to coral reefs. After wading and boating in estuaries at day and night, I have grown to respect their temperamental nature! I think of estuaries as challenging places where rivers merge with the tides. The marine environment receives inputs of freshwater, nutrients, organic matter, and sediments from the terrestrial environment, while sharing its own blend of such environmental factors. In this place, great terrestrial and marine systems are connected, share their biota, and form a new environment with unique properties.

What is an estuary? There are different ways to classify estuaries, but there is one definition that seems to separate estuaries functionally from nonestuaries. An estuary is now generally considered an ecosystem in which terrestrial runoff enters into a semi-enclosed arm of the sea and measurably dilutes seawater (e.g., Pritchard 1967; Odum 1971; Smith and Smith 2001). However, this simplistic definition says little about the great diversity of estuaries, which also have been further classified by geomorphology, hydrography, and energetics (reviewed by Odum 1971). Hydrographic differences primarily due to water density and circulation patterns also can be used to separate estuaries into those that are highly stratified due to a salt wedge, those that are partially stratified, and those that are completely mixed.

Four types of estuaries are commonly recognized, based on geomorphology: (1) coastal plain or drowned river valley estuaries, (2) fjord estuaries, (3) bar or barrier-isolated (i.e., semi-enclosed bay) estuaries, and (4) tectonic estuaries (Pritchard 1967; Odum 1971; Nybakken 1997). The most productive estuaries exist in the coastal plain where inundated river valleys are influenced by freshwater runoff and tidal action (e.g., the Hudson River). Fjord estuaries are caused by glacial erosion and are very deep, for example, those in Norway or Alaska. Barrier island or bar-formed estuaries are the familiar semi-enclosed shallow bays or sounds of the Carolinas, which are oligohaline estuaries that are semi-isolated from the sea and protected by barrier islands such as Hatteras and Cape Lookout. Estuaries also can be formed by tectonic forces shaping the continents—an example is San Francisco Bay. Finally, two other types have been considered. Odum (1971) suggested another type of estuary, the river delta, such as the Mississippi River delta, a place where a sediment-laden river has produced complex estuarine systems by direct encroachment into the marine environment. The hypersaline lagoons of Texas and Mexico also can be considered to be special types of an estuary (reviewed by Odum et al. 1969).

Two of my early mentors, Tom Odum and B. J. Copeland, were principals in an entirely different conceptualization of coastal ecological systems (Odum et al. 1969). These authors used ecosystem energetics to classify systems and included salt marshes, grass beds, kelp beds, oyster reefs, mudflats, and fishes as subsystems of temperate estuaries. From a fish perspective, system energetics tell us two general things: (1) estuaries have harsh physical conditions (stressors) that fishes must cope with, and (2) an incredible amount of food supplies energy for those tolerant of conditions. Unfortunately, there is continuing disturbance to estuaries due to the actions of humans. These impacts include damming rivers, bulkheading shorelines, discharging wastes, and dredging channels. In some locations, including the Cape Fear River estuary where I used to work, humans are in a continuous struggle to maintain deep shipping channels in naturally shallow river estuaries, with resulting dredging and filling, construction of weirs and jetties, and the incidental resuspension of anoxic sediments. Also, most people do not know about the many waterways and ship channels along the coasts, such as the Atlantic intracoastal waterway. It exists all along the Atlantic coastline so that smaller vessels, such as barges, and recreational boats can travel north and south along the Atlantic coast without having to brave the open ocean. Maintenance of this waterway is the responsibility of the U.S. Army Corps of Engineers, and it is a continuing process of dredging and waste disposal to keep the channels open. Boat traffic constantly disrupts sediments in the channels and associated estuaries by churning and wake.

DROWNED RIVER ESTUARIES

Structure

This coastal plain or drowned river estuary is presented here because of its extreme importance to fishes and humans. As an example, two of the largest and most studied estuaries in the world are the Hudson River estuary of New York (e.g., Smith 1988) or the huge Chesapeake Bay system of multiple rivers and associated estuaries, where ecological studies were documented as early as 1912 (Hildebrand and Schroeder 1928). The richness of these systems is hard to imagine today, but amazing historical accounts are provided by several contemporary authors (e.g., Franklin 2007; Roberts 2007).

In simplest terms, a river estuary is formed when a river enters the sea. But riverine and marine forces are not very compatible, and the river estuary can be imagined as a great struggle between opposing forces: The estuary is semi-enclosed, and the river is able to dominate the system seasonally with large flows of freshwater; the ocean exchanges blows daily with huge tidal inputs of salt water. Other factors such as wind complicate this picture.

River waters flow downhill and transport inorganic and organic sediments, dissolved nutrients, and freshwater organisms into the estuary. Rivers are relatively shallow and fluctuate in temperature according to the ambient air temperature. The magnitude and extent of freshwater flows change due to the amount of seasonal runoff and more local precipitation events, but the flow is almost entirely due to constant gravity. Riverine flows are usually of relatively high velocity and run in channels until the gradient becomes very low.

Seawater flows in and out of the estuary (about twice per day in most places) due to the increasing and decreasing water levels that result from tidal forces. Coastal sea level changes occur around the world due to the rotation of the Earth, its moon, and its relative position with respect to gravitational pull of the sun and the moon. Tides are very high when these two celestial bodies are together and very weak when they are apart. Tides also vary in magnitude and develop tidal currents according to topographic features (e.g., tidal bore). Tides can vary in amplitude from less than 0.4 m in the northern Gulf of Mexico, from 2 to 2.5 m in the southeastern estuaries, and possibly more than 16 m in the Bay of Fundy (Herke and Rogers 1999). Ross (1997) also presented tidal amplitude for other locations in the Atlantic, Pacific, and Gulf coasts.

With changing tidal amplitudes, estuarine organisms can experience wide ranges in conditions, especially exposure to sun, rain, snow, and ambient air temperature. At high tide, the estuary can be dominated by salt water. Tidal flows from large oceans do not fluctuate as greatly in water temperature as river flows, and the flows are usually of lower velocity than rivers. The low-velocity tidal flow must oppose the river, and it tends to move estuarine waters laterally. River flow is especially dominant at low tide, when riverine flow travels in dendritic tidal channels, also called tidal creeks, which may flow through exposed mudflats and marshes.

Wind also is a factor affecting water depth and flow in estuaries. Water depth in an estuary can be increased or lowered as wind blows in the tide and moves river water upstream, or blows with river flow and against the tide, forcing water out to sea. In large, wind in shallow sounds and bays can have a considerable effect, especially when pushing an incoming tide. I have seen boaters, perhaps more accustomed to lakes, leave their vehicles parked near a boat ramp at the water's edge of a large shallow estuary. After fishing all day, they return to a vehicle inundated with salt water. A lesson was no doubt learned, but what a hard price to pay!

Estuarine structure is further complicated by the relative density of fresh and salt water. Freshwater is less dense than sea water (about 1.0 compared to 1.03) and, in the absence of turbulence, tends to float on top. This creates a "salt wedge" in which seawater, usually of 35 ppt (or practical salinity units, psu) salinity, moves up the river channel overlain by freshwater that is almost always less than 0.5 ppt. Wind and currents, especially the force of fast incoming tides, can disrupt this stratification; otherwise, the lighter freshwater can even flow out into the ocean.

Salinities vary seasonally in river estuaries according to the amount of freshwater flow and daily according to the tide. As seawater moves upriver, it will reach a point at which it can no longer move against the flow and instead is mixed with freshwater. In this fashion, salinity gradients occur from upper reaches, where salinities may be approaching that of freshwater, to the lower reaches, where salinity may be near that of the sea.

As the flowing river water reaches the wider estuary, it slows, and its capacity for transporting sediment is decreased. Sediment and larger suspended matter (e.g., detritus) can no longer be transported. It mixes with the heavier seawater and soon falls to the bottom. As freshwater mixes with seawater, dissolved organic and inorganic nutrients (especially phosphates) and suspended fine organic matter also are concentrated in the more dense seawater. Various ions in the salt water will cause silt to flocculate into larger particles that settle to form the characteristic mud bottom. As water continues to flow from the river, it seeks the path of least resistance, and at low tide, the previously deposited shoal forms an obstruction that is overcome by eroding channels through the shoal and by flowing around the shoal. Water that flows around the shoal tends to erode soft shorelines and form a characteristic estuarine basin (Figure 15.2).

A very productive system is thus formed from riverine sources of organic matter and inorganic nutrients (especially phosphates), which are combined with nutrient-laden salts, organic matter, and sediments swept from oceans into estuaries by tidal action, with organic matter produced within the estuary. All contribute to the productivity of the resulting system, which forms a nutrient trap in which dense salt water transports this rich material up into the estuary with each tidal cycle, as the lighter freshwater, largely depleted of sediment, gradually mixes with salt water at the surface and moves seaward.

Function

Inputs of organic matter and nutrients produce estuarine food webs based on detritus and algae (plankton and attached). Primary productivity is derived mostly from phytoplankton, marsh vegetation, seagrass flats, and algae attached to rocks and plants, all of which consume dissolved organic matter and inorganic nutrients. Plankton supports herbivorous fishes in larger estuaries, and many

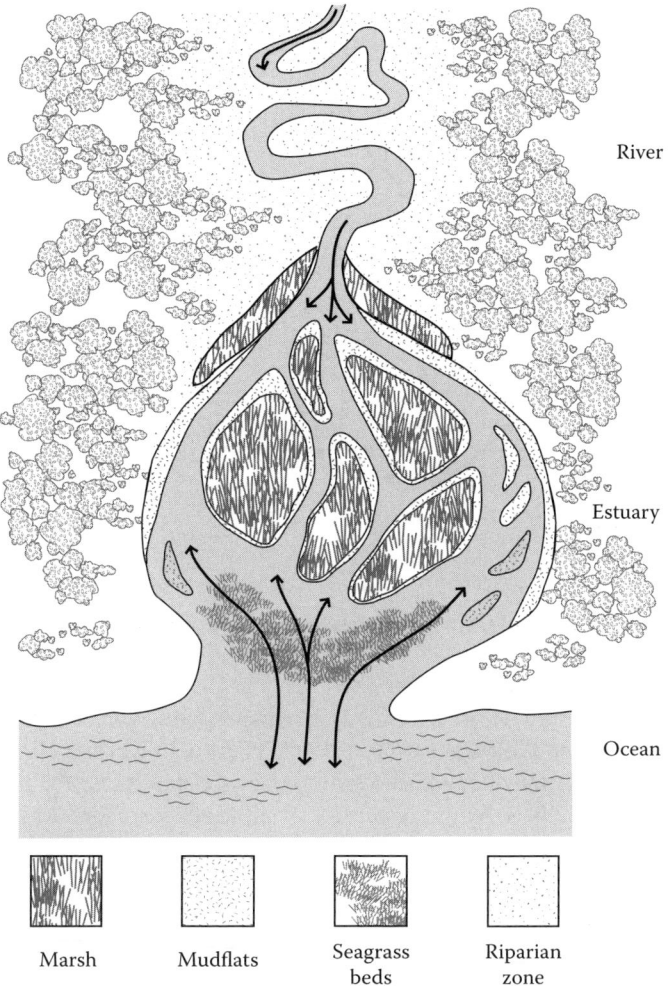

Figure 15.2 A hypothetical river estuary that includes several estuarine systems.

species use the estuary as nursery habitat. Detritivores, such as mollusks and mullets, are able to consume organic matter directly. However, it is likely that many detritivores may be stripping off the coating (i.e., biofilm) of bacteria and other microorganisms from the detrital particles and digesting it without degrading the detritus any further (Nybakken 1997). Dominant predators in estuaries include fishes and birds, which consume invertebrates and fish. A typical estuarine food web is presented in Figure 15.3.

Associated with and in river estuaries are plant and animal communities that may be recognized as ecosystems in their own right. Along the Atlantic coast of North America, these include mudflats, subtidal grass beds, salt marshes, oyster beds, and mangrove swamps (e.g., summaries of these systems are usually provided in general ecology texts, such as Smith and Smith 2001). Briefly, sediment deposition in the estuary can provide a perfect habitat for a number of colonizers. Shallow mudflats tend to be colonized by burrowing worms or clams and transient arthropods, while deeper ones can be colonized by mussels and oysters. Further away in more sandy bottoms, subtidal grass flats may provide food and shelter. In more protected shorelines, there are stands of irregularly flooded and

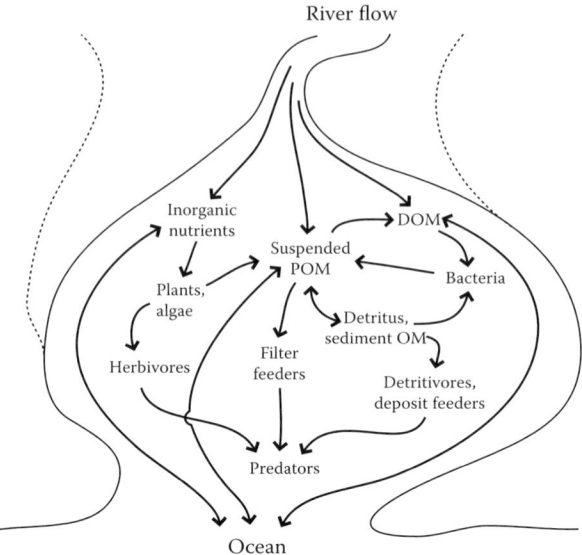

Figure 15.3 An estuarine food web. OM, organic matter; DOM, dissolved organic matter; POM, particulate organic matter.

flooded tidal marshes (e.g., *Spartina*), including low marshes, which are flooded each day, and high marshes, which are infrequently flooded (Figure 15.4). Finally, in warmer climates of the subtropics and tropics, tidal marshes may be replaced with mangrove stands, especially in locations where there are highly organic mudflats. Larger estuaries of all kinds also support phytoplankton-based systems.

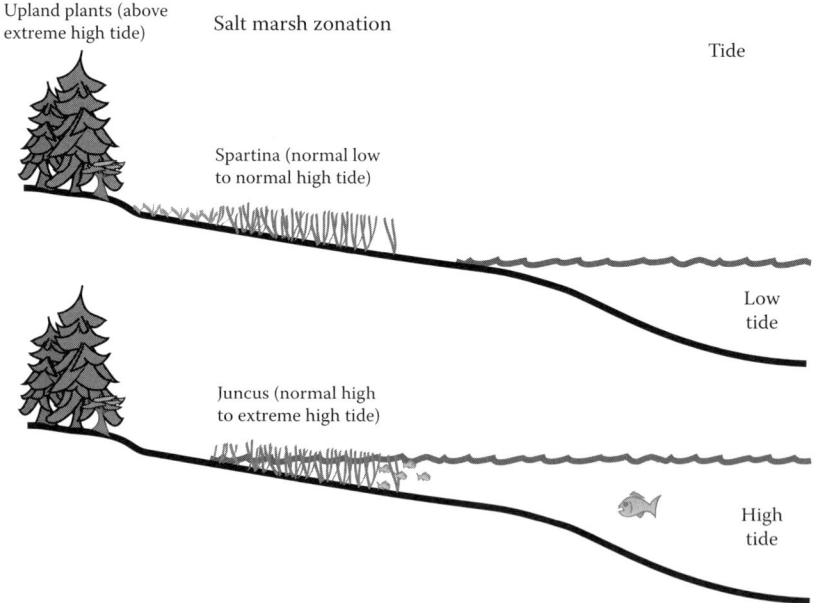

Figure 15.4 Tidal flooding in a salt marsh. (Courtesy of McCutchan, J. H., Jr.)

ESTUARINE FISHES

Our knowledge about estuarine fishes has been greatly increased by a few in-depth publications such as Allen and Cross (2006) for west coast species and Able and Fahay (2010) for east coast species. Freshwater, estuarine, and marine fishes occur in estuaries, and they may be either transient or permanent residents (e.g., Pacheco and Grant 1965; Herke and Rogers 1999). Estuaries provide extremely important nursery habitat for fishes and shellfishes, supporting commercial and recreational fisheries. West coast estuaries support about 70% of the juvenile fishes and provide habitat for about 50% of fishes harvested. Atlantic coast estuaries support about 80% of the catch, and Gulf of Mexico estuaries support almost the entire fishery (reviewed by Ross 1997; Herke and Rogers 1999). Of 552 fishes captured by Hoese and Moore (1998) in the upper Gulf of Mexico (Texas and Louisiana) over more than 20 years, a full 21% were estuarine species (including some freshwater visitors).

All fish that occupy estuaries have to cope with constantly changing conditions, most of which are summarized in Table 15.1. Those species that can adapt to such conditions benefit from the great productivity of the estuary, which generally exceeds that of fresh or salt water. The availability of food is an important reason why estuaries are used as a nursery area. Another benefit is a reduced level of competition or predation during the young (small) life history stages, due to the availability of shallow vegetated or reef habitat that can be used as cover.

In spite of the harsh and stressful conditions of the estuary, its advantages are compelling, and many fishes have evolved by using various strategies for taking such advantage. The most important is movement. Fish move in and out of the estuary depending on food availability and tolerance to the prevailing conditions, which may occur seasonally or with the tides (Figure 15.5). Fishes also move around locally as estuarine conditions change. Another strategy for life in an estuary is to reduce osmoregulatory stress by seeking the most favorable salinity regime: for freshwater species, this would be toward the upper end of the estuary, and for marine forms, it would be the lower end. Finally, some fishes develop physiological adaptations that allow them to be permanent residents in estuaries, but not many. More permanent residents are favored by the very large estuaries, while small estuaries may have few year-round residents.

It has long been recognized that estuarine fishes are a mix that changes constantly, and there have been efforts to classify them according to life history traits. For example, Dionne et al.

Table 15.1 Factors Affecting Fish in River Estuaries

Factors	Problem and Effect on Fishes
Temperature	Variable—due mainly to solar heating at different water levels and the seasonal effect due to river flows. Fish may move in response.
Salinity	Variable—due to river flows, wind, and tidal action; excludes stenohaline fishes or restricts them to one or the other end of the estuary.
Oxygen	Naturally abundant in shallow estuaries, may be a problem with nutrient enrichment (pollution) by humans.
Exposure	Wetting and drying of littoral habitat can kill organisms or expose them to terrestrial and avian predators.
Cover	Marshes, grass flats, oyster reefs, etc., can provide fish protection from aquatic and terrestrial predation.
Siltation	Naturally high levels build mudflats and carry nutrients and organic matter. But dredging can destroy habitat and resuspend toxic chemicals in the sediments.
Predation	Can be seasonally high in nursery areas with inshore movements of predaceous marine fishes.
Competition	Schools of estuarine-dependent fishes can consume large amounts of plankton and benthic invertebrates.

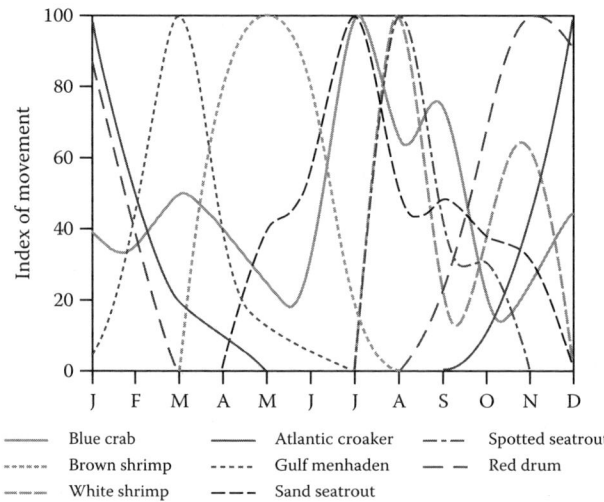

Figure 15.5 Seasonal use of estuaries by fish. (From Herke, W. H., and Rogers, G. D., 1999. *Inland Fisheries Management in North America, 2nd Ed.*, American Fisheries Society. With permission.)

(1999) classified 47 native estuarine fishes captured in salt marsh estuaries of the Gulf of Maine into the following: 10 resident, seven migratory, and 28 marine transients. These represented five categories: residents—small species of hardy killifishes, silversides, and sticklebacks; migratory—mostly anadromous and one catadromous; and transients—further divided into fishes that spawned in the estuary, those that used the estuary as a nursery habitat for young spawned in the ocean, and species that visited the estuary only as adults. The overwhelming majority of these fishes were non-residents (37/47 = 79%), an observation that has been widely accepted for estuaries in general (e.g., Sette 1971; Copeland et al. 1969; Moyle and Cech 2004). Another interesting observation is that all of the transient organisms do not occur at a single time of the year; instead, there seems to be a sort of partitioning out of species that share the estuary at different times of the year.

An extensive study of newly hatched fishes in the Middle Atlantic Bight region (located from Cape Hatteras, North Carolina, to Cape Cod, Massachusetts) by Able and Fahay (1998) used spatial and temporal patterns of estuarine fishes during their first year of life to assign fishes of this region into 11 categories of which seven included transients partitioned into various categories, three were residents that spawned there, and one unclassified. Although there was great variation in the location and timing of the spawning of these fishes, the most common juveniles were classified as transients (58) or residents (18), and the remaining (four) were unclassified. These proportions are very similar to California coasts.

Moyle and Cech (2004) divided estuarine fishes into five categories that vary somewhat from the above, but whose terms seem to be more widely accepted: *freshwater fish* that live in the upper, oligohaline portion of an estuary, such as sunfishes and white catfish; *diadromous fish* that only pass through an estuary, such as anadromous river herring, shad and salmon, and catadromous American eel; *true estuarine fish* that are permanent residents in an estuary, such as white perch, spotted seatrout, mullet, and croaker; *nondependent marine fish* that enter the lower, higher-salinity reaches of an estuary but do not complete their life history there, such as pinfish, pigfish, and weakfish; and *dependent marine fish* that complete at least one of their life history stages there, usually as a nursery area, such as menhadens and croakers. Important for commercial fisheries in the Atlantic and

Gulf coast of the United States are two species of menhaden (*Brevoortia tyrannus* and *B. patronus*), whose young use estuaries as a nursery habitat.

The population dynamics of fishes in small estuaries and the seasonal occurrence of Atlantic menhaden, one of the most important estuarine dependent fish—to be discussed later—were studied by Pacheco and Grant (1965). These authors reported the monthly abundance of fishes in eight locations of varying salinities in a shallow tidal creek associated with Indian River, Delaware. They collected 58 fish species (representing 48 genera in 32 families) and divided them into two groups, nonmigratory and migratory. The nonmigratory group consisted of only seven species of brackish water fishes that hatched there and may remain there during life, and a few incidental freshwater species. The migratory group was more complex, consisting of spring–fall migrants that used the creek as nursery habitat, diadromous fishes, incidental summer species, and marine wanderers. Seasonal use of the creek by spring–fall migrants and residents from April 1957 to June 1958 is shown in Figure 15.6. The influx of juvenile menhaden caused a great peak in fish diversity and abundance in May 1957 (19 species, $N = 22,617$), when 78% of the catch was composed of only three species: menhaden (61%) and two residents, rough silversides (*Membras martinica*, 17%) and mummichog (*Fundulus heteroclitus*, 13%). This changed in November when the menhaden moved out, reducing fish abundance by about one-half and changing the dominance (17 species, $N = 13,198$). At that time, 83% of the catch was composed of only three resident species: Atlantic silversides (*Menidia menidia*; 42%), sheepshead minnow (*Cyprinodon variegatus*; 26%), and mummichog (15%). By February, only the resident species remained, as reflected by lowest species diversity and abundance (5 species, $N = 192$). The catch was dominated (i.e., 79%) by two species: tidewater silversides (*M. beryllina*; 44%) and sheepshead minnow (35%). Fish abundance in November ($N = 192$) was only 1.2% of the peak in May.

Fishes occupy different habitats within estuaries, where they prey on various food sources, including detritus, marine invertebrates, and other fishes. Minello (1999), in a synthesis of 22 studies of estuaries in Louisiana and Texas, reported that fish species varied among habitats, including *edges of low marsh (S. alterniflora)*—most abundant fishes are gobies, pinfish, and gulf menhaden, also used by spotted seatrout and southern flounder; *submerged aquatic vegetation* (subtidal grass beds)—pinfish and gobies, also used by red drum and sand seatrout; *shallow*

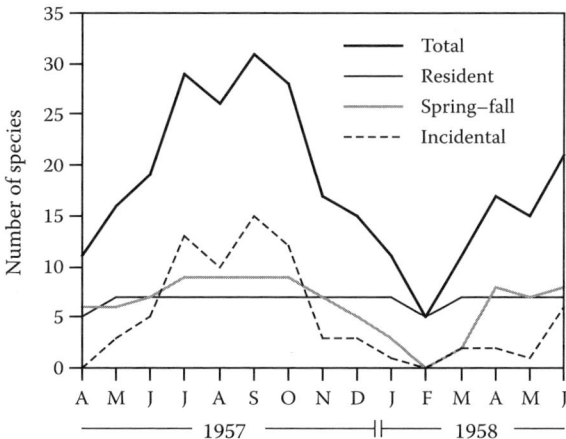

Figure 15.6 Fish movement in an estuary. (Redrawn from Pacheco, A. L., and Grant, G. C., 1965. *Seasonal Occurrence of Juvenile Menhaden and Other Small Fishes in a Tributary Creek of Indian River, Delaware, 1957–1958. Part 1, Studies of the Early Life History of Atlantic Menhaden in Estuarine Nurseries*, U.S. Fish and Wildlife Service.)

nonvegetated bottoms—gulf menhaden; *inner marsh*—killifishes and striped mullet; and *oyster reefs*—bay anchovies, gobies, and silversides. Commercially important species such as brown and white shrimp, blue crab, spotted seatrout, and southern flounder also occupied the edges of low marshes, while subtidal grass flats were used mainly by pink shrimp, red drum, and sand seatrout. Joining all of these systems are networks of tidal creeks. In fishing in such habitats with a gill net, I can attest to the importance of tidal creeks as modes of entrance into estuarine habitats by larger fishes. In North Carolina estuaries, I frequently set large mesh nets by blocking tidal creeks and captured mostly large striped mullet (a detritivore) and, infrequently, Spanish mackerel (its predator).

Plankton-Based Systems

In some cases, it is difficult to determine the roles of estuarine fishes because many systems, even large ones, have been greatly altered. In very large estuaries, such as Chesapeake Bay, pelagic systems can support a great biomass of plankton. Principal among the fishes that utilize this plankton are clupeid fishes, including anadromous herrings and shad, and estuarine-dependent menhaden. Along the Atlantic coast of the United States, great masses of young and adult Atlantic menhaden historically filtered the waters for plankton, and their decline due to commercial fishing has illuminated their keystone role in such systems. A recent popular book by Franklin (2007, p. 21) summarized much data and information and makes four points, which I paraphrase: (1) Menhaden are the only clupeids that feed almost entirely on phytoplankton and other tiny bits of vegetation. (2) Vast amounts of phytoplankton are produced by Atlantic and Gulf coast estuaries, where they were once the most abundant fish. (3) Menhaden was a food item for most large predatory fishes, birds, and marine mammals in estuaries. (4) Humans do not subsist on menhaden as food because it is too oily, but humans have decimated populations of the fish to make commercial products because the fish is relatively cheap. Perhaps they are not really so cheap considering ecosystem damage due to their declining numbers.

To illuminate the points given above, there now are restoration efforts underway in most if not all the large estuaries in the United States, but there have been problems. As an example, shellfish beds in Narragansett Bay have had difficulty making a comeback due to anoxic bottom conditions that resulted in a loss of clams, fish, and 4 to 5 billion mussels. The problem has been oxygen depletion linked to nutrient overloads that stimulate great summer algal blooms. Absent are millions of menhaden that historically foraged on the algae, due to commercial fishing. Menhaden also served as a buffer for other clupeid fishes, such as the anadromous river herrings, and shad, whose populations now have declined precipitously in some areas due to predation from fishes that used to consume menhaden. In the absence of these fat clupeid fish as prey, top predators such as striped bass have declined and some have apparently gotten sick from switching to a poor diet of invertebrates. In the Gulf of Mexico, two species of jellyfish have expanded greatly; these ecological equivalent species of menhaden apparently benefit from menhaden overfishing (Franklin 2007).

Oyster Reefs

Odum and Odum (1976) compared the energetics of an oyster reef with that of a city. As might be imagined, both are great users of energy, but the reef is much more efficient at the process. Oyster reefs are among the most productive systems on the planet. However, this system is more difficult to study, and the ecological significance of oyster reefs has not been as widely appreciated as that of tidal marshes and mudflats. Oysters along the Atlantic coast occur as intertidal and subtidal reefs of varying size, dependent on several factors of which the most important are perhaps low salinity to limit marine predators, a rich supply of food and nutrients, and a hard substrate. These reefs are extremely important as a resource in their own right, including ecosystem services and commercial value to humans. Oyster reefs also are considered essential fish habitats under the Magnuson Act (PL 104, et seq.), in part because of their great value to fishes for feeding grounds, reproduction, and nursery habitat. Coen et al. (1999) listed 79 fishes that used oyster reefs in seven areas from Maryland to Texas. These fishes included 12 reef residents (permanent and facultative) and 67 transients that included freshwater, estuarine, and marine species. Even more fishes were observed to use waters around the reefs, but these were not captured in the studies. Permanent residents (seven) were gobies (three), clingfish and blennies (two), and toadfishes (two).

ECOLOGICAL CONCEPTS

- *Estuaries as Nutrient Traps*—Estuaries tend to be more productive than the freshwater and saltwater systems that create them. They receive sediment supplies from rivers, the sea, and also from surface runoff from lands around the estuary. Rivers tend to provide vast supplies of organic matter and nutrients in spring runoff. Tidal action helps entrain matter in the estuary and also contributes dissolved salts and organic matter from the marine environment. A considerable amount of organic matter also can be produced within the estuaries by subsystems of marsh, grass flats, plankton, and so on. Estuaries can vary widely in such inputs, and as an example, river flow provides the majority of rich sediments for the Apalachicola River estuary of Florida, but in the Tay estuary of Scotland, the marine systems provide 70% of sediments. For most estuaries, it is a mix of inputs, such as in Chesapeake Bay, where 56% of sediment occurs from river input, 31% from shoreline erosion, and 12% from tidal action (McLusky 1989).
- *Estuaries as Nurseries*—Estuaries provide juvenile habitat for about two-thirds of the recreational and commercial fisheries of the Atlantic coast, where young organisms experience high survival and fast growth. Interestingly enough, spawning of these organisms usually occurs frequently upstream in freshwater (e.g., by anadromous species such as herrings, salmonids, and striped bass) or downstream in the marine zone (e.g., from marine adults of penaeid shrimp, blue crab, and menhaden). Spawning and rearing areas in freshwater rivers are likely to be features of the spring floods and are thus ephemeral, while the estuary is a better place for juvenile fish than the marine system because the young fish are protected from wave surge and are provided with shallow turbid water, subtidal grass beds, and marshes, all of which can provide protection from predators.
- *Remane's Curve*—A plot of species diversity versus salinity shows that freshwater and saltwater portions of estuaries support more species than brackish water portions, producing a characteristic curve that reaches a low point (minimum diversity) at about 5 ppt NaCl. (Attributed to Adolf Remane; McLusky 1989.)
- *Critical Salinity Hypothesis (Remane's Minimum)*—This is attributed to V. V. Khelbovich, who suggested that lowest diversity of estuarine organisms at about 5 ppt is due to the changing chemical properties of water below that level. This hypothesis has been replaced, and low diversity is now attributed to increasing physiological stress (see next paragraph).
- *Stability–Time Hypothesis*—This concept was proposed by H. L. Sanders and was related to estuaries by McLusky (1989) in which resident species abundance of estuarine organisms is limited by great physiological stresses in an environment of relatively recent origin.
- *Migrating Subsystems*—Copeland et al. (1969) originated this term to explain the function of animal populations that move from one ecosystem to another. These subsystems take advantage of specific components of several systems as part of their life history strategies. In coastal systems, migrating subsystems connect freshwater and marine systems with estuaries and thus increase energy availability. As an example, great masses of young menhaden were an extremely valuable component of Atlantic coast estuaries historically, enhancing system function by consuming enormous quantities of phytoplankton. The energy thus consumed is then cycled up through estuarine and inshore marine food chains (e.g., see Franklin 2007).

CASE STUDY: ALEWIVES AS MIGRATING SUBSYSTEMS

Background

Many of the transient fishes in estuaries are diadromous, and adults of those species migrate from the ocean to spawn in freshwater (anadromous) or from freshwater to spawn in the ocean (catadromous). Some of the most important transients along the east coast are members of the family Clupeidae (herring family), which includes herrings, shads, sardines, menhaden, and other related fishes. In this case study, we will discuss the importance of river herrings, specifically alewives, and the function of these fishes as migrating subsystems. Clupeids include many fishes that support commercial fisheries in various parts of the world.

The alewife (*Alosa pseudoharengus*) is an anadromous fish that is native to the Atlantic coast of North America. It is a northern fish, and its distribution overlaps with the blueback herring to the south (Loesch 1987). It also occurs as landlocked populations in freshwater lakes, including the Great Lakes. Adults are schooling fish of nearshore marine waters that feed primarily on zooplankton. In spring or early summer, depending on latitude, alewives begin their spawning migrations by moving toward the coastline where they enter into river estuaries and migrate upstream into freshwater. The fish is also called the "branch" herring because they seek smaller streams and spawn in shallow sluggish headwaters or in freshwater ponds. Mortality of adults during these spawning migrations can be high, especially in warmer climates. In the cooler waters of Maine, adults first spawn at about age III and some continue to spawn until age VIII (Havey 1961). The adults place their eggs in habitats that are flooded in the spring and produce productive blooms of plankton. As these flooded areas begin to contract in size and warm up, the juveniles swim downstream and feed in estuaries for a time until they finally move into the nearshore environments that they prefer as adults. These locations are proximal to their natal estuaries and within about 80 miles of land (Mansueti and Hardy 1967).

Alewife migrations (also known as "runs") seem to announce the coming of warmer days of spring in many locations, where local folk consider them an important part of their environment. Naturalist John Hay (1959) wrote a very compelling book about an alewife run in Cape Cod. This book was inspirational to me because I read it when I was doing a graduate study of the fish.

The alewife was important to North American natives long before the arrival of European settlers. Their use of the fish for food and fertilizing crops is common knowledge, thanks to William Bradford's *Of Plymouth Plantation* (1621), wherein he wrote about a native American who taught them how to use the fish (Hay 1959). The alewife was very important to the colony, and the first fishery law passed in the New World was the Plymouth Colony Fish Law, enacted in 1623 for the protection of alewife runs (Belding 1921).

Significance

As stated above, alewives have long been used by humans as food in several ways: scored (cut in intervals through the meat to the bone), deep fat fried to crispy, and bones consumed; canned as meat by cooking the bones to insignificance; or as the famous "pickled herring"—mostly a breakfast food. But the ecological value of alewives as a migrating subsystem is much more interesting.

Alewives migrate into freshwater systems and spawn their adhesive eggs over vegetation that is flooded by spring runoff. The runoff also floods dry land, releasing nutrients from the soil and contributing organic matter to the aquatic environment—in this case, small shallow ponds. By the time alewives arrive, the nutrients and detritus have provided a rich soup of plankton, which is the favorite food of the developing fry, and larger organisms that can be eaten by juveniles. These rich ponds do not occur at other times of the year and thus cannot support fish year round. By migrating and spawning there, the alewife has taken advantage of an underutilized food source and an ephemeral and shallow habitat that does not support many predators. In turn, the young alewives provide a rich protein source for larger stream fishes, such as freshwater bass, pickerels, and sunfishes. As the ponds begin to dry up, the juvenile alewives are large enough to swim into other habitats and eventually enter into the estuary, where they take advantage of food and the nursery habitat for as long as a year, finally returning offshore to begin the cycle all over again. Offshore, the adults are eaten by larger inshore marine fishes, such as striped bass and bluefish, thus supporting important recreational and commercial fisheries. In addition, there are land-locked populations of river herring in reservoirs, such as Santee-Cooper in South Carolina, where they are fed upon by striped bass.

Alewife stocks have been heavily fished since the late 1800s. North Carolina was historically the leading alewife-producing state, producing over 20% of the national catch and supporting a fishery of over 35 million lb in 1896 (Smith 1907). However, this catch declined to 30 half of that in 1900 and then to only 6 million lb in 1937 (Woodward 1956), dwindling since then to less than half of that annually.

Alewives have always been considered an important resource in New England, and many steps, including legal protection and refuge, have been taken historically to preserve their runs. However, dams, drainage, pollution, and other factors have reduced alewife stocks in New England as well, as in the case in Maine. Saunders et al. (2006) argued that diadromous fish populations have been drastically reduced from most Maine rivers. They provided a comparison of historic and recent catches of river herring (mostly alewife, with some blueback herring) for certain rivers in Maine. In six rivers that have both historic and present catch data available, the catch has been reduced from 5.3 million fish (1867–1899 data) to only about 650,000 fish (2000–2005 data)—an 88% decline. The decline of the river herring there also has been linked with a corresponding decline of Atlantic salmon. The herring presumably served as food for the salmon and also served as a prey buffer to protect salmon fry and smolts from native predators (Saunders et al. 2006). Estuarine function depends on a coevolved system, and restoration of the entire system would be needed to restore its parts.

Alewife Run at Mattamuskeet

The coastal plains of southeastern United States are low-lying expanses of shrub bog (i.e., pocosin) and associated shallow coastal bay lakes. Lake Mattamuskeet is the largest of these lakes, located near Pamlico Sound, a wide oligohaline estuary formed by barrier islands that partially enclose the sound from the Atlantic Ocean, which is about 30 miles away. The lake has an interesting history. Beginning in 1914, it was partially drained by pumping lake water via a canal to the sound and was then farmed. Evidently, farming was not profitable, perhaps due to wet years and inadequate pumping, so the practice was discontinued. The area was known as a haven to Canada geese and other water birds; in 1934, the lake was sold to the U.S. Fish and Wildlife Service (USFWS), and it became the Mattamuskeet National Wildlife Refuge. Since then, the lake has been maintained as a bird refuge, and management practices have favored the development of freshwater marshes and farming to produce food for geese and ducks. Lake Mattamuskeet also was repeatedly stocked with recreational sport fish from a federal fish hatchery, and a sizeable population of largemouth bass, crappie, and bluegill resided in the lake along with native longnose gar, bowfin, bullheads, and introduced carp. Striped bass also are reared at the hatchery and were likely introduced as well.

To drain the lake, canals have been dug to Pamlico Sound and equipped with water control structures. These structures are wooden, gravity-operated flap gates that allow the lake to drain in the spring but limit inflows of salt water to a trickle in the summer. There is evidence that the original drain of the lake was Lake Landing canal, Figure 15.7 which enters the sound at a small boat harbor used by commercial fishermen.

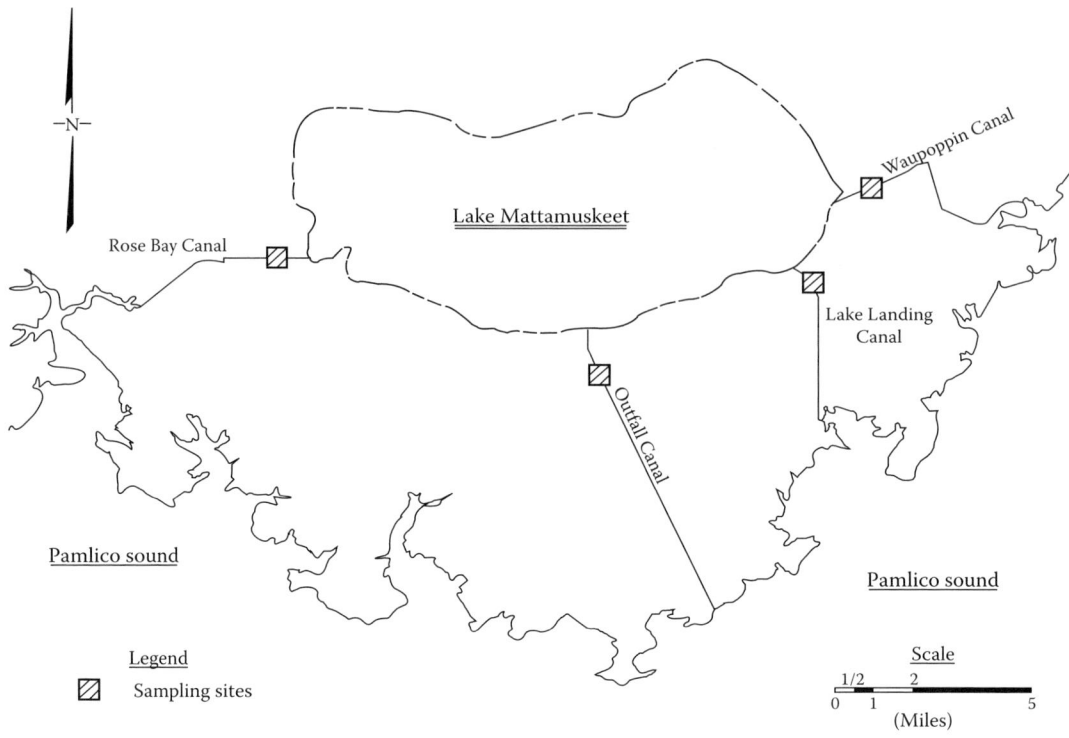

Figure 15.7 Lake Mattamuskeet is connected to Pamlico sound by four canals that were used to sample alewife runs (From Tyus 1971). Roads and other details omitted.

The Cooperative Fisheries Research Unit at N.C. State University received a small contract to investigate fisheries at Lake Mattamuskeet during the time I spent there as a graduate student. My interest in the lake came about while I was working with other students on some fish surveys. While talking to locals, I was told not to miss the herring run, or I had to see the "hurrin dippers." I found out that in early spring, local people with very large handmade dip nets congregated at these canals during the warming days of spring (late March or early April) and then they would walk out on the water control structures to search the water for fish. One day, a dipper would detect and catch alewives, and the word would be spread: "the hurrin are runnin'." Soon, there would be scores of people dipping shoulder to shoulder and making huge catches for so small a stream (Figure 15.8). I had never seen anything like it, and I believed that the refuge was a perfect place to study the migrations of the alewife. There were obviously thousands of the rather large fish entering the lake to spawn, and this had to be an important component of the Mattamuskeet ecosystem. After reading John Hay's book *The Run* (1959), I was hooked on the alewife!

A Study of Alewives

As I begin to investigate alewife life histories in coastal North Carolina, I found that very little was known about them, even though much was known about historic runs in more northern waters. With little to go on, a pilot study was used the first year to design a comprehensive 2-year study. Sampling indicated that most of the migrating fish entered the lake through the Lake Landing canal. Using nonuniform probability sampling, efforts were allocated in favor of the most used canals. All were sampled, but more efforts were allocated on Lake Landing, where a permanent trap was built

ESTUARIES AND COASTAL ZONE

Figure 15.8 Herring fishermen dipping their nets at Lake Landing canal, Mattamuskeet National Wildlife Refuges (From Tyus 1971).

downstream of the water control structure. Sampling at all canals was done immediately downstream of the fishing area. When the fish were sampled, a creel census also was done to determine how many fish were taken by dip nets.

When the first year of the comprehensive study was completed (1970), I found that the run peaked on April 5–6, when it was estimated that 199,600 fish (108,400 lb) had entered the lake, principally through the Lake Landing canal. Dip-net fishermen had spent 2233 dip-net hours and captured 88% of the prespawning fish. Aging of a subsample of 48 fish in 1970 indicated that the spawning fish ranged in age from II to IV, with the majority of spawners having an age of III. The mortality rate from III to IV was an alarming 84% (Figure 15.9) (Tyus 1971). Based on this information, the USFWS decided that this catch rate, which was mostly due to commercial fishermen who used the fish for crab bait, was too high. To protect the run while allowing almost unrestricted recreational fishing, it was decided to reduce the number of days allowed for fishing from 7 to 5. The following year, we found that the fishermen caught only 69% of the run of 167,300 fish, which was accepted as a safe limit. Further aging of 229 fish in 1971 (peak of spawning, April 2–3) produced almost the same results: same age for spawners, with an age of III to IV and a mortality rate of 86% (Figure 15.9). Alewives live longer in more northern climates, but perhaps due to the fast warming and high temperatures of the Carolinas, their existence there seems precarious. The caution exercised to restrict fishing was logical since removal of a large number of the prespawning fish could decimate the run.

Admittedly, presence of the water control structures tended to concentrate the fish and increase fishing success. But the alewives were experts at slipping through the large flap gates. Anyway, the large gates were always to be open in spring due to the accumulation of water in the lake from its drainage basin, and limbs, sticks, and other debris get caught in them so some water flows back and forth via the gates at all times of year. The ability of the alewives to sniff their way back to natal spawning grounds is described by Hay (1959). It is this uncanny ability to find historic routes by olfaction (Thunberg 1971) that perhaps guided 97.5% of the fish into the Lake Landing canal in deference to the three other ones, because the earliest record that I could find indicated that in 1909, the only connection between Lake Mattamuskeet and Pamlico Sound was the Lake Landing canal. I assume the fish never forgot!

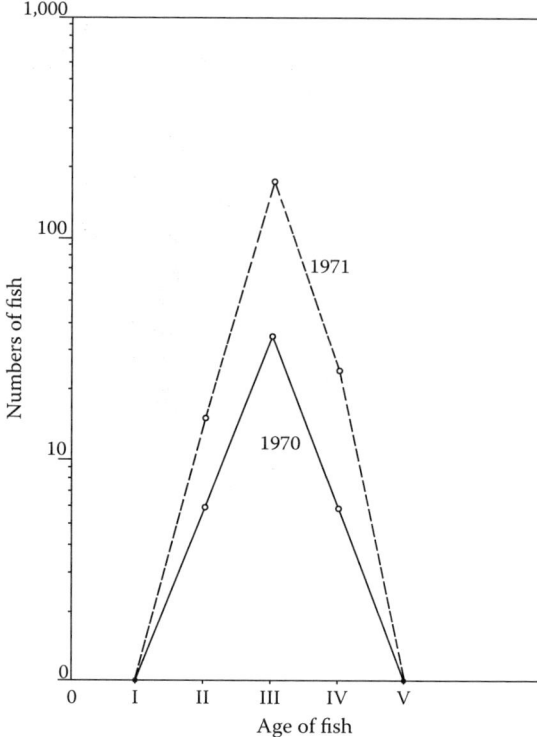

Figure 15.9 Catch curve of alewives captured at Lake Mattamuskeet (From Tyus 1971).

Tragedy Strikes and a Lesson Learned

I was proud of my work at Lake Mattamuskeet and believed that I had helped protect a valuable natural resource. As years passed, I would think about those days and how great it was to be involved in successful management decisions. Ironically, my success was short lived. I received a phone call from the USFWS in 1994 inquiring if I had any idea why the alewife run at Lake Mattamuskeet had disappeared! A copy of my thesis was found at the refuge headquarters, and they wanted to know what I thought. In 1989, the old wooden flap gates were replaced by smaller but heavier stainless steel gates that required a great deal of pressure to open. No testing had been done to see if the alewife could use these gates and navigate the high-velocity water that was released. Evidently, the fish could not. In 4 years, the run had declined to only a few incidental fish. To give credit where it is due, the refuge workers had tried propping some gates open when it seemed clear to them that the fish were denied access.

The tragedy that happened at Lake Mattamuskeet was very troubling, especially since the lake is in a national wildlife refuge. To its credit, the USFWS soon took measures to ensure fish passage by installing slotted fish weirs in 1996 that were opened manually. Unfortunately, these also were found unsatisfactory and, in 1999, were replaced with wooden gates of the original design at Lake Landing and another canal. The modified 1996 gates and these old-style gates were studied by Rulifson and Wall (2006), who found that high water velocity precluded fish passage through the stainless steel gates. The slotted fish weirs were more efficient than the steel gates but were not satisfactorily due to turbulence. Wooden gates of the original design were the best.

The alewife run at Lake Mattamuskeet collapsed and has been in danger of extirpation. This is evident in the numbers of fish present in the spawning runs: although I reported historic runs of

170,000 to 200,000 fish, the population sizes in 1997 and 1998 were estimated at only 178 and 454 fish (Rulifson and Wall 2006).

However, there is hope. The last report that I know of was that of a marked increase in the number of alewives in 2001, which amounted to an estimated 8836 fish (Godwin and Rulifson 2002).

I am amazed that there were any alewives left at all at Mattamuskeet, and I can only imagine that the adaptable alewife population has persisted because alewives trapped in the lake were able to survive and reproduce. I captured young alewives year-round in the lake. But not only the alewives have been affected, blue crabs and young marine fishes (including anchovies, croakers, pinfish, needlefish, puffers, and menhaden) also have been captured moving into the lake (Tyus 1971), and these organisms also would have been blocked. Hopefully, the USFWS has learned an important lesson and will continue monitoring this situation. Blockage of fish migration routes must be considered every time construction affects water flow in coastal areas. Federal agencies are required to complete an environmental assessment under the National Environmental Policy Act (NEPA) provisions and regulations, especially in wetland or floodplain habitats.

Passage of alewives at Mattamuskeet at times conflicts with the management objectives of the waterfowl refuge, which relies on the flap gates to reduce the amount of brackish water that enters the lake. In early spring, with freshwater runoff, the adult alewives can slip into the open gates. However, later in the season, when salt water flows back into the lake, the flaps close, and the emigration of late adults and juveniles is blocked. Historically, many of these fish died in the narrow channels that become stagnant, hot, and low in oxygen.

Alewives are especially capable of passage through more natural channels, such as "nature-like" fishways with a wide, low gradient (1:20 slope or less). Placement of boulders allows the fish to make the passage aerobically (i.e., at sustainable speeds). In testing such fishways, Franklin et al. (2009) found that 95% of alewives were able to ascend them in less than 22 min, and they suggested that such nature-like structures be constructed at low-head dams. I have seen concrete ladder-type fishways in use for alewife passage elsewhere, but I would like this more natural and less costly design implemented on at least one of the canals at Mattamuskeet, preferably the original one at Lake Landing. Perhaps this passageway could not be operated all year, but it might be extremely useful during spring, with its operation enhanced by water level control.

SUMMARY

Costal zones are transitional areas that include the inshore marine zone and estuaries. These aquatic systems are constantly affected by freshwater runoff, tides, ocean waves, and wind. Four types of estuaries are recognized, based on geomorphology, and the coastal plain or drowned river valley type is the most common type in coastal United States. In simple terms, a river estuary is formed when freshwater of a river is diluted by the sea. However, these estuaries are dynamic, and mixing can occur for miles in a semi-enclosed basin. As fresh and salt water mix, riverine sources of organic matter and inorganic nutrients are combined with nutrient-laden salts, organic matter, and sediments swept from oceans into estuaries by tidal action. Productivity of the resulting system is also enhanced with organic matter produced within the estuary and by formation of a nutrient trap. This trap forms and acts as a dense wedge of salt water that transports nutrients and organic material up into the estuary with each tidal cycle. It is covered by a layer of lighter freshwater that loses velocity and drops suspended material into the dense water below. Associated with and in river estuaries are plant and animal communities that may be recognized as ecosystems in their own right. Along the Atlantic coast of North America, these include benthic-oriented systems such as mudflats, subtidal grass beds, salt marshes, oyster beds and reefs, and mangrove swamps. Freshwater, estuarine, and marine fishes, such as seatrouts, mullets, croakers, pinfish, and striped bass, may be transient or permanent residents in these estuaries, which are important nursery habitats for fishes and shellfishes. All fish that occupy estuaries have to cope with stressful conditions, and most do so by moving in and out seasonally or with the tides. Stress and energy expense are compensated by the great availability of food. Fish also benefit through a reduced level of competition or predation during young (small) life history stages, because the young can make use of shallow vegetated habitats and reefs as cover. Large estuaries in eastern United States support a great biomass of plankton that is utilized by clupeids, including anadromous herrings and shad and estuarine-dependent menhaden. Along the Atlantic and Gulf coasts, great masses of young and adult Atlantic menhaden historically filtered the waters for plankton, in turn providing high-quality food for large predators. Other planktivores include several anadromous clupeids such as the alewife, which is discussed here in a case study. Alewife adults feed on zooplankton in the marine inshore zone and migrate all the way into freshwater systems, such as the marshy shallow lake (Mattamuskeet) in the study, where they spawn. Spring runoff also floods dry land, releasing nutrients from the soil and contributing organic matter to the aquatic systems. By the time alewives arrive, the nutrients and detritus have provided a rich soup of plankton that is the favorite food of the alewife young. These rich ponds do not occur at other times of the year and cannot support fish year round. The alewife has taken advantage of an underutilized food source and an ephemeral and shallow habitat that does not support many predators. In turn, the young alewives provide a rich protein source for larger fishes, such as freshwater bass, pickerels, and sunfishes. As the ponds begin to dry up, the juvenile alewives are large enough to enter into the estuary, where they take advantage of food and nursery habitats for as long as a year, finally returning offshore to begin the cycle all over again. Offshore, the adults are eaten by larger inshore marine fishes, such as striped bass and bluefish, thus supporting important recreational and commercial fisheries. The case study also relates tragedy on a national wildlife refuge, where the alewife run was blocked by the construction of water control gates. The lesson: many resources are lost due to poor management and carelessness rather than intention.

CHAPTER **16**

Marine Environments, Intertidal Fishes, and Sharks

OCEANOGRAPHY AND MARINE ECOLOGY

Oceans and connected seas cover more than 70.8% of the Earth with salt water to an average depth of about 3700 m, a volume that contains 97.2% of all the water on or near the surface of the planet (Thurman and Trujillo 2002). Oceans are so extensive that evaporation of seawater greatly exceeds that of land. As a result of wind transporting rain clouds inland, continents receive much more water than they lose in evaporation, resulting in an abundance of freshwater (e.g., see Smith and Smith 2001). Also, the separation of continents on the globe produces a moderating effect on their climate because water has a higher heat capacity than land. The present distribution of continents also produces much continental shelf habitat, which greatly enhances global productivity.

Marine environments are so obviously different from terrestrial ones that little discussion is warranted here. However, what is obvious to one person is not to another, so an example is given: Large flowering plants dominate productivity in terrestrial systems. Even though the largest biomass of green plants occurs in oceans, it occurs mainly in the form of microscopic algae (phytoplankton). In turn, the herbivores also are very small, resulting in at least five links in the marine food chain from autotroph to apex carnivore, while there may be as few as three in the terrestrial or freshwater system (plant–deer–cougar; algae–sucker–pikeminnow). With plants so small, herbivores also must be, so almost all large animals at sea are carnivores. An interesting reversal of life histories also results, with long-lived terrestrial autotrophs and short-lived carnivores, while at sea, the opposite is true (Nybakken 1997). This suggests that lessons learned from terrestrial or even freshwater systems may not be appropriate here. For example, it is a mistake to even consider oceans as large lakes (e.g., Magnuson 1988), an admonition that will be supported by the remainder of this section.

Seawater is in constant motion due to extraterrestrial gravitational forces, rotation of the earth, wind, temperature differences, and changes in salinity. Such factors result in tides, waves, surface currents, and deep density flows of water in the oceans and seas. These forces also affect fish habitat because of the mixing of nutrients, oxygen, and heat, and they affect living conditions and food supply. Currents also provide for the transportation of various life history stages of organisms.

Unlike the lithosphere, water exists mostly in liquid phase, and a vast quantity of water is affected by the Earth's rotation, which produces a bulge around the equator. Oceans also are affected by two extraterrestrial objects, the moon, which is close, and the more distant but larger sun, which produces about half of the gravitational pull of the moon. The gravitational attraction of the moon and the sun produces tidal effects, which are amplified (spring tides) when they are lined up in space and their attractions are combined (during the new and full moon stages).

The Earth rotates once every 24 h, but the moon makes one orbit in 27.5 days. As a result, there is one tidal cycle (one high and one low tide) at a given point on the Earth every 24 h and 50 min (a lunar day). But it is not this simple. Tidal height and frequency depend on the location and shape of landforms. While there are semidaily (two high and two low) tides on the Atlantic east coast, tides in the northern Gulf of Mexico occur only daily (one high and one low), and there are mixed (unequal) semidaily tides on the Pacific west coast (Sumich and Morrissey 2004).

Oceans have other attributes as well, including a global system of currents, at the surface and at depth. Surface currents such as the important Gulf Stream on the Atlantic coast and the California current on the Pacific coast have a great effect on fish populations as well as on humans. Such currents are mainly caused by the rotation of the Earth, locations of obstructions (continents), and wind. Deeper currents are more affected by water density.

Wave action is generated by wind and, to a lesser effect, by surface currents. Wind produces offshore waves that transmit energy by wave shape, not by moving individual water molecules across any great distance. Locally, water molecules tend to move in a circular fashion, and only the energy is transmitted across any distance. Waves mix the water locally, perhaps to depths as great as 50 m, but when waves contact with shallow water, the lower part of the wave slows; eventually, the top outruns the bottom of the wave, and the full force of its stored energy crashes down. When this occurs, the energy can be substantial, eroding shorelines by suspending sediment and organisms and translocating them, usually in alongshore drift. At sea, waves may move in different directions and produce unusual patterns and, occasionally, what is known as "rogue" waves, which are unpredictable superwaves of enormous size and force.

Surface currents are mostly due to the rotation of the Earth and the prolonged effects of wind blowing from one direction. In this case, water is moved at the surface by friction, so there is a real movement of the water mass horizontally. Prolonged winds are produced from the eastward rotation of the Earth, and wind tends to move water to the west. However, movement of water can produce unanticipated patterns as we view them on the surface; this is due to the Coriolis effect (named after G. G. Coriolis, a seventeenth century mathematician), which results in a polar deflection of water, that is, northerly in the Northern Hemisphere and southerly in the Southern Hemisphere. The Coriolis effect is nil at the equator and intensifies toward the poles (reviewed by Sumich and Morrissey 2004; Castro and Huber 2007).

Ocean currents are extremely important for fishes, as we will discuss in detail later, so more explanation is needed here about how these currents form. The equatorial region is subjected to great solar warming, and this results in warm surface air rising and being replaced by cooler air that flows toward the equator from the north and south. However, at the same time these flows occur, the planet is rotating. Earth's spherical surface is traveling at about 24,000 mi/day (1600 km/h) at the equator, its greatest diameter, and surface speed would decrease steadily as you travel toward the poles. For example, at 45°N (the latitude of Minneapolis, Indiana), the surface speed would be 17,000 mi/day (1133 km/h) (Colinvaux 1993). Air that would travel south to replace rising equatorial air would thus be deflected by the more powerful flow, which is, in this case, to the right or slightly north. The winds thus produced are known as the northeast and southeast trade winds.

The trade winds about the equator also affect the surface of the oceans down as deep as 100 m. Again, on the rotating Earth, water flow will be subjected to the Coriolis effect, but with a lag time. Surface flow from wind friction also will be transmitted by friction down the water column, with each increment responding to the Coriolis effect and bending slightly north (i.e., the Ekman spiral), thus resulting in north and south equatorial currents around the globe. Continents complicate the matter, however, because they present a barrier to global current flow, resulting in huge water circulation patterns known as *gyres* (Figure 16.1) that occur due to the combined forces of water flow and wind at about 30°N/S latitudes. As an example, equatorial flow results in the piling up of water 1–2 m in height in the Caribbean Sea. This shallow and very warm water is forced north by the Coriolis effect along the Atlantic coastline of the United States as the Gulf Stream. This flow encounters colder water and westerly winds and continues to bend in a circular fashion as it meets

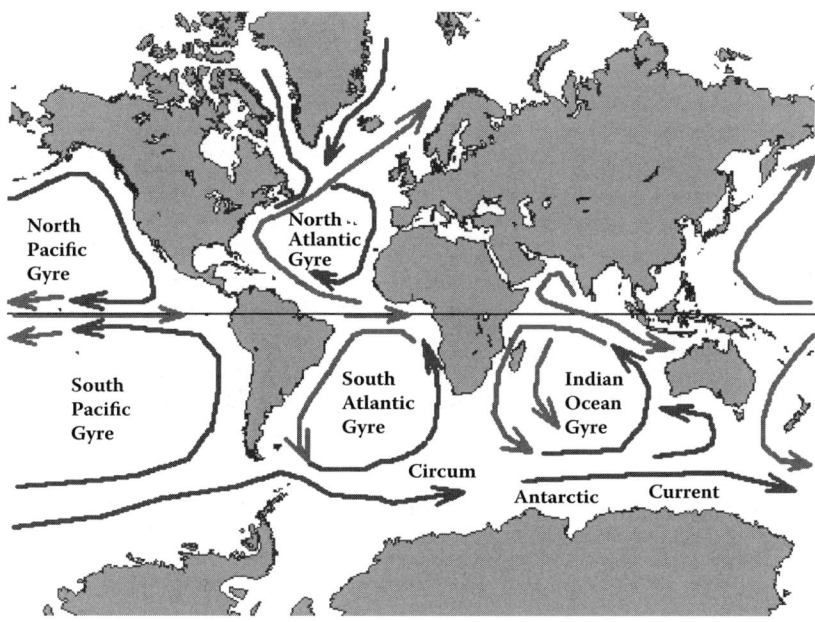

Figure 16.1 Formation and persistence of gyres. (Courtesy of the U.S. National Aeronautics and Space Administration (NASA), http://search.usa.gov/search/images/query=ocean+gyres&locale=en&m=false&commit=search.)

western continental flows; it is then once again entrained by equatorial currents, completing its cycle. A similar gyre also can be observed in the Pacific Ocean, where the warm flow north and east is known as the Japan Current (Figure 16.1).

Off the coast of California, the prevailing winds tend to move water across the Pacific Ocean, resulting in a somewhat net western displacement of surface water (about 2 m in height). An important consideration for fisheries occurs with *upwellings* of deep nutrient-laden water along the California coast that flows up to fill in the void as water is blown westward (the La Niña effect). As surface currents encounter the continental shelf of Asia, they are deflected northward along the Asian continent as offshore continental (Japan) current. These waters then encounter westerly winds that drive the current easterly once more, setting up another gyre.

Vertical water movement in the oceans also can occur due to differences in temperature and density (thermohaline circulation) (Nybakken 1997). Seawater comprises about 97% of water on the planet, and of course, it is salt water. Salt water can be diluted to under 1 ppt in estuaries and can exceed 40 ppt in the Red Sea, but in general, seawater in the open ocean averages about 35 ppt and varies only within the range of about 34–37 (Nybakken 1997). Seawater does not display the same properties as freshwater, and its density does not reach a maximum at 4°C. Instead, as it gets colder, it is denser and sinks or remains at the bottom. Thus, the oceans do not turnover like lakes; they are too deep and remain permanently stratified. This means that nutrients and organic matter that sink into depths tend to remain there. As a result, the upper waters in the open ocean are characteristically low in nutrients (especially phosphorous), and they are analogous to a desert in the terrestrial system. Ironically, the deep sea underlying these waters may be nutrient rich. It is only because of deep currents that these rich waters can be circulated to the top as upwellings, as mentioned earlier, and significant upwelling can occur in deep continental margins such as the Pacific coasts of California and Peru.

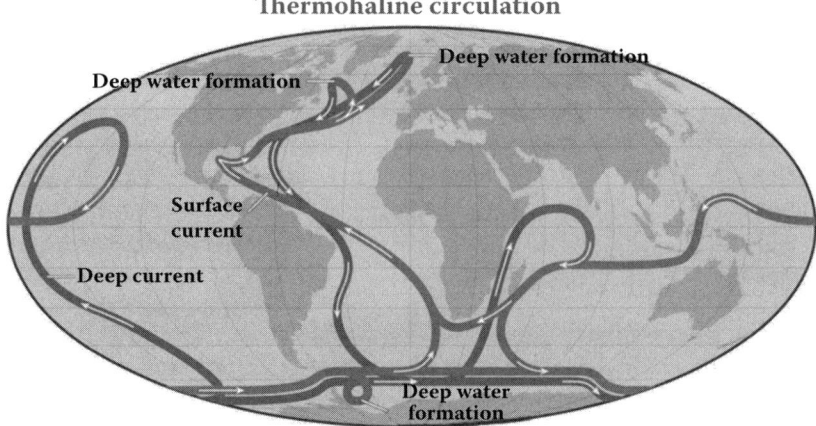

Figure 16.2 Ocean Conveyor. (Courtesy of NASA, http://search.usa.gov/search/images/query=thermohaline+circulation&locale=en&m=false&commit=search.)

Global thermohaline circulation presently occurs due to the Ocean Conveyor System (reviewed by Castro and Huber 2007) (Figure 16.2). This circulation occurs because surface water is heated in the tropics, where it also is subjected to a high evaporation rate. In the Atlantic Ocean, some of this light but very saline water is transported northward by gyral circulation of the Gulf Stream, eventually meeting frigid water of the Labrador Current, which moves south. As the tropical saline water is cooled, it becomes extremely dense and sinks, joining a very deep ocean current that flows from the northern to southern hemispheres. Cold, dense water warms enough to rise to the surface near Antarctica in about 1000 years (Thurman and Trujillo 2002), and the conveyor makes one complete circulation in about 4000 years (Castro and Huber 2007). This is an important concept, because oxygen is taken to depths as the deep sea waters are slowly recirculated.

Weather can also produce other phenomena such as storms, which can produce wind currents (directional movement) and turbulence (nondirectional). Wind can have some extreme effects on plankton, as they can be dispersed both horizontally and vertically. Turbulence also makes it difficult for organisms to feed on zooplankton. Lasker (1975) found that periods of calm were very important for the growth and survival of larval fishes. These calm periods have been named in his honor as "Lasker events," which are periods of calm (wind speed < 5 m/s that lasts at least 4 days). This also is known as the "stable ocean concept."

Oceans are vast, and marine habitats are many. There are, no doubt, more distinct ecosystems than have been recognized to date, and there are already too many for convenient memorization. However, the complexity of the system makes it necessary to partition the faunal environment into convenient portions that are easy to understand. Thus, it is customary to divide the oceans into just two realms, the pelagic realm and the benthic realm, because organisms can be classified as benthic or pelagic. The biggest problem is the *intertidal* zone (also called the littoral or coastal zone) which is dewatered with low tide. This zone is exclusively benthic, because the pelagic realm must always be wet. The next zone also needs clarification, because the *neritic* (or sublittoral) province includes the mass of water that lies above the continental shelf to the continental break, where the shelf slopes down at a greater angle, and the benthic zone that is included beneath. The *oceanic* province then includes all of the pelagic zones (water) and benthic zones (seafloor) of the open ocean (Figure 16.3). We recently completed our discussion of the estuary, so we will continue to the closest marine environment: the inshore marine systems of the intertidal zone and the fishes that live there.

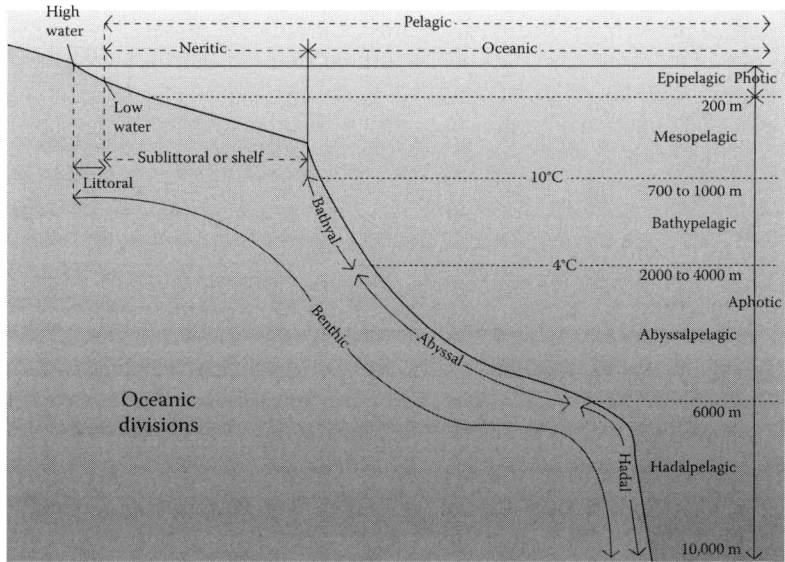

Figure 16.3 Ocean realms: pelagic and benthic. (Courtesy of the National Oceanic and Atmospheric Administration (NOAA), http://commons.wikimedia/wiki/File:Oceanic_divisions.svg.)

INTERTIDAL ZONE: STRUCTURE AND FUNCTION

We have discussed different plant and animal communities that are associated with estuaries, and this section deals with some marine systems that are associated with ocean coastlines. There are many ways to partition coastal systems, and only a few are discussed here. The intertidal (or littoral) zone is exposed to the force of waves and, as the tide goes out and the upper portions are dewatered, to terrestrial predators as well; then with high tide, it is exposed to marine predators. The zone can be a rich one, however, due to inputs of organic matter and nutrients that are produced both from terrestrial and aquatic sources. Detritus is mostly produced by the terrestrial runoff and is carried along the beach with alongshore transport, which also carries live, dead, and bits of dead organisms. Here, we recognize the two main inshore systems: (1) intertidal beaches of mud, sand, or gravel and (2) intertidal rocky shorelines and associated tidepools.

Beaches

High-energy beaches are dynamic systems that support similar invertebrate organisms throughout the world. Exposed to the full energy of the open ocean, wind- and tide-driven waves strike the coast and, where the surf breaks at an angle to the shore, produce alongshore transport of sediment, which can range from clay and silt (mud) to sand and gravel, depending upon the energy of the wave action. The sediments are not stable and mostly inhabited by organisms that can live in and beneath the disturbed surface (infauna). Residents of these soft bottoms are almost entirely marine invertebrates and algae. Larger organisms include mollusks that remain hidden, anchored by use of their muscular "foot"; crustaceans such as amphipods and mole crabs ("sand fleas") that dig with their feet; and lugworms that crawl and eat their way through the sediments. However, there are hundreds of species of very small organisms that live in the interstitial spaces provided by the sediment (i.e., meiofauna). On Atlantic beaches, predators, whether various birds or fish, make their living by digging these out of habitats that exist in at least three major areas: above high tide (amphipods, ghost crab), intertidal (mole crabs, lugworms, clams), and subtidal (fishes, blue crabs).

Fishes are not well represented in the upper intertidal zone of beaches, but further offshore, deep intertidal and inshore subtidal areas are patrolled by several species of fishes feeding on plankton, detritus, mollusks, and other fish. Whiting (kingfishes) find prey buried in the substrate, rays hunt for anything and often settle for cropping the siphon tubes of mollusks, roving schools of bluefish eat any smaller fish they can find—usually fishes of the herring family that are feeding on plankton, and juvenile fishes seek protection by moving along the coastline. Rays and sharks may wait for high tides to take their turn at feeding on almost anything, including the bodies of larger animals washed up with the tide. In addition to feeding, transient California grunion and sea turtles make legendary spawning forays into the beaches to spawn with highest tides.

Rocky Shorelines

Rocky shorelines and associated tide pools are most common on the Pacific and New England coasts of the contiguous United States. These shorelines are composed of exposed rock due to the crustal rebound of the Atlantic coast, or uplift due to geological forces associated with the active continental margin. In contrast to the exposed softer beaches, rocky beaches are occupied by resident organisms that live on the rocks (epifauna) instead of in the substrate. Due to the extreme turbulence of this habitat, large aquatic predators are scarce, and resident organisms have to protect themselves more from dessication and terrestrial predation than from underwater predators. Some invertebrates (crabs and snails) do this by mobility, hiding out in wet protected places usually in tide pools, while sessile forms inhabit wetter areas permanently by having impervious coatings and attachment organs (barnacles and mussels). Fishes respond to the potential for dessication with behavioral adaptations, mostly movement and timing (Horn and Martin 2006) (Figure 16.4).

Although species vary between Atlantic and Pacific coasts, in general, rocky shores are characterized by three zones: the upper zone, inhabited by algae, lichens, and periwinkles; the middle, dominated by barnacles and mussels; and the lower zone, dominated by seaweeds and surf grasses (Castro and Huber 2007).

Rocky shorelines are a challenge for fishes primarily due to risk of damage from mechanical wave action and risk of stranding. However, there are the rewards of abundant food and reduced risk of aquatic predation. As might be imagined, most fishes that occur in the area will only venture in rocky shorelines during high tides (e.g., flatfishes). However, some fishes are permanent residents and have evolved a special suite of adaptations in morphology, physiology, and behavior (reviewed by Gibson 1986). Resident fishes include blennies, gunnels, pricklebacks, clinids, gobies, clingfishes, and sculpins. These fishes are small in size (usually <20 cm), maintain neutral buoyancy

Figure 16.4 Rocky coastline of the NW Atlantic Ocean. (Courtesy of Miller, J.)

(reduced or absent gas bladder), have a compressed body, and have fins that are modified for clinging to rocks (and to life!). Physiologically, these fishes are able to withstand changes in salinity and temperature (euryhaline/eurythermal), obtain oxygen when out of water, and resist drying. Bridges (1993) found that tolerance to dessication was an important attribute of intertidal fishes and reported that mudskippers (*Periophalmus*) can survive dessication of more than 20% body weight. Intertidal fish also tend to restrict activity with outgoing tide to avoid stranding. In order to do this, they select low-tide refuges such as deeper tidepools and return to them as needed (Gibson 1986).

Evidently, rocky shorelines can be very attractive for fishes, because in some regions, all of the fishes in the rocky intertidal zone may be true residents. Prochazka (1996) captured 5409 intertidal fishes of five families and 20 species from 12–14 tidepools in four different vertical shore zones. All of the fishes were intertidal residents, no transients were detected, and there was no observable seasonal trend in habitat use. It appears that intertidal fishes become highly specialized for life among the rocks that experience pounding wave action. Fishes without this specialty are not able to live there.

MARINE FISHES

Perspectives

Marine fishes comprise about 58% of fishes worldwide, and they are not randomly distributed in the oceans and seas. Four ecological divisions were presented by Helfman et al. (2009): continental shelf (neritic) fishes, comprising 45% with about 12,600 species; epipelagic, comprising 1.3% with about 360 species; deep pelagic (meso- and bathypelagic), comprising 5% with 1400 species; and deep benthic (continental slope to abyss), comprising the remaining 6.4% with about 1800 species.

Marine ecosystems are very different from the freshwater ones we have discussed so far. As we leave the intertidal zone and move into deeper marine systems, it will be helpful to have placed marine fishes in perspective with several issues. There are two issues that I wish to introduce at this point, because these issues will surface many times in the future: (1) species diversity in comparison with freshwater systems and (2) declining abundance.

Diversity Scrutinized

In a general sense, about 60% of the recognized number of fishes (N = about 28,000) (Nelson 2006) are marine, and the remaining 40% are freshwater. This seems strange, because oceans and connected seas constitute about 300 times more habitat by volume than terrestrial regions (Helfman et al. 2009). Assuming that habitat is equitable among species, multiplying 11,200, the approximate number of freshwater species (i.e., 40% of 28,000), by 300 times as much habitat would amount to 3,360,000 marine fishes. In other words, assuming that marine fish would use the same amount of habitat per species on average as freshwater fish would, there is enough volume for over 3 million species. However, there is only about 16,800 marine species (i.e., 60% of 28,000). Does this mean that the marine habitat is so very much inferior to that of freshwater? Is there another explanation? Obviously, my assumption of habitat use is incorrect. Putting my numbers aside, it is clear that there are less marine fish species that might be expected from the number of fishes that occupy comparable volumes of freshwater. This does not seem intuitive, because freshwater fishes on average experience more stresses imposed by life in freshwater habitats (excluding the small intertidal zone) due to more fluctuating and stressful conditions.

Diversity of fishes can be addressed from several viewpoints, and four issues for consideration that occur to me are (1) the amount of geographic isolation, (2) habitat heterogeneity, (3) productivity and habitat quality, and (4) effects of interspecific competition. There are other factors that could also be considered (e.g., Krebs 2008), but we know enough about these four issues to apply them here:

(1) Freshwater fishes can be greatly affected by local changes in topography and isolated geographically by barriers, both within and between continents. Marine fishes live in a more three-dimensional and interconnected world even though continents can restrict distribution. On balance, more geographic isolation in freshwater fishes seems patent.
(2) Genetic divergence in allopatric gene pools is believed to be a major factor in speciation. Accordingly, speciation should occur at highest rates where there is heterogeneity in the environment because spatial isolation and divergence would be more likely. There are two major faunal elements in the oceans: those that are associated with the bottom and those that are pelagic. Of all marine species, only about 2% of these are pelagic. The sea floor is relatively monotonous, effectively placing fishes that forage there in sympatry. On the other hand, heterogeneity abounds in freshwater habitats.
(3) The largest habitat on Earth is the marine pelagic realm, most of which is very low in productivity. There are poor-quality terrestrial habitats as well, such as deserts, high altitude, great latitudes, and perhaps even north temperate regions. Once again, some judgment is required, but the great and unproductive pelagic realm (with only 2% of the fishes) places a restraint on species diversity.
(4) Most freshwater streams and lakes of the temperate zone are low in diversity and perhaps more affected by physical restraints of the environment than by interspecific competition (n.b., not considering invasive species). This seems to be supported by increased competition in the more stable tropics. However, even in the competitive African Great Lakes, microallopatry can produce speciation. I have to place interspecific competition in a lesser role.

All of the issues identified may have had an effect at producing a seemingly inequitable degree of species diversity between freshwater and marine systems. However, the most plausible explanation is trying to compare volume of habitat as a common denominator between the two systems. In this comparison, almost all freshwater systems are littoral, associated with the benthos, while almost all of the pelagic systems under comparison are not.

As in freshwater systems, marine systems have been altered by anthropogenic effects, and those effects also need to be examined in order to understand marine biodiversity. In this case, there are several issues to be considered: the effects of overfishing, pollution, acidification, eutrophication, physical habitat alteration, and global climate change.

Declining Abundance

It was once imagined that the oceans contained an inexhaustible supply of fishes that could supply all the food that humans would ever need (Daniel and Minot 1954). Sadly, this was a fairy tale (Crowder 2005; Roberts 2007). Even though "untapped fisheries" were still being identified as late as 1971 (Shapiro 1971, four papers therein), it is painfully obvious that by then, humans had caused great declines in formerly abundant fish populations and major world fisheries had already collapsed. In some cases where simple overfishing was the main cause, there has been some recovery, but in other areas, there has not. Due to the severe decline in marine fishes, Jackson (2008) voiced what many others have decided: there is evidence now that the synergistic effect of human actions in the marine environment is leading to an Anthropocene mass extinction in oceans. Helfman (2007) produced a thoroughly referenced volume that discusses virtually all of the pressing issues involved in this global decline, closing his book with some optimism—although he professes that he has bouts of depression about the decline. Accordingly, I have made a substantial effort to include information and case studies to illuminate these very important issues that may affect the future of fisheries and fishes.

ECOLOGICAL CONCEPTS

- *Seafloor Spreading*—Obliteration of the deep marine benthic zone by subduction was widely accepted as fact by the 1960s. Ancient fish habitats and fish fossils more than about 150 million years old do not exist, except for more shallow sea floors associated with continental shelves.
- *Ocean Conveyor System*—This is a pattern of global thermohaline circulation due to the sinking of dense water in polar regions. These cold dense waters provide an oxygen supply to the abyss.
- *Stable Ocean Hypothesis*—The open ocean can be very turbulent, especially during storms. Lasker (1975) found that stable conditions are needed for effective feeding by marine fish larvae preying on plankton. Recruitment of marine fishes is enhanced by stable conditions.
- *Vertical Zonation of Intertidal Invertebrates*—These zones are an indication of great physical stress, competition, and predation. Upper zones are more likely a result of physical factors, and lower zones are more likely influenced by biological ones.
- *Interconnected Coastal Systems*—Coastal ecosystems are linked to estuaries and offshore systems by the migrating subsystems of organisms as well as by physical and chemical processes.

CASE STUDY: THE ULTIMATE MARINE PREDATOR

How Do Ecologists View Sharks?

Sharks and their closest relatives, the rays, are widespread in marine systems. Rays, with some notable exceptions such as the eagle and manta rays, are benthic predators—and they are superbly adapted for that role. Sharks are the ubiquitous marine predator due to their presence in almost every conceivable habitat that is large enough for them. Their ability to consume almost any large prey, including humans, reflects their diversity in feeding adaptations (and a good reason to fear and respect these creatures). But there are more reasons than that for us to respect and admire sharks. As pointed out earlier, they have been around for a very long time—at least 400 million years—and Devonian sharks developed a very recognizable sharklike morphology long before the appearance of the earliest teleosts.

In numbers, there are less than 1000 species of elasmobranchs, comprising only about 4% of the total numbers of extant fishes. Sharks account for less than half of that percentage but comprise about 7% of the fish families (i.e., nine orders, 34 families, and 403 species of sharks) (Nelson 2006). However, comparing sharks with other fishes may not be practical because sharks in general are much larger than almost all other fishes, and with few exceptions, they do not feed at lower trophic levels.

How do ecologists view sharks? Sharks have proven their worth as an apex predator if not in species numbers. They are superbly adapted to the marine environment, where they occur in all oceans and large seas. They are the most successful large predator in almost every system, rivaled by only a few of the largest marine mammals and other fishes. Most shark species occur mainly in coastal waters, where food is abundant, but sharks also occur from open layers of the ocean to the deep sea realm.

Most of the attention in this book is directed toward teleost fishes. However, this additional discussion of sharks is placed here because they are pervasive and very important in marine systems. In radiating throughout the oceans and seas of the world, they solved the problems associated with living in the marine environment, but, in most cases, in very different ways than in the teleosts. Unfortunately, shark populations are declining because some of the adaptations that equipped them for a very long persistence may now be a disadvantage in this human-dominated world.

Sharks as Human Predators

Perhaps the first thing we should consider about the shark is its predatory adaptation. All sharks are predators, having a formidable array of acute sensing powers to detect prey. These senses include some powers that we can identify with, such as an excellent sense of smell (olfaction), taste (gustation), vision, and touch. But they also have senses that we can only imagine, such as the lateral line for detection of vibrations and electroreception for detecting the electric impulses from other animals. Couple this with a tough skin with little embedded teeth (denticles) for protection, a powerful tail for thrust, enormous fins for agility, and large pointy teeth—well, all of this is scary to humans, especially those swimming at ocean beaches. So, with all of these adaptations and more, should humans venture into the oceans at all?

Actually, there are very few humans attacked by sharks on a global scale. One could say that the number of attacks is insignificant (unless the one attacked is you!). There are many reasons why sharks do not pose much of a threat to humans. Sharks have great intelligence at recognizing prey and avoiding potential danger, most sharks are smaller than humans, almost all sharks recognize and feed on teleost fish or invertebrates, and very few large sharks patrol inshore beaches. Some do, however, and the great white shark, which patrols for and consumes various marine mammals is known as a man-eater. Stories about attacks from man-eating sharks used to be common in outdoor literature, and *The Sharks* by Robert Burgess (1970) provides a good accounting of attacks if you can find a copy of it. However, contrary to the movie *Jaws* and the comic strip *Sherman's Lagoon*, sharks are not plotting to kill hairless beach apes (humans). In truth, it is we who are killing sharks, and in great numbers.

Humans as Shark Predators

Sharks have declined worldwide as fishing pressure and catches increased (Figure 16.5). This decline is unprecedented, and it has been especially dramatic for large species. I remember when there was little concern expressed for killing sharks because there seemed to be so many of them. On a study trip aboard RV Eastward in the Caribbean, I witnessed shark vulnerability to baited long lines. This gear is called a long line because the line can be miles in length. Thick lines (with smaller baited lines attached with a clip) are released from a drum and dropped on the ocean floor or suspended above the floor with floats. After a period of time, usually overnight, the lines, hooks, and fish are all winched back on board the ship. In my case, we were only doing experimental fishing and had a small line out (n.b., a popular movie, *The Perfect Storm*, shows details of long-line fishing). Unfortunately, with all of the fishing operations going on, long lines and gill nets are lost at sea. Called "ghost fishing," this gear can continue to catch fishes and especially sharks for a long time.

Our large catch of sharks on the research trip was unanticipated, and we tried to release most of them. It is hard for a nonfisherman to imagine the difficulties in freeing a large thrashing shark from a hooked line; it is dangerous, and most sharks are killed. As bad or worse are the large gill nets that are suspended at sea. Larger sharks are especially vulnerable to them. And with the release of several *JAWS* movies, recreational pressures increased for shark fishing. Many sharks are killed accidentally. However, there also are markets for shark meat, and even the docile whale shark is not immune from commercial fishing (Helfman 2007). An especially cruel practice is the harvest of shark fins, with the mutilated shark being returned to the sea alive. Finally, some people place sharks into the category of "dangerous trash fish to be killed at every opportunity"—I have experienced this notion. I became unpopular on a deep sea fishing trip when I hooked a large shark but cut the line to free it instead of trying to bring it on board.

But fisherman cannot be blamed for all of the adverse effects on sharks. In general, all ocean fishes are being affected by the actions of humans. Overfishing, habitat destruction, pollution, and so on are now having synergistic effects (Jackson 2008). Perhaps this combination of effects

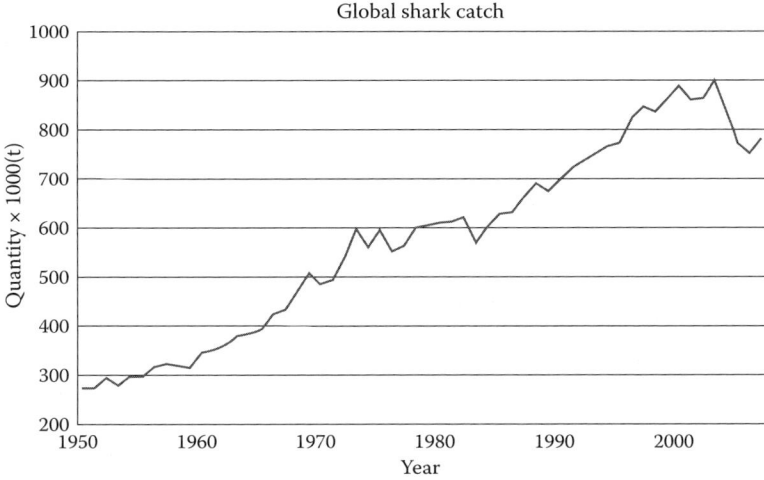

Figure 16.5 World catch of sharks. (Courtesy of Chris Huk, http://commons.wikimedia/wiki/File:Global_shark_catch_graph.png.)

resulted in the collapse of large shark species in the northwest Atlantic Ocean (Baum et al. 2003), where three species declined by 75% in 15 years. In a more recent study of 35 years of catch data, this decline was found to cause a top-down trophic alteration. With the decline of 11 large sharks, smaller species of sharks and rays (i.e., *mesopredators*) (Prugh et al. 2009) increased substantially in number, and their combined predation caused great declines in shellfish abundance. As an example, increased abundance of cow nose rays decimated scallop fisheries (Myers et al. 2007).

Sharks are very good at finding food because they are highly skilled and fully equipped predators. But they are especially vulnerable to net fishing because of their large size (half of all shark species are 1 m in length or longer) (Springer and Gold 1989), and some species are very large, reaching enormous sizes (as large as 18 m). They live long lives (i.e., 30 to 70 years), depending on the habitat and species. Such long life is necessary because they produce only small numbers of offspring. All sharks have internal fertilization, and the female protects the eggs or fetus within her body. Sharks also have a long gestation time, and live bearing sharks may require as much as 2 years to give birth. To sum this up, in nature, fully grown large sharks can expect to live long lives. They produce only a few large young sharks, which are already so large when they emerge that they are not very vulnerable to predation. Unfortunately, this formerly successful life history strategy, which evolved over hundreds of millions of years, is now contributing to their steady demise. It seems that humans and their tools of the industrial revolution are the ultimate marine predator.

Shark Swimming as Multitasking

The public in general views the shark as a relic of the past, and biologists might point out that shark roles are being taken over by the more advanced teleost fishes. To support this view, the need for constant swimming by the great pelagic sharks is held as evidence of inferiority. These beliefs are misguided. In truth, sharks are just as advanced as the more recently evolved fishes—but they have survived at least three periods of global mass extinction, which the teleosts have not. Yes, sharks must swim, but even with constant swimming, they have more efficient metabolism than that of the teleosts. How can this be?

The deceptively simple act of swimming is actually a well-integrated program of multitasking by the shark. The large heterocercal tail is capable of very sophisticated action as it propels the shark through the water, because it works in symphony with the large winglike pectoral fins. Acting together, they provide the stable platform needed for the proper functioning of sensory devices, but they also facilitate the launching of an attack on prey and enable exceptionally fine movements that may be needed to complete one. In addition, the fins provide hydrodynamic lift, an important part of swimming for a fish that has no gas bladder.

Associated with efficient and powerful swimming is the need to overcome gravity; otherwise, energy is wasted in an effort to avoid sinking. Instead, that energy is needed to push the fish through the viscous media. So the shark has developed and retained a cartilaginous skeleton to lighten its body—but that is not enough. It also needs more flotation, which it has obtained by storing oil in the liver. The oil makes it easier to swim against gravity, but it also is part of the multitasking because it provides storage for high-energy lipids and vitamins.

But overcoming gravity does not address the viscosity problem. That is addressed by having a thick skin with embedded little "teeth" (placoid scales) that reduce water turbulence as the dimples on a golf ball do in air. These placoid scales were a pioneering effort by the shark, and millions of years later, these were duplicated to a degree in teleost fishes. But there is more here, because the thick skin aided in another problem, which is reducing the entry of salt into the shark's body from seawater.

Swimming still requires significant expenditure of energy, perhaps the price of searching for food and obtaining prey. But here also is more multitasking. All sharks have platelike gills, with each slit promoting an efficient flow. By the simple act of opening its mouth while swimming, it obtains all of the oxygen it needs by "ram" ventilation. The transport and cellular supply of oxygen thus obtained in such large species require a circulatory system, which also is aided by swimming. Finally, muscular heat generated by swimming serves to warm the body, aiding digestion. Swimming seems to be a very good thing for a shark!

SUMMARY

The marine environment is immense, covering 71% of the Earth in seawater. Unlike the lithosphere, water exists mostly in liquid phase, and seawater is in constant motion due to extraterrestrial gravitational forces, rotation of the earth, wind, temperature differences, and changes in salinity. These factors result in tides, waves, surface currents, and deep density flows of water in the oceans and seas. These forces also affect fish populations, their food, and their habitats because the mixing of nutrients, oxygen, and heat affects living conditions and food supply. Currents also provide for the transportation of various life history stages of organisms. The equatorial region is subjected to great solar warming that results in warm surface air rising and being replaced by cooler air that flows toward the equator from the north and south, thus resulting in north and south equatorial currents around the globe. Continents present a barrier to global current flow, resulting in huge water circulation patterns known as *gyres*. For example, warm tropical water is forced north by the Coriolis affect along the Atlantic coastline of the United States as the Gulf Stream. This flow encounters colder water and westerly winds and continues to bend in a circular fashion as it meets western continental flows; it is then once again entrained by equatorial currents, completing its cycle. Associated with this flow are deeper circulations that are part of the Ocean Conveyor System that slowly circulates cold dense water though the ocean depths. From an ecological perspective, oceans are divided into two realms: the pelagic realm and the benthic realm. The *intertidal* zone is dewatered with low tide and is exclusively benthic. The next zone, the *neritic* province, includes the mass of water that lies above the continental shelf to the continental break and the benthic zone that is included beneath. The *oceanic* province then includes all of the pelagic zones (water) and benthic zones (seafloor) of the open ocean. Fishes are not well represented in the upper intertidal zone of beaches, but further offshore, deep intertidal and inshore subtidal areas are patrolled by several species of fishes feeding on plankton, detritus, mollusks, and other fish. Rocky shorelines present challenges for fish due to wave action and risk of stranding. However, more food and cover are present than on beaches, and some species are highly adapted for life among the rocks and tidepools. The greatest diversity of fishes occurs in the neritic province, and much lower numbers occur in the pelagic province. Although it was once believed that oceans contained an inexhaustible supply of fish, severe declines in commercial fisheries have occurred due to overfishing and habitat alteration. Fishes that have been affected by humans include large sharks. These interesting animals are presented in a case study that focuses on their adaptations. Sharks have proven their worth as an apex predator if not in species numbers. They are superbly adapted to the marine environment, where they occur in all oceans and large seas. They are the most successful large predator in almost every system, rivaled by only a few of the largest marine mammals and other fishes. Most shark species occur mainly in coastal waters, where food is abundant, but sharks also occur from open layers of the ocean to the deep sea realm. Why are sharks declining? Shark reproduction produces only a few young sharks, which are already so large when they emerge that they are not very vulnerable to predation. Unfortunately, this formerly successful life history strategy, which evolved over hundreds of millions of years, is now contributing to their steady demise. It seems that humans and their tools of the industrial revolution are the ultimate marine predator.

Recommended video: *Sharks!*, In: *Secrets of the Ocean Realm*, PBS Home Video, Howard Hall Production, 1997.
Further reading: Allsopp et al. 2007.

CHAPTER **17**

Neritic Province and Fisheries

INSHORE OCEAN IN PERSPECTIVE

The *neritic province* also has been called the neritic, subtidal, sublittoral, or offshore zone. This is somewhat confusing because this puts zones within zones, and nomenclature varies between general ecology and marine ecology/biology texts. Also, there are differences in how "littoral" is used in freshwater and marine ecosystems, so I use intertidal and subtidal to avoid this problem. The neritic province (sensu Sumich and Morrissey 2004) is the area of oceans where the pelagic realm lies over continental shelves, usually at depths less than 200 m, and it covers about 13% of the surface of the earth (Angel 1997). The neritic province extends from the subtidal coastline to the edge, or break of the continental shelf, where the *oceanic* province begins as the continents slope to the abyss. The pelagic part of the neritic is usually called the photic zone, and the continental shelf is part of the *benthic* realm (Figure 16.3). Most fishes can be placed in one or the other of these two realms (i.e., as a pelagic or benthic species) based on feeding adaptations and/or morphology.

The inshore part of the neritic province along a gradually sloping coast has been referred to as the inshore coastal zone, a relatively shallow subtidal area that extends from the low tide line (deepest part of the littoral zone) offshore to about 30 m. This area can be separated as the inshore neritic. Winds and currents mix these shallow coastal waters, constantly resuspending organic matter, nutrients, and oxygen. Much larger predaceous fishes occur here than in the intertidal zone, and there are many visitors from other zones. Fish communities in this zone may be different depending upon different habitats, which in general are soft or rocky bottoms, sea grass flats, kelp forests, and reefs.

The offshore part of the neritic province, the offshore neritic, extends from depths of about 30 m to the continental break. The outward edge of the neritic on the NW Atlantic coast is obvious most of the time, because it extends out to the inshore edge of the Gulf Stream where the water changes from a murky green to a deep blue. It is this relatively shallow offshore zone that produces some of the world's greatest fisheries. The offshore neritic can be of great size; for example, in the Gulf of Mexico, the extensive northwest Atlantic fishing banks off New England, Nova Scotia, and Newfoundland, and the Arctic Ocean off Russia (where it extends 800 miles). However, the neritic is narrow in other places, such as along the Pacific coast, and along shorelines of volcanic islands, the neritic province can be exceedingly narrow and sink precipitously to the continental slope. In the southern part of the Gulf of Mexico, the neritic sinks to a depth of 1 mile in a 2-mile distance (Idyll 1971).

Structure

The neritic province is composed of pelagic and benthic systems. Pelagic systems include an inshore continuation of the epipelagic (or upper pelagic) zone and, in some instances, an upper part of the mesopelagic zone; both of these will be discussed more fully in chapters on the oceanic province. The benthic zone of the neritic is the submerged edge of continents, the continental shelf. There are several neritic systems that are used by inshore marine fishes, including the open-water zone, benthic communities, and complex systems. Open-water areas are dominated by plankton. Subtidal bottoms of soft sand and mud mostly support populations of invertebrate infauna, and rocky bottoms are mostly inhabited by invertebrate epifauna. Also, there are other complex systems in this province, including seagrass beds, kelp forests, and coral reefs. Structure in this environment is influenced by currents, tides, and winds. Human activity in some areas has disrupted structure by pollution and bottom trawling.

Function

Productivity in the neritic province is based on two major features of this environment: (1) input of organic matter and nutrients carried into the water by surface runoff, wind, and stream flow via estuaries and (2) autotrophy from phytoplankton, algae, and grass flats. Strong tidal currents and winds circulate the shallow water and constantly suspend sediments, which include organic matter and nutrients. This can result in high productivity if solar radiation is not limited by the resulting turbidity. Productivity also varies with latitude and seasons, influenced by temperature, day length, and solar intensity. In some locations, the neritic zone also receives nutrients from upwellings, but upwellings in most cases occur or are linked with deeper waters, and this will be discussed more fully in subsequent chapters. Finally, it is very important to note that human occupation of the warmer coastal zones of continents is very high, and enormous quantities of detritus, hydrocarbons, and nutrients are disposed of as pollution into neritic waters. The term pollution is used here because not only "cultural enrichment" occurs, but also heavy metals and thousands of industrial, medical, and domestic chemicals are added. Many of these pollutants, such as mercury, are entering marine food chains and concentrated into the flesh of top carnivores. People who eat a lot of fish should check U.S. Environmental Protection Agency (EPA) guidelines. Mercury levels can be high in some fishes, such as swordfish and albacore.

PELAGIC SYSTEMS

The pelagic realm is represented in the neritic province by a photic zone, which is continuous with waters of the epipelagic zone, and sometimes the upper part of the mesopelagic zone. In contrast to epipelagic waters of the open sea, there are major differences in habitat. First, open waters of the neritic are direct recipients of substantial inputs of organic matter and nutrients from continental runoff. These nutrients and the bodies of dead organisms are essentially trapped by shallow expanses of continental shelf and suspended by wind, currents, and tides, so they can be accessed by decomposers and producers over and over again. The shallow shelf water also is well illuminated by the sun (i.e., photic), and this encourages much proliferation in phytoplankton throughout, and for seagrass and attached algae in the inshore areas. Planktivorous fishes abound in the neritic and include economically important clupeids, such as menhaden, herrings, sardines, and also anchovies (engraulids).

BENTHIC SYSTEMS

The continental shelf zone of the eastern United States is shallow and receives huge amounts of sediment transported by continental runoff, wind, and wave action. As might be imagined, most of the benthic habitat is thus composed of soft substrates, ranging mostly from mud to sand. Some areas, especially those affected by current, have pebbles and shells on the surface of mud and sand. Rocky substrates are much less available but as important as fish habitat when they do occur. Except for the mouths of large river estuaries, the Pacific shelf is much steeper and rockier.

Soft Substrates

Like mudflats in estuaries, most of the organisms that live in soft neritic bottoms are infauna. However, there are less shifting currents and tidal action to expose these animals, so the fishes and other organisms that feed on them the most expose them by digging them out in some fashion. Rays, skates, and some sharks are well adapted to do so, and even marine mammals such as porpoises have developed techniques to capture invertebrates and fishes that live in the sediments. On coarser bottoms, many of which contain old and broken shells, some fishes are lie-in-wait predators, such as flounders and rays. Other fishes such as cods are roving predators of the deep neritic and continental slope, taking all kinds of fishes, shellfishes, and other invertebrates, including squid (Bigelow and Schroeder 1953).

Seagrass Flats

Underwater grass flats (Figure 17.1) are important components of the inshore neritic province, especially in warmer climates, where productivity can be high. Eelgrass occurs on both U.S. coasts, and it is replaced in South Florida and the Caribbean by turtlegrass. Seagrasses are not grasses at all, but flowering plants that are adapted for a submerged life. Their long, strap-like leaves (blades) appear to be grass, but they are not. They have air channels within them for flotation. The grass needs to have some current to supply nutrients to them, and beds tend to accumulate fine sediments. Fishes, birds, and invertebrates graze the grass, some of which probably only consume the biofilm that lives on the blades. Much organic matter also originates from seagrass beds as detritus from dead leaves. Seagrasses provide nursery habitat for important invertebrates like penaeid shrimp and cover for many fishes. The young of spotted sea trout, an important inshore sport fish, typically use seagrass beds for cover (Hoese and Moore 1998).

An eelgrass fish community was outlined by Marshall Adams (1976) in coastal North Carolina, and the fishes consumed food produced in the eelgrass by more than 50% of their total weight. Pinfish, pigfish, spot, and filefish were considered the main consumers of eelgrass detritus, but at least 17 food items were detected. In all, three trophic groups comprised the community, with pinfish and pigfish as omnivores, spot and filefish as detritivores, and silversides, silver perch, pipefish, flounder, and gag (a grouper) as carnivores.

Figure 17.1 Seagrass bed in Florida. (Courtesy of the National Oceanic and Atmospheric Administration (NOAA), http://www.noaa.gov/.)

Rocky Substrates

Rocky bottom habitats are more stable and provide more substrate for organisms than soft bottoms; hence, they are rich in epifauna. These habitats include outcroppings of bedrock, reefs, and artificial substrates such as sunken vessels. Areas of such habitat are common on the offshore neritic from North Carolina to Florida and support excellent offshore fishing for sea bass, grouper, snapper, and many other marine sport fish.

Kelp Forests

Large brown algae are called kelps, and they occur attached to rocky substrate in colder waters of the world. These algae have no roots but use a holdfast organ to remain attached. Kelp beds and forests (Figure 17.2) occur in the north temperate zone on both coasts of North America. Kelps obtain heights of 20 m and more and remain suspended erect in the water by use of a float (pneumatocyst) attached on the upper part of the structure just beneath the blades (limbs). Kelps must have cool temperatures, plenty of light, and rocky substrates. The forests occur generally in a band alongshore, bordered on the shoreward side by unacceptable wave action, and offshore by depth and lack of light penetration. As might be imagined, these huge plants provide a variety of habitats and food choices for invertebrates, fish, and marine mammals.

The fishes of the Pacific coast kelp forests are dominated by rockfishes from central California north, and from that point south, they are dominated more by warmer species, including perches, damselfish and wrasses (Sumich and Morrissey 2004). Fishes may be active at different times of the day and occupy various habitats, divided more or less vertically. In addition, some larger fishes can be expected to cruise or lurk at will. Some rockfishes (Scorpaenidae) and kelp bass (Serranidae) feed on fish and invertebrates near the bottom, where sheepshead forage for and crush invertebrates. Surfperches, wrasses, and other fish feed toward the center of kelps, picking invertebrates off the plants. Topsmelt feed on zooplankton around the plants (Castro and Huber 2007).

Coral Reefs

Found in warm, shallow seas, coral reefs (Figure 17.3) are spectacular underwater systems, and they attract scuba divers in great droves. These reefs are one of the most productive systems on Earth, but corals are sensitive to water conditions, requiring stenohaline (seawater) as well as

Figure 17.2 Kelp forest. (Courtesy of NOAA, http://commons/wikimedia/wiki/File:Kelp_forest-blue.jpg.)

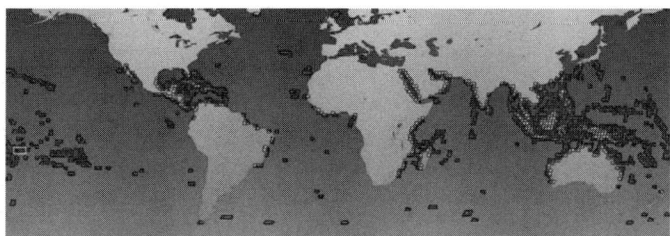

Figure 17.3 World distribution of coral reef. (Courtesy of the National Aeronautics and Space Administration (NASA), http:commons.wikimedia/wiki/File:Coral_reef_locations.jpg.)

stenothermal (23°C–25°C) environments. Coral reefs support hordes of interesting fishes. Thus, a comprehensive coverage of their adaptations to the reef, surf, channels, lagoon, associated open waters, and so on and also those of invertebrates would be outside the scope of this book. Only a brief summary is given here, with reference to more complete accounts.

Excellent discussions of coral reef structure and function, with emphasis on the important invertebrate faunas, are provided in general marine biology and ecology texts such as those cited here (e.g., Nybakken 1997; Sumich and Morrissey 2004; Castro and Huber 2007). For scuba divers who are interested in observing and interpreting the behavior of coral reef fishes, I suggest a copy of Wilson and Wilson (1992). A good start for budding ichthyologists or fish ecologists with some previous knowledge is provided by Sale (1991) and contributors, who cover every imaginable topic pertinent to bony fishes of coral reefs. For many of us that do not live in areas with tropical climates, a trip to an aquarium will undoubtedly find a coral reef display.

Coral reefs are structures of calcium carbonate, formed from biological rather than geological action. The reefs are neritic features, because they are formed on stable shelf areas at shallow depth. In reef-building corals, the coral polyps (i.e., the sessile stage of animals classified as Cnidarians) produce calcium carbonate skeletons, and billions of them are needed for a well-sized reef.

The ability to produce a reef is greatly aided by the presence of a symbiotic dinoflagellate (algae) called zooxanthellae that live within the tissues of the coral animal. Zooxanthellae are photosynthetic and perform two important functions: providing extra food for the coral and assisting in depositing the skeleton. Corals feed on zooplankton but greatly benefit from the zooxanthellae productivity during periods of low food. The zooxanthellae also benefit by having a place to live. They receive the waste products produced by the coral and use it as carbon and nutrients to produce more food. This is a win–win (mutualistic) relationship. Other algae also aid with reef building, including encrusting or coralline algae that grow in sheets, which aid in the incorporation of sediment into the reef structure.

Geologically speaking, coral reefs can grow rather quickly. The largest coral reef in the world (covering about 225,000 km^2) is the Great Barrier Reef of Australia, but curiously enough, it began to grow only about 9000 years ago with a rising sea level. The present reef is testimony to the speed at which the reef was able to "catch up" with the rising water level, which it did from 4000 to 6000 years ago (Sale 1991).

Thousands of species inhabit coral reefs, including about 100 different families of bony fishes (Leis 1991). Of these, eight families of higher teleosts are closely associated with, and thus characteristic of, coral reefs: three labroids (wrasses, parrotfishes, and damselfishes), three acanthuroids (surgeonfishes, rabbitfishes, and moorish idols), and two chaetondontoids (butterflyfishes and angelfishes). However, there are many more fishes of wider distribution. These include demersal blennies and gobies, which can be abundant. Other fishes include invertebrate predators (cardinal fishes and grunts), tetradonts (boxfishes, puffers, and triggerfishes), and larger predators (squirrelfishes, sea basses, snappers, and emperors) (Sale 1991). Also found on reefs are a number of roving sharks and rays, whose species depend on the geographic location of the reef. In all, 120 species of sharks and 112 species of rays have been identified in association with rocky and coral reefs (Michael 1993).

Most of the differences in coral reef fishes can be seen in feeding adaptations. Surprisingly, there are many herbivores, presumably due to the year-round warmwater environment, but most fishes are generalized carnivores on zooplankton, invertebrates, and other fish. However, parrotfish are highly specialized to feed directly on the coral. Having both jaw and pharyngeal teeth, parrotfish use the jaw teeth to bite off pieces of coral, which it grinds up with the pharyngeal teeth. The fish voids clouds of finely ground coral, which is recycled back into reef building.

A final note on coral reefs is that about one-fifth of coral reefs worldwide have been lost since the 1950s due primarily to direct human impacts. These include pollution, overfishing, and recreational use. Indirect effects of humans also have been implicated. Some reefs have been adversely affected by bleaching, which is expected to increase with warming climate.

NERITIC FISHES AND THE TEMPERATE ZONE

Neritic or continental shelf fishes inhabit the pelagic and benthic habitats of the shoreline and offshore continental shelf. This region can be visualized as a narrow band around continents and islands, which extends in depth to about 200 m. Neritic fishes are the largest component (45%) of the total marine fish fauna, comprising about 12,600 species (Helfman et al. 2009). Ocean fishes are affected by changing ocean climates, especially noticeable in the neritic province because of its shallow coastal waters.

In temperate oceans, the distribution of fishes along temperature gradients can be used to place them in faunal groups. Taking into account seasonal movements, Robins and Ray (1986) identified six landmarks along the northwest Atlantic coast: Cape Sable, Florida, southern limit for temperate fishes moving south from the Gulf of Mexico in winter; Palm Beach, northern limit for many tropical fishes; Cape Canaveral, southern boundary for many temperate fishes; Cape Hatteras, dividing cold and warm temperate fishes; Cape Cod, northern limit for many species, including those moving inshore from the Gulf Stream; and the Strait of Belle Island; southern limit for arctic species.

Fishes also are distributed along temperature gradients in the Pacific coast of North America, but the zoogeography of fishes there is influenced by at least four different factors compared to the Atlantic coast (e.g., Thurman and Trujillo 2002; Horn et al. 2006): (1) There is a surface current along the coast produced by the North Pacific gyre. It is similar to the Gulf Stream, but it is a coldwater flow south instead of warmwater moving north. (2) The Pacific coast is a tectonic one and the leading edge of the moving continent, not the trailing edge. Subduction occurs as the continent encounters the ocean floor, resulting in a continental shelf that is deep and narrow, instead of shallow and wide. Continental runoff can sink to depths, and nutrients are not circulated as much as in the shallow Atlantic coast. (3) Major upwellings are very important for food production, but they are variable (discussed in chapter 18). (4) Long-term warming/cooling trends on the order of 20–30 years produce differences in the fauna (represented by sardine anchovy shifts in abundance). Along the Pacific coast of the United States, it has been customary to recognize two different fish regions; a northern one from upper California, Oregon, and Washington (includes Puget Sound and inflows from the Columbia River) (Glude 1971) and a southern one that includes most of southern California and into the Baja of Mexico (Leet and Cramer 1971). In California, a cold temperate fish fauna dominates the coastal fish fauna north of Point Conception (at latitude 34.5°N), and a warm temperate/subtropical fauna occurs south of the point (Horn et al. 2006).

Identification of neritic fish faunas is also complicated by inshore movements of deep-sea fishes (residents of the ocean beyond the continental shelf) and by seasonal migrations along the coast by fishes seeking favorable water temperatures. One area that is appropriate for studying neritic fishes is in the northern Gulf of Mexico, where almost all of the water is neritic. Hoese and Moore (1998) studied fish records for more than 20 years and reported 552 species. Of these, 118 were captured in estuaries and are excluded here. Of the remaining 434 fishes, 171 were inshore demersals, 78 were inshore pelagic, 101 were offshore reef dwellers, and 84 were offshore epipelagic species. Estuarine and inshore species (376) amounted to 67% of the total. Of these, inshore fishes were 45% and amounted to 57% of all marine fishes.

Most inshore fishes of the neritic are well known to fishermen and coastal inhabitants, and about 80 of the 563 marine fishes that occur in North Carolina are often encountered by sport fishermen (Schwartz and Tyler 1970). Goldstein (1986) also identified coastal fish species commonly captured by recreational fishing in the inshore subtidal areas of North Carolina. Larger sport fishes of the beach zone (lower intertidal and upper tidal zone) include quite a variety: spot and croaker, kingfishes (locally known as whiting), flounders, bluefish, Spanish mackerel, little tunny, spotted and gray sea trouts, red drum, striped bass, and pompano. Common fishes caught in the surf include roving bluefish and more sedentary kingfishes (Figure 17.4). Fishes caught in deeper waters and around structures include the same fishes, with the addition of sheepshead, cobia, amberjack, jack cravelle, black drum, larger sharks, and a few less desirable fishes that I reluctantly admit catching by hook or crook: dogfish shark, clearnose skate, stingrays, menhaden, silversides, Atlantic needlefish, toadfish, hardhead and gaftopsail catfishes, lizardfish, sea robin, white and striped mullet, hogchoker, puffer, filefish, and so on.

The inshore marine fish fauna of California is composed of about 519 species, of which 225 occur in bays and estuaries and 289 are marine (Horn et al. 2006). About 200 fishes occur in the California current, but only 79 species are reported as epipelagic and 124 species were deep-sea forms (meso- to bathypelagic) (Allen and Cross 2006). Recreational fisheries (private boats) (Love 2006) in northern California target rockfishes, salmon, striped bass, ling cod, and white sturgeon, while south of Point Conception, they seek kelp and sand bass, rockfishes, barracuda, mackerels, yellow tail, Pacific bonito, white sea bass, California halibut, dolphinfish, and yellowfin tuna. Southern recreational fisheries exist with a continental shelf that averages <5 miles wide, but it receives cold, nutrient-laden water from the California current and experiences periods of upwelling associated with ENSO (i.e., La Niña). As might be imagined, fishing for large pelagic species such as tuna and yellowtail (Figure 17.5) is affected by migrations in response to water temperature and upwelling (food fish, e.g., sardines). As a side note, most Atlantic coast fishers do not eat jacks or bonitos because the meat is dark and considered to be strong. On the other hand, yellowtail (Figure 17.5) and Pacific bonito are sought after in California. Fortunate to have caught both, they are excellent table fare for those that enjoy eating fish.

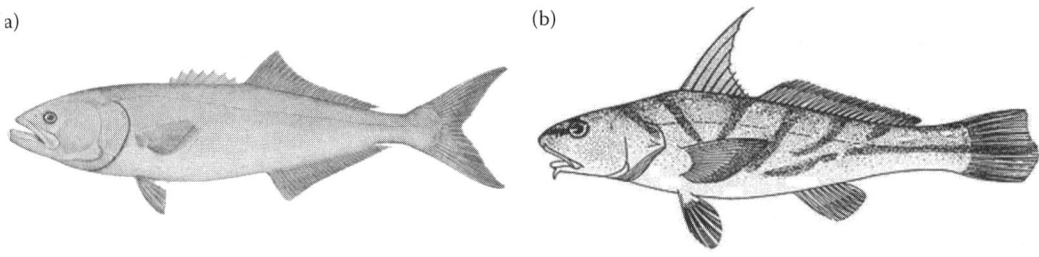

Figure 17.4 Bluefish (a) and kingfish (whiting) (b). (From the Duane Raver Fishfinder Collection, North Carolina Division of Marine Fisheries. With permission.)

Figure 17.5 Yellowtail jack (*S. lalandi*), San Diego, CA.

MARINE COMMERCIAL FISHERIES

The neritic province is critical to U.S. commercial fisheries, and these occur in both pelagic and benthic zones. Although the neritic province occupies only about 8% of the surface area of oceans globally, the fishes found there amount to as much as 90% of the world commercial tonnage. In the northwest Atlantic, two fisheries dominate the catch: One is pelagic and is composed mostly of Atlantic menhaden. This inshore fish spends its life cycle in the neritic and uses estuaries as a nursery area for its young. It is discussed in the previous chapter and again in chapter 31. The other fishery is benthic and is composed of "groundfish" (as in fishing grounds), the most important of which is the Atlantic cod. The cod is an offshore neritic fish, which also can occupy the inshore portion of the continental slope, where it may overwinter. We will discuss the cod here.

CASE STUDY: COD AND NORTHWEST ATLANTIC GROUNDFISHERY

America or Codland?

A great abundance of natural resources is often cited as the most compelling reason for the European settlement of North America. However, settlement was delayed for hundreds of years after the discovery of Newfoundland and the Canadian coast. A reason for this delay has been attributed to the aggressive nature of humans that already lived there, but another reason has recently been illuminated. Apparently, large stocks of Atlantic cod were discovered and exploited for hundreds of years in secrecy. Discovery of these fishing grounds by the English was the major reason for establishing colonies there (the cod was illustrated on the colonial seal of Massachusetts). According to Kurlansky (1998) and others (e.g., Jordan and Evermann 1923), there is ample information to prove that medieval Europe was fishing Atlantic cod from Newfoundland before Columbus set sail on his first voyage of discovery! What is this fish, why is it so valuable, and how does it still affect our civilization?

The Atlantic cod (*Gadus morhua*) (Figure 17.6) is unusual in appearance, having three dorsal fins and two anal fins, a single chin barbel, a pale lateral line, and a pelvic fin that is anterior in position to the pectoral. It is a benthic fish of deeper shelf waters. It is an omnivore, consuming many different food items such as almost any larger mollusk or crustacean, any fish small enough to swallow, squid, sea squirts, algae, and detritus. It is mostly a roving predator, feeding by sight and smell as individuals slowly move over the bottom, apparently making fast attacks on their prey. The fish can occur at almost any depth, and upon occasions, it feeds in surface waters, consuming sea ducks. Large fish have been captured at 3–5 fathoms (5.5–9 m) in winter and 7–8 fathoms (13–15 m) in the summer, but most cod were historically captured by commercial fishing at 20 to 70 fathoms (about 37–130 m). Young fish live more inshore, where they are bottom feeders in primarily the inshore neritic. The fish can reach a record size of over 2 m (6 ft) and 100 kg (211 lb), but 11 kg (25 lb) is considered a large fish. The cod is very fecund, averaging about 1 million eggs; however, a large female (34 kg; 75 lb) can produce over 9 million eggs. Some sharks, including spiny dogfish, are enemies of cod (Jordan and Evermann 1923; Bigelow and Schroeder 1953).

Cod have long been considered a valuable food fish. It has very flaky meat of white muscle, and it is easily dried and stored. This was extremely important historically because even in areas with cold climates such as Norway, the fish could be dried even in the frigid winter air. Once so prepared, it could be stored for long periods, and thus it was called "stock fish." And stocked it was. It was presumed to be the main food of Viking explorers in the tenth century and was widespread in medieval Europe (Kurlansky 1997). New techniques were perfected by Spanish (Basque) fishermen for preserving cod by using salt (not available in Scandinavia), which hastened the drying process, rendered the fish almost inert, and improved the flavor.

By the close of the middle ages (ca. 1450 AD), Icelandic cod were under heavy exploitation, and new grounds were searched for. English merchants discovered a new source of cod perhaps by 1481, but certainly by 1490 (Kurlansky 1997), presumably in the northwest Atlantic, but the location of their fishing grounds was kept secret. Secrecy ended with a 1497 voyage by John Cabot, who noted the rich cod fishing grounds in Newfoundland. Perhaps as important was his note that the "savages" call the cod "Baccalao," which is Basque for cod—obviously, the Basques had been

Figure 17.6 Atlantic cod *G. morhua*. (Courtesy of NOAA, http://www.photolab.noaa.gov/historic/nmfs/figb0314htm.)

there long before him (Idyll 1971)! Finally, Jacques Cartier arrived 37 years later in the vicinity of the St. Lawrence River estuary to note 1000 European vessels fishing for cod (Kurlansky 1997). By the mid-eighteenth century, Newfoundland cod were consumed in enormous numbers throughout Europe, and one French port fitted at least 4000 ships for the Newfoundland fishery from 1722 to 1792 (Roberts 2007). The race for cod was on.

For about 400 years, Atlantic cod were captured in the northwest Atlantic by bottom fishing with baited hooks. At first, cod were caught from small row boats (dories) using hand lines. Later, it also was common practice to set long lines (about a half mile in length or longer) with baited hooks attached every meter or so. These practices appeared sustainable because cod were captured in great numbers during this time, with little or no indication of overfishing.

New Technology: Dragging, Bycatch, and Bykill

By the mid-1800s, new methods emerged in Europe: sailing vessels were capturing fish by dragging a large weighted net on the bottom. The net was a beam trawl, in which heavy steel beams kept the net down and provided a horizontal opening into the bag. This device had to be used on the flat ocean floor, and sailing vessels dictated a small size for the nets because sail had little power to extract nets that became entangled. They were effective at capturing bottom fish, but they were not without problems. Roberts (2007) provides quotes from testimony given in an 1883 English court case about the adverse affects of trawls on fishing habitat. Testimony about the ill effects of trawling included killing of nontarget fishes and the young of target fish due to bycatch (like byproduct), and killing of fishes by net action (bykill). However, the most egregious impact was the destruction of bottom habitat by stripping the seafloor of invertebrate communities, seagrass flats, and so on and covering oyster reefs and other hard habitats with silt. Although the 1883 court decided that there was no evidence to conclude that trawling was damaging fish habitat, Roberts (2007, p. 157) cites an 1899 pamphlet published in resentment that said the following:

> Dozens of witnesses swore before the commission that the trawl destroyed the food of fishes, the spawn of fishes, and the immature fishes; many declared from their personal experience that it was enough to make an angel weep to see the awful signs of destruction brought up on deck by the trawl.

By 1889, there was little suitable habitat left in the North Sea that had not been trawled, and it was estimated that by the turn of the century, 260,000 km^2 were trawled twice each year (Roberts 2007). Trawls were taking in all of the fish, and other methods such as hook and line could not compete, so commercial fishers began switching to the trawl or going out of business.

The efficiency and destructive power of the trawl was very much increased with the emergence of steam power and a new type of net (i.e., otter trawl) that used wooden doors to keep the mouth open and huge links of chain to hug the bottom (Figure 17.7). The otter trawl could be used in uneven and rocky locations that were unsuitable for the beam trawl, and it was larger. Fishing yields of the

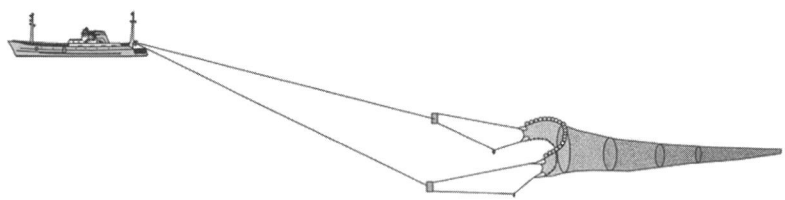

Figure 17.7 Otter trawl. (Courtesy of NASA.)

trawl was greatly increased with new technology and the larger size of the nets. Unfortunately, so was the damage. This process greatly increased yields, but it also disrupted marine communities that had developed over many thousands of years and destroyed habitat heterogeneity as thoroughly as in a ploughed field (for discussions about trawling and its impacts) (see Watling and Norse 1998; Jennings et al. 2001; or Helfman 2007).

Fisheries of the Grand Banks

As cod were fished in the northwest Atlantic, other species also were captured. These included other food fishes such as haddock, pollock, redfishes (three species), hake (two), and others. All of these fishes were not suitable for drying and salting, but as freezing emerged as a method of preserving fish, they also were sought, and the fishery was known as the NW Atlantic Groundfishery. As fishing and exploration continued, it was discovered that the fishing grounds were much larger than just the waters around Newfoundland. In time, a vast fishing ground consisting of 1.2 million km^2 was discovered on the continental shelf of southeastern Canada and the northeastern United States. Excellent coverage of the area is provided by Drinkwater and Mountain (1997). The shelf varies in depth, typically 100–200 m on the top of submarine plateaus that also may rise to 20–40 m and deeper (200–400 m) submerged basins. The shelf varies in width (about 500–50 km, north to south). Large and distinct areas of shelf have been named, including (from north to south) the Labrador shelf, northeast Newfoundland shelf, Grand Bank, Gulf of St. Lawrence, Scotian shelf, Gulf of Maine, and Mid-Atlantic Bight. A disjunct eastern plateau, Flemish Cap, was recently the subject of a popular movie (*The Perfect Storm*). The area is collectively known as the Grand Banks.

The abundant fishery resources of the Grand Banks developed principally because of the discharge of the St. Lawrence River into the shallow trailing continental margin. The river has an annual runoff of about 424 km^3, bringing large quantities of sediment, organic matter, and nutrients into the neritic zone. To place this runoff in perspective, it is greater than the combined runoff of the eastern coast of the United States (Drinkwater and Mountain 1997).

Commercial fishing continued in the northwest Atlantic by many nations, steadily expanding and developing new technologies. In the twentieth century, diesel engines replaced steam and even larger nets were dragged. Salt cod were exported from the United States well into the 1930s when technology was improved enough to begin machine filleting and freezing. In the 1940s, Canada allowed the use of the otter trawl. In the 1950s, larger diesel-powered ships were available from World War II, and they were used as mother (factory) ships to process fish at sea. Catches increased as ships grew larger, more powerful, and more numerous; stern trawlers were built; techniques were improved to allow pulling huge trawls with two ships; and fishing went faster—round the clock fishing was now possible. Europeans brought with them massive nets with stout steel towlines that caught underwater obstructions, including rock piles, boulders, and even wrecks, and dragged them, sometimes for miles, along the bottom. It was so bad that charts of underwater snags became useless (Roberts 2007).

New inventions also aided in opening more areas for trawling, such as rockhopping discs to allow trawling in rocky places and tickler chains to scare fishes such as flatfish out of hiding and into the water column where they could be swept into the net (e.g., Helfman 2007). New technology was also used to find schools of fish, such as sonar and spotter airplanes, and when fish concentrations were discovered, radios were used to call the vessels to the location (Kurlansky 1997). Catches steadily increased as fishing effort intensified. By 1965, the Soviets fished from Greenland to the Carolinas with 102 large and 425 smaller trawlers to supply 30 mother ships with fish for processing and freezing. Competing with them were vessels from Poland, Germany, Spain, Romania, Portugal, France, England, and, of course, Canada and the United States. By 1974, more than 1000 foreign vessels dominated the fishery (Roberts 2007). The United States had enough of this and increased its 3-mile territorial waters to a 200-mile economic zone—they kicked the foreign vessels out.

Unfortunately, the fishing industries in Canada and the United States soon doubled their fishing industry in response. In the 1980s, catches began to freefall, and fisheries scientist warned that a collapse in the fishery was likely.

On July 2, 1992, the Canadian government announced a 2-year moratorium on cod fishing, putting 40,000 people out of work and seriously affecting many more. After further investigation, the ban was extended indefinitely in 1993. With few minor exceptions, the ban continues (Roberts 2007). Also, after many years of warnings about pending collapse of groundfish in the U.S. portion of the Grand Banks, the U.S. Commerce Department responded. In 1994, fishing was closed in three large areas (11,220 km^2) of Georges Bank, the minimum net size for trawls in some other areas was increased to 6 in., and smaller mesh was prohibited in other areas to reduce bycatch (Ross 1997).

Why Did the Fishery Collapse?

Collapse of the NW Atlantic groundfishery (e.g., Figure 17.8) was a clear case of "recruitment overfishing," which means that the industry was taking fish before they had time to grow large enough to reproduce. This should have been relatively easy for fishery scientists to detect, but closure of the fishery was demanded from an unexpected sector: the fishing industry. Evidently, the trawlers were catching only small cod, and fishing became uneconomical. The fishing industry recalled their fleets and asked the government to close the fishery. Why wasn't this closure in response to government scientists? Martin (1995) provides a pointed essay on the subject and attributes this problem "corporate greed, union betrayal, political cowardice, and . . . a science that got lost somewhere along the way" (p. 5), and he provides us with three lessons to be learned. I condense and restate his remarks below:

- *Lesson 1.* There is a need for independent science. Researchers need to be free to make scientific conclusions, insulated to some extent from direct pressures of unions, industry, and government.
- *Lesson 2.* Fishing must be treated the same as any other industry and must require adequate environmental assessment of its actions. For example, the adverse impacts of trawling have not been adequately considered. Also, bycatch reporting has not been required, so catch of fishes were underreported. Bykill has not been considered.

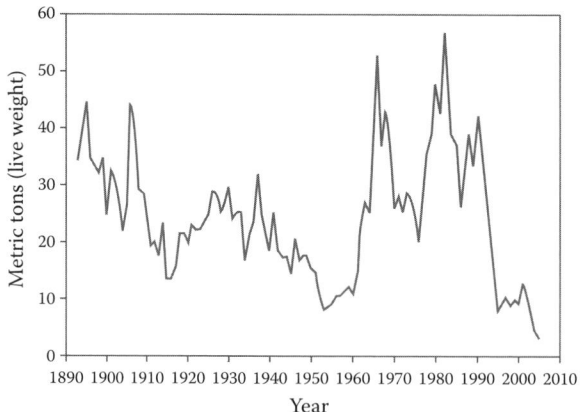

Figure 17.8 Commercial landings of Georges Bank cod. (Courtesy of New England Fishery Management Council/NOAA, http://commons.wikimedia/wiki/File:Global_shark_catch_graph_png.)

- *Lesson 3.* Fishery scientists and fishermen need to work in partnership. Scientists have one type of knowledge and fishermen have another. In one example where scientists would have benefited, it appears that fishermen had been increasing their use of electronics to locate concentrated schools of fish that retreat to deeper, warmer waters in winter. Fishing these concentrations produces a high catch per effort, but it does not represent fish density in the overall habitat. If taken to be representative, this would inflate predicted population sizes.

It has been about two decades since the Canadian closure, and I have not heard any substantive information that suggests recovery. Would there have been a collapse if only large hooks were used to catch the adult cod? We will never know, but 400 years of previous fishing with hooks and traps suggests that perhaps not. Contrary to trawling, fishing with hook and line does little to destroy habitat, and smaller fish cannot take larger hooks (i.e., recruitment overfishing cannot occur). How serious is the habitat destruction? There are many reports of the damage done by trawls and dredges (e.g., Engel and Kvitek 1998; Watling and Norse 1998). In most cases, productive ecosystems are trawled into a wasteland of rock and mud, with little left of the rich habitat that existed previously. Unfortunately, impact analysis has been hampered by a lack of controlled (never trawled) comparisons. The problem is that untrawled areas are hard to find (Engel and Kvitek 1998; Kaiser 1998). Thus, there is no environmental baseline upon which to make comparisons: we simply do not know what the neritic communities looked like centuries ago.

One interesting comparison has been made between clear-cutting forests and trawling (Watling and Norse 1998; Watling 2005); however, clear-cutting occurs in intervals of 50 to 100 years or more, whereas trawling is occurring in most areas on time scales of months to years. For example, in the North Sea, trawling covers the entire area about three to seven times each year, and some areas are trawled hundreds of times (Kaiser 1998). In this case, I think the trawling–clear-cutting comparison would be a more appropriate if an old growth forest were clear-cut and then plowed monthly for several years! Such loss of habitat is a serious ecological problem, and as incredible as it seems, such habitat destruction seems to have been effectively ignored in predicting stock biomass on the Grand Banks. Even without habitat destruction, there remains the problem of removing a tremendous number of adult fish. When the abundance of breeding adults is greatly reduced, it is possible that recruitment could suffer.

There are two final questions: (1) how many cod have we lost, and (2) will the fishery ever recover? There is a problem in evaluating the first question, because peak abundance probably occurred more than 100 years ago and records do not allow very good comparisons (Carlton 1998). However, some studies have addressed the issue. Using different methods (e.g., Rose 2004; Rosenberg et al. 2005), scientists have arrived at a figure: present stocks amount to only about 1% of the historic cod biomass—a loss of a whopping 99% for the Grand Banks fishery and about 100 billion breeding fish (Rowe et al. 2004). Regarding the second question, if the present situation is an indication of the rate of recovery, it will take a very, very long time. Why is this?

The slow or no recovery of cod is related to intensive overfishing by trawling. One problem is size selection of the fishing gear: in this case, it appears that larger, more aggressive, and fast-growing fish are harvested first. By removing these fish, there is selection for slower growth in the population and loss of aggressiveness, a behavioral trait (Biro and Post 2008). The larger fish feed on smaller predaceous forage species (known as *mesopredators*) (Prugh et al. 2009) that in turn feed on smaller items such as cod eggs, larvae, and young. Due to a low abundance of larger cod and other groundfish, recruitment now appears to be very low due to predation by these forage fishes (Walters and Kitchell 2001) and invasion by skates and sharks (Helfman 2007). Such a low population size may result in reduced survival and growth of young cod due to problems in reproduction (e.g., mating and low egg fertilization) (Rowe et al. 2004) as well as predation.

SUMMARY

The neritic province contains pelagic and benthic systems. Pelagic systems occur in an open-water photic zone, while benthic systems occur on the continental shelf. There are several neritic systems that are used by inshore marine fishes, including open-water areas that are dominated by plankton, subtidal bottoms of soft sand and mud that mostly support populations of invertebrate infauna, and rocky bottoms that are mostly inhabited by invertebrate epifauna. More complex systems in this province include seagrass beds, kelp forests, and coral reefs. Structure in this environment is influenced by currents, tides, and winds. Productivity in the neritic province is based on two major features: (1) input of organic matter and nutrients carried into the water by surface runoff, wind, and stream flow via estuaries and (2) autotrophy from phytoplankton, algae, and grass flats. Wind action also assists productivity in the relatively shallow neritic system because it can keep organic material and nutrients in circulation. Thus, high productivity results in a great number and diversity of neritic fishes, which are the largest component (45%) of the total marine fish fauna, comprising about 12,600 species (Helfman et al. 2009). Most inshore fishes of the neritic are well known, and about 80 of the 563 marine fishes that occur in North Carolina are often encountered by sport fishermen. The neritic province also is critical to commercial fisheries, producing about 90% of the world catch in only 8% of the oceans. In the northwest Atlantic, two fisheries dominate the catch: One is pelagic, composed mostly of Atlantic menhaden. This inshore fish spends its life cycle in the neritic and uses estuaries as a nursery area. The other fishery is benthic and is composed of "groundfish," the most important of which is the Atlantic cod—an offshore neritic fish and the subject of a case study. For at about 400 years, Atlantic cod were captured in the Grand Banks of the northwest Atlantic by bottom fishing with baited hooks. At first, cod were caught from small row boats (dories) using hand lines. Later, it also was common practice to set long lines with baited hooks. These practices appeared sustainable because cod were captured in great numbers during this time, with little or no indication of overfishing. By the 1800s, a new method was developed to catch bottom fishes. A large bag net (i.e., beam trawl) weighted down with steel beams was dragged by a ship. This was improved by the otter trawl, which came with wooden doors to keep the mouth open and huge links of chain to hug the bottom. The efficiency and destructive power of the trawl were very much increased with the emergence of steam and then fuel oil. The fishery declined and then collapsed. On July 2, 1992, the Canadian government banned cod fishing, putting 40,000 people out of work. The U.S. Commerce Department responded in 1994 by closing 11,220 km^2 of Georges Bank and restricting gear to reduce bycatch. The problem was "recruitment overfishing" or harvesting fish before they have lived long enough to reproduce. In addition, overfishing has removed the large predators, resulting in a proliferation of smaller predators (i.e., mesopredators) that consume young cod. The situation seems grim. Cod stocks are estimated to be only 1% of historic biomass, and after almost two decades, the cod have not recovered.

Recommended video: *The Perfect Storm*. Warner Home Video, 2000.
Further reading: Boreman et al. 1997.

CHAPTER **18**

Oceanic Province and Epipelagic Fishes

PROVINCE

The pelagic division of the marine environment includes the neritic province (all of the water over the continental shelves), and the oceanic province that includes all of the waters that are contained in the open, or deep seas, which lies offshore of the continental shelf break (Figure 16.3). This immense province is further divided into several zones on the basis of depth and other criteria.

Structure

There are several ways in which the oceanic province may be subdivided. The epipelagic zone is the upper, well illuminated layer of water that descends about 200 m. Deeper regions of the sea can be collectively called the abyss (Idyll 1971) and may be divided into the mesopelagic, the bathypelagic, the abyssopelagic, and the hadopelagic zones. Also in common use is the term "deep sea," which is given to the aphotic zone or for depths > 1000 m (Angel 1997).

The mesopelagic (twilight) zone, descends from 200 to 1000 m. Below the mesopelagic is the largest pelagic zone, the bathypelagic zone, which descends from 1000 to 4000 m. In very deep oceans there may be an abyssopelagic zone, which descends to about 6000 m. Even deeper, especially in ocean trenches is the hadopelagic (Figure 16.3). Not only does the amount of light vary in these zones, other physical, chemical, and biological attributes do also. Pressure increases greatly with descent due to high density of water. Temperature drops rapidly with distance from the surface, becoming very cold at depths, even in the tropics. Oxygen levels also decrease with distance from the top, and food becomes very scarce because it is eaten as it drifts down from above.

Oceans include waters of all latitudes from equatorial (very warm) to polar (very cold) regions. From an ecological standpoint, there appears to be more in difference than in similarity; however, broad principles apply to all. In general, there are two major systems in the oceans, an upper autotrophic system of relatively shallow circulating water that is sunlit, warm, and oxygen rich, and a lower heterotrophic one that is sunless, deep, and permanently stratified with respect to temperature and oxygen.

There are few large plants in oceans, and phytoplankton is by far the primary autotrophic organism (about 95%). Although the oceans are not very productive in comparison with inshore and terrestrial systems, they are vast in size and produce about half of the primary productivity and half of the oxygen on the planet (Castro and Huber 2007). For many years it was thought that larger net plankton could not produce this much productivity, and recent studies (with better technology) show that very small photosynthetic archea and bacteria are responsible for a large part of photosynthesis in the oceans.

Nutrients are more limiting in the open ocean than the inshore areas. Oceanic phytoplanktons tend to be small in size, and they are more concentrated in the upper layers of the ocean where there is sunlight penetration. Grazers on phytoplankton also tend to be small zooplankton. Zooplankton and small fish that eat them are at risk to predation in the open sea, and most of them are capable of vertical migration to deeper (darker) levels in day to seek some cover. Zooplanktons include organisms that spend their whole life as plankton (e.g., protozoa and crustaceans such as copepods and krill) and planktonic larvae of larger species (e.g., fishes). Small zooplanktons tend to be consumed by larger zooplankton, which in turn are consumed by small nekton such as herring. However, food chains can be short in the open oceans, and some of the largest nekton are planktonic (blue whale, whale, and basking sharks). Bottom communities in the abyss predominately feed on organic matter that is deposited from above. Surprisingly, the bottom supports a fairly rich community of organisms compared to the waters immediately above it. Decomposers predominate there, including invertebrate filter feeders, deposit feeders, collectors, and predators. Bacteria also are dense in numbers and are important food for deposit feeders. Deep water fishes probe the bottom or slowly glide along seeking prey. Diversity is exceptionally high at the location of hydrothermal vents, where chemosynthetic bacteria are able to make organic carbon.

Function

Oceans of the world are vast three-dimensional habitats with a low productivity per unit of area compared to terrestrial systems. Not including estuaries and the neritic province, phytoplankton of the open ocean only produces about 25% of the world's net primary productivity. This results from a small upper zone where there is enough sunlight for autotrophy but a low level of nutrients. In fact, most of the nutrients in the open ocean exist in living tissues, which is lost to the photic zone when they sink and are decomposed in lower levels. These nutrients are recirculated in temperate seas to some extent due to wind action, and are brought to the surface in areas of upwelling. In tropical seas there is a density affect that reduces circulation: surface waters are warm all year, and as a result, they are lighter than the deep layers. This results in a permanent density stratification and lack of nutrient recirculation, except for locations where upwelling occurs. Two types of upwelling, coastal and equatorial, predominate in the oceans. Coastal upwelling usually occurs because surface water is displaced by winds, and deeper nutrient-laden water rises to replace it. Equatorial upwelling occurs in the mid-Pacific due to the divergence of westward flowing water away from both sides of the equator and the rising of deeper water to replace it. Upwelling will be discussed in more detail later in this chapter.

EPIPELAGIC ZONE

Conditions

The epipelagic zone has been defined differently, most commonly as "the upper 200 m of the oceanic province" (Sumich and Morrissey 2004), but it also has been called the photic zone, or lighted area of the ocean in both marine and general ecology texts. Recently, the epipelagic has been defined as "the pelagic environment from the surface to a depth of 100 to 200 m" (Castro and Huber 2007), presumably including the zone in both neritic and oceanic provinces, a convention also used in some general ecology texts (e.g., Krohne 2001; Smith and Smith 2001). For our purposes here, I will include the inshore (neritic) portion of the epipelagic as the photic zone of the neritic province, and use the term epipelagic to refer to the upper 200 meters of the oceanic province.

Most of the pelagic fishes of the Oceanic Province are found in the epipelagic zone, even though the zone only amounts to about 10% of the open ocean by volume. Reasons for this include the availability of food and more desirable living conditions.

The epipelagic conditions are hospitable for fishes in several respects. First, there is a dependable supply of food, especially in the inshore regions that benefit from neritic inputs, and in areas of upwellings. Next, there are warm conditions due to sunlight penetration, which also aids fish in detecting food to eat. Another plus is the availability of sufficient oxygen for active life, including prolonged swimming.

Swimming is important in the epipelagic, because many resident fishes such as tuna, make long spawning and feeding migrations (e.g., oceanadromous), and part time residents include both anadromous (e.g., salmon), and catadromous (e.g., eels) species. Horizontal migrations are a characteristic feature of this zone that sets it apart from other zones of the deep sea.

However, there are some aspects of the epipelagic zone that make it difficult for fishes as well. One important aspect is gravity. In order to live there fish have to retard sinking in some way. Epipelagic organisms can reduce sinking by increasing their surface to volume relationship. Many plankers do this and even some fishes (such as young ocean sunfish) by developing a disk-like shape or by growing projections. Perhaps a better way is to develop buoyancy, which is not so difficult in the dense water medium. Pelagic fishes provide a sizable supply of lipids in the eggs that is maintained in teleost fry until it can develop a flotation organ, which we know as the swim (or gas) bladder. However, some fishes, especially fast predators may not have a functional swim bladder. Finally, some fishes have developed highly efficient fusiform body shapes with large crescent shaped caudal fins. In fast predators such as the tunas, jacks and mackerels, and billfishes, huge masses of red muscle are connected across a thin caudal peduncle with tendons to the large falcate caudal fin. In case of epipelagic predators that must swim quickly between various depths, efficient swimming in salt water proved to be superior to the swim bladder, which is lost in many of the fast predaceous fishes.

Once a fish became adapted for staying in the epipelagic zone, either by floating or swimming, there were other considerations. Staying alive was the most important. To escape detection, fishes became countershaded, that is, dark on top (dorsal) surfaces and light on ventral surfaces. In this way, a predator looking up might miss seeing the light fish in the light sky, or looking down, the dark fish against the dark sea below. But predators also hide with camouflage, and in many cases they find cover in floating weed, timber, lost cargo, wreckage (flotsam and jetsam), or trash. In fishing for large sport fish in the mid-Atlantic Gulf Stream, charter boat captains have learned that just about every large board potentially has a big tuna, wahoo, dolphinfish, or another predator hiding under it.

Many fishes of the epipelagic are schooling species, both predator and prey. Predators, such as tuna, school so they can find prey and then herd the prey toward the surface in a tight ball. Prey species also benefit from schools in several ways. It is safer to be in a school (or shoal), because prey distribution is thereby clumped in space, not uniform, making it less likely to be detected. Then, once detected the school can confuse the predator by making it difficult to focus only on one individual. Thus, a single predator is probably going to eat the slowest or sickest fish first. A good reason to stay healthy!

Living in the epipelagic zone also places constraints on reproduction, and almost all teleost fish that spawn there produce abundant numbers of floating eggs. Large tunas produce millions of eggs, and the ocean sunfish, the most fecund vertebrate, produces hundreds of millions of eggs. On the other hand pelagic sharks produce only a few offspring, but the young are large at birth and capable of foraging on most teleost fishes.

Fishes

There are about 325 fish species from 70 families in the epipelagic zone, which comprises about 1.3% of marine fishes (Helfman et al. 2009). For the most part they can be classified as plankton feeders and their predators. However, all fishes that occur in the epipelagic are not permanent residents. Many species find cover in deeper zones and make vertical migrations to feed or reproduce, and many deeper water residents have larvae that rise to the epipelagic zone. We will discuss them later.

Figure 18.1 Bluefin tuna an epipelagic fish. (Courtesy of National Oceanic and Atmospheric Administration (NOAA) Fisheries Collection, http://www.photolab.noaa.gov/bigs/fish2056.jpg.)

Bond (1996) identified 26 fish families that could be considered to be representative of the epipelagic zone. These included five families of sharks and 21 families of ray-fin fishes. It is interesting to note that of these 21 ray-fin families, 13 of them were higher teleosts in the order Perciformes. Thus, about 65% of the fish in the epipelagic were teleosts, and this number could be increased to 70% if the partial resident lanternfishes are excluded (they migrate up from the lower zones).

Large marine fishes that reside in the epipelagic zone are typically warmwater fishes that only access the colder temperate regions of the sea in summer. Most of these powerful fishes, such as tunas (Figure 18.1), wahoo, dolphin fish, marlins, jacks, and some sharks, are adapted to richer regions of the oceans that are located especially around the equator, where currents (equatorial divergences) result in upwelling of nutrient-laden waters (e.g., see Castro and Huber 2007). These waters are some of the most productive areas in tropical portions of the open oceans. Fishes from tropical seas can access temperate regions by entering in the great gyres located north and south of the equatorial zones, which are enormous eddies of circulating water. In the western Atlantic Ocean the Gulf Stream is included, and the North Atlantic gyre circulates around an area known as the Sargasso Sea (Figure 18.2). Important large fishes such as bluefin tuna, albacore, swordfish, and ocean sunfish take advantage of the Gulf Stream to move north as the water warms in late spring.

The Sargasso Sea is an area of very low plankton productivity, as evidenced by water clarity. Idyll (1971) gives a good idea of this by comparing the depth of water where light reaches a 99% extinction level: at Woods Hole, Massachusetts, this level was reached at a depth of 8 m (26 ft), in the Gulf of Maine at a depth of 32 m (105 ft), and in the Sargasso Sea at a depth of 150 m (490 ft). However, a large brown alga, *Sargassum* weed, takes advantage of the relatively warm, clear water there and grows profusely from the bottom. The plant has berry-like flotation chambers (pneumatocysts) on its blades, which serve to keep the plant afloat. During storms, the plants can be detached and still floating, they are entrained in the gyre and thus into the Gulf Stream. On fishing boats, we look for the lines of "weed" to tell us when we reach the area of the Gulf Stream, and if it is relatively calm when we arrive, you can almost always detect where the stream is: moving slowly, there is a line of demarcation; the front of the boat will pass from the cloudy greenish inshore water into the absolutely crystal clear waters of the stream.

In the barren open ocean, usually devoid of visual objects, suddenly you approach the Gulf Stream, and there appears a huge line of seaweed. As it drifts, the *Sargassum* weed transports its community of invertebrates and fishes with it, including two fishes endemic to Sargassa: the sargassunfish (a frogfish) and the sargassum pipefish. A recent study was done in the Gulf Stream using surface trawls in and around the floating weed, and videography shows the importance of the weed for animals, including fishes.

In an otherwise aquatic desert, Casazza and Ross (2008) captured 18,799 fishes representing over 80 species and 28 families in the weed, and 2706 fishes in open water samples near the weed. A very important feature is that 96% of the individuals captured were juveniles, which were very

OCEANIC PROVINCE AND EPIPELAGIC FISHES

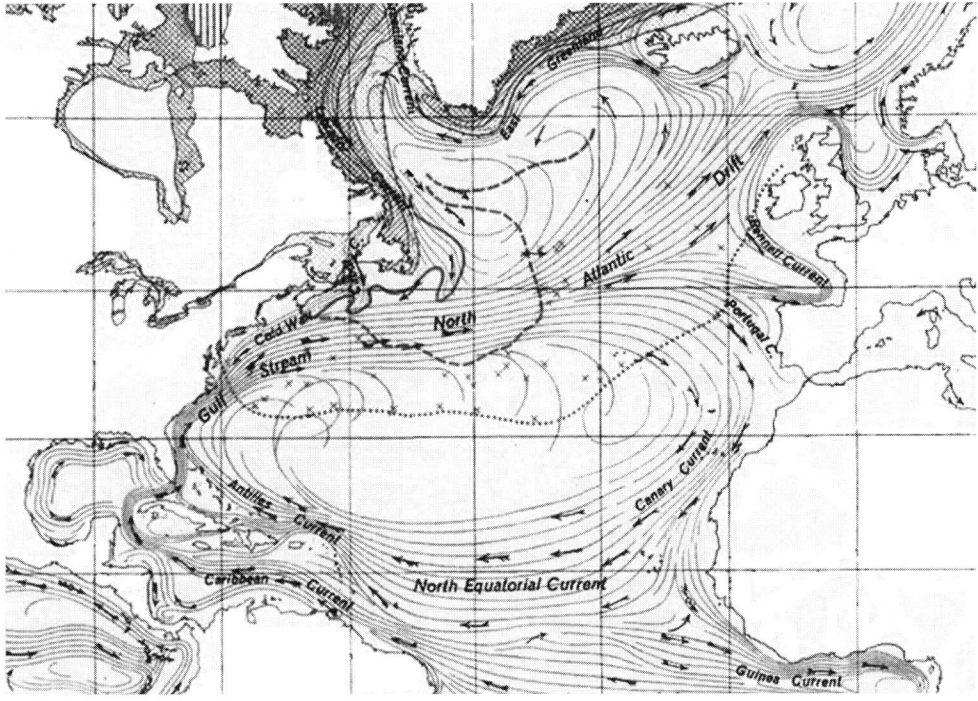

Figure 18.2 North Atlantic gyre. (Courtesy of NOAA, http://www.noaa.gov/.)

closely associated with the weed, presumably as cover and feeding. Of the total number, 93% was due to only nine species: three filefishes, three jacks, two flying fishes, and a triggerfish. The juveniles also included some commercially important fishes, such as dolphin fish, jacks, and amberjacks. The fate of these fishes is not completely understood, but some of them likely move inshore in summer, and others are transported, perhaps to Europe or even all the way round the gyre.

In the open ocean there are islands and reefs that add to the diversity of fishes in the epipelagic zone, but there are less visible sunken islands as well. Located at various depths, including a few meters below the surface in the epipelagic zone to thousands of meters in deeper zones, *seamounts* rise from the ocean floor. As currents sweep by them nutrients may circulate from deep layers to the epipelagic zone and increase its productivity. Large predators and other fishes such as eels may use seamounts for migratory cues and stopovers, or as feeding and spawning areas. Seamounts are deep-sea phenomenon and support deeper zones; they will be discussed in more detail later.

Closer inshore, along the continental breaks and where plankton is dense are concentrations of planktivorous fishes, including the herrings, sardines, anchovies, and the giant basking and whale sharks. Further offshore, plankton are not so abundant, and the few fishes there are unusual ones, such as the flying fishes, which can jump several meters high and glide long distances to escape predators. The flying fish generally can be divided into offshore and inshore species. Some of these fish are quite attractive such as the inshore band-wing flying fish of the Atlantic (Figure 18.3). Offshore forms share their habitat with the very large ocean sunfish, a tetraodont (Figure 6.16).

Ocean sunfish are so unusual that they deserve a few additional comments. The apex of fish evolution, they are said to be, in a sense, the world's largest plankton. The fish tends to drift along with little apparent movement, but it can scull along with its strange tetraodont swimming (i.e., with fins only). The fish consumes mostly jellyfish and other organisms that it can catch, growing in size to about 3 m, and it can be a ton in weight. The flesh is said to be flabby and parasite ridden.

Figure 18.3 Bandwing Flying fish. (Courtesy of NOAA, http://www.photolab.noaa.gov/fish/fish3023.htm.)

It usually has marine invertebrates, algae, and whatever can be imagined, growing on its surface. So perhaps the fish can be considered a community in itself. Little ocean sunfishes look so different from the adult that they appear to be different species. The larvae have a thorny suite of armor, which may function to retard sinking or could be used as an antipredator adaptation. They look like small medieval maces.

UPWELLINGS

Two types of upwelling predominate in the oceans, equatorial and coastal. These upwellings occur due to the rise of nutrient-laden water into the photic zone. Equatorial upwelling is a function of the Coriolis effect, which causes the equatorial divergence. A good example of this phenomenon occurs in the mid-Pacific due to the divergence of westward flowing water away from both sides of the equator by the Coriolis effect, and the rising of deeper, nutrient-laden water to replace it. This phenomenon produces a region of high productivity, and many fishes take advantage of it to feed or spawn. Skipjack tuna are one species that incorporates this region into their life history, making migrations from coastal areas in South America to spawn in one of these rich areas (see map in Sumich and Morrissey 2004).

Coastal upwelling in North and South America occurs on the western edges of the continents due to two mechanisms, water displacement from alongshore winds produce surface currents that are moved offshore by the Coriolis effect, and also by strong westerly winds that move surface water across the Pacific. In both cases deeper, nutrient-laden water moves up toward the surface, but the latter case, these upwellings occur irregularly. In normal or average conditions, alongshore currents cause water displacement and result in some amount of upwelling; however, along deep coastlines, there are two regions that receive occasional but substantial upwellings of nutrient-laden water. Those upwellings are caused by the same phenomenon, which is known as the El Niño–Southern Oscillation or ENSO (e.g., see Castro and Huber 2007). This event occurs in December and is locally known as El Niño.

Consider the Pacific Ocean as a large shallow pan of water that is almost full. If the pan is moved it sets up oscillations that slosh water back and forth. In normal years the trade winds blow westward and tend to pile up warm surface water in the western Pacific. This results in an upwelling of cold water that rises up on the eastern side to take the place of the surface water that has been displaced. This cold nutrient-laden water generally is found about 40 m below the surface on the west coast of North and South America. With El Niño conditions, which occur about 30% of the time, westward trade winds are weaker than easterly winds, which pile up water along the coast of the eastern Pacific, effectively burying the colder water and blocking normal upwelling. The reverse of this condition, called La Niña, which occurs about 20% of the time, results in very strong westerly winds that cause massive upwellings along the west coast of the Americas. La Niña results in cooler conditions and bountiful supplies of nutrients at the surface. This results in plentiful plankton blooms. Just offshore of the Pacific coast of Peru, an area about 30 × 300 miles is the richest area of upwelling in the world and significant community benefits. The keystone species is the Peruvian anchoveta.

CASE STUDY: PERUVIAN ANCHOVETA

When I was a child I helped my parents fertilize the garden with a commercial fertilizer called "guana." Little did I know that this fertilizer was produced by seabirds that roost and nest on islands off the coast of Peru. Those birds fed on anchoveta, a small anchovy (Figure 18.4), and over untold thousands of years their droppings (guano) accumulated in layer after layer. When discovered by humans, the layers exceeded 40 m thick. This guano was mined and supported a fertilizer industry, the world's largest organic supply. Millions of tons of this nitrogen- and phosphate-rich fertilizer were sold worldwide.

With discovery of the guano and publicity about the source, humans also began to exploit the anchoveta for use as fish meal. In the 1950s, Peru built a fish meal reduction plant, which encouraged expansion of the fishery. Large catches were made by purse seining vessels, which could make two or three hauls a day. Soon the vessels were taking 4 million tons annually, which was equal to about 1/7 of the world's catch in an upwelling area that is only 30 miles wide and 800 miles long (Marshall 1966). By the 1960s, the catch was an unbelievable 10 million tons, the largest fishery in the world. The nation of Peru, which was 27th as a fishing nation in 1957 rose to second place in 1961. Catches continued to increase, and finally, 12 million tons were taken in 1972, due to intensive purse seining by many nations and use of airplane spotters. After that, the fishery collapsed, and seabird numbers also crashed. Following another period of unfavorable El Niño years, the catch dropped to only about 94,000 tons. Extended poor conditions had concentrated the fish in a few areas where there was some food and they were very vulnerable. After that, there were periods of good years and bad years. Recently during El Niño conditions of 1997, the catch crashed 80% and then recovered somewhat in the twenty-first century. The fishery has had ups and downs and in the long term remains unstable (Figure 18.5).

Under natural conditions the anchoveta expanded during La Niña conditions and also supported expansion of the predaceous seabirds, mostly cormorants and boobies. In the good years, the bird populations rose to 25–30 million, which could consume 3 million tons of anchoveta in a year (Idyll 1971). During bad years, the plankton declined, and the fish population also was regulated by the birds, probably preventing a severe crash due to starvation. However, when El Niño occurred for two or more consecutive years, fish and bird populations starved, sickened, mortalities increased,

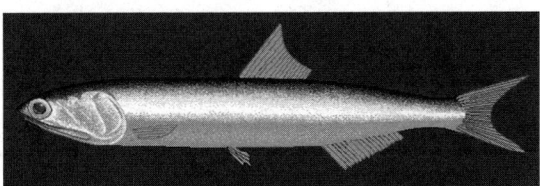

Figure 18.4 Peruvian anchoveta. (Courtesy of Cada, R., http://fishbase.org, http://commons.wikimedia/wiki/File:Enrin_u0.gif.)

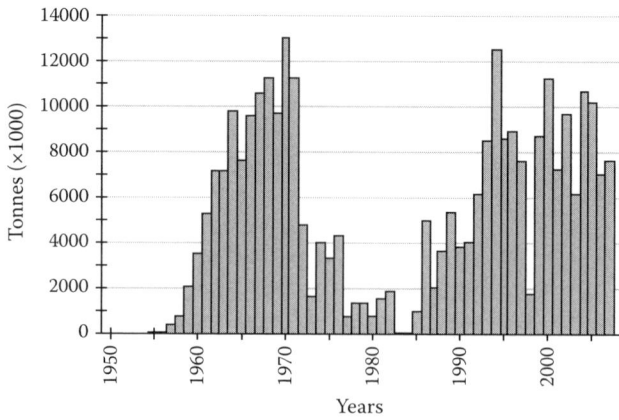

Figure 18.5 Anchoveta fishery. (From http://www.fao.org/fishery/species/2917/en.)

and reproduction failed. During average years, there were average conditions, and all was well. Now fishing has replaced the birds, but fishing continues during bad years. In this case, why hasn't the population been decimated like the northwest Atlantic groundfish? It was probably because of the different gear and techniques.

Purse seining is done with a very long floating net that is used to encircle a massive school of anchoveta. The net has a line looped through the bottom that is tightened, or pursed, so the fish cannot dive and escape. Smaller fish can swim through the mesh, and nontarget fish can be freed. Purse seining does not destroy the habitat, and bycatch is relatively low. Unfortunately, standard fishing models cannot predict climatic conditions very well.

Another fish that is subjected to similar conditions is the Pacific sardine (*Sardinops sagax*) (Figure 18.6). That species is also affected by El Niño conditions, but the picture is somewhat different due to exposure to cold currents in some years and competition by the northern anchovy, in general terms considered an ecological equivalent. The population has recovered somewhat after being decimated in the 1940s, and a major sardine processing operation in Monterey, California, was closed. There is a memorial there and a book written about the aftermath by John Steinbeck (*Cannery Row*). It has been argued that the California sardine was one of the most intensively studied fish in the world, and models have been constructed to evaluate various scenarios (e.g., see

Figure 18.6 Pacific sardine.

Moyle and Cech 2004). But the lesson is the same. Complex climatic conditions must be considered in regulating populations and caution taken not to overfish the resource.

SUMMARY

The oceanic province is an immense area of the sea known as the open or deep seas, which lies offshore of the neritic province. It is comprised of an upper epipelagic zone that descends about 200 m. Deeper parts of the province, collectively called the abyss, are divided into several zones: The mesopelagic (twilight) zone descends from 200 to 1000 m. Below the mesopelagic is a volume of complete darkness, including the bathypelagic zone (1000 to 4000 m) and the abyssopelagic zone (4000 to 6000 m). Even deeper, especially in ocean trenches, is the hadopelagic. These zones vary in physical, chemical, and biological attributes. As depth increases, pressure increases, temperature drops rapidly, oxygen levels decrease, and food becomes very scarce. The ocean province has two major systems, an upper autotrophic system of relatively shallow circulating water that is sunlit, warm, and oxygen rich, and a lower heterotrophic one that is sunless, deep, and permanently stratified with respect to temperature and oxygen. Oceans are not very productive in comparison with some inshore and terrestrial systems, but produce about half of the primary productivity and oxygen on the planet. The open ocean, specifically the epipelagic zone, produces about 25% of world productivity because nutrients are limiting. Nutrients and organic matter are lost to the photic regions as organisms die, sink, and are decomposed in lower levels. The deeper nutrient-laden water may be recirculated to the photic zone again by currents, wind action, and upwellings. Equatorial and continental upwellings are extremely important for commercial fisheries. Fishes representative of the epipelagic zone include about five families of sharks and 21 families of ray-fin fishes, of which about 65% (13 families) are higher teleosts. However, all fishes that occur in the epipelagic are not permanent residents. Many species find cover in deeper zones and make vertical migrations to feed or reproduce at night. Epipelagic fishes are mostly schooling species, both predator and prey. Predators, such as tuna herd the prey toward the surface in a tight ball. Prey species such as herring also benefit from schools because their distribution is clumped making it less likely to be detected. If detected, it is difficult for a predator to focus on one individual. Life in the epipelagic zone also places constraints on reproduction, and almost all teleost fish that spawn there produce abundant numbers of floating eggs. Large tunas produce millions of eggs, and the ocean sunfish, the most fecund vertebrate, produces hundreds of millions of eggs. Pelagic sharks produce only a few offspring, but the young are large at birth and capable of foraging on most teleost fishes. Large marine fishes that reside in the epipelagic zone are typically warmwater fishes that only access the colder temperate regions of the sea in summer. Most of these fishes (such as tunas, wahoo, dolphin fish, marlins, jacks, and some sharks) are adapted to richer regions of the oceans that are located especially around the equator, where currents (equatorial divergences) result in upwelling of nutrient-laden waters. These waters are some of the most productive areas in tropical portions of the open oceans. Fishes from tropical seas can access temperate regions by entering in the great gyres located north and south of the equatorial zones, which are great eddies of circulating water. In the western Atlantic Ocean, bluefin tuna, albacore, swordfish, and ocean sunfish take advantage of the Gulf Stream to move north as the water warms in late spring. In the Pacific Ocean, trade winds blow westward and tend to pile up warm surface water in the western Pacific. This results in upwellings of cold, nutrient-laden water that rises up on the eastern side, which is the west coast of North and South America. About 30% of the years have weaker trade winds and easterly winds block upwellings (El Niño effect). About 20% of the time, very strong westerly winds cause massive upwellings and plentiful plankton blooms (La Niña effect). Just offshore of the Pacific coast of Peru, a significant community benefits. Massive plankton blooms are exploited by the Peruvian anchoveta, which were fed upon by seabirds. The birds nested on islands and produced millions of tons of guano, used as a fertilizer. The anchoveta expanded during La Niña conditions, when bird populations rose to 25–30 million, consuming 3 million tons of anchoveta in a year (Idyll 1971). During bad years, the plankton and fish recruitment declined, with the birds cropping adult fish, and preventing widespread starvation. When El Niño occurred for two or more consecutive years, fish and bird populations starved, and reproduction failed. Now fishing has replaced the birds, but fishing continues during bad years. The fishery is unstable, characterized by boom and bust.

CHAPTER 19

Deep Sea: Twilight to the Abyss

FEATURES OF THE DEEP SEA AND ITS FISHES

It is difficult for humans to understand life in the deep sea. In the previous section, we discussed the upper layer of the oceans that we all recognize from personal experience or from media presentations. The surface has sunshine, it is relatively warm, and there are familiar fishes. Many humans swim, boat, and scuba there. But those happy outings provide misconceptions, because the real ocean is a far different environment, almost all of the hundreds of millions of cubic feet of it. To use a pun, we have just scratched the surface of the typical ocean, because about 97% of the ocean pelagic realm is deeper than 200 m, the lowest part of the epipelagic zone. That level is the average depth at which the penetration of sunlight is so weak that phytoplankton cannot make enough food to live. However, the abyss (deep sea) is a huge biome, the largest continuous volume of habitat on Earth.

We envision the coastal shoreline as the edge of continents. This is not so. The edges of the continental land masses actually comprise the continental slope zone and form the walls of the ocean basins. These walls descend rather precipitously in most areas to the ocean "floor." However, it is not really like a floor. Most of us have the idea that the ocean has a floor like the beach we wade in. But this is not true. The deep ocean floor is not featureless. It includes regions of flat muddy terrain, but over all, it is more rugged than the continents, with spectacular mountain ranges, valleys, basins, and deep submarine trenches. The rugged condition is due to a lack of exposure to the erosion that occurs on continents (Idyll 1971).

In addition to the major habitat differences noted above, there are environmental challenges to the physiology of marine creatures. Some of these are visibly obvious, like declining light levels with depth. Other features of the deep sea, such as great pressure, low oxygen, and extreme cold, are not. But more amazing by far than the hardships imposed on fishes is the way that thousands of species have adapted to these conditions. Fishes of the deep pelagic environment, which includes the mesopelagic and bathypelagic zones discussed here, comprise 5% of the number of marine fishes (about 1400 species), and fishes of the deep benthic zone comprise 6.4% (about 1800 species) (Helfman et al. 2009).

There is a lesson to this. Even though by human standards the deep sea has an exceedingly adverse environment, this environment presumably has not changed much in perhaps a hundred million years. From an evolutionary standard, we would assume that such a long time period would result in specialization in ancestral fish groups, and once such extreme specializations were attained, they would persist. This is just what we observe: Using data provided by Bond (1996), about 70% of teleost families representative of the epipelagic zone were the recently evolved perciforms. In the mesopelagic, only one perciform family is representative, resulting in 95% of the fish families (21) from lower teleost groups. This follows in the bathypelagic, with only one perciform family of

12 total representative or about 92% from lower groups. Even higher percentages of lower teleosts species would presumably result from a complete compilation.

Structure

The oceanic province can be divided into various levels by the use of depth and associated physicochemical parameters (Figure 19.1). Ecologically, however, there are three major zones (i.e., volumes of water) that are very different: the well-illuminated, oxygen-rich, circulating upper epipelagic zone; the deeper transitional mesopelagic zone; and a deeper region of cold, low-oxygen water that is permanently stratified and cloaked in perpetual darkness, which is commonly called the deep sea or abyss (Idyll 1971). These marine environments can be delineated by differences in solar illumination, temperature, oxygen, and pressure. They also are ecologically distinct. Food is produced by phytoplankton growth and reproduction in the epipelagic zone, where it is fed upon by resident and vertically migrating zooplanktons and larger invertebrates and fishes. Many organisms in the mesopelagic zone also share in this wealth by vertical migration because the deep sea is mostly heterotrophic. Organisms that live at depths feed on organic material that sinks from above and on some organic matter produced by the chemosynthetic activity of archaea and bacteria.

Deep-sea organisms vary in their adaptations according to the conditions that are largely related to depth. Life occurs even in the lowest levels of the abyss, because oxygen is never completely used up. The deep sea is continuingly supplied with oxygen from dense thermohaline flows that originate in the epilimnion and sink to depths at the poles (i.e., the Ocean Conveyor). Deep-sea fishes are able to respire at great depths, and they can theoretically occur horizontally and vertically throughout the abyss. Fishes are comparatively large residents and most of them incorporate vertical movement into their life histories.

The deep seas of oceans should not be thought of as huge basins. As mentioned earlier, there is much habitat heterogeneity in the oceans, which includes deep basins, submarine canyons, underwater mountains or ridges, and isolated groups or individual islands that rise all the way to the surface. Due to subsidence and sea level rising there also are sunken islands called seamounts. We know very little about most of the deep-sea habitats, but we are learning about the seamounts because they can support a diversity of fishes.

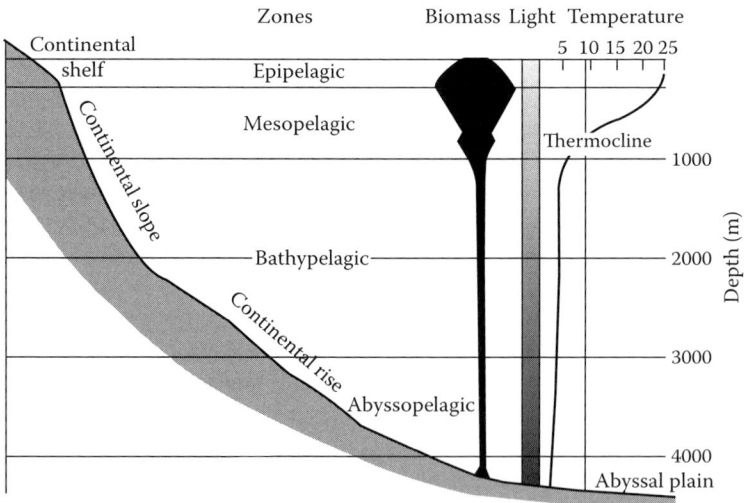

Figure 19.1 Zones of the ocean. (From Bone, Q., and Moore, R. H., *Biology of Fishes. 3rd Edition*, 38, Taylor & Francis Group, New York, 2008. With permission.)

Function

Food abundance in the oceans decreases as the distance from land and the depth increase. Except for microbial chemosynthesis that occurs on the sea bed (by archaea and bacteria), the deep sea is heterotrophic. Fishes of the deep sea consume matter from other organisms, from zooplankton to other fish. Life there is completely dependent on a supply of food in the form of living and dead organisms, fecal pellets, and detritus that sinks from the epipelagic zone (so-called marine snow). Most of this organic matter is consumed at higher levels, especially the mesopelagic zone; thus, food becomes scarce at greater depths.

Some organic matter remains and sinks to the bottom. This matter includes a digestible portion, such as the remains of organisms that are too large for complete consumption by most deep-sea consumers, and a nondigestible portion, such as chitinous exoskeletons, bone, cellulose, and so on. All of this matter will be subjected to slow bacterial decay, resulting in digestible protoplasm that also sinks to the bottom. As this matter accumulates, it continues to decay, and in turn, it supports more organisms than in the overlying deep pelagic zones. Also on the bottom, this decay process results in respiration, which requires oxygen. Thus, the pelagic zone, extending a few meters above the bottom can be reduced in oxygen to the lowest levels in the abyss.

The supply of oxygen in the ocean is regulated by atmospheric oxygen taken in at the water surface, produced as a byproduct of photosynthesis, and consumed by organisms in it by process of respiration. Fortunately, cold water can contain a large amount of oxygen and currents of cold, dense water flow along the bottom, such as the Ocean Conveyor System. These thermohaline currents transport oxygen from the surface water and allow respiration even at great depths.

MESOPELAGIC ZONE

Conditions

As stated previously, the mesopelagic zone lies between the surface and the deep sea (abyss). The zone lies immediately under the epipelagic zone at 200 m, the light compensation point for most phytosynthetic organisms, and extends down to 1000 m, where there is total darkness. About 20% of the food produced in the epipelagic zone drops into the mesopelagic due to gravity. There is little or no photosynthesis in this zone, but there is appreciable respiration as heterotrophic organisms, from bacteria to the largest fish, consume the food produced from above.

This zone is a transition area for several parameters, including light, temperature, oxygen, and pressure (Figure 19.1). Due to absorption and scattering, light is affected both in intensity and quality. Light is reflected at the air–water interface and attenuated exponentially with water depth. As light penetrates the mesopelagic, it gradually fades away into twilight, and at depths of 100 m, only about 1% of the light produced at the surface remains. Water also affects colors of the spectrum, and even in clear water, red is absorbed in the first few meters, yellow at less than 20 m, and orange gone at about 35 m. Blue penetrates more, and about 10% of blue makes it to 100 m. However, in turbid coastal waters, green can penetrate further than blue and results in the greenish cast of inshore waters as compared to the deeper blue of the open ocean (Sumich and Morrissey 2004).

Temperature does not fluctuate as much in the mesopelagic than in the epipelagic, but it experiences the major density thermocline there. A summer temperature of 22°C in the well-mixed surface zone would decline to about 5°C in the lower mesopelagic. Oxygen fluctuates differently: Oxygen levels remain near saturation at the surface due to mixing with the atmosphere and photosynthesis. However, once entering the mesopelagic zone, there is a great deal of respiration but not much photosynthesis. As a consequence, the oxygen minimum layer for the oceans is encountered in the mid-to-lower mesopelagic. Oxygen levels recover at greater depths due to less respiration and mixing of cooler waters below. Due to the very heavy weight of a column of water, pressure increases about one atmosphere (about 15 lb/in.^{2}, or 1 kg/cm^2 for every 10 m of descent; this means that at 1000 m, the pressure would be 1000 kg/cm^2).

Fishes

The mesopelagic zone is characterized by vertical feeding migrations of many organisms, including zooplankton, larger invertebrates (e.g., krill), and fish. Most of the larger organisms migrate upward at twilight to feed in the epipelagic zone at night, and return to the mesopelagic before day. In this manner, they take advantage of the more abundant food above, then retreat into cover—the protective darkness below. These migrating fish provide a sonar pattern that looks like a cloud, and is called the deep scattering layer. This cloud moves up and down daily and is produced by millions of mesopelagic invertebrates (copepods, krill, shrimp, jellyfish, squid, and fishes—mostly lanternfish). Filter-feeding invertebrates mostly remain in the mesopelagic as permanent residents, where they swim in search of organic matter (tissues or organisms, detritus, fecal pellets, etc.). Thus, the mesopelagic fauna is adapted for an active life in low-light conditions, and the fishes predominately feed on zooplankton and other fish. Except for the absence of countershading, fish are similar to those in the epipelagic: they are muscular, streamlined, and have large functional eyes and gills. Almost all fishes there have functional swim bladders.

Fish abundance is relatively high in the mesopelagic, about 750 species, almost all of which are lower teleosts (Bond 1996; Helfman et al. 2009). The most abundant fishes are the bristlemouths (Gonostomatidae) and lanternfishes (Myctophidae) (Figure 19.2), and these closely related families may constitute 90% of the fishes captured in trawls (Castro and Huber 2007). However, hatchet-fishes (Sternoptychidae) (Figure 19.3) also can be abundant, and many other species are common (see Table 18.1). Most fishes, such as hatchetfish, are relatively small and laterally compressed to hide their outline both from above and below. The fish are more uniformly colored and generally lacking in countershading, which would reveal their appearance from the side. However, they conceal their outline as viewed from below and the side with light organs called photophores that are scattered or placed in rows on the ventral and lateral parts of the fish. Viewed from below, the lights match the illumination from above and break up the outlines.

In addition to small fishes, there are some larger fishes that also are resident to the mesopelagic, and these include the piscivorous lancetfishes (Alepisauridae). This fish has a fierce appearance, and in pictures, it looks a little like the epipelagic sailfish. However, I have had the experience of handling a 1.5-m freshly caught fish (Figure 19.4), and it was very flimsy and gelatinous—a morphological adaptation to extreme pressure.

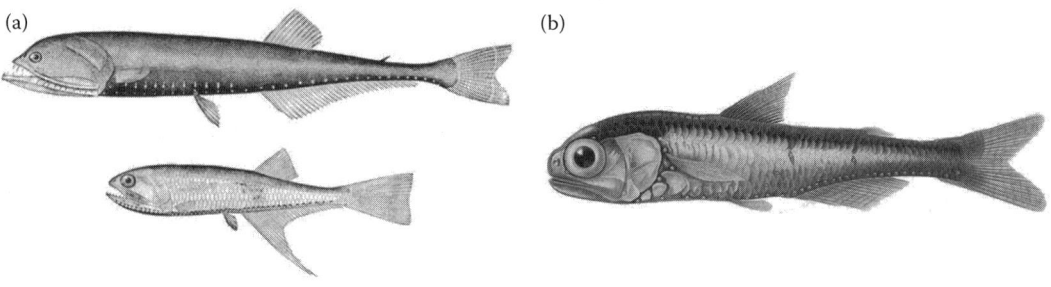

Figure 19.2 Mesopelagic fishes: bristlemouths (a) and lanternfish (b). ((a) Courtesy of Goode, G. B., and Bean, T. H., http://commons.wikimedia/wiki/File:Cyclothone_elongata_and_Bonaparte_pedaliota.png. (b) Courtesy of Kissling, E., http://commons.wikimedia.org/wiki/File:Myctophum_punctatum1.JPG.)

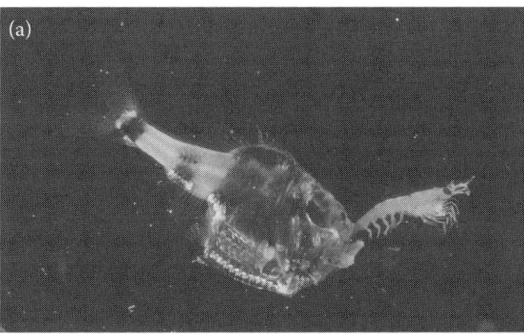 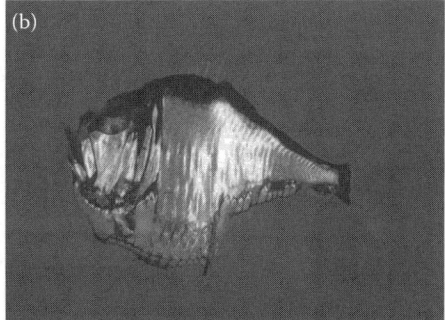

Figure 19.3 Hatchetfish, feeding on crustacean (a) and showing photophores (b). (Courtesy of Costa, F., http://oceanexplorer.noaa.gov/explorations/islands01/logsep23/media/hatchet_fish.html.)

Figure 19.4 P. Pate holds a Lancetfish *Alepisaurus ferox*, caught on *RV Eastward*.

Distribution of mesopelagic fishes in the oceans was first thought to be relatively uniform; however, recent evidence has shown that species composition can be related to food supply. Barnett (1984) studied mesopelagic fish geography in the central tropical and subtropical Pacific Ocean and found that low productive gyral faunas were different than those of upwelling areas. Of 301 fish species collected in the mesopelagic, the two great north/south Pacific gyres shared most fishes between them, but they did not share many species with the equatorial region between them. Presumably, the major factor influencing the distribution and abundance of mesopelagic fishes was food supply.

However, the distribution of mesopelagic fishes between regions of the oceans probably occurs from seasonal transport of larvae. Sassa et al. (2007) found that 31 species of larval mesopelagic fishes (representing 24 genera) were transported by currents into different regions, and accumulated in a transition region of the western North Pacific.

ABYSS

Below the mesopelagic zone, there is a sea of perpetual darkness that is called the abyss. This water has been divided into the bathypelagic zone, which extends from the mesopelagic at 1000 m to the average depth of the ocean floor at 4000 m. Deeper than that is the abyssopelagic, which extends down another 2000 m to a depth of 6000 m into sunken areas of the floor. The hadopelagic extends even more, to the deepest parts of the floor, which would include submarine channels. Although fishes have been observed at depths of about 8000 m (i.e., cusk eels), I emphasize the bathypelagic zone because it has a well-known fish fauna, while lower pelagic zones support few or no fishes. Table 19.1 outlines the changes in conditions from surface to abyss.

Food sources of the deep pelagic zones consist of "marine snow," which is particulate organic matter that is dropping from above. Large dead animals include dead fish or even whales that take a long time to eat and organisms that live in the deep sea. Marine snow consists of settling particles of inorganic or organic substances > 0.5 mm in diameter that sinks at an approximate rate of over 100 to 1000 m/day. Much of this material is covered with a biofilm of bacteria, and small particles are consumed by zooplankton while larger particles such as fecal pellets can be eaten directly by fishes (Gartner et al. 1997).

The bathypelagic zone (1000 to 4000 m) receives about 5% of the food produced from above; thus, food is very scarce. In addition to such scarcity, the conditions that fishes have to deal with are so bad that they are almost inconceivable in synergy: fishes live in the dark, with little food. They must tolerate very high pressure ranging from 1000 to 6000 kg/cm^2. They live in very cold, constant temperature of about 5°C or less, and the oxygen content of the water is low. However, in spite of such conditions, about 200 species are residents (only 3%–5% are perciformids) (Bond 1996; Helfman et al. 2009). Resident fishes include gulper and pelican eels (Eupharyngidae) (Figure 19.5), black swallower (Saccopharyngidae), viperfishes and dragonfishes (Stomiidae), and numerous species of deep-sea anglerfishes (Ceratioidea), among others. How can they survive deep-sea conditions?

Using an anglerfish as a model, we examine bathypelagic adaptations: Bathypelagic fishes are generalists and will eat whatever they encounter that is smaller than they are. We assume this includes living and dead organisms in most cases. So the fish have very large mouths and a gut that

Table 19.1 Surface to Deep-Sea Conditions and Fish Response

Light	Full sunlight at the surface to darkness where bioluminescence provides some illumination. Fish countershaded.
Temperature	Warm at surface (20°C–25°C) to cold at depths (2°C–5°C). Fish select temperature at surface by moving, but not at depths.
Oxygen	Near saturation at the surface and very low in abyss. Fish conserve energy by reducing activity at depths.
Pressure	Weight of water column increases 1 atmosphere (1 kg/cm^2) for each 10 m of depth. At 8000 m the pressure would be 8000 kg/cm^2 or about 12,000 lb/in^2. Fish replace air bladder at depths with fluids.
Food	Almost all food comes from the epipelagic zone. Food and fish abundance low in lower pelagic zones. Fish adapted for consuming large food items.
Volume	Very large volume of deep sea, but very low abundance of fish. Special sensory adaptations enable detection of prey and mates.

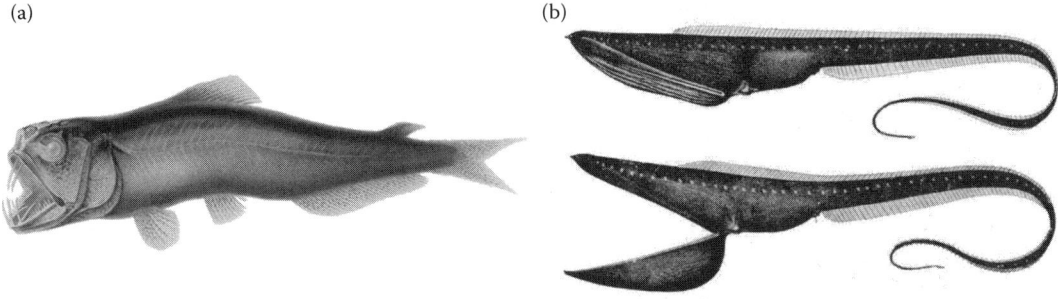

Figure 19.5 Sabertooth (a) and pelican eel (b). ((a) Courtesy of Brauer, A., http://commons.wikimedia/wiki/File:Coccorella_atrata.png. (b) From Goode, G.B. and Bean, T. H., *Oceanic Ichthyology, a treatise on the deep-sea and pelagic fishes of the world, based chiefly upon the collections made by the steamers Blake, Albatross, and Fish Hawk in the Northwestern Atlantic*. Special Bulletin of the U.S. National Museum, 1896.

can expand to accommodate large-size prey. Once they engulf an organism, they do not wish to lose it, hence the large recurved teeth. Anglerfish presumably consume mostly living organisms, so they must be camouflaged. It turns out that this is easy, as anglers are dark colored, mostly black. Due to the need to conserve energy, they do not stalk their prey; they float, wait, and attract it with lures or lights. Also, energy is conserved by floating along and being slowly propelled by the action of expelling water over the small gills. Since there is little need for prolonged swimming, the musculature is reduced. In order to deal with neutral buoyancy under extreme pressure, tissues are very watery (about 95% water), and the swim bladder is lost or rudimentary. Buoyancy is achieved by having reduced bones and protein. Structures that are unneeded are lost or reduced, and there are no scales or pelvic fins and very small eyes. However, bathypelagic fishes have acute sensory organs of the acousticolateralis (head and lateral line) system and either sensitive barbels or tails to detect movement in the dense water. Reproductive life histories in most of the fishes of the abyss include buoyant eggs and epipelagic larvae. Table 19.2 contrasts representative fishes from the epipelagic, mesopelagic and bathypelagic zone.

Table 19.2 Ray Fin Fishes Found in Three Pelagic Zones of the Oceans

Epipelagic Zone	Mesopelagic Zone	Bathypelagic Zone
Anchovies	Bristlemouths	Gulper/pelican eel
Herrings, sardines	Lanternfish	Black swallower
Salmon	Hatchetfish	Cusk eels/brotulas
Tunas and mackerels	Lancetfish	Deep sea anglerfishes
Billfishes	Dragonfish	Sea devils
Sword fish	Spiny eels	Grenadiers/rat tails
Dolphinfish	Snipe eels	Bigscale fishes
Remoras	Sabertooth	Fangtooth
Jacks	Lizard fish	Whalefish
Flying fish	Barracudinas	Sabertooth
Ocean sunfish	Orange roughy	Pelagic cod

Source: Bond, C. E., *Biology of fishes*. 2nd Ed. Saunders College Publishing, New York, 1996. With permission.
Note: Each entry represents one or more fishes in a family unless indicated as more. See Taxonomic Index for names of species and families.

DEEP BENTHIC AND BENTHOPELAGIC

Food in the deep benthic habitat comes from many sources. Very large dead organisms from the epipelagic zone down, everything that is eaten and re-eaten, and organic material that is indigestible for most organisms can be expected to fall to the bottom. Thus, the bottom of the ocean is rich in organic matter compared to the deep pelagic zones. Accordingly, deep bottoms of the sea (>2000 ft) are inhabited by many invertebrate deposit feeders, such as polychaete worms, crustaceans, mollusks, sea cucumbers, and brittle stars. Also, a large number of fish species inhabit the deep bottom or hover over it in search for food.

There are about 1000 fish species that are included in the benthic zone, of which about 13% are perciformids (data from Helfman et al. 2009). This higher number of perciformids may be due to the greater abundance of food than in the pelagic zones above and also is related to the ability to invade the bottom by stages. Fishes of the benthic zone are morphologically similar to those of the mesopelagic zone than the bathypelagic due to the greater activity that is made possible by a more abundant and dependable food supply. The fishes are relatively large (20–50 cm), streamlined, muscular animals that have well-developed gills, functional eyes (except for hagfishes), no countershading, and do not utilize bioluminescence very much. These fish also usually have sensory organs of touch, odor, and smell to detect prey, such as barbels, feelers, or long flexible tails. Grenadiers (*Coryphaenoides*) (Figure 19.6) are an excellent example of benthopelagic fishes. They live an

Figure 19.6 Deep benthic fishes: grenadiers. (Courtesy of the National Oceanic and Atmospheric Administration (NOAA), http://www.afsc.noaa.gov/abl/MESA/mesa_sa_gren.php.)

active life, aggressively searching for food and fast in finding it, as shown in studies using television cameras rigged at 4000–6000 m (reviewed by Hart 1997). Representative fishes are the tripod fish (Ipnopidae), cusk eels (Ophidiidae), the cod-related grenadiers (Macrouridae), and hagfishes. In addition, there are a few sharks that can be expected to show up at depths of about 2000 ft, and soon compete with the hagfish to make quick work of anything very large.

One type of deep-sea benthic community receives its energy from hydrothermal (hot water) venting on the seafloor. The first vent community was discovered by physical scientists in the famous deep-diving submersible Alvin in 1977, and it has received much interest. In brief, seawater seeps into cracks and fissures in the Earth's crust and is heated by magma to a high temperature. Because there is so much pressure, the superheated water remains a liquid, and temperatures greater than 300°C have been recorded (with special thermometers!) (Thurman and Trujillo 2002).

As the hot water is forced to the surface, it brings with it dissolved minerals such as zinc sulfate (such vents are called white smokers) and copper sulfides (black smokers). Chemosynthetic archaea and bacteria then oxidize the sulfur compounds to produce energy and to fix carbon into organic matter. The organisms multiply and are filtered from the water or eaten as biofilm by several larger organisms, including clams, mussels, crabs, and worms, of which most forms are new species. These invertebrates also consume other organic matter which may sink from above. Depending on the depth, benthopelagic fishes, such as the cusk eels previously mentioned, also may be found. The vents are oases of life on an otherwise barren seafloor; however, they are temporary and get wiped out by volcanic activity.

Another type of bottom community also has been discovered, hypersaline seep biocommunities. The seep water is not heated and microbial growth occurs on growth mats. It has been suggested that the larvae of sea vent communities spread with deep ocean currents to colonize new vents. A plausible explanation was provided by a 1987 dive with the famous submersible Alvin, in which vent organisms were found feeding on sulfur from decomposing whale bones (Nybakkan 1997). It is presumed that the larvae are aided by food in the form of large dead creatures that have fallen from above. This is known as the "dead whale hypothesis."

SEAMOUNTS

There are features of the deep oceans that occur across the entire pelagic realm. These include seamounts that rise from the ocean floor. Seamounts are underwater volcanic islands, found at various depths from tens of meters below the surface to thousands of meters down. These features have either sunken or been flooded by sea level rise. Large flat-topped seamounts called guyots (after geologist Arnold Guyot; Idyll 1971) are drowned islands that have subsided, and they are common in some locations. Few seamounts have been studied; however, these slopes can provide habitat for invertebrates and fishes.

Located in the open sea the substrate provided by seamount differs from the ocean floor, because water movement around the submerged island sweeps away fine particles and provides permanent residence for epifauna. In this case it can be compared to a coral reef, because habitat is provided and a supply of food also would be born to residents by the currents (Norse et al. 2005). Also, ocean currents striking the seamounts can create eddy currents and upwellings, bringing nutrients to otherwise depleted locations and supporting a great diversity of organisms. As an example, Castro and Huber (2007) reported that 24 seamounts in the southwest Pacific Ocean supported over 850 species of which one-third were undescribed. Because there are about 30,000 seamounts known and thousands more that may be discovered, biodiversity produced by their presence could be extremely valuable from an ecological point of view.

Seamounts also appear to be important as part of life histories of several migrating species. They have been identified near spawning sites of the Japanese eel in the West Mariana Ridge

(Tsukamoto et al. 2004) and seamounts may be involved with spawning of other eels as well. Large migratory epipelagic fishes such as tuna use seamount as stopping places, and sharks, large sea birds, and other organisms stop to feed in the sea around them. Unfortunately, the fishing industry, pioneered by the former Soviet Union has been trawling them with little or no oversight, disrupting systems and destroying habitats (e.g., Norse et al. 2005; Watling 2005; Roberts 2007). Seamounts are features of the open oceans and in international water. Thus, there are no laws to protect them and no conservation programs.

FISH ADAPTATIONS IN THE DEEP SEA

In General

Fish adaptations to upper pelagic and benthic conditions discussed so far are contracted in Table 19.3. More in-depth coverage of special adaptations follows.

Bioluminescence

There is light in the sea, from surface to great depths, and even at night. Anyone who has looked over the stern of a vessel at sea during the night has seen the greenish glow of light that people refer to as "phosphorescence." This light is actually produced in countless bodies of marine organisms, mostly dinoflagellates. Bioluminescence occurs widely in nature (i.e., in five terrestrial and 14 marine phyla, including over 700 genera), but the morphological diversity in bioluminescence is higher in fishes than any other group—and confined entirely to marine habitats (Suntsov et al. 2008). Many other species also are capable of producing light, and this phenomenon has been especially well studied in insects and squid. Light producing organs are collectively called photo-

Table 19.3 Conditions and Fish Adaptations: Upper Pelagic and Benthic Zones of the Ocean Province

Pelagic Realm

Epipelagic zone—Surface to 200 m. Warm, mixed, sunlight allows photosynthesis. Most fishes occur near continental shelves or in upwelling areas. Offshore oceanic is nutrient poor and supports only about 2% of fishes. Resident fishes are adapted for active life under lighted conditions. Food web based on phytoplankton produced in this zone. Fishes are visual predators; most occur in schools and prey on zooplankton and fish. Streamlined bodies and countershading predominate among fishes. Horizontal migration is typical. Nonresident fishes access this zone on a diel basis at night. Cover when present is important.

Mesopelagic zone—Depths 200 to 1000 m. Twilight zone. Light fades to extinction here. Thermocline and oxygen minimum zone occur. Little water mixing, low nutrients, no photosynthesis; pressure increases while temperature and oxygen decrease. Resident fishes are adapted to an active life, but under low light conditions. Most fish are predators with large eyes and camouflaged by photophores. Vertical migration to the epipelagic zone for food is common.

Bathypelagic zone—Depths 1000 to 4000 m. Water permanently stratified, perpetual darkness, low temperature and oxygen. Great pressure. Food is scarce because it has to sink into this zone from above. Fishes are "float and wait" predators and lead a sedentary existence with no sunlight and little food. Muscles poorly developed, mostly white. Organs and other structures reduced or absent. Photophores common but used mostly as a lure. Fish have small body size, large mouths, and occur in low density.

Benthic Realm

Deepwater benthic zone—This is the bottom of the ocean and exists below the continental shelf break. Depths decline to the abyss. Conditions are cold and dark, and pressure is great. Food is more abundant as it accumulates from the layers above. Some food can be big in the form of large dead animals. Some indigestible food is converted by bacteria, which then provide food for invertebrates. Fish adaptations differ from the above layer: fishes have more food and are more similar to fish of the mesopelagic. Large species are common.

phores, but there are two different mechanisms involved to produce light, intrinsic light produced by photocytes, and light produced by bacteria that are retained in special organs.

A fish may produce intrinsic light by sending a nerve impulse to a photocyte (cell) or group of cells (i.e., a photophore), located in the epithelium. The photocytes then release the enzyme luciferase into a chamber of luciferin (a light emitting pigment), which then is supplied with oxygen via arterial blood (Idyll 1971), completing a reaction that emits light and byproducts (such as carbon dioxide). This is essentially the same reaction that occurs in fireflies, but the composition of luciferin differs somewhat (Nybakken 1997). The color of light produced can be changed according to the kind of filter that is made by the overlying skin. It is these photophores that are used to line the sides of mesopelagic fishes, and to produce bright lights in some fish. Also, there can be different colors and locations that can be used as the fish determines they are necessary.

Bacterial photophores are special chambers in which symbiotic luminous bacteria are cultured by the fish. The chambers are initially open to the gut or sea and are invaded by different species of *Photobacterium* in the larval stage of the fish. Apparently these bacteria may be passed on from generation to generation because different species of anglerfish may have different species of bacteria (Bone et al. 1995). In practice, the fish provides a secretion that nourishes the bacteria and induces light formation. Apparently there are limitations on the use of bacteria, and fishes with such light organs usually have only one, and not more than four of them; perhaps this is due to cost of maintenance. These photophores are most commonly used as a lure.

Light organs are most abundant in deep-sea mesopelagic fishes and developed to perfection. Colors can be white, red, bluegreen, yellow, purple, and etc. Locations of photophores can be along the sides as "portholes" along the flanks, on the underside, on the tail, in the lure (esca), under the eyes, on chin barbels, internally, etc. In the case of the gulper, or pelican eel, its long whiplike tail has a red light on the end that may be used to attract a prey item that is close, but the attenuation of red color in water is so fast that predators at a distance would not be able to see it. Lights can vary in intensity, or be released in a pattern of flashes by intervals. They can be internal to produce a soft glow, or external and filtered for color, intensity controlled in various ways (e.g., to match ambient light from above), and probably controlled in ways I cannot imagine.

In summary, bioillumination can be used as a trap to lure fishes into the mouth, as a lure to attract them close, to see by as bright search lights, as signals for species recognition, in schooling, as a defense to distract or blind an attacker, as camouflage to break up an outline from below or the side, and I am sure there are other uses as well. Interesting stories of fishes that have been observed using light organs in various ways are provided in a chapter by Idyll (1971).

Buoyancy

Fishes of the epipelagic zone, with the exception of pursuit predators have morphological projections, supply of lipids, or gas-filled swim bladder to aid in attaining neutral buoyancy. Deep-sea fishes attain buoyancy by different means.

Below 1000 m, gas-filled swim bladders are uncommon, nor does hydrodynamic lift appear to be very useful. Instead, deep-sea fishes have reached a state of neutral buoyancy by taking on more water in the tissues, thus reducing tissue density. Even muscles become watery and some fishes are covered with a gelatinous layer. At depths, those few fishes that have retained the swim bladder have likely replaced gas with fats (Pelster 1997).

Sensory

Long-time evolution in the deep-sea environments has produced exceptional sensory physiology in the fishes, which includes ". . . function-specific olfactory and vision systems, elaborate tactile appendages, membranous lateral line canals, and stalked superficial neuromasts" (Montgomery and

Pankhurst 1997). Deep-sea fishes have more completely developed their sensory systems than have fishes from any other habitat.

DEEPWATER FISHERIES

We tend to think of deep-sea fishes as isolated in their own little worlds, away from the grasp of humans. This is changing. Modern technology has made it possible to commercially exploit fish at extreme depths. Two deep-sea fishes (Figure 19.7) that have been commercially exploited for human food are of interest here. One of these, the Patagonian toothfish (Nototheniidae), which is marketed as "Chilean sea bass," has been trawled and long-lined at depths of 400–1500 m. Another fish that has been exploited is the "orange roughy" (Trachichthyidae), which could be any of several species of fishes known as sawbelly, or slimeheads. Fishing was initially done in the late 1970s with deep trawls at 700 to 1000 m. This fish is a slow grower, matures after age 20, and live more than 100 years. Such fishes are obviously not suited to very much exploitation and their habitats are presumably fragile. Because fisheries target larger individuals in a species, they remove the older, most fecund individuals first, and in late maturing species this practice cannot be sustained (e.g., Heppell et al. 2005). As might be expected, wherever theses fishes are heavily fished, their populations have suffered severe declines and not recovered (Helfman 2007).

CASE STUDY: DEEP-SEA ANGLERFISH

We have been studying the constraints imposed upon fishes by different aquatic systems. Here we are going to look at fish adaptation to life under the environmental conditions of the deep sea. Our champion is a small family of deep-sea anglerfishes that have captivated the interest and imagination of scientists for over a century (reviewed by Pietsch 1976). The first report of these interesting creatures was recorded in Iceland by the Danish Navy in the 1830s (Idyll 1971).

These deep-sea fishes are members of the family Ceratiidae, commonly called sea devils (Figure 19.8; Table 19.4), and they are related to 17 other anglerfish families (order Lophiiformes), which include pelagic and benthic fishes that also angle with lures to attract prey. More commonly encountered fishes in the order are neritic frogfishes (Antennariidae), which are benthopelagic or pelagic predators, and goosefishes (Lophiidae), which are benthic, lie in wait predators (Nelson 2006). Having encountered a very large and alive goosefish close up, I can personally attest to their ferocious appearance, which is due in part to an impressive business end: a very large flattened head and cavernous mouth equipped with numerous long sharp teeth. But the more recently evolved deep-sea anglerfishes have adaptations that are even more interesting than their distant kin. Here

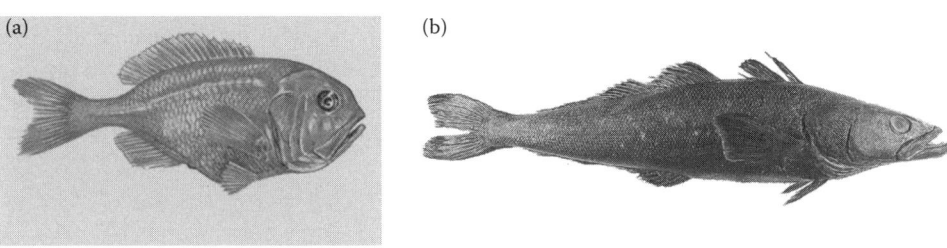

Figure 19.7 Orange roughy *Hoplostethus atlanticus* (slimehead; up to 60 cm) (a) and Chilean sea bass (Patagonian toothfish) (b). ((a) Courtesy of Cada, R., http://commons.wikimedia/wiki/File:Orange_roughy.png. (b) Courtesy of the USFDA, http://commons.wikimedia/wiki/File:Toothfisn.jpg.)

Figure 19.8 Black sea devil anglerfish. (Courtesy of Brauer, A., http://commons.wikimedia/wiki/File: Humpback_anglerfish.png.)

we will describe adaptations that have successfully enabled these creatures to survive in one of the most extreme environments on Earth.

Ceratoid deep-sea anglerfishes are mostly small, dark little fish, usually 10 cm or more in length, but at least one species can attain sizes of up to 1.2 m (Nelson 2006). Small anglerfishes had been dredged or trawled up from great depths, but are usually in poor condition when examined. But by 1922, it was noticed that some small fish were attached to large ones, and it was presumed that these were young attached to the mother. However, it was one of the large anglerfish, captured in Iceland and brought to the British Museum in 1925 that aided ichthyologists in detecting the truth. Ted Regan, Museum Director, realized that the tiny fish found attached to a deep-sea anglerfish was a parasitic male, and confirmed this by finding more attached males in the museum collection (Idyll 1971) (Figure 19.9). Once the male was described then it aided other workers to detect them from samples taken in surface waters. How did an adult fish that lives in the bathypelagic produce young in the epipelagic, and why?

Deep-sea anglerfish eggs are produced and fertilized by the adults in the deep mesopelagic or bathypelagic zone, perhaps at depths of 1000 to 2000 m or so. The eggs are buoyant and float up to the surface. Upon hatching the larvae are encased with a gelatinous envelope under the skin, presumably to aid in flotation (Idyll 1971) and perhaps the increase in size also helps to reduce the number of predators that can eat them.

Table 19.4 Deep-Sea Anglerfish

Life Cycle
Adults live and reproduce in bathypelagic zone (ca. 1600 m).
Buoyant eggs float to epipelagic zone and hatch.
Larvae feed on zooplankton.
Small male matures early; larger female takes longer.
Female develops esca (lure), illicium (pole), light, white muscle. Male has protruding teeth, red muscle, large tubular eyes.
New adults descend to depths, mature females release pheromone. Male has large olfactory lobes, follows scent, light, to female.
Male attaches to female (or dies), some organs lost, large testes.
Adaptive Benefits
Epipelagic larvae: rapid growth—food, light, warmth, oxygen.
Bathypelagic adults: low predation/competition.
Small parasitic male: more energy to make progeny.
Flabby body, little bone, oil filled bladder: buoyant/pressure.

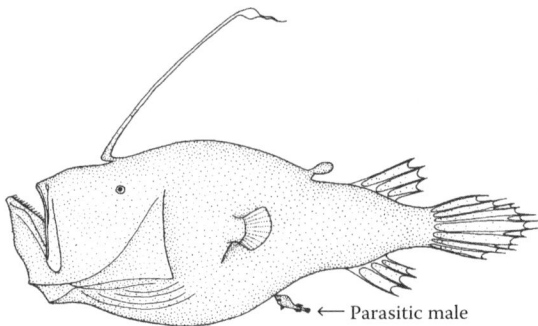

Figure 19.9 Deepwater anglerfish showing an attached parasitic male. (Courtesy of NOAA, http://commons.wikimedia/wiki/File:Cryptopsaras_couesii (triplewart_seadevil).png.)

The young appear identical at first, and feed upon copepods and other plankton. In about two months, gradual changes begin to appear as sexual dimorphism occurs. The females begin to sprout an illicium (fishing line) and develop a rounded, rotund appearance. Males begin to grow more slender and develop a large head equipped with curious denticles analogous with "buck teeth." The males mature at a smaller size than the females and after a few months begin a descent to greater depths. Females continue to grow until they have reached maturity and have developed a larger size, an esca (lure) with luminous bacteria and other features, when they also begin to descend.

By the time the male reaches the bathypelagic zone he is prepared to locate a female. This can be a difficult task, for the relative abundance of a female is about one female for every 800,000 m^3 of water (Helfman et al. 2009). Accordingly, the male has developed sensitive tubular eyes for enhanced vision, a large olfactory organ, and has red muscle for prolonged swimming. His life depends upon locating a female within a few months or, since he now cannot eat with protruding teeth, he will die. His search for her is intense. Presumably he searches for her light, and smells his way along pheromone fields laid out by her (reviewed by Montgomery and Pankhurst 1997). Evidently pheromone fields in deep-sea fishes exist in a pancake shape due to density stratification at great depths, where pancakes only 1 m thick can extend 1000 m horizontally (Jumper and Baird 1991). If he is lucky, he finds her and attaches to her flesh in one of several locations (e.g., belly, side, head, opercle). His digestive system, red muscle, and other organs begin to disappear as they are incorporated into the female's body, and he is fused with her as a lifetime mate. Although he is a parasite, his ration is a small one for such important contributions to her fitness. Relative size varies among species, but his weight is only a tiny fraction of hers and he requires very little food.

The young females that survive in the epipelagic finally mature and descend to the deeper zones to take up an adult life of a "float-and-wait" predator. The female develops an exceedingly large mouth for one so small and a fully functional esca. In addition, her flesh is now almost entirely white muscle, bone has largely been reabsorbed, and somatic tissues are about 95% water. The great pressure at depths will provide sufficient rigidity for her body. Other sensory organs develop in the lure, chin barbel, and on the body to detect the slightest motion in the dense water. At this point she is also capable of producing her own pheromone trail, and she begins advertising for her own mate. She moves along very slowly, aided by water that is pumped from her small gill opening. Her sense organs are fully functional and she is an efficient predator.

This reproductive life history reflects perfectly tuned adaptations to the adult life, which is a life of almost complete isolation in a low ration but also low predator environment. There is a need for epipelagic larvae because small food is obviously limited in the bathypelagic, and even if food was available the small larvae might take many months to a year to grow under such conditions of cold temperature. During that time they might all become prey to hungry mouths of many small predators. On the other hand, the epipelagic zone contains hosts of small organisms and even with predation pressure there the larvae swollen in its gelatinous capsule, is large enough to avoid the smaller predators. There is a great production of small zooplankton to be consumed up there, and growth can be fast in the warm water.

In the bathypelagic the female anglerfish grows to a size that food resources allow. She is a relatively large fish in that zone and equipped with highly efficient skills for attack and defense. Thus, she benefits by existing where there are low levels of other predators that are large and skillful enough to consume her.

Deep-sea anglerfish are not abundant and there is a great need to maximize fitness. This is done by conserving energy for use in producing a maximum number of eggs. In this case energy is allocated to eggs by converting somatic to reproductive tissue. This results in a relatively large size in females to escape predation and for egg production. This is where the parasitic male tactic is adaptive. There is no need for a powerful male to defend the young or guard the eggs. There is a need to have a ready supply of sperm available and to have as much energy as possible given to the female. Upon attaching the male is actually feeding the female, and reduced to a parasite, he requires only a small portion of her blood to sustain him.

ECOLOGICAL CONCEPTS

- *Deep Sea as a Desert*—This is not true. In fact, given the conditions, the deep-sea community is diverse. This is certainly true for fishes, which are highly diverse in the number of species, even though the abundance of any one species is usually low. Equally intriguing is the composition of the fauna, which is almost entirely dominated by the ancestral, lower teleosts. This diversity can be explained by somewhat conflicting hypotheses of Sanders (1968) and Dayton and Hessler (1972), which were reviewed by Nybakken (1997).
- *Stability–Time Hypothesis*—We have already discussed this hypothesis with respect to food rich estuaries and low resident species diversity. Here we reverse the direction and associate high deep-sea diversity under hostile conditions with the long-term stability (at least to Miocene in the Pacific gyres) (Barnett 1984). Persistence and high specialization among ancestral fishes seems to support this hypothesis. It is informative to contrast the food rich estuary with the low food abyss in making a comparison here. Food seems to be exploited due to specialization, but there is not enough for great individual species abundance. This idea is very palatable.
- *Food Cropper Concept*—Deep-sea fishes are certainly not crowded; thus, we do not accept that they are limited by competition for space. If competition exists then, it most certainly would be for food under the low food conditions of the abyss. This hypothesis, attributed to Dayton and Hessler 1972, suggests that all of the organisms of the deep sea are "croppers," consuming anything that is small enough, and assumes that there are no specialists. This would result in everyone eating everyone else with no competition. Unfortunately, there are great differences in morphology and behavior among fishes, including the deep-sea ones. We need to think about this one.
- *Dead Whale Hypothesis*—This is a way such as associated with vent explain how seafloor communities such as associated with vents, spread to new areas. In this case, organisms too large to be consumed on the way down aid larvae by providing food for them.
- *Millions of Eggs Hypothesis*—Okay, so this really hasn't been proposed as a formal scientific hypothesis, but it has been a working hypothesis of commercial fisheries for a century. Myers and Ottensmeyer (2005) coined this hypothesis, which can be restated as *high reproductive potential of marine fishes guards against extinction*. It has been assumed for a long time that the high reproductive potential and the great number of marine fishes will result in fast recovery from overfishing. Unfortunately these authors point out that even without fishing there is a maximum of about four to five progenies that survive from a spawn each year. Not only that, but fishing tends to crop the largest, more fecund individuals of a population, and fishing impacts on habitat and other species are not factored in, nor do modern fisheries leave any refuge areas. We will be happy to join with Myers and Ottensmeyer (2005) and trash this erroneous idea.

SUMMARY

The oceanic province includes the warm, illuminated, oxygen rich epipelagic zone, the deep transitional mesopelagic zone, and a deeper region of cold, low-oxygen water that is permanently stratified and cloaked in perpetual darkness, called the deep sea or abyss. Phytoplankton production is limited to the epipelagic zone, where it is consumed by zooplankton, larger invertebrates and fishes. Many organisms in the mesopelagic zone share in this wealth by vertical migration. Organisms that live at depths feed on organic matter that sinks from above. The deep sea is adverse, but relatively unchanged for perhaps a hundred million years. The fish fauna is mostly ancestral taxa that adapted and specialized. In the mesopelagic, 95% of the fish families (21) are lower teleosts, and 92% of representative bathypelagic fishes (12 families) are from lower groups. Fishes of the mesopelagic are adapted for an active life in low-light conditions—they are muscular, streamlined, and have large eyes. Bathypelagic and deeper pelagic fishes are generally smaller, have reduced musculature, watery tissues, tiny eyes, and well developed sensory systems. Fish in the benthic zone are relatively large and have adaptations more like mesopelagic species due to increased abundance of food. Bioluminescence is common in the deep sea and used by fishes as camouflage, species recognition, and as a lure. A case study of deep-sea anglerfishes is presented as an example of deep-sea adaptation. Adults live in the bathypelagic and produce buoyant fertilized eggs that float up to the epipelagic and hatch. The young feed on plankton, grow, and exhibit sexual dimorphism. The males mature at a smaller size than females and descend to the bathypelagic zone. They must locate a female within a few months or die. If a male is lucky, he finds her and attaches to her flesh. His digestive system, red muscle, and other organs begin to disappear as it is incorporated into the female's body, and he is fused with her as a lifetime mate. Upon attaching, the male is feeding the female, and reduced to a tiny parasite. He requires a small portion of her blood and supplies a lifetime supply of gametes.

Recommended reading: Randall and Farrell 1997 and Allsopp et al. 2007.

PART VI

Fish Adaptation

CHAPTER 20

Fitness, Morphology, and Ecophysiology

ADAPTATION AND FITNESS

In Parts IV and V, we considered freshwater, estuarine, and marine ecosystems as habitat for fishes, evaluating the characteristics of each system for potential benefits or constraints imposed on fishes that live in them. In Part VI, we will examine the inherent capacity of the fish to respond to those characteristics and to environmental changes. Specifically, we wish to understand how fish morphology, physiology, and behavior are adaptive to abiotic and biotic factors of the environment that affect distribution, abundance, and ultimately species persistence. We will focus on adaptations of the individual, but we also will consider certain attributes of populations.

In human perception, the ultimate goal of a species should be to persist for as long as possible. If it becomes extinct, hopefully, it will be due to pseudoextinction, so that its genes play a part in the evolution of future generations. As suggested by the Red Queen hypothesis, we presume that species are not static; rather, they are changing over time in order to persist (like the Red Queen, running just to keep in place). Individuals provide the genetic diversity to allow this to happen. Individuals must be well adapted to environmental conditions, perhaps by good fortune (e.g., by the persistence of rare genomes) for future conditions. Good fortune for a species is certainly enhanced by having an abundant and widely dispersed species population!

In our perception of life, the role of the individual is to maximize lifetime reproductive success, which is *fitness*. Adaptation must favor survival, growth, and reproduction if fitness is to be maximized. Herein is the problem: individuals can try to maximize fitness by expending energy in response to environmental pressures, but they will not expend energy on "super" adaptations that may never be needed. Most environmental factors are not static, so individuals will be able to respond to a range of conditions, usually over time. However, individuals cannot be adapted for every eventuality, such as a catastrophic event, and environmental factors can produce cumulative impacts and synergistic effects. Here is where distribution comes in focus, for widely distributed species have the best chance of avoiding local catastrophes.

One can assume that an individual is adapted to a general set of conditions but also may be confronted with some additional change in the usual factors. The response of an individual to such a change should depend on the severity of the change and its persistence. We have seen in estuaries that fishes tend to deal with changing seasonal conditions by movement into the estuary when it is favorable (profitable) to do so and out of them when it is not favorable. In this case, we have movement as a general adaptation to mitigate a very wide range of environmental perturbations that specific organs may not. Thus, local or short-term changes can be dealt with by a behavioral response, usually movement. If pervasive, then a physiological response (e.g., acclimation) might occur, and if persistent, then a morphological or life history change could occur. Finally, if the change continues for a very long time, then a genetic change can occur in the population due to natural selection, resulting in a change in gene frequencies.

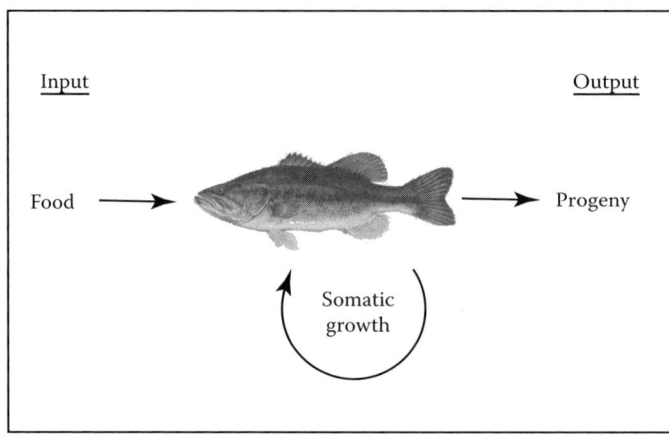

Figure 20.1 Using the fish as an input–output system for allocating energy to fitness.

As pointed out by Wootton (1990), the fish can be viewed as an input–output model (Figure 20.1) in which decisions have to be made regarding the allocation of energy (food) for various life history needs over time. From a fitness perspective, the fish receives inputs of food over its lifetime with the goal of converting that energy into the maximum number of progeny that will also survive to reproduce. To achieve this goal, the fish has a somatic framework (whole body less the germ cells), a life support system for the future progeny that may persist and reproduce over years of varying (i.e., good and bad) environmental conditions. By living long enough or expending a large amount of energy in the reproductive effort, it is hoped that a sufficient number of progeny will survive to ensure genetic participation in the future. First we discuss that framework here, and later we will consider how it operates. To encapsulate this discussion:

- The fish has to develop, maintain, and protect the framework (somatic body that transforms food into progeny) from larva to reproductive adult, a process that requires energy and time.
- The reproductive adult must survive as long as possible to ensure that the maximum amount of energy is transferred to the progeny, whose genetic complement will then add to the species gene pool.
- No organism can have specific adaptations for every possible environmental constraint. However, it can develop some general adaptations that can be put to work when needed to avoid unpredictable adversities. The most important of these is locomotion.

FISH MORPHOLOGY AND ECOPHYSIOLOGY

As we have studied aquatic systems, we have explored the wide diversity of habitats occupied by fishes and found that they are exposed to great differences in abiotic and biotic factors. In studying fish diversity, we have noted differences in morphology (structure) that are associated with life in various habitats. Now we also consider ecophysiology—how fish change in function; that is, the organism responds in physiology to the demands imposed by nature.

Integrated Fish Response

The individual fish is an integrated system of organs that support movement, foraging, digestion, metabolism, reproduction, and so on. We will consider the individual fish as being composed of a framework, or platform, which contains and assists the functions of various organs. We then will consider specific needs and address pertinent organs separately.

Skeletal and muscular portions of a fish with the external structures and coatings (the somatic tissues) can be considered as protecting and transporting the life support, sensory, and reproductive organs. It also serves to assist organ functions in various ways, as we have discussed in our treatment of swimming as multitasking in sharks (chapter 16). In an ecological context, the total integrated body is used to aid the fish in dealing directly with some physical environmental factors such as temperature regulation and salt balance, and biological threats such as predation. However, first we focus on its primary function: Locomotion, the ability to move from place to place in search of food or better conditions. Movement is absolutely essential for fish survival and is generally the preferred response to any type of environmental change, including predation and physical environmental stress. Movement is necessary to obtain food for supplying the body with energy for metabolism, for avoiding adverse local conditions, for regulating temperature, for avoiding predators, and for occupying spatially separated habitats needed for completing the life cycle (e.g., trout need four types of habitat; chapter 10).

We can learn much from the external anatomy of a fish. Two things stand out: the general body shape, and the number and position of the fins. Both of these features are adaptations to life in the dense and viscous medium that we know as water. However, shape and fins also are related to behavior. Behavior almost always is adaptive and reflective of the "profession" of the organism. Hence, the shape of the body, position and function of the fins, and other external things like the size of the eyes can tell us a great deal about the organism.

Locomotion, Shape, and Function

Fishes propel themselves through water in two very different ways, by use of the median and paired fins (MPF) and by undulation of the body and/or caudal fins (BCF) (Webb 1997). Different ways that fish swim are illustrated in Figure 20.2, which divide the two gaits just given into lift-based swimming and resistance based swimming, which then can be attributed to undulation of the fins (e.g., bowfin) or a more rapid oscillation of fins (e.g., tunas). Some fishes, such as tuna are pursuit predators. Their bodies are shaped for speed and have a torpedo-like shape. Of these fish, some swim by undulation of the body (e.g., from eels to salmon) and others swim by oscillation of the caudal fins (e.g., tuna). In case of the tuna and billfishes, speed is so important the fish has grooves (Müller 2008) to hide the dorsal fin and depressions into which the pectoral fins and other projections fit into. In these fishes oscillations of the powerful caudal fin has taken over most if not all of the effort! But great speed is not very helpful for foraging on plankton or invertebrates and many fishes are not shaped for speed. Some are laterally compressed (sunfishes) and others are flattened or dorsoventrally compressed (goosefish or rays). Using morphology we can decide that the fast swimming torpedo-form fishes likely are piscivores, the laterally compressed fish are made for quick turns in some type of cover (in bluegill sunfish, likely they lurk in vegetation to feed on insects), and the flattened ones are likely benthic hunters that need lift more than speed (such as rays). However, fishes can have a broad range of body shapes, such as trout, which are torpedo shaped in outline, but somewhat laterally compressed in cross section. Then there are pike, which also are torpedo shaped, but as ambush predators don't seem to swim very much, usually lurking around cover. In some cases, fish swim slowly along using only undulation of fins, such as skates and rays, or by fin oscillation, such as boxfish and puffers.

Fin position is another bit of morphology to consider in evaluating fish adaptation, and in the case of the pike mentioned above, the unpaired (median) fins are positioned far to the posterior (toward to caudal fin) to enable a powerful thrust forward. Its elongated shape and the location of fins give the profession of fishes such as these away. They are ambush (lie-and-wait) predators.

From the preceding, we find that the shapes of fishes can be related to physical features of the environment and feeding. But the shapes of fishes also can be related to other factors as well and it

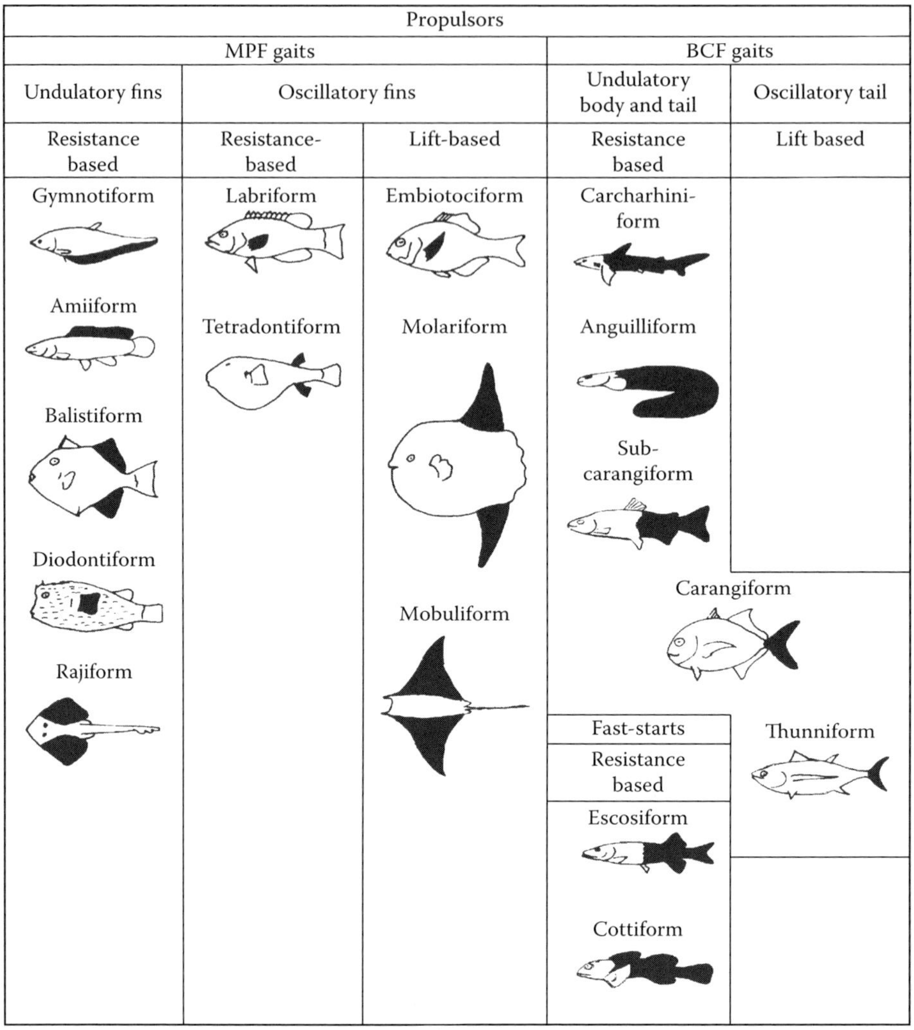

Propulsors				
MPF gaits			BCF gaits	
Undulatory fins	Oscillatory fins		Undulatory body and tail	Oscillatory tail
Resistance based	Resistance-based	Lift-based	Resistance based	Lift based
Gymnotiform	Labriform	Embiotociform	Carcharhiniform	
Amiiform	Tetradontiform	Molariform	Anguilliform	
Balistiform			Sub-carangiform	
Diodontiform			Carangiform	
Rajiform		Mobuliform	Fast-starts	Thunniform
			Resistance based	
			Escosiform	
			Cottiform	

Figure 20.2 Fish propulsion in water. (From Webb, P. W., *The Physiology of Fishes*. 2nd Ed., 5, CRC Press, Boca Raton, FL, 1997. With permission.)

would be a mistake to think all of the shapes we observe are due to movement through water. As an example, the fast growth of juveniles of many species is part of a survival strategy to become too large for others to eat. Of principal interest here is the concept of gape limitation (i.e., the limits on food size imposed by the size of the mouth opening), especially present in some taxa, such as minnows. A fish that grows too big to be eaten is comparatively safe in a gape-limited environment, and it may live to spawn many times, all other things being equal. Fishes use spines for this purpose and they also can have distorted shapes to avoid being eaten. This seems to be the case in some Colorado River fishes that have grown large humps (Portz and Tyus 2004). We will save this topic for a case study on predation when we will go into more detail.

We know from studying fish diversity that the location, shape, and size of fins have changed in evolution. The remora has an attachment organ made from a dorsal fin and the goby has one made from the united pelvic fins. Sharks have maintained the massive pectoral fins as a hydrodynamic lift for their heavy bodies. Changes in the fin sizes and location in higher teleost fishes are often associated with evolution of the swim bladder as a hydrostatic organ, which explains the diminution of the pelvic fins, their forward migration and use for other tasks, and, in extreme environment, their complete absence. Such changes also allowed teleost radiation into many niches.

In this chapter, we are interested in movement as a response to environmental factors. Most of this movement addresses two important environmental variables that limit growth, reproduction, and fitness: differences in the temporal and spatial distribution of food, and in the changes in temperature. Fish move to obtain food, and they also move to seek the best temperatures for processing food and to regulate metabolism. They also move to seek the protection of cover and thus to reduce predation. In the three-dimensional environment that they occupy, this movement is both horizontal and vertical. As we continue to study the responses of individuals and life histories of populations, we will find that movement plays a part in almost every aspect of the ecology of fishes. Fish move!

PHYSICOCHEMICAL ADAPTATION WITH ORGANS

Fishes, as well as humans, have special organs that deal with basic survival. For an aquatic organism the most important ones take part in obtaining food and oxygen, distributing oxygen and food to cells, disposing of wastes, regulating water and salt balance, maintaining heat for muscles, regulating the swim bladder (or lung), and the important function of avoiding predators. Most of these organs are familiar to us because the fish developed them and passed them along in evolution (Figure 20.3). Now we will discuss response of some important organs that deal with physicochemical factors in the environment, specifically how fishes deal with four important factors: temperature, oxygen, water density, and the salt content of water. In the following chapters, we will look at how fish respond to various stimuli as an integrated whole, and we will include other adaptations (e.g., for foraging and reproduction) as well. From an ecological point of view, the complexity and function of body organs are needed to deal with environmental variables, and the organs require a considerable amount of energy to operate.

Some organs that are absolutely essential for addressing physicochemical variables are discussed below, organized by the needs of the individual fish for movement, temperature, buoyancy and osmoregulation. More complex body functions, especially involving behavior, will be addressed later. Here I skim the surface of these adaptations, which are well known to biologists. More detail can be found in general ichthyology (e.g., Helfman et al. 2009; Moyle and Cech 2004), biology of fishes (Bone and Moore 2008), and physiology (Smith 1982; Evans 1997) textbooks.

Dealing with Temperature: Warm Muscle

The largest organ system in the fish body is skeletal muscle. It consists of about 60% of the total body weight, which is more in ratio than in humans (Smith 1982). We have discussed the importance

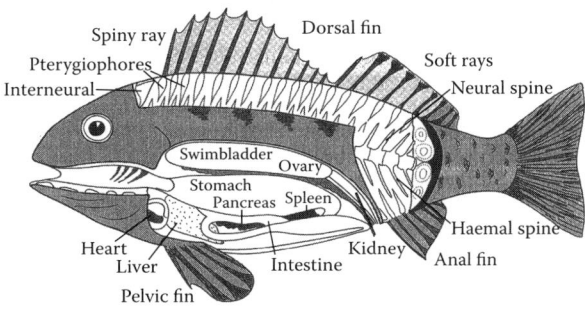

Figure 20.3 Internal organs of fish. (After Dean, B. Fishes, *Living and Fossil: An Outline of Their Forms and Probable Relationships*. MacMillan and Company, New York, 1895.)

of locomotion in fishes, and thus, it is highly important to keep the ability to move unimpaired. The skeletal muscle is not all the same: it consists of light (white) and dark (red) muscle. This can easily be observed in cleaning or eating flesh from fishes such as tuna or salmon. White muscle stores and uses mostly glycogen as an energy supply for high power output over short periods of time, and it is also known as fast glycolytic (FG) muscle. In contrast, red muscle produces a low power output, but over a longer period of time because it is nonfatiguing. Red muscle is powered by lipid in addition to glycogen and it is referred to as slow oxidative (SO) muscle (Webb 1997). Most fishes have mostly white muscle, and their life history indicates a need only for short bursts of movement. Fishes that are adapted for sustained swimming have the most red muscle, its long term action made possible by aerobic respiration.

As stated above, fishes have a combination of white and dark muscle. Sudden bursts of speed can be made possible with white muscle alone, but it does not have a great blood supply and it has to rely on anaerobic metabolism, which means it ceases to function as oxygen debt and lactate build up (we experience this as fatigue, pain, and cramps). Some explanation is in order here and we will discuss anaerobic metabolism in more detail later.

White muscle can be powered for 1–2 min using its stored energy supply (i.e., phosphocreatine) and then it has to utilize anaerobic metabolism of glycogen, which can extend muscle activity for 10–15 min (Jobling 1995). Unfortunately, the anaerobic process results in the accumulation of lactic acid (lactate, the end result of the anaerobic metabolism of glucose or glycogen). Lactic acid build-up in the tissue and blood not only results in fatigue, it results in an oxygen debt to convert the harmful substance back into glycogen. Exhaustion can occur in 2–15 min of extreme activity and take 24 h for recovery (Jobling 1995). Lactic acid levels of 150 mg/100 g of blood (accompanied by about 700–900 mg/100 g of muscle) can cause delayed mortality of fishes after they have been subjected to stressful activity (Smith 1982) such as handling, transport, or angling. Not only that, but anaerobic metabolism is costly. It requires oxygen to convert the lactate back into glycogen, and it requires the catabolism of about 1/5 of the lactate to convert the remaining 4/5. Not only is anaerobic metabolism very low in efficiency (it produces only about 5.6% of energy compared to aerobic metabolism) it builds up lactate which is toxic. It is very costly to excrete lactate, because this results in a loss of more than 90% of the original glucose (Smith 1982).

Obviously, anaerobic metabolism in white muscle is not metabolically very desirable for producing energy. White muscle is fine for short fast burst of speed, but efficient sustained speed cannot be accomplished with it. However, it is useful for quick power and for emergency response at activity levels approaching maximum sustained speed and above it. If white muscle can ensure the survival of the fish, it can be worth the energy inefficiency.

It is red muscle that is used for general movement and cruising at sustained levels. Red muscle seems to be located in areas of the body that are responsible to maintain constant movement, so in seahorses we would expect and find red muscle associated with the dorsal fin. In some fishes it powers the pectoral fins. In tuna we find it all along the side of the fish. Even present in small amounts it is a very important organ. Thus, it is critical to keep this organ system working properly.

Red muscle is powered by aerobic respiration and it requires a good blood supply, hence its red color. Also, it must have a constant temperature to operate at an optimal level. But blood coming directly from the gills will have a temperature that depends on ambient water temperature. This temperature will vary according to time of day, location from the surface, season, and so on. Also, activity can generate considerable heat within the red muscle tissues. Fortunately, the fish is able to maintain a relatively constant temperature in the muscle tissue.

Blood supply to red muscle far exceeds that of the white muscle, and it is in association with red muscle that large network of capillaries have been found. These networks are called the *rete mirabile* (pl. = *retia mirabilia*), or in English it is the "wonderful net." Within the net, venous capillaries move oxygen-depleted warm blood from the muscle back to the heart and gills, and arterial capillaries are bringing in a fresh supply of oxygen and nutrients in the cold blood from the gills. These beds of small vessels lie in a pattern where the incoming and outgoing blood vessels are close alongside each other; thus, the venous and arterial blood flow is opposite, or counter to each other, producing a countercurrent flow. In this fashion the net functions as a heat exchanger as heat is radiated across the thin membranes. It is amazing to note that heat transfer can be as much as 98% in this manner, and the muscles can be kept up to 21°C warmer than ambient water temperature (Bone and Moore 2008). Not only is the muscle warmed, the fish can exert some regulation and provide a steady state condition with respect to temperature within the red muscle mass. Of course, the temperature of the red muscle also can radiate to surrounding white muscle, but the function is not as well understood. Two main groups of fishes that use this mechanism are some recently evolved scombrids (albacore, bluefin, yellowfin, and skipjack tunas), and laminid and isurid sharks, including the great white shark.

As a final note, we assume this mechanism evolved in response to allow sustained movement in cold environments, and it also has been used to warm other important parts of the body, such as the brain, eyes, and stomach of some species. In this chapter we also will see how this countercurrent exchange system also is used for other purposes.

Obtaining Oxygen from Water and Air

Aerobic metabolism is not possible without a supply of oxygen. It seems simple to us. We obtain oxygen by breathing air. It is not that simple for the fish, because oxygen solubility is very low in water compared with air (about 1/20). Also, oxygen solubility is inversely related with temperature, so in warmer water there is less oxygen available. But wait! Metabolism of a fish, which is primarily an ectotherm, would increase with warmer water (it approximately doubles with a 10°C increase in temperature). So when the fish needs oxygen the most, that is, to feed an increasing metabolism, it would be less available in the water for them. So a fish needs to have an efficient way of obtaining oxygen from water, or resort to breathing air, which some fishes do. First we will deal with the problem of obtaining oxygen from water.

The fish has a sophisticated system for obtaining oxygen and for circulating it throughout the body (Figure 20.3). This system is different in fishes and tetrapods, so some explanation is in order. In normal conditions, depleted venous blood enters the heart and is pumped to the gills. There the blood flows through the tiny lamellae that are exposed to the flow of water. Here is the important point: *There is a countercurrent flow of blood and water.* This countercurrent flow through the very thin tissue of the lamellae results in the capability of the animal to withdraw as much as 90% of the oxygen, whereas concurrent flow can withdraw no more than 50%. The amount taken up by this osmosis is not enough, however, and large fish require more than what can be obtained in this fashion. What also occurs is binding of extra oxygen to the respiratory pigment, hemoglobin. Skipping the details, oxygen is bound to hemoglobin with higher pH levels (ca. 7.5), and disassociates at lower (acidic) pH (1/2 oxygen is lost at about pH 6) (Smith 1982). Practical application of this is the availability of oxygen at higher CO_2 levels and this is when cells need to obtain it. Venous blood at the gills would lose CO_2 to the water, raise the blood pH, and enable hemoglobin to bind oxygen once more.

Under conditions of stress, and especially with elevated water temperatures, gill function is affected by endocrine control and this will be discussed when we consider the effects of stress in the next chapter.

Many fishes have the capability to extract oxygen directly from air, and Helfman et al. (2009) listed 48 families of air breathing fishes that have several ways to do so. However, obtaining oxygen from the air also limits their distribution to the surface layers and interferes with other functions of the gills in obtaining nutrients and disposing of wastes, especially ammonia. We have already discussed the use of lungs in lobefin fishes, which in case of lungfish are used to survive during droughts. Some fishes such as bowfins and gars, are able to live in shallow swampy habitats that may have waters low in oxygen content. These fish deal with the hypoxic conditions by using a vascularized open swim bladder for absorbing oxygen from the air. However, in these examples, the fish is limited to an open swim bladder or lung. To avoid this problem, more derived fishes have developed other ways to extract oxygen from air. These are suprabranchial (above the gills) respiratory membranes that may occur in the form of labyrinthine plates, respiratory fans, or tree-like arborescent organs. Present within the gill chamber in association with the regular gills, these structures have stiffer supports so that they do not collapse in air like the regular gills (Helfman et al. 2009).

Buoyancy and the Swim Bladder

Water is so dense, especially salt water, that it is possible to float in it without much effort. By becoming buoyant a fish can avoid the energy costs associated with gravity, and they can remain in a selected level of the water column. Fishes such as sharks accomplish this by a combination of hydrodynamic lift, lightening the body, and static lift (i.e., buoyancy) due to storage of lipids. Teleost fishes remain buoyant due mostly to static lift from the swim bladder, also known as the air-float or gas bladder, which develops soon after yolk and oil globule depletion in most fishes (Govoni and Follower 2008).

The swim bladder requires energy to make and maintain, but it saves energy in swimming and provides a dependable way to remain at a certain level in the water column. As pointed out by Lanham (1962, p. 39) concerning the swim bladder: "It transforms a heavy mass of flesh and bone into a weightless craft that soars in the water, one that is instantly responsive to the slightest movement of fin and swimming muscle." This provides a responsive unit, and also a platform for the sensory systems, which function best when the fish is still. Swim bladder structure is variable due in part to its use for other purposes, such as hearing (Bone and Moore 2008), and there are two main types of swim bladders, the open (physostomous) and closed (physoclistus) types (Figure 20.4).

The open swim bladder converted a primitive lung into a flotation chamber that is connected to the gut via a pneumatic duct. The bladder is filled by gulping air and emptied the same way. This open float is filled at the surface and is found mostly in freshwater fishes, where depth adjustment is not very important. In many species this open float also can be useful in gulping air for its oxygen content, and some swim bladders are heavily vascularized for this purpose, especially in fishes that live in swampy habitats where oxygen in the water can be limiting (e.g., bowfin and gar).

The closed swim bladder is filled and emptied via the blood and has no connection with the gut. This bladder is found in many fishes that live at depths, because oxygen can be pumped into the bladder at great pressure. This bladder has a gas gland that is used to fill the bladder and an oval body that is used to deplete it (Figure 20.4). The mechanism works by countercurrent exchange of oxygen in a familiar structure of small veins and capillaries that we already know as the *rete mirabile* (wonderful net). But there is more to this. As arterial blood enters the gas gland it is acidified by lactic acid, which reduces the oxygen affinity of hemoglobin (Bohr effect). Oxygen also disassociates from the hemoglobin molecule because the carrying capacity of hemoglobin for oxygen is reduced with decreasing pH (Root effect). As a consequence, some hemoglobin (as much as 40%) is made into a deoxygenated state (T-state), and it can no longer bind oxygen (Pelster 1997).

As the small veins leave the gas gland, they lie among incoming arterioles. This allows oxygen and acid to enter the arteriole from the vein. As oxygen continues to exit the vein, the incoming blood also is acidified, enhancing the effect. The wonder net in this case is referred to as a secretory rete. Many thousands of afferent and efferent vessels are needed to produce the effect, and the degree of contact between them increases with length. Deep-sea fishes with swim bladders have large numbers of long retia.

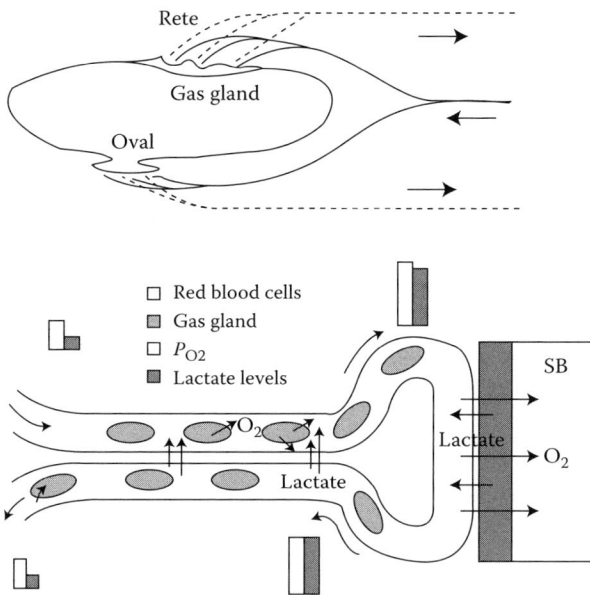

Figure 20.4 Swim bladder. (From Bone, Q., and Moore, R. H. *Biology of Fishes*. 3rd Ed., Taylor & Francis Group, New York, 2008. With permission.)

Osmoregulation

Fishes regulate their internal environment at some range of values, usually referred to as homeostasis. One aspect of living immersed in water is the need to regulate the amount of water retained in the tissues and its concentration of dissolved substances, especially salts. Obviously this concentration changes a great deal from salt water to freshwater, and almost all fishes live in water that has a different concentration of salt than that of their tissues. The skin of fishes is mostly impermeable to ion transfer, but there are locations where it is unavoidable to expose epithelium directly, such as the mouth and gills.

Marine fish live in water that is hypertonic to them, which has the effect of extracting water from the fish by osmosis, mostly through the gills—thus dehydrating them. Freshwater fish have the opposite problem (Figure 20.5). Water they live in is hypotonic to their tissues, with the result that water is drawn into the body by osmosis—the resulting dilution causing swelling. Diadromous fishes must contend with both conditions and be able to adapt to them. The end result is expenditure of metabolic energy and the development of organs to maintain a proper range of dilution for the body fluids. This process is called osmoregulation and it is accomplished differently in some major fish groups.

The kidney is the principal organ in the body that removes excess waste products (such as ammonia from protein breakdown), salts and water from the blood. This function is shared by the functional unit of the kidney, the nephron, and millions of nephrons filter water through a structure known as a glomerulus, which conceptually resembles a kitchen colander. The large particles (cells and large molecules) are separated just like noodles would be through the colander, and the filtrate is processed in a connected tubule to remove desirable metabolites and water back to the circulation system, leaving urine as the final product. The monovalent (small) molecules of NaCl are troublesome because the kidney cannot concentrate NaCl in the urine higher than it occurs in the blood. This problem has been dealt with differently among fish groups.

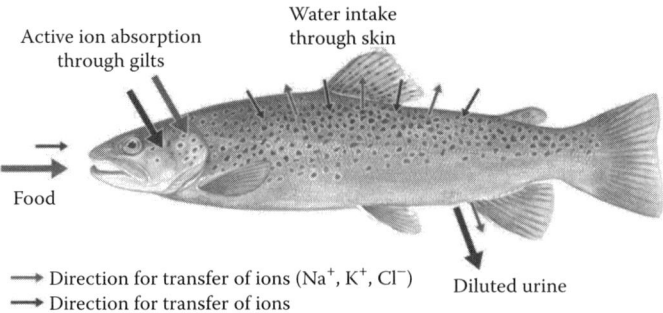

Figure 20.5 Osmoregulation in a freshwater fish. (Courtesy of Raver, D., USFWS, http://commons.wikimedia/wiki/File:Bachforelle_osmoregulation_bw_en.png.)

The long surviving hagfishes solved the problem by having their blood almost isotonic with seawater, thus effectively avoiding osmoregulation of small ions altogether, but they do regulate divalent (larger) ions. Sharks and their relatives solved the problem by concentrating a waste product, urea, in the blood. A molecule of urea is too large to pass out of the body (gills) by osmosis. Thus, the blood is slightly more concentrated in solutes than seawater. However, the fish needs water and does attain it, from seawater that enters via gill membranes and perhaps by drinking. The excess salt that is taken in must be excreted, and this is done via a special organ, the rectal gland, which actively secretes the salt.

Teleost fishes deal with osmoregulation in an entirely different way (Figure 20.5). Marine teleosts lose water passively across the thin gill membrane, which increases the internal salt concentration. Water replacement has to be obtained by drinking seawater, which is high in NaCl, and must be eliminated. The marine teleost rids itself of some NaCl by the kidney, but not enough. It excretes the rest with special cells that are located in the gill filaments. These special cells are the α-chloride cells, and actively pump chloride ions out of the body, followed passively by the sodium. Freshwater teleosts have no need to drink water because it enters into the body passively, mostly across the gill membrane. The kidney rids the body of the excess water by producing copious quantities of urine, in which the small ions, including NaCl are lost. To replace these ions the freshwater teleost also has developed chloride cells, but these are β-chloride cells and they are able to pump chloride ions from the water back into the body. The ontogeny and physiology of chloride cells are presented by Bone and Moore (2008) and Kaneko and Hiroi (2008).

Fishes that live in both freshwater and saltwater during their lives, such as anadromous species, must be able to make morphological and physiological changes to accomplish this. The most studied group of fishes is the anadromous salmonids. Of note are the functions of the chloride cells and other features of young fish as they change from freshwater to saltwater life (Steffansson et al. 2008).

MORPHOLOGY AND SENSORY SYSTEMS

Feeding Morphology

In our section on the African Great Lakes, we read about the great feeding diversity in the cichlids that inhabit them. It is critical for fishes to feed, because that is the only way they can assimilate energy. Fish cannot survive without food and they cannot grow, or reproduce without metabolizing more energy than is needed just to survive. Feeding requires special adaptations such as proper body shape for the task of seeking, arrangement of fins for power, and very importantly the shape

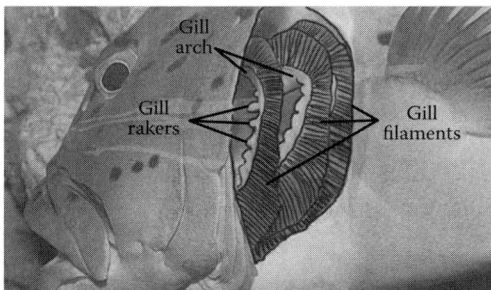

Figure 20.6 Fish gills showing rakers and filaments. (Courtesy of Richardson, M., http://commons.wikimedia/wiki/File:Fish_Gills_Labeled.jpg.)

and function of the jaws and teeth for catching and eating. However, once captured the prey must be retained and prevented from damaging the sensitive gill filaments, and this is done with the gill rakers (Figure 20.6). Some prey, such as invertebrates or vegetation need to be fragmented, pulverized, or ground—a job for pharyngeal teeth. And the processed prey also must be retained in the gut long enough to extract the nutrients. Each of these requirements is addressed below.

Body Shape

Even though propulsion through water requires some fundamental shapes in fishes, there are some patterns that tell us something about feeding. A few examples will get to the point. The torpedo-form streamlining and large powerful fins we associate with the impressive pelagic sport fish, such as tuna is recognized as the "need for speed" in chasing down and consuming other fish. Lateral compression and agility sunfish, butterfly fish and spadefish is associated with a life in or around cover, and hunting among the vegetation or reef for invertebrates or other small fish. The flat pancake like bodies of rays, flounders, soles, goosefish, and other bottom dwellers speaks of deception, and a patient game of hiding and waiting. Finally, the long graceful body of the pike, with its powerful unpaired fins clustered toward the rear, suggests a sudden attack from ambush.

Mouth

The mouth of teleost fishes is quite adaptable, and it may be different in its morphology, position, and size for three major uses: for feeding by suction, ram ventilation, and manipulation (Liem 1980). Mouth morphology includes the degree of protrusion, useful in quickly reducing the distance from the fish to its prey, and by reducing the size of the opening. Position refers to whether it is terminal, which is an attribute of most predators, or subterminal as occurs in suckers in which it is useful for benthic feeding. Opening size in concert with a flexible head, powerful muscles, and effective sealing of the opercle can serve to produce a very powerful suction. Size of the mouth also is related to the prey, and small mouths usually indicate feeding on invertebrates. A very large mouth is most useful to a predator, and much has been written about "gape" limitations—to be discussed later in chapter 25.

Foraging by ram action is common in planktivorous fishes of all sorts, from paddlefish to herring. These fishes generally lack protrusible mouths. Some of our largest fish, including the huge whale and basking sharks, as well as many clupeids, use ram feeding to capture zooplankton, larvae of marine invertebrates, krill, and other items.

Capture of larger food items may require some dexterity in handling them, especially those that can be dangerous. Manipulation of prey can include biting, scraping, rasping, and so on (Wootton 1990). This is obvious in case of puffers that feed on crabs, or largemouth bass that feed on crayfish.

Teeth

In most fishes, teeth occur not only on the jaw bones, but also elsewhere, including the tongue and pharynx. In piscivores, the teeth are usually long and very sharp, many of which also are angled outward. The teeth of molluscivores are usually made for crushing, and the teeth of herbivores are made for grinding. The presence of both jaw teeth and pharyngeal teeth also can provide a measure of diversification. For example, some fish pluck food with jaw teeth and then crush or grind it with the pharyngeal teeth.

Gill Rakers

These structures serve to separate the food from the water passing over the gill filaments. Generally, piscivores have a low number of shorter and stout rakers while plankton feeders have a greater number of fine, longer rakers to filter out the plankton.

Gut

The type of diet is related to the complexity and relative length of the gut. In higher teleosts the carnivores tend to have short guts (equal or less than body length) and the presence of stomach. The stomach helps to dissolve the food. In herbivores there is no stomach or a very small one and the guts are longer (from several to many times greater than body length) (Bond 1996).

Sensory Systems

Fishes constantly monitor their environment through an impressive array of sensory organs, as illustrated for a shark in Figure 20.7. Some of these, such as the eye, are very similar to that of land vertebrates. Other organs are difficult to appreciate, like hearing, because we have sound perception, but it is not nearly as sensitive. Some organs, like the lateral line system and electroreceptors, we can only guess about what their use might reveal. But all of these sensory systems are used to find the most suitable conditions, to detect prey or predators, to tell direction and time of day, find mates to locate spawning grounds, interact in social behavior, and a host of other things. These senses allow the fish to interact with the biological as well as physical and chemical components

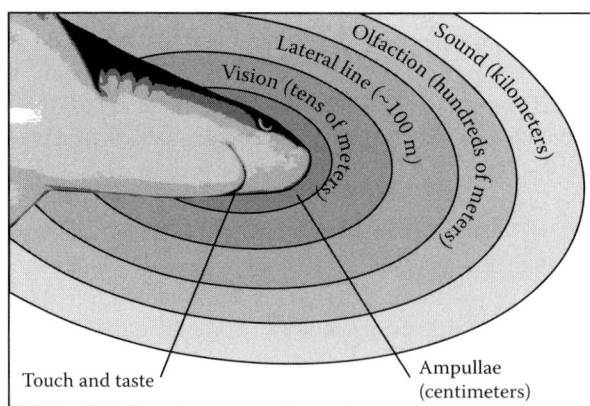

Figure 20.7 Sensory abilities of a shark. (From Springer, V. G., and Gold, J. P., *Sharks in Question: The Smithsonian Answer Book*, Smithsonian Institution Press, Washington, DC, 1989. With permission.)

Vision

Fishes have very good eyes, thankfully, because we inherited them. They are constructed very much like the human eye, except they have an added capability to see in very low light levels. They do this by having a reflecting layer or *tapetum lucidum* (curtain + reflective). The tapetum is reflective and it reradiates the light to the retina to enhance the image. This works very well in low light conditions. In bright light the retina occludes the tapetum by a layer of pigment that migrates over the reflective layer. This occlusion serves two purposes: it aids vision in brighter light and perhaps as importantly, it hides the reflective layer for better concealment. The reflection just described can be observed in deer and other nocturnal creatures as "eye shine." In addition to rods and cones, some fishes have special cells to detect ultraviolet light, and receive ultraviolet from other conspecifics as signals. Some fishes also have tubular eyes, to allow binocular vision and a large lens to gather light (Bone and Moore 2008).

Hearing

What terrestrial creatures consider to be "sound" is really just a neurological sensation that we perceive from interpreting vibrations traveling through air. Such vibrations are transmitted faster in water, allowing fish to quickly detect sound as to direction, intensity, and frequency. Sound reception in fishes also is aided by structures that aid in transmitting and amplifying the sound, which usually involve the swim bladder. Sound amplification is very effective in ostariophysian fishes due to the Weberian apparatus, which connects the ear to the swim bladder. Sound vibrations also are detected by the lateral line (Schellart and Wubbels 1997).

Chemosensory

Fishes have three chemosensory systems—olfaction, gestation, and solitary chemosensory cells (SCC) (reviewed by Døving and Kasumyan 2008).

Olfaction; Smell

The olfactory organ of fishes, usually paired, has numerous receptor cells. Water flow is circulated through the receptors, which are usually found in platelike lamellae. Chemoreception that occurs there is transmitted directly to the olfactory part of the brain, and it is very sensitive in all species. This is due to the solubility of most chemicals in water, whereas terrestrial animals first have to intake air into the nasal cavity where particles in the air are then dissolved in water.

Gustation; Taste

Taste and smell are hard to separate, so we generally refer to both as chemoreception. Taste buds of fish can occur almost anywhere, but they are found outside as well as inside the mouth, mostly in mainly cutaneous organs like the "whiskers" of catfish and carp. These cells aid in discriminating between "good and bad" foods.

Solitary Chemosensory Cells (SCC)

These cells can be found in many locations on the body of a fish. Their function seems to be mysterious, but they respond to complex substances such as mucus.

Mechanoreception

Neuromast cells and the lateral line system (better known as the acustico lateralis), can detect changes in water pressure and vibrations through hair cells located in canals along the side of the fish and elsewhere that act as receptors. Receptors in the head and rostrum aid in detecting movements or vibrations of prey, while receptors along the side of the body pick of swimming organisms.

Electricity

Electroreception occurs in many fishes, but not many teleosts (von der Emde 1997). A well known array of electroreceptors occurs on sharks (Ampullae of Lorenzini), and is used in perception, such as in navigation and locating prey.

Magnetism

Magnetic crystals have been found in special sensory cells that occur in the olfactory tissues of trout. It is presumed that this arrangement enables directional finding along the magnetic lines of force that circle the earth. Tuna and other fishes also are thought to have such skills. Sensory perception of magnetic lines of force also seems to be related to electroreception (Figure 20.7).

SUMMARY

An individual is tolerant of a general set of environmental conditions, due to morphological, physiological, and behavioral adaptations. It also may be confronted with some additional changes. The response of an individual should depend on the severity of the change and its persistence. Movement as a general adaptation can mitigate a very wide range of environmental perturbations that specific organs may not. Thus, local or short-term changes can be dealt with by a behavioral response, usually movement. If pervasive, then a physiological response (e.g., acclimation) might occur, and if persistent then a morphological or life history change could occur. Finally, if the stress of change continues for a very long time, then a genetic response can occur in the population due to natural selection, resulting in a change in gene frequencies. This chapter discusses major adaptations, but no organism can have specific adaptations for every possible environmental constraint. In this case, a likely general response is locomotion. Fish movement is possible using a combination of white and dark muscle. Sudden bursts of speed are possible with white muscle alone, but it has to rely on anaerobic metabolism and ceases to function as oxygen debt and lactate build up. However, it is useful for quick power and for emergency response. Red muscle is used for general movement and cruising at sustained levels and powered by aerobic respiration, which requires a good supply of oxygen and constant temperature. Fortunately, the fish is able to maintain a relatively constant temperature in the muscle tissue, which generates heat from metabolism. Fish have a sophisticated system for obtaining oxygen and for circulating it throughout the body. There is a countercurrent flow of blood and water that allows the animal to withdraw as much as 90% of the oxygen from the water. Large fish requires more than can be obtained in this fashion. And extra oxygen is obtained by binding to the respiratory pigment, hemoglobin. To partially offset energetic costs, most fishes have a swim bladder that uses atmospheric gases to provide static lift (buoyancy). The bladder may be open or closed, but allows a fish to remain at the same level in the water column. But life in water has fundamental problems, such as osmotic pressures, requiring osmoregulation. Hagfish blood is almost isotonic with seawater, but all other fishes need osmoregulation. Freshwater teleosts must rid themselves of excess water that enters their bodies and marine teleosts have the opposite problem. Salt regulation in both depends on the kidney and chloride cells. These physiological needs all require energy and fishes have to obtain energy from food, which requires adaptations for obtaining and processing food (e.g., body shape, type of teeth, gill rakers) and acute sensory systems to detect food and danger (e.g., vision, hearing, chemoreception, mechanoreception, and electroreception). Most of these adaptations we have as well, because the fish passed them along to us.

CHAPTER 21

Energy, Metabolism, and Growth

ENERGY BUDGETS

All fishes are consumers. They must obtain energy in the diet in order to live and carry out activities necessary to complete their life history needs. Bioenergetics, in part, is the study of how much energy is taken in and what happens to it, that is, how it is partitioned and used to provide for the needs of the organism. The amount of energy available in the food varies among food items and within the fish, primarily due to differences in costs of metabolism and fish activity. As previously discussed, the relative amount of energy available for growth and reproduction can be critical in considering fitness. In this section, we will explore assimilation and partitioning of energy for various body functions. Later we will study how energy is obtained (i.e., feeding) and used (i.e., activity, growth, and reproduction).

In the last chapter, we considered the fish as a "black box," a theoretical input–output model, where the input of food was converted to an output of growth in somatic and reproductive tissue. Now we will explore how energy is used by the fish in reaching the goal of fitness, and how the distribution and abundance of some foods also may affect the fish distributions.

The energy budgets of heterotrophic organisms have been of great interest in the study of animals in nature and in culture. Such a budget is defined as the amount of food consumed (total intake over time), which is then partitioned by the body as undigested material that is ejected (feces), material that is absorbed and then excreted as wastes (urine), material that is incorporated into the body as growth (somatic and reproductive) in individuals or populations, or energy lost in respiration. Thus, the food consumed is lost in feces, urine, and respiration and gained in biomass. The trick is how to estimate food consumption and use. Windell (1978) identified five methods that have been used to estimate the intake and use of food in fishes, including the Winberg equation, balanced energy equation (bioenergetics), nitrogen balance, radioactive isotopes, and digestion rate (food progression). Of interest here is bioenergetics. Windell (1978) reviewed the concept and provided historic examples of its development. Adams and Breck (1990) provided an extensive list of the application of bioenergetics in fisheries and fish biology. Wootton (1990) and Diana (2004) do a good job of reviewing energy equations and fish bioenergetics.

The Russian physiologist Winberg is given credit (Windell 1978; Adams and Breck 1990) for developing the basic energy budget equation in 1956. In this case, three variables were considered with the idea that if two are known the third can be calculated. In general, the digestible portion of food consumed would be spent in metabolism and growth. In a simplified way then:

$$p \cdot C = M + G \tag{21.1}$$

In the Winberg equation, p is the portion of food that can be assimilated, C is the food consumed, M is metabolism (catabolism), and G is growth (anabolism) (Figure 21.1). The coefficient p needs

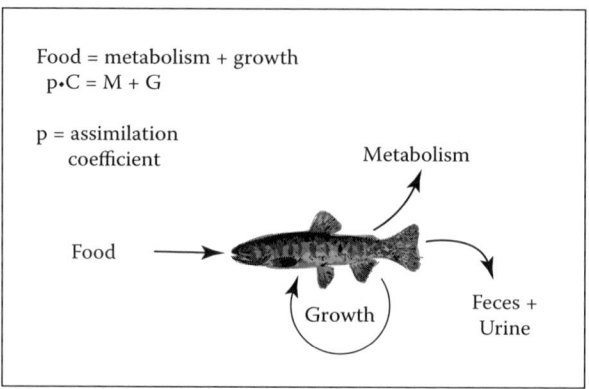

Figure 21.1 The fate of food energy assimilation in a fish.

some explanation. In case of most fishes, which eat other fish, the food is mostly protein and about 80% is digestible. In this case, $p = 0.8$, with 20% voided as feces and urine (Figure 21.1).

It is more instructional to view an expanded equation, which can be used to explore various relationships among energy allocation in the body. Warren and Davis (1967) did this by separating M (metabolism) into the standard metabolic rate, the increase in metabolic rate attributed to activity, and the metabolic rate increase due to the energy required to digest food. Growth was then separated into somatic growth and reproductive growth (reproductive tissues). Doing away with the assimilation coefficient, they added two new terms for waste on the right side of the equation for feces and urine. Unlike a homeotherm, fish do not have a basic metabolic rate; instead, the rate is more influenced by its medium. In this equation the metabolic rate is indicated as standard metabolism, which reflects that fish metabolism varies according to the water temperature. For most purposes, the standard metabolism is usually 20°C, but not always, so the temperature must be provided for comparison and interpretation.

$$C = (Mr + Ma + SDA) + (Gs + Gr) + (F + U) \qquad (21.2)$$

As indicated above, C is the rate of energy consumed, Mr is standard metabolic rate, Ma is metabolic increase due to activity, SDA is metabolic activity due to digestion (includes peristalsis, assimilation and storage, and detoxification of ammonia), Gs is growth of somatic tissue, Gr is growth of reproductive tissues, and F + U is feces and urine.

It has been recognized that energy cannot be destroyed, but it may be changed from one form to another. In this case, conversion of energy is governed by the laws of thermodynamics, and as energy is changed there will be a loss of heat energy that cannot be avoided. This fact and the universal use of Q to represent a quantity of energy prompted the detailed form of the energy equation presented below (Webb 1978; Diana 2004), which I have reorganized:

$$Qr = Qs + Qf + Qh + Qn + Ql + Qg + Qp \qquad (21.3)$$

In Equation 21.3, Qr is the amount of energy in the food ration, Qs is standard metabolic rate, Qf is feces, Qh is waste heat (SDA), Qn is nitrogen excretion, Ql is locomotion, Qg is somatic growth, and Qr is reproductive growth. Ration refers to the amount of food eaten over some time period. In aquaculture it is generally computed in the per cent of food given relative to body weight (BW). For a 1 kg fish, 100 g food per day would be a ration of 10% BW/d.

What does an energy budget look like for a fish? Details were worked out for several carnivorous fishes by Brett and Groves (1979). In this case, the prey were composed of almost all protein, yielding to about an 80% degree of assimilation (energy that can be digested). with about 20% passed as feces. After subtracting for metabolism (7%), feces (20%), heat loss (14%), and nitrogen excretion (7%), there was 52% remaining that was used for activity (23%) and the remaining 29% was available for growth (somatic and reproductive). Equation 21.3 looks like this:

$$100\% = 7\% + 20\% + 14\% + 7\% + 23\% + 29\%$$

We note that about 48% of the energy consumed is lost to digestion, heat, feces, and nitrogen excretion. Regarding fish survival and reproduction, the last two numbers are telling. A carnivorous fish generally has about 23% of the food energy available for locomotion or other activity, which assumes feeding, moving to better conditions, escaping predators, and so on. The 29% energy represents the amount available for growth in young fish, or in case of old fish, most of that number presumably would be allocated mainly for reproductive costs (either in building reproductive tissues or in reproductive activity).

Diana (2004) prepared similar information about an herbivorous fish, grass carp, for comparison. He reported that if grass carp eat a diet entirely of vegetation, that the diet was only 19% capable of being metabolized, due mostly to cellulose. Of this 19%, only about 3% was left for growth and reproduction. I wondered about how much energy might be left for activity, so using his figure of 16% for activity, heat, and standard metabolism combined, I reasoned that an approximation (somewhat educated guess) could be made using values from the carnivorous fish, recognizing that there is about 1/4 the amount of food to be digested. Q_n should be very low because the protein level in plants is low, so I allowed 1%. Q_h should be about 1/4 due to the lower food volume, so I calculated 3%, and Q_s I left at 7%. This total is only 3%–4% of the energy that can be allocated to activity such as feeding.

Comparison of energy obtained in the diet only tells part of the story. In nature, the amount of energy that can be obtained is related to the time available for feeding, the activity needed to feed or escape predators, the availability, quantity and quality of food stuffs, and the capacity of the gut to digest food at the prevailing temperature. Compared to the 23% available for activity of carnivores, there is not much energy available for herbivores to forage or escape from predation. But there would be problems with such a comparison. Carnivores eat a large meal and rest for digestion. Most herbivores have no stomach for food storage and their gut is long. However, some herbivorous cichlids (e.g., haplochromines and tilapias) have a large stomach where extreme acidification can occur to hydrolyze many refractory forms of plant tissue. In that case, the gut is short and mostly absorbs the digested portions. In both cases, herbivores eat more or less continuously and in this fashion they can greatly increase their energy uptake each day. So the trick for herbivores is to be in a place where there is plenty of vegetation to eat, and to be able to eat all year. Such a place is the tropics. It is no coincidence that this is where almost all of the herbivorous fish live.

There are more problems. Grass carp, can and do eat other food items, such as worms, and no doubt epiphytic organisms such as a biofilm on the plants they eat. As calculated by Diana (2004) a grass carp fed tubifex worms could assimilate a whopping 60% of the worm biomass, and even if their digestion is not fine-tuned to be a carnivore, they can increase their energy input by eating miscellaneous organisms. Regarding the value of herbivorous fishes, even though they may not be highly efficient in extracting energy from plant food, they also act as shredders, reducing the vegetable matter into smaller pieces that can be eaten directly by others or support bacteria and other decomposers. Of course, we must always consider that ontogenic shifts occur in the foods eaten by fishes, and herbivores almost always start life as carnivores, usually on zooplankton. In the preceding exercises all of the fishes compared were adults.

The two main uses of energy in carnivorous fishes are for activity and growth. Thus, it is interesting to compare the distribution of stored energy between a sedentary and very active fish, for they should show opposite values in these energy compartments. Kitchell (1983) compared the relatively sedentary adult largemouth bass with the very active skipjack tuna. Using his raw data, Adams and Breck (1990) improved its presentation, and Figure 21.2 shows the values that might be expected from maximum rations. The bass with a small expenditure of energy for activity allocates a great deal of energy to growth, while the reverse occurs in the tuna. More about this study shortly.

As fish mature, growth is switched from somatic to reproductive tissues, and reproductive costs also have to be balanced with metabolism. A good example is reproductive costs in American shad. This anadromous fish ranges all the way from Florida to Canada, and allocates energy costs differently in northern versus southern races. In northern climates the fish uses about 40% of its energy in up-migration and 15% for down-migration and lives to spawn more than once. In southern areas (Florida) the fish expends more energy at warmer water conditions and dies at spawning, spending 80% of its stored energy. The increased energy used is mainly for the increased cost of metabolism (additional 10%), but it also switches 30% of its energy into making more gonadal tissues (Glebe and Leggett 1981). This is likely a response due to increased fitness in fishes that can reproduce only once.

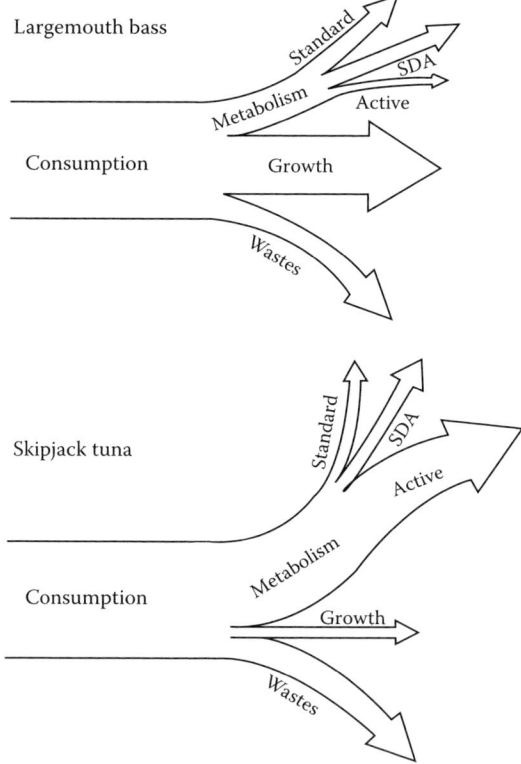

Figure 21.2 Allocation of energy in largemouth bass and skipjack tuna. (After Kitchell, J. F., *Fish Biomechanics*, 312–338, Praeger, New York, 1983; Adams, S. M., and Beck, J. E., *Methods for Fish Biology*, American Fisheries Society, Bethesda, MD, 1990; Adams, S. M. (Ed.), *Biological Indicators of Stress in Fish. Symposium 8*, American Fisheries Society, Bethesda, MD, 1990. With permission.)

FISH ENERGETICS

For a good historic review of the application and use of fish energetics in ecology and management, Hansen et al. (1993) provided much information, guidance, and many pertinent citations. In addition, use of bioenergetics models has greatly expanded in the past 10–20 years, and Hartman and Kitchell (2008), introducing a special session of *Transactions of the American Fisheries Society* dedicated to bioenergetics modeling, reported that articles on fish management and ecology have been published in at least 35 journals since 1994. The most common application was to estimate the amount of food consumption for various fish species under different conditions, and the models have been applied to individual fish, larvae and juveniles, habitat use, and for refining and testing other models. These authors stressed that there is still a great deal of uncertainty in model inputs and that testing outputs may not be possible. They recommended that models be kept simple and cautioned that models are useful tools, they need to be used appropriately. In this context, Chipps and Wahl (2008) analyzed 32 bioenergetic studies for model performance and found the differences between model performance and field data ranged from –84% to 770%: 82% of the studies that estimated food consumption were higher than data collected in field studies. They found that even though bioenergetic models were poor in predicting outcomes when compared with field and lab data, modeling was still being used to make important management decisions. They recommended adherence to a framework for model development and application in order to reduce uncertainty and to facilitate continued use of models to better understand fish physiology and feeding ecology.

Kitchell (1983) addressed the above problems in his excellent chapter on fish bioenergetics which I paraphrase to close this section: He identified the issues many years ago.

The use of energy models can help to identify the important factors to consider in the survival, growth, and reproduction of individuals and species. Models are especially useful because it is difficult and perhaps impossible to study all components of the energy budget in a natural setting. However, many bioenergetic studies are done in the laboratory using a single natural food or an artificial one and thus are "dangerously simple" (Kitchell 1983). In some cases, juveniles or subadults are used as surrogates for old adults and the possibilities for ontogenetic shifts in resource use and behavior should be obvious. The well-known and widespread degree of plasticity in fish diets (i.e., trophic adaptability) is often ignored as too complicated to consider in doing studies. Metabolism in fishes is dependent on their size, the temperature, and their activity, and only surplus energy is available for growth.

Growth is the most sensitive variable component in the energy equation, and thus, growth is the standard for assessing likely values of other components in an energetics exercise. However, small affects of temperature and other environmental variables also can have a large affect on growth, and this affect may be hard to detect. Kitchell (1983) used an example of growth of perch under three close temperature regimes to illustrate how small temperature affects can produce large changes.

Different life history strategies also may be evaluated in fish energetic and we have seen the differences in alternative energy budgeting strategies by comparing a "sit-and-wait" predator (largemouth bass), representing an "energy saver" strategy with that of a highly active pursuit predator (skipjack tuna), representing an "energy speculator" strategy. Energy requirements for the tuna were five times more at maintenance ration and eight times more at maximum ration than the bass. However, it is not this simple. The bass wastes a lot of time waiting, and the tuna expends a lot of energy catching. In the open ocean food is patchy. Once found, however, there is a potential for gorging. Both strategies obviously work, and we can learn much about fitness from them both.

METABOLISM

In Perspective

As ecologists, we are interested in energy budgets because they can provide a way to evaluate the effects of environmental variables on the physiology of organisms, that is, on their function and activity. Basic to an understanding of how energy is produced in the body is an understanding of metabolism. A few definitions might help in understanding terms.

The word *metabolism* means *change*. It is the way organisms derive energy and new compounds from more complex ones and also how they change those simple compounds into specific ones needed for maintenance of the body. This change involves *catabolism*, a degradative process that changes large compounds into simple ones with the production of energy, and *anabolism*, a constructive process to form more complex compounds from simple ones and to store energy. Metabolism aids to maintain homeostasis in the body by providing needed materials and by generating energy used for heat. It is important to understand that metabolism generates all chemical compounds used in the body for respiration, activity, bioluminescence, and so on.

Metabolism requires other bodily functions to support it such as feeding, digestion, and respiration. Feeding requires work by the organism in order to energy in the form of food. This seems clear, although behavioral aspects of foraging is complex and will be discussed later. However, metabolism is often confused with *respiration* and *digestion*.

Respiration is a process that supports metabolism, and aids to produce metabolically usable energy, mainly from the oxidation of compounds in food. The word means to breathe air, and historically, the biological definition included the external process of obtaining oxygen and removing carbon dioxide from the blood and the internal task of supply oxygen for catabolism (Ganong 1963). The term fermentation was used for the process of anaerobic respiration. Present usage requires a modifier: either *aerobic* or anaerobic respiration (Lawrence 1989). Aerobic respiration is a process that furnishes oxygen to fuel catabolism. Anaerobic respiration is a chemical reaction that occurs in the absence of oxygen, and produces energy at the expense of an oxygen debt needed to rid itself of lactic acid, a toxic byproduct.

Digestion is a process in which food is transported in the gut and converted into nutrients. It includes mechanical and chemical degradation, absorption, and the elimination of indigestible parts in the feces. The digestive system is usually thought of as the gut (esophagus, stomach, intestines, caecae, etc.), but there are other organs that are integral to the process. They include the teeth, spiral valve (if present), liver, gallbladder and pancreas. The system produces acids and enzymes needed for the breakdown of food and also functions to absorb the products.

Energy Source and Quality

Most fishes are carnivores and their diets contain mostly protein; however, prey items can vary greatly in fat content, and herbivores may consume substantial amounts of carbohydrates directly. The proximate composition of a diet is "approximately" what is in it. In this case it is the percent of macronutrients: fat, protein, and carbohydrate. This composition is usually provided on sacks of fish chow that are used in hatcheries. The following data was provided by the U.S. Fish and Wildlife Service (Piper et al. 1982). Dietary requirements of trout (t) are contrasted with that of channel catfish (c):

Fat (t)	9.4 kcal/g	85% digestible	8.0 kcal assimilated
Fat (c)	9.4	90	8.5
Protein (t)	5.6	70%	3.9
Protein (c)	5.6	80	4.5
Carbohydrate (t)	4.1	40%	1.6
Carbohydrate (c)	4.1	70	2.9

The preceding demonstrates that fishes vary in their ability to digest foods; in this case, the comparison was made between coldwater and warmwater species. Trout, which are carnivores, show a great capacity for digesting fat and to a lesser degree protein, but do not get much energy from carbohydrate. On the other hand, the omnivorous catfish has a higher efficiency in digesting all of the nutrients.

Metabolic Output and Rate

Metabolic rate of mammals, such as basal metabolic rate (at rest) can be measured directly by the amount of heat produced. Fish metabolism is only about 1/20 that of mammals, and they live in water. In this case metabolism is usually measured indirectly by sealing the fish off in a tank of water and measuring oxygen depleted over time with a probe. Since fish metabolism also varies with water temperature, then metabolism should be measured at a standard reference temperature, or at least the temperature has to be stated. The resulting measurement would be standard metabolism at a given temperature. In general, we have to remember that the metabolic rate of most fishes will approximately double if the water temperature is raised 10°C. In general, there is an increase of 1.65–2.7 × standard metabolism for each 10°C increase in temperature, depending on the species and temperature range (Jobling 1993).

As suggested by the energy equations, activity also has a great effect on metabolism. In this case, elevated activity such as searching more for food or escaping predators can be so high as to reduce the amount of energy normally allocated to other functions, such as reproduction or growth. An indication of just how costly elevated activity can be in an individual fish has been provided by numerous studies of prolonged swimming and the rate if oxygen consumption (a measure of metabolic rate). In general, an increase of speed of one body length per second (BL/s) requires 2–2.5 times the amount of oxygen. Compared to standard metabolic rates, maximum prolonged swimming may require four to seven times the amount of oxygen, and high speed (unsustainable) swimming may require 10–15 times the resting level (Jobling 1993). It is clear that high speed swimming is very costly to the fish and to be avoided if possible due to exhaustion and the buildup of an oxygen debt in the muscles.

Fishes are composed mostly of red and white muscle as previously discussed. Red muscle is greatly infused with blood and is capable of sustained speed. Red muscle is fueled by aerobic metabolism, which converts glucose into carbon dioxide and water within muscle cells. The aerobic process is about 62% efficient in producing energy and results in harmless byproducts. The amount of energy produced is constant, but relatively unresponsive and limited in power. On the other hand, in an emergency situation where burst swimming may be needed, white muscle is very responsive at short notice and can generate a great amount of almost instantaneous power using anaerobic metabolism. This extra energy is additive to the power of the red muscle. Unfortunately, the yield of energy from stored glucose under anaerobic conditions is not as efficient and the chemical reaction involved results in formation of lactic acid, which must be replaced by converting it aerobically. More explanation is warranted. In early states of metabolism (glycolysis), the glucose molecule is split into two molecules of pyruvic acid with the release of energy (2 ATP). In aerobic respiration the pyruvate is further metabolized (oxidative phosphorylation) in the Citric Acid Cycle to produce another 34 ATP for a total of 36 ATP. However, in anaerobic metabolism only the first state is available for producing energy (2 ATP). Thus, it is only about 5.6% as efficient as the aerobic pathway (Jobling 1995) and it produces a harmful end product (lactic acid) that must be removed from the body.

One-fifth of the lactate produced has to be converted with oxygen to carbon dioxide and water to produce enough energy to convert the remaining 4/5 of the lactate back to the storage compound (glycogen). This produces an oxygen debt that must be paid back with more energy. If the fish is so fatigued that this debt cannot be paid back soon, it is likely to go into shock and die. On a practical side, it may be fun to enjoy a long fight "playing" a hooked fish, but if the fish is to be released alive, perhaps a shorter fight and faster release would be more humane.

Metabolism can be substantially affected by several abiotic factors, some acting independently and some in synergy. We have discussed previously the need for obtaining oxygen, regulating salt balance, temperature effects, activity demands, and so forth. Fry (1971) identified the effects of abiotic factors on fishes, providing a conceptual way to evaluate these effects on metabolism. He named those factors as lethal, controlling, limiting, masking, and directive. The concept and factors are reviewed in more detail by Wootton (1990) and Diana (2004). Environmental factors, such as temperature, can work as more than one of these factors, in this case lethal, controlling, and limiting.

Lethal factors such as temperature and oxygen can disrupt the metabolic machinery so much that the individual cannot tolerate the situation. As an example, if temperature is too high or oxygen is too low, the fish will eventually die. A controlling factor affects metabolic rate. Temperature changes are a good example in fish. A limiting factor limits the supply or removal of metabolic products, such as a high level of carbon dioxide in the water would depress respiration. A masking factor modifies the effect of other factors. A directive factor is a stimulus that elicits a response from the fish (zeitgeber).

GROWTH AND AGING

As pointed out by Kitchell (1983), growth is the most variable energy component in the energy budget. A fish has to allocate energy according to the need to maximize fitness, and so standard metabolism and SDA are first because of the need to maintain the life support system. Waste products also must be eliminated, from two perspectives: one is that the fecal wastes need to be voided to make room for more digestion, and the other is the problem of being carnivorous and the need to rid the body of poisonous nitrogenous wastes. Activity also comes before any energy can be allocated to growth. In this case, the need to avoid being killed and eaten, the need to forage for food and seek refuge to rest and metabolize food, and the need to make spawning migrations or other movements are important for survival and reproduction. So we must view growth in the energetics equation as the place where surplus energy is stored. Viewing this from another perspective, energy allocated for growth also may be needed for other purposes. A good way to visualize this is to construct a simple budget as a pie diagram (after Wootton 1990) as shown in Figure 21.3. If more time or energy is needed for foraging or resting and evading predators, it will be taken from growth and reproduction. On the other hand, fish are indeterminate growers, and if conditions for reproduction are not favorable, they can store energy and use it for somatic increase.

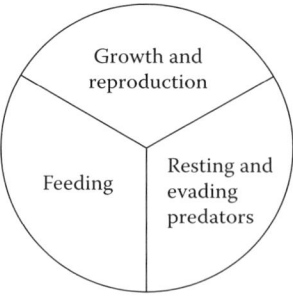

Figure 21.3 Allocation of time or energy in a fish.

It follows that growth is an indicator of internal (physiological) and external (environmental) conditions. For growth to occur metabolism, osmoregulation, feeding, activity, and so on must be already accomplished. Obviously, growth will be better with better environmental conditions. As growth increases, the fish will be larger and less vulnerable to predation; thus, growth is related to survival. As growth increases, the fish can produce more offspring; thus, growth is related to fitness.

Growth can be considered as an increase in length, biomass, or energy content. However, it must be given as a growth rate or change of size over time. For fishes, the growth rate can be expected to decrease with age, as more energy is allocated to reproductive growth, so size at a known age can be informative. The amount of biomass also can be seen as a variable per length of the fish. This observation is most useful in evaluating the relative condition of fish, because of the attribute of indeterminate growth. In this case, fish condition has been determined as a parameter called the *condition factor* K, first used by Frederick Heincke in 1908 (see Nash et al. 2006 for the history), in which K describes a relationship between fish weight and length (Table 21.1), namely, K = weight/length^3. In order to provide a K factor that ranges from 0 to 1 the following formula is used:

$$K = \text{weight} \times 10^5/\text{length}^3,$$

in which weight is given in grams, and length is in millimeters. By cubing the length it provides a volumetric variable to divide into weight, and multiplication by 100,000 (as a scaling factor) results in a K (g/mm) that ranges from 0 (very thin) to 1 (very fat). There is an equivalent condition factor C (lb/in) when using English units.

The condition factor as index of the relative degree of plumpness usually ranges in value from about 0.3 to 0.7. It has been used as a standard in fisheries science (e.g., Lagler 1956) to evaluate environmental conditions for stocks of a species inhabiting different locations, or to evaluate growth increments produced in different years. Again, the condition factor is most useful when it is used for comparing fish of the same age. Thus, it is necessary to understand how fish are placed in age groups.

There are several methods for estimating the growth of fish in nature (Jearld 1983) and Busacker et al. (1990) provided a detailed key for the selection of various options. The options include cohort identification, length frequency analysis, back calculations, and mathematical models. These authors provide a detailed explanation of aging techniques and various associated methods for analysis.

By far the most common method of aging fish is the use of body parts that retain annual growth marks (Table 21.2). There are several body parts that can be used, including scales, fin rays, spines, vertebral centra, and otoliths. Scales are widely used for this because they do not require sacrifice

Table 21.1 Condition Factor

Condition refers to the relative degree of "plumpness" in a fish and it is usually reported for fish that are aged in some way.
—Condition is a way of relating mass and length.
—Condition allows a way to assess environmental conditions.
—Condition factor is species-specific and quantifies the amount of energy that is stored.
—Condition factor can also be used to evaluate allocation of energy to reproduction and to detect when spawning occurs.
Condition factor is mostly used in the metric system. It describes the relationship between weight (mass) and length (grams and millimeters) as follows:
Condition factor K = weight/length^3 × 100,000 or K = $W(10^5)/L^3$
Numbers obtained range from K = 0 (emaciated) to K = 1 (very fat)
Usual range of numbers ranges from 0.3 to 0.7

Table 21.2 Evaluating Growth in Fishes

Growth can be expressed as
—An increase in length, mass, or energy content
—A rate such as increase in length or weight per year, such as
$Y_2 - Y_1 / t_2 - t_1$, in which Y = length or weight, and t = time
A standard tool in fisheries is to compare size at annual ages
—Age and growth require aging fish and obtaining lengths
—Aging techniques
  Known age
  Length frequency
  Growth rings in hard structures
    Scale method: back calculation where age of fish at year i = TL_i/TL of fish = distance of annulus at year i/total distance of scale
Factors influencing growth: endogenous vs. exogenous
—Endogenous: yolk quantity and quality
—Exogenous: ration, activity, temperature, oxygen, salinity, etc.

---

of the fish. In this case, annual growth is determined by "back-calculation" in which a simple ratio is used in equation form: the total length of the fish (TL) is known, the total distance across the scale (Ds) is known, and the distance from the center of the scale (the focus) to each annulus (Dan) also is known:

$$Dan_i / Ds = \text{fish at age}_i / TL$$

Solving for the unknown $age_i = Dan_i / Ds\ (TL) + \text{Intercept}$.

To be precise, an intercept is added to add distance on the scale that would be missing when it is formed. Of course, this method does not work on fish without scales and scale regeneration can be a problem. Also, annular marks result from periods of fast and slow growth that are associated with seasons in the temperate latitudes, so this would not occur in the tropics.

Another way to age fish, especially younger ones, is to use cohort analysis, in which a certain cohort is identified by size and followed over several years, or "length frequency" curves that include several cohorts. Large samples of fish populations will show clumping by lengths of successive age classes. This method is not very useful in very old fish, because fast and slow growers will tend to obscure the separation.

It also is possible to maintain fishes in ponds or to tag or mark them. In either way the identity can be determined and the age will be known. This method can be questionable if the goal is to determine age and growth of wild fish, because cultured or marked fish may grow at a different rate.

Finally, it must be understood that growth in fishes is controlled both by endogenous and exogenous factors. Endogenous (internal) control is primarily related to switching growth from somatic to reproductive as the fish matures, but also can come into play by fishes that energy costs, or lack of environmental cues result in less than annual spawning. Exogenous controls result from ration, temperature, oxygen, salinity, and other factors. Ration is usually expressed as the amount of food per day or as a percentage of body weight eaten per day. The ration can range from nothing to satiation (maximum). Temperature can have an effect on feeding (food consumption) and the efficiency of metabolism, both affecting growth. Oxygen can be as important as food, and fishes of swampy habitats may have to rely on air. Other factors such as salinity (osmoregulation) or other stressors can increase metabolism and reduce the amount of energy available for growth.

METABOLIC STRESS

Stress is due to an exogenous (outside) force (stressor) that disrupts normal metabolism. These forces can be physical, chemical, or biological, and they occur separately or in combination. Such stressors elicit a compensatory physiological response from the fish that is initially adaptive, but it can be detrimental depending upon the severity of the stress and its persistence.

Stressors are usually physicochemical, such as exposure to temperature extremes, low oxygen concentration, extreme salinity fluctuations, high acidity or alkalinity, and combinations of the above. Even more insidious may be the pollution, which can result in waterborne contaminants and poor water quality (e.g., Brown 1993). Combinations of these stressors can be synergistic and result in much greater stress, such as elevated temperature with low oxygen availability.

Biological stressors also can disrupt normal metabolism, such as vigorous exercise needed to escape predators. Often biological stress results in "fright" producing a fright reaction. Examples are handling stress from anglers or due to confinement or transport. The fright reaction can result in an "emotional meltdown" is fish as well as in humans. The results can be serious and has been called the "fight-or-flight syndrome," in which the body prepares to escape or if this is not possible, to fight for life.

In a number of published papers spanning from 1936 to 1973, Hans Selye (1950, 1973) was a pioneer in the study of stress. He defined stress as the sum of physiological responses to a stressor, which strive to maintain a normal metabolism and he developed the *General Adaptation Syndrome* (GAS) to explain the stress response. Also, he documented three stages of physiological changes to stressors that were associated with the fight or flight response (reviewed by Smith 1982 and Wedemeyer et al. 1990, and found on dozens of Web sites):

- *First stage*: Alarm—This stage is an endocrine response that occurs in seconds due to the release of catecholamines (e.g., epinephrine, from the sympathetic nervous system) to minutes due to the release of corticosteroids (e.g., cortisol from the head kidney). This first stage prepares the fish to fight and/or flee, and also supports resistance.
- *Second stage*: Resistance—This stage is associated with acclimation, which may be achieved by action of various organs, and allows the organism to physiologically adapt to the stressor. This acclimation may require a costly expenditure of energy and it can result in a lower level of activity (e.g., susceptibility to early fatigue) or even repress growth and reproduction, but temporarily at least it wins the "fight."
- *Third stage*: Exhaustion—This stage occurs when stress occurs for so long that the compensatory mechanism is no longer able to maintain itself. At this time, the body has exceeded its capability for adaptation and its resistance has become harmful. Death will occur if this stage is allowed to continue.

It is obvious that changes will occur to the individual fish in response to the GAS. Using results from the studies of a host of other authors, Wedemeyer et al. (1990) prepared an outline that addresses the progressive and increasing changes that occur in fishes due to stress, using a conceptual framework. They have organized the changes that occur in the fishes to three responses, which I will group as "changes" to avoid confusion with the above GAS "responses":

- *Primary changes*: Endocrine—The stressor stimulates the pituitary gland to release a message (ACTH) that results in the release of cortisol. Also, the sympathetic nervous system releases epinephrine. In this phase there is an acute stress and a fast response.
- *Secondary changes*: Blood and tissues—Presence of hormones in the blood cause, among other things, an elevated level of glucose in the blood, and a decrease of glycogen in the liver. But soon there is depletion of glycogen, overworking of secretory organs, and osmoregulatory problems due to increased permeability of the gills for obtaining oxygen. A continuing response becomes chronic and drains the body of energy.

- *Tertiary changes*: Individuals, populations and ecosystems—Growth, survival and overall fitness are reduced due to energy drain that lowers disease resistance, reproduction, and recruitment. Ecosystems may be affected due to altered energy flow and disruption of species populations. The bottom line here is that the energy cost of maintaining homeostasis is so great that energy is not available for other purposes.

STRESS IN FISHES

Stress poses a serious problem for fishes and their populations. An in-depth treatment of those problems, how to identify stress in fish, and how to reduce the problem, is beyond the scope of this book, whose purpose is mainly to promote awareness. There are obvious stressors on fishes that we can identify and easily avoid, such as stress imposed by "playing" a "catch-and-release" fish to exhaustion. Hatchery workers also have learned to reduce transport stress in hauling fishes by adding common table salt to the freshwater and thereby reducing osmoregulatory stress. But most environmental problems are much more difficult to deal with.

I attended a presentation on fish diseases by Dr. Stanislos Snieszco, eminent scientist and former Director of the U.S. Fish and Wildlife fish disease laboratory at Leetown, West Virginia. Dr. Snieszco pointed out that diseases in fishes were influenced by three conditions: resistance of the host, virulence of the pathogen, and environmental conditions. Stress in fishes obviously reduces host resistance and makes them susceptible to disease, including some that are usually present in water and cause no harm (Snieszko 1974). He also observed that an outbreak of one disease can result in the infection by another, leading to an epizootic. Wedemeyer et al. (1990) identified 12 causative agents that produced diseases in fishes due to stress, and Piper et al. (1982) listed 11 stress mediated diseases of Pacific salmon, trout, catfish, carp and shad, including furunculosis, bacterial gill disease, columnaris, and others.

The American Fisheries Society hosted a symposium (i.e., Symposium 8) on biological indicators of stress in fish (Adams 1990), which remains a valuable addition to the bookshelf. Individual papers deal with most of the problems likely encountered, and the summary paper (Heath 1990) provided a synthesis of ways that might be used to measure stress in fish, which include biochemical and molecular, physiological and morphological, performance capacity, and ecological variables. However, the author points out that it is difficult to diagnose the cause of stress once detected, and the next step, that is, to alleviate environmental stress may be difficult or impossible. Although we must leave this subject for now, we will also address the topic of stress in chapter 29. This is a case study of fish salvage and imposed stress at the U.S. Bureau of Reclamation fish facility at Tracy, California.

CASE STUDY: MEASURING GROWTH AND AGE IN HARD TISSUES

Age determination in fishes seems to be rather straightforward—just get a few scales and count the rings. Unfortunately it is much more difficult than that. Some fishes have no scales and have to be aged in other ways. Other fish that have scales do not form them until after the fish is hatched. Some fish shed scales easily and replace them with new ones that have no early rings in them. Some fishes have scales that are most difficult to age. Tropical fish may not produce annual rings, and most importantly ages need to be validated in some way, such as determining growth in recaptures of previously aged fish.

Age and growth in some common fishes, usually sport fish that are intensively managed, have been studied by many persons all over the continent. Their growth rates are well known for widely separated geographic areas and the ages have been repeatedly validated. However, age and growth in most endangered fishes are rarely known before they are listed, and then they are difficult to study because of rarity.

Colorado pikeminnow historically grew to perhaps 2 m TL, but their maximum longevity in years is unknown. Aging these fish by scales has been problematic due to one or more missing annuli in young fish, indiscernible annuli, possible erosion of the edges of scales in old fish, and no age validation for large (old) fish. Also some studies suggested that scales are not reliable for pikeminnow older than age 10 (Osmundson et al. 1997). After talking with Gerry Smith at the University of Michigan, I also realized that there were quite a few museum specimens of Colorado pikeminnow around the country that might be used to compare historic growth rates with present condition if ages could be determined for some hard structures such as vertebrae, and also there was a need to compare different ages obtained from a variety of structures and techniques. So we rounded up all of the specimens we could find and tried to get scales, otoliths, and vertebrae from them.

We had some fish that were frozen, some collected many years ago and some recently and we were able to extract pairs of otoliths from them. I got one of Gerry's students (Dave Schultz) to age one of the pair using thin sections, and W. L. Minckley aged the other one by grinding. We also took some vertebrae from fish and had many scale samples. Unfortunately, after moving to the University of Colorado I did not have the time to complete the study and the samples sat in a desk drawer for a few years. One day out of the blue, a student knocked on my door and said, "Hi, I am Lorraine Hawkins and I heard that you worked with fish. I would like to do some statistical work on some fish data. Do you have anything I could do to help you with some studies?" To make a longer story short, I gave her the samples and she attacked them with a purpose. Between the two of us we soon added ages estimated from vertebrae and scales to the otolith results.

We were able to obtain and age all four structures from only eight fish (Table 21.3), but the results show the close agreement between vertebral centra and sectioned otoliths and the underestimation of older fish by scales. Polished whole otoliths produced the most variable ages. These findings were supported by the larger samples (48 fish) of unpaired structures (Figure 21.4). Ages obtained from vertebral centra (ages 2–23) were more precise for aging Colorado pikeminnow ($r^2 = 0.73$) as otoliths (0.61) and much better than scales (0.59) or whole otoliths (0.33). This was important, because museum specimens usually have been soaked in formalin and this can ruin small delicate structures like otoliths. You can read more about this study in Hawkins et al. (2004b).

Table 21.3 Ages of Colorado Pikeminnow Estimated from Scales, Sectioned Otoliths (s), Whole Otoliths (w), and Vertebral Centra Removed from Each of the Eight Fish

Fish Number	Total Length (mm)	Vertebral Age	Otolith (s) Age	Otolith (w) Age	Scale Age
1	324	5	5	5	5
2	445	4	5	5	6
3	453	5	4	6	7
4	471	7	5	6	5
5	478	8	7	7	6
6	492	5	6	8	6
7	518	8	9	5	6
8	602	10	10	10	7

Source: Hawkins, L. A., et al., Southwestern Naturalist, 49, 203–208, 2004b. With permission.

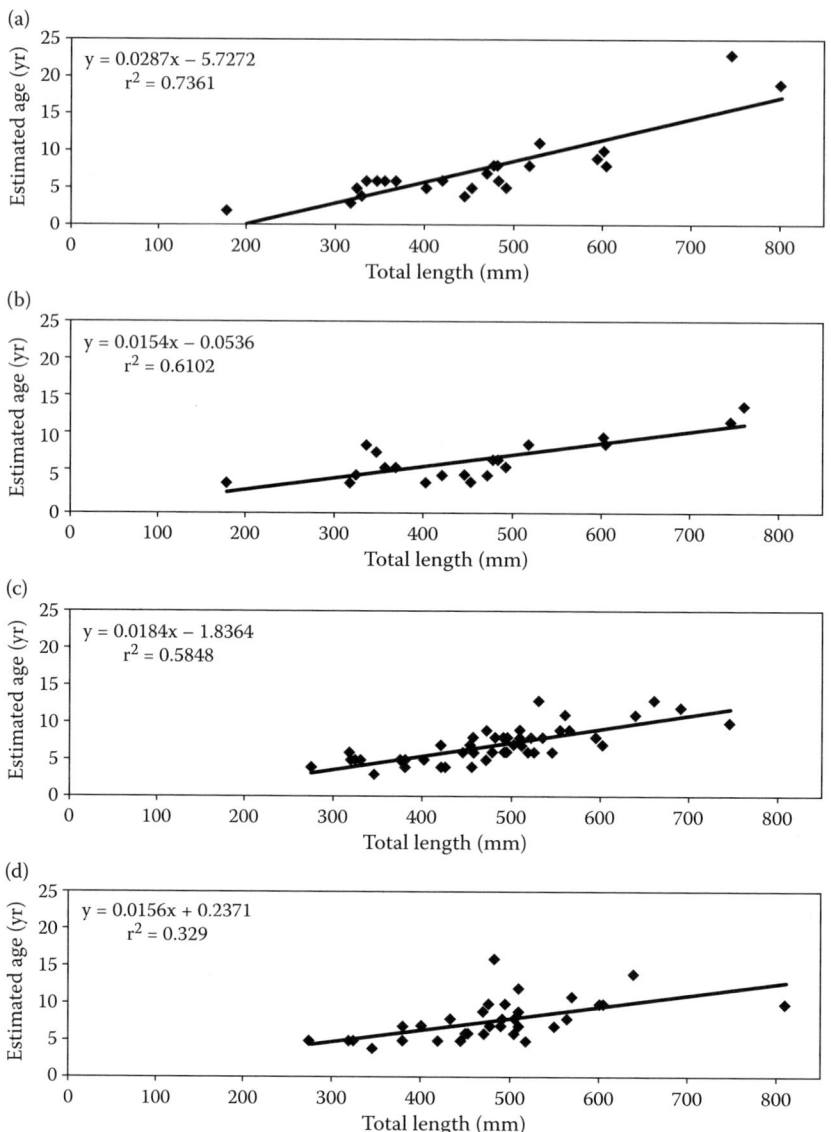

Figure 21.4 Colorado pikeminnow total length vs. estimated age. (a) Ages from vertebrae, (b) from sectioned otoliths, (c) from scales, (d) from whole otoliths. (From Hawkins, L. A., et al., *Southwestern Naturalist*, 49, 203–208, 2004b.)

SUMMARY

Fish are consumers and must obtain energy in their diets. Energy budgets are helpful in determining how the energy inputs and yields affect fitness. Growth and reproduction are the last yields in the energy equation and can be curtailed by increasing energy costs associated with activity and stress. Calculation of condition factors are based on relative plumpness and are helpful in evaluating environmental conditions. Fishes vary in their ability to digest different foods and they have indeterminate growth, which can be a good indicator of environmental quality. Annual growth rates can be calculated from use of hard structures or from holding or tagging individuals. Fish are very susceptible to stress and the GAS is useful in understanding the mechanism that produces primary, secondary, and tertiary changes in fish as individuals and populations. Age determination in hard structures is explained in a case study of adult Colorado pikeminnow.

CHAPTER 22

Adaptation, Niche, and Species Interactions

ADAPTATION

All students of biology and ecology should be familiar with the concepts of adaptation and the ecological niche. However, many students that enter senior level ecology course seem a somewhat vague about adaptation and the niche, and the significance, meaning, and application of such concepts. Accordingly and at the risk of oversimplification and redundancy, I will attempt to provide a better baseline here that the student may build upon: first, that the niche concept is all about adaptation and tolerance; second, that there are at least three ways in which the niche concept has been applied.

Historically, the term "adaptation" has had two closely related but different meanings in evolutionary biology, one that relates to evolutionary processes, and the other has to do with features of an organism (Burian 1992). The first meaning considers the alterations in organisms from generation to generation, which allow them to improve their ability at reducing the adverse effects of some environmental factor. In this case, natural selection would dictate that those organisms best "adapted" to solving or reducing the effects of those environmental problems would have the highest survival rate. The second is a trait or capacity of an organism that is produced by adaptation.

Most recently, the term has been used in a more functional and specific way. Adaptation is considered a process rather than an alteration or a trait per se. It is "the evolutionary modification of a character under selection for efficient or advantageous (fitness-enhancing) functioning in a particular context or set of contexts" (West-Eberhard 1992). Thus, a character produced or modified as result *of* adaptation, would be *an* adaptation.

Obviously, it would be difficult to test each character of an organism to see whether it is due to natural selection, but we can safely assume that organisms have "inherent qualities" that make them able to tolerate some environmental conditions and that these qualities can improve fitness. These inherent qualities are synonymous with adaptations, which are produced by the process of evolution. In summary, an adaptation is a heritable response to abiotic and biotic factors in the environment and the response in an individual may be morphological, physiological, and/or behavioral. As noted by Odum (1971), the most important abiotic factors affecting organisms in the terrestrial environment are likely to be differences in temperature, light, and water, while in aquatic systems, they are more likely to be temperature, light, salinity, and, in some cases, oxygen. Adaptations to biotic factors are much more complex but, in most cases, are related to interactions at population or species levels.

Adaptations to one set of environmental conditions, as judged by the degree in which overall fitness is improved, may not allow the organism to do as well under another set of conditions. Thus, these adaptations are presumably working under some sort of compromise as previously discussed for metabolism using the Fry paradigm (Fry 1971) in the previous chapter. A way of looking at this in general is to consider the concept of *tolerance*.

In an ecological sense, tolerance generally refers to the ability of an organism to exist within a range of one or more environmental factors, such as temperature or oxygen concentration. In this case, existence may be divided into survival, growth, and reproduction. Also, there will be a lower limit (e.g., cold) and an upper limit (e.g., hot) for each factor, beyond which the organism cannot survive. These values delineate the "limits of tolerance" for an individual or species. V. E. Shelford (1911) used tolerance to develop the "major conceptual tool for physiological ecology" (Krebs 2008, p. 36), which is known as *Shelford's Law of Tolerance*. Krebs (2008) succinctly states the law as follows: The distribution of a species is controlled by the environmental factor for which the species has the narrowest tolerance. However, Allaby (2005) places Shelford's 1911 proposal in a wider context in which the presence and success of an organism depend on a complex mix of environmental factors, and the deficiency or excess of any one factor may approach the limits of tolerance for that organism. Figure 22.1 provides the classic view of tolerance, in which populations are most abundant in conditions where environmental factors occur within an optimal range, rare due to physiological stress in areas where factors are less than optimal, and absent in areas in which the factors are outside (i.e., below or above) the limits of tolerance.

Tolerance provides a way of looking at adaptation under different conditions and specific circumstances. Given some time (e.g., as resources are depleted or change), an organism may adjust metabolism and acclimate to conditions that might otherwise be intolerable. In addition, individuals can show different ranges of tolerance if they are living in different geographic regions (acclimatization). Also, a major adaptation (such as may occur in ontogeny) may require changing tolerance limits to several environmental factors at once. In this context, the ability of an organism to adjust the phenotype in response to environmental influences is known as adaptive flexibility, or "phenotypic plasticity" (e.g. Allaby 2005), as previously discussed for the Devils Hole pupfish (Lema 2008). Next we consider the sum total of all the adaptations of individuals and species, and the integration of all tolerances, which, shape the *ecological niche*.

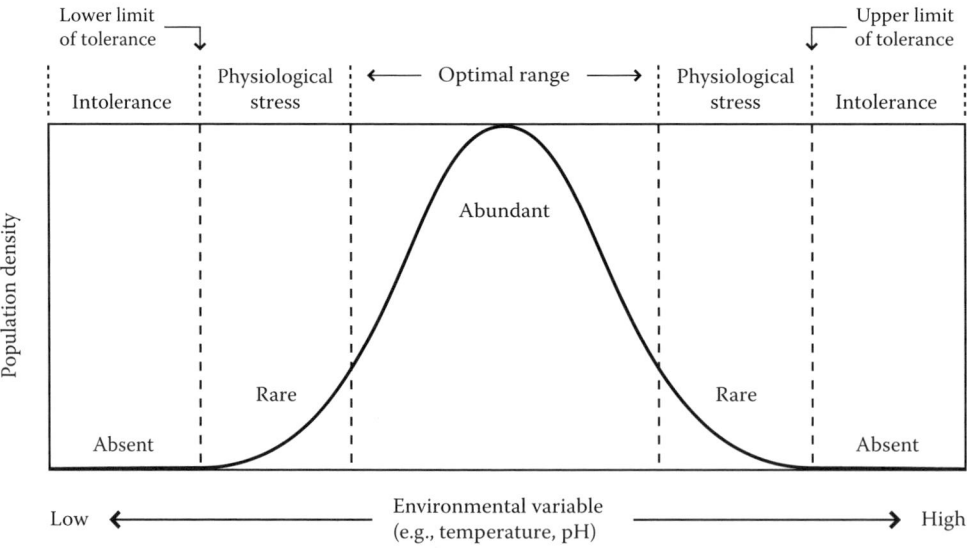

Figure 22.1 Concept of tolerance, as proposed by V. E. Shelford (1911).

THE NICHE

As a concept, the ecological niche (to ecologists) has been as elusive as the concept of species (to systematists), and it is arguably just as important. On the other hand, some important scientists have considered the concept unnecessary or better left undefined (e.g., Margalef 1968; Bell 1982). Why such differences of opinion? In some instances, it is a result of different interpretations of the niche concept. Colinvaux (1993) identified subtle differences in three compatible interpretations of the ecological niche, all of these concepts being linked in various ways:

- *Eltonian niche*: After Elton (1927)—functions within a community. In this case, a niche could be filled with *ecological equivalent species* such as the niche of a large carnivore (occupied by lions or tigers):
- *Species niche*: After Grinnell (1904)—properties of a species population. A niche can be occupied only by individuals of the same species:
- *Multidimensional niche*: After Hutchinson (1957)—qualities of the environment. A niche has two components. A fundamental niche of a large size that results from the resource requirements (abiotic and biotic factors) of an organism, and a realized niche that is reduced in size by sharing resources with others.

I have found that students like confessions from professors so I will relate some of my previous experience (and shortcomings). Early in my education, I learned (in my first ecology course) that an organism had an address, which was its habitat, and it also had a profession, which was its ecological niche. Also, I read in my textbook that the niche was "inherited and/or learned" (Odum 1959, p. 27). I was never happy with these explanations, and to be truthful, I just could not figure out how habitat and profession could be so easily separated— or how you learned a niche! I can only blame myself now for not having read a little further, because on page 231, Odum (1959) cleared up (or muddied?) the situation to some extent, by stating that an ecological niche "involves not only the physical space occupied by an organism but also its place in the community, including its energy source, period of activity, etc." This seemed like an untidy hodgepodge of things to have in a niche, and the process of adaptation was not offered up as a way organisms developed a niche. However, it was made clear that once a niche was defined, then an organism or population occupying that niche could be fatally affected by a competing organism with an overlapping niche. It is confusing (and perhaps a little melodramatic), and more background is needed.

The term "niche" in general usage (e.g., in my pocket dictionary) and in older dictionaries is given as an alcove or recess in a wall or room in which a curio or statute could be placed, or something (such as a position) for which a person is especially suited. It took some digging, but I located a recent source (Dictionary.com, © Random House, Inc., 2009) that defines the niche in two newer ways: As a segment of the stock market and, related to ecology, "a position or function of an organism in a community." Synonyms include a calling or vocation, or a berth, cubbyhole, and so on. There are similarities between the general use of the term "niche" and the term "ecological niche," but making such comparisons has been difficult. The term "ecological niche" (simply "niche" used here) was first used at least a century ago and different concepts have been developing since then. In piecing together the story and philosophies concerning the niche, I have greatly benefited from accounts provided by Hutchinson (1978, 1979) who knew many of the players and was a part of the historic process, by Griesemer (1992) and Colewell (1992), who have thoroughly investigated the development of the concepts and also benefit from hindsight, and Schoener (1989) who also provides a historical perspective. However, after 20 years of reflection Schoener (2009) somewhat abandoned the historic path and provided a more general perspective and useful basis for understanding concepts.

In discussing his development of the multidimensional niche concept, Hutchinson (1979, p. 243) relates the importance of competitive exclusion in development of the niche concept. He considered competitive exclusion principle to be a "simple, powerful, and sometimes incorrect generalization . . . but it is actually the theoretical basis of a very large part of modern ecology." The competitive exclusion concept was mentioned by several scientists between 1857 and 1924, but Hutchinson (1978) gives Volterra and Haldane credit for developing the concept. Gause demonstrated it in experimental work and he stated that due to competition, "two similar species scarcely occupy similar niches, but displace each other in such a manner that each takes possession of certain kinds of food and modes of life in which it has an advantage over its competitors" (Gause 1934, cited in Hutchinson 1979). Joseph Grinnell (1904) also developed this idea and used the term niche in 1914 to represent the space "in which two closely allied species did not co-occur" (Hutchinson 1979, p. 244).

However, Hutchinson (1978) credits R. H. Johnson as the first person who used the term niche (in 1910) in an ecological sense. Shortly thereafter, J. Grinnell used the term "ecological niche" in his 1913 thesis. Taylor, who worked with Grinnell, also published a paper in 1916 using the niche concept, and Grinnell (1917) then published a paper with niche in the title. Soon after, Charles Elton (1927) addressed the niche concept in his famous book, *Animal Ecology*. So the niche was off and running! These early concepts of the niche were similar, but not exactly the same.

Griesemer (1992) points out that Grinnell focused on distribution and evolution, envisioning niches relating to abiotic and biotic factors—each niche the property of but one species. On the other hand, Elton was focused on trophic (food) relationships as the structure of animal communities he envisioned the niche in terms of animal relationships to food and enemies, but he also was conscious of many factors that influenced niches. Thus, both early proponents of the niche concept believed in a niche shaped by biotic and abiotic factors, and they thought of the niche as a place, or a role in the environment that a species occupies. The two differed, however, on the uniqueness of the niche, with Grinnell believing in one niche–one species, and that species with similar adaptations to food might interact to the detriment of one by the more locally adapted one—effectively excluding the less favored. Elton believed that niches could be occupied with different species and did not mention the concept of competitive exclusion. (As a side note, Elton's view persisted long into the middle of the twentieth century, even after Hutchinson's concept was unveiled, e.g., see Margalef 1968, p. 6.)

Controversy about the finer points of the niche concept finally reached a pivotal point with the emergence of mathematical modeling of two species by Volterra, which demonstrated the mixed effects associated with predator–prey relationships, and testing of those interactions in laboratory experiments by different approaches used by Gause (1934) and Park (1948, 1954). Essentially, these experiments upheld the tenant of the competitive exclusion principle, which states that two species with the same food habits are not likely to remain in balanced numbers in the same region, and one will usually crowd out the other. These studies demonstrated that species with similar niches can have an effect on one another. However, outcomes indicate that slight differences in climate may favor one or the other of two species placed in competition.

The ecological niche concept continued to develop. Due to his ability to incorporate the fuzzy, theoretical concepts into a mathematical model, a revolutionary new approach to the ecological niche was proposed by G. Evelyn Hutchinson in 1957. The new model was constructed by incorporating earlier concepts of the ecological niche and by describing the niche in mathematical terms. In my opinion he did not refute earlier concepts, rather he seems to expand them. In reviewing his seminal 1957 paper, I will attempt to condense, simplify, and make a few observations as to the use of his model.

Hutchinson (1957, p. 416) used an opportunity to make concluding remarks at a symposium as a vehicle to present his concept of the niche, which he described as a multidimensional hypervolume of resource axes. In a nutshell, he used "the limiting values permitting a species S_1 to survive and reproduce" for two independent environmental variables (X1 and X2), which would plot an area (two dimensions) that shows the environmental conditions under which S_1 can exist (e.g., the tolerance range). Then he adds another variable, X3 whose range of tolerance is introduced as a third axis, to result in a volume (three dimensions) of all possible conditions under which S_1 can exist. He continues to add variables (X4 – Xn), which are all of the ecological factors affecting S_1. "In this way an n-dimensional hypervolume is defined, in which every point corresponds to a state of the environment which would permit the species S_1 to exist indefinitely. For any species S_1, this hypervolume N1 will be called the fundamental niche of S_1."

Hutchinson (1957, p. 416) further states that this fundamental niche is "merely an abstract formalization of what is usually meant by an ecological niche." In addition, he proceeds to define a realized niche, which is further clarified in a later publication (Hutchinson 1978): Two fundamental niches can be separate or overlap; in this case, they are called a preinteractive niche, or a postinteractive niche. In case of the latter condition, a second species may or may not interact with the first, but in case of overlap competitive exclusion would occur with either one or both of the fundamental niches constrained by the other.

One of the most compelling demonstrations of fundamental and realized niches was presented in a study of two species of barnacles on a Scottish coast. Connell (1961) did the study and found that young of a species may have one niche and adults another. Also, settlement of the larvae of one species represented the fundamental niche, while the surviving adults showed the realized niche.

Hutchinson (1957, p. 417) recognized some restrictions imposed by his model, and I paraphrase some relevant to our purposes: The described niche will have optimal and suboptimal regions. The model is instantaneous, representing a single instant in time. The model is devised to clarify niche specificity (one niche per species). Fundamental niches may have points in common (they intersect) or not (they are separate).

In reviewing Hutchinson's contribution as a "geometrical abstraction of the niche," Griesemer (1992) gives Hutchinson credit for a radical change in the way the niche is viewed: Instead of a place or role of an organism in a community, the niche considers how individuals of a species utilize environmental properties. Thus, the niche, once considered to be an attribute of the environment, has been redefined as an attribute of the species (or population of a species in a specific location) that is related to its environment; in other words, an ecological description of the population phenotype (Schoener 1989; Colewell 1992). MacArthur (1968) provided a helpful analogy between ecology and genetics. As a conceptual aid, we can associate the genotype with the fundamental niche and thus the phenotype would be likened to the realized niche.

Emergence of the Hutchinsonian concept has provided a powerful tool for conceptualizing the ecological niche. However, as a model Hutchinson (1957, 1978) clearly stated that the multidimensional niche approach has restrictions, and its early evaluations were preoccupied with competition. Colewell (1992, p. 247) reflects that "niche theory has become synonymous with competition theory." He made an earlier plea for broadening the niche concept to include other species interactions (i.e., predation, parasitism and mutualism).

Major physical forces have resulted in fish extinctions on Earth; however, biotic factors are extremely difficult to assess over geologic time scales. In paleoecology, we considered massive extinctions as the reasons for radical changes in biodiversity. However, early niche concepts considered abiotic factors and the niche to be stable. Although the basic concept of the niche considers all of an organism's adaptations, the interest and most progress early on were in evaluating the biological component of the environment. No doubt, adaptations produced as a result of species interactions (including but not limited to competition) have been a major evolutionary force, but there would be more comfort if niche investigations had placed the same amount of effort in considering abiotic and biotic factors as causes.

Schoener (1989) reviewed and discussed the development and historic concepts of four concepts of the niche, and made a case for overcoming some of the constraints imposed by the multidimensional model. In his latest paper (Schoener 2009) he reiterated some points, but he also developed new ways (and new terms) in which to consider the niche. He has combined the contributions of Grinnell and Elton and provided three concepts of the ecological niche:

(1) *The recess/role niche* considers the niche as the role, place, or recess occupied in an ecological community, thus combining the habitat and trophic components mentioned earlier. Basically, this is the historic niche attributed to R. H. Johnson (in 1910) and J. Grinnell. In this case, there may be species that occupy different locations and perform the same ecological role, which are "ecological equivalents," and there may be "vacant" niches.
(2) *The population/persistence niche* is the multidimensional niche of Hutchinson (1957). It is a quantitative description of the range of conditions that allow for the persistence of populations of a species in an n-dimensional hypervolume. The conditions are best studied in laboratory conditions. This concept focuses on species; there are no ecological equivalents nor vacant niches.
(3) *The resource/utilization niche* is an operational construct of the Hutchinsonian model, and it is quantitative and multidimensional. It differs in that it considers what organisms actually do, or what resources they utilize, rather than having to determine a range of values that allow for population persistence. Attributed to Robert MacArthur and Richard Levins (in 1967), it describes the natural history parameters of an organism, which are precisely the data collected in field studies.

Finally, Schoener (2009) discusses recent techniques, or environmental niche modeling (ENM) which can combine all of the three niche concepts thus presented. Using computer aids such as geographic information systems, species can be compared according to their realized niches, and related to evolutionary processes. When we use the statement "ecological niche" we need to be aware of the various concepts and to communicate which concept we are using. There is no ecological niche. There are concepts of the ecological niche.

NICHE OVERLAP AND RESPONSE

Fish are very plastic in the use of resources, especially food sources, as we will study later. Thus, it is not uncommon to find more than one fish species eating the same thing. This would constitute *niche overlap* for that resource; however, if it is plentiful and not limiting, it would not constitute competition. Fishes move and access different foods, and usually forage on the most abundant items. In theory, niche overlap of this sort might be considered to be competitive exclusion, but in actuality it happens very frequently. Using Hutchinson's (1957) concept of the niche, two species might overlap in fundamental niches with little problem. Since the niche concept is for only an instant of time, we could imagine that feeding at different places or times of day also might produce momentary overlap with little consequence over the long haul. However, in cases of extreme overlap there may be a response over time.

Responses to niche overlap for a resource can be detected in the *niche width* (i.e., niche breadth) of a population. In this case, a simple graph of the use of a resource along a resource gradient will provide the visual. A narrow distribution indicates a trend toward specialization, and a wide distribution indicates generalization. Resource use thus outlined may overlap between two or more species, but since overlap means less available for both species, the degree of overlap is usually decreased over time (if it is extensive) by natural selection.

Niche shifts tend to reduce interspecific competition. An ecological response can occur short term to reduce interaction, usually by behavior, but an evolutionary response also can occur due to a change in morphology or behavior. Werner and Hall (1976, 1977) noted a shift in behavior between three sunfish species that were stocked separately, and also together in ponds. Bluegill (*Lepomis macruchirus*) seemed to be the most flexible species, changing its feeding habits and becoming more pelagic when placed in the same ponds with pumpkinseed sunfish (*L. gibbosus*), or with green sunfish (*L. cyanellus*). In another study where bluegills and pumpkinseed lived together (Robinson et al. 2000), bluegill removal resulted in pumpkinseed expanding their feeding range by moving into pelagic areas previously used by bluegill and also changing their morphology. In a niche shifts such changes, are called resource polymorphisms, and they also occur in ontogeny (development of fishes from embryo to the adults), to be discussed in detail in "The niche revisited" section of chapter 28.

Niche compression is a related term, for the contraction of habitat use. It applies to situations where a competitor invades a broad and patchy niche and forces the original inhabitant to utilize less space (in case of fish this is usually intraspecific competition and serves to reduce the amount of energy needed to defend a territory that is larger than required).

On the other hand, *niche expansion* may occur when competition is reduced. Also called ecological release, as in case of birds that invade islands that lack competitors, but it could occur if a competitor species is removed due to disease or eradication.

Finally, a long term adaptation called *character displacement* can gradually separate species by a shift in a character trait. In fishes, such a change has been observed in a shift in the number of gill rakers present in whitefish (*Coregonus clupeaformis*), which was related to the presence or absence of a competing cisco (*Coregonus sardinella*). In presumed competition with the cisco, whitefish populations are dominated by fish with fewer gill rakers (Lindsey 1981; Diana 1995). In another example, alewife competition on a whitefish (*Coregonus hoyi*), reduced the number and length of gill rakers, and resulted in a habitat shift (Crowder 1984). Character release also may occur with removal of a competitor, such as in the case of pumpkinseed above (Robinson et al. 2000).

SPECIES INTERACTIONS

Intraspecific Competition

Species can interact within a species population or between species. It seems that most texts I have read emphasized interspecific interactions, and as a result, most students do not consider intraspecific competition to be a serious issue. In reflection, however, they soon realize that intraspecific interactions can be a powerful force. In fact, intraspecific competition is of the worse sort, because it is between individuals living in the same space and eating the same food. If any one of the necessary resources is limited, and there usually are limitations, there is an intense competition for it, with a major influence on the population.

In subpopulations of easily studied species, such as insects, it has long been discovered that intense intraspecific competition can result in the death of every member. In an early experiment on flies, Nicholson (1954) noted that if there was not enough food and it was spread out evenly, the flies scrambled for it, ate it all up, and the entire population died because each competitor received insufficient food for survival. He called this "scramble," or *exploitative competition*, (reviewed by Parker 2000) in which all exploit the remaining resources. Another type of competition is contest, or *interference competition* (Park 1954) in which there is division between individuals in such a fashion that the resource is not shared equally. The most able or higher ranking members obtain more, such as in hierarchical (e.g., size dominance or territorial) situations.

The influence of intraspecific competition can result in an increase in niche width (i.e., a tendency toward being generalists) if there is a wide range in characteristics of an important resource. This tendency is produced by size difference in many fishes, in which intense competition for an optimum resource depletes it, and resources at the ends of the gradient become more attractive. On the other hand, if there is competition for a resource with a comparatively narrow range of characteristics, then intraspecific competition could result in reduced growth and condition, with a reduction in niche width.

Interspecific Interactions

It is fashionable to use the term "species interactions" instead of "symbiosis" due to problems in definitions. Symbiosis means living in the same place. It implies no specific interaction, but the word has often been used in an obligatory sense or used instead of "mutualism" (historically, Odum 1971, p. 229, used obligate symbiosis with respect to mutualism). The current use of symbiosis implies two species that have been living in close association for a long time, which suggests that an obligate condition might exist (e.g., see Smith and Smith 2001). Reluctantly, I am going to avoid the word, instead inserting "sympatric" where possible to replace symbiosis.

Populations of two species, S_1 and S_2, living sympatrically may affect each other in various ways. Using (+) to indicate positive effects, (−) to indicate negative effects, and (0) to indicate no apparent population effects, they are depicted as follows:

Interaction	S_1	S_2
Mutualism	+	+
Competition	−	−
Predation	+	−
Parasitism	+	−
Commensalism	+	0
Amensalism	−	0
Neutralism	0	0

I will mention a few cases of such interactions in this chapter, and we will visit many more situations in the remainder of the book.

Mutualism has received much recent attention by evolutionary biologists who study coevolution. In fishes, there are numerous examples of situations among, or between fishes, and other organisms that benefit both species. Many fishes school with other species as an aid to finding food and to escape predation. An obvious mutualism exists with fish species (such as gobies, angelfishes, butterflyfishes, and wrasses) that specialize in "cleaning" other species by removing and feeding on parasites and dead skin. Cleaners (goldsinny wrasse *Ctenollabrus rupestris*) have even been recruited by the aquaculture industry to remove sea lice and necrotic tissues from salmon (Bone and Moore 2008). Another mutualism includes gobies and shrimp that live in burrows: the goby uses burrows but cannot dig them, whereas the shrimp has poor eyesight and needs the goby to aid it back to its burrow in case of danger (Preston 1978). Finally, everyone recognizes the colorful clown fish (AKA Nemo) that lives among the poisonous nematocysts of the sea anemone. The clownfish feeds on scraps of the anemone's prey, while enticing new victims within range. Fishes also can have mutualistic associations with other organisms, such as luminescent bacteria.

Competition has been discussed to a great extent in this chapter (Table 22.1). In general, it means that one or more individuals, or species in this instance, use a resource that is limited. In terms of the niche, this means that the fundamental niches of two or more species are overlapping. Contrary to intraspecific competition, the population size vs. resource use graph will show that the niche will be affected by moving the population curve to the most optimal resource use, perhaps resulting in a tendency toward specialization. In most cases such interactions will be done without aggression, and it involves sharing of food or space. However, regarding mechanism, interspecific competition has been placed in six categories (Schoener 1985, reviewed by Stilling 1999):

- *Exploitative* or consumptive, which refers to competition for resources. This is the most common type of competition.
- *Preemptive*—competition for space and *overgrowth*. In which one organism is deprived of light or resources by another, and this usually occurs only in sessile plants and animals.

Table 22.1 Exploring Competition: Niche, Resources, and Tests

Competition: over resource use—food, space, and mates.
Intraspecific—tends to broaden resource use away from optimum.
Interspecific—tends to restrict resource use toward optimum.
Competition occurs due to the following:
    Interference—when resource access is limited directly or indirectly.
    Exploitation—when resources are limited.
Competition is difficult to prove:
    Hard to determine in nature; lab studies may not be credible.
    Resource sharing is not competition.
    Individual fish of the same species may eat different foods.
    Must consider different life stages for ontogenetic niche shifts.
Tests of competition:
    Competitive inhibition—introduced organism changes resource use in original residents due to constraints on the niche.
    Competitive (ecological) release—removal of suspected competitor changes resource use due to expansion of the niche.
    Character displacement—change in morphology due to the presence or absence of a potential competitor.

- *Chemical*—one organism displaces the other by use of toxins, mostly in terrestrial environments, presumably, due to difficulties in concentrating chemicals in water.
- *Territorial*—one organism defends space against another.
- *Encounter*—organisms interact directly for specific resources.

Competition has been reported in perhaps 40%–75% of species that have been studied. This large range of values was reviewed and discussed by Stilling (1999), who reported valid reasons why competition might be over- or underestimated. Competition may be overrepresented due to difficulty in publishing negative results, and by bias of people who wish to study competition. It may be underrepresented because organisms have evolved to avoid competition or it may occur only during years when resources are limited. Competition has been studied intensively from a theoretical standpoint, as we will see in the next chapter when we discuss the logistic model.

Competition among fish species is difficult to observe and more difficult to prove. The reasons for this are many, but associated with the difficulty of studying aquatic organisms, the great mobility of most fishes, trophic adaptability, phenotypic plasticity, and ontogenetic separation of life stages. Most fishes are generalists and consume a wide variety of prey. They also are able to feed at different trophic levels. As with most animals, fish are very active and forage to obtain the most energy; thus, when one food item becomes low in supply they switch to something more abundant. For these reasons it is not unusual to find fish consuming the same foods, but not actually competing. The other reason is that different foods and resources may be used by the postlarvae, juveniles, subadults, and adults. For example, most sucker species start feeding on zooplankton when young and switch to other foods, such as algae and detritus, when adults. Competition has been observed in laboratory studies, and indirectly by character displacement. It is my belief that a potential competition for food has been investigated too much, while competition for space may be the most important factor for most of the large piscivores. After an extensive and thoughtful review, Matthews (1998) concluded that, although interspecific competition occurs in fishes, it is more likely that fishes share common foods when they are abundant. He believes that resource sharing is not usually a regular or highly competitive situation, and he did not consider competition to be the dominant factor in structure or function of stream fish communities.

Detecting competition is difficult, but theoretically possible, and Diana (2004) has provided three "tests of competition." First: Competitive Displacement. In this case, introduction of one species into the habitat of another can result in a change in resource use by the original inhabitant.

Such change is the use of different habitats or consumption of different foods in the presence of a competitor. Second: Character Displacement. Perhaps the easiest to detect is the change in physical appearance due to competition or predation (inducible defenses) (discussed in chapter 25). A change in growth rate in a different habitat, or change in the number of gill rakers due to a change in food eaten (e.g., alewife effects on a whitefish, *C. hoyi*) (Crowder 1984) are examples. Third: Ecological Release. The antithesis of the first test, here a competitor is removed and the affected fish changes resource use. In practical application the easiest and most reliable test would be the second. Diana (2004) also points out some conflict with these tests and the compression hypothesis of MacArthur (1972), which holds that competition forces organisms into smaller space but they consume a wider variety or food. Perhaps both are right depending on the organisms and habitats.

Fishes are consumers, and predation (carnivory and herbivory) in fishes is common. Predation can have a great influence on prey fishes, affecting populations through induced local (ecological) and evolutionary change. Predation will be discussed more fully in the chapter on foraging.

Parasitism is common among fishes, and they are host to many organisms. Also, some fishes such as the sea lamprey are themselves parasites. Most of the larger parasites of fishes are worms and fishes transport roundworms, tapeworms, and flukes. After cleaning at least a ton of marine sport fishes including billfishes mackerels and tunas, I can attest to that fact because I have seen heavy infestations in muscle tissues (no sushi for me!).

A well-known cestode is *Diphylloborthrium latum*, the broad tapeworm, which can be transmitted to humans in undercooked freshwater fishes. A similar tapeworm (*D. pacificum*) occurs in marine fishes. The invasive Asian fish tapeworm (*Bothriocephalus acheilognathi*) can cause death in carp and has infested endangered Colorado River fishes (U.S. Geological Survey, NWHC Information sheet, August 2004). Trematodes also can be a problem, especially the fluke *Nanophyetus salmonis*, which is carried (cecaria) by salmon and trout, and transmitted also to wild mammals, dogs, and humans. Fishes also are hosts to many diseases (e.g., Post 1987), such as whirling disease caused by a myxosporidian (*Myxobolus cerebralis*). This disease originated in Europe and brown trout presumably evolved with it, because the fish is very resistant to it. Other trouts such as rainbow trout and most Pacific salmon can be devastated by infection. Damage due to infection mostly occurs in young fish because disease spores migrate through the tissues and along nerves, eating cartilage. Fish disease specialists (e.g., Dr. Snieszco) believed that whirling disease, formerly absent in the United States, would never be a problem due to federal quarantine laws and ease of detecting it. However, relaxation in inspections resulted in the importation of diseased fish eggs with subsequent spread throughout most of the country. In many cases, hatchery stockings have shifted from Western salmonid species to the resistant brown trout. Another parasite of salmon, several species of sea lice (copepods and especially *Lepeophtheirus salmonis*), is causing problems in salmon farms and can kill juvenile salmon (Hume et al. 2004).

Special forms of parasitism also occur in fishes, such as nest parasitism by several species of minnows, and between minnows and sunfishes (Matthews 1998). Brood parasitism also occurs by catfish (*Synodontis*) and at least six species of mouth brooding cichlids (Barlow 2000) are so parasitized in the East African great lakes.

Commensalism is widespread among fishes. Remoras are well-known hitchhikers of sharks and other large fishes where they benefit from free transportation and a free meal cleaning up scraps. A similar situation is caused by large suckers that suspend invertebrates as they feed on the bottom, attracting other small fishes that benefit from consumption of the leftovers. Small schooling fishes of several species also may take advantage of the protection afforded by a large fish, especially slow moving sharks, most of the time at extreme annoyance to the shark.

Amensalism can occur in fish due to various types of poisonous substances. Toxic byproducts from algae can results in incidental death of many species (e.g., red tide). Amensalism also can result from the presence of larger or more aggressive species, such as crayfish. The larger organism is avoided by smaller ones, thus depriving them of resources, yet there is no effect on the larger.

SUMMARY

An adaptation is a heritable trait that improves fitness, or modification of a character due to its fitness-enhancing role under a certain set of conditions. In aquatic systems and fishes, adaptations are likely to be related to temperature, light, salinity, oxygen, and interactions at individual or species level. The sum total of all organisms in a species is generally regarded as the ecological niche. In the past, the niche concept has been related to an organism's habitat use or food consumed. A new paradigm was advanced by Hutchinson in which the niche was defined as an "n-dimensional hypervolume" of resource axes, including all of the adaptations of an organisms (fundamental niche) and a niche that is constrained by encroachment of others (realized niche). There are at least three main categories of the niche considered today: *the recess/role niche* considers the niche as the role, place, or recess occupied in an ecological community; *the population/persistence niche* is the multidimensional niche of Hutchinson (1957); and *the resource/utilization niche* is an operational construct of the Hutchinsonian model (or ecological niche). It differs in that it considers what organisms actually do, or what resources they utilize, rather than having to determine a range of values that allow for population persistence. In some cases, environmental niche modelling combines all of the three niche concepts thus presented. When we use the statement "ecological niche" we need to be aware of the various concepts and to communicate which concept we are using. Niche overlaps are common in fishes with abundant food and likely not competition. Niche shifts are usually behavioral and are most common in the ontogenetic development of fish. Competition can occur in fishes as intraspecific and interspecific. Intraspecific include scramble or exploitative competition and interference competition, which are the worse kind because of the same resource base between individuals. Interspecific interactions can vary from mutualistic, such as cleaner fish, to predation or parasitism. Interspecific competition tends to promote more specialization and has been reported in 40%–75% of species; however, it is difficult to observe or to prove. Predation is typical because most fishes are carnivorous. Parasitism is common and can occur in many forms from direct to brood parasitism. Commensalism (symbiosis) also is common among fishes such as remoras, and between suckers and minnows. Cases of Amensalism include toxic effects of red tide and loss of resources due to larger or more aggressive organisms.

CHAPTER 23

Populations, Growth, and Regulation

FISH POPULATIONS IN GENERAL

Fish experience great differences in conditions in the wide array of habitats they occupy. In the case of freshwater fishes, we have seen that they may occur in isolated single species populations such as desert environments (e.g., Devils Hole) or in mixed populations of hundreds of species, such as in the African great lakes. On the other hand, marine fishes are seldom completely isolated, but they do occur in locations, such as the epipelagic zone of the open ocean, which have low fish species diversity per volume. They also occur in mixed aggregations of fishes in such species-rich systems as coral reefs. Factors affecting the distribution and abundance of these species obviously cannot be precisely the same.

In the short term (dozens to hundreds of years) somewhat isolated fish species in temperate climates are affected mostly by physical changes in the environment, while multispecies aggregations in stable tropical environments are affected mainly by each other. In the long haul (thousands to millions of years) we have noted major waves of fish extinctions whose cause is likely due to a predominating change in abiotic conditions, perhaps affecting various habitats and fish communities differently (and some permanently!).

In this chapter, we consider the basic knowledge about the growth and possible regulation (i.e., changes within limits) of animal populations in general. We will then interpret this information in the context of what we know about fishes, and further apply population ecology to fishery exploitation.

PRESENT STATUS OF FISH POPULATIONS

The science of Population Ecology has been concerned historically with the forces affecting the growth and maintenance of natural populations. Fish populations have been regulated by natural forces, both abiotic and biotic, for millions of years. Fish populations have responded in various ways, and barring extreme events, their populations are well adapted to the limitations imposed by nature. As a result, their population attributes are generally understood and their fluctuations somewhat predictable. On the other hand, anthropogenic forces have increased over a few hundred years to the point where the actions of humans have added a whole new level of complexity.

In freshwater systems, anthropogenic regulation of native fish populations has been mostly unintentional. Such affects have been related to a perceived need to "control" nature. An example of this is the widespread conversion natural stream systems to regulated networks of impoundments and fragmented sections of rivers. Historically, the need to regulate or to maintain native fish populations was not a priority and not seriously considered in almost all cases. On the contrary, native riverine fish and natural lake populations were further disturbed by nonnative, invasive species, mostly introduced for sport fishing purposes or by accident. Many of these invaders were aggressive and hardy predaceous species and have persisted. Due to these invasions, native populations of freshwater fishes have generally declined in response. After these physical and biological alterations, extreme environmental changes result, making further attempts to increase populations, such as endangered species, largely fruitless. The changes just mentioned have continued to increase to the point that no major river or lake system in the United States has been spared.

In marine systems (and in large lakes such as the Laurentian Great Lakes) there have been concerted efforts to manage and regulate populations of many native fishes subjected to commercial and recreational exploitation. In this case, humans have devoted considerable effort to understand fish population dynamics and after many years of decline in fishing success, sustainable fisheries management has been attempted. This ostensibly has failed even with increases in fishing technology and fishing effort. To date, ocean fisheries, including historic and recent, have been depleted, suffered distortions in abiotic and biotic ecosystem structure, and most remain in a protracted state of decline, as we have seen in preceding chapters. Presently, the only option for *stock* (part of a species population in a defined area) recovery seems to be a cessation of fishing effort entirely, and even then recovery of major fisheries is either much slower than predictions would suggest, or nonexistent.

Great lakes fish communities have suffered similar fates, although mostly due to invasions that have devastated the natives. Newly constructed great reservoirs are characteristically blessed with unstable fish populations, even when "preadapted" nonnative fishes are continually stocked into them. Fish declines and extinctions in Laurentian, African, and Asian Great Lakes are now legend.

It should be painfully clear that all citizens should be concerned with the adverse effects produced by such alterations in environmental forces. Fish ecologist need to be especially aware of all aspects concerned with the regulation of fish populations due to the combination of natural and anthropogenic forces. Throughout most of the world, natural freshwater and marine fish ecosystems are being changed.

POPULATION CHARACTERISTICS

Populations are considered either to be a species population or a semi-isolated part of a species population referred to as a subpopulation or a stock. Freshwater fishes cannot be spread across the landscape and may occur in semi-isolated stocks. For example, coldwater fishes, such as trout, may occur in separate headwater streams that are semi-isolated by warmwater sections below them. These stocks might only be able to exchange individuals during winter conditions or between long intervals of time. Resident estuarine fishes and tidepool residents also might conceivably be considered in this light. Conservation biology is especially concerned with such situations, and the stocks may be considered to occur as part of a *metapopulation*.

Fragmentation of stream habitats by anthropogenic actions has stimulated a greater interest in the metapopulation concept, so an example of the concept is in order. Humpback chub (*Gila cypha*) were historically distributed throughout the mainstream Colorado River and larger tributaries. As insectivores, they are adapted for living in canyon-bound habitats that have large boulders and deep eddies, where they prey on aquatic and terrestrial insects. Disjunct populations were found in the lower Little Colorado River and in Grand Canyon, Cataract Canyon, Black Rocks, Desolation and Gray Canyons, Flaming George, and in Yampa Canyon. All of these canyons are separated (i.e., hundreds of miles in most instances) by long alluvial reaches that are not considered suitable as humpback habitat. However, the canyons were connected by flowing water, and there was opportunity for some interchange of genetic material, perhaps due to upstream movement of subadults or downstream displacement of young. Some of those populations have been lost due to reservoir creation, and presently most of those subpopulations are blocked by huge dams, including Glen Canyon, which blocks the largest population of fish in the Grand Canyon reach from the smaller populations in the upper basin. Meanwhile, upper basin populations have continued to decline and may be in an extinction spiral, perhaps due to adverse impacts from introduced species that prey on and compete with them. Once a metapopulation of at least 6 subpopulations, the natural metapopulation has been lost and with it a more diverse gene pool.

Populations are characterized by the sum total of its individual adaptations as acted upon by the environment. Populations also have other unique properties that are not found in individuals. Examples of those properties are size, density, a birth and death rate, a sex ratio, an immigration or emigration rate, and others. A population also can be viewed as a gene pool, and in case of the above example, the amount of genetic diversity would include all of its subpopulations.

A population would have a size that would be the sum total of all of its individuals, however it also would have density. The number of individuals per area inhabited would give a simple density, but this is not used much with fish. Instead, the density usually is given as the number of fish per river reach, or in the case of lakes, the number per surface area at a certain level (minimum, average pool, etc.). For fishes that are clearly restricted by habitat, such as intertidal, or herbivores of the riparian zone, an ecological density could be calculated using the amount of fish per linear reach.

In general terms, a population exists because there is at least an equal birth and death rate, and it expands due to a higher birth rate than its death rate. But it is not quite that easy, because fish populations can be maintained by strong age classes that occur infrequently, in case of some long-lived fishes, such as sturgeons and Colorado pikeminnow, a strong (successful in having large numbers of offspring that survive to recruitment) year class every 10 years could ensure survival. Many fishes display less than annual spawning of individuals, such as the common white sucker (*C. commersoni*), sturgeons, and pikeminnow, and it is necessary to know the age structure of each subpopulation in order to understand population dynamics. Such information can be used to construct age pyramids, which can portray how sensitive a stock might be to fishing pressure or other perturbations.

Another valuable bit of information is the sex ratio of a stock or population. This is extremely important information to know about a rare, threatened, or endangered fish population, because it allows the computation of effective population size (Ne), which is a reflection of the actual number of males and females that are contributing genetic material to future generations. To use an isolated example from my experience, Colorado pikeminnow are long-lived fish, capable of living perhaps 50 years in the wild. My crew and I were able to sample aggregations of spawning fish in two different spawning grounds in two different rivers during an eight-year period by locating them with radiotelemetry and then netting them.

We found ripe spawning fish ($n = 208$) with averaged sizes of 555 mm for males, and 654 mm for females (Tyus 1990b). Females were significantly larger than males (t-test, $P < 0.05$) and presuming equal growth rates, they would need to live longer than males in order to spawn. This raises the possibility of deferential mortality of females. This mortality seems evident because only 14 ripe females were present in 208 ripe fish, the rest (194) were males. Two problems emerge: a low number of females, and a departure from the 1:1 male–female sex ratio. Such a low number of females in these spawning aggregations would mean that not all males presumably would be able to fertilize eggs, and use of a 1:1 ratio would overestimate the amount of genetic diversity in the population. Using methods to be explained later in the conservation section, the 208 fish that were captured have about the same genetic diversity as only 52 fish spawning with a 1:1 sex ratio! As an aside, a "better" sex ratio might be derived by counting all fishes collected in various locations during those years even though many had no sex products that could be "milked" from them. These additional fish (67) were composed of suspected males, which were smaller fish with some breeding coloration, and suspected females, which were larger fish with little coloration. Inclusion of these fish would result in a "better" sex ratio. However in my opinion, this is poor science, because these fish obviously were not part of the spawning effort documented, and repeat spawning by the males at a later date (should some of the females produce viable eggs) was not certain. Further compounding this problem is the observation that not all adult pikeminnow spawn each year.

POPULATION GROWTH

The number of individuals in a population increases because of survival and reaches a stable point or declines only by death of individuals. Thus, survival rates (the reciprocal of mortality) are of interest in predicting population growth. There are differences in the survival rates of individuals over time, and four curves are widely accepted (e.g., by Slobodkin 1961). Figure 23.1 illustrates four types: Type I is typical of species that have high survival until old age, when there is high mortality. This curve occurs in some plants and large mammals. Type II represents a constant survival rate and represents some plants and sharks that produce live young. Type III shows a higher mortality in young and less in the very old, and it is characteristic of some birds. Type IV is representative of most fishes and insects that produce large numbers of small young. It demonstrates that large fish are relatively immune to mortality until very old, a function of being too large for others to consume. An important point is the change in some of these survival curves due to the influence of humans. For example, infant mortality in humans, especially in the U.S. has been greatly reduced, as reflected in an increased infant survival rate in the first year of life. On the other hand, human influence has resulted in increased mortality of very large fishes that are highly vulnerable to fishing gear such as gill nets.

Growth in animal populations is complex, including not only how many offspring are produced by an individual, but also including factors such as litter size, and generation time, which is related to age at first reproduction. The study of factors that influence the growth of populations, i.e., increases in number of individuals or collective growth in biomass, is called population dynamics. Growth in the change in individuals present in the population (ΔN) per time elapsed (Δt) and expressed as $\Delta N/\Delta t$. This expression is produced by the number of births (b) minus the number of deaths (d), each of which are related to population size (N) and this results in the following equation:

$$\Delta N/\Delta t = bN - dN \text{ or } \Delta N/\Delta t = (b - d)N. \tag{23.1}$$

The expression ($b - d$) is the instantaneous rate of increase r, and expressing the preceding equation as an instantaneous rate results in

$$dN/dt = rN. \tag{23.2}$$

When this equation is plotted arithmetically, the familiar J shape curve is produced (Figure 23.2). This curve tells us that in the absence of constraint, the larger a population the faster it grows, and

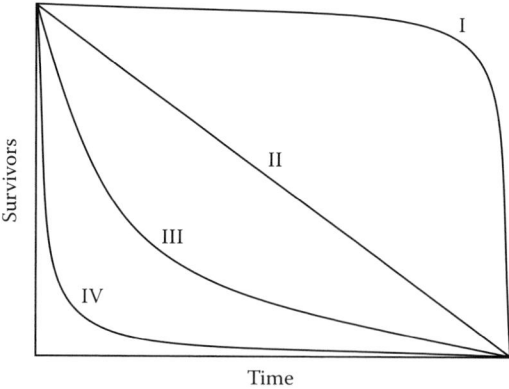

Figure 23.1 Examples of survivorship curves.

POPULATIONS, GROWTH, AND REGULATION 279

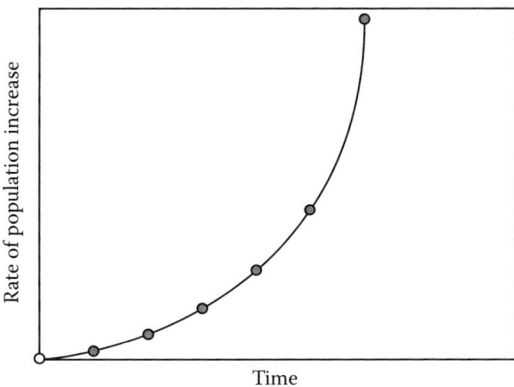

Figure 23.2 The exponential or J-shaped curve.

the curve thus generated is considered to be the maximum rate of increase—an indication of the biotic potential.

Although it is possible for a population to increase at such an exponential rate, it would soon outgrow finite resources such as food and space. Its very numbers would work against it as intraspecific competition, thereby reducing food levels, increasing starvation and risk of disease, etc. Poor condition of its individual members would then result in a lowering of fecundity, and an overall slowing and perhaps reversing of population growth. Theoretically, as resources become limited, the rate of population increase will begin to slow until the demands of the population are balanced with the resources available, when further growth would not be possible. Constraints thus imposed by the environment are called the *environmental resistance*. A notion of the response of an animal population to this resistance can be plotted by letting K equal to the absolute environmental limit of resources (i.e., the *carrying capacity*) and rewriting the above equation as

$$dN/dt = rN\,(K - N)/K. \qquad (23.3)$$

In this equation, when N is small the population rate is high, when N is large, the rate decreases, and when N approaches K the rate of growth becomes very small or none at all. This is a *logistic model* of population growth, developed by Pierre Verhultz in 1845–47. It produces the well-known sigmoid, or S-shaped curve (Figure 23.3). Construction of this curve requires accepting some

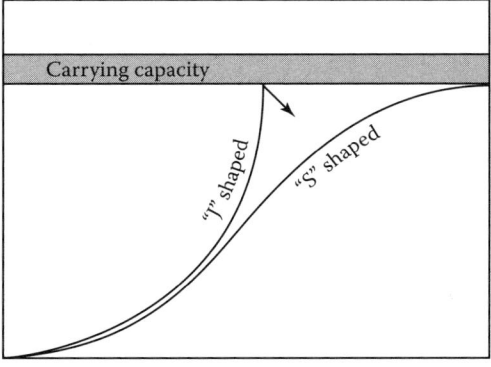

Figure 23.3 Theoretical J- and S-shaped curves and carrying capacity.

assumptions: every individual has the same ecological properties such as identical ages, mortality, and birth rates; all animals instantly respond to changes in their environment, and the rate of change in population size is a function of the present population size; and there is a constant limit to increasing population size, and the rate of population increase is linearly proportional to the difference between the population size at that time and the upper limit.

The complex adaptations of species and the vagaries of nature should make it obvious that one perfectly balanced curve cannot represent the population growth attributes and regulatory mechanisms for all species. Although development of this model is helpful in concept, it has long been noted that, except on very simple organisms, the logistic curve does not approximate animal growth patterns. This was recognized long ago by Slobodkin (1961, p. 102): "All of the metazoan examples we have discussed above differ from each other, but none of them can be considered to approximate the logistic curve in their growth pattern except in a very crude way." As pointed out by Colinvaux (1993), such curves have been generated in laboratories where populations of simple species have been kept in controlled environments. Even under laboratory conditions, more complex species have failed to show the logistic effect nor have populations in the wild exhibited it. Like most models developed for natural phenomenon, the logistic model is a mathematical simplification of nature that can be used as a paradigm to generate and test additional hypotheses.

POPULATION REGULATION

There are only two types of population regulatory mechanisms, density-independent mechanisms and density-dependent mechanisms. Both of these mechanisms cause increased mortality at higher population densities, but always work together. When population growth rate is plotted against population density, the difference in population growth is apparent. As population density increases there is no increase in growth rate in case of density-independent growth; instead, the population is controlled by abiotic factors, and when those factors are finally exceeded, the population crashes, as shown by line 2 of Figure 23.4. In line 1, there is a declining rate of population growth as density increases due to density-dependent mechanisms. However, there is another form of growth in which population growth rate can be highest at intermediate population densities (line 3). In a general sense, this effect is due to a declining fitness in individuals as population size decreases, and it is called the Allee effect, in honor of W. C. Allee, a strong proponent of the concept (Allee 1931, 1941; Allee et al. 1949). The effect can be observed as population size reaches a critical mass

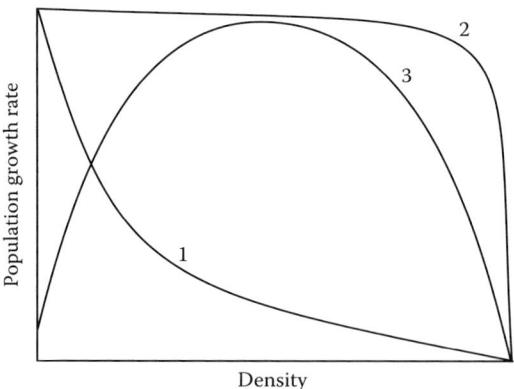

Figure 23.4 Population growth rate vs. density. Line 1 = density dependent growth; line 2 = density independent growth with limits; and line 3 = highest growth rate at intermediate densities.

at which the population growth rate is negative (an Allee threshold). This can be imagined in cases of island nesting seabirds, in which birds on touching nests can fight off predators such as gulls, but the success goes down when nests are too far apart, or when all nest space is utilized. This concept presumably applies to nesting fishes as well, as discussed for cichlids in the African Great Lakes (chapter 13). Stephens et al. (1999) reviewed the Allee concept and defined two separate cases: component Allee effects that are due to a component (i.e., natural or human induced) of fitness and demographic Allee effects that are related to total fitness. Allee effects also can be related to the tendency to become extinct, which is increased with multiple Allee effects that can occur when more than one component affects growth, reproduction, or survival at the same time (Berec et al. 2006).

Density-independent regulation occurs mainly due to forces such as climate that are not influenced by the size of the population. In this case, some factors may only affect populations at climatic extremes (Figure 23.5). In other words, population birth and death rates do not change with populations size (the value of N above). However, density-independent force can abruptly cause changes that can significantly increase or reduce population density regardless of population size. Such effects on fish can include positive effects due to increased upwelling of nutrients, or negative affects due to the appearance of cold currents. For fish larvae in both freshwater and marine systems, climate can result in very high levels of mortality especially during critical periods.

On the contrary, density-dependent regulations are those affects due entirely to the size of the population. In this case, birth or death rates are affected by the value of N. Because larger N causes a slowing of population size and smaller N causes an increase in mathematical models, density-dependent effects are thought to regulate populations within some bounds, in most animal species. Density-dependent regulation includes intraspecific and interspecific competition for food and space, dispersal, predator prey interactions, and other regulatory actions. Fishes are able to reduce intraspecific competition by ontogenetic niche shifts, which result in use of different habitat and foods by the young. In riverine systems, this also is aided by dispersal from spawning areas, and most riverine fish larvae enter into the drift as a transport mechanism. Interspecific effects by two or more competing species on the population attributes have been the subject of much study.

Effects of competition between two or more competing species can have an effect on population growth of either or both of them. The interaction between two species has been thoroughly investigated from a theoretical perspective, beginning a century ago. By use of logistic Equation 23.3, two

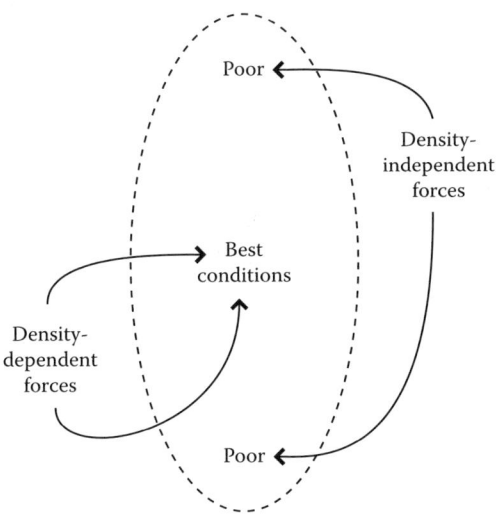

Figure 23.5 Effects of species range on population regulation.

individuals are given credit for deriving a way to describe the interaction between two species that are utilizing the same resource: this is called the Lotka–Volterra model in honor of the two scientists who independently derived it. The Lotka–Volterra model consists of two equations representing each species. Here is how they did it. Recall the following logistic equation:

$$dN/dt = rN(K - N)/K \tag{23.4}$$

It is possible to rewrite this equation to represent two species by using N_1, N_2 for two populations, and to represent the carrying capacity of each by using r_1, r_2, and K_1, K_2 for each population respectively. This results in two equations that are mathematical expressions that pertain only to those populations. The next step involves the need to interpret the competition of one on the other, and this is done by adding a coefficient that accounts for the competition in terms of the resource, of each on the other. If α = the competitive effect of species 2 on species 2, and β = the effect of species 1 on species 2, then two equations can be generated:

$$dN_1/dt = r_1 N_1 (K_1 - N_1 - \alpha N_2)/K_1 \tag{23.5}$$

$$dN_2/dt = r_2 N_2 (K_2 - N_2 - \beta N_1)/K_2 \tag{23.6}$$

The logistic growth Equation 23.3, which addresses only intraspecific competition, now has been altered to also consider interspecific effects. It can be observed that the pair of equations each contains a new component that interprets the numbers of one in terms of the numbers of the other. In other words, an increase in the number of one species negatively affects population growth of the other.

The growth of each population is thus considered in terms of the other (i.e., competition coefficients α and β). This is done by determining the number of individuals in each population that are equivalent in resource use to numbers of individuals in the other population. For population 1, the term α is the per capita effect of species 2 on 1. For population 2, the term β is the per capita effect of species 1 on 2. The carrying capacity for both species is reached when the same resource (in experimental studies it is a food, such as flour or a culture media) is depleted. Since the numbers of individuals required to reach K will be different for each species, say beetles. Starting with 50 beetles for N_1, with a carrying capacity (determined in pure culture) of K = 1000, then the competition coefficient for N_2 will be $\beta = 50/1000 = 0.5$, and the total effect of species 1 on species 2 at that instant, would be $0.5N_2$.

The results of competition under laboratory conditions that included differences in temperature and humidity have resulted in three outcomes: competitive exclusion, unstable equilibrium, and stable equilibrium. Gause (1934) tested the theoretical outcomes of the Lotka–Volterra equations with a laboratory experiment using yeast and protozoans. His work was later expanded by others with tests of more complex organisms, which are more compelling and easier to describe. Using flour beetles (Park 1954; table also presented by Krohne 2001), the results were not complete exclusion of one species over another all the time. Instead, competitive exclusion in hot/moist conditions excluded species 2, under cold, dry conditions though, species 1 was excluded. In all other cases, the species coexisted in stable or unstable equilibrium. All of the general ecology texts I have seen go into greater detail with diagrams and provide further clarification over this sketch (e.g., Smith and Smith 2001; Krohne 2001; Begon et al. 2006).

The preceding study by Park (1954) demonstrated that the results of competition depended upon the adaptations of the two beetles to climate (simulated by temperature and moisture). It also demonstrated that outcomes are not certain even under such simple conditions. Other experiments conducted afterward further stressed that complex factors such as parasites, host resistance to diseases, types of habitats, and attributes of life cycles may exert considerable influence.

Experiments between two species are useful; however, we must understand that this is a gross oversimplification of nature, where several species may interact simultaneously or seasonally. In addition, the role of humans in spreading species more or less indiscriminately around the world also has made the matter of species interactions more difficult to assess. Also, human-induced changes in habitat invariably results in large areas or pockets that are more suitable to some invasive species. Also, we are learning that fishery-induced trophic cascades can alter ecosystem function to the point of creating alternative states (Salomon et al. 2010). In that case, target species can be negatively affected by predation from small species and the Allee effect.

EQUILIBRIUM/NONEQUILIBRIUM

Are there forces on the planet that result in populations that are self-regulating and able to avoid population crashes? There are at least two schools of thought about this (reviewed by Krohne 2001). One supports the concept that populations fluctuate at random over long time scales (nonequilibrium theory), and another supports the concept that populations are not random and thus regulated to some extent (equilibrium theory).

Nonequilibrium (Density Independence)

Abiotic factors that affect population growth are mostly attributed to daily changes in weather and seasonal and annual changes in climate. Over geological time scales, however, climatic factors can have extreme consequences, the most recent being Pleistocene glaciation, and we are apparently experiencing some consequences of global warming. These density-independent factors and phenomena can cause great shifts in abundance and even extinctions of organisms, and they are indicators of nonequilibrium population regulation. The concept here is that major factors in the weather reduce population sizes of organisms to such low levels that density-dependent factors are less important. Regarding fish populations, Cushing (1995) convincingly shows climatic changes have produced "dramatic and extensive change" in marine fish stocks. In freshwater fishes, Matthews (1998) proposed that stream fish communities may experience very high mortalities from overwintering conditions and stream flooding. Such mortality would serve to keep populations low, which would reduce competition due to more abundant resources to share.

As we have seen, fish populations worldwide have experienced declines due to the actions of humans and these actions are mostly not related to population densities. Some fishes that we have studied, such as Atlantic cod, have not responded to the concept of density regulation after fishing pressure has been removed, and this nonequilibrium is apparently related to habitat destruction.

Equilibrium (Density Dependence)

Density-dependent population controls affect birth and death rates as the population density changes. It is thought that such control results in populations that fluctuate around certain levels due to both extrinsic and intrinsic biological factors. Extrinsic factors, such as the supply of food occur outside of the population in question, and intrinsic factors include mechanisms that are internal to the population such as stress, territory establishment, and movement.

Several extrinsic biological factors have been shown to regulate animal populations due to top down (predation) and bottom up (food supply) affects. Among these, food supply has been accepted as such a factor for a very long time, as has predation. Freshwater pond fish, such as bluegills (Centrarchidae) may exhibit stunting, in which a large number of small but mature individuals are present. This stunting is related to inadequate food supply and absence or low levels of predators.

Introduction of predators can soon result in the opposite effect, with a smaller number of large individuals present.

Parasitism and disease also can influence populations. Ich, or "ick" (*Ichthyopthirius multifilis*), is an example. Young of this parasite erupt from infected fish, swim to another fish, and bore into the epidermis. The parasite is more successful in infecting other fish when fish density is high.

Intrinsic factors are a form of self-regulation. Stress can result from density-dependent effects and can regulate populations, increasing aggression and reducing fecundity. Delineation and defense of territories, and movement of individuals out of the population (dispersal) also can affect population density. Movement is very common in fishes as a means of reducing stress.

Complexity of Regulation

As might be anticipated, equilibrium and nonequilibrium effects interact in complex ways. Earlier arguments that one or the other was the controlling mechanism missed the point. It should be clear now, that both mechanisms exert an influence. As suggested earlier, individuals of a species that are living at the northern or southern, higher or lower, and other fringes of their ranges are likely affected by density-independent effects of climate, while individuals in the center of their range might be affected more by density-dependent factors (Figure 23.5).

Historically, fishery management biologists who are concerned with models of exploited fish stocks have worked their models in accordance with density-dependent attributes, accepting abiotic factors as constant. The term "carrying capacity" usually did not play a part in calculations and the concept is usually not even used, even in calculating maximum sustained yield (e.g., see Ricker 1975; Van Den Avyle and Hayward 1999 for information on population dynamics of exploited fish populations). Also, there apparently was little concern about the impacts to habitat associated with fishing.

Out of curiosity, I scanned through four fishery management books (published in the 1980s) looking at the section on gear. I found no discussion about the impact of fishing gear (including bottom trawls) on habitat. In the past 20 years, it has become so painfully obvious (e.g., see DeAlteris and Morse 1997; Helfman 2007) that you wonder what the authors (unnamed) were thinking, because damages have been documented for at least 150 years! Severe impacts have occurred, world class fish stocks depleted, and carrying capacity of many areas have been depressed by commercial fishing. Hopefully, we finally understand that human impacts and other density-independent factors need to be fully considered in stock management. Hopefully, effective models incorporating habitat considerations, climate change factors, and ecosystem impacts (such as sustained by trawling) will soon be implemented.

CARRYING CAPACITY PROBLEM

The term "carrying capacity" was introduced earlier in this chapter as the absolute limit of resources that are available for individuals of a species population. In the Lotka–Volterra models, carrying capacity K is the maximum number of individuals that can be sustained with the resources available, and further population growth beyond that point is impossible. In the real world, this is not true. Animal populations can exceed the carrying capacity of the environment for a time before their numbers come crashing down (AKA boom and bust), as shown in Figure 23.6. Compared with the theoretical logistic equation, populations are likely to cycle around the carrying capacity, and lower it due to habitat destruction. This phenomenon has been well documented for highly visible ungulate populations, such as the famous 1928 Supreme Court case, where federal agents reduced the population size of deer that were overbrowsing their habitat in the Kaibab National Forest in Arizona (Bean 1983). This deer population greatly expanded in the absence of natural predators

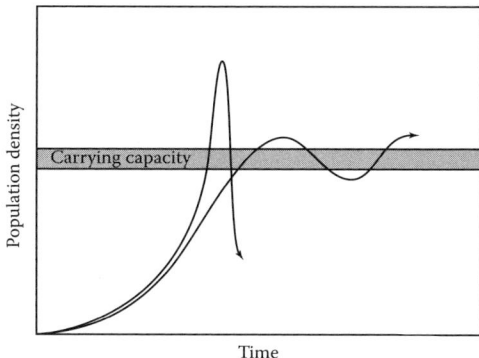

Figure 23.6 Growth of populations in nature showing likely response to carrying capacity.

and hunting, damaging vegetation on the range and lowering carrying capacity. As a result, much damage was done to vegetation, and the deer experienced poor condition and reduced fecundity.

In my experience, most young biologists and all laypersons are incredibly naive about carrying capacity. Carrying capacity is not a magic number, a cure-all or static. Carrying capacity changes with a great number of factors, most of which are climatic but which also include biological interactions. The concept of carrying capacity is crucial for understanding the logistic model, and instructional in concept. However, in an applied sense, we must consider the carrying capacity to be always in the past; because by the time we have sufficient data to calculate it, then the data are old and potentially outdated. This certainly has been the case with native riverine fish populations due to changing physical conditions and increasing numbers and species of nonnative fishes. Although we have stated above that natural fish communities do not appear to be regulated by interspecific competition, this is likely due to coevolution among native species. Such coevolution has not occurred with invasive species from other geographic areas and continents; thus, competition cannot be ruled out. Competition for resources such as food or space in this case would also reduce the carrying capacity.

COMMERCIAL EXPLOITATION

Overfishing

Historically, commercial exploitation predictably results in overfishing. In the remainder of this chapter, we will explore reasons why overfishing occurs, and it is helpful to consider that there are different ways in which overfishing affects fish populations. Three types are common in commercial fisheries:

(1) Growth overfishing, in which fishing pressure has resulted in catches of smaller-sized fish than would be captured with application of an appropriate maximum sustained yield program.
(2) Recruitment overfishing, in which spawning adults have been so reduced in numbers that they cannot produce enough young to replace themselves; thus, the population is declining.
(3) Ecosystem overfishing, in which the structure and function of an ecosystem is under change. This type of overfishing has resulted in the decline of several species, smaller sizes of fish, and ecosystem instability, and population impacts are synergistic with habitat alteration and destruction by fishing gear.

In the following section, I will attempt to explain a management process that has led to overfishing and eventually loss of profitable fisheries.

Fallacy of Maximum Sustained Yield

The logistic model is helpful in explaining some concepts associated with the management of fish stocks for commercial exploitation. If you observe the symmetrical plot of the sigmoid curve (Figure 23.3), as the growth rate of a theoretical population increases toward the carrying capacity, (K), it begins to slow as it encounters the environmental resistance, and the population ceases to grow when population size $N = K$. It is instructional to note that the environmental resistance is first experienced at the point on the curve that corresponds to the value of K/2, which is the point at which the exponential growth rate ceases. In fishery terms, K/2 is a magic number, for below this point, population growth is speeding, and above it, the rate decreases with every new individual in the population. As K is approached, the rate is interminably slow. From a fisheries perspective, fish that you remove near K are slow to be replaced due to the environmental resistance (e.g., low food and space). However, if the fish population is cropped to a size that occurs just above K/2 (i.e., the *surplus*), these individuals would be quickly replaced. Theoretically, a *maximum sustained yield* (*MSY*) would be possible in this fashion. Jobling (1995, p. 409), in a section on exploited fish populations, did a good job of explaining this wishful thinking, and at the risk of redundancy, I quote his concise description:

> In the majority of wild fish populations, food availability might be expected to be the overriding factor limiting individual growth rates and determining the timing of the onset of maturity of individuals. In other words, most if not all individuals would be expected to have the potential to reach maturity at an earlier age if they were given access to increased food resources. A reduction in fish population density would increase the food available to the remaining members of the population, with the consequence of increased rates of individual growth and lower age at maturity. This could not, however, continue indefinitely, and at some stage, the fish would reach the point at which they were maturing at the minimum possible age. Any further increase in food resources, due to the continued reduction in population density, would not result in a lowered age at maturity but rather would yield an increased size at maturity.

This seems to be a case of having your cake and eating it too: the more you fish, the same amount of fish you get, and they are bigger ones too! There are a few problems with this logic. Firstly, Jobling (1995) was discussing a "previously unfished population," which would be exceedingly difficult to find!

Secondly, the fishing effort would be maintained at the same level (is this really possible?). What generally results is a different situation. The successful fishing is shared by others and the "tragedy of the commons" scenario emerges. Fishing pressure usually increases, but even if it could be maintained at the precise MSY level, individual fish become smaller, not larger at maturity (ages determined from hard structures). This is due to Lee's Phenomenon, named after Rosa M. Lee, who wrote about it in 1912 (fully explained by Ricker 1975) and the effect is primarily due to increased vulnerability of larger fish in a year class.

Unfortunately, the "devil is in the details." Different versions of sustained yield management depend on the logistic model in which the surplus is quickly replaced by density compensation. In this case, a reduced population size due to removal of adults would result in an increased per capita population survival of young and an increased growth rate. However, no proof of this apparently exists, and in some cases, a reversed compensatory effect (depensation) (Ricker 1975) has been detected.

In implementing the MSY approach, it is assumed that the carrying capacity is constant. We know (from chapter 17) that fishing with huge trawls removes the large target fish and at the same time it disrupts bottom communities, destroys habitat, and kills target and nontarget species. Also target species of one fishery can be captured by other fisheries (bycatch), as it injures and kills its own target fishes (discards and bykill) (DeAlteris and Morse 1997; Helfman 2007). Without a doubt the catch is underreported, the populations decline and the carrying capacity also is affected—it is reduced by the fishing effort, which is repetitive, potentially lowering the capacity each time. This results in the exploited population being reduced in size and establishes a new lower equilibrium, with a greater potential to fall to very low, unsustainable levels. More sophisticated technology to detect fish concentrations can conceal a decline if fishes continue to congregate (the schools are less frequently encountered, but catch/effort can remain high where fish are). Furthermore, no correction has been made to account for species interactions, such as increases of other fishes that share and expand to the point of depleting the same resources. Also, smaller populations are more affected by short-term changes in environmental conditions that could have been tolerated by a much larger stock, such as in the case of the Peruvian anchoveta. The results of sustained yield management and various models developed from the concept have decimated fish populations and the only recourse has been to abandon the practice and to stop or greatly limit fishing on general principles.

Fishery-Induced Depensation

In fishery management, it is expected that fishing pressure on adults will result in recruitment compensation, that is, an increase in growth rates and recruitment of progeny. This is expected because there should be a greater availability of resources (i.e., food and habitat) due to a reduction in intraspecific competition. In other words, as the number of large fish are reduced by fishing the per capita number of progeny surviving increases (a negative correlation) (Ricker 1975). However, reverse compensation or depensation has been suggested as a possible reason for the decline of some stocks and a failure of them to rebound after fishing pressure is removed. Perhaps the best case of this has been the Atlantic cod declines in the North Sea and Grand Banks.

In the Northwest Atlantic cod fishery, overfishing and habitat destruction has resulted in a collapse of the large (apex) predators and a rebound of forage fishes (Frank 2005) followed by an increase of rays and sharks (e.g., spiny dogfish). These changes have altered the structure of the ecosystem and perhaps placed it in an alternative state. Walters and Kitchell (2001) suggested a "cultivation/depensation" mechanism and advanced the *cultivation hypothesis* as an explanation.

In the coevolved, cod-dominated system, adult cod "cultivated" the environment in preparation for the emergence of their offspring by eating smaller forage fishes that might prey on the eggs and young. By cultivation, the adults reduced predation and competition, and presumably increased their fitness. However, in the present system, overfishing and habitat destruction have increased the populations of other fishes that prey on young cod, thus producing a depensatory effect in response. In addition, the low abundance of cod apparently has resulted in reduced egg fertilization and mating difficulties, which are producing Allee effects (Rowe et al. 2004). Continued reduction in survival and reproduction could reach a critical level that ultimately could lead to stock extirpation.

A Look at Fisheries Yield Models

In closing this chapter, I decided that some additional information about the traditional fishery yield models might be helpful in understanding overexploitation. This is a historical reflection and not intended to be a complete review of fisheries models, which could be multispecies as well as the single-species models given here (for a review of fishery models see Jennings et al. 2001). Regulation of the commercial fisheries industry is accomplished mainly by using the success of the fishing effort in some way. In general, catches are monitored and annual statistics are obtained to allow calculation of the catch and age composition. Using this information, mortality rates and the amount of recruitment are determined. Natural mortality is considered fixed and fishing mortality is the remainder. With these data, the yield is determined and fishing effort permitted for the next year (total allowable catch) is calculated according to some target for fishing mortality. In general, three different types of models have been used to determine mortality and to set optimal levels of fishing: surplus production, dynamic pool, and stock recruitment models. A few words about each are summarized from a more detailed presentation by Rothschild et al. (1997).

Surplus production models equate fishing yield to effort. They assume that abiotic and biotic conditions that affect the target population do not change, and that size of the stock and its yield will be constant if fishing effort is constant. Allowable fishing mortality is calculated at a target MSY.

Dynamic pool models are concerned with "yield per recruit," which is an indication of fishing mortality. A practical consideration is that the maximum yield allowed with this model can exceed MSY, thus an artificial correction factor is used to avoid this problem. This model also has no component for including changes in environmental conditions.

Stock recruitment models consider stock size versus recruitment, intended to show that recruitment is dependent upon the size of the stock. Good and poor year classes can be predicted and it is possible to determine the minimum size of a spawning stock that would prevent population collapse. Unfortunately, these models only predict average recruitment and they do not incorporate corrections for changing conditions.

SUMMARY

Populations have unique properties not found in individuals, such as size, density, a birth and death rate, a sex ratio, an immigration or emigration rate, and others. A population can be viewed as a gene pool, and the amount of genetic diversity would include all of its nonisolated subpopulations. In general terms, a population exists because there is at least an equal birth and death rate, and it expands due to a higher birth rate than its death rate. But it is not quite that easy, because fish populations can be maintained by strong age classes that occur infrequently. Populations of some long-lived fishes, such as sturgeons and Colorado pikeminnow, may survive due to a strong (successful in having large numbers of offspring that survive to recruitment) year class every 10 years or more. Also, many fishes display less than annual spawning of individuals, such as the common white sucker (*C. commersoni*), and it is necessary to know the age structure of each subpopulation in order to understand population dynamics.

The logistic model was developed in an attempt to portray population growth in nature, producing the S-shaped or sigmoid growth curve, and incorporating the concept of environmental resistance and carrying capacity. The logistic concept is often used to describe natural population growth, but its use has mainly been theoretical. Population regulation has been related to density-independent (mostly climatic) and density-dependent (mostly population density) mechanisms. Two schools of thought consider population fluctuations, the concept that fish regulate their population size (equilibrium or density-dependent concept) and the second one considers fish populations to be random over long time scales (nonequilibrium or density independence). Fish management in the past has been concerned only with density-dependent effects to calculate expected fish yields, ignoring abiotic factors, which were considered to be stable and not affected by fishing. This thinking resulted in the concept of maximum sustained yield, in which fish would be harvested in the exponential growth phase of the logistic equation before the environmental resistance slows the growth rate. Based on catch statistics, harvests almost always resulted in overfishing. Also, intensification of the fishing effort can result in extreme habitat alteration and community disruption, thereby resulting in rapid and persistent population declines. Overfishing of Atlantic cod has likely resulted in depensatory response caused by predation on the young and low abundance of breeding fish. This situation is apparently limiting stock recovery.

Further reading: Recent: Boreman et al. 1997 and papers therein.

CHAPTER 24

Instinct, Learning, and Social Behavior

WHY THE INTEREST IN BEHAVIOR?

There are great differences in the behavior of fishes, but in general, fishes are complex and intelligent organisms. Humans tend to think of them as "just fish" because most freshwater fish (the ones we see as kids) look so similar to the untrained eye. This notion is a great mistake, because how fish appear on the outside has everything to do with aquatic adaptation, but says little about behavioral capabilities. There are so many species and so much difference in their behaviors that it is truly mind-boggling, and not many biologists are expert on the behavioral adaptations of more than a few groups—mostly aquarium varieties. In my professional experience, most nonbiologists lump fish together and cannot imagine why fish require anything other than some food and water. Brown (2004) was compelled to ask, "So why does everyone think fish are plain stupid?" In truth, fish have good memories, recognize individuals, make choices, have complex relationships, and some even use tools. It is fish behavior that informs us so much about their intelligence and their needs. Their behavior is a direct link to fitness, and any behavior that requires more than a small energetic cost is adaptive, significant, and deserving some special consideration. It is by their behavior that we know them.

Behavioral ecology, in this text, considers how the actions of fish function as part of integrated systems. We view their actions or behaviors as response to environmental factors, with the goal of maximizing fitness. Fishes must first survive in nature by feeding and avoiding predation. Reproduction is then possible after enough energy is obtained to produce growth of somatic and reproductive tissues. But successful reproduction and recruitment are usually accompanied with some form of sophisticated behavior. Fishes, as complex organisms, fit into their world and adapt to changes in it with innate and learned behaviors, which in turn are guided by social factors and environmental cues. Behavior is thus of paramount interest to ecologists.

INSTINCTIVE BEHAVIOR AND INNATE MECHANISMS

Basic behaviors required for survival are likely to be instinctive and linked with innate mechanisms. These simple types of behaviors would include movement away from unfavorable physicochemical factors (such as heat), ways to detect food, or to avoid predators. Such fundamental and necessary traits are deeply ingrained because they are needed at the emergence of free-living life stages. Other behaviors such as the direction of migration in trout and social behavior in guppies also have been linked with genetic control (e.g., Noakes 1986). Learning, we shall see later, provides a way to improve the response. Innate behaviors include orientation, kineses, reflexes, and taxes.

Orientation

The sense of proprioception, in which the body "knows" the position of its parts, and vision are used in instinctive responses of fish to adjust body position, just as we balance ourselves to avoid falling. However, we are responding to gravity and most fishes are less concerned with gravity because they are neutrally buoyant. Instead, an overriding factor is vulnerability to predation, and most fish, especially pelagic ones, are camouflaged by countershading. A dark back and light belly aid in hiding them from predators viewing from above and below them. However, countershading only works when you are properly aligned. Thus, pelagic fishes respond to light by aligning the body to light intensity even if it is directed at an angle (e.g., Frankel and Gunn 1961; Bond 1996). This is called the "dorsal light reaction." It is a *primary orientation* because it is used to maintain body posture. A so-called *secondary orientation* (Bone and Moore 2008) would be displayed by a fish also maintaining a position with respect to its surroundings, or habitat. An example is the instinctive ability to float a certain distance from the bottom, or to maintain a position in a school.

Instinctive responses also can be very complex, and done in a series of steps. Such responses have been found throughout the animal world, and in fishes, it is very common in elaborate courtship behavior, mostly in fishes that reproduce in pairs. Such behavior may be sequential, and require appropriate colors, signals, or other cues that produce a fixed action or "stereotyped" behavior. This also is very common in the territorial defense of birds, and a good example in fishes is in sticklebacks (Fitzgerald and Wootton 1986) and especially the three-spine stickleback (Bone and Moore 2008).

Kineses, Reflexes, and Taxes

Fish generally react in some fashion when they are exposed to an environmental stimulus, and the response might occur as an orientation, an activity, or a directed movement. Kinesis is random or orientation movement to a better environment not to or from the stimulus per se. In kinesis, the response depends on intensity instead of the direction of the stimulus. A common example in fishes is photokinesis in lampreys. Placed in a tank, the lamprey response is to move away from light and they will accumulate in a darker area (Frankel and Gunn 1961). Reflexes are involuntary responses to a stimulus that are common in vertebrates and easily understood in case of the "knee jerk" reaction in humans. For example, a sudden flash of light or attack by a predator might elicit a random flight response (i.e., in fish, a C-start, discussed later). Environmental changes can elicit stress reflexes also, which can be detected by a sudden increase in opercular rate in fishes. However, directed, nonrandom movement in response to a stimulus is different, and it is called a taxis. Such responses are widespread in fishes (Bond 1996); examples of these follow:

- *Phototaxis*: Response to light. Larval fishes of many species are attracted to light, and a way of capturing them is to use a light trap. Bathypelagic fishes also use light as a lure or a recognition signal. They also use it for camouflage. In epipelagic fishes the *dorsal light reaction,* discussed previously, is used to maximize the effect of countershading.
- *Thermotaxis*: Response to temperature. Very common for fishes to seek out a favorable temperature for resting and metabolizing food.
- *Geotaxis*: Response to gravity. Fishes are able to detect the force of gravity with the inner ear even when they are neutrally buoyant, thus remaining upright in the water column.
- *Electrotaxis or Galvanotaxis*: Response to magnetic or electric fields. Many fishes locate prey with electroreception and they are attracted to direct current, the basis for electrofishing. Sharks and other fishes presumably detect and align with magnetic lines of force that circle the Earth, and use this method to navigate. Studies have shown that sharks also use this sense to detect prey (Keenleyside 1979).

- *Thigmotaxis*: Response to touch: Fishes have a strange sense of "touch" using their lateral line system to detect changes in water pressure. Fish are able to detect structure and orient to it, and they also can detect prey items by this means.
- *Rheotaxis*: Response to flow. Water current elicits a strong orientation by fishes of all sorts and many fishes move upstream by instinct. A very interesting study by Lyon in 1904 (reviewed by Frankel and Gunn 1961) used fish enclosed in a sealed bottle of water. When placed into flowing water, the fish would align with the flow even though they could not feel it (e.g., presumably they used optical stimuli). I also have noted that native fishes of river backwaters would move out of them at the slightest indication of water flowing out—presumably an adaptation to prevent stranding (Tyus et al. 2000).
- *Chemotaxis*: Response to chemicals. The ability of sharks to detect and follow a very small blood trail is well known. Fishes presumably use a combination of gustation (sense of taste) and olfaction (sense of smell) to do this. Fishes use chemotaxis to detect and orient to their prey, and antipredator responses are now well documented, even in larvae (e.g., Hawkins et al. 2004, 2007, 2008). Fishes also use pheromones in mating and produce it for a warning (e.g., shreckstoff).
- *Biotaxis*: Response to other organisms as a stimulus. The response depends on many features of others and the situation, as influenced by such things as color, size, movement, aggression. This includes behavior resulting from the presence, actions, and behaviors of other fishes, such as in schooling behavior, predator prey interactions, feeding, hierarchies, etc. As an example, some fishes display kin selection, sharing high quality feeding territories with siblings (Griffiths and Armstrong 2002), and so on.

BIOLOGICAL CLOCK (BIORHYTHMS)

The concepts of an internal clock and circadian (24-h) rhythms ostensibly involve both innate and extrinsic components, and I have placed this section here, between instinct and learning. I have always been fascinated by the concept of "the biological clock" and eagerly read through the little book with that title when I was in graduate school (Brown et al. 1970). As pointed out by Palmer (1970), the capacity for an organism to respond to the rhythms of nature has been observed and debated for a very long time. As for me, well I dutifully set my alarm clock and then always awaken just before it rings.

A very famous experiment on the biological clock was conducted by Brown (1970). He studied shell opening in oysters shipped from Connecticut to Illinois that were cultured in nonfluctuating water. In the first 2 weeks, the oysters opened their shells at the same time as high tide occurred about 1000 miles to the east. Thereafter, they shifted this response to one that would have resulted in high tide if Illinois was a coastal situation. Although the oysters remembered the time for a while, they used environmental cues, perhaps the moon, geophysical forces, etc. to reset their activity. This experiment and many others he referenced indicated that organisms reset their activity to some environmental cue. This does not disprove internal timing, but it indicates that some rhythmic activity is not dependent on it.

Fishes also have a biological clock and this can be noticed by observing sunfish in aquaria. We had a building full of fishes in aquaria at North Carolina State. You could walk around apparently little noticed by the fish until lunchtime when they crowded on the glass to receive food. They ate once a day, and if you fed them at other times, they seemed to almost eat hesitatingly, as if afraid that the wrong time might be associated with some dire consequence! In addition, there is widespread use of dawn and dusk to tell time among many different organisms. In this case, the organism is perhaps remembering that things are supposed to occur with the different light levels, or some internal biorhythm is alerted with the stimuli. However, there clearly are examples when fishes are not exposed to a stimulus, but still produce a rhythmic response based on some internal guidance. Such a case would occur with bluegill sunfish that increase their activity an hour or more before dawn each day, or in the case of the blenny (*Coryphoblennius galerita*), which lives in a tidal regime. When the fish is removed to static conditions in the laboratory, it continues to show rhythmic activity every 12.5 h, the length of the tidal cycle, for several days (Adler 1975). The above shows that an internal clock exists and that it can be reset by environmental factors.

It is evident that both endogenous (innate) and exogenous (environmental) mechanisms are involved in circadian rhythms. In humans, rhythms like the sleep/wake cycle are controlled in the hypothalamus of the brain. In fishes, the pineal gland is involved (Hazon and Balment 1997).

COGNITION AND LEARNING

Cognition is the ability to know things. It includes awareness, perception, and reasoning. It provides the means for learning and solving problems for such important things as foraging, orientation and navigation, recognition of individuals, kin, and predators and antipredator responses. Unfortunately, most people think that fishes are capable of little more than consuming food and responding to stimuli. As indicated by Brown et al. (2006), fish have been underrated because of their low position on the vertebrate "evolutionary ladder," a concept untenable with natural selection. In truth, fish evolution has advanced right along with the other vertebrates, exhibiting the ability to learn and developing highly sophisticated behaviors. They recognize one another as individuals and relatives, and adhere to social structure accordingly (Griffiths and Armstrong 2002). They also cooperate in various tasks such as feeding and predator evasion. Some of the human impacts on fish populations include removal of larger (and older) individuals. This is more damaging than we can imagine. Removal of the older and most knowledgeable fish can disrupt social structure.

Learning is a process of receiving information from others or the environment and incorporating it into the individual by modifying behavior. The fish obtains experience, remembers, and acts in a different way. There are five principal ways (Bone and Moore 2008) in which this experience is obtained and incorporated:

(1) *Imprinting*: The fish obtains a memory that is stored for future use. At a proper time, the memory is recalled, usually due to a stimulus or endocrine action. In case of fish migration, the young may be imprinted to its surrounding by odor, and then the memory is activated due to endocrine action and/or upon detecting the odor again. Imprinting is widely accepted in salmonids, suspected in more families and will be discussed in our chapter on migration (chapter 26).

(2) *Habituation*: In this case, experience results in ignoring a stimulus that occurs in the fish's habitat, but that poses no threat. As an example, a brightly colored organism that has a warning color or design may be ignored if the fish is "habituated" to it. Fish reared in hatcheries or fish farms are habituated to many stressful factors that do not occur as frequently or at all in nature. Survival and better growth of those individuals are desirable and thus selected for in breeding with the usual end result being a loss of "wildness."

(3) *Conditioning*: Also called "associative learning," in which a fish can be trained over time to interpret the action of another as a signal. For example, learning by experience that a certain behavior in another species, normally not of interest, has a novel meaning, perhaps a warning or a signal that food is plentiful. Fish are easily conditioned to feeding because it has an immediate reward component. Fish in raceways can identify fishery workers that feed them, and, if sampled often they learn to avoid a person with a net. More to the point, repeated exposure to a predator also improves recognition and skill at avoidance. Thus, conditioning provides useful information and can contribute to fitness.

(4) *Trial and error*: Fish learn from mistakes. For example, it is important to judge the distance that is safe when near a predator. If caught, but escape, the same error will not occur the next time. Also, fish learn to avoid a hook if they escape from it (as suspected by legions of anglers).

(5) *Reasoning*: AKA "social learning," in which fish can learn from others and communicate with them. Documented examples include learning antipredator responses from others (e.g., Mathis et al. 1996). Fish also can remember things for at least 11 months (Brown 2004). We discuss more about this later in the chapter.

The ability to learn certain things has been linked to the state of development in fishes just as in humans. In this case, "sensitive periods" have been identified in some fishes in which they possess enhanced capability to learn relevant cues (Bateson and Hinde 1987). For example, in some salmon, imprinting is most effective in smolts or fry, depending on the species.

Specific learned responses may build on innate mechanisms to improve them, as in the case of the imprinting just mentioned. In homing migration, fishes can take a wrong turn and enter the improper tributary where odor is not detected, and then backtrack downstream to find the proper one using reasoning. A student of behavior would need to consider this complex possibility and separate innate and learned mechanisms experimentally. Otherwise, it would be difficult to fully understand the origin of a behavior.

NONREPRODUCTIVE SOCIAL BEHAVIOR

Social interactions in fishes can occur for several reasons, but the most frequently seen nonreproductive social behaviors in fishes are communication and the use of space. Communication is some sort of signal, display, or conduct that serves to reduce aggression and injury yet allow differential access to resources. In extreme cases, the encounters can be agonistic and result in physical pain or mental distress. Examples are given below for communication and spacing behavior.

Communication

Individual behavior is normally associated with some type of communication. Fishes may wish their presence known by others to avoid conflict. Communication also can be some form of alarm, or distress. Social status also may be recognized. In reproduction, there are many mechanisms for communication as well, and those will be discussed later. Nonreproductive communication can include the following:

- *Physical contact*: Fishes in an enclosed area or when placed in reproductive or territorial dispute may make contact, usually as a last resort. However, some fishes are aggressive, such as fathead minnows, especially in protecting eggs. Contact may include bumping, nudging, nipping fins, and biting. One of the participants may be displaced.
- *Visual display*: Usually involves some alteration or movement of the body. Can include flaring of the fins or gill covers to look larger, and this can be associated with coloration or shading to enhance the effect. Movement can include rushing, or approaching from an unusual angle or direction.
- *Light reflection*: Teleost fishes in several groups have silvery layers of reflective substances on various parts of the body and use it to catch light, reflect colors, and provide signals. Such signals can be used to increase the efficiency of schooling, communicating location, and providing information about orientation and movements (e.g., mackerels). Some fishes (e.g., guppies) also can distinguish ultraviolet light (UV) and use it as a criterion in mate selection. But even more amazing is the ability to flash UV signals to indicate territorial presence, such as eye roll signaling and UV reception in razorback suckers (Novales Flamarique et al. 2007 and references therein).
- *Chemical signals*: May be detected by smell (olfaction) and/or taste (gustation). Fishes have very sensitive chemoreception, and produce chemicals, such as pheromones for signals. Shreckstoff, used as a warning, has already been discussed as an alarm substance. Fish also can recognize odors particular to their species or to recognize dominant fish by smell. Even more impressive is an ability to recognize odors of parents and siblings (Griffiths and Armstrong 2002).
- *Coloration/shading*: It is important to note that fish can change color. Dominant fish are usually darker. Colors can be bright red, yellow, or with pigment. Some fish may be subdued in color to match the habitat, and usually for the purpose of hiding (absence of communication), such as in flounder. Color patterns can be informative, such as poster colors for warning. Fake eyespots can confuse predators; fake egg spots can confuse females.

- *Sound*: Not well understood. May involve air bladder for amplification. Can be used in reproduction as well. May be used to warn of danger.
- *Electric signals*: Can be used for locating prey or for species recognition, to frighten, or for navigation.

Spacing Behavior

Spacing behavior also is prevalent in fishes and occurs in a great many different situations:

- *Personal space*: This space is carried around with the individual. Very prevalent in humans and violated on elevators, spacing violation causes stress. The elevator response is easily recognized by fixed stares at the ceiling, at numbers of floors displayed, or at nothing.
- *Schooling space (shoal)*: Most fishes occur in a conspecific (one species) school at some time and about 25% of fishes stay in a school throughout life. A school is the most stereotyped spacing behavior in fishes. Once formed, there are no aggressive interactions and a school can be viewed as a beneficial social aggregation. A social group of fish is most appropriately called a "shoal" whether schooling or not (Pitcher 1986). Shoal behavior serves to benefit the schooling fish in at least four ways:
  (1) To reduce the risk of predation. The space occupied by a compact school is less than that used if all the fish were spread out. Thus, the probability of encountering a predator is less. Also, there are many more sensory organs in a school, so a predator can be detected more quickly. For a predator, it is difficult to detect a single prey due to the fountain effect (Figure 24.1).
  (2) To increase the efficiency of foraging. The school is larger than an individual so that prey is more easily located. When prey is detected, the entire school may benefit. However, predators such as tuna also have learned to school and they typically approach from underneath to force the prey to surface, which concentrates the prey into more of a 2-D environment.
  (3) To increase reproductive success. Mates are easy to find in a school. Synchronous spawning also is an aid to dampen predation.
  (4) To increase the efficiency of swimming. Mucus from fish ahead tends to reduce vortices and make swimming more efficient for fish that follow.
- *Social hierarchy* is not a space per se but affects the use of space. Easily seen in trout streams as a dominance hierarchy that is a function of size. Big fish get the best spot and so on, in alpha–omega sequence.
- *Home range*: An area that is not defended but used in regular activities, so this habitat is familiar to the fish. This area minimizes effort and may be avoided by smaller individuals of the same species due to size dominance.

Figure 24.1 Dense schools of fish can confuse predators. (Courtesy of the National Oceanic and Atmospheric Administration, http://commons.wikimedia/wiki/File:Sixfinger_threadfin_school.jpg.)

INSTINCT, LEARNING, AND SOCIAL BEHAVIOR

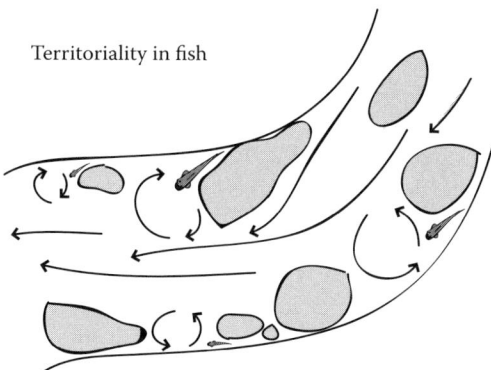

Figure 24.2 Trout have size-related dominance in feeding. (Courtesy McCutchan, J. H., Jr. With permission.)

- *Territory*: May be defended by threat, attack, or advertisement. It is a physical area that is defended from others of their own species, but perhaps not all species. Territories are usually related to reproduction and presence of a nest or spawn. The size of the area is usually correlated with food density. In trout, feeding territories may be established and defended (Figure 24.2). However, this also is related to dominance hierarchy based on size, as presented above. Kin selection though can reduce aggression, size, and dominance.

Multispecies Groups

Fishes can occur in mixed groups with individuals of other species as well as in groups of their own. Some of these associations are well known, such as the tendency for individuals of smaller species to cluster around a larger one for protection (Figure 24.3), or a larger fish can sometimes seek camouflage in a school in order to sneak up on its prey. However, many other such multiple species types of interactions are known as well. Matthews (1998) conducted an extensive review of such "mixed-species phenomena," which revealed that fishes are especially adept at participating

Figure 24.3 Small fish cluster around a big one for protection. (Courtesy of Diliff, http://en.wikipedia.org/wiki/File:Georgia_Aquarium_-_Giant_Grouper_edit.jpg.)

with other organisms. Marine fishes also may participate in the seemingly strange co-occurrence in schools with invertebrates! Mixed species schools may benefit from a broader range of experiences in avoiding uncommon predators. Other interactions of marine fishes include cleaning interactions between various species, and mutualism with invertebrates.

Matthews (1998) also reported on an early report (in 1920) of freshwater association in which feeding northern hogsuckers (*H. nigricans*) that were rooting along the bottom, were accompanied by small schools of shiners (10–12), presumably feeding off fragments of food thus dislodged. This behavior also was reported by Pflieger (1997), who noted that northern hogsuckers usually occur alone or in small groups, whose foraging was closely followed by bass, sunfish, and minnows. Such mixed species aggregations are clearly related to enhanced feeding, but a case of mutualism also may be involved. Similar feeding behavior also occurs with white sucker (*Catastomous commersoni*) and logperch (*Percina caprodes*), and in this case the suckers seemed to obtain some measure of safety when accompanied by the smaller species (Matthews 1998).

COOPERATION

The occurrence of fishes in multispecies groups as described above suggests that fishes derive benefit from such symbiotic associations, either as a commensalism (+/0) or mutualism (+/+). Alfieri and Dugatkin (2006) listed four nonmutually exclusive categories that explain how such cooperation may have originated: Byproduct mutualism, reciprocity, trait-group selection, and kin selection. These authors review numerous cases to illustrate various points, which are summarized for each category as follows:

- *Byproduct mutualism* may occur when cooperation by a group produces a more beneficial outcome than would occur by single individuals. In an intraspecific case of a "harsh environment" such as an area under territorial defense by damselfish (*Stegastes dorsopunicans*), a single individual blue tang (*Acanthurus coeruleus*) feeding on algae may not be able to ward off attacks of the defender, whereas the defender will not attack or be able to overcome blue tang cooperating in a group. In an interspecific case, cooperation between moray eels and groupers occurred in a case of cooperative hunting for prey. Evidently, the groupers actively recruit the eels to hunt and the two different species assume different roles.
- *Reciprocity* is exhibited in mutualistic cleaning relationships between a cleaner fish, such as the cleaner wrasse (*Labroides dimidiatus*) and various species such as groupers (Serranidae). In this case, the cleaner may attract clients by behavior (inviting movements) or the clients may move into special locations known as cleaning stations which are used repeatedly. Clients may encourage cleaning by opening mouths, flaring gills, etc, and also allow the smaller fish to clean inside their jaws without harming or eating them. Reciprocity also may be similar to trait group selection, but in this case, it is the individual that is of interest.
- *Trait group selection* is a form of cooperation in which some individuals cooperate in a group to benefit a larger population. In theory, sacrifice of a few to protect a larger part of the species population would be adaptive if loss of a few resulted in greater benefits to the species. *Predator inspection* is an example given, in which a small group of "inspectors" investigated the behavior of a potential predator to determine its intentions. There is some information that suggests that groups with inspectors are actually attacked less frequently than those without them, perhaps to the predator's disadvantage due to the "smoked me out/blown my cover" effect.

MACHIAVELLIAN INTELLIGENCE

In his 1513 book, *The Prince*, Niccolo Machiavelli considered the social and political interactions of humans. Franz de Wall applied these concepts of social intelligence to primate sociology in 1982.

This form of social behavior emphasizes certain forms of political behavior, such as making and breaking promises, rules, and alliances through lying, cheating, blaming, and other shady dealing. As dishonorable as it sounds, this is a clear case of extremely complex thinking, and an indication of advanced cognitive abilities. For example, the skills required for initiating such behavior would have to include recognition of individuals and their roles in society, and the presence of conflict within a social group.

It seems all of the prerequisites for Machiavellian intelligence have been met in fishes, and this cunning level of intelligence occurs in many fishes (Bshary 2006). Examples include individual recognition, understanding relationships of others, cheating the rules, and planning behavior. Cleaning mutualism in fishes is instructional because almost everyone has seen this cooperation on television. The following summary was obtained from a more detailed review by Bshary (2006). He reported that cleaner wrasse (*L. dimidiatus*) can recognize perhaps as many as 100 of their regular clients. They also discriminate among their nonpredaceous clients, "cheating" them by cleaning visitors before residents (NB, visitors do not wait if they are not cleaned quickly). This forces the resident clients to chase the cleaners around, demanding service. Chasing (i.e., punishment) by the client is effective, because it gives them a higher priority for cleaning in absence of visitors. In contrast, predatory clients are not cheated. When removed to an unfamiliar new site, the wrasse does not initiate cleaning, instead it associates with the new resident fish cautiously, riding on their backs and stroking them to reduce their aggression and only later begins cleaning them. Furthermore, it appears that prospective clients hanging around are "eavesdroppers" and also learn from the attentions given by cleaner fish to other clients, avoiding cleaners that are chased or whose client flees—thus avoiding selection of a biting cleaner.

Actions of fishes involved in trait group behavior also suggest Machiavellian intelligence. As an example, if a pair of fish moves away from their school toward a potential predator, it is assumed that they are exposed to the "prisoner's dilemma" situation from game theory reviewed by Alfieri and Dugatkin (2006), in which the greatest benefit is to defect, or betray your partner. If one stays out of harm's way, its partner is placed in greater risk—a case of amensualism (–/0). However, if both stay together there is a chance that neither will be harmed. In observed situations in which both remain at risk, this is a clear example of mutualism.

Machiavellian intelligence in cichlid fishes is evident in the clear waters of Lake Tanganyika. The account is long and only a few examples will be given, but a high level of social intelligence is obvious in the video *Lake Tanganyika, Jewel of the Rift*.

Conflicts occur over several things in the African lakes, including the need for defense against predators or competitors, use of shelter such as snail shells, and various aspects of reproduction. Cichlids that dwell in the lakes are coevolved and are able to understand individual and group social dynamics of their own species and others, and they adjust their behavior accordingly. Cheating to increase fitness is clearly evident in the behavior of some haremic ostracophylous (harem-like shell loving) cichlids (Barlow 2000). In one of these cichlids (*Neolamprologus callipterus*), only the females are small enough to fit inside the snail shells, with which the large male accumulates and tries to attract females. However, instead of spending time carrying shells and courting females, it is much faster to build a harem by stealing your neighbor's shells with females already inside (when he is not looking!).

HOW TO STUDY BEHAVIOR

Individual fish may be just as different as individual humans. If controlled experiments are needed, there will be enough differences among individuals to require a suitable number of replicates. Also, to reduce the number of confounding elements, it might be best to use lab-reared fish that have all been exposed to the same environment than to use wild fish or mixed stock. In this context, a study of learning should consider using fishes from the same parents, or at least the same genetic line, and of the same age to reduce differences of sensitive periods.

Other problems include "tank effect," due to unknown factors, and the likelihood that wild fish behave differently in the lab. Because fish behavior can be affected by the presence of a human or by handling stress, and remembered for some time (months), it is important to observe fish undetected. This is done in laboratory situations by using one-way glass, mirrors, and sound protection.

Fish are usually held for a time to acclimate them to laboratory conditions after transport. During this holding period, things like feeding them should be approached with caution because you would be teaching them to respond and congregate. This might really mess up a predator prey experiment!

Do not underestimate the sensitivity of fishes to environmental cues such as time givers (zeitgebers). My experience at Utah State University several years ago might be helpful in understanding this. The reason for my trip was to observe valuable fish that might be used for future broodstock. The fish laboratory there was indoors and subjected to artificial light conditions. A large tank was occupied by subadults of Colorado pikeminnow that were supposed to be too young to spawn. However, something caught my eye and it suddenly appeared to me that the fish were in breeding condition. I just reached in with my hand and touched one of them to discover that, sure enough, it had breeding tubercles. The lab director assured me that there was no way that the fish could have detected that it was the proper time for them to reproduce, because there was no difference in temperature (constant all year) and photoperiod was 12-h light and 12-h dark. However, after a bit of work, I found tiny high windows in the next lab and a faint amount of light was barely apparent around the closed door of the room in which the fish were kept. The fish were able to observe the differences in light duration and they responded to it. Perhaps there also was a slight difference in temperature as well, as indoors will be a little cooler in winter and warmer in summer. A controlled study of hormonal effects on reproduction using that lab could have produced erroneous results.

It is absolutely essential for any type of management or recovery program to understand movement patterns and migrations. Sometimes movements are deduced from capture sites of marked fish, but fish can move seasonally and several fish species have been recorded as nonmigratory due to repetitive captures during the nonbreeding season. I remember that pikeminnow were thought to be nonmigrants due to repetitive captures of the same fish from one year to the next. Unknown to the researchers, the fish was probably far away during a short breeding season and then returned to its home range each year to be counted. Also during our studies of Colorado pikeminnow, daily movements of one fish were monitored for 24 h a day for about a week, the fish presumed dead, and workers were complaining of boredom. Subsequently, the fish took off in the middle of the night, swam 3 miles up a tributary stream to a small inflow, presumably fed for a few hours, moved back downstream, and were home safe and sound at dawn.

The bottom line on movement studies: biotelemetry has been increasingly useful, but not intensive enough. Fish can be stationary for days, moving only at night and returning back to their preferred hiding spot during most of the time. Also, signals from telemetry must be verified, usually by displacing the fish. Much time and money has been wasted studying the behavior of a dead fish or an expelled transmitter.

The study of fish behavior requires preparation and alertness for the unexpected. Being prepared is to know everything you can about the life histories and requirements of the fish you will be using. Hopefully the following will be helpful.

CASE STUDY: BEHAVIORAL INTERACTIONS

The U.S. Fish and Wildlife Service started its Colorado River fish project in the late 1970s to determine how much water was needed to recover four big river fishes. The work required finding where the target fish lived and spawned, and physical characteristics of their habitats. However, after a few years of work we found that most important habitats were also occupied by nonnative fishes, including exotics (common carp from Europe) and fishes introduced from other basins. The effect of these fishes was unknown, but it was suggested that the small fish represented no harm.

We were able to identify the parameters that defined Colorado pikeminnow nursery habitats and began to monitor them, first with respect to flow effects. However, larval and juveniles of pikeminnow and other native fishes were vastly outnumbered. In general, young pikeminnow constituted less than 10% of the density of fishes, with 80% or more of the fish numbers composed of nonnative fishes. Most of these fishes were small species of minnows, such as red, sand, and redside shiners, and fathead minnow. These fish seemed innocuous enough, but there were also green sunfish present, which I knew quite well were very aggressive, and we also had bullheads that we suspected. I assumed most of these fishes would eat fish eggs and fry, but we lacked proof they were a threat to endangered fish. However, there was information to suggest that lack of razorback sucker recruitment was due to nonnative predation on the young. There seemed to be three things to consider: (1) did these nonnative fishes affect spatial use of habitat, or (2) were they affecting the pikeminnow by aggressive behavior, and (3) how predaceous were they?

I received verbal approval to do a small off-season study if we could do it within the present budget. We had an extra storage room that we cleaned out, moved in six 110L aquaria, installed red lights to make observations at night, shielded the tanks from one another so the fish could not see fish in another tank, placed grids on the tanks to record fish location, collected the fishes (of about the same total lengths) we needed from the Green River, and started to acclimate them to the tanks. In the meantime, we requested razorback sucker larvae from Dexter National Fish Hatchery, Dexter, New Mexico.

We decided to evaluate the behavior of juvenile Colorado pikeminnow in a tank with a native coevolved fish, flannelmouth sucker, and contrast this with its behavior in a tank with nonnative (noncoevolved) fish. So we placed juvenile Colorado pikeminnow (average size 50 mm) in each of six tanks with equal numbers and approximately the same sizes of native flannelmouth sucker, and nonnatives of fathead minnow, red and redside shiners, green sunfish, and black bullhead. We had what we needed to do the first controlled study of the behavioral interactions of Colorado pikeminnow, and to see if the fish we used would prey on razorback sucker larvae. The results follow (for more information, see Karp and Tyus 1990):

- *Use of space*: This part of the study was not revealing. Pikeminnow dispersed throughout the tank regardless of the species in it. There were differences in the vertical distribution of other fishes, but it was difficult to draw conclusions.
- *Aggressive behavior*: Nips, chasing, and threat display that resulted in displacement of the recipient were observed and judged to be agonistic acts (Table 24.1). Only two individuals were observed to receive injury, a green sunfish attacking a conspecific, and a fathead minnow attacking a pikeminnow. All nonnative fishes exhibited a great degree of interspecific aggression. Green sunfish, fathead minnow, and red shiner initiated the greatest number of aggressive acts, which were three times greater than those of pikeminnow. We were greatly surprised to find that >90% of pikeminnow aggression was intraspecific, while nonnatives displayed much more interspecific aggression: 80% of redside shiner, 63% of green sunfish, 50% of fathead minnow, and 42% of red shiner aggression was directed toward pikeminnow.
- *Predation on larvae* All fishes with the exception of the sucker and bullhead attacked razorback sucker larvae. Nonnative fish did so aggressively, and green sunfish and red and redside shiners consumed 90%, 20%, and 10% of the larvae, respectively, within 4 min. Colorado pikeminnow consumed 50% of the larvae, but it took 20 to 40 min to do so. It is noted that this aggressive feeding is not due to a lack of food, because all fish were fed 2 times daily with brine shrimp.

The results of this study indicated that small individuals of three of these nonnative species (green sunfish, red shiner, and fathead minnow) were more aggressive and displayed more

Table 24.1 Agonistic Behaviors in Mixed Species Pairs

Species Pairs	Chases	Threats	Total
Tank 1			
CP–CP	1.8	6.0	7.8
CP–FM	0.0	0.0	0.0
FM–FM	0.0	4.8	4.8
FM–CP	0.0	0.0	0.0
Tank 2			
CP–CP	1.2	1.2	2.4
CP–FH	0.0	0.0	0.0
FH–FH	12.0	2.4	14.4
FH–CP	10.8	3.6	14.4
Tank 3			
CP–CP	0.0	3.6	3.6
CP–RS	0.0	1.8	1.8
RS–RS	6.6	6.6	13.2
RS–CP	6.6	2.4	9.0
Tank 4			
CP–CP	0.6	10.2	10.8
CP–RSS	0.0	0.0	0.0
RSS–RSS	0.6	0.6	1.2
RSS–CP	0.0	5.4	5.4
Tank 5[a]			
CP–CP	1.2	2.4	3.6
CP–BB	0.0	0.0	0.0
BB–BB	2.4	0.6	3.0
BB–CP	3.0	0.6	3.6
Tank 6			
CP–CP	1.2	1.2	2.4
CP–GS	0.6	0.6	1.2
GS–GS	13.2	1.2	14.4
GS–CP	19.2	6.0	25.2

Note: First fish in a pair initiated the actions. CP, Colorado pikeminnow; FM, flannelmouth sucker; FH, fathead minnow; RS, red shiner; RSS, redside shiner; BB, black bullhead; GS, green sunfish.

[a] Fish densities in tanks 5 and 6 are half of those in tanks 1–4.

interspecific competition than did similar sizes of Colorado pikeminnow. Aggression of the nonnatives such as green sunfish has been observed by others. The aggression of the nonnative minnows was more unexpected and no doubt aids their invasion and proliferation in new systems. Such aggression by fathead minnows was surprising, and linked with fierce territorial defense.

Colorado pikeminnow were slower in their feeding responses to introduced fish larvae than were three of the nonnatives. Differences in feeding behavior during normal feeding with brine shrimp also indicated that red shiner and fathead minnow were more aggressive than pikeminnow, and the territorial behavior of fatheads may have inhibited the pikeminnow in that tank.

This study demonstrated that young Colorado pikeminnow are likely to be negatively affected by the presence of small aggressive nonnative fishes, especially during periods of overcrowding and resource limitations. It also suggests the nonnatives that ate or killed larval suckers might kill and consume pikeminnow larvae just as well. This study has been followed by many studies conducted by others who have confirmed the findings reported here and extended the work to other species. One recent one (Carpenter and Mueller 2008) reports that predation by small nonnative fishes, including four of the species that we used, on razorback sucker larvae could be limiting its recovery.

SUMMARY

Fishes are complex and intelligent organisms, and fish behavior is often the first response of a fish to environmental change. Behavioral ecology considers how the actions of fish function as an integrated system in response to environmental factors, with the goal of maximizing fitness. In essence, fishes survive by feeding and avoiding predation. But successful reproduction and recruitment of offspring also require highly sophisticated behaviors by the fish. Fish behavior is a combination of innate and learned responses that are guided by social and environmental factors. Fish react when exposed to a stimulus by changing their orientation in space, by reflex action, and by a number of taxes, a directed response to a particular stimulus. Fish also have a biological clock, are cognitive, and learn by imprinting, habituation, conditioning, trial and error, and reasoning. Fishes also communicate with one another by means of contact, display, odor, sound, electricity, etc. Because most fishes occur in schools for protection and foraging, they are experts at the use of space for schooling, home range, and territories. Fishes also occur in multispecies groups for protection or feeding. Fishes also cooperate with one another in mutualism, reciprocity, and trait group selection. There is a dark side to fish as well, and some fishes exhibit Machiavellian intelligence, a form of social behavior involved with cheating on the rules that is an indication of complex thinking. In studying fish behavior, it is clear that workers have not been thorough enough because fish have not been given enough credit. A case study on behavioral interactions among Colorado River and introduced fishes produced some surprises, with most interspecific aggression displayed by nonnative fishes. Aggressive predation on native sucker larvae by small nonnative fishes exceeded that of juvenile pikeminnow, top predator in the system.

CHAPTER 25

Trophic Concept and Feeding

TROPHIC CONCEPT

Fishes have a role in all of the major aquatic ecosystems on Earth. An important part of their function in those systems is the food they eat and how they do it. By now, it should be evident to readers that fishes are rarely very specialized; on the contrary, they are rather diverse in feeding adaptations between and among species. This is due to several factors, including size differences that occur as fish grow, the type of food available in different habitats, and changes in the abundance and distribution of foods due to seasonal and longer-term changes.

Most teleost fishes start life as small larvae that do not resemble the adults, and ontogenetic niche shifts occur as the larvae grow and attain maturity. Most larvae begin exogenous feeding on zooplankton, but the adults generally feed on very different food items ranging from plant matter to other fish. Even adult fish may consume a wide variety of foods, and fishes rarely are highly specialized in their food choices. It is common for two or more fishes to eat the same food items when those foods are abundant.

Ecology includes the study of energy flow through ecosystems, and an important concept is the transfer of the energy in food through a series of levels. This is the *trophic* (food) dynamic concept first proposed by Lindeman (1942): organisms are linked with inorganic materials that produce them, and energy consumed by herbivores is passed through an ecosystem in a stepwise fashion, with energy loss visualized as a function of food conversion from one level to the next (about 10% is lost at each level). Based on this concept, we place organisms at various positions based on productivity and consumption; these positions are called levels, and a diagram of them is called a *food chain* or, if interacting between several levels (food chains), a food web (Figure 25.1). In simplest terms, a food chain would consist of *autotrophs* consumed by *herbivores*, which are in turn consumed by *carnivores*.

Lindeman (1942) recognized four trophic levels. The largest level in biomass and energy, the photosynthetic producers, formed Level I, the foundation of aquatic systems (e.g., phytoplankton). This first level is preyed upon by herbivores, or primary consumers that constitute Level II (e.g., zooplankton). That level is preyed upon by the first level of carnivores at Level III (e.g., planktivorous fishes such as clupeids), and the highest level was made up of the large carnivores (Piscivores, e.g., tuna) that consumed the smaller fish at Level I. The figure formed by the process results in a pyramid as the biomass at that can be supported at each level decreases with energy loss, and it is referred to as the "trophic pyramid" (Figure 25.1). Lindeman had a good idea, but he left some things out of his model, most importantly, the lack of a decomposer level and interaction of *decomposers* in food webs. As we learned from studying the situation in Lake Victoria, the decomposer-fishes component of the food web can be of critical importance.

Lindeman provided a basis for understanding the structure of aquatic ecosystems. However, it was an oversimplification of nature due to complexity produced by fishes that feed at more than

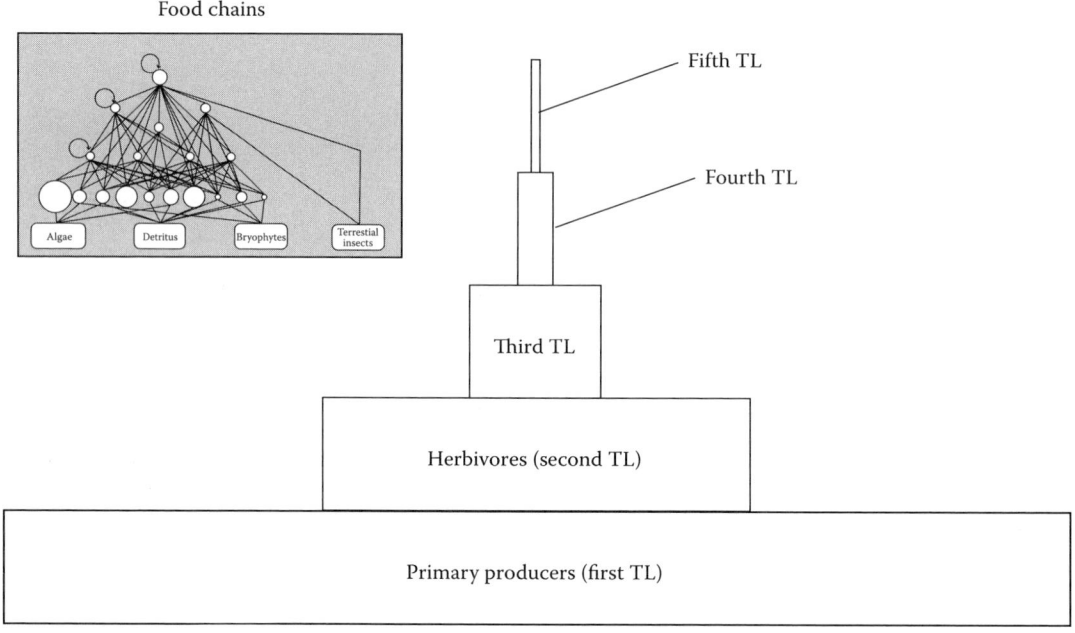

Figure 25.1 An aquatic food web and trophic pyramid. (Courtesy of McCutchan, J. H., Jr.)

one trophic level. We presently recognize two major food chains as important in aquatic ecosystems, the grazing food chain and the decomposer or detrital (small pieces of dead and decomposing plants and animals) food chain, which will be linked in food webs as shown previously (e.g., Figure 15.3).

Unfortunately, I have seen much evidence that makes me think the average person does not understand the basic trophic dynamic concept. At risk of oversimplification and redundancy, I take this chance to set the record straight regarding the erroneous concept of "trash fish": Big suckers eat living and dead plant and animal tissues and recycle them into thousands of little suckers. The little suckers are eaten by predaceous sport fish that are sought by fishers that kill the big suckers and throw them up on the bank. There is something wrong with this logic!

TROPHIC CASCADE

Effect: Bottom-Up and Top-Down

If we assume that the trophic pyramid is controlled in some way, it is easy to envision a system that is supported with food produced at Level I, and to assume that changing environmental factors, such as climate would result in varying levels of productivity. In that case, Level I would fluctuate in volume and produce more or less food for Level II. In a good year, there would be abundant food and level would increase, providing more food for Levels III and IV. In bad years the reverse would occur. This is called a "bottom-up" effect, which in theory should control the amount of energy available throughout the system, reducing or expanding biomass throughout the trophic pyramid.

A "top-down" effect also might occur if the numbers of the apex carnivore, perhaps tuna, were diminished by fishing. In this case, predation pressure would be reduced on Level III (say, herring), and its numbers would expand. However, all levels would not expand, for increasing numbers of

Level III, given time, might feed so heavily on Level II (zooplankton) as to *reduce* its numbers. A reduced level of zooplankton might then result in *increased* numbers of Level I (phytoplankton). This "ripple" effect of alternating increases and decreases throughout the trophic pyramid is analogous to the ripple effect of casting a stone into a calm lake. It is known as the cascade hypothesis, or more recently as the *trophic cascade concept* (reviewed by Colinvaux 1993; Gerking 1994; Salomon et al. 2010).

The trophic cascade concept is of great interest in considering community or ecosystem regulation and structuring. It also, like many concepts, is a gross oversimplification of aquatic ecosystems, primarily due to the extreme complexity of fish behavior, which suggests that such structuring would be ephemeral. There are several reasons for this statement provided here in no special order: Fish communities and ecosystems they occupy are not closed entities, and movements from other areas can and do occur. Fish do not feed at only one trophic level, but may forage at all levels. Predator increase results in changes in the behavior of prey species that make them less vulnerable, such as hiding. Decrease of a predator species may result in the increase of another, and decrease in numbers of one prey can cause predator switching. Population resilience in fishes can result in fast rebound in numbers. Ecologically equivalent species may expand to fill a void created by the population decline of another, such as with the California sardine and northern anchovy.

There are cases, however, when trophic cascading has been observed in fishes. The concept has been confirmed by numerous experimental studies (e.g., Carpenter et al. 1985), and confirmed by studies mainly in lakes. A well-known paper by Brooks and Dodson (1965) showed that fish could reduce the abundance of larger zooplankton in a New England lake. Higher up in the food chain, introduction of Nile perch, a nonnative piscivore, most certainly has resulted in decimating planktivores in Lake Victoria, which has resulted in a super abundance of algae (chapter 12). From the proceeding, it is easy to imagine stocking fish into a fishless lake with the results of decreasing the standing crop of large zooplankters and increasing algal biomass. On the other hand, it is intuitive that increasing nutrient levels might result in a greater abundance of algae in lakes, which effect would be increased by large numbers of planktivorous fishes. In a compelling review of 28 studies that evaluated the effects of fishes on trophic levels Matthews (1998) found ample evidence of multilevel trophic cascades that were presumably caused by piscivorous, insectivorous, or planktivorous fishes, with a three-level effect produced in 24 of the studies. Cascading effects were more prevalent in lakes than in rivers (average 3.6 levels affected vs. only 2.6 levels).

Fishery-Induced Trophic Cascades

As pointed out by Frank et al. (2005) trophic cascades have been documented mostly in freshwater systems and are characterized by small size of the systems, simple food webs, and low species diversity. Marine, open ocean systems, with their large sizes, high diversity, and complex food webs have typically not shown large trophic cascades. However, there is evidence that fishery-induced trophic cascades (FITC) have resulted from the removal of large (apex) predators (Salomon et al. 2010). FITC in open-ocean systems have been extensive, affected four trophic levels (Frank et al. 2005), and profoundly altered ecosystem dynamics.

Evidence for a large-scale trophic cascade was obtained by Frank et al. (2005) from the huge Scotian Shelf ecosystem, which lies near Nova Scotia, Canada. This FITC was a result of overfishing and removal of large apex predators, mostly Atlantic cod and other groundfishes. The apex predators had no doubt structured the ecosystem for thousands of years. Removal of these top predators released their forage fishes from heavy predation, resulting in a great proliferation of smaller fish species that placed heavy predation and competition pressure on the young life stages of larger predators. In the case of Atlantic cod, which are broadcast spawners, predation on eggs and larvae was evidently so high that populations of cod have not been able to compensate for the loss of adults. Instead, the stocks are remaining at a low density and exhibiting depensation (reduced per capita growth rate). As discussed in chapter 23, depensation has been linked with increasing predation of young cod by fishes whose populations were kept low (cultivated) by adult consumption (Walters and Kitchell 2001), but expanded greatly when the predation pressure was removed. Kept low in abundance, the cod populations have been unable to compensate due to the predation and difficulties with reproduction.

In this case, an FITC has resulted in restructuring of an entire ecosystem and perhaps the production of an ecosystem state shift due to internal feedback mechanisms (Salomon et al. 2010). Thus, even with a cessation in fishing the system has not recovered. Here we envision that the ecosystem has followed an unexpected path in recovery and is following an alternative one. Depression of cod abundance has led to depensatory recruitment caused by overfishing, then by habitat destruction and proliferation of small predators. A new system has developed, and it also will be affected by the inevitable invasion of roving predators such as sharks and rays. These changes may have stabilized the new system, accounting for the lack of response to the removal of fishing pressure. This scenario may persist for a long time, perhaps indefinitely.

FEEDING ADAPTATIONS

Trophic Categories

Application of the trophic concept to fish communities is not straightforward. In making the effort, Keenleyside (1979) recognized the limitations of the practice, due to complexity and omnivory in fish feeding, but he was able to place most fish in major feeding categories, although the categories deviate a little from a strict interpretation of the trophic level concept. For example, using present trophic concepts, the detritivores and scavengers would be combined as decomposers. I use his accounting to describe various levels (with the addition of omnivores) in the following discussion of fish as detritivores, scavengers, herbivores, omnivores, carnivores, and parasites.

Detritivores

These fishes feed upon bits and pieces of once living organic matter, or detritus. However, such indiscriminate feeding also results in the consumption of many more living items such as algae, bacteria, and protozoa. Examples of detritivores in freshwater include previously mentioned cichlids in African lakes, many suckers, some minnows, and others. Marine detritivores include mullets (Figure 25.2) in estuaries and especially rich salt marshes, and gobies on coral reefs. Mullets are sturdy fish and the adults have a scooping mouth to plow through sediments. Detritivory is an important mode of feeding and is included in the decomposer food chain.

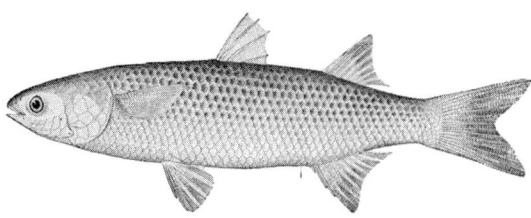

Figure 25.2 A detritivore: striped mullet. (From Goode et al. 1986 http://www.photolib.noaa.gov/htmls/figb0441.htm. With permission.)

Figure 25.3 A scavenger: channel catfish *Ictalurus punctatus*.

Scavengers

This also is a mode of feeding that includes fishes that feed on larger dead (or dying) organisms and some inanimate things as well. Note that many scavengers attack fish that are dying (e.g., caught in nets, bycatch, and bykill), while others consume long dead remains of large organisms. Technically, most scavenging could be placed into either decomposer or carnivore trophic levels.

The lowly hagfish is a specialist in scavenging, and a pretty successful one at that. Many fishes and perhaps the majority of sharks will scavenge on sick or dying fish, but some sharks and catfishes are consummate, and in some cases indiscriminate scavengers. Sharks have been found with every conceivable eatable item in their stomachs and a lot of uneatable items (e.g., cans, jugs, clothing, and driftwood) as well. Not to be outdone, the channel catfish (Figure 25.3) has a pretty good reputation as an omnivore, but it consumes so many things that it also can be labeled an opportunistic scavenger. In a diet study, Neal Nikirk and I (Tyus and Nikirk 1990) found that 575 channel catfish consumed aquatic invertebrates (31% of stomachs), vascular plants (28%), terrestrial insects (22%), algae and detritus (10%), fish (7%), and mice (1.5%), clearly an omnivorous fish. Here is where the scavenger part comes in. We discovered some unusual items (whole and pieces of) that were consumed by catfish in Dinosaur National Monument, to which I add a few additional items that I have seen since then: the plant horsetail (*Equisetum*), juniper berries, seeds, leaves, sticks, earthworms, fish, mice, corn, peas, charcoal, tallow, cheese, bologna, Mormon crickets, chironomids, mayflies, earthworms, etc. Just 15 years later I inspected about 50 catfish stomachs captured in the monument and found that every one contained pieces of a recently introduced crayfish, none of which were present in the original study.

Herbivores

As Gerking (1994) has indicated, fish species are not exclusively plant consumers. For one thing, all larval fish are carnivores (usually on zooplankton) and even "herbivorous adults" are not obligate plant eaters. They also eat animal foods, either inadvertently or on purpose. Do herbivorous fish require animal food in their diet, or can they exclusively live on plants? It appears from a long list of studies that some fishes can live exclusively on plants, but some do not grow, and it seems all can benefit from the inclusion of animal tissues. Lumping the fish that consume plant material of any kind is too broad-brush an approach, and Keenleyside (1979) divided these into three types, to aid in interpreting their foraging behavior: Phytoplanktivores, browsers, grazers, and I have added frugivores as a fourth type.

Phytoplanktivores

Phytoplanktivores feed on many species of phytoplankton included as diatoms, dinoflagellates, and algae, by filtering. Food consumption is aided by fine gill rakers that strain out organisms from water that is expelled via the gills. No doubt, other organisms such as zooplankton and larvae of other taxa are captured as well.

Browsers

These include fishes that bite emergent plants to obtain pieces. Many of these fishes live in tropical climates and have specialized teeth, such as leaf chopper cichlids and plant cutting characins. Some fishes that have no jaw teeth also are effective browsers such as several species of carp (grass *Ctenopharyngodon*), and also big head, and silver carps (*Hypophthalmicthys*), which have been introduced into the United States to control aquatic nuisance plants.

Grazers

Grazers scrape or rasp algae from rocks and other substrates. Many North American suckers have special scraping mouthparts that enable them to do this, including the bluehead (Figure 25.4) and mountain suckers of the west (Catostomidae) and minnows, such as the ubiquitous stoneroller (*Campostoma*) because it moves stones in its feeding activities. Well known marine grazers include parrotfishes (Scaras) that rasp algae and other organisms from coral reefs.

Frugivores

These consume reproductive products of trees and other vascular plants. Gerking (1994) reported that characins in the Amazon River of South America feed on fruits and seeds of 17 genera and 89 families of plants. These food items ripen and fall into the water, or they can be plucked off low branches during flooding. One of the most well known of these fruit eaters is the pacu (*Myleus pacu*) (Figure 25.5) a commonly imported fish found in aquarium (it can attain lengths of 70 cm or so). My friend Minckley had one of these, and it was especially fond of chunks of banana and grapes.

Omnivores

Omnivory is a mode of feeding that combines herbivory with carnivory. The term is used so frequently that I have added it here. Most, perhaps almost all fishes are omnivorous to a degree,

Figure 25.4 A grazer: bluehead sucker *Catostomus discobolus*.

Figure 25.5 A frugivore: pacu. (Courtesy McCutchan, J. H., Jr.)

consuming both plant and animal matter. Some omnivory is accidental or serendipitous. All herbivorous fish can be expected to obtain some animal life from invertebrates that live in association with plants, or if nothing else, certainly due to the biofilm that accompanies aquatic plants. Carnivorous fishes obtain plant material in a similar way, such as algae consumed with aquatic invertebrates. Omnivory is so common in fishes that the concept seems to be taken for granted and the term seldom used in classifying feeding (e.g., Keenleyside 1979; Wootton 1990). However, there are cases in which some fishes (e.g., common carp) are true omnivores (e.g., Bond 1996). Suckers also may be omnivores, especially those that are grazing on algae, because they also consume animals that live in the algae. Both of these fishes may consume large percentages of both plants and animals, and have the digestive system to metabolize it as well. In that case, the term is meaningful. However, omnivory should not be confused with other aspects of feeding by fishes.

Carnivores

Most of the foods eaten by carnivores (predators on animals) do not remain stationary to be fed upon as all of the above. Instead, carnivores have to have special adaptations to capture as well as find their prey. Keenleyside (1979) provided four categories: consumers of benthic invertebrates, zooplanktivores, aerial feeders, and piscivores.

Benthic Invertebrate Feeders

Foraging fishes that feed on benthic invertebrates are adapted for detecting and capturing arthropods and mollusks in several ways. Detection of prey may be aided by visual, mechanical, electrical, or chemical sensory systems.

- *Scan-and-Pickers*—These include fishes that swim over the bottom and locate, then pick up invertebrates by sight. These include many fishes, such as darters, cichlids, salmonids, and flatfishes.
- *Disturb-and-Pickers*—These are fishes that expose concealed prey in some fashion and then rapidly pick them up. Exposure can result from blowing streams of water, disturbing the sediment with fins or body movement. Some minnows allow suckers to do the hard work of exposing prey; the minnow follow and eat smaller particles that are suspended as the sucker moves along.
- *Substrate Sorters*—These suck up substrates and filter out the invertebrates and eject the inedible material. These include several kinds of cichlids (e.g., *Geophagus, Lethrinops*).
- *Graspers of Large Prey*—These include fishes feeding on large benthic invertebrates such as crabs, crayfish, sea urchins, and mollusks. Puffers approach crabs and bite through shell and all with platelike teeth and powerful jaws. Many species eat sea urchins using the puff-it-over and kill tactic (triggerfish), biting off pieces or beating them against rocks (Labridae). Molluscivores include fish

that consume them and crush the shells with pharyngeal teeth, such as shellcracker (redear) sunfish, and freshwater and marine drums, and some fishes bore through shells, others pull out tissues with the foot, and so on.

Zooplanktivores

These include fishes that feed by filter feeding in which the fish swims along with mouth open and water is strained through the gill rakers, or particulate feeding in which a few individuals are gulped in without sustained swimming.

- *Filter Feeding*—This occurs in adult fish, in several taxonomic groups; most fish larvae also are filter feeders. The most prevalent and widely known group that filter feed as adults are the Clupeiformes, which include herrings, shads, sardines, anchovies, and menhaden. All of these groups are schooling fish when feeding on smaller size prey. However, these fishes also are capable of consuming larger prey, such as in the alewife.
- *Particulate Feeders*—Particulate feeders such as alewife exhibit some complex feeding strategies. If alewives are feeding in schools and encounter larger prey, there is somewhat of a feeding frenzy with every fish out for itself, resulting in disruption of the school formation. Various prey are captured either by approach and sucking them up, by faster darting and sucking, or by swimming and gulping, depending upon prey density and swimming performance.

Aerial Feeders

These include insectivorous fishes that employ different strategies. The archerfish (Toxotidae) squirts water from the surface to knock down insects on branches. Arawana (*Osteoglossum bicirrosum*) can jump considerable distances vertically and out of the water to capture prey. Even trout can get into the aerial act by jumping out of the water to nab flying insects.

Piscivores

Piscivores are usually secondary or tertiary carnivores and near the top of the trophic pyramid. Piscivores can ambush prey, attract them with lures, stalk and attack them, or chase them down with speed. Many sight-feeding predators can be recognized by their very large eyes. Small eyes indicate that a piscivore relies primarily on other senses.

- *Ambush or Lie-in-Wait Predators*—These are usually sight-feeding fishes that hunt by day. They usually hide, use camouflage, or lie motionless and attack their victims with a powerful strike. If they miss the prey, they may strike again, but they do not try to chase the prey and catch it that way. Flatfishes are ambush predators, as are some groupers and sharks. Pike can assume the role of ambush predator also.
- *Lure Predators*—Lure predators include goosefishes and anglerfishes of the order Lophiiformes. Goosefishes (Figure 25.6) twitch their lure as they hide on the bottom. Deep-sea anglerfish, as studied previously, float in darkness and fish with a glowing lure.
- *Stalkers*—Stalkers detect their prey with one or more sensory organs and track them down with stealth. Once within range they can then attack with a vicious strike, especially in case of gar, pikes, and pickerels. Trumpetfish (Aulostomus) may disguise their approach by mingling with a school of small fishes or by hiding alongside a larger fish.
- *Chase Predators*—These catch their fish on the run, perhaps by a combination of stalking and chasing, but the end result is chase, overhaul, capture, and ingestion. Fast ocean fish such as tunas (Figure 25.7), billfish, jacks, mackerels, and sharks fit as this type of predator. The great white shark

TROPHIC CONCEPT AND FEEDING

Figure 25.6 A lure predator: goosefish. (Courtesy of the National Oceanic and Atmospheric Administration, http://www.noaa.gov/.)

Figure 25.7 A pursuit predator blackfin tuna. (Courtesy Raver, D., and U.S. Fish and Wildlife Service (USFWS)).

and other top predators may feed on dangerous game, such as sea lions. Sharks can easily outswim if not outmaneuver them, but the shark does not attempt to kill and consume the lion outright. Usually the shark will inflict a killing bite and wait until the lion has bled to death before returning to a meal.

Parasites

Males of ceratoid anglerfishes are true parasites on females of their species. However, other obligate parasitic fish do not exist. However, there are several groups that parasitize other organisms, mostly other fish, such as the parasitic sea lamprey that has devastated most of the large native fishes of the Great Lakes, catfishes (Trichomycteridae) that parasitize the gills of other fish, and scale eaters and fin biters among the cichlids.

Resource Sharing and Trophic Adaptability

Use of some resource by two or more species is not competition if there is plenty to go around, rather the resource can be shared in an overlap of some component of fundamental niches without affecting a realized niche. This concept may apply to *dietary overlap* in fishes.

The great variety of foods consumed by fishes and a high degree of dietary overlap between species has been noticed for a great while. This behavior has been referred to in various ways as trophic flexibility, resource sharing, feeding plasticity, etc. Most recently, the concept has been identified as *trophic adaptability*, and this term is defined as (Gerking 1994): "The ability to take advantage of the most profitable food sources at a particular time. . . ." The concept of trophic adaptability in fishes has been discussed for at least 60 years (Hartley 1948), but it has gained slow acceptance in the ecology of fishes.

Most of the evidence for the concept has resulted from diet studies of freshwater fishes, especially in streams, and the cause of the behavior has been related to adaptation to changing environments. The concept was presented earlier (chapter 11) with respect to the utilization of Mormon crickets by all species of riverine fishes captured. Another example, such an abundance of food and trophic adaptability involves carp and channel catfish, whose omnivory and scavenging is outlined above. Considered a bottom feeder, both species, but especially channel catfish, can exhibit unusual behavior in the spring by gathering at the surface in hordes to feed on emerging aquatic insects. I observed this behavior for many years in rocky river canyons. This behavior also was used to hold contests by drifting rafters, to see who could scoop up the most fish with a dip net in a given time. Although they could be caught at the surface with flies, the carp were too wary to be netted, but not the catfish. Winning numbers of catfish captured in this way might reach a hundred fish per hour, depending on the stealth and quickness of the dipper.

The trophic dynamic concept is helpful in considering the feeding relationships in fishes. However, as stated previously, fish usually feed at different trophic levels during their life, and adult fish frequently feed at more than one trophic level. As in African cichlids, even specialized fish can assume the role of a generalist when inundated with an abundance of food. Life is full of problems and opportunities. Those who can take advantage of those opportunities are most likely to survive.

Resource sharing of an abundant food cannot last, and when an abundant food becomes limited, prudent species will switch to another dietary item or be faced with a lower resource base due to competition. In the Mormon cricket example, fishes that usually lived in different habitats and/or ate different foods all congregated at the surface and in eddies to consume the crickets, then no doubt returned to their separate locations and professions when the glut ended. This is called *diet switching* (or in this case, diet switching back!). It is presumed that the native riverine fish community will return to its former food and habitat partitioning to obtain an efficient use of the resource base.

FOOD AND SELECTIVITY

What do fish eat? This has been a continuing question asked by generations of the fishing public. One concept used in fly-fishing is to "match the hatch," meaning to select a fly that approximates the most prevalent aquatic insect in color, size, and morphological appearance. Other types of fishing use lures of all types, having spinning components, other do-dads, spots, shading, lines, colors, and various shapes—many imitating animals (e.g., frogs or fish) and some not. Then there are scents that can be applied to the lures, and artificial baits that stink. Of course there also are live and dead baits of fish, worms, crickets, shrimp, sand fleas, etc., and cut-bait of squid, clams, and fish. All this is good for the fishing supply places, and potentially expensive for the user.

How do scientists figure out what fish eat? They read fishery science books and cut open fish stomachs. According to a legendary fisheries scientist, Karl F. Lagler (1956), freshwater adult fishes do not just eat one another; instead, they primarily consume aquatic invertebrates, including annelid worms, mollusks, and arthropods (crustaceans and insects). These foods are not uniformly distributed in aquatic habitats, and although likely locations to find these food items vary, they occupy five major zones. Lagler (1956) provides a nontechnical description:

(1) Bottom or benthic zone, on or in the substrate, debris, or rooted aquatic plants (most immature stages of insects, worms, mollusks, sowbugs, and crayfish)
(2) The open-water pelagic zone (plankton and plankton feeders, including larvae of some fishes)
(3) Just beneath the surface film (mosquito wigglers)
(4) The water's surface (water striders, fallen terrestrial insects, birds, frogs)
(5) Just above the water's surface (adults of many aquatic insects such as mayflies, caddisflies, and mosquitoes and other insects)

Foods consumed by fishes are sometimes evaluated by pumping their stomachs; however, this is difficult if not impossible for small fishes and fish larvae, and some fishes have no well-defined stomachs. So the most universally accepted method is to sacrifice the fish and remove their gut contents. Using only guts that have food, diets are reported as the percentage of the stomachs (or of guts) that contain each food item (frequency of occurrence), the number and percentage of each item in the stomach (stomach contents), and the percentage of stomachs with no food. In this way, the number of individuals that consumed each food item can be evaluated and the relative volume, weight, or number of each food item eaten can be calculated. The validity of this traditional approach is dependent on detecting different types of prey and being able to identify them. For invertebrates that have hard exoskeletons this is a reasonable expectation. However, for soft-bodied prey this is more difficult.

Fishes are vulnerable to predation by other fish, mostly when young, and it can be very important to understand the effects of piscivory on fish populations, especially when nonnative predators are introduced. Unfortunately, it can be difficult to obtain an accurate accounting of predation by various species. Although most fishes consume their prey whole, others chew their food before or during ingestion. Once eaten, digestion can be swift for delicate prey such as fish larvae. For example, Schooley et al. (2008) found that mastication of sucker larvae by fathead minnows and red shiners resulted in unrecognizable remains, whose identification was further compounded by very fast (<30 min) digestion. Even in sunfishes (*Lepomis*) that swallowed prey whole, digestion rendered larvae unidentifiable in about 60 min. Similar results were obtained by Legler et al. (2010) who worked on different species. They found that larval guppies fed to bluegills and yellow perch quickly lost morphological characters needed to identify them (i.e., fins and fin rays, heads, pigments, etc.). Furthermore, they found it unlikely that researchers would even detect the presence of larval fish in the gut after 2–4 h, much less be able to identify the larvae to species. The studies cited here and others show that gut content analysis can be unreliable for assessing predation on fish larvae or other delicate organisms.

Also, we might assume that fish would eat the most abundant food items. However, fish may exert a choice and prey on one organism over another. In this case, a far better way to evaluate diet is to determine what foods are available in the environment and to compare this with what food items are actually consumed. This gives an indication of food preference, or food selection. There are several ways to calculate selectivity and an early method illustrates the point. *Ivlev's index of electivity* was an early measure proposed by V. W. Ivlev (1961):

$$E = (r_i - p_i) / (r_i + p_i),$$

in which E = the index of electivity, r_i is the abundance (per cent) or (%) of prey item (i) in the gut, and p_i is the abundance or prey item (i) in the sampling environment. When calculated, E = –1 to +1. Positive values indicate selection greater than the occurrence in nature and negative values indicate avoidance or inability to utilize the prey. Zero suggests a random choice or no selection. There are problems relating to the use of this index since it depends on the behavior of the forager and on the numbers of each food present. Kohler and Ney (1982) and Wootton (1990) provide reviews of methods for determining food selection in fishes and suggest indices and statistical tests applicable for different situations. It is noted that selectivity depends on prey availability, experience of the predator, location of foraging, and other factors.

FORAGING BEHAVIOR AND THEORY

Optimal Foraging

Foraging means to search for food, and behavior is how it is done. We understand why fish forage, i.e., to obtain food in the short term, and to maximize fitness in the long term. However, to

Table 25.1 Optimal Foraging Theory

Prey model: Should a forager attack a prey or wait for a better one?
Focus: Individual prey
Benefit: Energy input if successful
Cost: Energy and time spent in pursuit and handling
Elements considered: Search time, net energy of prey, rate of encounter of prey, probability of a successful attack when encountered
Patch model: How long to feed in a patch?
Focus: Time spent feeding in one location
Benefit: Energy input over time
Cost: Energy spent in feeding and searching for a better patch
Elements considered: Encounter rate for patches, patch residence time. Net energy gain per unit of time

Note: Additional concepts: Decisions, patch depression, give-up time, risk-sensitive foraging, travel time, predator risk.

understand when, how, and where fish forage, we must understand their behavior and be aware of the advantages and limitations imposed on fishes as aquatic ecotherms (although some are heterotherms to a degree). As an ectotherm, fish can forage in good conditions and rest at low energy costs in bad conditions. From an energetic standpoint, this is very important because they do not have to waste energy as metabolic heat or excess activity when it is not necessary. Also, as we have discussed, trophic adaptability means that they are opportunists, and can utilize a broad range of foods if it is advantageous. Finally, it is important to remember that fishes continue to grow as they age and a species will likely utilize different trophic levels as they grow. This suggests that different foraging behavior is likely for different sizes of fish.

A foraging strategy is the total foraging behavior inherent in a species, and it presents in a way to maximize fitness for that species through the process of natural selection. In this case, the more energy extracted from food and available for reproduction, the greater the potential for maximizing fitness. However, the element of time also is important, because time also can be interpreted as energy expenditure. Swimming fish use a considerable amount of energy, and even immobile fish must account for standard metabolism. Thus, the overall strategy must be to maximize energy consumed while minimizing the amount of time that it takes to do it. Tactics are variations of foraging behavior that compensate for local conditions and differing adaptations of prey (Hart 1997).

The optimal foraging theory predicts how a fish can obtain the greatest net yield of energy from foraging, with the goal of maximizing lifetime reproductive success (Gerking 1994). Two principal models are pertinent: the *prey model*, which involves a predator that kills and consumes a prey, and the *patch model*, which considers a predator that grazes its prey—usually harming but not killing. These models consider three requirements of foraging: (1) a predator must decide whether or not to attack a prey, (2) what are the energy costs and gains, and (3) what are the constraints, or how do intrinsic and environmental factors affect the relationship between (1) and (2). Pyke (1984) provided a critical review of the basic concepts, and hundreds of papers incorporating and developing the concept have been published since then. Optimal foraging is outlined in Table 25.1.

Predatory Behavior and Prey Response

Before a predator can obtain energy by prey consumption, energy must be expended to find the prey and to deal with it. Predatory behavior is complex and may go through a number of steps that are mostly sequential:

- *Step 1. Search*—This can be an active search in which the predator swims or floats along looking for prey or an inactive searching by remaining in one spot and watching for movement of the prey. The important foraging concept is the amount of time between encounters. Active or swimming searches are done by pelagic fishes that can easily overtake its prey when detected, such as menhaden or herring searching for plankton. Lie-in-wait searches involve prey that may swim fast enough to escape. Fishes that use lures are considered ambush predators because of the escape problem, whether they float in darkness or hide on the bottom.

- *Step 2. Prey detection*—This depends almost as much on the behavior of the prey as that of the predator, because the predator has to know when and where prey are likely to be available and vulnerable, and what to look for. Prey also have some protection due to countershading and camouflage, and hiding or foraging among cover. Detection also has much to do with the experience and sensory capabilities of the predator. As illustrated by Springer and Gold (1989) sharks use their sensory organs in concert with one another, mostly depending on range. Sharks can detect sound vibrations from kilometers away, olfaction is useful at hundreds of meters distance, the lateral line is most useful at 100 m or less, vision works well at tens of meters, electroreception is very good at centimeters, and then there is contact and gustation. Contact not only tells the shark the consistency of the object, but the rough skin can easily cause bleeding, and sharks know very well about that.
- *Step 3. Attack*—At this point the predator has to make a decision whether or not to commit energy to catch the prey. This depends on whether the prey has seen the predator first or vice versa. If an intelligent prey (fish) detects the predator first, they are likely to signal that they are aware of the danger, and some may even taunt the predator by swimming toward it. With the element of surprise gone, and based on earlier failures, the predator may not commit. Nor is the predator likely to commit if it knows the prey is too large, does not taste good, or if it has color patterns or shapes of dangerous or poisonous organisms.
- *Step 4. Capture*—Even if an attack is made there is a good probability that it will fail due to prey evasion or an inaccurate trajectory on the part of the predator. Analogous to the coil of a rattlesnake, it has been observed that piscivorous fish place their body in an S-shape alignment in preparation to strike. Once the strike is released (an S-start), then it is hit or miss. If the predator is unseen, there is a good chance that the prey will not have a chance to respond. However, some of the time, the prey will have time for a reflex action that results from a rapid C-shape alignment (a C-start), and this direction of fish movement is random (compare with a startle reflex in humans) (Bond 1996; Diana 2004). However, if the prey detects the predator, it will gradually swim away and the predator is much less likely to capture it, because the prey can escape by moving at an angle to the attack direction.
- *Step 5. Handling*—or what do I do with this prey—is an important part of the energetics of predation and reflects prey defenses. I mentioned that white sharks that attack sea lions usually make a vicious bite then back away and let the lion bleed to death, thus avoiding potential injury before consuming the meal. Puffers that consume crabs do not prey on the very large ones due to likely injury. Largemouth bass consume all sizes of crayfish when they are soft from molting, but only select smaller and less dangerous ones when the shells are hard.

Prey Defense

Fish have many ways to avoid being eaten, such as the C-start reflex that can place them out of danger, the ability of some to detect a predator at a distance, to escape by swimming or flying through the air, and can discourage the predator with poisonous (urchins) or sharp spines (spiny rayed fishes). Other fishes, including puffers have special defense mechanisms that discourage (puffing up into a very large and prickly ball) or kill predators that consume poisonous body parts—just in case individuals learn to get past the first line of defenses. One special category that will be presented here as a case study includes changes in morphology that can be induced by predation.

Prey defenses also include chemical means of prey detection, such as shreckstoff. It has recently been discovered that early life history stages can detect chemical cues from predators. In one study (Jones et al. 2003), exposure to odors from a predaceous fish (burbot, *Lota lota*) delayed fry emergence while exposure to a conspecific trout (*Salmo trutta*) resulted in earlier emergence. These cues are innate, species specific and may be enhanced by learning (e.g., see Hawkins et al. 2007, 2008).

Basic Prey Model

The concept of optimal foraging has been applied to various aspects of feeding behavior. The *basic prey model* has been used to evaluate predatory behavior and prey selection. It is used to predict whether a forager is likely to attack a prey of avoid it in hopes of finding a better one. Like all models it has some assumptions that must be considered, and the model makes three predictions which I have summarized (Hughes 19, 137): Foragers should eat the most profitable prey type; foragers should eat less profitable prey only when the amount of energy consumed over time drops below average; and foragers should reject less than optimal prey. The model has five elements (Gerking 1994):

(1) Search time between prey encounters
(2) Handling time for attacked prey
(3) Net energy gain
(4) Rate of encounter
(5) Probability of attack for specified prey

The first and now classic test of the basic prey model was conducted by Werner and Hall (1974). Test results had some surprises, but after reflection, the experimental results matched the assumptions above. In their study, bluegill sunfish (*Lepomis macrochirus*) were given three sizes of *Daphnia* and allowed to forage on them. When large prey were most abundant, the fish ate the largest almost exclusively. When prey density was moderate, the fish ate the two largest-size groups according to the encounter rate. At lower prey abundance, the encounter rate was equal and the fish ate all sizes.

The results of bluegill foraging mostly satisfied the predictions of the model. However, even when prey abundance was highest, the fish still ate some of the small prey. Diana (2004) reviewed various aspects of bluegill foraging and clarified this apparent problem. He pointed out that bluegill moved from place to place where they waited to look for prey. Associated with their behavior are three hypotheses that apply to their prey selection. First, we would assume that the largest Daphnia would be eaten if there was a choice between two of different size (optimal foraging model). Second, the fish may be selecting prey based on *apparent* size (apparent size model). That is, a *Daphnia* that was closer than another would appear larger even though it was the same size or smaller. In this case, the intention was to eat a large prey. Third, if a smaller prey comes so close that little energy is expended to eat it (almost bump into each other), then it would also meet the less time aspect (first seen hypothesis), and associated with the hypothesis is the possibility of predator fixation that might occur due to a food item being so close.

Perhaps the greatest utility in developing models is not proving they are always correct, but in determining the reasons why outcomes in nature do not approximate the theoretical limitations. In this case, hunger and experience can be expected to be powerful forces. Hunger would drive a fish to consume prey no matter what the size, experience might teach the predator that a certain species, although smaller than another, is so much easier to catch that they are the best choice. In all of these examples, though, one could argue that optimal foraging is still being maintained.

Patch Model

The prey model is concerned about the predator–prey relationship, i.e., whether a forager should attack a prey item or wait in hopes of obtaining a better meal. The *patch model* is used to predict how long a forager should feed in a patch; in this case, it could be a patch of plankton, epilithic algae, seagrasses, coral reef sections or anything else that appears as patches across the aquatic environment.

Predictions about the outcome of this model were addressed by Charnov (1976) and Hart (1986), and presented by Wootton (1990, p. 55): "A solution to the problem, the *marginal value theorem*,

TROPHIC CONCEPT AND FEEDING

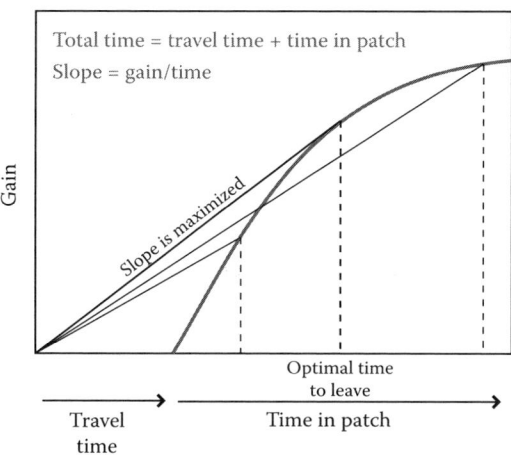

Figure 25.8 Marginal value feeding. (Courtesy of McCutchan, J. H., Jr.)

is that the animal should leave a patch when its rate of capture of prey immediately before leaving equals the average rate of capture for the food patches in the habitat." Thus, the marginal value theorem (Figure 25.8) suggests that an animal can calculate a benefit/cost evaluation based on the energy available between patches. Accordingly, a forager should feed in the most productive patches, stay in those patches until the foods obtained per unit of time declines to a value equaling the average condition in the environment, leave a patch when it declines below the average, and ignore patches of low food value. This presumes that a forager can do two intelligent things: it must be able to weigh the value of food in patches against one another, and it also must be able to keep track of time spent foraging.

In the patch model, a predator may be a carnivore, omnivore, or herbivore. A patch is considered to be composed of food organisms of the same type. Three elements are needed (Gerking 1994):

(1) Patch encounter rate
(2) Patch residence time of the predator
(3) Net energy gained per unit of time

As a forager works in a patch, the prey population is grazed and it will become necessary to move to another patch. This decline has been referred to as *patch depression*. The time expended until a patch is vacated can be called the patch resident or giving up time; however, the *give-up time* is defined as the point in which the average amount of food intake in that patch declines to or below the average value of all patches in the environment, at this point that there is no incentive to stay (as pointed out in the marginal value theorem).

The patch model has been studied under experimental conditions for many years, and substantive research has investigated the relative ability of fish, and animals in general, to determine patch quality. This can be demonstrated by studies that monitor fish behavior and show that fishes remain longer in patches of higher productivity. DeVries et al. (1989) were able to confirm this ability in an experimental study of bluegills, in which the fish tended to minimize time spent in poorer patches. Another study of bluegills also confirmed longer resident times with higher prey densities (Wildhaber et al. 1994). Finally, Milinski (1994) found that three-spine sticklebacks remembered and used patches that were productive in the past, preferring areas that recently produced foods over areas of more historic interest.

Patch residence time can be affected by risk-sensitive foraging. This is an economic risk that is related to the abundance and distribution of food. The decision that a forager has to make in this case is whether to stay in or return to a known patch or take a chance in finding a new, unknown patch. Predator risk is another variable on patch use, and fishes in general will not choose to increase predation unless a new patch has a great deal of food that justifies the potential risk. For example, creek chub will switch to a hazardous feeding site only when prey density is very high (Gilliam and Frazer 1987).

CASE STUDY: PREY RESPONSE—A MATTER OF HUMPS?

The native fish community of the mainstream Colorado River is composed of seven large fishes, and known as the "big river fish community." These fishes, four minnows and three suckers, have coevolved for many thousands of years. In addition to their relatively large adult size, some of these fishes have nuchal (nuch = back of the neck) humps and some do not. The nuchal hump is an interesting morphological feature that is seen as an abrupt bulge on the back immediately behind the head and over the opercle, and is very pronounced in two species (as their names suggest), the humpback chub and the razorback sucker (e.g., Figure 25.9). Bob Miller (1946) described the humpback chub, and speculated that large humps would aid a fish to stay on the bottom as water flowed over the hump because it would push the fish down. However, Miller did not observe the fish swimming and its habitat use was hypothetical at the time. Later, LaRivers (1962), perhaps after reading Millers' account, hypothesized that the hump in the razorback sucker might serve a similar purpose. This "nuchal hump hydrodynamic advantage hypothesis" (NHHAH) was around for many years, and with the passage of time (and the absence of any experimental tests to prove or disprove it), the hypothesis was treated as fact by many authors.

I had been interested in testing the NHHAH for several years, and fortunately for me, Don Portz, a graduate student in my Ecology of Fishes class, agreed to devote his master's study (Portz 1999) to experimentally test the hypothesis. Don decided that tests of the drag and lift components of large humps would provide the answer. He made whole body casts of adult humpback chub and razorback sucker from museum specimens, and then removed the humps by sanding them off and retested the same casts without the humps. The casts were placed in an experimental flume and subjected to high and low water flows. A force beam and foil strain gages were attached to each cast and trials were run to evaluate fluid-induced drag and lift forces (Figure 25.10). Results demonstrated an appreciable drag associated with the hump and no negative lift (toward the bottom) (Figure 25.11).

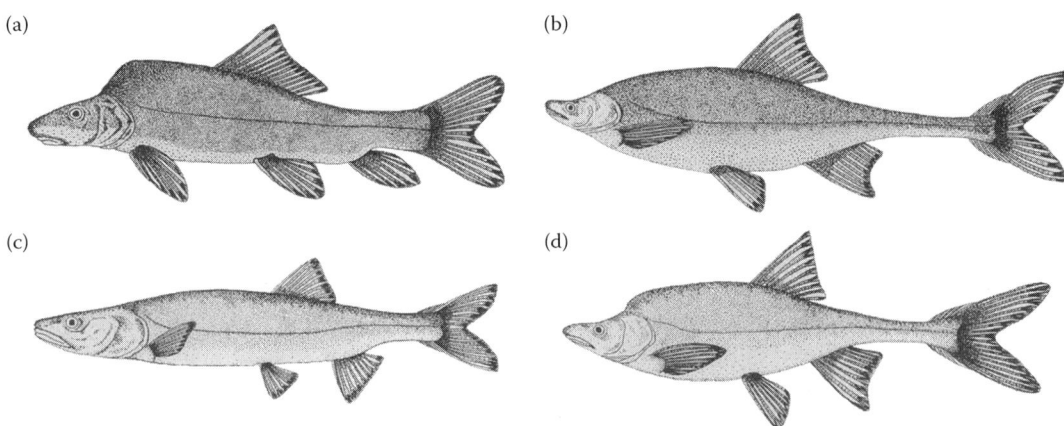

Figure 25.9 Four Colorado River endangered fishes. (a) a razorback sucker, (b) a bonytail, (c) Colarado pikeminnow, (d) a humpback chub. (Courtesy of J. Beard, USFWS.)

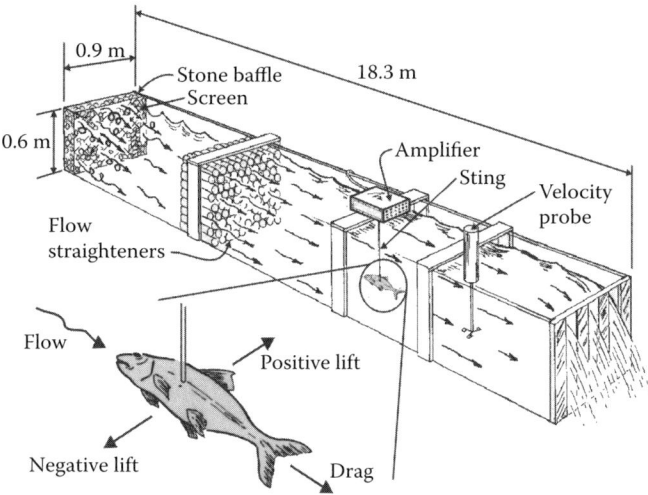

Figure 25.10 Experimental flume used to determine drag and lift. (From Portz, D. E., and Tyus, H. M., *Environ. Biol. Fishes*, 71, 233–245, 2004. With permission.)

Instead there was a lift component like an airplane wing, that would tend to make the hump (with fish) rise upward. In other words, presence of humps made staying on the bottom more difficult in fishes that had them than in fishes that did not. Also, increase in the drag component would require more energy to swim in fast currents. Also catch records and radiotelemetry indicated that two species are not fast water dwellers anyway, both preferring eddies and bottom habitats. Other fast water forms of chubs and suckers had no humps, presumably due to higher energetic costs that such a shape would produce. Thus, the NHHAH was rejected.

In science, it is not customary to discard one hypothesis unless there is a better one to offer. So we started working to build a case. First, we had to consider whether these were relict structures with no purpose—highly unlikely we figured, because there was an energetic cost associated with maintaining humps in the population. Another thing was troublesome also. The humps on the two fish are analogous, but not homologous: they look similar but the chub has a hump of muscle, and the sucker has one of bone. We pondered, is this convergent evolution? We thought so. Now the final clue: Both of the humped fish lived in eddy habitats while fishes of runs and riffles had no large humps. What else preferred eddy habitats? It was the Colorado pikeminnow, a predator with a relatively small mouth and no jaw teeth. They lived sympatrically with the predator! It was time for more work.

Don and I drove to Dexter National Fish Hatchery in New Mexico to measure hump sizes and gapes of live fish, thanks to the kindness of hatchery manager Roger Hamman. When the data were analyzed, Don found that humps provided substantial protection from predation by the pikeminnow (Figure 25.11). About 73% of suckers and 83% of chubs with no humps could be consumed by very large pikeminnow (805 mm TL), but only 55% of suckers and 71% of the chub with humps could be ingested. The differences would be even greater for the average size pikeminnow. Based on gape size alone, the pikeminnow piscivory would be constrained by the presence of enlarged humps. With no jaw teeth, pikeminnow have to suck down the slippery prey head first and cannot bite into it as an aid in swallowing it (Figure 25.12).

After obtaining the gape restrictions with and without the humps, we then had to relate these experimental findings to the natural river system. Based on field sampling we discovered that the average size of Colorado pikeminnow was about 536 mm TL. Based on body depth at hump, humpback chub, >21 mm, and razorback sucker, >220 mm, would be immune to predation from an

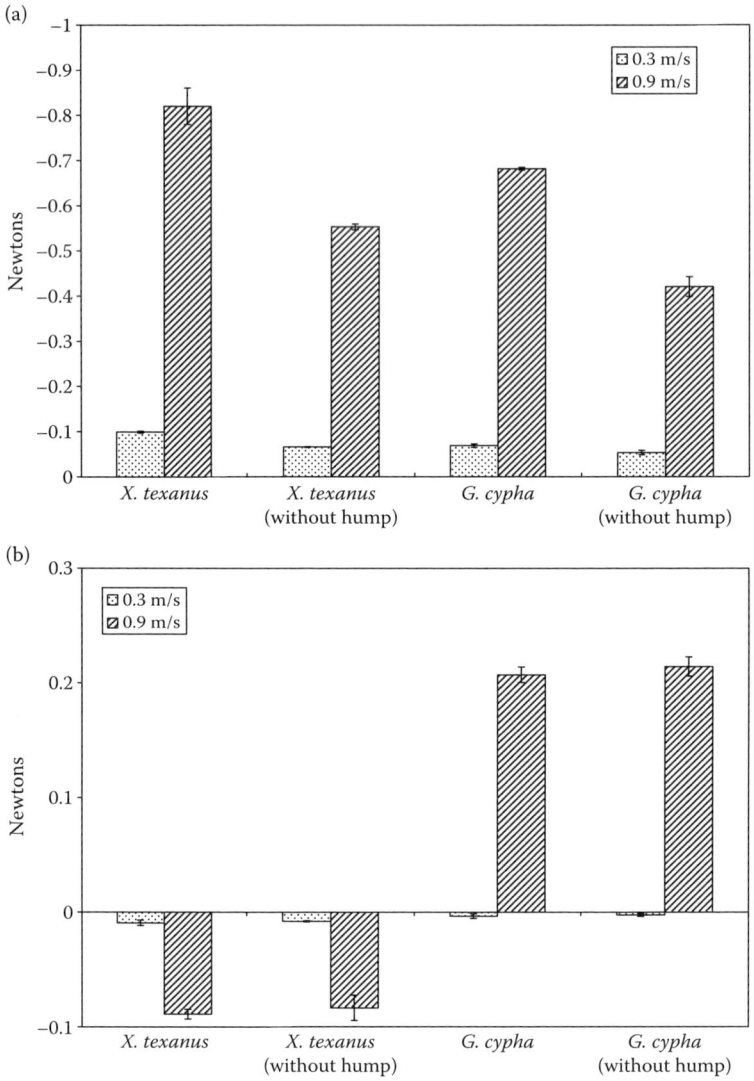

Figure 25.11 Drag (a) and lift (b) components associated with humps. Data exhibited by body casts of *Xyrauchen texanus* and *Gila cypha* in the experimental flume under water velocities of 0.3 and 0.9 m/s. Casts without hump are the same as those with the hump, except the hump has been removed by sanding. (From Portz, D. E., and Tyus, H. M., *Environ. Biol. Fishes*, 71, 233–245, 2004. With permission.)

average size pikeminnow. Furthermore, the young of both prey species utilize shallow habitats until they reach about this size, when they begin to move into deeper water where they are most vulnerable to large pikeminnow. The adults of both species are even larger, and they would be safe from the largest pikeminnow in the present system (Figure 25.12).

Was there more information about these fish to support our claims? Further support exists from other morphological characteristics of the prey that appear to be predation related. They have smooth bodies and a lack of or reduction in the size of scales, which would make them slippery. They also have laterally flattened bodies; with large paired fish that would make them highly maneuverable in eddy habitats. An almost scaleless, slippery, living fish would be difficult for a pikeminnow to

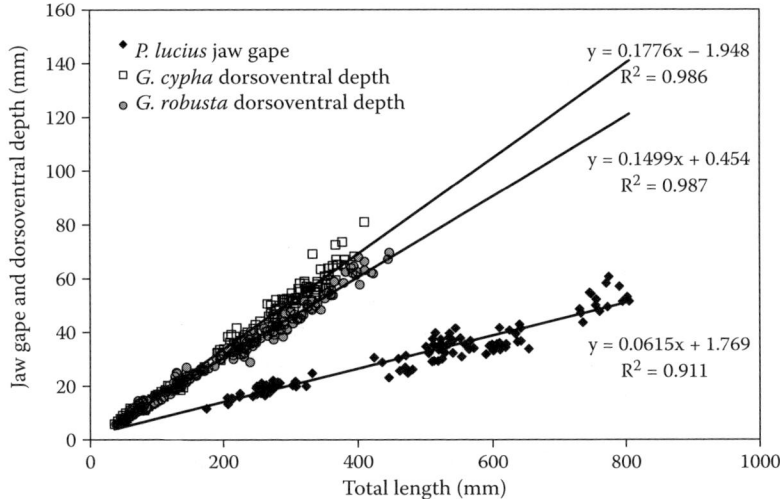

Figure 25.12 Jaw gape of predator and hump depth of prey. (From Portz, D.E., and Tyus, H. M., *Environ. Biol. Fishes*, 71, 233–245, 2004. With permission.)

capture and manipulate for effective feeding and prey body shapes would facilitate complex and evasive movements.

We also looked elsewhere for support, and found that there are long lists and books written about morphological structures that have evolved to provide a defense against predation. It also appears that another cyprinid fish, crucian carp (*Carassius carassius*), increases its body depth when living in habitats occupied by a predator, northern pike. The change in morphology can occur in a few months and it is so extreme as to lead to its mistaken identity as another species. This change increases handling time (making escape easier) and also produces gape limitations for the pike (see Portz and Tyus 2004 for a better discussion and references). After completion of our study, several other investigations have been published on Crucian carp (e.g., Holopainen et al. 2005; Vøllestad et al. 2004). Also, Kishida et al. (2007) reported phenotypic plasticity in tadpoles (*Rana pinca*) as a predator-induced defense, and discussed its genetic basis. Robinson et al. (2008) reported on pumpkinseed sunfish ecomorphs, which increased body depth and dorsal spine length when exposed to walleye. The whole concept of "the ecology and evolution of inducible defenses" can be reviewed in a book with that title by Tollrian and Harvell (1998), and Benard (2004) provided an informative review of the subject.

As a postscript, such a remarkable adaptation may have contributed to a degree of predator naiveté in these species due to immunity from ingestion and even recognition of large body depths by the coevolved predator, instead of more evasive behavior (e.g., see Johnson et al. 1993). This could be a very serious problem for them in an environment that now has been stocked with toothy and large-mouth predators such as northern pike striped Bass smallmouth and largemouth basses, walleye, and flathead catfishes.

SUMMARY

The trophic concept demonstrated how energy in the form of food could pass through different levels from photosynthetic producers (e.g., phytoplankton), to herbivores (zooplankton), to first level carnivores (planktivorous fish, e.g., clupeids) to large carnivores (piscivores, e.g., tuna). Also included are decomposers (detritivores). This concept has been referred to as a trophic pyramid, but the pyramid can be distorted due to changes in a trophic level, usually phytoplankton (bottom-up effect) or an apex predator (top-down effect), either of which can cause a ripple or trophic cascade through the food pyramid.

Fish do not fit into the trophic pyramid very nicely because most fish will feed at more than one trophic level. However, fishery-induced trophic cascades can occur with the removal of apex predators. Fishes may be herbivores, carnivores, omnivores, or detritivores, but they also may be classified more specifically under these categories and the categories can be subdivided to be more specific about the mode of feeding. There is a great deal of resource sharing among fishes, including dietary overlap, explained by the concept of trophic adaptability. What do fish eat? Just about anything alive or dead, consisting mostly of zooplankton, aquatic and terrestrial insects, worms, mollusks, arthropods, other fish, etc., but food for fish is not uniformly distributed. We understand why fish forage, i.e., to obtain food in the short term, and to maximize fitness in the long term. However, to understand when, how, and where fish forage, we must understand their behavior and be aware of the advantages and limitations imposed on fishes as aquatic ecotherms (although some are heterotherms to a degree). As an ectotherm, fish can forage in good conditions and rest at low energy costs in bad conditions. From an energetic standpoint, this is very important because they do not have to waste energy as metabolic heat or excess activity when it is not necessary. Also, as we have discussed, trophic adaptability means that they are opportunists, and can utilize a broad range of foods if it is advantageous. Finally, it is important to remember that fishes continue to grow as they age and a species will likely utilize different trophic levels as they grow. This suggests that different foraging behavior is likely for different sizes of fish. Optimal foraging theory predicts that fish will seek to obtain the maximum amount of energy at the lowest cost, and cost may be equated with time. Two models are presented, the *basic prey model*, which involves a predator that kills and consumes a prey, and the *patch model*, which considers a predator that grazes its prey—usually harming but not always killing. These models consider three aspects of foraging: (1) how does a predator decide whether or not to attack a prey, (2) what are the energy costs and gains, and (3) what are the constraints? In the prey model, foragers are expected to maximize energy input while minimizing the time spent searching for prey. In the patch model foragers have a residence time when the feed in a patch. When the food intake is about average for patches in general, then foragers are expected to "give up" and move to another patch. The chapter ends with a case study on a morphological structure (i.e., development of a hump), testing and rejecting an earlier hypothesis of hydrodynamic advantage and advancing a new one that suggests that the humps evolved as a defense against predation.

Further reading: Godin 1997.

CHAPTER 26

Reproductive Ecology and Life History Patterns

REPRODUCTIVE PROCESS

Reproduction provides for the origin of new individuals. Reproductive biology is the study of the reproductive organs and biological processes that tell us *how* organisms reproduce. Reproductive ecology of fishes on the other hand, is most concerned about *where* and *when* a fish reproduces and how much energy is allocated to the process. First, we must understand the process of reproduction, then we can determine how environmental factors influence the process, and finally we can explore how the fish responds to the environmental factors in a way that maximizes fitness.

Fishes are extremely complex organisms, a very diverse and ancient group, and consequently they exhibit a wide range of reproductive behavior, strategies, and tactics. As an example, Patzner (2008) summarized reproductive strategies in 55 orders of fishes. Here we will outline the process of reproduction for a generalized teleost fish in a probable sequence, emphasizing the main considerations of reproductive ecology, which are timing and energy allocation (Wootton 1990). The following outline considers major reproductive requirements in teleosts:

- *Events preceding reproduction*—A combination of endogenous and exogenous conditions must be met for reproduction to occur.
  - The fish must be healthy and in good condition.
  - It must have at least an adequate amount of high-quality food, preferably rich in protein and fat.
  - The fish must be physiologically ready to reproduce. Minimum age at first breeding varies among fishes, and the capability to reproduce must be met. Readiness is under genetic control influenced by complex factors of age, condition, stress, and social behavior.
  - The proper seasonal stimuli are required. Reproduction may be arrested at some stage and require an environmental signal (zeitgeber) for completion, such as providing yolk (process of vitellogenesis). Timing signals usually include attaining a certain photoperiod (light:dark ratio), water temperature, odors such as pheromones (e.g., presence of conspecific males or females in breeding condition), a certain substrate, vegetation, salinity, etc.
- *Prespawning condition and behavior*—If the above conditions have been met, some changes in morphology or certain behaviors may occur due to physiological stimuli, which may be under hormonal control:
  - Secondary sex characteristics may become evident, such as brighter or different coloration, breeding tubercles, hooked jaw, etc.
  - The desire to move inshore, offshore, upstream, or downstream, perhaps due to an awakened memory, such as in imprinting, to find a spawning location.
  - Identification of spawning areas by imprinted behavior, presence of conspecifics, odor of reproductive byproducts, etc.
  - Mate selection, which may involve courtship, a ritualized display, and other activities that may be influenced by size, behavior, coloration, odor of sex products, presence, and construction of nest.

- *Spawning and egg fertilization*—Final egg maturation and ovulation into the body cavity (or lumen of the ovary) should now have occurred and the eggs must be fertilized within hours in most cases.
  - Mating can now occur with the release of female and male gametes.
  - Most fishes are oviparous with external fertilization of the released eggs.
  - Usually mating occurs with cross-fertilization but gynogenesis (egg development initiated by penetration of sperm from male of similar species but not fertilized) is known.
  - A variety of mating systems occur. Promiscuity (several males one female) is ancestral. Monogamy is occasional.
  - Some fishes are ovoviviparous with internal fertilization of the eggs and eggs hatched in utero with no placental involvement.
  - A few fishes are viviparous (about 6%) (Patzner 2008), and eggs have placental attachment (sharks, guppies, surfperch).
- *Parental care*—May or may not occur. When it occurs, one or both parents can be involved.
  - No parental care is provided in most pelagic fishes.
  - Some benthic spawners may guard a nest or bear the oviparous young in the mouth or in pouches.
  - Viviparous and ovoviviparous species bear the young internally.

The preceding account and emphasis here is reproduction in teleost fishes. Parsons et al. (2008) provides an excellent review of the reproductive ecology of sharks, and provides the reproductive mode for eight orders and 34 families of sharks. Although sharks can be oviparous, ovoviviparous, or viviparous, the point is that most of them give live birth to a fully formed juvenile that are so large that they are not very vulnerable to predation, and capable of feeding on most adult-size fishes. Furthermore, Parsons et al. (2008) reported that 60% of elasmobranchs are viviparous and hypothesized that two pelagic traits, obligate ram ventilation and vivipary, may have coevolved in pelagic sharks.

LIFE HISTORY PATTERNS

The preceding outline of the reproductive process in teleost fishes shows that there is a great diversity in fish reproduction, but also that reproduction requires numerous decisions that need to be made. These form the basis for development of different life history patterns. The evolution of these life histories is a result of the goal to obtain an optimal resolution of conflicting environmental demands placed on an organism in an attempt to maximize fitness (e.g., Ricklefs 1990). In other words, a life history pattern includes major features of the life of an individual that are the parts of a strategy to improve survival and reproduction. For fishes, these features, or traits, are major adaptations that include variations of sexual reproduction, mating systems, sexual selection, site selection, and the reproductive effort. Frimpong and Angermeier (2009) provide a database of the ecological and life history traits for 809 freshwater fishes found in the United States.

Reproduction is energetically costly, and decisions must be made in the allocation of parental energy to the young (e.g., amount of yolk), to select the spawning site (e.g., migration), and whether to provide care, and how much (e.g., nest, protection). Throughout these decisions, there is a need for proper timing of events. First we discuss timing, and then we will evaluate different life history traits and energy allocations.

Timing of Reproduction and Environmental Cues

Reproduction begins when a fish become sexually mature, and reproductive ecology is concerned about whether and how environmental factors influence the decision to reproduce and subsequent events associated with reproduction. Age at first maturity is a decision that affects future growth, reproductive output, and survival. Thus, there is great variation in the age of first reproduction in fishes and this variation is molded by life histories in different environments. For example, fish that inhabit very stressful and sometimes ephemeral locations, like pupfish and other cyprinodonts, can reach sexual maturity very early (weeks to months), while other species like paddlefish and sturgeons of large and variable rivers may not attain maturity for several years. Even within the same species, age (and size) at first maturity will vary according to environmental conditions. Stunting occurs in several freshwater fishes usually due to density-dependent population regulation and can result in sexual maturity at a much smaller size and age than normal.

Fishes have an endogenous mechanism, including perhaps a "biological clock" that is sensitive to endogenous conditions and exogenous cues. There is evidence in fishes that such an endogenous cycle is the basis for an annual rhythm resulting in sexual maturity (Bye 1984). Old timers used to link biological events such as timing of fish "runs" (migration) to other biological events that occur at about the same time. I looked for these and noticed that razorback suckers started to spawn at about the same time that local apricot trees began to bloom. I had a harder time figuring out the Colorado pikeminnow, but can do that one a little more scientifically; they spawned with the summer solstice when days and nights are of equal length.

There is ample evidence that egg maturation proceeds to a certain point, but that ovulation requires an external stimulus. The stimuli associated with ovulation have been linked with photoperiod, temperature, spawning substrate, visual and chemical stimuli from conspecifics, and other environmental factors (reviewed by Stacey 1984). However, there may be striking difference in the factors and how they affect gonadal maturation and provide spawning cues among different fish taxa. Two of the most studied groups, salmonids and cyprinids, show such differences. The spawning act also may require that certain conditions be present as well, and as an example, some fishes spawn at night and/or with high tides.

It is thought that riverine fishes generally respond to photoperiod and temperature to obtain environmental cues for migration and spawning. But of course, that logic has to be a little backward. For oviparous fishes, the important thing has to be timing of larval emergence so that the proper temperature, foods, and physical habitat conditions are present for survival when larvae are ready for exogenous feeding. So at first, environmental cues might be more general ones, then natural selection, i.e., survival and then recruitment of the young, could result in survival of fish that spawn according to more site-specific cues, such as flow levels (usually associated with the spring hydrograph in temperate regions).

Sex and Mating

Patzner (2008) conducted an extensive review of the modes of reproduction in fishes. He reported that sexes are separate and genetically determined (gonochorism) in almost all fishes (88%). However, some fishes also are successive hermaphrodites and change from male to female (protandry) or female to male (protogyny), and a few are synchronous hermaphrodites, in the latter case they may mate with either or both sexes. Finally, gynogenesis is known in at least one fish (Amazon molly) in which all fish are females.

Fishes are highly variable in their mating systems (reviewed by Taylor and Knight 2008). *Monogamy*, or pair bonding, is a feature of fishes that are associated with substrates and defend a territory. Centrarchids, ictalurids, cichlids, serranids, and so on exhibit social monogamy. Largemouth bass are good examples of a monogamous fish, and guard the nest and larvae for as long as a month after spawning. *Polygamy*, in its various forms, is the rule in most fishes, and polygamy includes repeat spawning or spawning aggregations of more than one mate for males, females, or both. All pelagic fishes, freshwater and saltwater, can be included in this group, and cyprinids are a good example. Broadcast spawning is common, and dual polygamy is common in pelagic eggs, and females mating with multiple males (polyandry) is common in demersal eggs. Finally, there are two types of polygyny that can be important in fishes, resource defense polygyny, in which more than one female lays eggs in a nest, such as in three-spine sticklebacks, and female defense polygyny in which males defend a group of females (harem), such as in shell nesting cichlids.

Sexual Selection

Almost all fishes reproduce sexually, a process which produces genetic variation and aids in preventing a buildup of deleterious genes. However, each parent loses one-half of its genetic material in the process. This costly exercise makes it imperative to choose the best mate possible. There are two ways to do this: spread the eggs around to a number of males and hope for a good one, or select a superior one based on some criteria. This process can involve intrasexual competition between likely candidates, or intersexual selection by male or female.

Intrasexual selection usually occurs as males compete with one another for the privilege of mating. Size dominance is common in fishes, and usually the largest male will control spawning sites or nests and be selected by the female or females. However, males of some species will turn to aggression in competition for space and females, for which the males of Siamese fighting fish are famous. However, most fishes mate by intersexual selection. This is usually a clear female-dominated resource selection process, the male displaying its markings, building nests, or collecting shells to entice females to spawn with them. However, females also will select a mate based on "genes only" selection (no territory, nest, or other resource a factor), and this type of selection may be influenced by apparent size, brilliant coloration, or the ability to successfully complete a complicated ritualized display. Extremely colorful or highly active males also may be preferentially selected, perhaps a case of the "handicap hypothesis" if such coloration or behavior makes them vulnerable to predation. In this case, the fish that can survive with the greatest handicap must be of superior qualities to those that cannot.

Alternative Breeding Tactics

Usually some type of intraspecific or interspecific selection of males is made in fish reproduction, but alternative ways to achieve male fertilization of the eggs are relatively common. Gross (1984) considered these behaviors as tactics, and provided an excellent discussion their evolution as related to overall population fitness. There are two basic types of alternative reproductive tactics related to fitness. These are subordinate tactics, which may reduce fitness but make the "best of a bad situation" (BBS), and equivalent tactics which are of equal fitness.

Ovipary and external fertilization of the eggs creates an opportunity for rivals to deposit sperm as eggs are being deposited. A dominant male cannot always be on guard, and may be busy defending the eggs from another, he might be at one end of the nest or territory and cannot defend the other end, and also he must take time to release his own sperm. In this case, smaller males may dart in and fertilize eggs. These small males, known by various names according to different species and their mode of action, may be known as sneakers, streakers, satellites, peripherals, female mimics, jacks, helpers, accessory males, etc. To understand their prevalence and need, consider three situations: (1) Reckless courtship of a large brightly colored male might attract predators, whereas drab and smaller males are less vulnerable. (2) Large males can guard the female while fighting off other males—in this case sneakers may fetilize more eggs than the dominant male; and (3) Jacks may not be inferior, but just early maturing.

Well-known examples of alternative male mating systems occur in many fish groups, including sunfishes (Centrarchidae) and salmon (e.g., Keenleyside 1979; Gross 1984). Bluegill sunfish have three different males, a large dominant that pairs with the female and defends the nest, sneakers that dart in during fertilization, and satellites that mimic females and overtly join in matings. Pacific salmon have two sizes of males as a rule, the large hooknose that migrates back after 2 or 3 years in the ocean, and smaller Jacks that return after only a year at sea. Hooknose males breed by winning a fight with another male, but the jacks are sneakers.

Spawning Site Selection

About 1/3 of fishes are pelagic spawners and 2/3 are demersal spawners. Some degree of spawning site selection occurs in both situations. The common notion of pelagic reproduction is a random spawning act by a school wherever they may be. However, this is generally not true. Pelagic fishes do not appear to spawn at random locations as might be supposed. Some species, like menhaden, move into locations where prevailing currents can transport the young into estuaries that are used as nursery areas. In freshwater species, spawners may use riverine currents to disperse the eggs into downstream habitats. An example is the Rio Grande silvery minnow, in which an upper section in the river is chosen, and spawning occurs at highest flow. The male wraps himself around the female in the upper water column and the pair release and fertilize their eggs directly in the flow.

Demersal or substrate spawners characteristically choose spawning areas carefully and prepare the site for the deposit of eggs. How sites are selected is a mystery for most species, but in some, there is an identification and orientation to natal areas presumably in association with olfactory cues and presence of reproductive byproducts. Demersal spawners of inshore marine zones use preselected areas to deposit eggs (Potts and Wootton 1984). North sea herring move to, and spawn in stock-specific areas, and homing of salmonids, cyprinids, and catostomids to natal areas are well known.

Parental Care of Eggs and Young

Balon (1984) divided teleost fishes into reproductive styles and guilds depending on their reproductive habits. In this system, he divided all fishes into three major behavioral (ethological) groups:

(1) Nonguarders—Fish that do not guard their eggs or young. This behavioral group includes fishes that are broadcast spawners in pelagic and benthic habitats, and those that hide their brood.
(2) Guarders—Fish that guard eggs or young. These fishes that choose to spawn in certain substrates such as rocks or plants, and those that use nests.
(3) Bearers—Fish that protect the young with their bodies. These include external bearers, such as mouth or pouch brooders, and those that bear young internally, such as viviparous fishes.

REPRODUCTIVE EFFORT AND ENERGY ALLOCATION

Lifespan and reproductive effort tend to be inversely related in fishes, and critical decisions must be made about how to allocate the available energy acquired by the fish to the reproductive process (reviewed by Kamler 1992). This can be divided into two main categories: (1) how much energy is allocated to each of the young individuals (e.g., in the amount of yolk (food), and (2) what proportion of available energy resources are allocated to the reproductive effort (e.g., movements, courtship, site preparation, mate selection and breeding behavior, and shelter and protection for the young).

To the Young

The vast majority of fishes are oviparous, with the release of unfertilized eggs in 94% of species (Patzner 2008). An important consideration for the welfare of their young is the amount of energy to be allocated to each individual in the form of yolk. Fish eggs are highly variable in the amount of yolk provided among taxa, with smaller eggs < 0.3 mm in diameter and largest eggs > 90 mm. As an indication of the amount of yolk, largest eggs can be millions of times as heavy as the smallest eggs (Kamler 1992). The amount of yolk supplied per egg is a major reproductive tradeoff, because the same amount of energy can produce thousands of small eggs or only hundreds of large ones. For example, cod produce millions of tiny eggs, trout lay hundreds of eggs, and sharks and rays may lay only one or two fertilized eggs. The obvious denominator in the size of the egg is the amount of yolk, from a tiny amount in the cod egg, to a glutinous amount in the shark egg. (NB: As astonishing as it may seem, Kamler (1992) reported that the largest egg produced by fishes weighs 34 million times heavier than the smallest egg!) What are the tradeoffs that need to be balanced in number-of-eggs-versus-amount-of-energy dilemma? There has to be a balance between quality and quantity. This issue is related to survival prospects of the mother and the young.

The achievement of maximum fitness is a function of the survival of the mother and the progeny. If the mother will have an opportunity to spawn another season or another year, there would be a potential benefit. The mother might enhance fitness by spreading the reproductive effort in a changing environment. In this case, it would be best to conserve her energy for survival rather than maximizing eggs and/or yolk and jeopardize her life. On the other hand, if there are not enough offspring to survive physical conditions or predation, or they do not have enough energy to mature, then her fitness is zero. In this section, we address why varying amounts of energy are allocated to the young. Later we will incorporate the rest of the dilemma by addressing energy needed by the females.

There is a great difference in the amount of yolk provided to developing larval fishes. The amount fluctuates between two extremes of a spectrum, those larvae that receive only a small amount of yolk, which are known as *altricial* young, and those that receive a large amount of yolk, which are known as *precocial* young. Given that the size of an egg mostly reflects the supply of food, teleost eggs can range from about 0.25 mm to about 7 mm, and comparison of 101 marine and 33 freshwater fishes from Europe demonstrated a tendency toward smaller diameters (Wootton 1990). The largest of these are eggs of salmonidae, an ancestral group, which suggests two things: an evolutionary trend toward more and smaller eggs, and production of larger eggs in cold-adapted groups.

Blaxter (1988) investigated the relationships between producing a large quantity of small eggs versus fewer but much larger eggs in a number of species. Not surprising, production of a large quantity (fecundity) is inversely related with egg size, and increases with age and size of a female. Given the same temperature conditions, large eggs take longer to develop than small ones, presumably due to sequential steps in the process of development. Fecundity is higher in fishes that are marine pelagic spawners and lower in freshwater fishes and in fishes that provide care to the young.

Altricial young are true larvae and little resembles the adults. Upon hatching, they are incomplete in development and nearly helpless. Development of some of the major organs is incomplete, and these may include such important structures as the eye, jaw, and gut needed for feeding, and the lateral line and fins, also needed for feeding and locomotion. The completion of these organs may occur after swim-up as *sac fry* and it is dependent on the energy contained in the yolk sac (i.e., endogenous feeding). In some fishes (Scheidegger and Bain 1995), such development may occur after emergence, entrainment, and transport to downstream areas, presumably where there is food for consumption and exogenous feeding is necessary for survival. Nature generally has provided enough yolk for altricial larvae to develop mouths and gut, but as inefficient swimmers as they are, they must be delivered to a place of high food density in order to survive. Many are pelagic feeders and their first live foods are usually small zooplankton or other small invertebrates.

Precocial teleost young are more fully developed at hatching, and by yolk depletion, they are more like juveniles than are altricial fry. They are larger, have sensory organs, mouth, jaws, and digestive systems that are capable of detecting, catching, and consuming prey. They are usually benthic predators, and trout young begin feeding on smaller stages of aquatic insects.

We will discuss ontogeny and early life histories of both types of larvae in our larval fishes chapter (Chapter 28). Extremely large and precocial young can be borne by viviparous fishes, including sharks, and they may receive additional internal nourishment from the mother's body.

To the Reproductive Adults

The reproductive effort in fishes can require considerable energy to be expended by the adults both before and after spawning occurs. Major needs that require energy are summarized here in approximate order, but the list is not complete:

- *Secondary sex characteristics*—As suggested in the above outline, many species develop morphological characters associated with reproduction. Most species develop a bright coloration that is different from the nonreproductive condition, and special characters may develop, such as breeding tubercles, which are common in cyprinids and ictalurids. In salmonids, the development of a hooked jaw occurs in male salmon and can be used as a weapon against rivals. Other characteristics may include pheromones produced by males and females, and other secretions such as sticky mucus from sticklebacks that is used to build their nests.
- *Migration*—Almost all fishes select and congregate at spawning areas, thus requiring energy to travel to them. These costs can be very great in case of migratory species, and especially for diadromous ones that have to make physiological changes in their metabolism to survive movement between saltwater and freshwater. The well-known travels of salmon and eels stand out, but many freshwater fishes also perform spawning migrations, such as the Colorado pikeminnow, which may make round trip migrations of 200 miles. Such travels can be hazardous due to physical damage, exhaustion, and predation.
- *Courtship*—Successful mating requires investment in a specialized courtship behavior, which serves several functions, including proper species identification, mate selection, ensuring reproductive synchrony among the sexes, and to establish a pair bond for protecting or caring for the young. Once mate selection is initiated, usually by a female swimming in close proximity to a male, then courtship may begin. The behavior may be ritualized as in the well-known three-spine stickleback, in which correct visual display and stepwise procedures must be followed, or it may be more generalized territorial displays and swimming behavior such as in the cod (Keenleyside 1979). Although courtship may not require vast amounts of energy, Wootton (1990) points out that the fish are not foraging while courting, and the displays can make them vulnerable to predation as well.
- *Spawning site preparation*—Many fishes provide shelter for their young after they hatch, even if the adults do not remain to guard them. The degree of attention given to providing shelter is related to several variables, one of which appears to be whether the eggs are adhesive or not. Trout, which lack adhesive eggs, usually dig a depression to hold the eggs and clear away fine sediment so the eggs can fall into interstitial spaces. Other species with adhesive eggs may clean stones, leaves, etc. by scraping the surfaces clean so the eggs will stick to them (Keenleyside 1979). Some species, including lampreys, dig pits or shallow depressions and cover the eggs. Some species of minnows carefully remove and sort through different size gravel or stones which they will replace into the nest and spawn on them or use them to cover the spawn. Nest builders, such as sticklebacks, do even more work to construct a spawning site. Mouthbrooding cichlids shelter their young internally, as do pouch brooders, such as seahorses. Fishes with internal fertilization provide the shelter of their bodies.
- *Cultivation*—Adults may prepare the environment in which their young emerge by reducing the abundance of other fishes that would prey on or compete with them. They can do this by eating them or as discussed below, by aggressive behavior. This is the *cultivation hypothesis* (presented in Chapter 25) as proposed by Walters and Kitchell (2001), and it is a way of maintaining ecological dominance. It is presumed that this behavior is net yielding in energy and not a cost.
- *Defense of territory or nest*—The value of the nest occupied by the young is indisputable, and the nest and area it occupies may be defended with force if needed. Large tropical fish, such as the emperor cichlid, can be an effective defender. Salmonids may defend or dispute an area that is being excavated for a nest, and then several nests that have been constructed. Some cichlids and sunfishes

are colonial nesters, and the entire area of the nesting colony will be defended by any and all nesting fishes.
- *Parental care of young*—Few teleost fishes provide care for the developing young after the eggs are produced and placed into the nest, other than perhaps covering them. However, males of some fishes like mottled sculpin may defend the young of several females for at least as long as they are on endogenous food. Mouthbrooders and pouch brooders and other bearers obviously provide care for their young. Cichlids, stickleback, and other fish fan the developing eggs to provide oxygen and remove debris that falls into the nest.

Fish that continue to provide an energetic investment to their young may be reducing the success of the next reproductive effort. This tradeoff between the reproductive effort and future reproduction is the *cost of reproduction*, which has been defined as ". . . the loss in expected future reproduction . . . that is attributable to present reproduction" (Sargent and Gross 1986). This concept has been called *Williams' principle* after G. C. Williams (1966). In connection with this concept, it is common in fishes for the male to care for the developing offspring, which allows the females to grow larger (due to indeterminate growth) and thus be able to produce more eggs. The male spends very little energy in producing sperm in comparison, thus the cost of reproduction for the female can be reduced by substituting male care.

TWO LIFE HISTORY STRATEGIES

Semelparity

Costs of reproduction can increase to the point that the survival of the adults may be at risk. This is especially true in salmon and eels, fishes that have to undergo the exertion of long diadromous migrations and undergo physiological changes to deal with osmotic regulation. Such fishes are *semelparous,* they store a large amount of energy and expend it all (60% to 85%) in one great reproductive effort to maximize fitness. This seems like a crazy kamikaze type of strategy, referred to as the "big bang" strategy (Gadgil and Bossert 1970). However, there are perfectly sane reasons for it. It is no secret that many salmon die upon spawning, using 80% or more of their somatic energy (Diana 2004) in the reproductive effort. The question has to be why. As first invaders into streams that drained Pleistocene glaciers, Pacific salmon were able to travel in large rivers and spawn in predictable coldwater habitats that contained no other fish competitors. By producing eggs with large supplies of yolk, the precocial young produced were large enough to feed on stream invertebrates virtually undisturbed. Thus, good recruitment was produced every year. In this case, the maximum amount of precocial young can be produced, and in dying, the adults bring high-quality nutrients into the stream where their young are foraging. In summary, there are high energetic cost and likelihood of dying for the adults; good recruitment of the young in a dynamic, but relatively predictable environment; and few competitors and predators.

Iteroparity

Most fishes spawn more than once in their lives, reproducing repetitively either seasonally and/or annually. These *iteroparous* species are on the opposite end of a spectrum of reproductive strategies from semelparous species; however, as more is being known, the propensity for either strategy is not absolute. Some fishes, like American shad are semelparous in the southernmost extent of their range and iteroparous in more northern latitudes. In Canadian waters, they expend about 55% of their stored energy in migration. In warm Florida waters, they expend about 50% for inshore migration, but also expend 30% of their energy producing larger gonads while migrating (Glebe and Leggett 1981). In anthropomorphic terms, if you are going to die anyway, why not make more eggs before you do. In the ecological world, we anticipate that greater fitness would result from more eggs, and the trait be selected for.

Iteroparity is related to unpredictable environments. It is associated with less allocation to reproductive tissues (25%–50%) (Diana 2004), and iteroparous fish have a tendency to produce altricial young. There may be other tradeoffs as well, as in the case of large river fishes, such as paddlefish, sturgeons, pikeminnows, and white sucker. These fish are all migratory and may move long distances in freshwater migration. Females display less than annual spawning; presumably, this is a survival issue. Due to high cost of reproduction, there may not be enough resources to support a large fish and allow for the substantial energy costs of migration and growth of reproductive tissues. Of course, less than annual reproduction also may be related to the absence of needed environmental cues.

REPRODUCTIVE TRADEOFFS: r AND K SELECTION AND A 3-D CONTINUUM

Adaptation can be seen as a result of compromises in allocation of time and energy to various life history needs, and it is adjusted by natural selection. It also is clear that different life history strategies are favored by the size of the species population. If a population is sparse then natural selection would favor those genotypes that result in rapid population growth. On the other hand, a population with a high density would soon be constrained by carrying capacity. These considerations were noted by MacArthur and Wilson (1967) in discussing the theory of island biogeography. They noted that islands tend to be invaded by species dispersing from other locations, and that these species were early colonizers that were able to rapidly increase at low population densities. However, as islands become more populated those rapid colonizers evolved into populations that grew more slowly, but were more competitive at higher population size.

These concepts are referred to as r and K selection for the two ends of a spectrum. Species which are r (for exponential growth curve) strategists are opportunists. They have the ability to disperse and to quickly multiply due to their high fecundity. They are more successful at lower population densities. On the other hand, K (for carrying capacity in the logistic growth curve) strategists are equilibrium species; they take energy away from reproduction to build structure and to persist to reproduce over longer time scales. For this reason, they are more competitive at higher population densities.

Few species are "pure" r or K strategists because the ends of the spectrum were used to characterize the two strategies: tradeoffs are the most risky in an evolutionary sense. Let us look at the tradeoffs:

Tradeoff	r Strategists	K Strategists
Age at first reproduction	Young (small)	Older (large)
Number of spawning efforts	Few	Many
Offspring at each spawning	Large/body size	Fewer/body size
Length of reproductive life	Short	Long

My sense of this is that r-strategists do well in habitats that have environmental problems, but they need to reproduce in many locations (be able to disperse) to find favorable ones. However, they are small and can use smaller habitats, such as ponds or marshes. Some species (like catfish) can walk away to another pond or sit out unfavorable periods. Temperate regions seem to be ideal for these species because they are able to cope with catastrophes that are not related to adaptation.

In larger, more stable and connected habitats such as rivers, the K-strategists do well. In the presence of many competitors, reproduction may not occur until late in life to allow growth in size, which reduces vulnerability to predation and aids mobility. Because they have long life, they can sit out less than optimum conditions and continue to grow, amassing a size that can produce a great many young when conditions are better. Tropical conditions are good for these species because competition and efficient resource use are selected for.

It is unfortunate that we tend to think of life history patterns as two ends of a diverse spectrum. Nature does not always exist at extremes, and conditions change as noted by MacArthur and Wilson (1967). Instead, most species can be expected to participate in "bet hedging," and combine various elements of the r and K models. In what is known as "Grime's triangle," an alternative to the r and K concept in which there are three strategies: R (ruderal plants tolerant of disturbance), C (competitive), and S (stress tolerant) (Grimes 1977). Winemiller and Rose (1992) have provided additional insight into life history strategies of fishes by placing life histories into three strategic groups, and fitting them into a 3-D continuum (reviewed by Bone and Moore 2008). The axes used are age of adult maturity, fecundity, and juvenile survivorship. This provides for three classifications, and two of these are similar to r and K strategists. The concept is presented here by describing the extreme situation in each of the three dimensions:

(1) Opportunistic spawners with small body size, early maturation, multiple spawning episodes each season, and little parental energy investment. Examples of these fishes are the anchovies and Cyprinodonts. This strategy favors colonizing in environments that undergo frequent abiotic change in space and time.
(2) Periodic spawners have larger body size, delayed maturity, high batch fecundity, little parental investment, and long generation time. Examples include most salmonids and clupeids. This strategy is favored where there are large spatial variations and cyclical changes.
(3) Equilibrium spawners have variable body sizes, low reproductive effort but high parental involvement, and moderate/long generation times. Examples are cichlids, centrarchids, and catfishes. This strategy is favored in environments that are dominated by biotic interactions and have relatively stable physical conditions.

CASE STUDY: TIMING OF SPAWNING

We tend to think of spawning as a spring occurrence because most fishes do spawn in the spring. Some fishes like char spawn in fall in response to very short growing seasons. But why would a fish spawn in the middle of the year? That is a question that has been asked for the Colorado pikeminnow. It was previously thought that the fish spawned earlier in the year due to the capture of fish in breeding condition at or slightly after the peak of spring runoff. But this is not the case.

Thanks to the efforts of the Larval Fish Laboratory at Colorado State University, identification of larval pikeminnow from other cyprinids has been a great aid in determining life history needs. A further step was made in determining and ground-truthing the rate of development of larvae, so that researchers could estimate time of hatching by the size of the young and water temperature. Using these techniques, and adding to this captures of spawning fish, it was possible to determine the spawning of the fish on a spawning reach for several years. Figure 26.1 shows the average annual distribution hydrograph for the Yampa River, Colorado for three years selected from eight years of data (Table 26.1) to show low, average, and high water years. Included in the diagram are spawning periods. It can be seen that spawning varies depending on the timing of spring flooding.

The spawning periods all occurred around the summer solstice and the relative length of the estimated optimal spawning period lasted about 26 days. Spawning occurred earlier in low water years and later in high flow years. Water temperature and discharge also varied during these years. The average minimum and maximum water temperatures were 19°C and 24°C, and discharge ranged from about 25 to 108 m^3/s (Table 26.1). What does all this mean? In general, spawning occurred during a 26-day period of declining flows at almost equal day lengths and increasing water temperatures following most of the peak in spring runoff.

REPRODUCTIVE ECOLOGY AND LIFE HISTORY PATTERNS

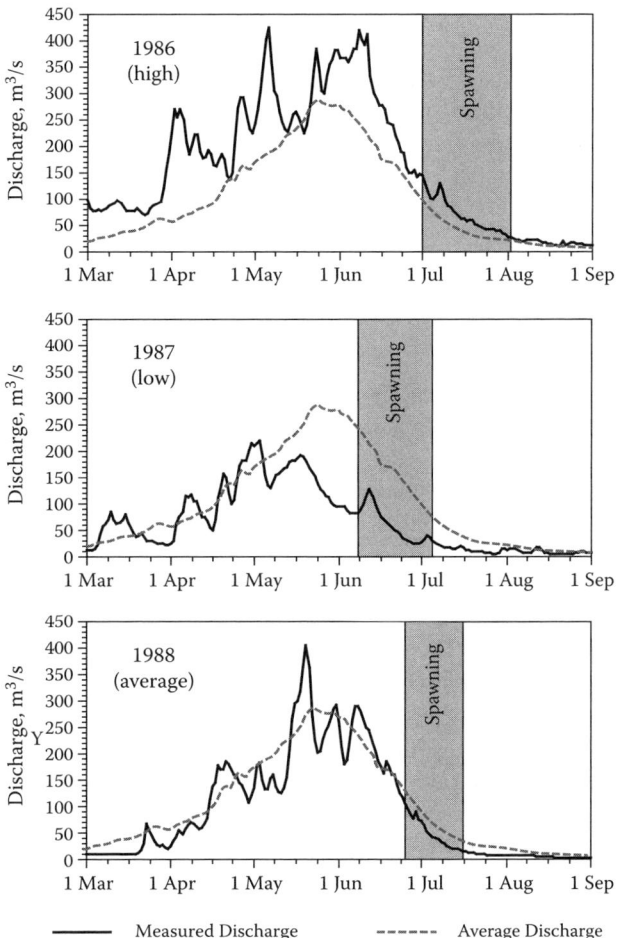

Figure 26.1 Spawning of Colorado pikeminnow, Yampa River, Colorado. (Courtesy of Tyus and Karp, USFWS Biological Report, 1989.)

Table 26.1 Conditions during Colorado Pikeminnow Optimum Spawning Periods, 1981–1988, Yampa River, Colorado

Year	Flow[a]	Spawning[b]	Mean Flow (m³/s)[c]	Temperature Range (°C)	Mean Minimum (°C)
1981	Low	6/23–7/13	25.27	18.0–25.5	19.3
1982	Average	7/8–8/1	108.25	16.5–27.5	19.5
1983	High	7/15–8/10	86.17	18.0–27.0	21.0
1984	High	7/16–8/15	71.74	20.0–24.0	20.3
1985	High	6/25–7/17	64.02	14.5–25.5	17.8
1986	High	7/1–8/2	69.82	18.5–23.0	19.5
1987	Low	6/8–7/5	58.69	16.5–24.5	17.9
1988	Average	6/25–7/16	49.14	18.0–25.0	19.5

Source: Tyus, H. M. and Karp, C. A., *U.S. Fish and Wildlife Service Biological Report*, 89(14), 1–27, 1989.
[a] Based on average annual flows 1922–1987.
[b] Optimum period based on age of larvae, collections of ripe fish, radiotracking.
[c] U.S. Geological Survey flow data, sum of gauges for Lily Park (1981) or Deerlodge Park gauge (1982–1988), Yampa River, Colorado.

Using statistical techniques (regression analysis), the amount of peak discharge preceding spawn, and the occurrence of mean minimum conditions of temperature were highly correlated ($r = .84$ and $.88$; $P < .05$). This suggests that a flow level and minimum temperature are sought for spawning; however, we have to be cautious because flow, temperature, and photoperiod are correlated. So these factors should not be separated without very good reason. But why try? It looks like a higher cold flow is not warm enough so high flow years are later. In low water years, the temperature warms more quickly, so low flow years are earlier.

But none of this tells us *why* the fish are spawning here at this time of year; we need to think about natural selection to figure that out! It turns out that the sac fry emerge at the spawning site and are transported downstream and out of the whitewater canyons and into the alluvial reaches of softer sediment, meandering channels, and semi-isolated backwaters that they use as nursery habitat. The backwaters are formed by the lowering water levels. Thus the location and timing of spawning is programmed so that the larvae swim up and are transported in the drift downstream to perfect conditions for them to begin exogenous feeding. To maximize their fitness, spawning adults must make the right decision about when and where to spawn. Under natural conditions, they seem to do a good job of it! However, when dams alter flow regimens there can be problems. We will talk more about pikeminnow migration as a life history trait in the next chapter.

SUMMARY

There is great diversity in fish reproduction and in the process numerous decisions will be made. These decisions collectively form the basis of different life history patterns. The evolution of reproductive life histories should help obtain an optimal resolution of conflicting environmental demands on an organism in an attempt to maximize fitness. Fundamental to this are strategies that improve survival, reproduction, and recruitment. Fishes have major adaptations that include variations of sexual reproduction, mating systems, sexual selection, site selection, and the amount of reproductive effort. Reproduction is energetically costly, and decisions must be made in the allocation of parental energy to the young (e.g., amount of yolk), to select the spawning site (e.g., migration), and whether to provide care and how much (e.g., nest, protection). Throughout these decisions, there is a need for proper timing of events. Reproduction begins when a fish becomes, sexually mature, and reproductive ecology is concerned about whether and how environmental factors influence the decision to reproduce and subsequent events associated with reproduction. Age at first maturity is a decision that affects future growth, reproductive output, and survival. Thus, there are great differences in the onset of reproduction among fishes. Breeding in fishes is mostly monogamous or polygamous, and results in the release of eggs that are externally fertilized. Energy is allocated to the reproductive process before, during, and after spawning, depending on the species. Semelparous fishes put all of their effort into a single spawning, but iteroparous fishes spawn multiple times. Prespawning fish may expend energy in migration, courtship, and/or site preparation, or in allocating various amounts of yolk to the egg. Parental care of the eggs after spawning and fertilization includes guarding, nonguarding, and bearing of the eggs in pouches, mouth, or internally. Young can be precocial, altricial, or intermediate in development depending on the amount of yolk or nourishment provided. A case study of the timing of spawning is presented as an example of how environmental signals can be incorporated in the reproductive effort.

Further reading: Rocha et al. 2008 and Classic: Breder and Rosen 1966.

CHAPTER 27

Migration

FISH MOVE, DISPERSE, AND MIGRATE

Humans have been intrigued by migrations of animals for centuries. We see migrations of songbirds as they seasonally appear in our yards and feeders. We also hear birds overhead as geese and cranes fly in passage. We learn about the herds of caribou, wildebeest, and other mammals that seek vegetation and water. We see nature presentations that show arctic gray whales wintering in the Sea of Cortez. However, few birds and mammals can surpass the migrations of tunas, salmon, eels, and other fishes for their endurance, hardships, and sacrifice. Bluefin tuna have been documented in a migration pattern from Florida to the coast of Norway, at least 10,000 km distant, with individuals averaging 5 km/h for a passage of 2–4 months (McKeown 1984).

The above examples are just the beginning of long lists of migratory species. But why do so many animals migrate? Migration is energetically costly, and any ecologist will point out that if it is costly behavior then there is a very good need for it—and that has to be associated with fitness, which is a product of natural selection. The reasons for migratory behavior are many; not all migrations are for the same purpose. Some are for feeding, others for better climatic conditions, and of course many are for reproduction. Migration can allow access to more food, different foods, faster growth, quicker maturity, increased fecundity, lower intra or interspecific competition, escape from predation, access to better and safer reproductive sites, and more.

Feeding migrations are common among marine fishes that use estuaries at some time in their life, such as flounders, Spanish mackerel, and menhaden. Adult flounders and Spanish mackerels move in to feed with the tides, a regular movement in most cases. Menhaden adults spawn in currents near the estuary and their young enter the food rich water for a nursery area. They return to sea for adulthood, thus the feeding and reproductive migration is separate. Fishes of floodplain rivers in tropical and temperate climates also make feeding migrations, in this case lateral ones in spring or rainy season to exploit terrestrial food sources in the flooded lands. Mesopelagic fishes make vertical feeding migrations each day.

Climatic migrations occur seasonally in a number of fishes along the Atlantic coasts, including tuna, swordfish, striped bass, and others. Fishes of warmwater reservoirs and lakes will make vertical migrations to cooler water in order to digest food more efficiently.

Spawning migrations are extremely common in fishes and usually need little explanation. This type of migration clearly is programmed to maximize fitness by selecting a spot that is perfect for reproduction and provides access to feeding grounds for the young. Spectacular spawning migrations of salmon, shad, and eel are well known, but a great many other species also make spawning migrations in oceans and freshwater systems.

We must assume that migration in fishes emerged because of some need, and movement to a new area increased survival and growth. But what was the origin? Tsukamoto et al. (2009) presented the *random escapement hypothesis*, which suggests a behavioral origin based on a "psychological repulsion" to conditions of stress that results in a motivation to escape. A stimulus, such as crowding—a violation of personal space—or adverse conditions of the physical environment (e.g., low oxygen, high temperature, hunger, predation) can result in "jumping" behavior, especially noted in wild fish confined to aquaria. Fish also may respond by swimming into flowing water in which better conditions (e.g., food, more oxygen) could be detected upstream. A pattern can thus be developed in which seasonal movements can be established. The rest is history.

It is no wonder that migrating fishes have evolved general and specific morphological, behavioral, and physiological adaptations to allow their purpose.

But it is only in the past few decades that we have come to appreciate the amazing ability of fishes to overcome enormous obstacles in finding their way to specific locations. What is it, why do they do it, and how do they do it? In this chapter, we explore these questions and consider the environmental factors that require, are effected by, and facilitate migration as a life history pattern. We begin with definitions:

- *Movement* is a generalized term that signifies some activity on a local scale. It can be a shifting position, it implies no starting or stopping point, and it requires no rate of return.
- *Dispersal* means to move away with no intent for a return. It is ostensibly a one-way trip. It is common in mammals when the mother kicks out the young adult to find its own territory.
- *Migration* is a specialized behavior that has its origin in natural selection. In simple terms, it is a round trip, a seasonal movement that implies some element of return to an initial starting place. As alluded above, fish migration can be a feeding migration, climatic migration, or a spawning migration. Also, there are several attributes of migration, which include a direction, a periodicity, a distance (usually fixed), a degree of return (usually high), and homing (usually a high degree of accuracy to a predetermined location).

When I began studying migration in fishes, I was amazed to find that most humans accepted fish migration as some "darned fool thing" that fish did. There was little or no credit given to the notion that the fish had any control over what they did. Baker (1982) convinced me long ago that in order to study migration you have to accept the concept of a fish as an intelligent being, acting as an individual. He pointed out that the study of migration was at one time torn between those (ethologists) who believed fish migrated and found their way entirely due to some programmed reflex responses and others (population ecologists) who worked on mathematical models of dispersion. Baker applauds behavioral ecology, which is supported here: the study of behavior from an evolutionary perspective.

Using this as a metric, it is not difficult to assume that fish find their way in many aspects just like humans do—without the global positioning system of course!

FINDING THE WAY BACK—HOMING

Home Stream Concept

The ability of adult salmon to return to the stream of their birth has been known for a very long time. Montgomery (2003) identified several prominent persons who expounded on this idea historically, including Alexander Fraser, a Scottish naturalist who studied the life history of Atlantic salmon. Evidently, Fraser performed studies and also had access to experiments done by others, in which wires or morphological marks were placed on juvenile Atlantic salmon and discovered years later on adults returning back to the same stream where they were born (Fraser 1833, cited by Montgomery 2003). By 1860s, the life history of Atlantic salmon and their homing behavior to natal streams was well accepted in the British Isles, and some naturalists believed that this behavior took the fish to the very gravel bar from which they emerged as hatchlings (Montgomery 2003). This "parent stream theory" was extended to Atlantic salmon in the New World, where homing of marked juveniles was confirmed by captures of these fish as adults as early as 1653 (Cooper and Hirsch 1982). However, there were doubts about Pacific salmon in the western United States.

Dr. David Starr Jordan, a monumental figure in ichthyology and the leading expert on Pacific salmon in his day, at first believed that the salmon entered steams at random and stated in the first edition of *American Food and Game Fishes* (Jordan and Evermann 1902) that the authors did not believe in natal homing. This changed in the revised edition of the same book (dated 1923), in which the authors gave two explanations for the movement of adults back into natal streams: "Some thought that the salmon return to their own stream because they possess a marvelous geographic or homing instinct," while others believed the young reared to maturity close inshore and entered the same stream because it was "nearest at hand." The reason for this change of mind was presumably related to the findings of several scientists, which occurred as early as 1906 (no references given). One of these was Dr. C. H. Gilbert, whose ability was obviously held in high esteem. Due to Dr. Gilbert's influence, Jordan and Evermann (1923) supported the concept of "homing instinct" and Dr. Evermann suggested changing the name "parent stream theory" to a more appropriate "home stream theory," the reason for suggesting this change will become clearer later. The point is that by the 1920's there was an endorsement of homing behavior by the most respected and famous team of ichthyologists in the country.

The ability to "home" to and locate natal areas for reproduction is a powerful one because two important processes are produced by it. One is an aid in survival and thus fitness—the spawners are living proof that young produced at that site have survived to recruit. The other is more subtle, but perhaps just as important—survivors can perpetuate successful genotypes, thus producing locally adapted gene pools. Both of these processes are well supported in the literature, but often not succinctly stated.

Homing has been most extensively studied in some Pacific salmon, principally coho, as will be described shortly. However, homing also has been noted in other fishes for centuries. In North America, the anadromous alewife has been discussed as an important resource to early European settlers. Many early anecdotal accounts of its homing and its wonderful ability to find spawning areas can be found (e.g., Hay 1959; Thunberg 1971). Migrations and homing of North American suckers also have been studied (e.g., Raney and Webster 1942; Dence et al. 1948; Werner 1979). Similar homing migrations have been reported in Asia for various minnows (Nikolskii 1961; Breder and Rosen 1966; Tyus 1990b). Evidence that has been building suggests homing and olfactory imprinting are not as rare or as restricted to a few taxa as once thought.

Mechanisms for homing are still being investigated. Such studies are difficult, but what has emerged is that the phenomenon is not stereotypic. A great deal of flexibility seems to be emerging in which different taxa recognize signals from different sources. Skill for recognizing spawning, feeding, and nursery areas could involve the ability to detect many cues and the skills needed for direction finding, location of general areas, and site-specific recognition. This is all very exciting to an ecologist once we understand some of the concepts, behaviors, and mechanisms.

A Few Terms

In understanding migration and how fishes find their way, we must first consider that in some cases the individual already has made a trip away from the site to which it now wishes to return (i.e., to "home"). As simple as this may sound, it appears that most people fail to appreciate this point. Now the fish must use its instinct, learning, and decision-making ability to find the way back. Even more interesting is the ability of some fish to use cues in locating sites that they have never before experienced. We proceed by considering a few related concepts and then by looking at probable mechanisms:

- *Orientation*—We do this by looking or moving our body, i.e., we orient to that location. In this case, we must establish a position to a point in space. If the orientation changes with time, then we have to consider space and time. When you feed them at the same place each day, they will be waiting at the same place facing the same direction.
- *Piloting*—Many people do not think about this, but an airplane pilot is not an explorer. The pilot is trained to travel over a route, thus gaining knowledge or becoming familiar with the area. Thus, piloting is to find your way using knowledge of a familiar area, or sensory cues. A pilot boat has a pilot that travels the channel each day and records changes so it is easier to guide in a larger boat or ship. Bees learn the surroundings of the hive and return to it. When you put fish in an aquarium, they will explore it to learn where everything is located.
- *Navigation*—This requires some thinking and adjustment. In navigating, we travel to a certain point in space. We navigate with a map. In the mountains, we can navigate by sight if we follow a distant mountain. We can navigate by following the sun, but we have to know that the sun's apparent location changes predictably with time. Thus, if I know the time I can tell you where the sun will be located (E, S, or W). When you feed fish in an aquarium at the same time each day, they will be waiting at the same place at that time. If you have the aquarium in the sun, the fish will learn to be in the same relative position, even if you rotate the aquarium 180 degrees.

Orientation Mechanisms

Celestial (sun)—Most fishes have excellent and sensitive eyes. Presence of light in water provides a good cue for determining position in a water column; however, it does not help at night. Fish that are near the surface or that inhabit shallow freshwater rivers and lakes have a dependable daytime cue. Light is polarized as it travels through water, aiding in its use for direction finding. Several studies have demonstrated that fish can be trained to orient to a sun compass and to polarized light (reviewed by Smith 1985). Hasler (1971) set up a complicated experiment using a rotating tank with hiding places arranged in a circular central part. A fish was taught to use only one slot of many such hiding places, and then the tank was rotated. It was demonstrated that fish could be trained to orient with the sun. With an overcast day, the fish could not make the proper choice, and when artificial light was used to replace the sun at the improper time of day the fish oriented to the light when finding a slot that would be appropriate if the sun had moved (Figure 27.1).

Celestial (moon)—There is little evidence for exclusive use of the moon to navigate in fish migration, although some fish migrate at night. Brannon et al. (1981) demonstrated that orientation of sockeye salmon fry was aided by moonlit nights.

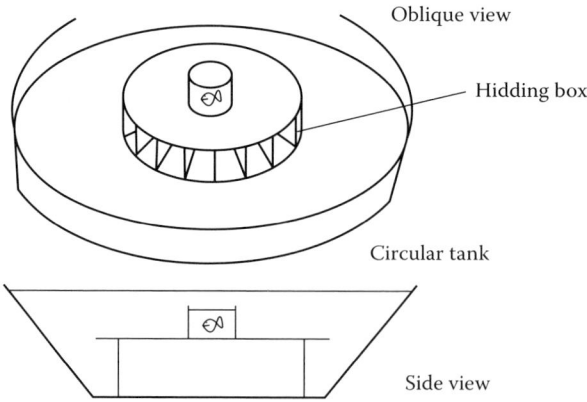

Figure 27.1 Experimental results: fish can tell sun time. (From Bone, Q., and Moore, R. H., *Biology of Fishes*. 3rd Ed., 427, Taylor & Francis Group, New York, 2008. With permission.)

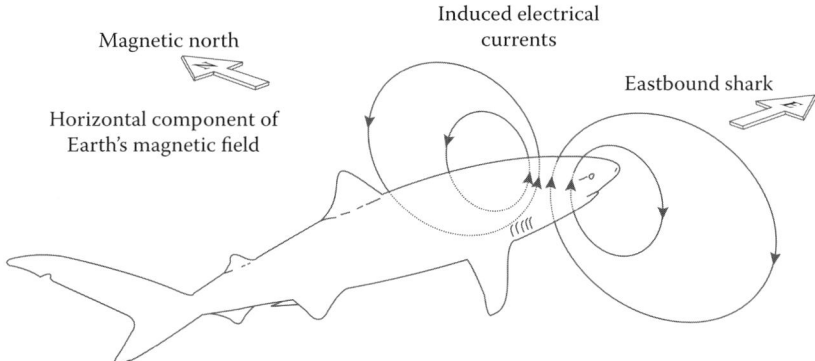

Figure 27.2 Sharks can detect magnetic lines of force. (From Springer, V. G., and Gold, J. P., *Sharks in Question: The Smithsonian Answer Book*, Smithsonian Institution Press, Washington, DC, 1989. With permission)

Magnetic fields—Smith (1985) reviewed several studies that demonstrate that some fish, including salmon, have the ability to detect magnetism, and to acquire a compass reading from the magnetic lines of force. Magnetic substances capable of inducing currents have been found in the tissues of several fishes. One of these, rainbow trout, has magnetic receptors containing magnetite crystals in the olfactory lamellae (Lucas and Baras 2001).

Electrolocation—As discussed in our section on feeding, many fishes (mostly nonteleosts) and especially elasmobranchs and even coelecanths have electroreceptive organs (Figure 27.2) that they use to find food, locate objects, and detect electrical impulses. They also can receive and measure the Earth's magnetic field (Figure 27.3) and obtain compass readings (von de Emde 1997).

Water flow—Fish can certainly tell upstream from downstream, and larval fishes of several species have an active response to detect and move up into the flowing water so that they will be transported downstream and away from a spawning area. It is presumed that fish who wish to enter a river know the difference from upstream and downstream.

Chemoreception—Karl von Frisch created a great stir with his discovery of *shreckstoff* in 1938, as did Hasler and Wisby (1951) with their olfactory hypothesis for homing. The bottom line is that fishes have extremely acute abilities to detect odors through olfaction and/or gustation. By the late twentieth century, numerous studies had confirmed that chemoreception was used in orientation by many fishes (Kleerekoper 1982). As examples of sensitivity, detection of odors present in concentrations of only 10^{-16} is possible (Lucas and Baras 2001). Chemoreception is extremely important, because (unlike celestial cues) it can be used 24 h a day. It is important to note that sensory perception of odors and the use of them as environmental cues can be related to general physical and biological components of the

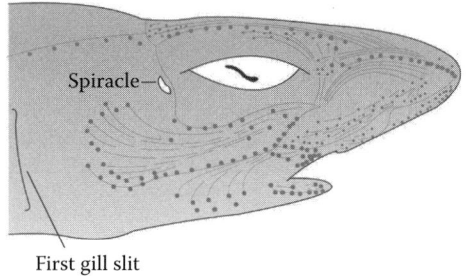

Figure 27.3 Electroreceptive organs in a shark. (Courtesy of Huh, C., http://commons.wikimedia.org/wiki/File:Electroreceptors_in_a_sharks_head.svg.)

environment. Fishes can detect and avoid the odor of predators, be attracted to the odors of conspecifics, display fright when exposed to alarm pheromones, and home to areas due to local odors they have learned and/or by conspecific odors including pheromones and reproductive products which they may recognize innately. Chemoreceptive organs develop early in the life of fishes.

Vision—Sight perception is important in determining the proper orientation to several cues, such as light, sun, moon, etc. Vision also is used by fishes to locate reference points used in piloting.

Gradients—Smith (1985) cited several studies that demonstrated that fishes orient to temperature gradients, and fish also orient to salinities as well.

Random search—No doubt, fish sometimes search at random as do humans; however, this type of orientation mechanism would be costly. It also is difficult to characterize what is meant by "random" search. Fish that are orienting to a spawning area sometimes overshoot the mark by traveling past a proper tributary, and then backtrack, locate, and follow the proper one. Some might consider this to be a "random" event, but I consider it a mistake corrected. On the other hand, there is usually some small percentage of "straying" among migrating fish; whether this is due to a physiological defect, a random search, or some other factor is unknown.

SPAWNING MIGRATIONS

We may not be sure about the intent of some migrations, whether they are taken for feeding, to avoid unfavorable conditions such as climate, or for both. However, spawning migrations are unambiguous to humans and important to fish! There are three major classifications of spawning migrations: oceanodromous, diadromous, and potamodromous. There also are some subdivisions. The common feature of all of these types of spawning migration is the location of spawning sites previously used. The use of natal areas is advantageous because the returning adults were recruited there and they have survived to spawn again. Homing to natal sites for reproduction also aids in maintaining locally adapted genotypes.

Oceanodromy

Oceanodromous (sea + runs) migrations usually combine wintering (seasonal), feeding, spawning, and nursery (in various combination) migrations in a large circuit instead of an upstream–downstream fish migration that we are so accustomed to learning about. This circuitous pattern is usually developed in accordance with oceanic circulation patterns, such as gyres (e.g., bluefin tuna in Atlantic and Pacific, skipjack and bigeye tunas in the Pacific). However, in order to navigate the oceanodromous fishes would require some timing device or be able to recognize some type of environmental cue. Harden-Jones (1981) developed the groundwater seepage hypothesis in 1980 to do just that. Hypothesized as a marine analog to the olfactory hypothesis envisioned by Hasler and Wisby (1951) for streams, Harden-Jones suggested that spawning of many marine fishes that spawn above the continental shelf zone (such as Atlantic herring) are able to recognize and "home" to areas of groundwater seepage. In temperate zones of the ocean, migratory routes of fishes often take a triangular shape with adult migration to spawning, feeding, and wintering areas, and an associated drift of larvae from spawning to nursery area, feeding, and to adult wintering areas.

Diadromy

Diadromous (double + runs) fishes have received much attention, which is surprising, because they include only about 1% of fishes. Their migrations are even more complicated. McDowall (1987) identified 32 families and estimated that about 160 fishes have diadromous migrations. He also defined three different types:

- *Anadromous (up + runs) migrations*: Adult fish migrate from ocean to spawn in freshwater. Feeding and growth of the adults occur in the sea while reproduction occurs in freshwater. Anadromous species include lampreys, sturgeons, salmon, clupeids, striped bass, etc.
- *Catadromous (down + runs) migrations*: Adults feed and grow in freshwater and migrate to the sea for spawning. Catadromous fishes include a strange assortment of fishes, but include *Anguilla* eels, several mullets, and others.
- *Amphidromous (double + runs) migrations*: Larvae hatch in freshwater and quickly return to saltwater to feed and grow, then juvenile fish return to freshwater to grow and reproduce. Amphidromy is rare and technically not for the purpose of spawning (McKeown 1984), so we mention it for completeness of record. These strange fish include some galaxiids, sleepers, gobies, etc.

Potamodromy

Potamodromous (river + runs) migrations occur only in freshwater. This type of spawning migration is probably the most common type in fishes. Due to the short distances involved in many fishes it is not considered as glamorous as diadromy; however, some spectacular migrations are known. In the United States, the most well-known migrations occur in cyprinids and suckers. Some salmonids also make freshwater migrations that may include rivers and lakes.

EXAMPLES AND DESCRIPTIONS

Anadromous Migrations and Homing of Pacific Salmon

Anadromy in salmonids is well known, partially due to important food and sport fish. We know their life histories due to countless studies. Also, we understand their migrations because A. D. Hasler was interested in them and he studied the work of Karl von Frisch and Konrad Lorenz. I never met Dr. Hasler, but had the great privilege of meeting two of his students: Peter Johnson and Al Scholz, both of whom I pestered about Colorado pikeminnow migration. My thanks go to Dr. Scholz for coauthoring a wonderful account of "olfactory imprinting and homing of salmon" with Dr. Hasler (Hasler and Scholz 1983), from which I freely draw in outlining the historical perspective and other particulars about salmon imprinting and homing.

The life history of various salmon species is well known. Pacific salmon feed and grow in the ocean, and at sexual maturity, fish move perhaps hundreds of miles to the mouth of a river that enters the ocean. The mature salmon ascend this river, and (coho and chinook) swim upstream, rejecting tributary streams until they reach the one where they were born. They then ascend this stream and reach a suitable spawning area where they spawn and die. The fertilized eggs are placed in a nest (redd). The young hatch out, absorb the yolk, and begin feeding (alevins). They grow and become juveniles (parr), and inhabit the stream for months to a year, when they begin to lose the parr marks, undergo smoltification (a process that prepares the freshwater fish for a marine life) (Steffansson et al. 2008), and begin to move downstream. The smolts enter the ocean where they feed and grow to mature and begin the long migration home.

How do the adult salmon remember the natal stream? How do they even find the mouth of the river? How do they know where to deposit the eggs? Hasler and Wisby (1951), after careful reasoning, developed this hypothesis: The young fish become imprinted to the distinctive odor of their natal stream before they enter salt water and mature. Adult salmon have this information stored in long-term olfactory memory as a cue. Upon reaching sexual maturity, the memory becomes vivid, and they use it to relocate the stream during spawning migration. This is now known as the olfactory hypothesis for salmon homing. The authors developed and published their hypothesis without testing it, because they had a problem. They were located in Wisconsin and the salmon were 3000 miles away. They assumed that someone else would be able test it for them.

Hasler was able to perform some experimental studies, which demonstrated that salmon could differentiate different rivers by smell, and they conducted a field study that showed salmon with plugged nostrils could not detect their own stream. However, this did not establish a long-term memory was present from the smolt, and they tried another approach. They began to look for a synthetic chemical that they could use to imprint the young and then test for as the young mature into adults. Wisby was able to find a suitable organic compound, morpholine with the aid of faculty at the university.

The years passed and still no one tested the olfactory hypothesis. Then in the 1960s, Hasler had some luck. State fish agencies began introducing coho salmon into the Great Lakes. The salmon were moving to him!

Hasler and his students imprinted coho salmon smolts to two different chemicals in a hatchery as a substitute for home odor: morpholine and phenethyl alcohol. Then the fish were stocked into Lake Michigan. In subsequent tests during spawning season 18 months later, they introduced the imprinting chemicals each into a separate stream. They captured 95% of the morpholine and 92% of the phenethyl alcohol imprinted fish in their respective streams. The fish were able to learn (imprint) the synthetic chemicals as smolts and to remember them later. They also conducted other experiments with similar results. Salmon can find their way by using odor.

The olfactory hypothesis was a wonderful contribution, but it had to be adjusted by a few minor changes (reviewed by Lucas and Baras 2001). One is the "pink problem". Although Hasler and his students imprinted fish as smolts, some salmon, such as pink salmon, also shows homing behavior but it migrates to the ocean immediately after hatching and does not undergo smoltification. Then it was noted that hatchery raised salmonids released in a presmolt stage were capable of imprinting to and homing back to their release site instead of the location where they underwent smoltification. This was once associated with straying, but perhaps not! It now seems that the imprinting (which is a learning process) can occur very early in the lives of fishes, and the learning can be associated with layering at ontogenetic events. This has led to concepts of sequential imprinting, producing multiple memories. Such problems and their explanation have broadened the olfactory hypothesis to include more fishes than just salmonids.

After many years, the olfactory hypothesis has been expanded and modified. An updated version is recapitulated:

Every stream has its own distinctive odor from the special combination of plants and soils in its basin and from the organisms and detritus in its bed. This combination of chemicals is known as the home site odor bouquet (HSOB). Some salmon species (e.g., coho) are imprinted to this HSOB as smolts, and some salmon that migrate directly to sea (pink and chum) are imprinted in the egg or as fry. This imprinted or learned memory remains with them throughout their lives, to be remembered when reproductive hormones become activated. There is some indication that the smolts also can have a memory of several sites in sequential pattern.

In the ocean, the salmon range far over the North Pacific in search of food. They may roam for 2 to 3 years until full grown and mature. At maturity, sex hormones activate an endogenous signal—they remember the HSOB and perhaps a sequential pattern of odors as well. They cannot detect HSOB in the ocean and begin to seek the coastline with exogenous stimuli, probably a sun compass orientation or magnetic lines of force or both. Most salmon arrive about 40 miles or less from the parent river. At this time, HSOB is probably detected.

After entering the main stream, the fish swims upstream with a strong rheotactic urge related to the level of sex hormones. Fish may make some mistakes and pass by a tributary, backtrack, and reenter the correct one. When they near the natal area, odor of conspecifics in spawning condition may invigorate them, and fine-tuning, perhaps to buried remains of old reproductive byproducts, may lead them to the exact spot where they hatched. Nests (redds) are constructed, eggs are laid and fertilized, and the cycle begins again. The adults die and their bodies provide additional high-quality nutrients into the stream. Their lives and death have been spent to produce young.

Catadromous Migrations of Anguillid Eels

McDowell (1987) identified 11 families of fishes that had catadromous species, but by far the best-known catadromous fishes are 15 species of anguillid eels. These fish were very important to Europeans who considered eels to be a delicacy. However, their ribbonlike larvae were mistakenly identified as an organism named *Leptocephalus* (small + head) by Linnaeus who thought them to be a different species. It was not until the 1890s when the Leptocephali were reared in aquaria and developed into young (glass) eels. Once the larval stage was confirmed, the hunt for their origin began. Danish professor Johannes Schmidt set out to find where the eels spawned by finding larvae of the European eel. He began in 1904 looking for smaller and smaller sizes and, in 1922, found the smallest larva south of Bermuda in the Sargasso Sea (Baker 1978). Presumably, they spawn there, but eggs or even adults have never been captured in that location. It is believed that two species, the European and American eels, spawn in the same region but in different spawning grounds (Figure 27.4).

The small eel larvae are entrained in the Gulf Stream and carried both to the North American coast (about 1 year or less) and to northern Europe (2 years) (McKeown 1984). When the eels enter freshwater, they are clear and called glass eels, but they transform into pigmented elvers in freshwater. The elvers are very clever and can move overland when it is rainy to find new habitats. The elvers grow into 60 to 80 cm adults called yellow eels. These adults mature in 11–20 years (Helfman et al. 1987) and move into rivers and thence to the ocean. When they enter the ocean, they change into silver eels and have large eyes. Presumably, the silver eels then migrate to the spawning grounds in the Sargasso Sea. However, there is some speculation that the European eels die and the entire population is composed of ecophenotypes of the American eel. This has not been proven and is probably incorrect (McKeown 1984).

There are only three spawning areas known for anguillid eels, the Sargasso Sea in the Atlantic Ocean, off northeastern Australia in the South Pacific, and south of Japan in the North Pacific. All of these areas have different eel species. It is assumed that eels migrate to these natal areas and spawn there at depths, but this has not been observed. Information about the return migration of adults is scanty and nothing is known about their use of environmental cues.

McCleave et al. (1987) studied the Sargasso Sea, and suggest that the migrating eels arrive at an extended frontal zone, which extends across the sea at latitude 24° to 29°N. This frontal zone is within the North Atlantic Subtropical Convergence, and it separates a warm permanently stratified

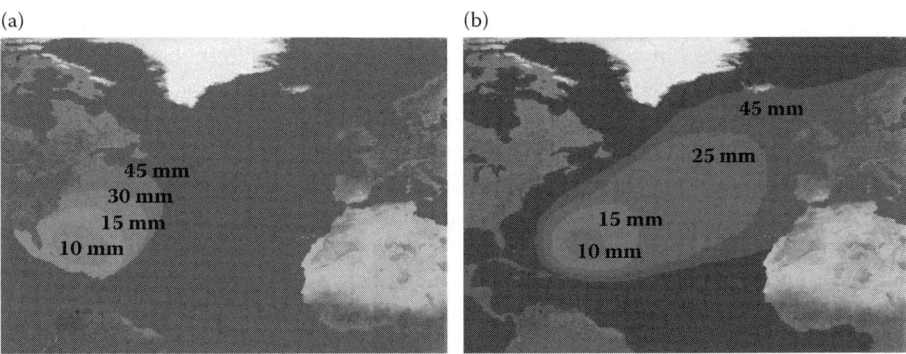

Figure 27.4 Spawning areas and movement of eels (a) American eel (b) European eel. Showing movement away from Sargasso Sea with age (size), (Courtesy of Kils, U., http://commons.wikimedia/wiki/File:Rostratamuk.jpg, http://commons.wikimedia/wiki/File:Anguillamuk.jpg.)

mass of water in the southern Sargasso Sea from the cooler northern Sargasso Sea, in waters 5000 m deep. They sampled both areas and found that very small leptocephalus larvae were abundant only in the frontal zone and to its south. This led to the hypothesis that some environmental cue, most probably odor, in the frontal zone serves to attract the adult eels, which stop migration and spawn there. This hypothesis was further supported by discovering the early development of olfactory organs in the eel larvae. Also, male eels are attracted to females, and this is an indication that a species-specific sex pheromone exists. It is well known that eels have a very acute sense of olfaction. The American eel can detect phenethyl alcohol (an imprinting chemical) in concentration of only 10^{-18}, which would be about one molecule in the olfactory chamber (Bone and Moore 2008).

A search for the spawning areas and environmental cues associated with the Japanese eel has been ongoing for many years and numerous publications have emerged (e.g., Tsukamoto et al. 2003). It also is thought that olfaction is a cue that attracts the Japanese eel to the vicinity of three seamounts that are associated with the North Equatorial Current of the Pacific Ocean. So there is much similarity between the present concept of eel migration in the Atlantic and Pacific oceans.

The last part of this story is the selection of rivers to ascend by the young eels. It was once thought that these eels were homing to some environmental cue but since they had never been in the stream before it could not be natal homing. It is probable, as in the case of natal homing in sea lampreys, that pheromones from upstream immatures are the signal used for stream selection.

Eels are greatly sought for food and Europe consumes million annually. Unfortunately, eel abundance dropped 80% the last 60 years for unknown reasons and they have been placed on the IUCN red list (www.IUCNredlist.org/apps/redlist/details/60344/o/full).

Oceanadromous Migrations of Atlantic Herring

While migrations of marine fishes are well known and historically documented, concepts of homing, home site fidelity, and natal imprinting have been less documented in marine systems, primarily because studies would be quite difficult to do. One species that seems to be a good one for study is Atlantic herring, because there is at least a century of data available from commercial fishing in many different locations.

It has been universally accepted that Atlantic herring exist in many local populations that have their own attributes and tendencies. Those populations have a separate spawning area from all other populations and the fish of each population migrate and spawn with a very high degree of homing fidelity. Using tagging data the homing rates vary, but can be as high as 90% (Wheeler and Winters 1984). Given the problems of using tagging data to make such a determination, it is probable the homing rate is even higher. Perhaps, the explanation for Atlantic herring population richness (multiple numbers of subpopulations) is due to genetic semi-isolation that would result from natal home-site spawning fidelity. This phenomenon is well reported in the literature for other fish species (reviewed by Smith 1985).

However, such a homing hypothesis has not been generally accepted by marine biologists, and in its place, some rather complex hypotheses have been developed (see chapter 28). Curious about this, I explored the literature for some answers and the best discussion that I could find about the homing concept in marine fishes (included were plaice, cod, and herring) was presented by a well-respected authority on fisheries ecology that I choose not to identify. This source pointed out that homing in salmonids is peculiar to streams and that smoltification was required for imprinting to occur. He continued with a question, whether the fish larva is competent to acquire an imprint, and that the larvae hatch from pelagic eggs. There are at least four issues that need to be considered further:

(1) *First*, all salmonids do not undergo a protracted period of smoltification. On the contrary, there is the "pink problem" mentioned earlier, and it is well known that pink and chum salmon do not undergo smoltification but do imprint. These species move out of freshwater as the yolk is depleted and immediately enter estuaries and the sea.
(2) *Second*, it would be more appropriate to compare fishes with other members of their own family (i.e., other clupeids) instead of placing salmonid requirements on fish of different families. The homing ability of shad and river herring to their natal areas with great temporal and spatial precision has been widely noted. Natal homing has been supported by three tests: (1) meristic and morphometric differences between river systems, (2) establishment of new runs where old runs were depleted or in new systems by stocking them with spawning fish, and (3) recognition of home-site water in olfactory experiments (reviewed by Loesch 1987). Is it not likely that Atlantic herring share some of these attributes with other members of their family? In any event, it would be prudent to do some controlled experiments to find out.
(3) *Third*, as will be mentioned in chapter 28, olfactory organs develop early in fish larval development, and prolarval Atlantic herring seemed to be equipped for chemoreception at an early stage. They can detect odors from prey even before they begin exogenous feeding (Døving and Kasumyan 2008).
(4) *Fourth*, although plaice and cod have buoyant eggs, Atlantic herring are demersal spawners and deposit adhesive eggs on the sea floor. To be fair, perhaps the anonymous author really meant to refer only to plaice and cod.

Some additional background and information is warranted before I proceed further. Atlantic herring populations of several European stocks may mix during feeding but spawn at widely separated spawning grounds (Baker 1978). The fish are benthic spawners over substrates described from four different sites as loose stones and rock, flint and gravel, gravel and small stones, and shingle and gravel (Harden-Jones 1968; McKeown 1984). Thus, the spawning area is specifically located in space, permanent, porous, and thus available for storing sensory cues.

Herring eggs are demersal, occur in ribbons, and are adhesive. Spawning beds are relatively small (one given as 300–360 m × 3200 m) (Harden-Jones 1968). The eggs hatch in about 2 weeks, larvae remain on the bottom about 4 days more, then they enter the drift. It also is noted that the larvae drift in the same mass of water that they were hatched before it is measurably diluted (Harden-Jones 1968). Also, it is well documented that the young herring make extensive movements as juveniles, then after a few years (Baker 1978) enter and complete the same migrations as their parents and return to their natal areas as spawning adults. If the fish are imprinted as early larvae, then dispersal would not be a problem because in other fishes, mature imprinted fish would seek the natal site for spawning no matter where they were located at the time of reproductive maturity.

One of the advantages of site fidelity in spawning migration is the development of locally adapted gene complexes, such has been documented extensively in anadromous fishes, including other clupeids (reviewed by Smith 1985; Loesch 1987). A problem with earlier studies was the expectation of major changes in genetic composition. However, major advance in techniques allow detection of subtle but significant genetic differences, which have been used to compare genetic diversity among clupeid stocks (e.g., in American shad; Waters et al. 2000).

Harden-Jones (1981) proposed the groundwater seepage hypothesis in 1980 as a possible olfactory homing cue for marine fishes, including herring. Because herring spawning areas are used over and over again, it is also possible that an accumulation of reproductive byproducts such as egg cases (e.g., see Foster 1985) could accumulate and provide a powerful olfactory cue in the porous substrates and deep sites used for spawning (40–200 m) (Baker 1982; McKeown 1984). Whatever the cues might be, the spawners from separate areas would likely be able to orient, navigate, and find the appropriate natal area. Other clupeid species such as alewife and shad can exhibit homing preciseness of well over 90% (e.g., 97% in American shad) (Waters et al. 2000). I think we need more studies of Atlantic herring to confirm or disprove imprinting to natal spawning areas as the driver for the persistence of separate stocks.

Potamodromous Migrations

Spawning migrations of freshwater fishes are very common, but only a few of these migrations have been studied in detail. Fortunately, we now have an entire book addressed to the subject: *Migration of Freshwater Fishes,* by Martyn Lucas and Etienne Baras (2001). These authors bring to light many of the problems that have plagued the study of freshwater fish migrations. They point out that the awe, conspicuousness, and commercial importance of anadromous fishes (mostly salmon) have been responsible for so much study of so few fish migrations worldwide. To be more specific and make up a few descriptive quotes: The awe of seeing large salmon jumping out of the water to get over waterfalls ("how *do* they do that?"), the amazing numbers of marine fishes that show up in narrow streams ("where do they come from and why did they pick this place"), and the opportunity for commercial exploitation ("well someone needs to profit from all these fish!") have resulted in countless studies and markets. On the other hand, freshwater fishes seem less important. After all, these are not exotic species; they are just local fish. Also, most of their migrations are less impressive in distance and danger, and many are not of much commercial importance ("we can catch these any time, and what good are those suckers anyway?"). As a result of all the above, the term "migratory fish" generally means diadromous species (salmon, striped bass, sturgeon, shad, eels) to fisheries management agencies.

There have been some problems in scientific perception too. An early fish ecologist, Nikolsky (1963, p. 249), contrasted "migratory fish" (i.e., diadromous ones) with "marine and freshwater fish." In the United States, fish biologists developed perceptions that led to the "restricted movement paradigm" for freshwater fishes in general. In the 1950s, there was a great interest in stream dwelling sunfishes, and these resulted in findings of little movement and small home ranges. This perception was reinforced by publications that characterized freshwater fishes as having restricted movements (e.g., Gerking 1959), even though there was strong evidence for migration and homing in some of the more common species (e.g., white sucker) (Raney and Webster 1942; Werner 1979).

Many freshwater fishes undertake regular potamodromous migrations for feeding, climatic adaptation, and spawning. Many of the best examples are fishes of large river systems around the world, including tropical and temperate regions. It is not surprising that we do not understand many of these migrations because it is very difficult to work in large rivers. Unfortunately, most of those large systems are now dammed and migrations have been blocked, so we will never know the full extent of freshwater migration with these conditions, and disruption of the migratory component in the life history of many fishes is resulting in restricting their distribution and abundance and leading to endangerment.

Lucas and Baras (2001) provide a long list of fishes that have freshwater migrations, including many species of cyprinids, catostomids and characins, to name a few. Fortunately for me, they included Colorado pikeminnow (formerly the Colorado squawfish) as "perhaps the most celebrated example of migration in cyprinids" (p. 159) and "possibly the best documented example of homing in nonsalmonid fishes" (p. 20). So we end this chapter examining pikeminnow migration.

CASE STUDY: MIGRATION OF COLORADO PIKEMINNOW

I saw my first Colorado pikeminnow in 1979, when I was selected to lead a U.S. Fish and Wildlife Service field research station on the Green River in Utah. The Green River is the largest upper basin tributary to the Colorado River, larger than the present upper Colorado River, which was formerly named the Grand River, and renamed the Colorado mostly for political reasons. No matter, it is the Green River that supports the largest wild populations of the native Colorado big river fish community. It was our job to find out everything we could about the endangered fishes that lived there. Little was known about the fishes at that time; however, it was believed that the apex predator, Colorado pikeminnow, undertook migrations (accordingly, locals called it Colorado River

Figure 27.5 (a) Colorado pikeminnow in spring. (b) Female pikeminnow in spawning condition.

salmon, or white salmon), but it was unclear just why they migrated and where they migrated to and from. Young of the year (age 0) pikeminnow had been captured and it was presumed that the adults spawned in sections of the river where the young were found.

We went to work looking for sexually mature Colorado pikeminnow (ripe, with expressible gametes). We intensively sampled the whole mainstream river and especially locations where the young were found. After 2 years, we found no spawning fish, but the young appeared each year. This seemed hopeless; however, we were told that a new developing technique, radiotelemetry, was considered an option, and we would get some transmitters to see if they would work in the high-conductivity water. We hoped to get a better idea about the specific habitats, including depths and velocities used by the fish, but also with the possibility of locating a spawning area. I learned how to surgically implant the transmitters into the body cavity and we started implanting and monitoring the fish.

To make a long story short, we captured and implanted several large pikeminnow (Figure 27.5) and tracked them in the same locations repeatedly. Then we lost most of the large ones and one was tracked for a while moving upstream toward Dinosaur National Monument, where the river entered Split Mountain Canyon, and rapids. Surely, the fish did not go there, but we had to find out. We were working with the park service and had a ranger available to help with tracking. A field crew went into the monument in rafts at upper Yampa Canyon and discovered several of our telemetered fish together in an aggregation at river mile 16.5. They contacted me as soon as they were able with the news and we set out the next day with nets to see if we could capture some of the fish. We did, they were running ripe, presumably spawning (Figure 27.6), and we had

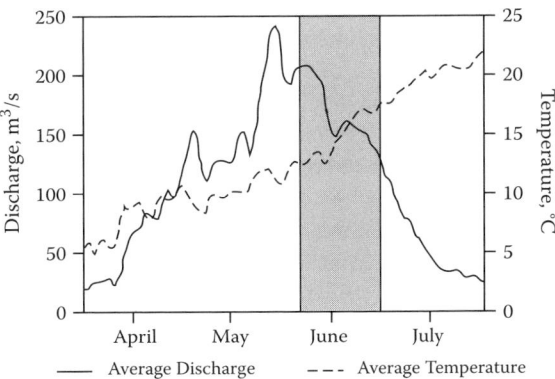

Figure 27.6 Flow and temperature conditions associated with Colorado pikeminnow migration. (Courtesy of Tyus, H. M. and Karp, C. A., *U.S. Fish and Wildlife Service Biological Report*, 89(14), 1–27, 1989.)

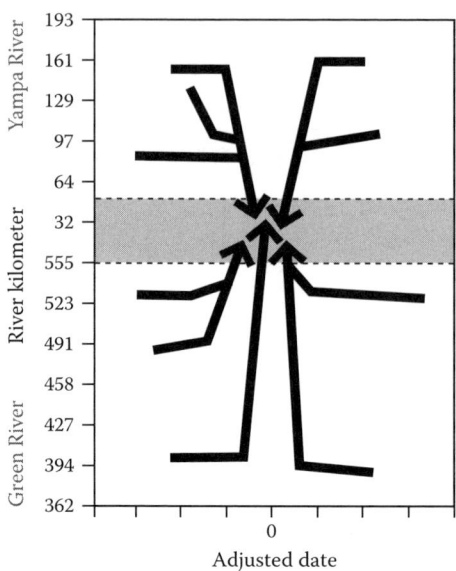

Figure 27.7 Homing migration pattern of Colorado pikeminnow in the Yampa River Colorado. (Redrawn from Tyus, H. M. and Karp, C. A., *U.S. Fish and Wildlife Service Biological Report*, 89(14), 1–27, 1989.)

finally confirmed the first known spawning area of the Colorado pikeminnow. We studied movements of the fish for several years in the Yampa and Green rivers and were amazed to find that fish converged from above and below the Yampa spawning area, displaying an incredible homing pattern as provided in Figure 27.7.

After several years of work throughout the system (see Tyus 1990, 1991), we located only one more spawning area (in lower Desolation and upper Gray canyons of the Green River, and to this day no more have been discovered in the Green system. We also found that pikeminnow in the White River did not spawn there, but moved downstream, entered the Green River and spawned at either the Yampa or Desolation sites (this pattern has been confirmed by Irving and Modde 2000). Sampling below both of these spawning sites (Figure 27.8) revealed the presence of larval Colorado pikeminnow shortly after adults were detected in spawning reaches, but not immediately above the sites.

To summarize the spawning migrations: Colorado pikeminnow adults have home ranges throughout much of the Green River mainstream and some of the larger tributaries, especially the Yampa and White rivers. In mid- to late June (average June 15 in the Yampa River and June 21 in the Green River and at water temperatures of 13.7°C to 15.8°C. Most of the larger fish migrated to the Yampa or Desolation canyons and spawned in about a month after they initiate their movements. Fish that did not migrate remained in their home range and presumably did not spawn that year. The migrating fish moved either upstream or downstream to reach these spawning areas, depending on where they have established home ranges. After spawning, they return to the general location, and perhaps the exact same place, where they were implanted with radiotransmitters. Sampling below the spawning areas yields very young larvae, which enter the drift and are transported out of the rocky canyons. Their destination is the downstream alluvial reaches that have good shallow shoreline areas and backwaters that they utilize as nursery areas (Figure 27.9). As summer progresses, these fish tend to move downstream slightly. Thus, we have covered the cycle of migration for the adults, and the outmigration (or one-half of the cycle) of the young. As indicated, the young fish are displaced downstream over time, and when they become large subadults, they tend to move upstream, presumably in search of a home range. They complete the journey back to their natal

MIGRATION 349

Figure 27.8 Sampling for larvae above and below spawning areas, Yampa River, Colorado.

Figure 27.9 Larval fish nursery areas, Green River, Utah.

location upon sexual maturity. It is likely that the spawning sites are located by olfaction, since the very young larvae have developed olfactory organs at swim up.

The homing mechanism is probably imprinted in the larvae, as in pink salmon. The adult fish will be able to detect the home site odor and migrate toward it, or, if it is not detected, the fish will know to migrate downstream (it could not then be upstream of them!). Fish are intelligent and quite able to detect or not detect odors and make an appropriate response, as Peter Johnson determined in his PhD research (Johnson 1978). The presence of imprinting odor resulted in positive rheotaxis but absence of the odor resulted in a negative rheotaxis, this acting as a sign stimulus as suggested earlier by Harden-Jones (1968). In Colorado pikeminnow that were tagged in the upper river, 76% of the migrations were downstream.

Colorado pikeminnow are large, torpedo-form, powerful fish, perfectly suited to long distance migration. Distances migrated in this example ranged from 32 to 373 km, depending on where each fish had their home range. Those who do not know the habitat assume that the pikeminnow is forced to migrate to these locations to find rocky substrate in which to lay their eggs. This is not true, for upper Yampa River fish pass through many miles of rocky riffles and rapids and fish moving up from the middle Green River pass though Split Mountain and Whirlpool Canyon with major rapids and riffles. Figure 27.6 reveals the answer. The swim-up fry are transported downstream and out of very favorable spawning habitat and very unfavorable nursery habitat, arriving downstream into warm, shallow, and productive alluvial reaches with much nursery habitat (shallow alongshore ephemeral embayments and cut-off side channels).

There are two considerations in spawning for the Colorado pikeminnow: (1) temporal, so that flows are declining and backwaters are being produced, and (2) spatial, to select a location that places the drifting larvae (sac-fry) the right distance away so that when they swim up and drift downstream, they are placed in nursery habitat by the time that exogenous feeding is necessary. It is natural selection at work placing the postlarvae into the nursery habitats, because the adults that spawn in the wrong place would have no young that survived to make the same mistake. Unfortunately, this life history pattern has worked against those fish in the dammed and diverted mainstream Colorado River. Now an endangered species, it has been extirpated from 80% of its native range.

SUMMARY

Fishes use movement for many purposes, including temperature regulation, feeding, escape from predators or stress, for reproduction, and for many other reasons. Migration is a specialized behavior that has its origin in natural selection. In simple terms, it is a round trip, a seasonal movement that implies some element of return to an initial starting place. Fish migration can be a feeding migration, climatic migration, or a spawning migration. Also, there are several attributes of migration, which include a direction, a periodicity, a distance (usually fixed), a degree of return (usually high), and homing (usually accurate) to a predetermined location. Those that study migration understand that fish are intelligent beings, and capable of making choices. Although ethologists once believed that fish found their way entirely due to some programmed reflex responses, the phenomenon of migration is now firmly under the purview of behavioral ecology. Fish find their way in many aspects just like humans do—with orientation, piloting, and navigation. Fishes can orient to many stimuli, of which the sun and moon and chemoreception are commonly used. A. D. Hasler and students worked out the olfactory hypothesis to explain how Pacific salmonids found their way: Young salmon are imprinted with odors present in their home stream, perhaps sequentially as the fish move downstream. This imprint, or learned memory, remains with the fish as they mature and move into saltwater. Activation of reproductive hormones awakens this memory and the fish begins to move to the coastline, perhaps by sun compass or electromagnetic fields. After finding the home stream the fish move up with a rheotactic urge, smelling their way to the site where they were raised. At that point, redds are constructed and spawning commences. Soon the semelparous spawners die, leaving their bodies in the stream to provide a source of high-quality nutrients. Spawning migrations of other anadromous fishes, such as American shad, may be semelparous or iteroparous depending on latitude, but most fishes are iteroparous and survive spawning. Migrations of many fishes such as the anadromous alewife and potamodromous pikeminnow share behavioral attributes of migrating salmon.

Fish migrations can be expensive in energy cost, but are an important tactic in the reproductive ecology of many freshwater, marine, and estuarine-dependent fishes.

Further reading: More recent: Dadswell et al. 1987.

CHAPTER 28

Larval Fish

INTRODUCTION AND IMPORTANCE

When humans consider other animals, they usually envision the adult life stage. However, ecologists must consider all life stages, especially early developmental intervals that we refer to as "young." Adequate numbers of young must survive or there will be no adults, no reproduction and in time, no species. Perhaps we should even be more concerned about young fish because they are usually more sensitive than adults to changes in physical, chemical, and biological conditions of their environment. And for most fishes, the young are indeed important regulators of the fish population. Even a small variation in mortality rate, growth, or stage duration can result in a "tenfold or greater fluctuation in fish recruitment" (Houde 1987, p. 17).

For the vast majority of fishes, the earliest young (i.e., embryos and larvae) are very different from juveniles and adults in morphology (Figure 28.1). More to the point here, they also are ecologically different in fundamental adaptations such as foods eaten, habitat occupied, and behavior (Snyder 1990). As exciting and potentially revealing as such studies could be, the needs of young fishes were not adequately considered in historic conservation and management efforts, which usually focused on the habitat requirements and limiting factors of adult fish. In most instances, the reason for this bit of apparent shortsightedness has been due to a lack of information about the needs of young fish—in many instances due to difficulties in identifying and sampling them. As a result, the role that the young can exert on regulating the size and other attributes of fish populations and, consequently, important fisheries has not been fully appreciated.

The larvae of most important commercial fishes were described by the late 1970s, but only about 10% of marine and about 15% of freshwater and anadromous fishes of North America were known at that time (Snyder 1983). Passage of environmental statutes such as the National Environmental Policy Act of 1969 and the Endangered Species Act of 1973 provided the stimulus to study larval fishes and learn how they might be affected by anthropogenic actions. Today we know much more, thanks to the efforts of many fishery scientists, government agencies, and supportive organizations such as the Early Life History section of the American Fisheries Society. However, in the United States and Canada, only about 40% of freshwater fishes have been at least minimally described, and only 25% have been described adequately. In contrast, the ontogeny of more than 70% of fishes in the western North Atlantic Ocean is now known (D. E. Snyder, personal communication 2010).

When I received my graduate training there were few larval fish experts, and most of them were marine ichthyologists. Furthermore, texts on the distribution and identification of fishes mostly ignored fish larvae, except in the cases of well-known transformations in the development of fishes such as anguillid eels and flatfishes. However, as early as 1927, Hildebrand and Schroeder (1928) included drawings and descriptions of larval fishes in their monumental *Fishes of Chesapeake Bay*. These were mostly larvae of commercially important species (anchovies, eels, menhaden, alewives, white bass, striped bass, porgies, and tautogs) and other previously described larvae (silversides and

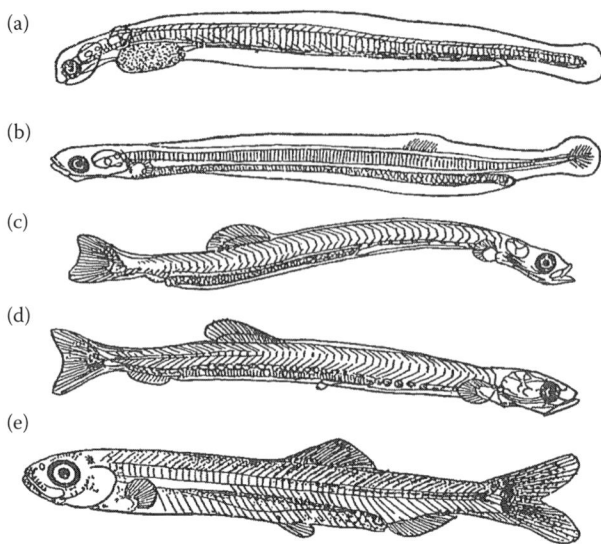

Figure 28.1 Larval development of the Atlantic herring: (a) newly hatched 7-mm larva, (b) 10, (c) 19, (d) 29, and (e) 41 mm. In (e), at about 3 months, they are beginning to look like adults. (From Bigelow, H. G., and Schroeder, W. C., *Fishery Bulletin*, 74, U.S. Fish and Wildlife Service (USFWS), Washington, DC, 1953. With permission of the Smithsonian Institution.)

sticklebacks). Bigelow and Schroeder (1953) in *Fishes of the Gulf of Maine* included more discussion about fish larvae, with illustrations of 40 species (including all of the above fish plus more). A more ecological approach was provided in a short section of Nikolsky's (1963) text, and within 4 years, an impressive publication on larval fishes by Mansueti and Hardy (1967) appeared. At the same time, Marshall (1966) provided entire chapters (i.e., 15 and 16) on the early life of marine and freshwater fishes. By the present century, almost all textbooks on fish contained a major section, if not chapter, on early life stages. Fuiman and Werner (2002) even published an entire text on "the unique contributions of early life stages" to fishery science.

Emphasis on the early life of fishes was historically concentrated on marine fishes in an attempt to understand recruitment processes in commercially valuable species (e.g., see Blaxter 1974; Lasker 1981, 1987; Chambers and Trippel 1997; Moser and Watson 2006 for review). More recently, significant works have included estuarine larvae (Able and Fahay 1998) and monumental undertakings in selected marine systems. As an example, Moser (1996) described larvae of 467 fish species using 1171 illustrations. With few examples (e.g., Fish 1932), progress on the identification and ecological attributes of freshwater fish larvae generally stagnated until it became necessary to comply with environmental laws and regulations of the 1960s and 1970s. While comprehensive works on continental fish larvae similar to that of the marine system have not emerged, many publications, including larval identification manuals, have been produced for lakes, rivers, drainage basins, and regions (e.g., see Snyder 1983 for a list of North American manuals 1932–1982). Annual larval fish conferences, which began in 1977 and the Early Life History section of the American Fisheries Society, established in 1980, have provided avenues for communication and research. Study and recovery efforts for native freshwater fishes in the United States, including the threatened and endangered fishes of the Rio Grande, Colorado River, and other systems, have been aided by work done by the Larval Fish Laboratory at Colorado State University, founded in 1978. The lab produced a video on larval fish ecology in 1983 that is still available and is referenced at the end of this chapter. Other centers for work on young freshwater fishes include federal and university laboratories around the country.

I have given this chapter the title of "Larval Fish" because the focus is on the ontogeny and ecology of true (i.e., free-living) larvae, not embryos or juveniles, although we will mention all of these life history stages here. At some risk of being redundant with information already presented in chapter 26 and, to some extent, chapter 27, we will review some basics about fish reproduction in this section to foster an ecological perspective for fish larvae. We also will consider historical viewpoints and present ecological concepts about larval fishes in light of recent developments.

REPRODUCTION AND EARLY LIFE

In contrast with human reproduction and that of other mammals, and excluding a few ovoviviparous and viviparous fish taxa, almost all fishes (94%) (Patzner 2008) are oviparous, producing eggs that are fertilized outside of the body. As we have already discussed in chapter 26, these eggs and the young produced from them can be protected (guarded) for a while by the parents in various ways (especially in benthic spawners). However, for the vast majority of species, especially pelagic spawners, there is no parental energy investment provided after the eggs are released and fertilized.

After fertilization, fish embryos soon hatch (in a few days to several weeks depending on temperature) and universally remain protected to some degree by the substrate (e.g., trout), vegetation, or even a thick case (e.g., rays and most sharks) for a period of time when the energy provided by the mother (i.e., yolk) is metabolized. In precocial young, this period can be a lengthy one (months), and when the generous supply of yolk is depleted, the young fish, although small, transform and closely resemble the adults. However, in the altricial egg found in most teleosts, the supply of yolk is smaller, and it may be consumed entirely in a week or so. It is these altricial hatchlings that we are mostly concerned about in this chapter. Compared to the adults, they have different adaptations expressed in morphology (i.e., such as lack of the full complement of adult fin rays), physiology, and behavior. Compared with precocial young, they assume entirely different life history patterns appropriate of free-living organisms. A classical example of development of altricial larvae is the northern anchovy *Engraulis mordax* (Hunter and Coyne 1982). True larvae, they are not fully formed, having only fin folds, lacking an efficient digestive system, lacking functional gills (obtaining oxygen from cutaneous respiration), having poorly formed eyes and mouth, and lacking other organs. Thus, a small larva with poor vision, inefficient fins, poorly formed mouth, few sensory organs, a primitive gut, and other incompletely developed organ systems must have a difficult time in catching and assimilating food. With such small amounts of energy supplied in the yolk, most of the larvae are not very efficient as predators, even on plankton.

But the supply of yolk is not the only contribution of parental energy given to aid the young. I have pointed out previously (chapter 26, etc.) that production of an altricial egg and lack of parental care of eggs after spawning may not be the only parental contributions to the reproductive effort. Rather there may be a great amount of parental involvement into the effort *before* the eggs are released to insure that the young will have a hospitable environment for their development. In freshwater streams, altricial eggs are mostly adhesive and placed in a good hiding location. After the young develop into sac fry, they swim up and may be transported downstream in the drift to more productive nursery habitats. Taxa include North American catostomids, cyprinids, percids, and centrarchids (e.g., Scheidegger and Bain 1995). Not as common, are some freshwater fish species that produce pelagic or semibuoyant eggs that drift in the water column (e.g., emerald shiner, Rio Grande silvery minnow, plains minnow, some clupeids and striped bass, freshwater drum, etc.) or along the bottom (e.g., sturgeon) and are thus transported to disperse them.

Also, there is widespread dispersal of the young fish in marine environments, including capelin and herring (Fuiman and Higgs 1997). In the case of near shore pelagic spawners such as Atlantic menhaden and plaice, adults also spawn in areas where tidal currents can transport their offspring into estuarine nursery areas. In case of offshore spawners, dispersal may be due to transport of eggs and larvae in surface currents as in red snapper (Johnson et al. 2009) or due to deeper currents (e.g., gyral) as presented earlier for anguillid eels.

DESCRIPTION AND TAXONOMY

Larvae are the young of animals that undergo metamorphosis from a free-living immature form to an adult. In this case, fish larvae are morphologically different in comparison with the adult and will undergo metamorphosis, that is, "change in form and structure" (Lawrence 1989) before becoming an adult. Almost everyone is aware of metamorphosis in insects, in which a caterpillar develops into a butterfly or a maggot-like larva becomes a very different adult, such as a honeybee. Metamorphosis in some fishes is just as interesting. Although not discussed here, the early development of fish and the role of the endocrine system in metamorphosis were presented by Hall (2008) and Power et al. (2008).

Adults and larvae of a fish species are adapted to a life in water, but that can be where similarities end. In general, there are major changes that occur over time in the life of a fish: one is growth or change in quantity (biomass), and the other is change in development that affects form (through change in morphology, physiology, and/or behavior). Larval developmental occurs with changes in the amount and location of pigment, the proliferation of true fins with rays or spines, maturation of organs, and many other characteristics used to identify species or diagnose the stage of individual development. Thus, in order to determine the ecological requirements and other aspects of the developing fish it is helpful to group the early life stages into conveniently defined developmental intervals. Unfortunately, terms used in characterizing intervals of larval development can be confusing because more than one system is in use (Snyder 1976). Fishermen and early fish culturists used the terms sac fry as loosely equivalent with yolk-sac larvae, fry for larvae, and fingerlings for juveniles. Kendall et al. (1984; reproduced in Bond 1996) compared several developmental interval terminologies in use for the entire life cycles of the fish or portions thereof, based on selected developmental events. Snyder (1983) similarly presented three terminologies for the larval period and phases (subintervals) thereof that are widely used in North America and emphasized the utility of such for taxonomic purposes. To the latter, I have added an earlier terminology and summarized all as follows:

(1) Lagler et al. (1962), using terminology proposed by Hubbs (1943) presented within-species concepts in use at that time: Young fishes were classified as either *prolarvae* (sac fry) or *postlarvae*. Prolarvae are identified by presence of the yolk sac. Postlarvae have no yolk sac but are advanced fry with adult features. Fish in the late postlarval state appear as diminutive adults (e.g., salmonids).
(2) Mansueti and Hardy (1967) and Hardy et al. (1978) incorporated earlier criteria and identified three phases of development in association with feeding: *yolk sac larvae*—a phase that occurs between hatching and exogenous feeding; *larvae*—a phase between yolk depletion and presence of adult fin rays; and *prejuvenile*—a phase between presence of adult fin rays and adult body form.
(3) Another system is based on a combination of Ahlstrom et al. (1976) and Kendall (1984). In present usage (e.g., see Moser 1996), yolk sac larvae are defined as hatchlings with yolk present, and larvae are defined as a stage after the yolk has been used up and the fish is in the process of transformation into a juvenile or adult. The transformation period was divided into three substages: *preflexion, flexion*, and *postflexion* according to the degree of flexion in the end of the notochord in formation of the caudal fin. This system applies only to fishes with homocercal tails and is believed to be associated with differences in swimming ability. The transition from flexion to post flexion also serves as the transition for measuring standard length from the posterior end of the notochord to the posterior-most margin of the hypural plates in the tail.

(4) Snyder (1983; and references therein) developed a different system for describing larvae and formulating keys to species. Phases used are *protolarvae*—a phase between hatching or birth and the presence of the first fin ray or spine in the nonpaired (median) fins; *mesolarvae*—a phase after the formation of the median fin ray or spine and the appearance of both the pelvic fin or fin buds and all of the principle soft rays of median fins; and *metalarvae*—a phase covering the period of development from mesolarvae to a fish that possesses the adult complement of spines and rays if characterized by such, some segmented rays, and loss of all finfold tissue.

Snyder and Muth (1988, 1990) suggested that terminologies 2, 3, and 4 given above are not competing alternatives but rather complementary options with phases defined for different purposes. They can be effectively combined as needed. They also offered a "combined developmental interval terminology" in which the mesolarval phase could be subdivided according to notochord flexion (e.g., preflexion, flexion, or postflexion mesolarvae—protolarvae understood to always be preflexion and metalarvae always to be postflexion) and the presence or absence of yolk denoted with appropriate modifiers (e.g., yolk-sac or yolk-bearing protolarvae, prolarvae without yolk, metalarvae with yolk).

LARVAL ONTOGENY

The above terminologies for phases of larval development are aids in tracing ontogeny, or changes in growth and development of organisms (and their capacity for adaptation). Some of the changes are gradual, occurring by a continuous accumulation of small changes, but some tend to be salutatory (stepwise) (Balon 1979) with short episodes (thresholds) of great morphological change. Human children develop gradually and we tend to think of changes occurring in most organisms as gradual processes (identified as a "gradualistic bias" by Gould 1977). Thus, historic naturalists, including Charles Darwin and his many followers visualized ontogenetic development as a gradual process. Consequently, small, free-living fish larvae that are radically different in appearance from the adult form have been classified as entirely different species, such as the leptocephalus larvae of eels, ammocetes of lampreys, and tadpoles of amphibians.

Although normal growth in development (epigenesis) can be expected to display different rates of growth due to the rate of food intake, temperature, activity, and so on, ontogeny is measured by change in a character state, the appearance of new characters, or loss of characters (Ditty et al. 2003). In addition to changes produced by variable growth rates, an organism may function as a free-living form with certain adaptations for a time, but then undergo a more or less striking change. For examples, flatfish have eyes located in normal bilateral position as early pelagic larvae, then one eye shifts to the opposite side (Figure 28.2) and the fish takes up a benthic life style; razorback sucker hatchlings first become pelagic and feed on plankton, then the terminal mouth rotates to an inferior position and it begins feeding on the bottom, and of course there is the often-used example of the eel leptocephalus that becomes a glass eel, then an adult. These changes during growth can result in size structuring and niche shifts in fish populations (e.g., Werner and Gilliam 1984; Claessen and Dieckmann 2002).

The development of a fish may occur in different ontogenies depending upon how many developmental intervals may occur (e.g., see Balon 1985). An example is given in anadromous salmonids, but this is not typical for most fishes. In general, it is best to define the larval period to begin upon hatching or birth, when the young fish are exposed to its environment. Although there are good reasons why ecologists might wish to also consider its source of nutrition (i.e., endogenous or exogenous), fish larvae may begin exogenous feeding while there is a remnant supply of yolk left. This behavior is likely an evolutionary "bet-hedging," in case of undesirable environmental conditions.

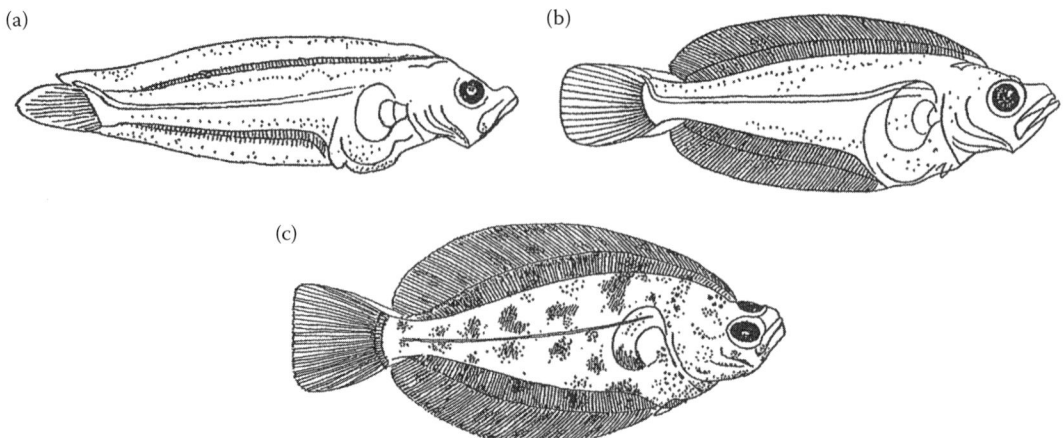

Figure 28.2 Atlantic halibut young of year showing migration of the left eye: (a) 16.2, (b) 22, and (c) 34 mm. (From Bigelow, H. G., and Schroeder, W. C., *Fishery Bulletin*, 74, USFWS, Washington, DC, 1953. With permission of the Smithsonian Institution.)

In this case, storms or other stochastic events can adversely affect food supply, feeding conditions, rate of development, or timing of emergence.

The larval period can last weeks to years, but ends with completion of metamorphosis. This is a period of rapid change in which adult features, such as true median and paired fins, are attained. When completed, the young fish resemble adults in morphology.

The adult period is a reproductive one in which growth of somatic tissue is reduced in favor of energy allocated to producing gametes and supporting reproductive behavior (e.g., migration, parental care). The senescent period may or may not be present in all fishes, but when it is, it is characterized by decline in or negative growth and a reduction of gametes produced.

With over 30,000 fish species, there are usually exceptions to every paradigm, and one is progenesis, or precocial sexual maturity in larvae. Progenesis is common in meso- and bathypelagic marine zones, presumably due to life in an impoverished environment. The advantage of progenesis is linked with bioenergetics. Presumably, energy is saved by producing a mature but otherwise less developed organism in less than 1 year and by avoiding costly migrations (Marshall 1984).

Eugene K. Balon authored and coauthored numerous papers on fish ontogeny, and he explained discontinuities in fish development by reference to the theory of saltation. This concept was first applied to the ontogeny of fishes in two 1953 publications by Russian scientists (Vasnetsov; Kryshanovsky, Disler, and Smirnov; both publications cited in Balon 1979). Saltation as used here refers to changes in the development of an organism that occur in a series of steps. Quantitative changes are due mostly to longer periods of growth separated by shorter events called thresholds. In an ecological perspective, a threshold is "a rapid transition from one state to another of the organism–environment interaction" (Balon 1984). Accordingly, it is during these threshold events that structural and functional adaptations (i.e., ontogenetic shifts) can emerge and be critical in shaping the course of ontogeny (e.g., see Claessen and Dieckmann 2002). A well-known example of such a shift occurs in the metamorphosis of amphibian larvae and enables the animal to abandon life in the aquatic environment in preference for a terrestrial existence (e.g., Rowe and Ludwig 1991). Major ontogenetic shifts also occur in fishes, and in many cases, the metamorphosis is just as striking. Examples of saltatory ontogeny in which development of a fish is characterized by periods (steps) and brief intervals of change (thresholds) have previously been discussed in the case of the European eel and the deep-sea anglerfish (e.g., see chapter 19), both of which exhibit striking differences in more than one step before they are transformed into adults.

Descriptive morphology of developmental change in larval fishes has been extremely helpful in many ways, but it has limited application in others. Taxonomic and descriptive systems presented above have even been criticized for being arbitrary and contributing to "ignorance of comparative ontogeny" (Balon 1979) and confusing matters from an ecological perspective. It seems that ecologists are caught in the middle. We must have ways to identify larvae to species and to be able to trace their development. However, as ecologists we need to recognize that larvae of one species may have major differences in their sources of energy compared with sympatric species. As an example, the event of hatching or parturition marks emergence into nature, but it may not be very helpful from an energetics perspective. This event may signal the beginning of exogenous feeding or only represent delayed endogenous feeding, such as with yolk in fish or glandular input of milk such as in marsupials. It is the ecologist's role to sort these things out.

LARVAL FISH ECOLOGY

Ecological Interactions

The larval stage is a free-living organism but it has many handicaps compared to adults. Thus, survival, distribution and abundance of larval fishes can be greatly affected by changes in abiotic and biotic factors of the environment. Larvae are especially vulnerable due to small size, and incomplete morphological and physiological development. As in adults, their best adaptive responses to such changes are likely related to behavior. However, behavioral responses such as fast and efficient swimming may require a degree of developmental sophistication that will not occur until later in larval development. In this section, I focus on some major environmental influences that affect fish larvae and adaptations related to habitat use, feeding, and predation. Important differences in marine and freshwater systems will be identified in the process.

Habitat Selection

Most larvae of marine fishes (important exceptions include some species with demersal eggs, such as Atlantic herring and cod) are produced by pelagic eggs. These larvae spend most, if not all, of their lives in the upper pelagic zones of the neritic and oceanic realms. They also have specific adaptations for buoyancy, such as a large surface to volume relationship, presence of oil droplets, a high degree of hydration, and so on. Also, there are many larval species that rear in the epipelagic, but are not actually spawned there. Instead, they originate in deeper zones and float to the surface. Habitat selection in this case is by vertical movements into the upper zone of the ocean in order to exploit plankton (dinoflagellates, copepods, larvae of marine invertebrates, etc.). Food selection by larvae is specialized for different sizes of plankton and many fish species exploit the same resources.

Once in the upper pelagic, habitat selection by small organisms is difficult, and chance events can result in transporting the larvae vicariously. Fish larvae can be transported great distances by currents produced by wind, tidal action, and ocean circulation patterns such as gyres. Also, productivity in the marine environment is not uniform, as we have learned. Areas of greatest plankton biomass are those influenced from continental runoff, upwellings, and ocean circulation such as the Gulf Stream, the latter being warm and productive. Such areas tend to result in patches of plankton that are not uniformly distributed (i.e., patchiness). Also, even when larvae are dispersed into areas of food and acceptable temperature (i.e., "good habitat"), storms and wind action can produce turbulent conditions that make it difficult for larvae to obtain food, resulting in "bad habitat." In addition to patchiness, productivity varies with season. In the temperate zone, it is affected by temperature which also drives continental runoff. In tropical climates, it is affected by seasonal weather patterns such as monsoons.

In contrast, freshwater fish larvae are primarily produced by demersal (nonbuoyant) eggs. Most of the freshwater systems are relatively shallow and usually vegetated, resulting in a greater variety of habitats suitable for larval fish. Consequently, freshwater fish larvae are more general in trophic adaptation and diverse in habitat use. Larval dispersal and habitat exploitation is likely more horizontal than vertical, especially in lotic habitat. Riverine habitat is an exceptionally diverse mixture of lotic (channels) and lentic (alongshore embayments, backwaters, and oxbows). As shown earlier, larvae of stream fishes (which primarily exist in cool- and warmwater habitats) can and do take advantage of downstream flow to disperse into nursery areas. These habitats also tend to be ephemeral, varying in presence size due to river discharge. They also are exposed to terrestrial climate, where ambient conditions, especially temperature, can greatly affect small fishes.

Feeding

The feeding ecology of fish larvae, that is, the kinds and sizes of their prey and their behavior preying upon them, is so different from the adults that the larvae are more like "separate species" (Gerking 1994, p. 139). The main food of the larvae in both marine and freshwater systems is zooplankton, but small larvae and smaller larval species also may consume algae (such as dinoflagellates, diatoms, and green algae) and then switch to animal food (copepods, cladocerans, and larvae of marine invertebrates) as they grow (reviewed by Gerking 1994). Many freshwater fish larvae consume cladocerans and rotifers, and shallow water environments promote more demersal feeding on such foods as chironomids. Also, larvae of piscivorous fishes may consume smaller fish larvae. For example, at 6 mm total length, blue marlin larvae (Figure 28.3) begin to consume fish larvae and feed on them exclusively when the marlin reaches about 12 mm (Gorbunova and Lipskaya 1975).

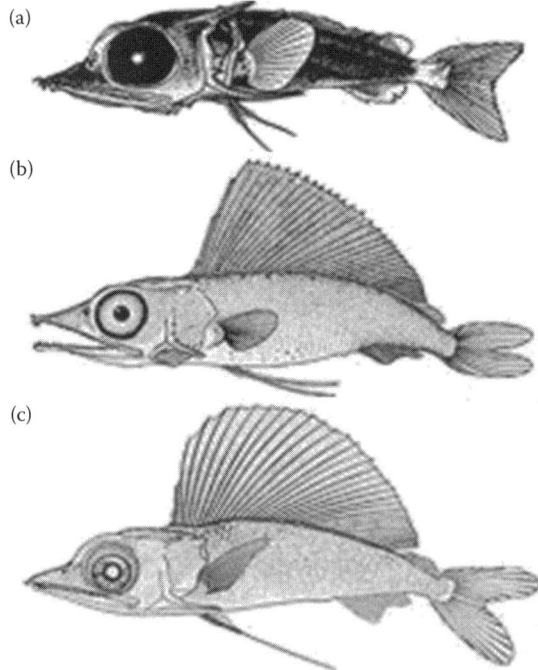

Figure 28.3 Blue marlin larvae. (a) = 12.6 mm, (b) = 21.0 mm, (c) = 27.1 mm. (Courtesy of USFWS, http://commons.wikimedia/wiki/File:Blue_marlin_larvae.jpg.).

As this suggests, fish larvae increase in feeding ability as they age due to growth in size (e.g., larger mouth and more powerful musculature), the appearance of new characters, and experience. Such change in feeding as larvae mature has been recently reported in estuarine fishes by Able and Fahay (2010) who detected ontogenetic feeding change in 23 species from 18 families.

Fishes may commence exogenous feeding before the yolk is depleted, but there will come a time when they must find food or starve. The length of time a young fish can survive without food is related to temperature and the degree of activity. However, Hunter (1981) provided a synthesis of data that indicated larvae from larger eggs had greater energy reserves and more time to find food before a period of "irreversible starvation" set in. He pointed out that larvae produced from small pelagic eggs may have only 2 days or even less after yolk absorption before starvation was irreversible. Demersel larvae of herring and plaice (flatfish) were able to persist for three times longer (6 days). When larger (just before metamorphosis), herring can last 15 days without food.

Feeding success of larvae requires searching, detecting, and capturing food. All of these activities require movement and sensory perception. Swimming speed is related to body length, water temperature, turbulence, and fish condition. Such small organisms are constrained more by viscosity than inertia. Hence, movement is undulatory (details provided by Müller 2008). Larger larvae (>5 mm) have an advantage due to the nature of water viscosity, which affects both cruise ability (detecting prey) and burst speed (capture of prey). With their limited ability, fish larvae need to be close to their prey to capture them, for example, small larvae such as anchovies capture 95% of their prey in less than ½ body length away (Hunter 1972, 1981). Sensory perception occurs very early in larval ontogeny and functional eyes and chemosynthetic organs occur early in the larval stage.

Prey detection by vision and olfaction are primary means of sensing prey by fish larvae. This is because young larvae with indirect ontogenies typically have only cones in the retina, and they feed visually during light hours. In some larvae, vision is enhanced by stalked eyes, which increase the field of vision and place the eye closer to the prey (Moser 1981). However, fish larvae also feed at night, and studies on several species have shown this to be aided by chemoreception and mechanoreception. Newly hatched larval herring respond to odors of amino acids and extracts of their prey organisms (Døving and Kasumyan 2008). Although the lateral line system typically appears late in larval development, newly hatched fish larvae have neuromast cells located in various places on the body and can detect hydrodynamic disturbance with those cells (Pankhurst 2008).

It has long been noticed that the distribution of plankton in the ocean occurs with patches of high density against a much lower average background. Small fish larvae can survive and grow only with high density of prey—so high that the fish are practically bumping into plankton. Upon first feeding, larvae must consume enough food to grow 15%–20% per day, which requires a high prey density of about 1000 microcopepods per liter (reviewed by Moser and Watson 2006). However, estimates of food abundance by various workers have shown that food densities needed by larval fishes exceed the average densities present in the oceans (reviewed by Pitcher and Hart 1982; Mullin 1993). These facts have lead to the "patchiness" concept, which suggests that high larval fish densities are a result of larvae locating and utilizing these food patches, which are affected by disturbances such as storms (Lasker 1975, 1981; Hunter 1981).

The concept of patchiness in the marine environment has its counterpart in freshwater systems, especially in rivers. There, productive larval habitat is concentrated in areas that are the least disturbed by river flow. These areas may be totally dependent on water level (stage), such as in quiet ephemeral alongshore embayments, eddies, or partially dewatered side channels.

Predation

As might be anticipated, such diminutive prey as fish larvae are vulnerable to predation from a variety of sources, and predation is considered a major source of mortality (Rice et al. 1997). However, the degree of predation that will occur on larvae is a function of habitat selection and timing by the parents. To some extent, vulnerability to predators may be decided even before the eggs are released or larvae emerge (e.g., see Pope et al. 1994).

As in real estate, everything seems to depend on three things, location, location, location! The same might be said for fishes. In most fishes, the spawning location is different from the nursery site. In pelagic spawners, and especially in rivers, the eggs may be released in a water mass that is transported to other locations and habitats by currents. This transport can be beneficial to larvae because planktonic predators would be placed at a disadvantage in fast flows and turbulence. In demersal spawners, the eggs are protected by placing them in gravels or attaching them to the substrate. This also allows a measure of protection from pelagic predators as the embryos mature to the swim-up stage. (Note: some fishes like grunions and leaf spawning characins even remove the embryos from the aquatic habitat entirely.) Floating eggs or emergent embryos also may be transported by water currents such as alongshore, tidal, or river flows to locations in which they may hide from predators by moving them into patches of vegetation, rough substrate, or shallow water.

An equally important component in the spawning event is timing. Many species have planktonic eggs and larvae, including marine invertebrates as well as fishes, and synchronous spawning can result in predator swamping. Synchronous spawning or the synchronous emergence of benthic young is common by many species in marine (e.g., coral reefs) and lacustrine systems (e.g., in the African great lakes). Finally, adults usually select the most productive time of year, when temperatures are optimal and food is abundant to release their eggs. If growth is maximized, then the vulnerability of planktonic larvae would be reduced over time as they outgrow their predators.

Many fishes guard the nest and protect the young from predation. This protection even extends to mouth brooding and vivipary. However, large predators also may "cultivate" the environment to the advantage of their young by removing (eating) mesopredators that would prey upon or compete with eggs, larvae, or small juveniles. This is called the cultivation hypothesis and was discussed previously (chapters 23, 25, and 26).

Once placed in the natural world, fish embryos and larvae are exposed to direct predation. In the marine system, planktonic marine invertebrates can consume large numbers of fish embryos, including sac fry. However, marine invertebrates of the plankton are very gape limited, and larger stages from larvae to post larvae are consumed primarily by fishes. Clearly, rapid growth leading to an increased body size could improve survival in the plankton, and this is the "bigger is better" argument. But wouldn't larger larvae be consumed by larger prey anyway, such as other fishes? In fact, some studies suggest that increased body size of larvae can make them more vulnerable to predation than smaller larvae (Cowan et al. 1997; Houde 1997). Cowan et al. (1997) reviewed the subject, performed simulation experiments, and noted that as larvae grew they escaped predation from one organism but became vulnerable to predation from another. There comes a time however, when fish are large enough to escape predation from all but the largest predators, which are present in few numbers. Thus, faster growing prey will have the edge in survival. In most cases, species- or individual-specific attributes of the predator and prey, coupled with previous exposure to the prey and different environmental conditions, were judged to be more important than size (Cowan et al. 1997).

Fish larvae also protect themselves from predators directly. Pattern development for pigment cells occurs early in embryonic development and pigmentation patterns of larvae are no doubt used for camouflage. This type of defense helps disguise larva, but it also can produce an effect called "flicker fusion." This occurs principally due to postanal melanophores in teleost larvae and tends to startle or confuse predators similar to the effects produced by snakes. As the larva moves, bands of pigment tend to attract attention to the rear of the fish and by the time the predator focuses on the patterns the fish is gone (Moser 1981). There are six types of chromatophores (special pigment cells of different colors) in fish that may be present at hatching (Kelsh and Parichy 2008).

Another type of camouflage is produced by bioluminescence, which occurs early in ontogeny. Larval fishes usually have intrinsic bioluminescence that is produced chemically. Light also can be produced by bacterial luminescence, but it is less common (see taxonomic review provided by

Suntsov et al. 2008). Larval photophores occur first on the head, and are used for attracting prey and feeding as in deep-sea lanternfish and/or as counter illumination for camouflage as in hatchetfish (Moser 1981; Suntsov et al. 2008).

Some fishes produce large spiny projections (e.g., serranids), or a covering of prickly projections (e.g., in tetraodonts). These antipredator features also aid in retarding sinking in pelagic larvae (e.g., Moser 1981; Helfman et al. 2009), but they are complex structures and require energy to make.

Larval fishes can have innate behavioral responses to avoid predators or even predator odors. Larval fishes typically produce a "startle response" of rapid escape movement when they detect the presence of a predator, either visually or through mechanoreception (Pankhurst 2008). However, they characteristically use chemosensory systems to detect predators (Døving and Kasumyan 2008). Many cyprinids react to alarm pheromones and odors of conspecifics, as early larvae and reaction to predator odor also is common in fishes. Atlantic salmon have recently attracted research attention and they can recognize predator odor (northern pike) at hatching and also they can distinguish between pike odor and that of harmless minnows (Hawkins et al. 2004, 2007). Some larvae can detect predator odor before hatching, and larval emergence in Atlantic salmon can be delayed by exposure of the embryos to predator odor (Jones et al. 2003). A longer period of development can aid in locomotion and sensory perception, thus resulting in a greater ability to avoid predators.

Why Larvae?

Larvae are the first free-living stage of almost all fishes and most of them are on their own in a relatively hostile world. It is quite remarkable that these small, delicate, and presumably vulnerable organisms hold the future of species. Their ability to survive, grow, and transform into juveniles and adults depends largely on abiotic and biotic environmental factors for which they have no or very little control or defense. However, as important as they obviously are, human awareness of their very existence and the ability to unravel their almost secret lives has occurred very recently.

By the close of the nineteenth century, it was becoming obvious to marine fisheries biologists that the maintenance and profitable existence of major commercial fisheries depended on an understanding of the factors influencing the distribution and abundance of fish larvae. We will discuss more about larval recruitment and commercial fisheries concepts shortly. But for now, we need to address a fundamental ecological question: Assuming precocial young are larger, more capable, and less prone to predation, why do fishes produce small and relatively helpless (altricial) larvae? More to the point, why does nature maintain such a life history pattern and why haven't more precocial forms evolved to replace them?

Internal fertilization and retention of the embryo inside the mother's body are ancient reproductive tactics. Certainly, ovovivipary and vivipary existed in placoderms, coelecanths, and ancient elasmobranchs, as proven by the fossil record, yet the teleost fishes have dominated the fish world by producing mostly altricial eggs and larvae. Persistence of larvae in such a large number of species supports the evolutionary value of this life history pattern. The logical answer to "why larvae" is this: Fishes produce altricial eggs and larvae because the larger number of smaller young results in a greater reproductive success (fitness) than producing a much smaller number of larger, more developed ones. The question that remains is "how do small life forms result in maximizing fitness?"

In major commercial fisheries (Atlantic cod and other gadoids, plaice, and Atlantic herring), the mortality of adults is about 5%–10% per year, while the mortality of larvae is about 2%–10% per day (Pitcher and Hart 1982). Obviously, it would be beneficial to increase the survival of larvae. Fishes have two options as an evolutionary response: to increase parental care, or to increase fecundity (the number of eggs). Let us see how this could work.

There are several factors related to fish larval success: (1) sheer force of numbers which can result in a diversity of individuals and invasion of a great many habitats; (2) parental involvement in their life history *before* eggs are released, as an aid to obtain suitable habitat; (3) egg and larval dispersal to distant locations can aid in spreading the resources; (4) exploitation of resources not available to the parents, including different habitat and foods; (5) separation of life history states, which reduces or avoids cannibalism and intraspecific competition; and (6) exploitation of seasonal food in abundance. These factors are somewhat related. However, it appears that the ultimate success of the altricial tactic is due to the huge number of larvae produced, which makes it impossible for the parent to produce enough yolk to enable development all the way to the juvenile period. Instead, nature is relied upon to provide food for the successful larvae, sometimes in very large numbers.

The parent's role in protecting their young in the pelagic ocean is constrained (e.g., they cannot build and defend a nest), but their role remains just as critical in two important requirements: (1) a spatial requirement to place the embryos in a location where a large number can find habitat with adequate food, or be dispersed into such habitat, and (2) a temporal requirement to spawn at a time when food is or will be available to the larvae when they begin exogenous feeding.

Pitcher and Hart (1982) pointed out that abiotic and biotic factors working together are the source of mortality on early life stages. Figure 28.4 provides a visualization of factors that affect four early life history stages. We will explore these points in more detail as we look at recruitment in fisheries. However, they provided an interesting analogy between a game of darts and larval recruitment that bears mentioning here. Abiotic factors are stochastic ones, largely due to climatic events and difficult to accurately predict. They inserted such factors into a game of darts: The object of a dart game is to hit the bull's-eye with a single dart. Let us give players two choices of propelling darts. A player might opt to use a super accurate crossbow to propel a single dart or have an alternative of using 20 darts. No matter how accurate the bow, some stochastic factors might ruin the aim, whereas it is more probable that one of the 20 darts will strike it. However, now we insert further complication to liken the game to conditions affecting survival and growth of larvae. In this case, stochastic affects are like a large and powerful steel magnet appearing at random intervals to deflect the path of the steel darts. In this case, it is much better to have 20 chances and to get some right, or close to the bull's-eye, than to take the chance and lose everything. For given enough time the single shot will certainly be lost.

As an endnote here, this discussion would not be complete without reminding the reader of the present situation with large pelagic sharks. Those fish produce a very few large young, and

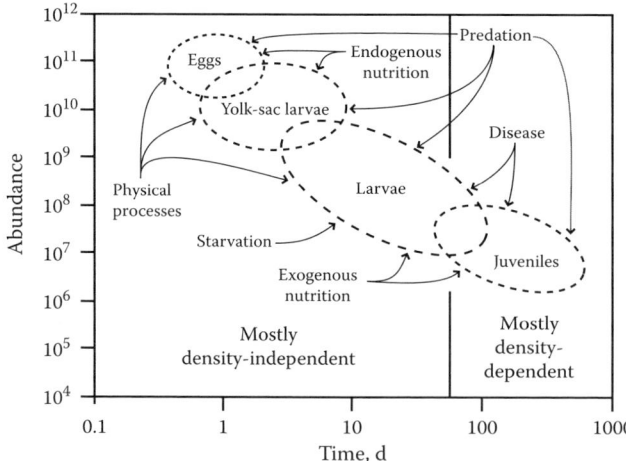

Figure 28.4 Fish larvae are vulnerable to abiotic and biotic factors. (From Houde, E. D., *10th Annual Larval Fish Conference.* American Fisheries Society Symposium 2, 17–29, American Fisheries Society, Bethesda, MD, 1987. With permission.)

as mentioned earlier, this life history strategy that served so well for millions of years has been compromised by humans. This large size now makes them and their embryos vulnerable to fishing gear. As inferred from Dollo's Law, evolution is not reversible, and this specialization is leading to unsustainable mortality rates in some of our largest sharks.

THE NICHE REVISITED

Are sweeping ontogenetic changes compatible with the ecological niche concept? The ecological niche of Hutchinson (1957, 1978) is an attribute of populations, and subjected to genetic changes and phenotypic plasticity (e.g., Colewell 1992). Growth, usually an extended gradual process, is an integral part of life and in this case quite necessary for larvae to become reproductively mature. Growth can alter habitat use and trophic relationships, but it does not alter genetics in any direct way. However, saltatory ontogeny can result in abrupt and sweeping changes in development because the organism makes a change in its quality (something new) rather than simply increasing in its quantity (somatic growth). As presented by Balon (1979, 1984), a change in quality provides a threshold that allows the organism to have an entirely new relationship with the environment through new adaptations that potentially affect habitat use, feeding, predation, and so on. In this case, ontogeny proceeds in response to DNA programming—the individual has the same DNA as before, but adaptations change over time as a function of the genome and the environment. In turn, the species niche widens due to adaptations of individuals, which exhibit differences in growth, sex, and ontogenetic life stage.

Unfortunately, the emergence of new characters can have a dark side, and rapid change during thresholds can be affected by environmental stressors such as changes in temperature or oxygen. Browman (1989) cautions that critical periods can occur as changes occur in embryology, neurobiology, and ethology, and these periods can occur both separately and in combination. Such developmental critical periods can then produce ecological critical periods in response to environmental influences, which may lead to death of individuals.

Dr. Hutchinson (1978, p. 176) acknowledged two evolutionary processes that act to widen the niche: "one leading to more polymorphic, less individually adaptable organisms, the other to less polymorphic, more individually adaptable organisms." Ontogenetic shifts would be included in the former process, and the latter process can broaden individual adaptations such as resource polymorphisms (e.g., trophic adaptability) (Smith and Skúlason 1996). In continuation of this observation, he stated that "though usually one species lives in one niche, there are cases where one species, by sexual or other types of polymorphism, does in fact occupy two or more niches. . . . In some fishes nonsexual polymorphism of a profound kind, putting the various forms of a species in quite different trophic niches, has recently been described" (p. 211). Thus, Hutchinson's concept of the niche was dynamic. He recognized niches within niches and looked at portions of the niche of interest. The practical matter of concern here is how to recognize and consider the needs of early life history stages of fishes in conservation and management. Although a fundamental niche for a species with a great degree of ontogenetic change would theoretically exist, it would be several times greater with respect to resource axes and unnecessarily complex to deal with for most purposes. In this case, an ontogenetic niche that included only a few of the most important resource axes could be very helpful in visualizing the adaptations, roles, and needs of a fish over its lifetime. This, the resource/utilization niche concept (chapter 22), would be a practical way to present the ontogenetic niche of a developing fish larva.

In the decades that have passed since the publication of the multidimensional and resource utilization niche concepts, there have been many studies illuminating ontogenetic niches of fishes. These studies have focused on habitat or trophic niches by ontogenetic life stages (e.g., review by Werner and Gilliam 1984; a series of papers included in Balon 1985), and more recently investigators have looked at evolutionary branching related to ontogenetic niches (Claessen and Dieckmann 2002). A few general ecology texts also have addressed the issue, such as in Smith and Smith (2001, p. 258), who stated that "the fundamental and realized niches of an organism can change with its growth and development." However, most general texts do not discuss the importance of niche shifts in fishes.

FISHERIES ECOLOGY AND RECRUITMENT CONCEPTS

Historic Concepts

Ocean fisheries have been important in supporting the wealth of nations. Those nations have responded to declines in commercial fisheries by hiring scientists and providing funds for study. Major breakthroughs in science have resulted from studies on the great marine fisheries and some historic review is warranted. In our contemporary civilization, overfishing has resulted in the loss of major ocean fisheries. Recent studies have indicated that young organisms are very sensitive to environmental conditions and we understand that the size of fish populations likely fluctuate due to variable rates of young mortality (i.e., recruitment failure). This has not always been the case, however, and there was little concern about overfishing for centuries.

As strange as it may seem today, historic declines of fisheries in Europe during the eighteenth century and even later were not attributed to overfishing or recruitment failure. Instead, a popular "theory" was the *polar migration* hypothesis attributed to Johan Anderson and published in 1746. The notion is reviewed and discussed by Sinclair (1997, 2009). I note main points here: According to the "theory," the home of the herring lies under the north polar ice cap at great depths. This home is peaceful, has plenty of food, no predators, no fishing, and continuous reproduction. Herring disperse south from this home where they are affected by predation, the major cause fluctuations in their abundance. The main feature of the migration concept was an inexhaustible supply of herring for humans, due to an unceasing migration south from the polar region. Variations in the supply were attributed to predation (Mother Nature has failed again) and not overfishing. The theory was accepted for over a century.

Basic to the migration concept was a renewal of herring stocks by relatively constant production of new individuals and the theory was later expanded to include other fishes as well. The notion was well supported until the close of the nineteenth century, when dissatisfaction grew as more was learned about reproduction and real migration patterns of several species. In time, oceanographic conditions were suspected as causes of variable migration patterns, which resulted in good or poor fishing. Finally, the theory was undermined by Professor Heincke in 1882, when he discovered that two herring stocks in the Baltic Sea were actually separate races, presumably induced by different environmental conditions affecting larvae. His work was followed by studies of speciation, egg and larval fish abundance, and distribution, which laid a foundation for modern scientific approaches.

By 1909, Johannes Schmidt (later famous for his work on eel migration) hypothesized that spawning cod were sensitive to conditions of temperature, salinity, and depth, with the result that spawning was restricted to areas that had the right mix of conditions. This was known as the "spawning sensitivity" hypothesis. His findings were soon supported by the work of others. In addition, it was recognized that fishing yields were linked with environmental conditions that affected larval survival and also with different drift patterns affecting larval dispersal. Another famous biologist, Norwegian Johan Hjort (reviewed in Allen 1914), put these details together. Working with widespread egg and larval studies, personal observations, and discussions with his staff (see Solemdal 1997), he developed a new paradigm in fisheries ecology: Fluctuation in the fisheries of northern Europe was due to variable recruitment in populations. Hjort also supported the "critical period" hypothesis for fisheries and generally is given credit for its development. The critical period hypothesis linked recruitment failure with starvation of young fish. In this case, larval survival depends on the timing of spawning, which had to coincide with food abundance. Hjort also hypothesized that a critical period coincides with the first exogenous feeding of the larvae, a time when high food density is imperative (Bradford and Cabana 1997). Thus, for a very long time abiotic changes due to unstable conditions in oceans have been implicated in the abundance and location of food for fish larvae.

Recent Concepts

Timing of spawning and the critical period hypothesis have been accepted as a factors in recruitment variability in some fisheries (e.g., sole, sardine, anchovy, and herring) but not for all. There has been some evidence that a critical period may occur later in some species than in the early larval stage (e.g., Atlantic mackerel) or not at all (reviewed by Lasker 1987). Part of the problem may exist in data quality and being able to detect the difference in mortality curves with and without a critical period (Pitcher and Hart 1982). Also, potentially compounding the problem are critical periods that may occur in several levels of larval development affecting embryology, neurobiology, and ethology. Browman (1989) related those critical periods to the early life history stages of fish, in which massive die offs can also produce an ecological critical period. Thus, developmental critical periods also exist and likely are related to the ecological critical period concept.

The critical period hypothesis has been extended by Lasker (1981) and others to suggest that food limitations do occur, but even then some food exists for larval fishes—as evidenced by some larval survival during bad years. In an intensive study involving lab and field observations, Lasker (1975) found that an unstable ocean reduced larval feeding success, mostly due to turbulence, although prey size and species also were important factors. Ocean instability was linked with areas of patchiness in larval food, which could in turn be disrupted by upwelling and storms. Lasker (1975) found that plankton used as larval fish food can be dispersed horizontally with currents and vertically by wind. He also found calm periods resulted in good feeding conditions for larvae (4 days or more of wind speed < 5 m/s). In his honor, these conditions are called "Lasker events" (Pauly 1989; Cushing 1990). In northern anchovy, larval survival was linked with higher numbers of Lasker events per month (Peterman and Bradford 1987), and Lasker events can be characterized by mathematical notation. An i/j Lasker event would be a calm period lasting i days with wind speed of j m/s.

Patchiness in larval food distribution offers an explanation for larval survival at sea due to small-scale events (Mullin 1993). Given the low average density of food at sea, such patches could provide for growth, whereas average food density in the ocean is so low that only larval survival can be maintained (Smith 1981b). Plankton patchiness and larval drift patterns can be disrupted by storms and turbulence, and dense layers of dinoflagellates used as food by fish larvae can be scattered by storm winds (Mullin 1993). Thus, density-independent factors have been implicated as important regulators of larval survival, and this is recognized as the "stable ocean" hypothesis and attributed to Reuben Lasker and his coworkers.

In general, continued heavy fishing of a stock after several poor years of recruitment (e.g., in the Peruvian anchoveta, Atlantic herring, and Pacific sardine) has resulted in lowering fish populations to a size that fisheries have a difficult time recovering, pointing to density-dependent factors. However, there have been some cases in which fisheries have recovered quickly from very low population size (e.g., the Japanese sardine fishery of the 1970s) (Lasker 1981). Such a recovery has been associated with changes in density-independent (abiotic) forces, and also attributed to stable ocean conditions.

Another extension of the critical period hypothesis was made by D. H. Cushing, who developed this concept by the early 1970s and expanded it in the 1990s (Cushing 1990, 1995). In essence, his "match/mismatch hypothesis" links the amount of larval recruitment to the production of plankton (larval food). The hypothesis holds that spawning must occur at the proper time of the year so that the first exogenous feeding occurs at a time of high productivity of larval food. Cushing (1990) went into great detail explaining his original concept and evidence for it, and also extended the concept from temperate fish stocks to include tropical and subtropical climates as well.

Jobling (1995) explained how the match/mismatch concept relates to the timing of spawning so that the onset of larval feeding would coincide with food production, and provided insight about how small populations might be especially affected due to unpredictable climatic events. Such events were illustrated by Lasker (1987) as related to Northern anchovy (Figure 28.5). In essence, a large population of spawners would be expected to exhibit some diversity in the onset and duration of spawning so that a period of relatively small food production might still be utilized. However, small populations of larvae could miss most or all of food production during a bad year. In addition, it is reasonable to expect that several environmental factors might affect the timing component. Pope et al. (1994) suggested that spawning periods in high latitudinal seas should occur at high productivity, but also occur before the responding predator wave can develop. In this case, rapid growth of larvae could (surf the wave of food) outdistance predation by rapid growth. Hutchings (1997) also suggested that in low latitudes, tropical fish will tend to multiple spawning during the year, perhaps in an attempt to match food production in a more subtle and structured environment.

We now understand that larval transport regularly occurs in oceanic and riverine environments. In such cases, transport of larvae into favorable areas would require dependable current patterns for dispersal and proper timing. If match/match is extended to riverine systems (as discussed for pikeminnow), proper timing may not only include photoperiod and temperature, but also with runoff patterns.

J. T. Anderson (1988) developed a growth–mortality concept that suggests small individuals have a lower probability of survival than large ones (reviewed by Hutchings 1997). This is an important idea, because predation on larvae can be just as important as starvation (Mullin 1993). Two closely related hypotheses have been developed, partitioning the effect. The "bigger is better" hypothesis gives a size advantage to faster growing, larger individuals that are more capable of avoiding predation, foraging, and resisting starvation. Certainly larger and more developed larvae would have an advantage in intraspecific interactions as well. However, the concept does not consider all aspects of gross vulnerability and has been questioned (e.g., see Litvak and Leggett 1992). A related "stage duration" hypothesis posits that faster growth reduces the length of time that an individual spends

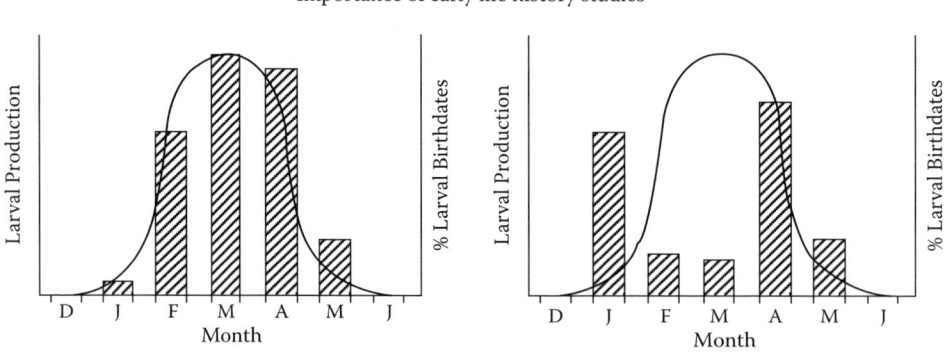

Figure 28.5 Environmental perturbations and northern anchovy. Left = larval production of northern anchovy by month (histogram) assuming no environmental perturbations, and a curve fitted to the distribution. Right = what larval production might look like with perturbations, compared to the curve generated for ideal conditions. (From Lasker, R., *10th Annual Larval Fish Conference*, American Fisheries Society Symposium 2, Bethesda, MD. 1987. With permission.)

in the early life stages in which mortality rates are very high. Both of these hypotheses have been supported by field research (e.g., see Hare and Cowen 1997). Another related hypothesis is the "point of no return" concept, presented by Blaxter and Hempel (1963), in which larger eggs (more yolk) extended the point of no return, a point in which starving larvae could not survive even when given food, from 15 d after yolk depletion for herring larvae hatched from small eggs to 28 days for larvae hatched from large ones.

A good summary of the recruitment issues thus discussed was presented by Houde (1987), who conceptualized sources of food and mortality in fishes, plus likely population regulatory mechanisms by life history stages. At any time, there may be both abiotic (density-independent) and biotic (density-dependent) forces affecting recruitment. As he points out, fishes exhibit indeterminate growth, and marine fish typically emerge from altricial eggs. The size increase from egg to adult may commonly range 10^5 to 10^7 times in weight, and in the first year of life, increases of 10^3 in cod and 10^5 in tuna are usual. Thus, at these size ranges it is difficult to imagine the plethora of affects that might be caused to the larva by abiotic and biotic factors. Houde (1987) also effectively demonstrated that mortality of fishes during the larval stage can be much higher than in the juvenile stage. In this case, reduction of time spent as a larva would result in an increased level of survival. Thus, for survival, the attainment of a certain size may not be as important as a faster growth rate.

Another issue in larval fisheries ecology has to do with population richness in some commercially fished species, such as Atlantic herring. As previously mentioned in this section, it was determined before the arrival of the twentieth century that Atlantic herring exist in smaller, localized subpopulations (also referred to as demes or stocks). Each of these smaller populations have different attributes such as spawning at different times of year, characteristically different levels of abundance, different spawning areas and feeding distributions (reviewed by Harden-Jones 1981; McKeown 1984; Sinclair 1988). In an effort to explain the existence of these separate populations existing in overlapping ranges and sizes, Iles and Sinclair advanced a "herring hypothesis" in 1982. These authors postulated that the larvae are able to maintain their presence in a diffuse environment by behavior and the physical features of the geographic area. This hypothesis was subsequently extended 6 years later by the "member/vagrant hypothesis," which proposes that the number of populations of a species depend on the presence of physical geography that provides for retention and survival of the young following hatching (Sinclair 1988). "Members" are individuals that remain in the population and contribute to the local gene pool. "Vagrants" are those that are lost by geographic displacement or biological factors like predation or starvation and do not contribute to the gene pool. As in the herring hypothesis, the larval "retention area" is a geographically stable physical space. Another hypothesis was more recently offered (adopted/migrant) (McQuinn 1997) in which young fish learn to find spawning areas by following older fish and may spawn in more than one spawning site. In contrast, other workers have found that spawning site switching is less than 10% and homing rates to spawning sites in some populations are as high as 90%. Based on this, some authorities believe that further work is warranted to explore mechanisms that facilitate natal homing in Atlantic herring (Brophy et al. 2006).

In general, hypotheses are much easier to produce than experimental data, and some hypotheses are difficult if not impossible to test. Yielding to Ockham's Razor (that the simplest hypothesis should be chosen), it appears that the simplest hypothesis is one of natal homing involving learning (including imprinting) rather that more complex ones like the member/vagrant hypothesis that would be difficult to test. As discussed in chapter 27, there are aspects about homing and imprinting that need to be reevaluated in marine fishes, and especially Atlantic herring. In my opinion, the most logical explanation for Atlantic herring population richness (multiple numbers of subpopulations) could be due to genetic semi-isolation that would result from natal home-site spawning fidelity. This phenomenon is well reported in the literature for other fish species (reviewed by Smith 1985). Hopefully, I have given a lucid argument for this in chapter 27.

Marine versus Freshwater Environments

Concepts about the role of early life history stages in marine fish recruitment discussed above were developed during a period of over 100 years. These concepts have been subjected to intense scrutiny and continue to be refined and extended. In the past few decades, increased efforts to determine factors influencing recruitment in native freshwater fish communities have created an interest in applying concepts learned from marine fisheries to freshwater systems. This is potentially useful, but differences and limitations of such an approach have to be understood. For example, we have identified major differences in the reproductive biology of marine and freshwater fishes and gross differences in oceans compared to freshwater systems.

Some hypotheses developed in marine systems, such as critical period, including its logical extension to the match/mismatch hypothesis have to do with general principles and can be applied to freshwater situations. I believe the match/mismatch concept is very helpful in understanding the intricacies of Colorado pikeminnow reproduction and larval survival in the Colorado River system (chapter 27). Also, it was useful to have an existing concept as a point of departure. However, in every case that I can imagine, major differences in freshwater and ocean systems would require careful examination to see if the concepts really are compatible. While both systems are aquatic and exist in dynamic equilibrium with environmental influences, the environments, dynamics, and equilibria are different.

Environmental conditions in freshwater systems are extremely variable, especially for small larvae. In riverine systems, years with more stable flow regimens will likely expose larvae to higher predation risk. Years of greater flow variability can result in flooding or dewatering nests or nursery habitat, causing abandonment or displacement of fish eggs or larvae (e.g., Harvey 1987; Jennings and Philipp 1994). Lakes appear deceptively similar to marine environments, but there have been severe warnings about analogy.

John Magnuson (1988) in a paper titled "Two Worlds for Fish Recruitment: Lakes and Oceans" pointed out that principles involved were common to both freshwater lakes and oceans. However, the systems, although having some basic features in common, were different in some fundamental ways. His bottom line was this: fish recruitment in lakes (which are like islands) are primarily influenced by factors that promoted extinction, while fish recruitment in oceans (which are like continents) are primarily influenced by factors that promoted colonization. Clearly, the shallow freshwater systems would be affected greatly by terrestrial influences, while deep waters of oceans would not. Comparisons of these systems can provide some incredible insight, but gross application of marine concepts to freshwater systems, even lakes, should be avoided.

CASE STUDY—LARVAL FISH MOVEMENT

It has long been known that fish larvae can maintain their vertical position in a column of water (presumably by responding to light and gravity) and use this positioning while being transported by river flow (e.g., Gerlach and Kahnle 1981). Such dispersal is common in some North American fish families, including Catostomidae, Cyprinidae, Percidae, and Centrarchidae (Scheidegger and Bain 1995). We have previously discussed the drift of larval Colorado pikeminnow from spawning sites in rocky canyons to shoreline habitats in downstream fluvial reaches. In the historic Colorado River system, unpredictable changes in river stage could occur on short time scales, but in the present system, dams can produce more frequent changes even on smaller scales. As a result, the present system can affect backwater nursery habitat by frequent flooding and dewatering. Can small fish larvae deal with these changes and how? General features of the larval transport in rivers are known, but some questions remain. For example, do the larvae respond only to high (i.e., fast) flows or will they travel on lower flows? What flows are needed to transport the fish but also allow them to move into the backwater nursery areas? Also, do some flow regimes result in stranding them in backwaters that become disconnected?

Although I was almost exclusively a field researcher, I knew that answers to the larval movement and transport questions would be hard, if not impossible, to obtain from nature. In order to get the answers we needed we would have to do some controlled tank studies. The chance came when we learned that excess larval fishes might be available for such study from a national fish hatchery, so we got busy in a hurry. First, we needed to design and construct a special tank. We were able to utilize our technicians to build a large experimental tank of marine plywood coated with epoxy resin and painted with nontoxic latex paint. The tank had seven compartments and water was circulated with an electric pump situated in a separate container. Water circulation in the tank flowed linearly from one compartment into another via small ports that were cut in the partitions (Figure 28.6). We adjusted the flow with a valve to obtain the low flow rates that we wanted.

Water circulation was observed by using florescein dye and measured with a flow meter. Water moved along the tank walls and at the ports, but velocity was too low to measure in the middle and bottom of each chamber. This provided a quiet area (backwater surrogate) for the larvae in which they might rest, or be entrained. Movement through the tank, especially at very low velocity (2–3 cm/s at ports), would require a directed, rather than a passive effort by the larvae!

We obtained Colorado pikeminnow (Cpm) and razorback sucker (Rz) larvae of less than 1 week old from Dexter National Fish Hatchery for the study. And we also were able to obtain a few 36-week-old Cpm juveniles from the wild. The larval fish were used only once, and all fish were fed with live brine shrimp and pellets of larval fish food.

Trials were run in simulated daylight (five white incandescent 25-W bulbs) and dark conditions (one red bulb). Five replicates of five test fish each were used for varying conditions of fish age, light, and flow rate. At the beginning of each trial, the test fish were isolated in the middle chamber of the tank and allowed to acclimate for 15 min before the trial began. The distribution of larvae in the chambers was obtained after 5, 15, 30, and 60 min by counting the number of larvae. Two measures of movement were then calculated: total activity in any direction, up or down flow, which had a maximum of three; and directed activity, either upstream or downstream (negative values upstream, positive values downstream). Figure 28.7 shows the average total and directed activity for each species of larvae and size group (1-, 3-, 6-, and 36-week-old Cpm and 2-week-old Rz), day (a) and night (b).

All ages of fishes were able to explore their surroundings and to locate the small ports connecting the chambers. Average total activity of the juvenile Cpm was high for day and night, but the smaller larvae of both species were most active in daytime for all flows. Directional movement was different: At night, smaller larvae tended to move downstream with low flow, and all fish moved downstream with increased flow. The larger 36-week Cpm were extremely mobile, and their activity usually peaked in 15 min or less, even with no flow conditions.

Although the results of this study might suggest that the larval fishes would be swept downstream at night and perhaps out of the river system entirely, such is not the case for these fishes in nature, as indicated earlier (chapter 27). Although the swim-up behavior of larvae would result in

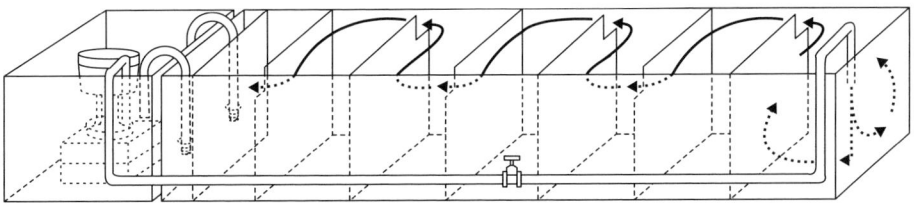

Figure 28.6 Experimental tank for movement study. (Courtesy of Tyus, USFWS.)

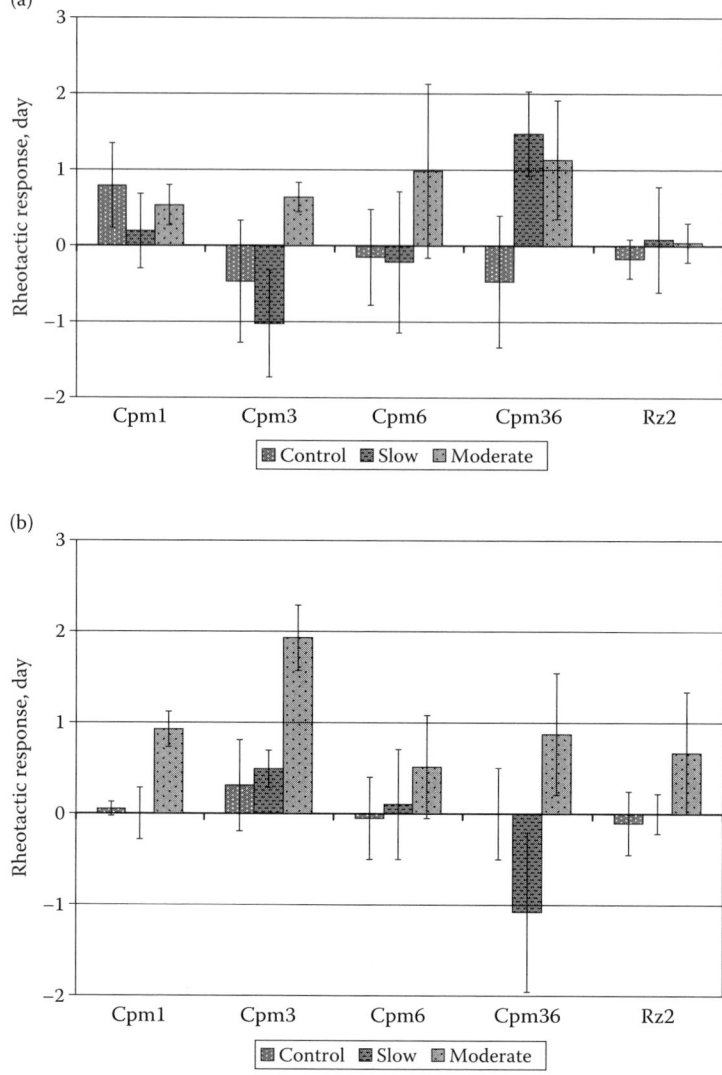

Figure 28.7 Average total and directed movement. (Courtesy of USFWS.)

an initial period of rapid transport out of canyon bound areas, the larvae are able to find and enter slower transport along shallow shorelines of the alluvial nursery reach. At some point they are able to detect and enter ephemeral (depending on river stage) alongshore embayments referred to as "backwaters." Thus, we assume that their movement in the tank reflects the same behavior in an interconnected system of riverine backwaters. In that case, moving downstream with the flow could be an adaptation to take them into a backwater that is filling and out of one that is draining. This offers an explanation why these native fishes are not usually found stranded in isolated backwaters, in which deteriorating habitat conditions could exceed thermal limits, limit oxygen, produce desiccation, and lead to death.

The results of this study also compares favorably with an earlier one that tracked movement of larger juveniles in river backwaters. In that study (Tyus 1991), almost all of the juvenile pikeminnow (85%) moved among backwaters and other habitats with changes in flow, and established a diurnal pattern of movement from backwaters to main channel habitat that was linked with water temperature.

We have shown that young Cpm can detect very subtle water velocities and become entrained in downstream flows. We believe that this offers a plausible explanation why stocking of juvenile Cpm Trammel et al. (1973) failed to establish a reservoir fishery (in Kenney Reservoir, Colorado). In a very short time, the abundance of stocked fish decreased to the point that they were exceedingly difficult to find and subsequent sampling indicated all had moved downstream and over the dam.

As we have learned, sensory systems of early altricial larvae are poorly developed, but they do have reasonably good vision and use it in orientation, detecting food, and so on. Our study showed that small larvae are much more active during the day than at night (i.e., indication of visual rather than mechanoreceptive sensing). This would make the larvae more vulnerable to introduced (nonnative) sight-feeding predators, which now dominate the fishes in backwaters. Also, we demonstrated that the larger 36-week Cpm were very active all the time, presumably engaged in a searching pattern (see Tyus et al. 2000 for more details). Behavioral studies have demonstrated a relative "naiveté" of native fishes in the presence of nonnative predators (Johnson et al. 1993) and a basic lack of aggression compared to the same size of nonnative fishes that are common in the system now (Karp and Tyus 1990). Our study suggested that the behavior of the young fish we studied would have different consequences in the present system than historically. The movements would surely increase their chance of encounter with the aggressive introduced fish that now live in the system, and increased death by predation is the anticipated result.

SUMMARY

Larval and postlarval fishes (early life history stages) are usually very different in morphology and physiology compared to adults of the same species. They also are ecologically different in fundamental adaptations such as foods eaten, habitat occupied, and behavior. However, the needs of young fishes were not adequately considered in historic conservation and management efforts, which usually focused on the habitat requirements and limiting factors of adult fish. In most instances, the reason was lack of information about the needs of young fish and difficulties in identifying and sampling them. As a result, the role that the young can exert on regulating the size and other attributes of fish populations, and consequently important fisheries, has only been recently appreciated. Most larvae of marine fishes (important exceptions include some species with demersal eggs) are produced by pelagic eggs. The larvae spend almost, if not all, of their lives in the upper pelagic zones and have specific adaptations for buoyancy. Food selection by larvae is specialized for different sizes of plankton and many fish species exploit the same resources. These larvae are greatly affected by storms. In contrast, freshwater fish larvae are primarily produced by demersal (nonbuoyant) eggs. Most of the freshwater systems are relatively shallow and usually vegetated, and freshwater fish larvae are more general in trophic adaptation and diverse in habitat use. These habitats also tend to be ephemeral, varying in present size due to river discharge, and exposed to terrestrial climate, where conditions can greatly affect small fishes. Fish larvae can be precocial and use a large supply of yolk to emerge as juveniles, which are analogous to small adults, or altricial with a much smaller supply of yolk. The altricial young are true larvae and do not appear as little adults. They are not fully formed. Instead, they are lacking in fin development and some of their organs may not be functional at swim-up. Development of altricial larvae to the juvenile stage will require exogenous feeding and they may pass through stages of development known as salutatory ontogeny before they undergo metamorphosis into an adult. Fish larvae may pass through several periods when they eat different foods and occupy different habitats, experiencing one or more ontogenetic niche shifts. These altricial larvae have to be presented with an environment that has a large amount of easily captured food for them to survive and grow. Why would fish produce such delicate young? Several reasons include sheer force of numbers to increase diversity and exploit different habitats, exploitation of habitats and food not available to the parents, exploitation of seasonally abundant food, separation of life stages to avoid cannibalism and intraspecific competition. In this strategy, the adults must place their eggs in the right place at the right time to enhance larval survival. The study of larval fish ecology was promoted with the recognition that variable recruitment was the cause of fluctuations in stocks, which lead to the spawning sensitivity and critical period hypotheses in ocean fisheries in the early 1900s. Soon to follow were the patchiness, stable ocean, match/mismatch, and other hypotheses concerning larval fishes. However, some mechanisms by which larval fishes cope with environmental conditions are difficult, perhaps impossible to learn by studying the fish in nature. A case study of larval fish behavior under controlled conditions is provided to demonstrate this point.

Further reading: Recent: Finn and Kapoor 2008.

Recommended video: *Larval fish ecology: A critical management concern (24 minutes).* Produced by D.E. Snyder 1988. Video catalog number TVO4735 (English) or TVO 6104 (Spanish). Office of Instructional Services, A-71 Clark Building, Colorado State University, Fort Collins 80523.

PART VII

Applied Ecology: The Human Factor

CHAPTER **29**

Exploitation and Fisheries Management

INTRODUCTION

Applied ecology considers the role of humans as a major factor affecting the distribution and abundance of species, the quality of habitats, and function of ecosystems. As an early example, Hinckley (1976) addressed the basics of this field in a nontechnical course called "Ecology and Man." This chapter traces the progression in human use of fish and wildlife resources beginning with early and sometimes relentless exploitation. In time, humans realized the need to reduce their adverse impacts, and they established management practices such as natural resource conservation. More recently, concepts of sustainability and ecosystem management have emerged. Applied ecology is of worldwide interest, but here we focus on North America with consideration of worldwide problems.

Human impacts on North American ecosystems vary from "not much" to "a lot," but these terms are not very informative. A better understanding of the degree and characterization of human influence is essential to understanding exploitation and natural resource management philosophies. Descriptors that address different types of impacts are needed, especially in consideration of the severity of change. In the following, I identify various abiotic and biotic changes in terrestrial and aquatic ecosystems as an aid in developing a perspective of degree of human influences. Ecosystems, communities and habitats can be described with the following:

- *A pristine* system (or systems) exists in a pure, unspoiled state. There is no trace of modern humans (postindustrial revolution) or their activity. All native species are present. There are no introduced species. Examples include all ecosystems of the North American continent before about 1500 (pre-Columbian), including terrestrial and aquatic systems, and Rocky Mountain streams, which could be considered pristine before about 1840.
- *Wilderness* is unsettled by modern humans, thus uncultivated and undeveloped. Most, if not all, native species characteristic of systems are present. An occasional introduced species may be present, including humans. Examples are some presently designated and more remote wilderness areas, and aquatic habitats of northern Canada.
- *A natural* ecosystem is produced by nature. Such systems appear as wilderness in general, but with some human-induced effects such as grazing by domesticated animals. All native species are present. Some introduced species can occur. There is a presence of humans and domesticated animals, but few artificial (human made) structures. In general, I place most national parks and some less disturbed warmwater streams of the eastern United States in this category.
- *Human-influenced* systems are considered to be "disturbed." In these systems, a low density of modern humans causes an obvious visual impact. Some native species are absent. Many nonnative species are present. Habitats are fragmented to a degree. I include environmental-friendly (green) housing developments and most channelized streams in the eastern United States in this category.
- *Human-dominated* systems have a high density of humans, or they are extremely affected by human activity. Many native species are extirpated. Nonindigenous species are common. Suburbs, intensively cultivated farmland, maintained intracoastal waterways, and reservoirs are examples.

- *Human-disrupted* systems are artificial, not natural. These systems replace natural systems and have little or no wild, native species. They are dominated by introduced pests and domestic species. Included are cities, concentrated rearing facilities such as chicken farms, or aquacultural facilities such as hatcheries.

This attempt at classifying the degree of human influence on natural systems has not been met with total satisfaction by all, mostly due to sociopolitical disagreement with the use of terms. However, it can be used as a starting point for further discussion. I believe that all informed persons recognize there are human effects that alter the structure and function of nature in various ways, and that it is important to qualify the magnitude.

An extreme view of anthropogenic change is that humans are "not only controlling nature, but wholly transforming it . . . consuming the wildness of nature" (Wapner 2010, p. 4). Thus, McKibben (1989) considered the present situation as the "end of nature," because of the pervasive effects of humans in altering the natural world. In this case, nature can be defined as the result of billions of years in which the world has acted "by itself." Historically, "nature" is an idea, a social construct to recognize the nonhuman world (Wapner 2010). Consequently, here I use "natural" in the context of the processes, phenomena, and systems that have occurred over millions of years and use the term "artificial" to identify those changes in nature that are an outcome of human influence.

Management is the process of exercising some control over nature, and rather than an exhaustive accounting of the management process, the goal here is to build a foundation for understanding various concepts common to fish management—a field that developed with a scientific basis deeply rooted in application of ecological principles. To do so we explore historical aspects leading to the present state of aquatic ecosystems and fish communities, review management practices and concepts for fishes in fresh, estuarine and marine systems, introduce emerging concepts in sustainability, and discuss concepts associated with ecosystem management.

Basic to the notion of fisheries management is the use of fish as a commodity, and human use can affect fish populations in various ways (e.g., see Scalet et al. 1996). In this context, some clarification of terms will be helpful: Someone who is a *direct* user of the fisheries resource for example, might fish a pond, or benefit from those that do, whereas an *indirect* user might graze livestock in the shoreline or take water from the pond. Furthermore, *consumptive users* are those that catch and eat fish while *nonconsumptive users* might only view or take pictures of fish. As might be imagined, there are gray areas in such classifications, and one might argue that all human use of natural resources is consumptive to a degree that fish behavior and habitat is usually affected by the activity. In general, however, consumptive users of fish (i.e., that catch and kill fish) are divided among subsistence, recreational, and commercial activities (Ross 1997). Nonconsumptive users are really "less consumptive." These users generally include boaters or hikers that view fishes in an ecological perspective as they experience outdoor recreation and education as in *ecotourism*. Having said this, ecotourism can result in wounding or killing large fish and marine mammals (e.g., paddlefish and manatees) by boat propellers, and habitat damage increases as numbers of recreational users increase.

HISTORIC PERSPECTIVE

Exploitation

All organisms exploit nature to some extent, in an ecological viewpoint to use resources in an attempt to maximize fitness. Such use also can be viewed as selfish use. However, nature tends to balance such use because organisms are exploiting one another. The meaning of exploitation by present standards involves an unethical destruction of resources by some unsustainable practice.

In order to understand the present problems it is necessary to trace the causes, which are rooted in history and dominated by socioeconomic forces.

Humans were once part of nature—just another animal. But our clever ingenuity has made us special and dangerous. Unlike other animals, humans have developed complicated languages for complex communication; domesticated beasts of burden such as oxen and horses; enlisted the aid of other skilled animals such as dogs for protection, hunting and herding; cultivated, raised and stored crops; and used stone, brass, and iron tools and weapons. Most recently, humans harnessed the energy of fire, water, wind, fossil fuel, and the atom to do work.

Early humans apparently formed groups for safety and worked collectively as scavengers, gatherers, and hunters. They also developed agriculture and domesticated stock, which made them less dependent on wildlife and allowed more permanency, resulting in settlement. This provided more time for civilization, including development of culture and technology. They developed metal tools and weapons, diversified into trades, and multiplied in numbers. Radiating out of Africa, they entered more temperate environments and continued to change nature to meet their purposes. Forests were cleared and plowed; marshes drained; streams were blocked, impounded and diverted for agriculture. Wildlife populations, including freshwater and anadromous fishes were extirpated in and near areas of human settlement.

Steel tools and social cooperation allowed construction of boats, vessels, and ultimately ships for local, then oceanic transport. This facilitated trade between geographic locations and increased exploration. In time, ocean fisheries were established with hooks and crude traps. Fish were valuable and avidly sought to augment failing supplies of fish from depleted streams and rivers. It was at this stage that the new world of the Americas was rediscovered by Christopher Columbus and European cod fishermen (ca. 1500 AD) (Kurlansky 1997). Unlike the Spanish who were mostly interested in precious metals, the cod fishermen immediately struck it rich harvesting a highly sought and valuable commodity. Consequently, the fishermen had little or no reason for leaving the comparatively safe and well-supplied ships to invade a dense wilderness occupied by millions of Native Americans.

Native Americans were well adapted to the North American continent and had a sophisticated culture. They were nomadic descendents of hardy Asian hunters/gatherers that moved into Beringia and from thence into the North American continent at the end of the Pleistocene. Moving south in waves, they colonized the entire continent in less than 1000 years. They also have been implicated in the extinction of many large land animals (i.e., overkill hypothesis) (Colinvaux 1993). These invaders had language skills and tools. However, they lacked large domesticated beasts and their civilization knew only Stone Age technology. Even so, their success is history, and by about 11,500 AD, people of the ancestral Clovis culture were numerous (about 40 per km^2) (Gray 1993). These people presumably resulted in or hastened extermination of North American megafauna and their ancestors were capable hunters and fishers. Some tribes were greatly dependent on salmon in the Pacific northwest and all tribes used freshwater, anadromous, and estuarine fishes for subsistence where they occurred. Tribal skills at fishing and using fish for other purposes, including agriculture, are well documented (e.g., interesting reviews by Hay 1959; Crowder 2005; Roberts 2007).

Tales of the richness of the New World and promises of land and freedom finally resulted in European invasion of North America. Spanish conquistadors invaded the southwest in the sixteenth century looking for riches, and northern Europeans invaded at the beginning of the seventeenth century for various reasons, producing large permanent settlements at Jamestown (1607) and Plymouth (1620). These settlements occurred late enough in history to leave a decipherable account of the ecological impacts caused from the invasion of relatively advanced humans into a temperate wilderness.

Europeans brought steel tools and weapons, horses, oxen, plows, wheels, and firearms. They also brought simple necessities and trinkets that amazed the natives. Natives no doubt saw advantage in trading with these new invaders and they would not have imagined the hordes of invaders that would follow. The native people even taught the Europeans how to fish for food and fertilizer, and how to plant crops (documented in William Bradford's *Of Plymouth Plantation*; written in 1621 and referenced by Hay 1959). Europeans were amazed at the richness of the fauna, especially cod, and the sudden appearance of incredible numbers of anadromous fish, some (e.g., alewives), arriving at their very doorsteps.

Moving forward in time the numbers of invading Europeans with their livestock, pets, pests, and diseases, steadily increased, wars with the natives were fought, and forests were cleared for more and more settlements (e.g., see Gray 1993; Moulton and Sanderson 1997). As civilization and law reached these settlements, political problems emerged. Sick of European laws that restricted harvest of fish and game to the privileged (landowners, and "qualified persons" of wealth and nobility), the colonists wanted the control of natural resources to reside with all citizens (Bean 1983). This concern prevailed in the formation of the United States and no natural resource protection of any kind was considered in the new constitution. Later, common law would award the ownership of fish and wildlife resources to the states to be held in trust for the citizens (Bean 1983).

What did the average citizen think about the great North American landscape? Early colonists, including Captain John Smith, were favorably impressed by the natural resources present and lauded the richness of the waters (review by Roberts 2007, p. 45). However, their response to the heavily vegetated interior (wilderness) was different. Nash (1982) considered the reactions of Europeans to the forests, which I summarize. William Bradford, upon landing there in the *Mayflower* wrote of a "hideous and desolate wilderness" and those that followed him continued with defiant hatred of wilderness, celebrating the advance of civilization (i.e., clearing of forests and establishing settlements) as a great blessing from God. Much later, traveling to Michigan territory in 1831, a Frenchman, Alexis de Tocqueville, wrote in his journal that American frontiersmen considered him mad because he wished to travel in the wild forests for pleasure. He also wrote that the Americans only appreciated the works of man, and were insensitive to natural wonders. The philosophy of the frontiersmen was recorded as insensible to nature, felling trees, draining swamps, diverting and blocking rivers, ending solitude, and disrupting nature (Nash 1982).

With this lack of care and little understanding of the function of nature, it is not surprising that the wealth to be made from commercial trade (mostly with European nations) resulted in great exploitation of wildlife, including fishes in the American colonies. Gray (1993) provided a good account, noting that as early as 1748 South Carolina alone shipped 160,000 deer hides to England. Furbearers were ruthlessly exploited by settlers, and traders enlisted the aid of natives to whom they furnished goods, trinkets, and firearms. By mid-eighteenth century, flintlocks began to replace the cumbersome and inadequate matchlocks, fowling pieces (shotguns) became effective, and the deadly American long rifle was well established by the close of that century. In aquatic systems, anadromous fishes took an immediate hit due to the predominately coastal settlements. Early regulations, such as the Plymouth Colony Fish Law of 1621 (chapter 15), were needed to prevent complete eradication of anadromous runs. In the Great Lakes, exploitation and mismanagement is legend (chapter 13).

Westward expansion in the nineteenth century found millions of passenger pigeons, which occurred in flocks of many thousands in the Midwest; further on, herds of 50–60 million bison were found, and then herds of 40 million pronghorn antelope. They would not persist for long. In a scant 50 years, the herds were greatly reduced, as were all species of flocking birds such as curlews, plovers, ducks, and geese (Gray 1993). Although naturalists raised a cry of protest, little was done to stop the slaughter, and the degree of carnage was astounding. By the beginning of the twentieth century, the passenger pigeon and heath hen were extinct (1914 and 1932, respectively), bison, and pronghorn were almost extinct, and other formerly abundant species such as elk were extirpated from much of their range and greatly reduced in numbers. Depletions in marine fishes are more difficult to assess. However, the harvest of fish can stagger the imagination. An example is provided by a chinook salmon fishery on the Sacramento River, which produced 2500 cases of canned salmon in 1874. The business was profitable, and the fishery expanded to 1500 boats supporting 21 canneries.

In only 8 years (1882), the fishery produced 200,000 cases (Nielsen 1999)—an increase of 8000%. The fishery declined after 1882 and never recovered.

Water played an important part in expansion of human populations into the interior of the nation, and supported trade back and forth between the coast and inland settlements. Such transportation was aided in the mid 1800s by public works, including the Army Corps of Engineers, which is responsible for maintaining navigation as a national interest. The Corps "improved" the streams and rivers. They dug canals, opened and deepened channels, installed riprap, training dikes, levees, and other improvements in large rivers such as the Ohio and Mississippi (Nielsen 1999). Not as easy to observe, and in some cases thoroughly unpredicted, were associated effects on fish habitat and communities. Damage to fish populations occurred from dredging and siltation. Changes to fish habitat occurred due to loss of associated floodplains and structural changes. Changes in fish communities occurred due to the invasions of marine organisms such as the parasitic sea lamprey and alewife. Native freshwater fish populations also were reduced by fishing, and early declines in Great Lakes fishes and in native western trouts (e.g., greenback cutthroat trout) have already been discussed.

The discovery of gold in the mid 1800s also resulted in great losses of stream habitat due to mining impacts. Large valuable fishes such as lake trout, salmon, sturgeons, and later, paddlefish came under assault in lakes and slower moving rivers such as the lower Mississippi River. Even in fast and difficult river systems such as the Missouri and Colorado rivers commercial fishing activities resulted in large catches, especially during low water periods when the rivers could be seined. Declines of native freshwater fishes did not first result in conservation efforts. The losses were interpreted as loss of fishing opportunity, and they were "mitigated" (i.e., the loss of fishing reduced) by stocking of other fishes to increase the yield to fishermen.

Beginning in 1871, the new U.S. Fish Commission would focus on fish culture. One of their first acts was to transport American shad to California and Pacific salmon to the east coast. By 1873, the commission was stocking fishes via railroad (including many species of game fish and the highly desirable but exotic common carp) in accordance with requests received for fish from all across the United States. Trout and salmon are always popular with fishermen. German brown trout were introduced in 1883; by 1896, many eastern states supported the reproduction of populations of rainbow trout, and Alaska advocated sockeye salmon hatcheries on all major streams (Nielsen 1999). By the close of the nineteenth century, the predominant job for fish biologists was hatchery related. However, it was soon discovered that stockings of invasive fishes created new problems, and continuing habitat deterioration resulted in decreasing fish populations, making it clear that stockings were not a cure-all.

Natural Resource Conservation

Decline and losses of wildlife and fish populations in the United States during the nineteenth century were met with mixed reaction by different states. At first there were few established laws protecting animals, thus there were no seasons or limits imposed on exploiting them. Then, harvests were regulated to protect the diminishing stocks by setting creel and bag limits and establishing open and closed seasons. But too often, the regulations were ignored and commercial trade continued by poaching. According to Gray (1993) the first game wardens were hired by Maine in 1852, and the first hunting season established by New York in 1864. Other states slowly followed, but even so, most species were not protected at all on public land—especially plants and nongame animals. Although the United States had no express authority to manage wildlife, the establishment of Yellowstone National Park in 1872 was done in an effort to preserve the Yellowstone wilderness, and hunting was not allowed in it. Curiously, this prohibition was never challenged in court. Just as important, beginning in the 1870s, public opinions were changing: sportsman groups and periodicals advocated abolishment of commercial trade that was having a devastating effect on wildlife populations (Bean 1983). However, it was not until Theodore Roosevelt became president of the United States in 1901 that a national conservation ethic became a reality.

Theodore Roosevelt, a Harvard graduate, rancher, outdoorsman, naturalist, big game hunter, writer, and politician, was an advocate for "wise use" of natural resources. He was a strong supporter of "conservationists," who advocated for natural resource conservation: the use of public land for economic gain, but only if such use were compatible with nature (i.e., the health of nature was maintained for future generations). This generally means that the natural resource being used was not depleted. On the other hand, the "preservationists," supported by the Sierra Club, believed that public land was to be protected, and they wanted large tracts of public land set aside as wilderness in which animals were not exploited in any way (Gray 1993). Both viewpoints exist to this day.

Roosevelt made good on promises to conservationists. In addition to national parks, he also set aside public lands as refuges for wildlife, with a focus on migratory birds. The first national wildlife refuge was established at Pelican Island, Florida, in 1901, and he added 31 additional ones in 6 years. Although it was once judged impossible to develop a body of national wildlife law, in subsequent years the federal government entered into fish and wildlife management to protect its property, to assist states in commerce, and to make treaties with foreign governments (i.e., the property, commerce, and treaty-making clauses in the U.S. Constitution) (Bean 1983). The property clause needs further explanation: With parks and refuges, it is the property clause that grants the federal government the right to protect and manage its property—including fish and wildlife. In this case, to manage means to direct activities or control the property for the benefit of some interests.

MANAGEMENT PRACTICES

In General

The term "management" is a broad and somewhat nebulous term. Historically, to manage, means to use the hand (= *mano*) in the process of handling, as in handling (training) horses (e.g., Wikipedia.org). To manage means to control or to take charge. Management then, is the act or process of managing. It also refers to regulation, administration, treating, conducting, or dealing with people, places, or things, such as managing a household or a business. Drucker (1973) pointed out that it is the tasks to be performed in meeting a purpose or objective that really defines management, in which a human is in control of something for some purpose that would be beneficial to them or others. Thus, natural resources can, and has been managed for a variety of purposes, ranging from preservation (e.g., a wilderness) to total exploitation (e.g., a bombing range or parking lot).

Historically, humans have exploited nature to obtain some type of harvest. In ecological terms, we might say that early "management" was nothing more that human interference with ecosystem function with the goal of obtaining socioeconomic benefits. This is easily seen in the case of forest "management." Humans might wish to harvest some trees for fuel and lumber, thus replacing a forest with fields used for domestic livestock grazing, plots and wood for building, and cultivated land for farming. Remaining forest could be kept for future use, and, in association with clearings, it functions as wildlife habitat. However, logging requires roads for access and hauling logs. Building roads may cause serious impacts due to siltation, bridging streams, changing drainage patterns, and by providing easy access for human exploitation of game. As time goes on, land is further fragmented with the forested areas reduced in size so that interior species are lost, or the forest may be removed entirely. This is exploitation. It is not natural resource management.

Natural resource management is a special term that applies to how humans deal with nature to meet some need or acquire some benefit. The benefit would be lost if exploitation resulted in eradicating the resource yielding the benefit. Thus, successful management in this sense would need to be *sustainable*—a requirement hard to judge due to the complexity of nature. For instance, completion of a forest management plan would have to address the issues indicated in the previous paragraph. Such a management plan also might include different provisions for different parts of the area being managed. Some part of an area might be left as is ("preserved"), some areas could be allowed to grow naturally (or be "restored"), and others used for some new purpose ("reclaimed"). Thus, natural resource management plans can become extremely complex and tedious.

Fisheries

Management of a fishery includes three interacting components: organisms, habitats, and people. In addition, management of fisheries is complex, requiring knowledge and consideration of at least a dozen fields (Figure 29.1). Ross (1997) pointed out that the first management controls were initiated by government agencies when it became clear that there were too many people for a limited resource, especially freshwater and anadromous fishes. This was followed by a quick denial and blame-placing on predation and competition by other fishes from suckers to predators. After removal programs (usually the use of fish toxicants) for nongame fishes proved too difficult, too expensive, or failed, humans then took charge of nature by raising and replenishing target species, or in most cases by introducing sport fish from other systems, with little interest or comprehension about ecological effects.

Modern fisheries management has emerged to include three different techniques to provide benefits to humans while striving for sustainability: regulation of human exploitation, habitat management, and manipulation of target organisms. The techniques pertinent to fisheries management are discussed below.

Regulation

Perhaps the most familiar management tool, regulations limit the ability of humans to exploit fishes. It is usually the first response to a declining harvest. In general, regulations (or rules) are

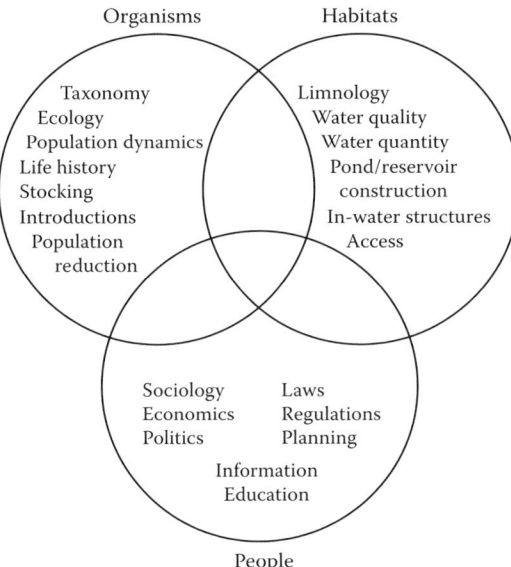

Figure 29.1 The basis of Fisheries management is depicted in three overlapping circles. (From Nielsen, L. A., *Inland Fisheries Management in North America*. 2nd edition, 3–30, American Fisheries Society, Bethesda, MD, 1999. With permission.)

codified into enforceable law by some government agency, usually a state wildlife department or federal fisheries authority. Regulations can be basic: you cannot fish unless you purchase a license, at the peril of a large fine or even jail time. Regulations also can be very detailed, such as setting bag limits (limit the number of individual fish of a given size for each species that can be taken), specify the season in which fishing is legal, remove areas from fishing activities by exclusion, or restrict the gear used for fishing. Regulations also can affect habitat by enforcing dredge and fill restrictions, monitoring and destroying invasive species, and other things. Regulations also have been recently passed that prohibit introduction of nonnative fishes.

Habitat Management

Fish habitat is dynamic and also susceptible to adverse effects associated with natural events and anthropogenic actions. Droughts and floods are prime examples of natural riverine events that can produce effects ranging from mild to catastrophic. Such effects, although anticipated, are hard enough to predict and assess. With the inclusion of the usual human related impacts such as increased grazing, timber cutting, clearing, farming, etc., it became clear that regulating take of fishes alone was not effective. Also needed is habitat, including feeding, spawning, rearing, nursery, and overwintering habitats for fish, all of which must be accessible and connected. Management of habitat may simply require fencing of riparian areas to restrict livestock use. It also can regulate habitat by modifying flow regimes. However, all fish have not been treated as equals in management programs. The funds for state fishery agencies are almost all obtained from the sales of fishing licenses and federal excise tax on fishing equipment. Thus, fish management (e.g., stocking programs and restoration projects of the twentieth century) were focused on sport fish management, with endangered species priorities appearing only at the end of that period due to federal funding available to states (under Section 6 of the Endangered Species Act of 1973).

Manipulation of Organisms

Overfishing of freshwater fishes in the latter part of the nineteenth century was generally met with stocking hatchery reared fishes to improve fishing harvest. In the eastern United States, construction of small impoundments referred to in general as "farm ponds" provided much opportunity for implementing controlled stocking of native fishes, while in the western United States, fishery managers generally believed that the fish fauna was not very diverse and could benefit from introducing more species from the east, which could use "vacant niches." Also, new aquatic systems such as reservoirs gave ample room for tinkering with nature by introducing game fish and even supplying nonnative fishes for a forage base. As a result of this, there was a preoccupation with fish stocking of hatchery fishes that began in the late nineteenth century and flourished for much of the twentieth century. One reason for all this interest in manipulating fish communities was the realization that it was easier to learn how to raise and stock fish than it was to understand population ecology and dynamics of natural communities. However, stocking of fishes was frequently unsuccessful for the same reason (e.g., see Ross 1997).

FISHERIES: PRACTICES AND PROBLEMS

A Scientific Approach

A "fishery" involves a mix of strange bedfellows. As indicated above, a fishery would include the fishes, fish habitat, and humans. Humans affected by fishing also are managed right along with the fish. Thus, management strives to manipulate aquatic systems and human activities to sustain socioeconomic benefits for people while sustaining the necessary environmental conditions to do so. One of my professors (Dr. Fred Barkalow) told our class that wildlife management is easy. You just manage the people and then the wildlife will manage themselves! Perhaps so, but dealing with all of the sociocultural, economic, and political components as well as ecological ones makes wildlife management exceedingly complex (e.g., see Krueger and Decker 1999). To compound the issue is the difficulty inherent in working with aquatic systems. In terrestrial systems you can visually observe the habitat, organisms, and results of the human actions (e.g., as in timber cutting). However, in aquatic systems, and especially large ones such as rivers, lakes, and oceans, you can only observe the habitat and changes in it with great difficulty. This problem has resulted in fish stock declines in oceans, due to socioeconomic and political pressures to exploit more and more marine fish stocks for food and economic value. Looking back in history, it is fortunate that the model for fisheries management incorporated a scientific approach that was way ahead of its time.

Development of a scientific basis for fisheries management was nicely discussed by Nielsen (1999). In the early twentieth century, fisheries were managed by catch statistics, primarily the concept of maximum sustained yield. However, it became clear that fisheries could not be sustained without having a scientific understanding of the organisms and their habitat. Development of the field of ecology provided this basis and fishery science and aquatic ecology grew in a mutualistic relationship. Nielsen (1999) listed three areas that were critical for effective fisheries management:

(1) Understanding the fishes and their invertebrate food sources
(2) Knowing the physicochemical attributes of waters
(3) Knowledge of the life histories and ecology of fishes

The need for such knowledge resulted in professional degrees in fisheries management for undergraduate and graduate programs in the United States, and government aid was furnished in the form of cooperative extension, including federal assistance to provide fishery research units at land-grant universities. As fishery programs grew, thousands of fishery managers were matriculated.

Armed with a scientific tool chest, fisheries biologists were ready to take on the daunting task of fisheries management. However, management of various aquatic ecosystems took different directions, required different skills, and produced different outcomes. In the following sections, we will compare management of freshwater, estuarine, and coastal (inshore), and offshore marine fisheries.

Freshwater Fisheries

Small freshwater impoundments and natural ponds have been used to raise fish since antiquity. Early humans managed irrigation works for impounding water and for raising fish. Presumably fish were native to a system being altered and were trapped in ponds or small lakes. They could have been used for food or other purposes and likely were captured by traps, spears, hooks, and nets, especially during dry periods when the pond levels were low. Later, constructed ponds were made to be drained, facilitating harvest, weed control, water quality, and so on. This type of exploitation altered species composition by introducing preferred species (food fish) usually larger predators. Then, as is the practice today, some ponds would have been constructed with the goal of raising fish in them exclusively. Right away, we see humans taking charge of nature on a small scale. Although the scale increases with technology and the construction of huge reservoirs, the ideas are the same. Streams are converted to "lakes" and stocked with introduced fishes. If the new habitat seems unproductive, more species are stocked, including forage fishes.

With government subsidies for pond construction, the number of farm ponds jumped from about 20,000 in 1936 to more than 2 million in 1965 (Swingle 1970). As a result, farm ponds are very common features of privately owned farms. Small impoundments and farm ponds may be constructed mostly for watering stock or irrigation, but they are almost always used to raise fish for personal use. This is largely a private endeavor in which management is mostly unregulated by fishery management agencies. The owner of the pond purchases fish from a private hatchery and manages them according to current practice. County agents or other extension persons are usually available to provide guidance to the owner with respect to the construction of the pond, fertilization (which can increase fish production 10-fold) (Swingle 1970), species stocked, harvest practices for species, and problems that may arise, such as nuisance aquatic plants, eutrophication, and summer or winterkill. Perhaps the most successful pond fish management is concerned with rearing largemouth bass and bluegill in warmwater situations. Maintenance of this fishery is dependent upon the owner diagnosing potential problems, and this can be done by simple seining and adherence to standard management practices (Table 29.1) (Swingle 1956, 1970; Flickinger et al. 1999). Cage culture of catfish and other species provides almost total control over pond culture, and caged fishes are usually fed pelleted food as in hatcheries. In that case, fishes are easily obtained and biological parameters such as growth and health can be monitored.

Pond culture also is an important part of the aquaculture in the United States, especially for rearing food fish (e.g., channel catfish and tilapia). However, ponds also are used for rearing baitfish and for rearing fishes for stocking. Such commercial interests are regulated by several agencies. Concerns are due to health issues and also there are potential problems due to wastes that drain out of the ponds, and escapement of nonnatives.

Larger impoundments and natural lakes are usually managed for multiple purposes and management becomes complex due to watershed features and public ownership or control. Recreational fisheries are almost always the main concern of fishery management in larger impoundments, and key factors influencing management are water temperature and nutrients. Even large reservoirs are susceptible to differences in productivity and affected by the eventual trophic downsurge and siltation (Figure 29.2). Thus, the environment for fishes changes over time, and fish abundance and community composition can be expected to change. Also different species may be introduced in

Table 29.1 Status of Bass and Bluegill Populations in Ponds, Determined by Seining

1. Seining produces no young largemouth bass and:
    a. Many young (recently hatched) bluegills but few bluegills of intermediate size *Conclusion*: Bass and bluegill populations in the pond are in temporary balance but bass are overcrowded by the bluegills.
    b. No young bluegills, but many of intermediate size. *Conclusion*: Pond unbalanced with overcrowded bluegills and too few bass to eat them.
    c. No young and few intermediate bluegills. Tadpoles, minnows, or crayfish present. *Conclusion*: Unbalanced with overcrowded bluegills and very few bass.
    d. No young and few intermediate bluegills. *Conclusion*: Unbalanced due to presence of fish competitive with bluegills.
    e. No young or intermediate-size bluegills. Presence of competitor species noted. *Conclusion*: Unbalanced, crowding by competitor.
    f. No recent hatch of young or presence of intermediate bluegills. *Conclusion*: No fish present; pond may be unsuitable for bass–bluegill reproduction.
2. Seining produces young largemouth bass and:
    a. Many young bluegills, but few of intermediate size. *Conclusion*: Balanced.
    b. Many young bluegills, very few or no intermediates. *Conclusion*: Balanced but slightly overcrowded by bass.
    c. No young or intermediate bluegills. *Conclusion*: Unbalanced due to lack of reproduction of bluegills.
    d. No young or intermediate bluegills. *Conclusion*: Temporarily balanced but possibly suffers from competition by another species.
    e. No recent hatch of bluegills, but many intermediate size fish. *Conclusion*: *Unbalanced. Similar to 1b.*

Source: Swingle, H. S., *N. Am. Wildl. Conf.*, 20, 298, 1956. Swingle, H. S., History of warmwater pond culture in the United States. Pages 95–105. In: N. G. Benson (Ed.). *A century of fisheries in North America*. Special Publication No. 7, American Fisheries Society, Washington, DC, 1970. Flickinger, S. A., et al., Small Impoundments. Pages 561–588. In: C. C. Kohler and W. A. Hubert (Eds.). *Inland fisheries management in North America, 2nd Ed.* American Fisheries Society, Bethesda, MD, 1999.

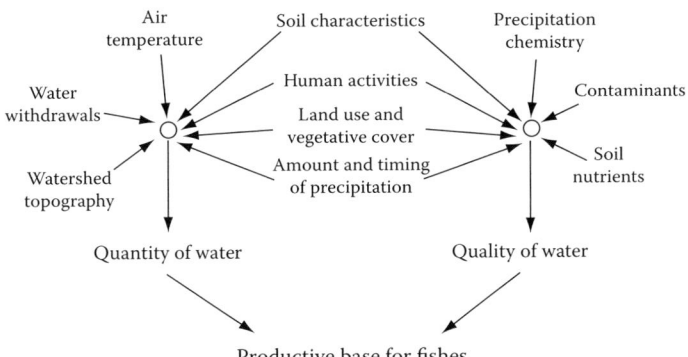

Figure 29.2 Water quantity and quality is influenced by watershed features, climate, and human activities. (From Hayes, D. B., et al., *Inland Fisheries Management in North America*. 2nd edition, 589–622, American Fisheries Society, Bethesda, MD, et al. 1999. With permission.)

response to a decrease in fishing, producing an increasingly complex and volatile system (Hayes et al. 1999). In an earlier case study (chapter 13), we read about management problems in the Laurentian Great Lakes. Management there, as elsewhere, has been difficult due to many physical changes in habitat, overfishing, and invasive species. On the other hand, some mitigation has occurred due to management action taken to control the numbers of parasitic sea lamprey, and the stocking of large Pacific salmon has been successful in the presence of abundant food (alewives) (e.g., see Jude and Leach 1999).

Streams and rivers are extremely complex, fluctuating systems that are more difficult to study and manage. In most instances, riverine systems were changed by the construction of reservoirs on almost all of the larger streams and rivers in the United States, and reservoir construction was viewed as a way to increase fishing opportunity. In the Pacific northwest, decline of salmon by habitat loss and stream blockage was met with increasing stocking programs to mitigate the loss. In addition, stream management included protection of riparian areas from channel degradation caused by grazing, timber cutting, and erosion. In many instances, previously channelized coldwater streams are being converted to a more natural habitat by restoring complex channels and providing within-channel structure. However, in warmwater streams, such as in the lower Missouri River, many natural streams have been converted to a system of interconnecting reservoirs. Recreational fishing there is big business to state game and fish agencies, whose primary concern is management of game fish. The development of fishery management practices that include native fish communities has been difficult, for many of the problems affecting native riverine fishes have been related to the game fish.

Management of stream fishes is presently concerned with maintenance or restoration of complex stream habitats (Table 29.2) (Orth and White 1999; Hunt 1993). Consequently, the complexity of large regulated streams and rivers dictates the need for an understanding of the functioning of a new ecosystem that is a complex combination of lake and river. Management also is divided over those who wish to catch predaceous gamefish in reservoirs and tailraces, and those who are interested in native stream fishes in remaining natural stream sections. Both interests require adequate supplies of water, and stream flow management is a major issue, hotly contested in courts. Much effort has gone into determining the stream flow needs of native, and especially federally listed fishes, so that management programs can be developed. However, in almost all developed riverine systems, it is difficult to partition out historic physical habitat needs of native fishes in the presence of introduced ones. Finally, while there are restoration projects on smaller streams, large rivers are used for multiple purposes, and once developed, restoration is difficult or impossible.

Table 29.2 Stream Degradation and Effects on Fish Due to Changed Conditions

Habitat	Channelized streams are lacking in habitat heterogeneity, usually due to a reduction in meandering sections, pool-riffle sequences, and loss of instream cover, resulting in ecosystem simplification. Also, forest removal can expose streams to solar radiation, change the temperature regimen, and increase harsh conditions. Loss of forest litter can alter the food web.
Flow regimen	Altered flow regimes due to peaking operations or base-flow alteration can disrupt fish life history patterns, interfere with resting and escape cover, reproduction, and rearing, and favor invasive species over natives.
Water quality	Increases or decreases in turbidity, suspended and bed load can reduce cover or make it difficult to forage. Changes in temperature regimen can alter fish responses to environmental signals (zeitgebers). Increased salts and contaminants can weaken or kill fishes and their food organisms.
Sediment	Whether from logging, road construction, or livestock grazing, increased erosion from stream banks and basins can result in smothering vegetation and stream invertebrates, and filling pools used by overwintering fishes.
Blockage	Loss of stream passage due to blockage interrupts fish movements and migrations. Spawning migrations can be destroyed in this way.

Note: See Orth and White (1999) for more details.

Estuarine Fisheries

As aquatic systems get bigger and more complex in structure and function, they are more difficult to understand and to effectively manage. Estuaries in general are harsh environments due to the different forces that act upon them, but they also are nutrient traps and their relatively shallow habitats usually have vegetative cover and ample food that make them extremely productive and desirable nursery and feeding areas for many fishes. For those reasons, larger estuaries usually support economically important commercial and recreational fisheries.

Unfortunately, estuarine fisheries have universally declined due mainly to upstream dams and diversions, human exploitation and estuarine habitat degradation (Herke and Rogers 1999). Thus, management efforts for estuaries have included regulating the take and focusing on protection of remaining habitat by refusing development projects that destroy or degrade structure, such as dredge and fill of marshes.

Very large estuaries have been most difficult to manage due to divided government controls, effects of increasing human population, and in some cases unpredictable changes in the ecosystem. Chesapeake Bay, the largest tidal estuary in the United States, includes portions of six states. Its productive waters have been greatly affected by a variety of changes (to be more fully discussed in the next chapter). Early signs of system changes were observed at the end of the nineteenth century with increasing plankton blooms, dead zones, and fish kills.

Management of Chesapeake Bay and other large estuaries includes laws restricting development and requiring mitigation of habitat destruction. The concern is not only on direct destruction such as bulkheads, draining and filling, and dredging new channels for navigation, but also includes indirect effects of upstream development and dams, and pollution due to sewage and chemicals. Restoration of estuarine habitat is extremely costly and not well understood. Consequently, agencies attempt to reduce and prevent more human-induced changes, while at the same time collecting information on abundance, distribution, and needs of fish and shellfish populations. Many other estuaries face similar problems that can cause such system-wide disruption in basic structure and function.

Marine Coastal Fisheries

Management of smaller freshwater systems placed humans at least in partial control of outcomes. On the contrary, marine systems are vast and management problems are compounded by many factors, including basic system functions as well as socioeconomic and political influences on dividing up and managing large areas of valuable marine resources. Of primary concern regarding management is a basic question of ownership: Who "owns" the resource, or better put, who has the right to manage it?

Marine systems can be divided into coastal (or inshore) and offshore zones for management purposes. Historically, coastal fisheries management was restricted to the appropriate state agency as per the "Public Trust Doctrine" in which wildlife and fishery ownership is held in public trust by the states, subject to rights surrendered in the constitution to the federal government (reviewed by Bean 1983). In other words, private ownership of marine resources is prohibited, including fishes on submerged lands and navigable waters of a coastal state's territorial water (extending seaward 3 mi). State agencies historically had the right to manage coastal resources, while the United States retained the right to regulate fisheries in "fish conservation zones," which extend offshore of the 3-mile mark (this will change, see below).

As might be imagined, historic management of the coastal zone was not uniform in any sense as each coastal state applied their own priorities based on multiple issues. Also, not all states had the same ability to regulate their resources due to a lack of funds and enforcement capability. Management was also complicated by right of land ownership resulting in trespass restrictions and seasonal (climatic) migration of some stocks along coastlines.

Principal inshore fisheries of the United States include food fin-fisheries, industrial fisheries, shellfisheries, and recreational fisheries. Historically, only a dozen fish groups accounted for 80% of landings (McHugh 1980). The industrial catch for fish oil and meal includes two species of menhaden (*Brevoortia tyrannus* and *B. patronus*) on the east and gulf coasts, and sardines (*S. sagax*) and anchovies (*E. mordax*) on the west coast. These fishes have made up the bulk of commercial landings. Recreational fisheries also are important components of U.S. coastal fisheries. This fishery exceeded the commercial catch for some species of food fishes, and it includes predominately bluefish, cod, two species of flounder, striped bass, and weakfish. Many other fishes also are captured.

Management of coastal fisheries has almost exclusively been through regulation of one sort or another, including setting seasons, restrictions on gear and landings, bag limits, fish size restrictions, geographic closures, etc. Some management activities also include restrictions on permits for coastal development projects such as bulkheads, small dams, dredge and fill, etc. Habitat improvement also has occurred by constructing artificial reefs used for recreational fishing and closure of some areas to all fishing. In general, it appears that state management alone through the 1970s was not very effective in preventing fluctuations and declines of important fisheries (McHugh 1980).

Offshore Marine Fisheries

For centuries, it was generally agreed that all nations had the right in international law to fish open oceans outside of territorial waters (Bean 1983). This fact was least partially related to the notion that oceanic fisheries were inexhaustible and that nations lacked the ability to exert much harm on fish stocks. However, by the twentieth century, advances in technology and sheer force of numbers made it obvious that offshore fisheries indeed could be affected by overfishing. In fact, yields from offshore commercial fisheries predominately are affected by cycles of "boom and bust" (e.g., Jennings et al. 2001).

The goal of marine fisheries management is to obtain a high yield while avoiding overfishing, which leads to decline and eventual collapse of a fishery. Collapse results in economic hardships to the fishing industry, which is often the prime concern of regulators. However, we are learning that disruption of ecosystems can occur with loss of habitat and trophic changes, which can result in unpredictable, often tragic outcomes (Hutchings and Reynolds 2004; Levin et al. 2006). Such outcomes can suppress fish populations for decades and more, and in case of some species, recovery may take human lifetimes to occur if at all. In view of this, it is interesting to note that most commercial fisheries have been subjected to so much fishing pressure that (according to fish recruitment models) they are fully exploited or overfished. Clearly, it would be in the best interests of coastal nations to avoid overfishing of offshore stocks, but no nation had that exclusive authority until the United States acted immediately after World War II.

Concerned with increasing numbers of foreign fishing vessels off the coasts of the United States, and the capacity of huge "mother" ships to process and store vast catches of satellite vessels without the need to return to port, President Truman issued Presidential Proclamation 2667 (59 Stat 884), known as the "Truman Proclamation" on September 28, 1945. The United States proclaimed that in addition to territorial seas, it would be appropriate for the United States to establish conservation zones contiguous to its coasts and to regulate and control all fishery activity in them. Truman's intent was to negotiate such zones with affected nations, but by 1958, it was an accepted part of international law and the International Conference of Law of the Sea: all nations unilaterally had this right. However, the conference provided no absolute boundaries. At that time, most nations had a 3-mile territorial sea and a 12-mile conservation zone (Bean 1983). Outside of this 12-mile zone, foreign fleets continued to build up and worry the U.S. fishing industry, which pressured the government for more action.

In 1973–1974, the international conference met again to consider the issue, and at that time most nations agreed on a 12-mile territorial and a 200-mile conservation zone. However, the conference reached no specific agreement. Finally, tired of negotiations and concerned with the impact of foreign fleets, the U.S. Congress passed the Fisheries Conservation and Management Act of 1976 (FCMA; also called the Magnuson–Stevens Act) that established a 197-mile conservation zone offshore of a 3-mile territorial sea. This act also set up a management program for these waters. It is noted that 200 miles is large enough to effectively manage fishing of the rich continental shelf areas of the eastern United States. This was not possible earlier due to lack of control over other nations.

Fisheries management under the FCMA would be different. The U.S. Department of Commerce was given the lead in the development and implementation of fishery management plans that would be completed by eight regional councils representing all coastal areas. Each council would be composed of the appropriate Regional Director of the National Marine Fisheries Service (NMFS), a representative from each state, and appointees by the Secretary of Commerce from a list provided by each regional state governor. Broad management measures would be approved and implemented by the Secretary of Commerce. Each plan would reflect a new policy of implementing optimum sustained yield (OSY) instead of maximum sustainable yield (MSY). OSY includes social and economic factors plus condition of the stocks as the criteria for fisheries harvest. Thus, with OSY, stocks might be underfished (or overfished) relative to MSY.

It is instructional to consider the nature of marine fish management. According to Jennings et al. (2001), marine fisheries are managed because uncontrolled fishing predictably results in the disaster of overfishing, which can result in increased costs and declining wages, or a fishery collapse can result in unemployment. Human effects also can be associated with slow or little recovery of the stocks, alteration of species composition, and loss of habitat. But it is even more complex. Management is a regulator of four overlapping and conflicting objectives: to derive biological, social, economic, or political benefits from a fishery, all accomplished while minimizing costs. These benefits are obtained from management by control of catch (i.e., number of fish harvested) and effort (number of people, boats, type of gear), plus technical measures (targeting a specific size and sex of fish with gear modifications).

Under the new FCMA, the councils were to replace foreign fishing fleets with U.S. vessels and to provide a system that identifies maximum sustainable levels of fish harvest for various stocks. However, there seemed to be little improvement in fish stocks in the years after the act was passed, and 57 of 80 major stocks were considered fully exploited or overfished in 1994 (Jennings et al. 2001). Although there was progress on increasing the U.S. fishing effort, no system for defining sustainability emerged. The U.S. Congress amended the act in 1996, specifying three goals: to prevent overfishing and rebuild declining stocks, to avoid and minimize bycatch, and to identify and protect "essential fish habitat" (i.e., the environmental features needed to support all life stages of target fishes). In performance of the goals above, councils were to use the "best scientific data available."

Unfortunately, decisions made under the FCMA have tended to favor short-term economic gain over long-term sustainability, presumably due to domination of regional councils by advocates for the fishing industry, which allowed user groups to control their own economics (Reiser et al. 2005). Progress toward meeting goals also were affected by continued litigation: Fishing interests delayed management plans in court, while conservation groups litigated NMFS to attend to its duties. In 2002, at least 41% of federally managed fish stocks were either overfished, or at levels too low to support sustainability (Reiser et al. 2005). On a happier note, the act has resulted in the Fishery Stock Sustainability Index (FSSI), a performance index developed by NMFS to assess the sustainability of 230 fish stocks at the national and regional level (reviewed by Jacobs and Jepson 2009). Another positive step is the concept of *essential fish habitat*, which government agencies must identify and consider in all programs of managed fishes. *Essential fish habitat* was the subject of a Sea Grant symposium funded by NMFS and published by the American Fisheries Society to further develop the concept (i.e., Benaka 1999 and authors therein).

CONCEPTS OF SUSTAINABILITY

Overfishing has been common throughout the commercial fishing industry. Also, virtually all fisheries have significant amounts of bycatch and some gear can destroy habitats. These attributes certainly make it difficult and in some cases perhaps impossible to support a sustainable fishery. However, here is a dilemma—the major tenant of the FCMA is the ability to manage fisheries in a sustainable way. The basic concept of sustainability involves providing long-term benefits to humans. This seems clear enough, but there are some interesting fine points. Hilborn (2005) considered the sustainability concept with application to commercial fisheries and evaluated three widely used definitions: long-term constant yield, intergenerational yield, and a system producing biological, social, and economic benefits. The discussion is encapsulated here:

- *Long-term constant yield*—is an inappropriate definition for sustaining fisheries because fishery yields are not constant. Natural fluctuations in fish stocks are difficult to predict and humans take the largest possible harvest, which makes yield fluctuations even greater.
- *Intergenerational equity*—this definition accepts fluctuations and overfishing if a stock can rebuild and continue to provide benefits within a human generation. Sustainability is judged patent even if yields vary over generational time scales. Presumably, a fishery would have time to recover from adverse impacts during generational time.
- *Maintaining a biological, social, and economic system*—this includes long-term exploitation of hundreds to thousands of years in which marine systems undergo changes induced by human and natural forces. Thus, transformations in ecosystems and even loss of some species (as long as they are not critical) would be acceptable as long as the system in question produced long-term benefits. This idea shows a potential conflict between social sustainability and the need to maintain biodiversity. It is pointed out that long-term human activity of all sorts historically has resulted in reducing biodiversity.

Many fisheries scientists believe that dramatic changes need to be applied to make fisheries management sustainable and quickly. For example, Helfman (2007) provided four points required to manage fisheries in a sustainable way, which I have paraphrased: First, the single species approach needs to be replaced by multiple species or ecosystem based management. Second, large scale physical factors such as climate cycles and so on affect fish stocks in a density-independent manner and need to be considered. Third, there is a need to shift away from common resource use to "resource ownership" (by limited entry) to replace competition-driven, short-term economic gain with long-term cooperation and yield. Fourth, the ratchet (incremental tightening) effect of increasing effort despite decreasing yield needs to be avoided.

Humans fish marine environments mostly for money, and there is an incentive to maximize net economic return due to the chance of striking it rich quickly (e.g., as in the highly publicized Alaskan crab fishery). However, net economic gain also can occur by reducing the fishing effort, but this will require incentives and understanding between agencies and users. Further related to sustainability is the need to avoid habitat destruction, the ability of industry to tolerate fluctuations in allowed yield, the use of nondestructive fishing gear, and ways to reduce bykill and bycatch.

Overfishing remains a serious issue and it must be stopped. However, the issue is complicated. There are many reasons for overfishing, and early cases of recruitment overfishing was allowed to continue in North Sea herring stocks for decades because of the notion that recruitment was independent of population abundance. The notion became scientific "dogma" and thus resistant to change even in the face of compelling evidence (Sinclair 2009). It took two decades and several additional fishery collapses before recruitment overfishing was accepted as the explanation. In marine fisheries, where there is no ownership and intense competition, the concept of the "tragedy of the commons" (Hardin 1968) no doubt applies to many situations of simple greed. However, in some situations, there also appears to be disbelief and denial mixed with hope that continued fishing will produce a lucky episode. Also, fishermen think they know more about fish than the scientists, and some situations have reinforced that notion. Some fish stocks that are declining still exhibit a tendency to congregate in large schools, giving the impression of more abundance than is warranted. Such congregations are picked up on sonar gear and efficiently exploited, for a while producing large catches. Finally, there is the problem of social and political pressures on scientists and managers, who feel that only with indisputable evidence can a fishery be closed. Unfortunately, by the time sufficient evidence is obtained, if ever, the fishery is collapsing.

ECOSYSTEM APPROACH

Management of natural resources appears to be moving away from demographics toward management of whole ecosystems with the goal of sustaining ecosystem structure and function while obtaining benefits to humans. In this case, an ecosystem approach might be used in a management unit, such as a watershed, lake, or estuary. The ecosystem approach also is considered a necessary part of fishery restoration projects (e.g., see Castagna 2010), which, according to Crowder (2005), "isn't rocket science, it's harder than that! But it is possible." To date, restoration has occurred in some overfished stocks like Pacific sardine and Peruvian anchoveta just by reducing fishing pressure. However, in fisheries that have suffered community to system-wide damage due to bycatch and habitat loss, such as the northwest Atlantic groundfishery, the prospects remain dim after decades.

Constituting a paradigm shift from single species management, ecosystem management is a recent "buzzword" for which obtaining a definition that fits all cases has been elusive. The concept is structured around three core and interacting elements: ecology, economics, and sociopolitical forces (i.e., also known as humanism). A problem arises because the term ecosystem management has a different focus and a different meaning depending upon which of many user groups is providing the definition. Two ends of a spectrum have emerged to shed some light on the problem (Stanley 1995): a *biocentric view*, with the goal of protecting long-term ecosystem integrity, and an *anthropocentric view*, with the goal of manipulating ecosystems to obtain human needs, with a lower priority given to maintaining ecosystem integrity. Another way to look at these extremes considers whether ecosystem management is "nonconsumptive use" to meet the goals of conservation biology (e.g., to protect or restore natural ecosystems and biodiversity) or "consumptive use" of commercial fisheries that changes natural ecosystems and may result in a loss of biodiversity. In both cases, the benefits to humans can be considered "sustainable" using definitions presented earlier. Notice that the focus in the first case is the system and its biodiversity, which are sustained to benefit humans, while in the second case the focus is sustained harvest even if ecosystem scale changes occur. The approach in both instances is based on the notion that humans have the knowledge and understand the composition, structure, and function of ecosystems well enough to obtain long-term benefits.

Meffe et al. (1997) provided an excellent discussion of the ecosystem approach, contrasting traditional management with ecosystem management. Those authors pointed out that ecosystems are dynamic, our knowledge of them is incomplete, and thus, management results are uncertain and full of surprises. As testimony to this, Pine et al. (2009) reported examples of management manipulations with fish populations in which the responses to treatments were counterintuitive: they were either much smaller or in the opposite direction than predicted by experience and population models. Of note were behavioral responses and juveniles adaptations that restructured ecosystems in even some simple models. Those authors concluded that our ability to predict ecosystem response to management is often not very good, no matter what the complexity is. The results of this work can be used to improve models, but points out that our knowledge of fish ecology is not very good, especially in understanding behavioral responses and early life history stages. The results also suggest that the ecosystem approach should incorporate long-term *adaptive management* with the goal of learning as you go, freedom to take risks, intensive monitoring, and the ability to make changes in direction. Adaptive management is a scientific approach that treats management actions like experiments. In this respect, Pine et al. (2009) viewed management failures as opportunities to learn and do a better job.

Commercial fisheries long depended on single-species models for calculating allowable yields (harvest). Those models were very limited in predictive capability and are being replaced by more complex ones, such as multispecies models that consider biological (competition and predation) and technical (fishing for more than one species) interactions in a single fishery. These multispecies models have been around for a while and they are data intensive and costly. As an example, "virtual population analysis" or VPA is an age-structured model used to obtain stock sizes and mortality rates based upon cohort survivors, which provide a minimum estimate of survival the previous year. Survival the rest of the year would depend on fishing and natural causes that can be estimated. In multiple species VPA (MSVPA), the natural causes are further partitioned into predation by other fish in the model plus other sources of mortality (Jennings et al. 2001). As an example, much effort has been expended in developing MSVPA for use in management of North Sea fisheries (e.g., Magnússon 1995). A well-known effort was coordinated by the International Council for Exploration of the Sea. Working with groundfish surveys, data on predator prey relationships were obtained from many thousands (>54,000) of fish stomachs (Jennings et al. 2001). Using MSVPA, workers derived total biomass for each of 10 fishes subjected to predation from 1974 to 1995.

Ecosystem models already have been designed to quantify energy flow in exploring trophic relationships (computer software such as Ecopath) and how climate or change in fishing can affect ecosystems (Ecosim), and spatial dynamics in predation, feeding, and fishing (Ecospace). Using another piece of software (Atlantis), Kaplar and Levin (2009) found that fishing can alter communities, resulting in replacement of long-lived species by more short-lived ones, and these results are similar to that of an earlier investigation by Levin et al. (2006). No doubt even greater sophistication will be incorporated in future models. The challenge that lies ahead for managers is being able to interpret these models as to their importance, relevancy, and accuracy, and to determine how to use them in making fishery management decisions about vague terms such as ecosystems, sustainability, ecological health, essential fish habitat, and so on. Not only do we need to do a better job of understanding ecosystems and fish ecology, but we need to be able to design more complex models.

I believe that we must consider ecosystem structure and function to the best of our ability before we commit to its potential disruption by human actions. However, some believe that ecosystems are too complex for adequate understanding and that management is presumptuous. This belief was evaluated by Stanley (1995) by comparing the assumptions of ecosystem management (three attributed to Cains (1990) and the fourth to Stanley 1995) with similar assumptions underlying humanist beliefs (Ehrenfeld 1981). Humanism has a central tenet that the natural world exists for the benefit of humans, who in part, have a great faith in the power of human reason to solve problems. The humanist assumptions are "(1) all problems are soluble by humans, (2) many problems are soluble by technology, (3) problems that are not soluble by technology alone have solutions in the social world, and (4) humans will apply themselves and work together for a solution before it is too late" (Stanley 1995). The comparison showed that the assumptions of ecosystem management were so similar to that of humanism, that ecosystem management appeared to be just another example of the "arrogance of humanism." Stanley (1995) found that, as currently stated, "ecosystem management is a magical theory that promises the impossible." This viewpoint also was given by Ludwig et al. (1993) who recommended a cautious approach to deal with the uncertainty that we must face in resource management, and that we cannot expect science and technology to solve every resource problem (known as *technoarrogance*). If we pay attention to these concerns, there is a potential for maintaining ecosystem integrity with human use. Unfortunately, increasing human populations can be expected to disrupt ecosystems in the long term (Stanley 1995) and there remains a great need to understand the total effects of fishing. For example, we now understand that fishery-induced trophic cascades can alter ecosystems so much that a new structure and state of quasi-equilibrium can emerge (e.g., Salomon et al. 2010).

Ecosystem-based management seems a logical shift from typical natural resource management to the management of natural systems—which in essence is a change from preoccupation with demographics to a viewpoint of sustaining ecosystems. It has long been argued (e.g., Leopold 1949) that ecosystem management will serve to protect the whole, thus saving the parts. Alpert (1995) pointed out the problems with that logic in the present world: The existing systems have been subjected to many changes for the worse, including habitat loss, alteration, and fragmentation; invasive species; and decreasing biodiversity. It is these "worn systems" that continue to be affected by human actions that we have to manage—not pristine ones. Also, management changes in these systems are further constrained by the socioeconomic policies that changed them in the first place. To date our successes in aquatic habitats have been limited.

CASE STUDY: FISH SALVAGE AT TRACY

Operation and Change

The Tracy Fish Collection Facility (TFCF), constructed in 1951 by the U.S. Bureau of Reclamation (USBR), is located near the town of Tracy, CA. Part of the Central Valley Project, the facility was designed to "mitigate" (to reduce the loss) project impacts to fish by rescuing fish from water that was to be diverted from the Sacramento–San Joaquin River Delta (henceforth referred to as Delta) and used for irrigation, industrial, and municipal use. The facility was innovative, a great idea, and the first such facility of its kind to use a system of louvers to separate and collect fish (see Figure 29.3).

The fish targeted for salvage in 1951 were juvenile chinook salmon (native) and striped bass (introduced), both valuable sport fish. However, other species would be rescued as well. In practice, the facility would capture fish entering the Delta–Mendota Canal via the Tracy water pumping plant and move them back to the Delta downstream of the pumping facilities. The facility tested various types of screening devices and by 1955 selected a louver, bypass, and holding tank system (F), considered the best available technology at that time, and tested to prove efficient in separating the target fishes, except when fouled by debris. The vertical louvers would serve to orient fish from the flowing water and direct them into bypasses which connected with large holding tanks (Figure 29.4). Collected fish would be transported via fish tanker trucks to release sites in the Delta, which are downstream of the pumps. All of the information presented thus far and much more is given by USBR (2001), and on the TFCF Web site.

Figure 29.3 Tracy Fish Collection Facility (a) and bypass louvers (b). (Courtesy of USBR.)

I visited the TFCF in 1995 to observe the physical plant and some of the fish studies in progress. In a few days, I was able to see the entire operation of the facility from diverting fish with the louvers into the bypass system, transfer of the captured fish into holding tanks, counting the fish by species, transferring the fish into the transport trucks, accompanying the fish truck to the release site in the Delta, and releasing them back into the Delta. I also was able to participate in capturing predaceous fish that were living in the louver–bypass system.

At first, operation of the passive louver system was simple, means were provided to keep them clean and efficient, and the years began to pass. In operation for more than 50 years, the facility has

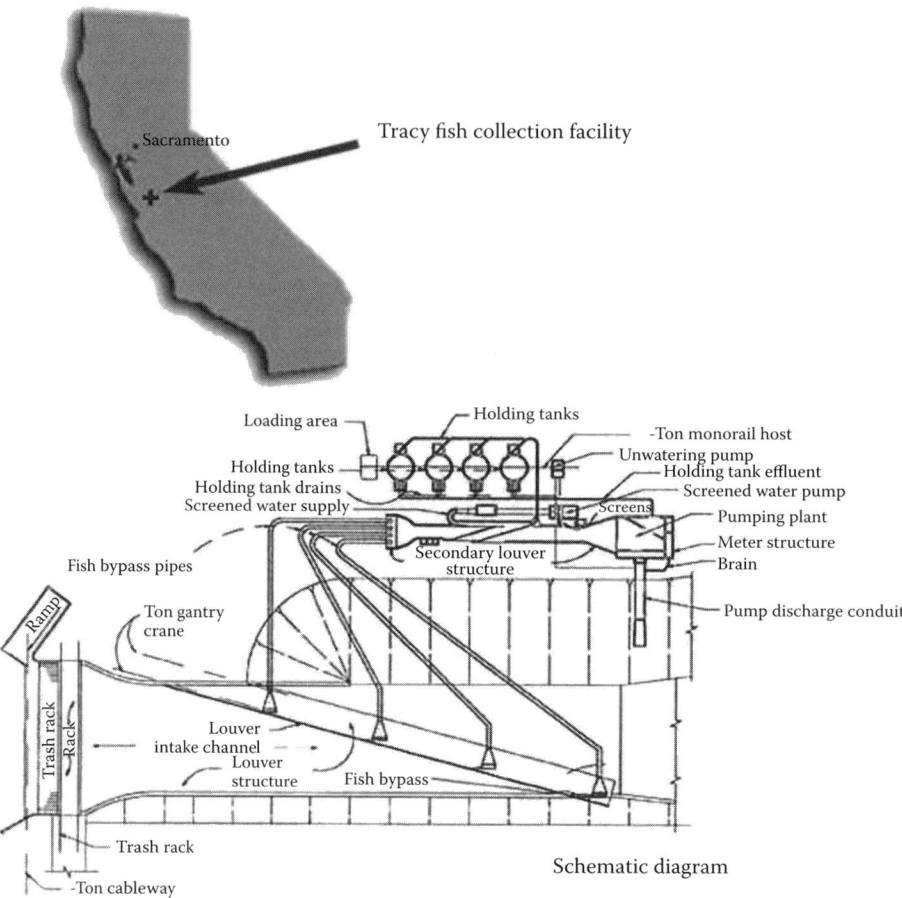

Figure 29.4 Collected fish are placed in holding tanks and then transferred to trucks. (Courtesy of USBR.)

endured many changes, from environmental to sociopolitical, which have caused problems. It also serves as a good lesson to us and shows why adaptive management practices are so desirable. I list six of these changes below, which all happened after the 1950s:

(1) Federal and state endangered and other special status species designations were passed, protecting other fish species, including delta smelt *Hypomesus transpacificus*, and splittail *Pogonichthys macrolepidotus*, in addition to chinook salmon already mentioned.
(2) Many nonnative organisms have been observed at TFCF. Fouling by these exotics (e.g., water hyacinth) (Figure 29.5) has been so adverse that the facility has at times been stopped. Also, masses of pest organisms (such as mitten crabs) have overcrowded and no doubt stressed fish in holding tanks.
(3) Predaceous fishes, including striped bass, have adapted to the facility, learned to live and maneuver there, and consume fish.
(4) Increased pumping rates due to water development have affected fish salvage in complex ways.
(5) Early life history stages (i.e., from eggs to larvae) of several protected fishes have been entrained in pump water and flushed through the facility into water export canals (Figure 29.6).
(6) Predaceous fishes and birds congregate at release sites for the salvaged fish and eat them.

Challenges such as the above and more have resulted in a response by the TFCF and improvement programs led by scientists and engineers at USBR's Technical Service Center, Denver,

EXPLOITATION AND FISHERIES MANAGEMENT

Figure 29.5 An infestation of water hyacinth is one of many management issues.

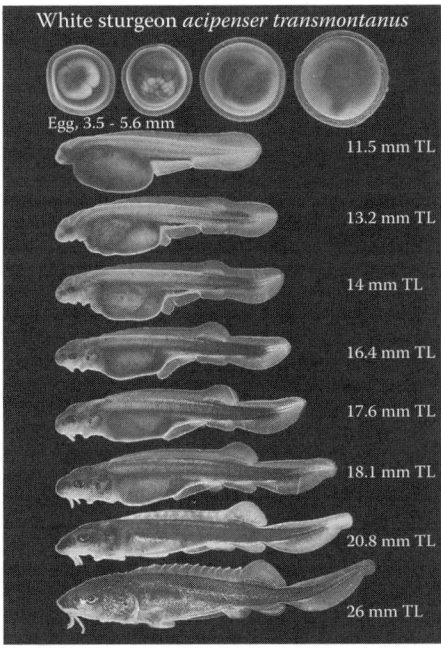

Figure 29.6 Entrainment of early life stages of fishes such as sturgeons remains a problem. Courtesy of the U.S. Bureau of Reclamation, Tracy Fish Collection Facility, Technical Services.

Colorado, have been implemented. In brief these include predator removal programs, better louver efficiencies, holding tank improvements, fish egg and larval entrainment studies, development of fish friendly pumping devices, mitten crab removal system, fish hauling improvements, and etc. These improvements have generated dozens of technical papers that I have used, in part, to further discuss changes made and increasing demands at the facility.

Demands on Operation

Endangered Species Acts

Federal and state endangered species designations have resulted in an increased effort to comply with the need to protect and restore listed fishes. A total of 53 fish taxa have been handled at TFCF, including taxa listed as threatened or endangered (TE). Fishes on the TE list included six runs (distinct populations) of chinook salmon and steelhead with critical habitat designated for three of these, delta smelt with critical habitat, and splittail (USBR 2001). Under Section 7 of the Endangered Species Act of 1973, all federal agencies are precluded from any activity or action that is likely to jeopardize the continued existence of a listed species, or to adversely modify or destroy critical habitat. Also, Section 9 of the act prohibits take of any listed species. In order to decide what effect the operation of the Jones Pumping Plant at TFCF would have on a listed species the USBR had to formally consult with the U.S. Fish and Wildlife Service. The service issued a Biological Opinion (December 15, 2008) that covered continued long-term operation of the Central Valley Project, and which affected the salvage operation of the TFCF. In brief, most of the concern was directed toward critical habitat and take of delta smelt, specifically larval and juvenile entrainment and rearing habitat, plus migration of adults. In general, there also were concerns about stress or mortality by predators or induced by handling. Monitoring presence of listed fishes was required and frequent reporting was also needed. However, the Tracy plant was not designed to efficiently capture small sizes (i.e., larvae and juveniles) or small species of fishes, thus there is a need for more study and refinement of facilities (USBR 2001).

Invasion of Nonnative Species

More than 100 nonnative species have been introduced in the Delta, of which about half have been observed at TFCF, such as the Chinese mitten crab. The catadromous mitten crab was first noticed at the TFCF in 1992. Thereafter their numbers increased exponentially, reaching 750,000 in 1996–1998 and over one million in 1999 (White et al. 2000). Such numbers were a strain on fish salvage operations and staff. The crabs produced high fish mortalities in the holding tanks and USBR had to install an expensive traveling belt screen. The screen reduced the number of crabs through the facility by about 90%. Since then numbers of the migrating crabs have been much smaller.

Smart Predaceous Fish

Predator response to the TFCF over time targeted the secondary channel where smaller fish congregate and are susceptible to capture. In 1991 and 1992, a removal program yielded 6549 striped bass (1805 lb) and 4800 white catfish (346 lb) Liston et al. 1994). This required 65 draw downs to catch and remove the predators. At present, there is no easy way to remove or discourage the predators, but sound, strobe light, CO_2, and electricity have been used as deterrents.

Increasing Demand for Pumping Water

Increased water diversions and demand has resulted in higher flows, which affect the efficiency of the louvers and fish bypasses of the TFCF. Some modifications have been made to the current facility, but to cope with the need for more experimentation and capacity, plans were made for constructing a new collection and test facility near the old one (USBR 2001). Although an environmental assessment has been made, no action has been taken in view of short budgets.

Entrainment of Early Life Stages

Studies at TFCF have determined that striped bass are spawning near the trashracks at the entrance to the facility and their eggs are immediately entrained and moved downstream, as were larvae of other species. More than 150,000 eggs and larvae were entrained from February 6 to June 6, 1994 (Sigfried et al. 2000). So far, there has been no solution to the problem.

Predation at Release Sites

A major objective of the TFCF is the transport and release of captured fishes back to the Delta. At present there are four release sites used where the tanker trucks back into a dock and affix their pump hose to a metal cylinder that allows the fish to be returned to the river, downstream and away from the influence of the pumps. USBR has determined that predators appear to recognize the release sites and congregate to feed on the released fish. The predators consist of striped bass, Sacramento pikeminnow, largemouth bass, black crappie, and other sunfishes (D. Portz, personal communication 2009). It is unknown just how many fish are being eaten at the release sites but it is substantial, especially when avian predation is included. Even though pipes are underwater, cormorants and gulls are able to locate and consume a great many of the released fish.

Stress

Holding and transporting wild fish is stressful and many fish are either too stressed to be able to avoid predation or they are stressed to the point of inevitable death. Stress is evidently increased by holding predators and prey in the same transport tanks, presumably due to species recognition by odor, behavior, morphology, or visual means. Also, there can be a considerable amount of debris and vegetation present. Pumping fish from one place to another also is stressful, and there is the prospect of replacing conventional pumps with those based on the Archimedes screw principle. USBR now has an in-house research physiologist involved in studies to reduce the stress factor.

FUTURE OF THE FACILITY

What does the future hold? The TFCF was a good idea and has proven its worth for half a century, during which the environment and political objectives have changed, and fisheries management has become more sophisticated. Consequently, this facility provides a good example of unanticipated problems that can occur in implementing a management strategy over time. In most cases, we have no such example, primarily due to the difficulty in working in aquatic habitats. In this case, a need to incorporate adaptive management is logical. However, adaptive management requires that problems be addressed as they emerge. Although this has been done to some extent, not all problems can be addressed because new facilities would be needed for an experimental approach. Such facilities have been proposed, but are doubtful with present budgetary restrictions. For the foreseeable future, USBR will have to continue working with the restrictions of the old facility or conduct experiments elsewhere as appropriate (NB: they already are doing this at their research facility in Denver).

SUMMARY

Applied ecology addresses the effects of humans on ecosystems. Human impacts can cause a progression of changes in natural systems. Thus ecosystems can be classified as pristine, wilderness, natural, human influenced, human dominated, and human disrupted, depending on the degree of anthropogenic change. Humans effects may be direct or indirect, and human use may be consumptive or nonconsumptive. Humans have a long history of exploitation, and starting with the American colonies, many laws regulating human use have been enacted. Westward expansion in the nineteenth century was devastating on wildlife with extinctions of passenger pigeon and near extinction of vast herds of bison and antelope. A new national conservation ethic of wise use emerged with the aid of President Theodore Roosevelt in 1901, which included numerous parks and wildlife refuges. Management of those lands sparked the development of natural resource management, which ranged from preservation to total exploitation. Fisheries management was guided by a scientific approach that stressed the need to understand the fishes and their food sources, know the physicochemical attributes of habitat and water, and know the life histories and ecology of fishes. Fish management includes attempts to control fish populations in freshwater systems (ponds, lakes, and streams), estuaries, inshore marine, and offshore marine zones. Sustainability is an integral part of marine fisheries management under the Fisheries Management and Conservation Act of 1976; however, sociopolitical influence resulted in overfishing most of those stocks. Ecosystem management offers a promise for improvement, but our knowledge of ecosystem function is incomplete, and future efforts will require an adaptive management approach to learn and formulate new management practices and tools, such as multispecies or ecosystem models. Unfortunately, we do not understand everything about ecosystems and how they work. Ignorance notwithstanding, humanism has resulted in technoarrogance, in which people believe they can identify and solve every problem that comes up. The case study of fish salvage at the TFCF, a study of unanticipated problems, exposes the folly of technoarrogance.

Further reading: Norse and Crowder 2005.

CHAPTER 30

Conservation of Fishes I: Crisis and a Response

INTRODUCTION

It should be clear from the preceding chapters that humans have had a big impact on natural resources, especially fishes and aquatic systems. In response, an ecosystem approach to management has emerged in an attempt to save the parts needed for ecosystem function. In this chapter, we discuss what those parts are, why they are important, and integrate some concepts of species conservation and recovery with ecological-based management. The plans and actions for conservation and recovery of species are clearly related to the ideals of ecological (ecosystem) management as presented in the preceding chapter.

BIODIVERSITY

In a general sense, *biodiversity* is the variation inherent in living things. This variation is found in the genes of each individual, in an aggregate group of individuals such as subpopulations, or within a species, community, ecosystem, or landscape (e.g., Noss 1997). Benton (2003) points out that biodiversity is a recent term (dating to 1988) and used interchangeably with "diversity." However, the term also has been used in a restricted sense. It was applied to the number of species in an ecological community, which is the same as "species richness" (Krebs 2008). Presently, the biodiversity concept has expanded so much that common usage applies to the genetic diversity within a species (an indication of its adaptations) or to the species diversity within a community or ecosystem (an indication of the degree of collective genetic heterogeneity).

A good understanding of biodiversity would be predicated on the knowledge of how many species are on the planet, which major groups they represent, and how all of them function in respective ecosystems. Unfortunately, we are a very long way from such an understanding. Some groups, like vertebrates, are well known; others are poorly known, such as insects, even though the most diverse organisms are insects. About 1.8 million species are known, and recent crude "estimates" (educated guesses) for the total number was 13–14 million (Krebs 2008) to 20–100 million (Benton 2003). Why do we not have better numbers? Benton (2003) pointed out that it took from 1758 to the present to document 1.8 million species and estimated that it would take 4000 years at that rate to document 30 million. We cannot wait that long!

We have previously considered differences in biodiversity across the planet in zoogeography (chapter 8), and identified six potential causative factors (Krebs 2008). In this context, biodiversity is not uniform and some "hotspots" occur where biodiversity can be very high, such as in tropical rainforests and coral reef systems. We have related this high diversity with favorable climatic conditions, a corresponding high rate of primary productivity and a long time of stable conditions needed to foster the evolution of specialized niches. But what is the value of high biodiversity?

Many would argue that ecosystems are like a piece of art and should be maintained for moral and ethical reasons. Perhaps, but from a utilitarian viewpoint the most important feature of biodiversity is to facilitate continued functioning of the life support system of planet Earth (long argued by Ehrlich and Ehrlich 1981), providing goods and services that humans depend upon as part of the biosphere (like oxygen and food!). Humans also acquire and use medicines and other substances produced by organisms. All this is contingent on ecosystem integrity, which is affected by survival of a number of species needed to have a functioning unit, and ecosystem stability to make the system function efficiently.

The genetic diversity of individuals and populations aid in maintaining ecosystems as environmental conditions fluctuate or change. Given time, natural selection will result in species adaptation for survival and functioning not only in the center, but also in different parts of their distribution. Thus, individuals in widely distributed species may not be genetically uniform throughout their range. In this case, genetic diversity inherent in widespread and especially in abundant populations can provide a safety net, such as in times of extreme climatic events.

Stability is another attribute of biodiversity, in this case it is generally accepted that community stability of complex natural systems is greater than in simple ones, especially monocultures. Simply put, diverse systems recover faster after environmental perturbations than do less diverse ones. In his informative book on wetlands, my friend Professor Bill Lewis addressed the diverse and valuable functions of wetland ecosystems and the role of biodiversity in ecosystem stability by providing an analogy between species diversity and a baseball team (Lewis 2001, p. 57):

> The team has a premeditated diversity of morphotypes that perform different functions. Some have the ability to deliver a long ball with uncanny accuracy, while others are as quick as rabbits over short distances, and so forth. The team as a whole can, because of its diversity, defend itself against virtually any kind of challenge.
> A very different sort of baseball team would be recruited under a peculiar set of rules allowing only the selection of catchers. The catcher's position would be filled perfectly, of course, and the team might do wonderfully at snagging low line drives. This team would not be competitive, however, because its excessive specialization could be exploited in a dozen ways by a more diverse team.

Baseball teams have functional diversity as do natural systems, they also have redundancy in the form of extra players to take the place of anyone that cannot play, or is removed. These "second stringers" will not be as efficient, and in cases where key players are eliminated, positions would be filled by less experienced players, and then players from other positions. Thus, the game would go on.

Natural systems also have redundancy; for example, there is some niche overlap between species. Trophic adaptability is a good example, in which several fishes can prey on organisms that are abundant for only a short time period (e.g., pelagic larvae). At such times, specialists (e.g., invertivores) may easily be outclassed by planktivores or generalists, but they can take advantage of the food source. In the case of the NW Atlantic cod fishery, overexploitation resulted in an invasion of spiny dogfish sharks into codfish habitats. These extra players may not do such a good job exploiting some species but they may be consuming more prey items, including young stages of target fishes. This could plunge the benthic community, already disrupted by overfishing and bottom habitat destruction, into a tailspin, with uncertain future outcomes.

BIODIVERSITY CRISIS

Human life on Earth occupies a time period characterized by increasing species diversity. However, proliferation of life on Earth during the past 540 my (the Phanerozoic, in which there is a good fossil record) has been greatly affected by episodes of extinction. In chapter 7, we studied fish extinctions that occurred up to the Cenozoic. We noted that some of those extinctions were called "mass extinctions" because the extinction rate was double or more the rate of 17 minor ones. Of special note are the "Big Five" extinctions in which a great deal of biodiversity was lost. A total of 20%–30% of families and 50% of species became extinct in four of them, and 60%–65% of families and 80%–95% of species were lost in the fifth one (the End Permian). From our human perspective, two of these stand out as exceptional: the End Permian (245 mya), and the well-known Cretaceous–Tertiary (K-T) extinction that ended the reign of dinosaurs (65 mya). These two events reorganized ecosystems in sea and on land, and changed biodiversity so much that the Paleozoic and the Mesozoic eras were ended by them. There also are indications that a third great extinction is in progress that will end the Cenozoic (Barnosky et al. 2011). It is identified by Ward (1994):

> I believe that the Third Event is well underway, having started at the dawn of the Ice Age . . . and since then accelerating in its rate of species destruction. In some ways, it is very much like the dinosaur-killing Second Event of 65 million years ago (mya), when a biosphere already stressed by rapid changes in climate and sea level was knocked into mass extinction by the impact of asteroids . . . A very similar scenario is currently unfolding. More than 2 mya, giant glaciers began to cover large portions of the earth, changing climate and sea level on a global scale in the process. And then, 100,000 years ago, another great asteroid hit the earth, this time in Africa. That asteroid is named *Homo sapiens*.

The Third Event also has been placed with the Big Five and called the Sixth Mass Extinction, or the Holocene Extinction. The potential magnitude of this present extinction is sobering. Wilson (2002) suggested that most species could be eliminated in 100 years! The cause of the extinctions is unlike any other—it is primarily due to the actions of a single species. This new wave also is proceeding at a very fast rate.

The background rate of species extinctions has been calculated to be about 99.9% (Raup 1991), meaning that life on Earth has been maintained by a slim surplus of speciations over extinctions. It would not take much to change the direction of this global survival curve, and the increasing extinction rate suggests it already has changed. Ehrlich and Ehrlich (1981) reported that a background extinction rate (before humans) was exceeded five to 50 times from 1600 to 1975. By the end of the past century, the rate was estimated at 100 to 1000 times background (Pimm et al. 1995), and the projected rate for the twenty-first century is estimated to increase to 5000–25,000 times background (Benton 2003).

Estimates of impending extinction on a global scale vary depending upon system and habitat. The 2009 International Union for the Conservation of Nature and Natural Resources (IUCN) Red List of Threatened Species (IUCN 2009) reported that 36% (17,291 of 47,677) of assessed species are presently threatened with extinction. By taxa, percentages ranged from 12% for birds to 70% for plants. However, in some ecosystems such as tropical rain forests and biodiversity hotspots (e.g., Pimm and Raven 2000) the estimates may be even higher. Also, *coextinction*, in which the loss of one species results in the loss of another that is dependent upon it, has not usually been considered. Koh et al. (2004) estimated that at least 6300 species are "coendangered" with host species presently listed as endangered. In case of the extinction of a *keystone species*, a species whose impact on the structure and function of a community or ecosystem is disproportionately large for its biomass or abundance (sensu Power et al. 1996), its loss might produce a cascade of extinctions (Ehrlich and Ehrlich 1981).

Freshwater fishes have been greatly affected by the biodiversity crisis, presumably because they exist in an insular state (poetically said: "an aquatic island in a terrestrial sea") and they are exposed to wide variations of climate, many are restricted to one drainage basin, and affected by large habitat changes due to dams, diversions, reservoirs, invasive species, etc. The statistics are scary: The IUCN (2009) Red List reported that 37% of freshwater fishes they assessed (1147 of 3120 species) were at risk of extinction (critical = 270, endangered = 241, vulnerable = 636). Harrison and Stiassny (1999) reported that 70 fishes went extinct in the last 100 years, noting that 93% of the extinctions were in the past 50 years. By adding recent reports of fish extinctions in Lake Victoria, they found that the number of fish vulnerable to extinction rose to 172, with 97% of these lost in the past 50 years. Fish extinctions were usually related to multiple human causes, and extinction of the 70 fishes they studied was associated with habitat alteration (71% of the fish), competition and predation by introduced species (54%), overfishing (29%), pollution (26%), disease and parasites (4%), and purposeful eradication (1%). Extinctions occurred mostly for fishes in lakes (61% of species), but also occurred for fishes living in rivers and streams (23%) and in springs and pools (16%).

Factors implicated in freshwater fish extinctions in North America were studied by Miller et al. (1989). During the 100 years they studied human activity resulted in the extinction of 27 species and 13 subspecies of native North American freshwater fishes ($N = 1003$), of which 30 were native to the United States (table presented by Ross 1997). Multiple agents are implicated as well. Physical habitat alteration was a factor in 73% of the total extinctions, introduced species in 68%, hybridization in 47%, chemical alteration or pollution in 23%, and overharvesting in 20% of the extinctions. However, for 30 fishes native in U.S. waters the ratios change. Nonnative (introduced) species becomes the most frequent factor associated with extinctions (80%, $n = 24$), followed by physical habitat alteration (63%), hybridization (47%), (pollution 23%), and overharvesting (20%). Also, one-third of extinctions were associated with introduced species in the absence of physical habitat alteration.

The Endangered Species Committee of the American Fisheries Society (AFS) has prepared three status reports of imperiled freshwater fishes over a 29-year period. The imperiled category includes taxa considered to be vulnerable to extinction, threatened, or endangered. The reports vary slightly due to incorporation of revised species concepts and a better understanding of biodiversity, but overall they are comparable.

The lists of imperiled species contained in the three reports have steadily grown. The first list (Deacon et al. 1979) contained 251 taxa, the second list (Williams et al. 1989) contained 364 taxa, and the third list (Jelks et al. 2008) contained 700 taxa—an increase of 179% since 1979, and 92% since 1989. Taxa on the most recent list included 133 genera and 36 families of fishes, or 39% of the North American fish fauna of which 230 were considered vulnerable to extinction, 190 were threatened, 280 were endangered, and 61 fishes were extinct. Regarding fishes on the 1989 list, there were few improvements in status, and 89% of the taxa were classified in the same category or worse in 2008 (Jelks et al. 2008). When imperiled fishes were plotted on a map of bioregions (Figure 30.1), the highest number of fishes was found in southeastern United States.

The preceding accounts should leave little doubt that, as remarked by Richter et al. 1997, there is a "quiet crisis" occurring in the freshwaters of the world, and various studies have implicated a number of "usual suspects." Our purpose in the remaining part of this chapter, is to investigate the causes for the decline and endangerment of freshwater fishes with the goal of understanding the reasons for their plight, how humans have responded with measures to reduce the damage, and the potential for future conservation efforts.

But what about marine fishes? We have already discussed the intense exploitation of several saltwater fishes. However, marine fish have not been as vulnerable to extinction as freshwater fish. The National Marine Fisheries Service (NMFS), Office of Protected Resources has the responsibility for conserving marine species under the Endangered Species Act of 1973 (ESA), including preparing marine lists. There are 72 listings given for the status of marine and anadromous fish (NMFS 2009), of which seven are candidates or proposed for listing and 30 are species of concern. The remaining 35 entries are threatened or endangered: and include 30 population segments of salmon, three sturgeons, smalltooth sawfish (*Pristis pectinata*), and totoaba (*T. macdonaldi*) (a foreign listing). Of the sturgeons, only the southern population of the green sturgeon (*Acipenser medirostris*) is listed as threatened, leaving only and the gulf and shortnose sturgeons (*A. oxyrinchus desotol* and *A. brevirostrum*) and smalltooth sawfish listed throughout their ranges. However, we must recognize that the family Acipenseridae is comprised of anadromous and freshwater fishes (Nelson 2006), thus they are not true representatives of a marine group any more than are the listed salmon stocks. This leaves the smalltooth sawfish as the only marine fish species officially listed in U.S. waters.

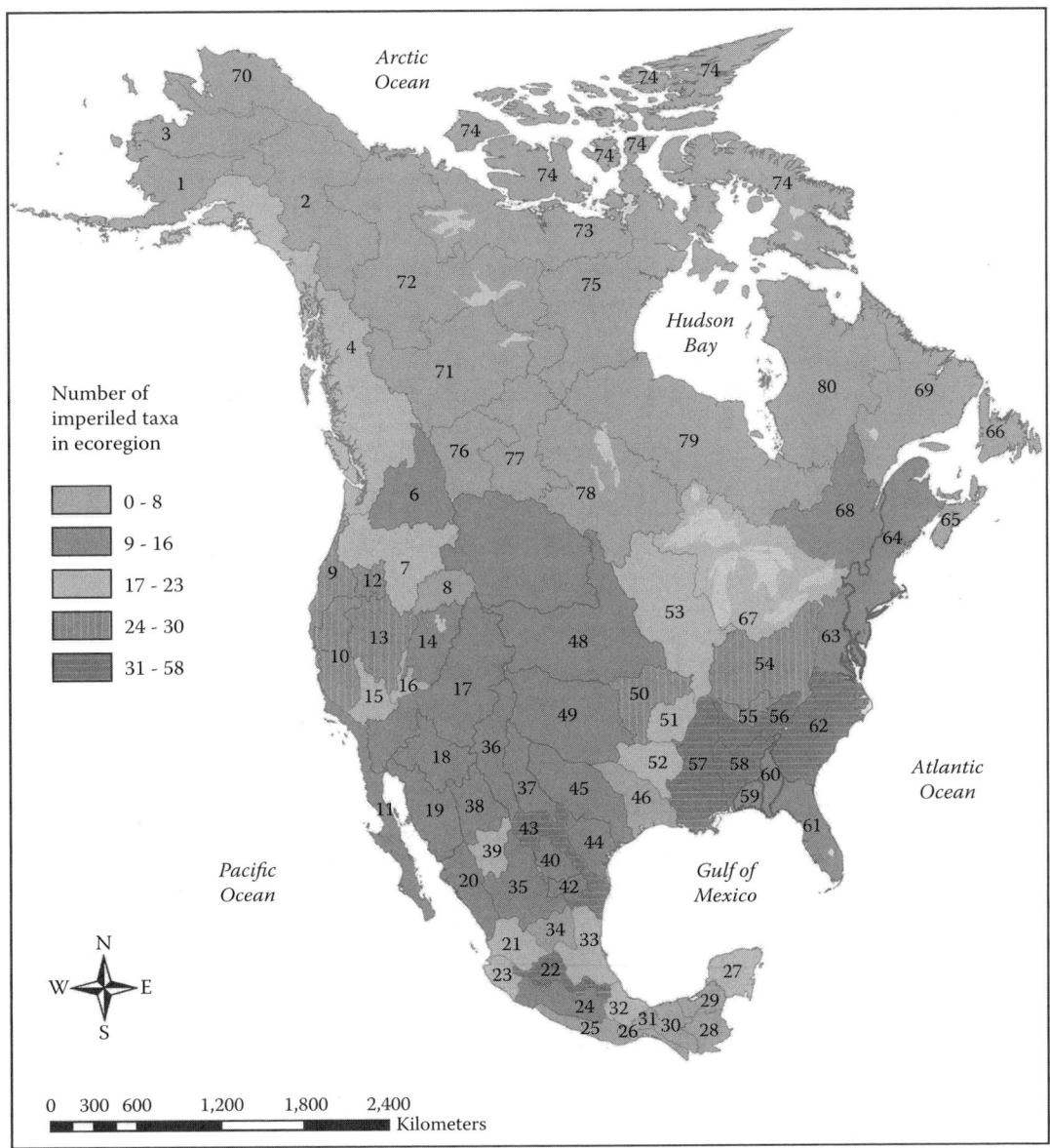

Figure 30.1 Bioregion map of imperiled species. (From Jelks, H. L., et al., *Fisheries*, 33(8), 372–407, 2008. With permission of the American Fisheries Society.)

Diadromous fishes have a complicated life history and in the process are exposed to more varying and/or different environmental conditions than are typical marine species—they encounter many of the same problems as freshwater fish with respect to anthropogenic habitat changes. For example, the abundance of 24 species has been reduced 90 to 98% of historic levels in the North Atlantic basin (Limburg and Walden 2009).

On the other hand, marine fish populations have suffered declines mainly due to overfishing, and most of the overfished pelagic populations have recovered after fishing pressure was reduced. Helfman (2007, p. 85) asks, "Are marine fish different?", pointing out that no pelagic marine fish is listed as recently extinct and only three of the 300 fishes he lists as extinct are marine: Texas pipefish and Galapagos damselfish (arguably extinct), and Banggai cardinalfish (functionally extinct). In this case, *functionally extinct* means they are not functional in their community, usually due to a lack of one or more life stages (e.g., not recruiting), and *arguably extinct* means that they are missing and presumed extinct pending a possible rediscovery (i.e., as *Lazarus taxa* for species, families, etc., that miraculously reappear after they are presumed extinct) (Benton 2003). The greenback cutthroat trout is an example of a Lazarus species.

Marine fishes, with the possible exception of reef systems, are not as restricted in range due to barriers as are freshwater fish, plus most have larvae that are very widely distributed by ocean currents. The *stable ocean* hypothesis holds that the large mass of water in marine environments tend to buffer the effects of weather and climate, and anthropogenic effects such as pollution. Marine fishes are free to move in all directions of the water column to seek favorable conditions—which is extremely important for early life history stages. Another argument is based on the great fecundity of marine fishes (i.e., the *millions of eggs* hypothesis), which has been discounted (Myers and Ottensmeyer 2005; Helfman 2007).

Some insidious changes are in progress in the oceans. One result of intense fishing has been change in genetic composition of stocks due to targeting larger, faster growing, and more aggressive fish. Other problems include less available habitat due to trawling and other direct effects of fishing that result in continual bottom disturbance, and community level distortions induced with bycatch of forage fishes and young of target species: all affecting ecosystem stability (Levin et al. 2006). Worm et al. (2006) found that declining diversity in marine systems was impairing the ability of the oceans to produce food, provide water quality, and to recover from disturbances. These factors are presumably related to findings of Hutchings and Reynolds (2004) in which 83% of over 230 populations of marine fishes had suffered a reduction of breeding population sizes, and although a few recovered rapidly, most of them show little or no signs of recovery even after 15 years.

Even though extinctions in marine systems are few compared to those in freshwater systems, there is a prevailing belief that marine declines and extinctions are an emerging problem (Myers and Ottensmeyer 2005). This is based in part by mounting indications that some marine fish species are recently in trouble, including the large pelagic and benthic sharks. According to the IUCN (2009) Red List, 20% of 547 sharks they studied were threatened with extinction from fishing activities. Other problems are emerging in fisheries that target large, very slow growing and late-maturing species, such as scorpaenid rockfishes of the Pacific coast, and fishes of southern oceans, such as orange roughy (*Hoplostethis atlanticus*), and Patagonian toothfish (*Dissostichus eleginoides*), mostly due to recruitment overfishing (reviewed by Helfman 2007). Unfortunately, problems with studying marine fish include the great costs in money and effort associated with studies, a general lack of baseline data, and poor controls.

WHY AND HOW ARE SPECIES GOING EXTINCT?

Past extinctions on earth have been linked with natural causes associated with a variety of environmental factors: climatic (resulting in glaciation and sea level change), tectonic (continental drift and volcanism), ecological interactions (e.g., predator–prey), extraterrestrial (asteroids), and others. However, the present wave of extinction is different. It is mostly due to the actions of humans. McKinney (1997) pointed out that fossil data may provide our only way of studying "natural" extinctions, because anthropogenic causes have been the major, if not the only, reason for the recent extinction of species: no conclusive example of nonhuman induced extinction has found in 8000 years! If so, then more humans should be associated with more extinctions, and that is just what we observe if we compare a curve showing extinction rates with another curve showing the increasing abundance of humans—the rate of increasing extinctions and growth of human populations over the same time are a close match.

There is no secret about *why* species are going extinct. All of the causes are presented above from studies of species that have recently gone extinct and those that are vulnerable to extinction. However, we also need to understand *how* they go extinct. There is something about human impacts that resemble a catastrophic event, except there is a temporal difference. Most mass extinctions apparently took place over many thousands of years. In contrast, human-induced extinctions are occurring over short time scales of perhaps a few hundred years or even less. It seems that humans can deliver a catastrophic blow to species, and in a very short time span. The rapidity is as important as the blow, because it limits the process of natural selection in producing adaptation. Two of the best examples have occurred in birds.

Two bird extinctions in the United States have been well documented. We know the *why* of their extinctions: both birds were very valuable to humans, apparently easy to kill, victims of the "tragedy of the commons," and exposed to multiple impacts. Now we will look at the *how*.

The passenger pigeon (*Ectopistes migratorius*) comprised about 40% of the entire bird population of North America (Day 1981). Flocks of the birds were huge and one flock passed overhead for 3 days at a rate of over 300 million birds per hour. Another flock was estimated at 2 billion birds (reviewed by Ehrlich and Ehrlich 1981). Typically, vast flocks nested in forest roosts of a few miles wide and up to 40 miles long, feeding on acorns and other forest products. The largest nesting flock (136 million) was found in Wisconsin in 1871. The location of the birds was sent out by telegraph and thousands of market hunters raced to the area. Millions of birds were killed, the remaining scattered, and the area was left a rancid wasteland (Gray 1993). The birds were subjected to a ruthless slaughter by large battery guns, by trapping, and by poisoning them with burning sulfur. The birds were sold in large quantities all over the eastern United States, and one merchant in New York sold 18,000 birds in 1 day (Day 1981). A market hunter shipped 3 million birds from Michigan in 1878, a last stronghold. It could not last. Only 11 years later (1914), the last surviving pigeon (Martha) died in the Cincinnati Zoo. Their huge population sizes and large flocks did not protect them. Final loss of the birds resulted from nesting failure due to human disturbance and predation.

The heath hen (one or more subspecies of the greater prairie chicken *Tympanuchus cupido*) also was an abundant and widespread species that ranged the entire eastern seaboard of the United States from Maine to Virginia. Early settlers loved it, because it was large, tasty, and easy to kill. It was native to the open dry land (eastern prairie), and it fed and mated in flocks of thousands (Day 1981). The ground-nesting bird was very conspicuous during mating when males made booming sounds. The bird also was killed for food by pioneers, settlers, and market hunters (and introduced dogs, cats, and rats). By 1791, its survival on Long Island was threatened as its habitat was converted into farms. By 1840, the bird had been extirpated from most of its mainland habitats, and by 1876, the last population remained protected on the island of Martha's Vineyard in Massachusetts. By 1900, there were fewer than 100 birds, and a 1600-acre refuge was established for them. After this, the population expanded to more than 800 birds (Shaffer 1981). Unfortunately, in 1916, a series of events reduced the population to 100–150 birds, and several factors led to their extinction (Shaffer 1981; Raup 1991): (1) Fire, spread by wind, destroyed much of the breeding area and reduced cover. (2) A cold winter followed, marked with the presence of migratory and predaceous goshawks that preyed on them. (3) Reduced population size promoted inbreeding and an unfavorable sex ratio. (4) Poultry disease from domestic turkeys killed many of the remaining birds. By 1927, there were 11 males and two females left. The last bird was seen in 1932.

Extinction of these two birds occurred in a similar pattern. First, they were exposed to catastrophic changes that they had not experienced before and had little or no ability to cope with. Such changes were a "substantial initial hit—a first strike" (Raup 1991). The First Strike was catastrophic due to its scale and the rapidity of its action. In less than a century, highly abundant and widely distributed species were reduced in population size and distribution to the point that they were vulnerable to changes in the environment that would have little affected them before. Extinction in this case would be caused by a case of "bad luck" followed by population declines due to "bad genes."

In a Darwinian sense, it would be bad luck that they perished because of a force never encountered previously and thus not available for the process of natural selection to provide some degree of adaptation in populations.

Are these bird extinctions applicable to fishes? In a general sense, they are. As a prime example, big river fish communities in North America (e.g., papers in Rinne et al. 2005) and throughout the world are in the process of suffering from a First Strike due to conversion of rivers to systems of multiple dams and reservoirs—blocking migrations and drowning spawning and rearing areas. Downstream discharges from large dams are not natural in regimen or water quality, and dams fragment the river ecosystem and isolate once interbreeding fish populations into smaller stocks. To make matters worse, a second strike follows quickly: alien fishes and other organisms that are well adapted to lakes, clear water, and less seasonal changes in water level have been introduced into habitats occupied by remnant communities of native fishes. These introduced organisms are almost always hardy, aggressive, and predaceous. The results of these invasions on native fishes are predictable: the ranges of native fishes are greatly reduced, long-term abundance decreases, and one or all life stages may become vulnerable to competition (food and space), predation, and in some instances, hybridization. Furthermore, such invasions also can result in additional strikes, because once stocked into altered reaches they can continually invade remaining unaltered river reaches, such as tributaries and affect remnant native fish populations there as well.

There is a point when abundant and widespread species cross a threshold from being almost immune to extinction to being the "living dead": these are not zombies, but living individuals of an endangered species. When populations become very small, they may enter into an *extinction vortex* from which escape is difficult, and eventually can result in their demise. That point occurs when population size drops below MVP, or "minimum viable population" size. At that size, extinction can occur due to four factors or uncertainties (Shaffer 1981; Meffe et al. 1997):

(1) *Random events* that work on survival and reproduction of individuals to lower the population size (AKA *demographic stochasticity or uncertainties*)—Problems include not enough mates, altered sex ratios (especially lower number of females), failure of reproduction due to change in weather, low or no survival of young, and other problems.
(2) *Environmental stochasticity* caused by change in weather conditions, or habitat factors, such as abundance of food, predator load, diseases and parasites, and competition—Compared with the "match mismatch" hypothesis, a large and widespread population could survive even with a mismatch, whereas survival of young in small restricted populations can be greatly reduced.
(3) *Natural catastrophes* that may occur at unpredictable times and produce changes in habitat—Examples are changes in river flow and water level due to storms (i.e., floods and droughts), and rocks falling from the sky (asteroids).
(4) *Loss in genetic variation* due to genetic drift and inbreeding (*Genetic deterioration*)—In this case, rare alleles are lost and reduce the capacity of the population to adapt to extreme or changing conditions. The occurrence of a population bottleneck, in which some genomes are lost due to chance alone, is a real danger.

Raup (1991) points out that behavioral mechanisms can affect small populations due to the inability to attract mates with low density of animals, or in herding animals in which diversity can be low (*social dysfunction*). Also, the Allee effect can result in a lack of protection in colonial nesting fishes when population size drops. Outside forces mentioned (e.g., fires, storms, disease, predation load, and other forces) affect all populations, but very small populations are exceptionally vulnerable. These may occur as natural catastrophes but not necessarily.

There is a notion that some groups are more vulnerable to extinction than others, a topic that was carefully considered historically by Lamarck, Lyell, Darwin, and Simpson. In general, past studies suggest that: "Species with large body size, high trophic level, specialized habitat needs, and poor dispersal are among the most consistently extinction-prone . . ." (McKinney 1997, p. 497).

However, large size in and of itself is not a good predictor of extinction, especially in the aquatic realm. McKinney (1997) concluded that individual traits such as specialization and low abundance are related to extinction vulnerability, especially with a synergistic combination of specialization, rarity, and fragmented range.

HOW MANY FISH DO WE NEED?

How do we determine MVP for a population? First, understand that calculation of MVP varies according to how much risk you are willing to take or, put in another way, how long you want a population to persist. One notion is to have a population size that has a 95% chance of persisting for 40 generations, or 100 years. However, for conservation of a species we might wish to have persistence in perpetuity and use a value of a thousand years! There has been some confusion here and a lot of debate. One early theoretical idea that included only genetics became known as the 50/500 rule, tentatively proposed by Franklin (1980) in which short-term effective population size should not be less than 50 (to prevent a loss of fitness by inbreeding depression) and long-term effective population size should not be less than 500 (to retard the effects of genetic drift, thus maintaining genetic variation). This rule is based on knowing the *effective population size* (Ne), which is an indicator of the potential genetic drift in a population of size *N*.

Effective population size of an animal population depends on several factors, but Ne is very sensitive to the sex ratio of breeding adults, which may be very different than just the number of organisms present. The effective population size of an ideal population assumes random breeding and an equal sex ratio (male–female, 1:1). If unequal, the effective population size (Ne) may be calculated as (Frankel and Soulé 1981, p. 38)

$$Ne = 4NmNf / Nm + Nf, \qquad (30.1)$$

in which Nm and Nf are the number of breeding males and females.

The value of this approach is easily seen in herding animals. In elk herds, one bull elk may have a harem of 20 or more females with which he breeds. Instead of having an effective population size of 21, Ne in that case would be calculated as 4(1)(20) / 1 + 20 = 80 / 21 = 3.8, or only 18% of the total number. Using my experience with the Colorado pikeminnow (Tyus 1990b), we collected 275 fish on known spawning grounds, of which there were 219 suspected males and 56 females. The effective population size according to the above equation would be 4(219)(56) / 219 + 56 = 49,056 / 275 = 178 or about 65% of the total number. However, although we may actually see and monitor the elk herd, it is vastly more difficult to do so with the fish. Thus, it is possible that females may not remain on the spawning grounds as long as males, smaller males may not be able to spawn, males and/or females may spawn several times during a few days, etc. In case of those problems, uncertainties must be acknowledged for endangered fishes by intentionally erring on the side of the species. We can use the fish study above in a different way: instead of using suspected males and females, one could use only fish that had sex products present, meaning they were captured in the very act of spawning. Then, the number of males shrinks to 194 and the females to 14. Now, Ne calculated for those fish would amount to only 10,864/208 = 52, or the effective population would be only 25% of the total spawners present. If the population of pikeminnow in a location were estimated at 5000, the effective size Ne would only amount to 1250 using these methods.

Conservation of endangered species is fraught with uncertainty, and the goal to preserve listed species in perpetuity and in nature (not in a lab) does not allow the use of the 50/500 rule; likewise an MVP determination can be used as a guide, but never as a precise rule in determining the fate of species. In this case, there have been several attacks on the concept of a "magic population size" that will allow an animal population to exist indefinitely. One problem is that the environment certainly has been changing globally, and this is not taken into account with the 50/500 rule. But there also are other problems. Thomas (1990) simulated fluctuations of three populations of different sizes (100, 1000, and 10,000) over 50 generations and found that about 5500 animals in undivided habitat were needed to insure that populations do not drop below MVP levels at some time in the future (in this case, 100 years). Lande (1995) attacked the rule directly, using spontaneous mutation rates to show that an effective population size of 5000 instead of 500 is needed to maintain adaptive genetic variance for the long term. Also, for endangered species populations in the wild, that are subjected to changing density-dependent and density-independent forces, actual population sizes of 10,000 or more may be needed to provide an effective population size of a few thousand.

Except for highly controlled environments such as zoos and controlled domestic stock the 50/500 rule is too risky. It needs to be classified as a "dead-on-arrival" approach for recovery of wild populations. This contention is well supported by others. Estimating MVP for 102 species by using the criteria of 99% probability of persistence for 40 generations, Reed et al. (2003) found predicted mean and median values of 7319 and 5816 adults were needed for MVP. They recommended that habitat needed for conserving wild vertebrates should be designed for at least 7000 animals to ensure long-term persistence. Looking at MVP estimates of 212 species over the past 30 years, Traill et al. (2007) found that a median of 4169 individuals ($P = 95\%$) was generally recommended. Finally, the point needs to be made that MVP is not a very good tool for protecting endangered fishes. It is usually calculated from the perspective of conservation genetics and the environment is assumed to be stable. This is usually a serious mistake, especially in freshwater systems and with populations that are small. (Most endangered species will have small populations because that is one of the requirements for listing them.) Fortunately, there are practical alternatives. Sanderson (2006) provided 18 different ways to set population target levels. He suggested the use of reference ecosystems for setting population levels and a four-tiered system to consider in increasing population sizes: (1) demographic sustainability, (2) ecological integrity, (3) sustainable use, and (4) restoration to historic sizes. I believe that this idea is a good one and it could take the focus off the continuing genetic debate. But if there are problems in accepting populations sizes needed for genetics, there will be more conflicts over the needed population size and habitats for maintaining ecological integrity and sustained use. Then, given the demands for more water for people, restoration to historic size populations in the presently altered rivers of the world will prove to be difficult if not impossible.

One way to address the likelihood of persistence for an endangered species is to use a risk assessment technique called *population viability analysis* (PVA). The analysis is done with computer models that predict how long a population being studied is likely to persist. The life history, habitat requirements, and environmental variability all go together to determine how long a population will persist and thus how critical it is. But perhaps the biggest advantage of PVA is in evaluating different management options. Unfortunately, PVA has been used inappropriately, usually due to a false assumption of ecosystem stability and inadequate data. One problem is the need for an extensive dataset to make accurate predictions, and it has been argued (e.g., Ellner et al. 2002) that PVA has little value because endangered species usually have poor datasets, and for many species, we even lack knowledge about basic life history traits. Because of continuing environmental change, PVA models should be run using different sets of assumptions about future conditions. Reed et al. (1999) also suggested caution in using PVA due to a past misuse of models and confusion in interpreting results. Those authors stressed that PVA should be subjected to external review and not used to determine MVP or the probability or reaching extinction.

SPECIES PROBLEM

Communities and ecosystems are often characterized by the species assemblages they contain, and biodiversity is calculated from the number of species and their relative abundance. Herein is the species problem: There are many concepts of species but no one concept or definition is completely satisfactory to biologists (Mayer 1964; Futuyma 1998; Williams 1992). Perhaps the most generally accepted one at present is the *biological species concept* in which species are interbreeding populations that are reproductively isolated from others (students seem to know this one). Unfortunately, this definition is limited because it cannot be used for asexual organisms nor for extinct ones. The *morphological species concept* (which predates Linnaeus) works to separate things that look different, including extinct forms, but phylogeny can be difficult. The *phylogenetic species concept* is useful, but we do not have all of the information to use it fully, and the species concept would be greatly different from historic (it would include more subspecies). The *evolutionary species concept* is compelling and thoughtful but difficult to use, as is the *ecological species concept*. These and more species concepts are discussed in biology textbooks and a good account was presented by Meffe et al. (1997).

There are two different perspectives on how to deal with the species problem. One is the *nominalist* (small) viewpoint that questions the reality of species and suggests that species are just artificial notions dreamed up by humans. The other is a *pluralist* (more than one) view that the species concept used should be the most appropriate one that fits the taxa and satisfies needs. I believe the vast majority of biologists/ecologists subscribe to the pluralistic viewpoint. This is logical, because the situation is similar to how we use other concepts, such as the niche: We use the concept that best fits the situation. There are several concepts of the niche and of species and they differ in a great many ways. However, both terms are quite useful as long as you clarify the way in which you are using them.

In addition to the above definitions of species concepts, I need to add another one, an *endangered species concept*. This concept and its definition have been codified into law. The ESA defines an endangered species as "any species which is in danger of extinction throughout all or a significant portion of its range" (excluding insect pests). Furthermore, for purposes of the Act, species are defined as "any subspecies of fish or wildlife or plants, and any distinct population segment of any species of vertebrate fish or wildlife which interbreeds when mature." Here we see the elements of more than one definition or concept previously discussed. I have heard this definition criticized because it gives more consideration to vertebrates, and also for including subspecies. But Congress got it right fundamentally. A species is an endpoint, while a subspecies would continue to evolve and perhaps attain the status of a new species.

The distinct population segment also recognizes that genetic differences (e.g., locally adapted gene complexes) may occur in populations or stocks. In practice, the "distinct population segment" has been replaced by the term "evolutionarily significant unit" (ESU), which is more explanatory. The ESU can refer to species, subspecies, races, or populations that have geographic separation, different allelic frequencies, or locally adapted phenotypic differences.

By recognizing subspecies and ESUs as threatened or endangered, the Act protects the process of evolution and not just the end products. Unfortunately, the concept of species in the minds of laypeople is likely in the context of "fixed" species such as familiar birds and mammals. But this mind set is now changing as we list more populations as ESUs. It is becoming clear that many organisms, including most fishes do not fit the "fixed species" category very well, thus subspecies (or race) is more appropriate. Interestingly, Charles Darwin (Darwin and Horan 1979, p. 107) considered this problem, pointing out that "no clear line of demarcation has as yet been drawn between species and subspecies. . . ."

A RESPONSE: THE NEW CONSERVATION

Concerned citizens, scientists, and government have responded to the biodiversity crisis by developing a new field of science (conservation biology, the development traced by Gray 1993) and the U.S. Government has decided that conservation of endangered species is a national policy (e.g., ESA) (Bean 1983). We previously used "conservation" in connection with "natural resource" conservation (i.e., wise use of resources). In this chapter, we use "conservation" in context of "saving." In this effort, we will be concerned with conservation biology (saving + life), which is a relatively new field of study that combines pure and applied sciences. For example, the Society for Conservation Biology (2009) has this goal: "to help develop the scientific and technical means for the protection, maintenance, and restoration of life on this planet—its species, its ecological and evolutionary processes, and its particular and total environment" (e.g., see www.conbiol.org). Krebs (2008) went further and defined *conservation biology* specifically as "the applied ecology of endangered species." As a subset of applied ecology, *conservation of fishes* means to preserve, guard, protect, or save fishes. This necessitates two things: (1) to maintain population size of an endangered species so that they do not enter an extinction vortex and (2) to reduce or remove threats that are adversely affecting listed species.

In the late 1960s and 1970s, the environmental movement in the United States resulted in the passage of several important laws related to the preservation and conservation of nature, including the National Environmental Policy Act of 1969, and the ESA. The first law aids mostly in protection of habitat, the second is to "conserve" species that are listed as threatened or endangered (and their ecosystems), to a point that they can be recovered to a nonendangered status and removed from the list. In deciding which species should be listed as a threatened or endangered species, the ESA requires the U.S. Fish and Wildlife Service (FWS) and NMFS to consider "listing" factors (A–E) provided in Section 4: (A) the present or threatened destruction, modification, or curtailment of its habitat or range; (B) overutilization for commercial, recreational, scientific, or educational purposes; (C) disease or predation; (D) the inadequacy of existing regulatory mechanisms; (E) other natural or manmade factors affecting its continued existence.

The ESA works in a stepwise fashion: candidate species are identified, their status determined by considering population size and threats, those populations or species that are determined threatened or endangered are placed on a list to receive protection by law, critical habitat is delineated if possible, and a recovery program is initiated with the goal of returning listed species to a less endangered status, the threat(s) of extinction is identified and removed, protective measures are implemented with other agencies and organizations, and the species is removed from the list and federal protection. As stated above, an *endangered* species is in danger of extinction in all or a significant part of its range (excluding insect pests), a *threatened* species is one that is likely to become endangered in the foreseeable future, and *critical habitat* includes specific areas, either in or outside of the occupied range of a listed species, that provide physical and biological features essential to the conservation of the species and areas that require special management consideration or protection.

At the end of 2009, there were 1898 U.S. and foreign species, subspecies, or distinct populations on the endangered species list. In addition, 88 species have been proposed for addition to the list, and 249 species are candidates for listing. At the most recent accounting to Congress (fiscal year 2005–2006), populations of 33% of the listed organisms were considered stable, 34% considered declining, 8% improved, and the condition of 23% was unsure (FWS 2009a).

ENDANGERED FISH RECOVERY?

Recovery of endangered species has been criticized as ineffectual (e.g., Tear et al. 1995). Using fish as an example, the endangered species list contains 138 listings for fish species and subspecies plus distinct population segments of salmon (26). Once listed, history reveals that these fishes are more likely to go extinct than to be recovered: only four fish have been delisted since the ESA was enacted into law. One was removed due to a taxonomic error in listing it (coastal cutthroat trout in 2000), and three were delisted because they were extinct (Tecopa pupfish in 1982 and blue pike and longjaw cisco in 1983). This is a poor prognosis for other fish. At least two plausible explanations exist: we do not know how to recover freshwater fishes or the task is too difficult to accomplish. In the best of situations, biologists can decide how many animals or resources may be needed and what habitat features are limiting, thus determining targets for recovery plans based on science, but they have little or no control over what happens next. Complex decisions have to be made that involve socioeconomic and political perspectives, often debating conservation against economics (Wilhere 2008).

The ESA seeks to keep endangered species from becoming extinct by protecting the organism and its critical habitat from harm by individuals and federal agencies, encouraging the use of habitat conservation plans (which allow incidental take permits for lawful activities that might result in loss of a listed species), encouraging states by providing matching funds for endangered species initiatives, and by developing and implementing recovery plans. Most species have recovery plans which contain a vast amount of information. Each plan is available free for public review and comment in draft form. When approved, every plan is available via the Internet at either www.fws.gov or www.nmfs.gov, depending upon whether it is a terrestrial, freshwater, or marine species. Recovery plans are excellent resources.

There is little wrong with the concepts developed in the ESA; however, there have been problems with its implementation. A video provides an example of my disappointment with implementation: *The Endangered Species Act: An American Legacy*, produced by the U.S. Department of Interior, Fish and Wildlife Service (N.D.), which described how the Act works by providing an analogy to a 911 call in which endangered species are taken care of and their "condition" stabilized. Yes, but recovery will require more that taking a patient into a hospital and keeping them alive, perhaps in a coma, with no or little prospects for improvement. This represents the 33% of endangered species whose populations are considered to be stable.

Recognizing that only 8% of all listed species are considered improving, one has to ask why the recovery rate is so low. The blame has been placed on recovery teams (Westrum 1994), recovery plans (Snyder 1994), and the conflicts with implementing recovery actions (Bower et al. 2001). Specific reasons are many. For one thing, the ESA is not permanent; it is debated, amended, and must be reauthorized over and over again. Thus, there has been hesitancy to push hard issues for fear of inciting a movement to not reauthorize it. Other problems include conflicts of interest between recovery of endangered species and development interests, disagreement in the use of natural resources, a lack of training in endangered species conservation, disinterest in endangered species recovery, unwillingness to take risks or apply adaptive management, focus on single species only, and a lack of funds and manpower.

We are learning the causes of extinction and reasons for the decline of endangered populations, which is imperative in making better management decision in the future. However, as time goes by we also find the threats to recovery of most species increasing, which makes the job even harder! In a recent study of 31 indicators, Butchart et al. (2010) reported that despite some local successes and an increased effort in general, global biodiversity continues to decline.

CASE STUDY: CAN SCIENCE SAVE THE SALMON?

Declining Pacific Salmon

Pacific salmon populations of the NW United States (i.e., in the contiguous 48 states) are declining in abundance and range. The severity of this decline was made abundantly clear in 1991, when the AFS published a report that identified 214 naturally reproducing populations of salmon and sea-run trout in the states of California, Idaho, Oregon, and Washington that were considered either at high or moderate risk of extinction, or of special concern (Nehlsen et al. 1991). Since that time, populations (ESUs) of chinook (*Oncorhynchus tshawytscha*), chum (*O. keta*), coho (*O. kisutch*), pink (*O. gorbuscha*), and sockeye (*O. nerka*) salmon, and distinct population segments of steelhead trout (*O. mykiss*) have been placed on the Federal endangered species list as threatened or endangered. Massive recovery efforts are underway for these fishes (e.g., see Good et al. 2007). However, this decline is historic: it has been in progress for over 160 years, and most wild runs of fish now represent less than 10% of their former population sizes (Good et al. 2007). The declining trend includes the Columbia River basin, which historically produced more salmon than any other

Figure 30.2   Map of Columbia River basin. (Courtesy of U.S. Geological Survey and U.S. Army Corps of Engineers, http://Vulcan.wr.usgs.gov/volcanoes/washington/ColumbiaRiver/Maps/map_Columbia_River_and_tributaries.html.)

river south of Canada. Columbia River salmon runs are now supported by large and very expensive hatchery programs. What is responsible for the decline of these attractive and valuable fishes? What is being done to save them? We will explore these questions by examining salmon recovery in the Columbia River basin, where massive recovery efforts are occurring.

The Columbia River is the fourth largest river basin in the contiguous United States, with a drainage basin of 260,000 square miles (Figure 30.2). Data and information provided by the Washington Department of Fish and Wildlife indicate that 10–16 million wild salmon entered the river predevelopment, but since 1970, their minimum estimate of returning adult salmon and steelhead peaked at 2.9 million in 1986 and dropped to a low of only 62,100 in 1995. Lackey (2009) provided a recap of some important events that surely added to the decline of salmon, for example, discovery of gold (ca. 1848), the Great Depression (1933), World War II (1942), floods (1948), and power blackouts (2001). Such events resulted in a steady procession of changes that negatively affected the salmon.

At present, 13 salmon and steelhead ESUs are listed under the ESA by NMFS in the Columbia River basin. There is strong public and government support to see the salmon recovered, not only for commercial value but also because of its important place in the stream ecosystem (e.g., see Gende et al. 2002). However, the Idaho Chapter of the AFS (Idaho-AFS 1995) reported that the millions of dollars spent and efforts of hundreds of fishery scientists working to restore salmon in the Columbia River basin has resulted in the most thorough documentation of extinction in history, but little else. This viewpoint is supported by a Government Accounting Office (GAO) study in 2002, which reported that federal agencies had spent $3.3 billion on salmon and steelhead recovery over two decades with little conclusive success to show for it. Do we know why the salmon stocks are declining? My answer is "Yes, certainly. Some excellent scientists have been working on recovery, and the problems are well known." Fishery scientists in public service and elsewhere have cautioned for years that unless the major problems are solved, salmon extinction is inevitable.

Four "H"s

In the northwestern United States, it has been customary to identify the factors associated with salmon abundance with words that begin with an "H." Four "H"s are traditional: *harvest*, *habitat*, *hydropower*, and *hatcheries*. Montgomery (2003) adds another "H" for *history*, because in his book, he stresses the importance (and notes the disregard) of learning from the past. As for me, I also would like to add another "H." We need an "H" for *hitchhiker* to represent nonnative species. However, we will follow convention and lump everything under the four "H"s. We also will explore reasons for the decline in Columbia River salmon and identify the most promising ways to aid restoration.

Harvest

Harvest regulation by limiting fishing is usually the first management action taken to protect a declining fish population. This is difficult for anadromous species because they require freshwater for reproduction and early life stages, and marine systems for growth of subadults and adults. Ocean catches are difficult or impossible to deal with in a management sense because the Columbia River salmon stocks are intermingled with fish from elsewhere, including large stocks from Alaska and British Columbia. Also, ocean fisheries are already heavily regulated, and catches are carefully monitored by the Pacific Fisheries Management Council. Recent restrictions have included closures, gear types, and harvest. Although fishing in the Columbia basin has harvested up to 90% of entering salmon, fishing is now more restrictive and listed stocks are now protected by the "take" provision of the ESA. Overfishing does not seem to be a main factor presently limiting abundance of the Columbia River salmon, although it undoubtedly is a factor. From a management perspective, there are also two sociopolitical issues: Tribal right to fish is guaranteed by federal treaty and recreational fishing as a desirable use of the resource.

Hatcheries

Hatcheries were considered in the 1870s, when it was becoming obvious that wild populations of salmon were declining in the western United States. Faced with the proposition of increasing the supply from nature, influential salmon experts supported hatcheries. "If you need more fish build a hatchery" seems logical and simple compared to other options. But simple solutions to complicated problems also seem counterintuitive! Many hatcheries were built and operated (e.g., by 1995, there were more than 90 state and federal hatcheries in the Columbia River basin; Busack and Currens 1995), while supporters and detractors argued, and more dams were constructed. Bowles (1995) pointed out that supplemental stocking in the Columbia River can never result in recovery of salmonids to historic levels because of the high mortality of smolts.

Hatcheries produced and stocked tremendous numbers of fry. During years of favorable conditions, the hatcheries were given credit for increased numbers of salmon, but during bad years, the salmon populations declined. It is said that hatcheries went for production (numbers of fish stocked) instead of quality, using the same broodstock in multiple locations that were already occupied by distinct populations. Furthermore, as revealed in Scarce (2000) hatchery workers would select, spawn, and rear the young over a few months to meet quotas, at which time they ceased to produce fish. The young of early spawners are fed longer, grow larger, and thus survive release in nature better. Hatchery fish produced this way turn out to be early spawners—not spread out during the season like wild fish, thus increasing the likelihood of reproductive failure and increased competition.

A total of 18 federal hatcheries produced 703 million salmon and steelhead that were stocked during 1980–1989 in the Pacific northwest. However, as the number of hatchery fish increased, wild stocks decreased (Ross 1997, figure 8.3, p. 199), possibly an indication of negative genetic effects on the wild population (Campton 1987, 1995), or other unknown factors. Heath et al. (2003) evaluated a supplemental breeding program in British Columbia where an average of over 550 million hatchery-reared salmon fry was released annually. They found a case of unintentional selection in which the egg biomass in hatchery fish decreased from 0.27 to 0.20 g/egg (1988–2001), representing a potential 24% decrease in survival if spawned in nature. Wild populations with a large percentage of hatchery fish showed decreases in egg size as well. In a hatchery where fry are fed regardless of size, there would be an advantage to having a large number of small eggs, which are fed as soon as they swim up. However, those eggs or small young would have a lower chance of survival in nature.

An NMFS team suggested that a 10-fold reduction in the density of wild spawning coho salmon in the lower Columbia River over a 30-year period might be related to hatchery practices. These practices were selection for early spawning, fry stockings of up to seven times carrying capacity, and stocking hatchery fry that were more than two times larger than wild fry (Flagg et al. 1995). As it turned out, carrying capacity for coho salmon was about 3 fry/m^2, but the hatchery fry were stocked at >22 fry/m^2. When stocked in May, hatchery fry weighed 1.85 g, and wild fry were only 0.75 g.

AFS sponsored a 1995 symposium on "uses and effects of cultured fishes in aquatic ecosystems" that addressed the ethical use of cultured fish. The keynote address (Radonski and Loftus 1995) pointed out that the historic decline of fisheries was not well understood, but public demand required some corrective action by agencies. The action was a hatchery program, which fueled increasing controversy, first between hatchery interests and biologists, and further heated up by concerns of the environmental movement. Brannon et al. (2004) considered the controversy surrounding the use of hatchery fish in Pacific salmon supplementation, and separated the problem into two areas: artificial propagation and fisheries management. Those authors advocated hatchery management reform to address such issues as genetic risk, and stocking of huge numbers of hatchery fish into wild populations. Pearsons (2008) discussed ecological interactions between cultivated and wild salmon.

Hatcheries have not turned out to be the cure-all once thought, because they address the symptoms of fish declines rather than the causes, allowing the causes to continue. But technological solutions are limited and inevitably exhausted as the causes increase in number and/or magnitude. Anadromous fish must have suitable freshwater and marine habitats, but Pacific salmon are losing habitat and access. Hatchery programs will fail if there is inadequate habitat for the fish. Perhaps hatchery failure is an example of attempting to dominate ecosystems instead of using ecosystem knowledge to manage the fish!

Habitat

Habitat is composed of physical (abiotic) and biotic components. But to use habitat, the fish have to have access to it via water, and presently, dams block about 1/3 of the historic fish habitat in the Columbia River basin. Conditions in shallow freshwater systems are highly variable due to climatic fluctuation, and fishes that live there have to be acclimated to them. On the other hand, life also is subject to changes in climate and food production in the ocean, but for the most part they are not as extreme and effects on fish are buffered by the great mass of water. There is little indication that salmon mortality has increased in the ocean in the past 100 years, and in fact, climatic changes in the North Pacific have favored salmon, producing high survival. Caution must be exercised in management, because favorable conditions in the marine environment can potentially mask degrading conditions in streams (Peterson and Schwing 2003; Schindler et al. 2008).

Salmon farming has some potentially lethal effects on smolts that may swim near them. Sea lice *Lepeophtheirus salmonis*, a parasitic copepod, are present on adult salmon, but the lice cannot tolerate freshwater; thus, salmon smolts are usually not exposed to them. However, sea lice flourish on salmon farms, and any juvenile salmon that moves in the vicinity of a farm is likely to be parasitized. The outcome is frequently lethal (Frazier 2008). There are no salmon farms close to the Columbia system, but sea lice do occur on salmon farms in Canada, and adult fish do escape from farms (Gudjónsson and Scarnecchia 2009) so the matter bears watching.

Human populations continued to increase in the Columbia River basin and land use change follows. Commercial logging on public and private lands and road building removes shade, increases stream temperature, and produces silt to smother habitat. Livestock grazing and clearing of trees and snags along shorelines reduce habitat heterogeneity. Mining, agriculture, urbanization, etc., all add to the problem. Urbanization has also resulted in more water pollution due to chemicals and sewage while agriculture has increased nutrient and pesticide loading and soil erosion into streams.

Habitat quality is important to the survival and growth of young fish. Paulsen and Fisher (2001) documented the highest survival of parr reared in wilderness areas, and the lowest survival in parr that reared in young, dry (mostly intensively managed) forests. In another study, Regetz (2003) evaluated degradation of chinook salmon spawning grounds, and found that three variables, representing urbanization, low water quality, and sedimentation, accounted for 60% of the variation in mean salmon recruitment. In areas where salmon have been excluded, habitat quality is degraded. According to Gende et al. (2002), salmon subsidize freshwater and terrestrial systems by several pathways, including a supply of high-quality nutrients taken up by riparian vegetation, stream invertebrates, fish, bears, and others.

If juveniles survive and grow in their freshwater streams they still have a hazardous trip downstream and into the ocean. Formerly a quick trip, now the fish must travel through reservoirs that are occupied with aggressive and predaceous fishes, including invasive species (Fuller et al. 1999) such as walleye, channel catfish, and smallmouth bass, because all of them eat young salmonids. The walleye is known to prey on salmon smolts in the Columbia River and accounting for about 1/3 of the predation (McMahon and Bennett 1996). Smallmouth bass predation also is a concern: daily consumption of salmonids by smallmouth bass near McNary Reservoir was more than twice that of the native top predator northern pikeminnow in May and June and exceeded that of pikeminnow in April through June in the Snake River, the bass preying especially on small sizes of salmon (reviewed by Zimmerman 1999).

There is another problem. When entering the lower Columbia River, salmonid smolts are exposed to predation by a colony of almost 11,000 breeding pairs of Caspian terns. The terns consumed about 7 million smolts in 2008. The problems and potential solutions to the tern problem are being investigated by a partnership of researchers from the government, academia, and industry (see www.birdresearchnw.org/project-information/columbia-river-estuary). A final environmental impact statement on management of the terns was released in 2005 (see www.fws.gov/pacific/migratratory birds/cate_background.htm). In addition, a colony of double-crested cormorants in the same area consumed about 9 million salmonids in 2007 and about 7 million in 2008 (unpublished data, 2008 final season summary, Oregon State University and Real Time Research, Inc. at www.birdresearch.org). Attempts are being made to relocate birds from the upper to the lower part of the estuary where it is expected that they will consume more marine fish and fewer smolts.

As mentioned above, the top native predator in the river system is the northern pikeminnow, a large minnow. The pikeminnow will feed on salmon smolts, but in natural river sections, smolts are not major prey items. It seems that dams provide atypical low-velocity areas where pikeminnow may lie in wait and feed on salmonids that are injured, dazed and confused by passage through the dam (Beamesderfer and Rieman 1991; Vigg et al. 1991). For this reason, there is a northern pikeminnow bounty program in the Columbia River. A total of 160,000 of the fish were captured and killed in 2008, and from 1990 to 2009, about 3.3 million have been killed for fertilizer. As it turns out the bounty program is a very popular way to make a little money and enjoy fishing as well; however the amount of benefit is difficult to quantify. The smolts are damaged or disoriented by passage through the dam and especially turbines where up to 70% of them can be killed. Since the pikeminnow have no jaw teeth, it may be more productive to select dead or damaged prey than to chase the lively, slippery smolts. At least some smolts that are eaten are already dead or would not survive. Also, pikeminnow predation on salmonids may have increased as nonnative predatory fish have proliferated and perhaps "outcompeted" pikeminnow for nonsalmonid prey (Zimmerman 1999). Smallmouth bass and channel catfish are interfering with Colorado pikeminnow recovery in the Yampa River of Colorado, and Vigg et al. (1991) noted that the mean daily ration of smallmouth bass was twice that of pikeminnow, channel catfish, or walleye.

The bounty program addresses symptoms associated with environmental change rather than the cause, and it sends a bad message. The pikeminnow is the native apex predator. I agree with Behnke (1992) that the northern pikeminnow has a rightful place in the Columbia River ecosystem and it coevolved with the salmonids for thousands of years. The pikeminnow has learned that smolts are easy prey in certain locations due to human intervention. Huge releases of hatchery fry, poorly planned dams, and lower than historic flow conditions provide easy prey (Brown and Moyle 1981). As popular as the bounty program is for sport fishing, it is perpetuating the "trash fish" myth instead of developing an ecosystem perspective. Persecution of the native apex predator does not seem like a good idea, because removal of apex predators has caused an increase of mesopredators (Prugh et al. 2009). Also, I do not think that protection or deferential treatment of walleyes, smallmouth bass, channel catfish, and northern pike is wise. They are invasive, aggressive, highly predaceous fishes that may cause more problems for the native fish community (including anadromous salmonids) in the future (e.g., see Sanderson et al. 2009). Warming conditions induced by the greenhouse effect could result in a decided advantage to the introduced warmwater fishes, especially in cooler upstream reaches. Petersen and Kitchell (2001) demonstrated that a 1°C increase in temperature will result in a 4% to 6% increase in consumption of salmonids by smallmouth bass and walleyes, and it is noted that the nonnatives prey on all sizes of salmonids.

Hydropower

Hydropower is a product of dams which have extremely negative effects on migrating salmonids. They pose total or partial barriers to the upstream migration of adults and the downstream migration of smolts. Dams also fragment river systems into a string of reservoirs that can be difficult for young fish to navigate. Dams can block access to upstream habitat. Dams can kill fish attempting to escape via turbines. Dams change physical habitat features by flow regulation, including reduction of peak flows, which can reduce downstream travel time and dams harbor invasive fish that compete with and eat small sizes of other fishes. Most or all of these things were known or suspected *before* the 18 mainstream dams in the Columbia River were operational (1938–1975). Montgomery (2003) provides a discomforting discussion: it seems that no one knew if the fish passages proposed would be effective. In the haste to construct dams, written documentation by the U.S. Commissioner of Fisheries stated that the effects on the fish were not understood, and protection of the fishes from "virtual extinction must at the present time be left to chance." The prevailing thought was the fish supply could be provided by hatcheries (Montgomery 2003, pp. 188–189).

Fish passages (ladders) were built on all but the upper five mainstream dams out of the 18 major dams (shown in Figure 30.3). However, there is some level of failure to be expected as a fish attempts upstream passage and this is about 5%–25% for each dam (National Research Council 1996). With four dams in place, survival of adults through the hydropower system ranged from 32% to 56%, and construction of four additional dams dropped the average survival to only 10%–30% (Williams et al. 2001). With changes in the system, the survival rate improved to 31%–59% for chinook and steelhead.

Loss of smolts in downstream migration is about 20% per dam (Raymond 1979), but this is compounded by increased migration time through reservoirs, dams, and tailwaters, which requires energy and induces stress. Also, loss of smolts passing through turbines is about 70% (Gessel et al. 1991), which also attracts predators.

A recent biotelemetry study of chinook smolts (McMichael et al. 2010) found a difference in passage time when three reaches were compared: (1) the Snake River from mouth to lower Granite Dam (110 km, 14 d, 14 km/d), (2) mouth of Snake River to John Day Dam (200 km, 5.5 d, 36 km/d), and (3) Bonneville Dam to estuary (200 km, 2 d, 100 km/d). Also, that study found that 41% of smolts survived their tests and made it to the estuary. Loss in the Snake River alone was 25% of the smolts. More passage time, distance, and slow water also increase exposure to predation.

Smolt-to-adult return in the present Columbia River ranges from only 1% to 4% (Gessel et al. 1991). In mitigation for the dam mortality, smolt transport has been slightly improved by increasing flows (spills) during their migration, catching and barging them downstream (at the expense of extra stress and disease transmission), and killing northern pikeminnow. Everything else pales in significance to the loss of migrating smolts in this system.

Breaching the Lower Snake River Dams

Salmon and steelhead losses are directly related to dams and these losses are sobering. This is no secret, and there is a movement to remove dams that are not deemed necessary or have outlived their usefulness. In the Klamath River basin of Oregon, a utility company has agreed in principle that four dams could be removed for salmon passage, pending government approval (Boxall 2009). There also is strong support to breach or remove the lower four dams on the Snake River. The Army Corps of Engineers has studied it, FWS and NMFS have supported the idea (Montgomery 2003), and the Western Division of the AFS has petitioned the National Oceanic and Atmospheric Administration (NOAA) (letter from R. M. Hughes to Jane Lubchenco, Administrator, NOAA, dated May 4, 2009) to breach them as well. NOAA has made the statement that breaching the four lower dams would result in a higher survival of smolts and a better prospect of recovery than any other initiative. There is much

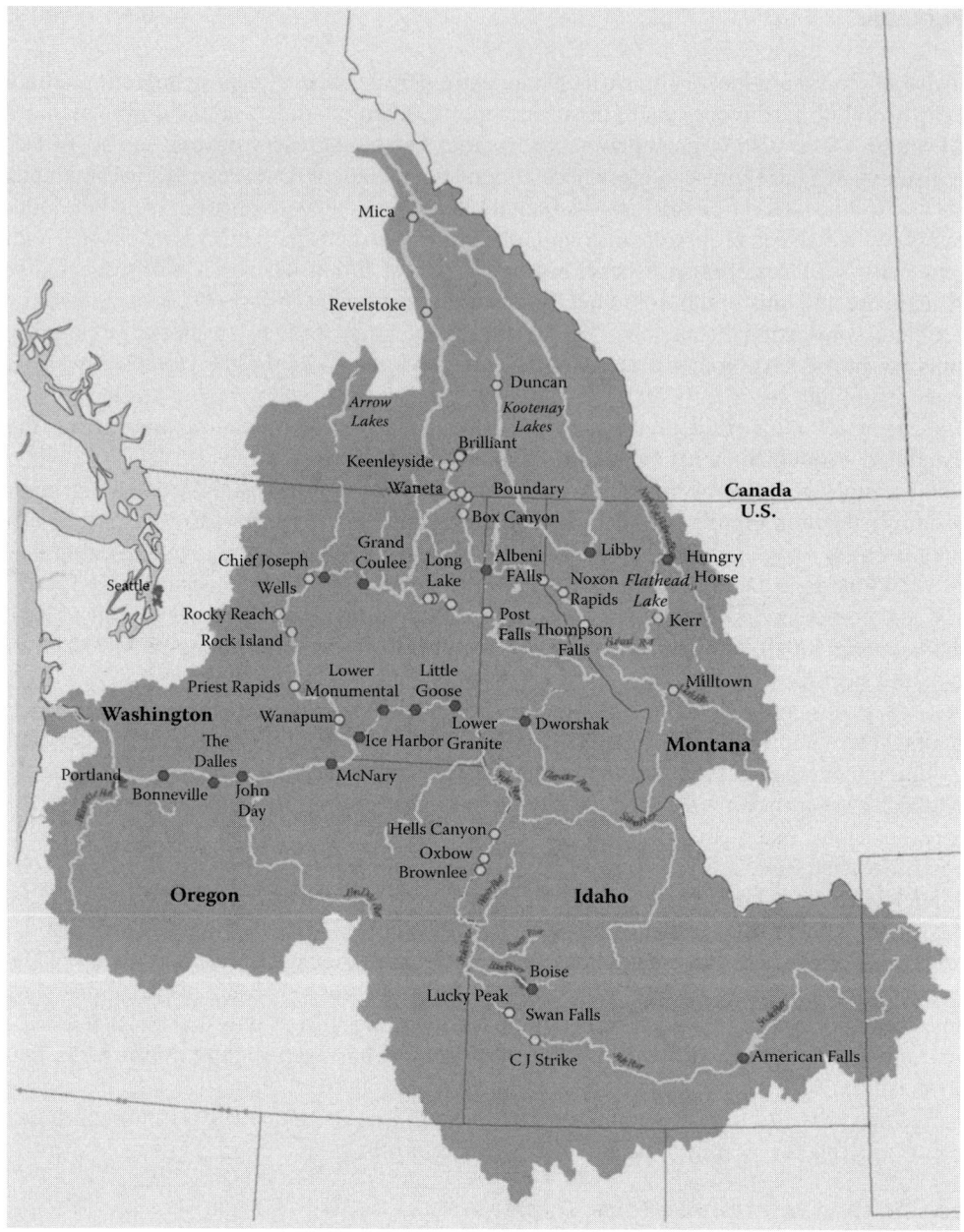

Figure 30.3 Location of 18 major dams. (Courtesy of U.S. Army Corps of Engineers, http://commons.wiki media/wiki/File:Columbiawdams.png.)

support for this, and Congress has passed the "Salmon Solutions and Planning Act" (HR 3503), which requires federal agencies to study dam removal and to produce a scientific analysis of federal efforts to restore salmon and steelhead. There are many tradeoffs to be considered, mostly economic, before the dams could be breached. Hopefully some progress can be made in the very near future.

The Snake River was the most prolific producer of chinook salmon in the Columbia River basin prior to dam building. It supported spring/summer/fall chinook salmon runs, coho runs, and a major run of sockeye. However, the chinook population is heading for extinction. It has suffered extinction of coho stocks, and only four sockeye survived the migration in 2009. Salmon that now descend into the Snake River tributaries have to pass through eight dams, thus only a few survive. The four lower Snake River dams are arguably the worst obstructions to their safe passage, resulting in the loss of at least 25% of the smolts, and adding considerable time and stress to the downstream passage (travel speed only 14 km/d) (McMichael et al. 2010).

Can Science Save the Salmon?

An awareness of declining numbers of salmonids was evident by the 1870s, when salmon hatcheries began to spring up with a goal to increase the catch. The commercial catch of salmon and steelhead subsequently declined from over 40 million lb to about 25 million lb by the pre-dam period in the 1930s, thence to the present catch of about 2 million lb post-dam building. In the past hundred years or so, $billions have been spent and the efforts of hundreds of fisheries scientists have worked on the problem. The result? According to the Idaho Chapter of AFS (Idaho-AFS 1995),

> We can confidently state that continuing the status quo of the lower Snake stream hydropower system will cause extinction of Idaho's salmon. . . . Until you are allowed to differentiate true concerns regarding the scientific merit of fish recovery from the stall tactics of those with economic and political agendas, no amount of science in the world will save salmon, bull trout, cutthroat trout, steelhead, sturgeon . . . .

Very early in the decline of Columbia River salmonid stocks, it became apparent that hatcheries were not providing a solution. But then, the perceived need for building dams to provide power and irrigation was the most important concept in the region. In this case, hundreds of dams, including 11 big ones on the Columbia River mainstream and four on the Snake River, suggest that dam building got out of hand. Now there are simply too many competing interests. It seems evident that the salmon solution is not going to be a technological fix because the problem is enormous. Dams have destroyed and blocked habitat and severely hampered migrations of adult and young salmonids. The remaining river has been so changed that the fish have a difficult time finding their way, and they also are more vulnerable to mortality—in passage and by predation from fish, birds, and humans.

Industry, agriculture, river transportation, and municipalities all have vested interests that are not in harmony with the needs of salmon. These groups evidently have resorted to arguments of bad science, inadequate information, and the need for technological fixes. I understand that many of the fixes were accepted in apparent "technoarrogance" by some fisheries scientists, further compounding the problem. No, science cannot save the salmon. Not alone. But now is the time for scientists to speak out, confront the causes of the problem, and work with engineers and planners to make a strategic plan for implementation. It is tempting to give my opinion on specific things that need to be done. However, I believe we already know what to do for salmon; it is all of the other interests that are troublesome. It does appear that a comprehensive program will emerge to remove some dams, make the rest more user-friendly, and to protect the remaining salmon habitat.

SUMMARY

A global biodiversity crisis and perhaps a mass extinction are in progress, caused solely by human-induced environmental changes. Even widespread and abundant species have become extinct following a decrease in numbers below MVP, when stochastic events result in extinction. Freshwater fishes have been greatly affected, and the causes of species extinctions included interactions with nonnative fishes, habitat alteration, hybridization, and pollution. In response to increased numbers of declining species, new fields of applied science, such as conservation biology, have emerged. In addition, the U.S. Congress passed the ESA, which made conservation of threatened and endangered species a national policy. This act is comprehensive and species protection is extended to subspecies or distinct populations (AKA evolutionarily significant units). Under the ESA, threatened and endangered species are recognized, placed on a federal list, and critical habitat is determined. A recovery plan is then written to guide efforts and a recovery team is appointed. Unfortunately, few species (and no fish) have been recovered to the point that they are removed from the list. A case study of salmon recovery in the Columbia River demonstrates the complexity of recovering endangered anadromous salmonids in a large regulated river. In this study, it is necessary to separate symptoms from causes and to avoid the quick fix and piecemeal approaches of the past.

CHAPTER **31**

Conservation of Fishes II: Understanding the Decline

FIVE CAUSES

In the previous chapter, we identified and discussed the global biodiversity crisis and how the United States has responded, principally through the Endangered Species Act (ESA). Now we will examine reasons for population declines and threatened or endangered status of federally listed species. Those reasons are identified as causes and provided in formal U.S. Fish and Wildlife Service (FWS) and National Marine Fisheries Service listings and recovery plans. More than one abiotic and biotic factor may be involved in the decline of a listed species, and several viewpoints have emerged regarding how these causes may be related.

Recent extinctions have been related to human-induced (anthropogenic) environmental changes, of two categories: direct and indirect human actions (e.g., Ehrlich and Ehrlich 1981). Direct endangerment results from exploitation, such as killing animals for food, fur, and other items, and indirect endangerment would include secondary impacts due to pollution, damming, and so on. Although these two categories are instructional, too many mechanisms are involved and further insight has resulted in dividing causes into other categories. In a well-known example, Jared Diamond (1989; reviewed by Pimm 1996; Moulton and Sanderson 1997) separated recent extinctions according to four causes that he called the "evil quartet," all related to human impacts. Chen (2005) characterized the evil quartet as the "four deadly horsemen of the environmental apocalypse" (the concept of the apocalypse—when civilization ends due to global catastrophes—is presented in the Christian New Testament). In Diamond's evil quartet, he presented causes of species declines as (1) overexploitation, (2) habitat fragmentation and destruction, (3) impacts of introduced species, and (4) secondary extinctions (i.e., the loss of one species resulting in the loss of others).

Separate categories for causes of species declines are useful; however, investigators almost always try to relate a species decline or extinction to a single, major cause. As Raup (1991) has shown, extinctions are usually related to more than one cause. In application of the evil quartet concept, Pimm (1996) noted that once a species declines, one member of the "quartet" is joined by another, confounding the effects and making it difficult to determine which of the causes started the process or will end it. It is important to recognize that extinction can be speeded up by synergism, such as the combined effects of nonindigenous species and presence of other anthropogenic stressors (Ruiz et al. 1999). Thus, the causative mechanism for extinction can be obscured by the impression that each of the factors involved appear to be too small to have caused an extinction. We have learned that a First Strike that results in major population decline may not be the cause that results in extinction of the survivors.

In evaluating the decline and endangerment of fishes, at least five causes are typically listed. I hasten to mention that (following the example of the "evil quartet") we could add a sixth factor to include secondary extinctions (AKA *coextinction*) caused by ecological imbalance (Terborgh and

Winter 1980). However, it is a difficult task to link extinction of one species to extinction of another, which could occur perhaps due to mutualism (Koh et al. 2004). For now, this cause has not been sufficiently explored; however, it is likely to become more important as species extinctions are more fully understood.

The five causes are the following:

(1) Physical habitat alteration (destruction, alteration, and replacement of habitats, or abiotic habitat alteration).
(2) Competition and predation from introduced species (also considered as biotic components of habitat).
(3) Hybridization (breakdown of reproductive barriers, reproductive loss, genetic introgression). Note that this category could have been placed under the introduced species one above; however, hybrids of native species also occur.
(4) Chemical alteration, contamination, or pollution of water (including poisons and nutrients). Note that this category could have been included in physical habitat alteration above, but this is not intuitive.
(5) Overharvesting (over exploitation).

All five of these factors have been implicated in the extinction of fishes globally and in North America and should be considered as "usual suspects." Recovery of endangered species will require action to mitigate or remove those factors. However, we must consider that any and all of these factors can and often do occur at the same time and affect different life history stages of each fish species. Also, what environmental changes will the future bring? We do not expect human populations to decrease in the near future and we are only beginning to understand how potential effects of predicted global climate change might increase or decrease the severity of causes.

In this section, we will address the effects of the five factors implicated as causes of species decline and endangerment in aquatic systems, individually and in combinations, recognizing that physical habitat alteration and introduced species have been major factors influencing freshwater fishes. Overfishing is of special concern because it is the factor most likely affecting species declines in marine systems. Mitigation of the effects associated with each of the five causes given will be discussed, and major problems and examples of affected systems presented. Many of these problems have already been identified earlier in this text, however here we strive for synthesis and focus on conservation biology. First, we consider effects of these factors in major aquatic systems, and then we will illuminate some of the major issues associated with each of the causes.

PHYSICAL HABITAT ALTERATION

Effects

Lakes

Deforestation of natural lake basins raises temperature, turbidity, and organic input into lakes. Draining and filling of marshes, isolation of wetlands, blockage of tributary streams, dredging of harbors and channels, digging canals, and so on all effect lacustrine fish populations. Natural lakes in the United States have already been settled by humans and much of the damage is done. Federal and state permits are now required in the United States for all of the impacts named above, especially for those that affect waterways and wetlands. Examples include the Laurentian Great Lakes and Lake Victoria, both studied previously.

Streams

Habitat alteration includes deforestation of the basin, grazing of riparian zone, road and building construction, dams, diversions, channelization, dewatering, artificial regulation of water flow, habitat fragmentation, and increased or decreased turbidity and sediment. All affect stream fishes. Headwater streams have been degraded by increasing temperature and siltation, while warmwater streams are being disconnected from floodplains and converted to systems of dams, diversions, reservoirs, and cold tailwaters. Examples are almost all large rivers, including the Columbia, Missouri, and Colorado rivers. Few natural large streams are left. Existing streams controlled by federal agencies are subject to re-regulation of flows (originally designed for hydropower, flood control, etc.) to protect endangered species. Extension programs strive to protect riparian zones.

Estuaries

Watershed development has changed the nature of water entering estuaries by increasing silt load, nutrients, and contaminants. Fishes and shellfishes also are affected by damming and diverting small coastal streams, bulkheading and filling of shorelines, and dredging of channels. Disposal of dredge spoil remains an issue in harbors and along intercoastal waterways (Figure 31.1). Examples: There has been a great loss of diadromous fishes, including clupeids along the eastern coast of the United States due to dams and diversions of small streams. Fish passageways have been constructed to aid fish movement but much more needs to be done to increase their effectiveness, and more are needed. Another problem is polluted and anoxic sediments and various techniques are being used to reduce their impacts, including upland disposal of dredge spoil. Many areas have been closed to commercial fishing due to habitat damage and changes to ecosystems. A major disaster is in

Figure 31.1 Dredge spoil disposal along the inland waterway.

progress in Chesapeake Bay (see case study below) due to land use change in the basin, pollution, overfishing, and habitat destruction in the estuary.

Marine Systems

Fish habitats and migration routes have been changed by the construction of harbors, waterways, jetties, wharves, bulkheads, piers, and so on. Increased beach erosion and dredging have increased sediment loading. Fish populations also are subjected to commercial and recreational exploitation in which physical and biotic components of habitat are affected by repetitive trawling, dredging, netting, and human presence (tourism and diving). All of these things change habitat suitability for feeding, growth, and reproduction of fishes. High technology has even permitted deep trawling of seamounts, and left few places untouched. Increasing pollution, especially by contaminants that also have degraded habitat. Examples: the North Sea and Northwestern Atlantic, where there has been closure of some areas to fishing based on vulnerability, degree of habitat damage, and depletion of stocks.

Issues

Anthropogenic habitat change, whether due to conversion of one type of habitat to another or due to system simplification is a major problem in aquatic ecosystems, especially streams and rivers, estuaries, and increasingly in the marine inshore and neritic zones. Humans are very good at converting natural systems into artificial ones and in the process, disrupting communities, ecosystem processes, and system stability. Habitat alteration in many instances was accomplished before the life history requirements of native fishes were well understood, and this lack of understanding has made recovery of endangered fishes difficult. Major projects such as large dams have been considered as permanent features in the past due mainly to sociopolitical issues, but also because they are difficult and expensive to remove, even after they outlive their usefulness. Nonetheless, dam removal is occurring as knowledge grows and attitudes change. Presently some mainstream dams in the Pacific Northwest are on the chopping block. But for the most part, large riverine dams are at least a semipermanent part of the environment. Dams present some big problems, and two habitat altering features of dams bear more scrutiny due to their threats on populations of native fishes: habitat fragmentation and flow regulation.

Habitat fragmentation is a physical habitat alteration that can cause a great many problems for fishes, including a loss of genetic variation by isolating populations. Terborgh and Winter (1980, p. 132) characterized the results of habitat fragmentation as "a deteriorative ecological chain reaction which begins with the stochastic loss of rare species" and may include top predators. Their loss can result in a "cascade" of coextinctions. Eventually the affected system is expected to reach a new equilibrium, but at the loss of diversity and efficiency.

An example of a fragmented system is given by the Colorado River system because there are dammed and undammed sections that can be compared. Figure 31.2 shows mainstream barriers consisting of dams, reservoirs, and cold tailwaters, which reduce native fish habitat and fragment their range. Remnant populations of the big river fish community are most abundant in the remaining connected and undammed portions of the Green and upper Colorado River subbasins, and an example of isolation is given by a population of humpback chub that is restricted to the Little Colorado River in the Grand Canyon. The chub is isolated from other chub populations by Glen Canyon Dam and isolated from warmwater habitat by discharge of cold water from the dam.

Severe impacts of habitat fragmentation have been studied in the endangered Rio Grande silvery minnow (*Hybognathus amarus*), a pelagic spawner that has lost about 95% of its range. A scientific advisory panel of five senior scientists reviewed all of the information about the research,

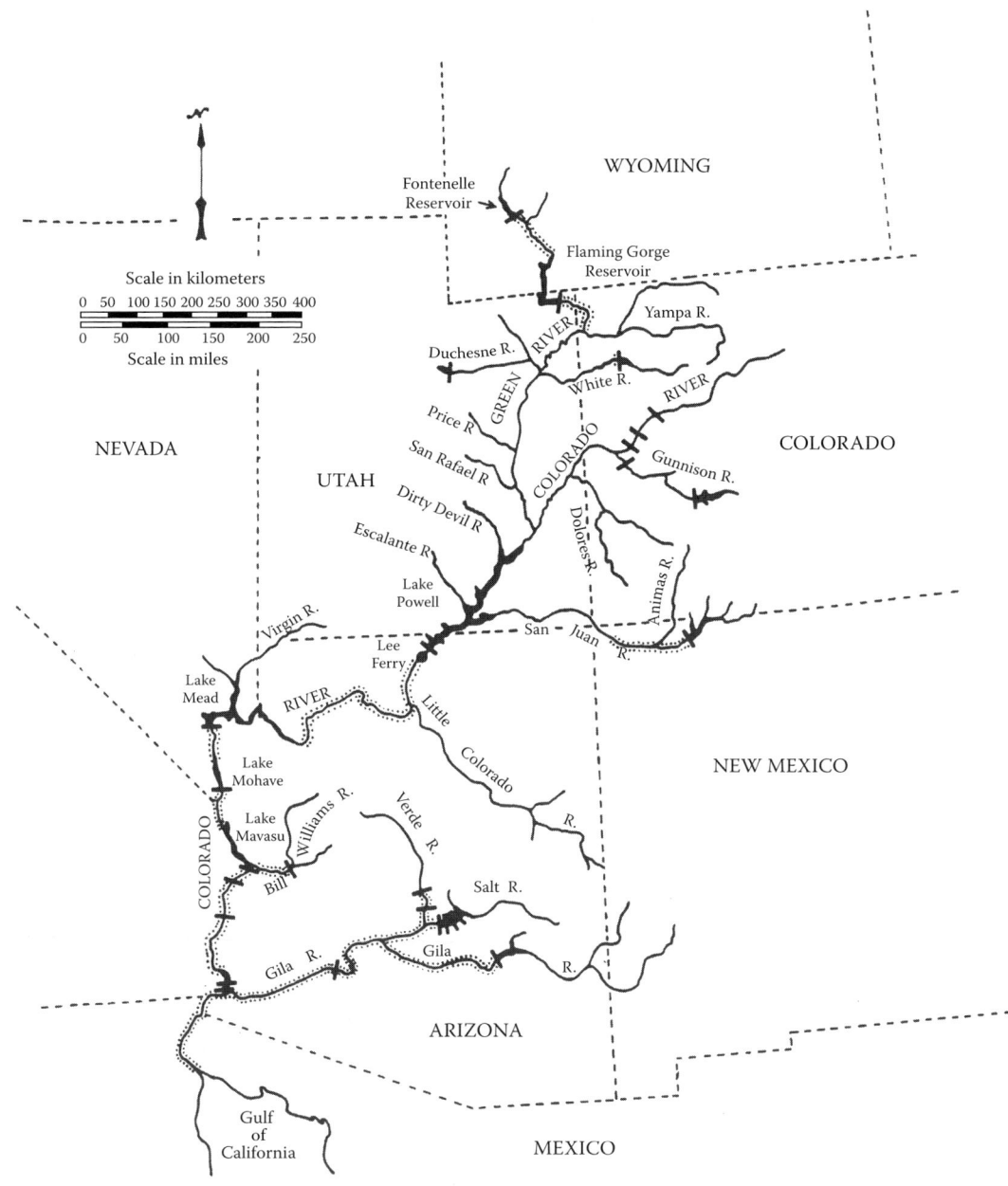

Figure 31.2 Fragmentation of the Colorado River. (From Tyus, H. M. and Karp, C. A., *U.S. Fish and Wildlife Service Biological Report*, 89(14):1–27, 1989.)

monitoring and management of the fish and produced a draft report recommending what seemed to be reasonable short-term (34) and long-term (21) recovery measures. The report was never finalized; the recovery program collapsed, and the recovery coordinator left (Martin 2005 provides an analysis of the recovery measures). In brief, the Rio Grande silvery minnow has a life history strategy that is not compatible with the heavily altered system, which includes five dams. Death of eggs and larvae in transit downstream produces heavy losses in annual reproduction and a large variance in reproductive success. In essence, fragmentation reduced effective population size (Ne) to a very low level and resulted in a "sweepstakes mismatch" phenomena in which reproduction is not well adapted to the environment. High reproductive success is possible, but only with exact conditions, resulting in the analogy of sweepstakes winning (Osborne et al. 2005). If the present situation continues, more than 500,000 adult minnows are needed to produce Ne of 500 (95% variation maintained), and more than 5 million adults will be needed for Ne of 5000 for which genetic variation would persist over evolutionary time (Alò and Turner 2004).

Alteration of the natural hydrograph by flow regulation for hydropower, flood control, irrigation, and other purposes is another pervasive problem caused by damming rivers. Large dams can store water during peak flows and release them in low flow (base flow) over the remainder of the year. What occurs is a more steady-state (unnatural) system. Dams and their operation can reduce or deprive native fishes of optimum or even satisfactory habitat conditions that they need at various times of year (e.g., spawning, rearing of young). Of course, dams also have other effects as previously discussed for reservoirs.

In considering the instream flow needs of fishes, it is clear that a flow regime that had natural seasonal fluctuations is necessary to maintain life history stages for riverine adapted fishes. Several techniques have been used to estimate flows needed by fishes, most requiring some judgment from fish biologists collecting data. In a move to become more quantitative, some modeling techniques (e.g., the instream flow incremental method and physical habitat simulation [PHABSIM]) have been seen as better ways to avoid subjective decisions and litigation. However, not only are such models rarely field-tested (i.e., to determine if their implementation results in an increase in fish abundance) (Armour and Taylor 1991), but their use must also be questioned in large warmwater systems. Furthermore, Williams (2010) stressed that PHABSIM outputs are not statistically valid in determining the amount of available habitat under varying flow regimes. In addition, our studies on the big river fishes (e.g., Tyus and Karp 1989) have shown that life history stages can be adapted to habitats that are seasonal features of the natural hydrograph. In one species, the larvae live in protected shoreline embayments, and the adults utilize eddies. Those habitats have little or no measurable flow, but both habitats require flows to produce and maintain them.

The bottom line: Fish habitats also must satisfy the needs of all life history stages, each of which may require appropriate flows at certain times of year. There is no substitute for understanding the ecological requirements of target species and there is no "quick fix" in complex systems. When flow requirements are determined, their implementation must be monitored to determine if they meet expected goals. The overriding goal is to produce more fish!

Attempts to mitigate the damage caused by building federal dams were first subjected to the Fish and Wildlife Coordination acts (1934 et seq). However, at first, mitigation was to lessen the loss of hunting and fishing opportunity, not to protect or manage ecosystems. This was done in most instances by constructing fish hatcheries, refuges, boat launching sites, or setting aside lands for hunting. After passage of National Environmental Policy Act (NEPA) (1969), all federal projects needed to evaluate project alternatives (which included no action as well as different sizes and locations for construction), and mitigation included selection of alternatives as well as reducing habitat losses. After passage of ESA (1973), conflicts with listed or proposed species had to be avoided and enhancement measures taken to aid recovery if possible.

Habitat alteration and destruction is also associated with some types of fisheries and gear. Although overfishing is given a separate category here, it consists of both population and habitat impacts. Impacts associated with mobile fishing gear on structured bottom habitats can be severe due to regional disturbances and loss of habitat complexity. In a review of the effects shown by 22 studies of the effects of mobile fishing gear, Austen and Langton (1999) found that the effects varied depending on gear and habitat types; however, all studies recognized a decrease in fish diversity and a reduction in habitat complexity over time.

INTRODUCED SPECIES

Effects

Lakes

The dominant sport fishes in reservoirs and most of the lakes in the United States (and throughout the rest of the world) are preadapted introduced species. Many of those cases are due to stocking various salmonids into coldwater lakes, of which many had no fishes. However, virtually all warmwater lakes and reservoirs are stocked with numerous introduced species, including exotics and nonnatives (i.e., nonindigenous). Declines of native fishes and proliferation of the nonnatives are usually the result. Effects have been most severe in older lakes that have endemic populations of lacustrine adapted forms. For example, Lake Victoria has been severely damaged by stocking of Nile perch, which presumably resulted in the extinction of hundreds of endemic cichlids. Introduction of lake trout into Yellowstone Lake has been received as a catastrophe that affects resident cutthroats and grizzly bears. The introduction of fish as top predators in lakes has resulted in cascade effects (Li and Moyle 1999). In the case of Yellowstone Lake, a fish removal program was initiated to reduce or remove lake trout. However, the well-adapted lake trout will be difficult if not impossible to remove. In another case, the parasitic sea lamprey devastated the lake trout in the Great Lakes, and larvicides are needed to control lamprey numbers.

Streams

Coldwater streams have been greatly affected worldwide by stocking various nonnative salmonids. In larger streams and rivers, most of the stocking was done to place gamefish into reservoirs. Examples: In coldwater streams and lakes, the threatened greenback cutthroat trout was beset by predation, competition, and hybridization with introduced trouts. One of the most studied large rivers affected by nonnative introductions is the Colorado River, where at least 67 species of nonnative fish have been introduced (Valdez and Muth 2005). Essentially the lower portion of the river below Lake Meade has been converted to a system of dams, diversions, and tailwaters, and the resulting reservoirs and remaining stream sections were stocked with predaceous fishes, including the voracious flathead catfish and the striped and largemouth basses (Mueller et al. 2005). The entire endemic fauna of the mainstream river is almost gone, replaced with a new fauna of about 44 forms (Minckley 1982), many of which were preadapted to the changed conditions in the river, and especially in reservoir and tailwater reaches.

The upper Colorado River system has been less physically altered, but its fish fauna has declined to the point that two of four listed mainstream fish are functionally extinct (maintained by hatcheries). Implicated in the decline of native fish are large coolwater and warmwater fish predators and competitors such as northern pike and smallmouth bass, omnivorous channel catfishes, and smaller fishes or young of large species that compete with and prey on young life history stages (eggs and larvae) of endangered fish (e.g., Carpenter and Mueller 2008; Karp and Tyus 1990). Olden et al. (2006) identified life history strategies associated with invasion and takeover by invasive nonnative fishes in that system. They found that nonnatives took advantage by filling niche opportunities provided by human alteration of the natural environment. Many of these nonnatives prefer slower flows, do not rely on fluvial conditions in reproduction, tolerate warmer waters, and are trophic generalists.

The impact of nonnative fishes on native fish faunas is finally being understood. However, there are hundreds of other organisms that also have been introduced into aquatic habitats, and their impacts on fish communities are less well studied. For example, introduction of nonnative crayfish into a system that had no native crayfish has resulted in competition with native fishes (Carpenter 2005), and trophic level changes in the system. In a study that I did with Neal Nikirk in the late 1980s (Tyus and Nikirk 1990), we found no crayfish in the stomachs of channel catfish in the Yampa River; however, when I returned to the river 10 years later (with an FWS team led by biologist Mark Fuller), every catfish stomach (n > 50) that we examined in Yampa Canyon (Dinosaur National Monument) contained crayfish body parts.

Efforts to control the numbers of introduced fishes that are established in large streams has generally failed, although in some species the larger size fishes (e.g., channel catfish) can be reduced in number. Some small streams can be poisoned with fish toxicants, first removing and then returning native fishes, such as in Rocky Mountain National Park.

Estuaries

Ballast water, fouling organisms, and discarded cargo have resulted in extensive introductions of exotic organisms into estuaries and bays. So many invasive species have been introduced into estuaries by shipping that it is tempting to say it is too late to do much about stopping the ballast water problem. In some ports, like the Great Lakes, where the zebra mussel problem started, ships have been forced to exchange freshwater ballast for salt water at sea.

The nonnative problem is most acute in locations where there is a busy seaport. Examples: About 250 invasive species have been captured in San Francisco Bay. A recent invader, the Asian clam (*Potamocorbicula*), was introduced after 1985, and by early 1990, the organism covered large areas of the bay. The European green crab (*Carchinus*) was introduced in 1989 and has been extending its range. The list of invaders is long and also includes seaweeds, comb jellies, sea squirts, and so on (Castro and Huber 2007). Introductions have been so numerous that some areas of the bay are called "introduced communities" because all of the inhabitants are nonnative, including filter-feeding worms and crustaceans, occurring in muddy bottoms in high densities, and bryozoans, sponges, and sea squirts that occur in fouling communities (Carlton and Ruiz 2005). Moving more inland into the Sacramento–San Joaquin Delta, Feyrer and Healey (2003) found the area dominated by nonnative fishes: captured were 33 fishes of which only eight were native, and the natives comprised less than 0.5% of the total number of individuals captured. Unfortunately, the ecological impacts associated with all of these invasive species are not well known, but the interaction of nonnative species with other anthropogenic stressors exacerbates the problem (Ruiz et al. 1999).

Marine Systems

The relative impact of introduced fishes on marine fish communities is less well understood than in freshwater. However, invasion of marine habitats by nonnative species has occurred throughout the oceans of the world. Vectors include ocean going vessels, canals, aquaculture, and so on (Table 8.1). Carlton and Ruiz (2005) have provided examples of such invasions for different marine habitats and included examples of ecological consequences. However, most of those invasions are unknown due to little study. Most invading species from 1800 to the present would probably be considered to be natives, although it is estimated that over 1000 fish invasions would have occurred from 1500 to 1800, and several thousands more could have gone unnoticed since then (Carlton 1998; Carlton and Ruiz 2005). Invaders include many marine groups such as crabs, clams, jellyfish, snails, sea squirts, diatoms, amphipods, and so on. Few fish are known, but examples are striped bass transferred from the Atlantic to Pacific oceans; goatfish from the Indian Ocean to the Mediterranean Sea; and snapper, tilapia, and grouper to Hawaii. In addition to fish, some small invertebrate invaders can cause significant damage to fish habitats when introduced into new systems, such as the case of two species of isopods. A burrowing isopod (*Sphaeroma quoianum*) from New Zealand has caused considerable damage and erosion to marshes in California, and an Indian Ocean boring isopod

(*S. terebrans*) works on taproots of mangrove trees in Florida, interfering with seaward progression of the trees (Carlton and Ruiz 2005).

The relatively low fish biodiversity of the North Atlantic Ocean has been associated with vulnerability to overfishing and collapse of some populations. Briggs (2008) has proposed a proactive approach to improve "poor" ecosystems in the North Atlantic by introducing nonnative fish from the North Pacific Ocean. This controversial approach is based on the premise that the North Atlantic fauna was impoverished and the addition of more species would produce a more stable and productive fishery. In a rebuttal article, Courtenay et al. (2009) pointed out that other nonnative introductions had not been positive and suggested that the cost outweighed the risk. They also stressed that the main problem in the North Atlantic was overfishing with attendant habitat damage and poor fisheries management. Until those problems were solved, other "solutions" seem moot.

Issues

The continued proliferation of nonnative species in the world is truly staggering. The U.S. Office of Technology Assessment (OTA) (1993) discovered that over 4500 exotic species have established populations in the United States. Such introductions are producing similarities in communities and systems worldwide, which has been called *biotic homogenization*, in which a few "winners" that do well in human-altered environments are replacing a great many "losers," producing a homogenous biosphere (McKinney and Lockwood 1999). This phenomenon has been considered an ecological holocaust because it lowers biological diversity (Williams and Meffe 1998). Unique faunas are disappearing, with the result that states are becoming very similar with respect to the number of fish species shared, as endemic faunas are invaded. For example, the states of Nevada, Utah, and Arizona now have fish faunas that are dominated by nonnatives (over 50%). The homogenization process is resulting in a disappearance of specialists and replacement by generalists.

Kolar et al. (2010) traced the history of fish introductions in the United States, which consisted of three waves and date back to the introduction of goldfish into the Hudson River in the late 1600s. Creation of the U.S. Fish Commission in 1871, led by Spencer Baird, resulted in national support for nonnative fish introductions. The United States received nonnative fish from other countries, cultured them, and distributed them across the country. These included "wonder fish" common carp and brown trout, which remain firmly established today.

Arguably, the introduction of nonnative species into habitats occupied by other faunas is just as important in producing fish extinctions as are physical habitat changes. These introduced species include alien (nonnative and exotic) species from other continents and species that are residents in one geographic area or basin, but not native in areas they are introduced. Intentional introductions of nonnative fishes and forage species (such as crustaceans) have been common in freshwater systems due to government programs and illegal stockings by bait buckets. Intentional introduction of nonnative and alien fishes of many species also occurs due to the Aquarium Trade, which is a growing problem (Strecker et al. 2011). Unintentional introductions occur by escapement from ponds and lakes, by contaminated stocking of other species, and so on. One of the biggest problems with nonnative introductions is the stocking of areas that have been converted from a natural to a human dominated system, such as a reservoir. In this case, the stocked species are usually better adapted to the new system than are natives, and they compete with or prey on the native (including endemic) species. Thus, even though some systems were altered in physical habitat it is probable that more native fishes would have persisted there without the introductions. Using Raup's (1991) concept, native fishes are thus hit with a 1–2 punch or double first strike!

Fuller et al. (1999), in the introduction to a monumental work on nonindigenous fishes, pointed out that some terrestrial introductions of nonnative species have provided socioeconomic benefits for humans, and some intentional introductions may at least be considered to be a "mixed blessing." However, beneficial aspects of introductions have been extolled by agencies responsible, while detrimental impacts have not been anticipated or studied. Curiously, no unintentional introduction of an aquatic species has been considered beneficial, and in general, they have been considered harmful and even catastrophic in some cases, "creating a growing economic and environmental burden for the country" (OTA 1993). Taylor et al. (1984, p. 352) stated that harmful effects to native fish populations should be a "foregone conclusion," and that the evidence was so compelling that a "no effects" argument would be implausible to the point of straining "one's confidence in ecological principles."

In a study of causes for fish endangerment in the United States, Lassuy (1995) found that adverse interactions with introduced species occurred in 70% of imperiled fishes. Furthermore, Ross (1991) reported that in 23 of 31 cases in which fish had been introduced into stream communities, declines were detected in native fish populations. Miller et al. (1989) listed 30 fishes that became extinct in the United States during the twentieth century. A total of 24 (80%) were affected by introduced species, and three fishes went extinct solely due to introduced species. In contrast, only 19 fishes were affected by physical habitat alteration (64%) (table presented by Ross 1997).

Why were intentional fish introductions made in the first place and why do they continue? An early investigator, Laycock (1966), started his book by invoking the "Song of Hiawatha" by Henry Wadsworth Longfellow, in which humans are not satisfied with life in a natural paradise. Laycock (1966, p. 223) ends his book with

> . . . we look away from our magnificent heritage of native wildlife and its habitat to cast covetous glances at strange creatures in far corners of the world. Then, ignoring lessons of the past we start another round of wildlife roulette.

Early stockings of nonnative fishes, such as common carp were done to introduce familiar fishes into unknown environments, and to fill what were considered to be "vacant niches" with no damage to fish communities expected (an erroneous notion, as discussed by Moyle et al. 1986; Courtenay 1995; Li and Moyle 1999). Instead, harmful effects to native fish populations of all sorts are so well documented in the literature (e.g., reviewed by Taylor et al. 1984; Fuller et al. 1999; Li and Moyle 1999) that adverse impacts are anticipated as an ecological reality. Many fish introductions were made by fisheries agencies in an effort to improve fishing in reservoirs or for other purposes, ignoring potential large-scale consequences. In trying to "improve nature," they have committed what Moyle et al. (1986) call the "Frankenstein effect," in which conditions in the new fish community "improved" by the introduction of the Frankenstein fish turn into a nightmare of unintended consequences. Poststocking studies have demonstrated inferior performance in survival and reproduction in hatchery produced fish compared to wild stocks (e.g., White et al. 1995). Increasing concern has been expressed over genetic risks and hazards from hatchery introductions, which can result in extinction, loss of within- and among-population variability, and domestication (Busack and Currens 1995; Leary et al. 1995). The American Fisheries Society (AFS) has developed guidelines for introductions of all aquatic species (Rasmussen 1997) in an effort to address these and other issues (e.g., to reduce habitat and trophic alterations, deterioration of wild gene pools, diseases, and interactions with endangered species). More pointedly, guidelines for conservation of freshwater fish in propagating and translocating them have been provided (George et al. 2009).

Introduction of alien fishes and resulting problems they cause in natural systems, have been reviewed by Courtenay (1995) and Kolar et al. (2010). Problems have been so great that invasions have been considered "biological pollution." For example, recent fish invasions of catastrophic potential have resulted in battles to control or kill two species of Asian carp (introduced for phytoplankton control) and prevent their access to the Great Lakes (Figure 31.3), and to control introduction of snakeheads (Figure 31.4) (an imported food fish).

Bighead and silver carp (weight up to 100 lbs) are considered a true menace (Finney 2011) because they compete with juvenile fishes for food and space. Apparently, two of the large carp passed through a $9-million electrical barrier constructed to exclude them from the upstream Des

Figure 31.3 Asian carp, an exotic pest. (Courtesy of USFWS, http://www.usfws.gov/AsianCarp/EricAsianCarp .jpg.)

Figure 31.4 Snakehead: "fishzilla." (Courtesy of U.S. Geological Survey (USGS), http://cars.er.usgs.gov/pics/snakehead/snakehead.html.)

Plaines River and the Great Lakes. The fish were found in the Chicago Sanitary and Ship canal in December 2009, where 2200 gal. of fish poison was thereafter dumped to kill them, and ship locks were closed to prevent their passage (Belkin 2009). However, a 19.6-lb bighead carp was subsequently captured in Lake Calumet in June 2010 (Woldt 2010). This lake is in the Chicago Area waterway and only 6 mi. from Lake Michigan. Regarding the snakeheads, by 2004, at least two species of these voracious fish had been found in California and the northeast. Discovery of northern snakeheads (*Channa argus*) in the Potomac River of Maryland caused a stir through graphic television coverage. By autumn of 2010, snakeheads were also reported from Arkansas, Florida, Pennsylvania, New York, and Massachusetts (Campbell 2010). Likely introduced by purpose as a lifesaving gesture by Orientals, live fish have been sold in supermarkets. Called "fishzilla" by National Geographic (Wikipedia, http://en.wikipedia.org/wiki/Snakehead_fish), these predators can breathe air and live up to 4 days out of water. They can breed several times per year, and can potentially disrupt aquatic systems. Efforts to eradicate them and others continue, but once established, eradication of invasive fishes is likely impossible (Kolar et al. 2010).

Native fish communities also have been affected by other taxa as well, some of which have made ecosystem level changes, affecting more than one fish community. The zebra and quagga mussels introduced into the Great Lakes are examples. The establishment and proliferation of these mussels have resulted in greater macrophyte growth in shallows, a new benthic zone due to their reefs and the deposit of feces and pseudofeces, lower standing crops of the plankton by consumption, and others. They have eradicated native mussels, increased populations of sunfishes, increased spawning habitat for walleye, provided food for some waterfowl, reduced food for planktivorous fishes, and so on.

Recovery of endangered species may require the reduction in numbers of nonnatives to a level that they are no longer an impediment to recovery action. There are at least two complications for fishes that are exposed to multiple predators: (1) the need to consider the potential for predation and/or competition as occurring at all life history stages and (2) the need to consider potential effects of invasive species in multiple aquatic habitats or ecosystems that may be used, especially for diadromous species (e.g., see Sanderson et al. 2009). It is truly amazing that aggressive and highly predaceous nonnative game fish are generally considered to have little or no affect on rare or endangered fishes. In fact, impacts of the nonnatives have not been adequately studied. In my opinion, the hoards of introduced fish such as walleye and smallmouth bass present in some mainstream reservoirs do pose a threat to recovery of listed species. As pointed out in chapter 25, predation on fish larvae is difficult to detect in nature (e.g., Schooley et al. 2008; Legler et al. 2010) and the potential needs to be evaluated in the laboratory as well as by careful field studies.

The control of nonnatives has already been presented earlier (chapter 9) as a major part of the recovery for the greenback cutthroat trout in Rocky Mountain National Park (i.e., use of fish toxicant to kill nonnative trouts in lakes that have barriers to further invasion). However, there are sociopolitical issues that make control projects even more difficult than they already are. Even though the public and even fish management agencies may support endangered species in concept, they may be unwilling to endorse programs to eliminate alien fishes. That is why the eradication of alien species has been called "the nasty necessity" (Temple 1990, p. 113). In truth, fish control measures have been a customary part of game fish management, and the removal of "trash fish" has been promoted for at least 100 years. Wydoski and Wiley (1999) provide a comprehensive review of techniques, which fall generally into three categories: mechanical, chemical, and biological. Mechanical means include nets, traps, and barriers, such as the electrical barrier mentioned above, and also more sophisticated measures such as the guidance devices mentioned at the Tracy facility in California. Chemical means include toxicants that may be selective for a life stage as in the case of the parasitic sea lamprey, or nonselective such as the fish poisons rotenone or antimycin, which have different effects on macroinvertebrates (Sanderson et al. 2009). Biological means might be to sterilize nonnative fish, or to introduce native predators that prey on nonnatives.

Impacts of introduced species on populations of native fishes are many, including interspecific hybridization, which has been placed in another category by itself. Unfortunately, introduced fishes are most likely impossible to remove and they are difficult and costly to control (Kolar et al. 2010).

OVERFISHING

Effects

Lakes, Ponds, and Springs

Native fishes in larger lakes can be quickly reduced to low numbers by commercial, subsistence, and sport fishing. In most cases, they have been almost entirely replaced with introduced species and fishing is now heavily regulated. Pond culture has expanded into a large endeavor in the United States and provides much fishing opportunity using nonnative species or stock. Overfishing for lake trout and Atlantic salmon quickly depleted stocks early in the development of the Laurentian Great Lakes, and other species (e.g., lake sturgeon) also were extirpated from some locations and reduced to very low numbers in others. Overfishing and introduction of nonnatives done together can be especially damaging as in the case of the parasitic sea lamprey in the Great Lakes, and extinction of the yellowfin cutthroat trout of Twin Lakes, Colorado. Large lakes are difficult to manage, however fishing restrictions can work. Ponds have been used to cultivate food and game fish for a very long time, and now they are also used to propagate and protect endangered fishes as refugia and to supply fish for restoration projects.

Rivers and Streams

Larger native riverine fishes are universally in danger of extinction. While it is difficult to prove overfishing as the cause of the decline of these fishes, most of them were fished commercially and for subsistence. As an example, in coldwater streams of the Rocky Mountains, overfishing combined with other impacts reduced populations of native cutthroat trouts to very low levels. Large rivers such as the Ohio, Missouri, and Mississippi were subjected to heavy commercial fishing that reduced the abundance of slow-growing, larger species. In the Missouri–Mississippi system, heavy commercial and illegal netting of sturgeons and paddlefish for caviar coupled with reservoir construction has driven pallid sturgeon to the brink and relegated many populations of paddlefish to hatcheries for survival. Records are sparse, but anecdotal evidence shows that western streams such as tributaries to the Colorado River were subjected to haul seining during vulnerable low-flow periods for Colorado pikeminnow and native suckers. Fish were salted and/or dried and sold for winter food. Strict laws and regulations now protect declining fishes from fishing pressure.

Estuaries

It appears that all of the larger estuaries have experienced overfishing in the recent past and many stocks have declined as a result. Large shallow estuaries, including bays and sounds along the eastern U.S. coast, are easy to fish and productive. Examples: Chesapeake and Narragansett bays have been heavily overfished for oysters, striped bass, menhaden and other target and bycatch species. (See case study that follows.) Better management is needed, especially for larger estuaries that are fished by trawling, dredging, and netting, because extreme damage can result in benthic habitats such as underwater grass beds and reefs.

Marine Systems

As mentioned above, overfishing seems to be the rule for marine fisheries. Hutchings and Reynolds (2004) reported a median reduction of breeding population size in 83% of the 230 populations they studied. Examples: Hundreds of marine stocks have been overfished in the North Atlantic Ocean from the North Sea to the coast of North America. A moratorium on commercial fishing of NW Atlantic groundfishes has been in existence for almost two decades with no recovery in evidence, and some Atlantic cod stocks have declined 99.9% in 30 years. No doubt, the genetic composition of many of these stocks has been compromised, and extensive damage has been done to some bottom habitats. Numbers of large pelagic fishes, including sharks and tunas also have recently been in decline.

Issues

Overfishing has been discussed in chapters 17, 18, and 23. It was previously thought that the oceans could provide an inexhaustible supply of fish for humans (Daniel and Minot 1954), and for at least 100 years, it was believed that fishing pressure had nothing to do with recruitment. How ironic it is that overfishing occurs where major fisheries exist, whether in great lakes or marine systems. This lack of constraint is a serious problem for the marine fishing industry. According to a 2008 report by FAO and World Bank (discussed on Wikipedia), fishing fleets are losing about $50 million each year due to poor fisheries management and depleted stocks. However, not only is there a socioeconomic problem, but also major ecosystems have been disrupted by overfishing. Historically, the industry has been in denial that fishing causes problems, even when there is clear evidence of change. Overfishing not only affects community demographics but also is associated with habitat damage caused by mobile gear such as flounder dragging, scallop dredging, and otter trawling. In an example in the Gulf of Maine, declines were accompanied with eight to 10 times an increase in dogfish sharks, skates, and stingray (Roberts 2007). The fishers began selling these pesky fishes for what they could get, and continued to trawl. But in a decade or so, numbers of these fish declined also. Now even greater damage had been done to bottom habitats and it is unknown when, or even if, the historic fish community will recover.

While it is doubtful that overfishing would be so thorough as to catch every individual and drive a fish to extinction, there is little doubt that overfishing can reduce the numbers of large populations to a great extent, removing certain genotypes, and reducing, degrading, and even destroying habitat. Thus, overfishing could be considered as an example of a "first strike" (sensu Raup 1991) affecting distinct populations (i.e., stocks). The notion that increasing economic costs will terminate fishing before biological extinction can occur could have merit; however, this does not apply to nontarget species or species of high value (Dulvy et al. 2003).

HYBRIDIZATION

Effects

Lakes

Coldwater lakes of North America support populations of various native salmonids, most of which are not native to the location they occur. Instead, the fish populations are managed by game and fish agencies, mostly by stocking rainbow, brown, brook, and lake trouts. Examples: Hybrid introgression (principally with hatchery strains of rainbow trout) has been common in western North America (Leary et al. 1995), and introgressive hybridization with rainbow trout has been implicated as the main factor in the decline of the many strains of pacific salmonids that developed there, such as interior cutthroat trouts (reviewed by Nelson and Soulé). Even established nonnative trout populations are affected by continual stockings of hatchery reared fish to provide recreational fishing, because the stockings result in steady outbreeding to hatchery stock. Finally, stockings can be a source of disease or parasites, as in the introduction to whirling disease to western streams. Almost all larger, warmer lakes of the United States are reservoirs, which are heavily managed by supplying nonnative and even native gamefish for recreational purposes. In this case, hybrids, such as wiper (white bass × striped bass) can be purposefully stocked. However, poststocking studies of hatchery fish have revealed that hatchery stocks in general have poor performance in growth and reproduction compared to the native stocks. Trends to grow bigger fish and to release them later have met with behavioral problems (White et al. 1995). Many states now strive to develop wild strains for propagation.

Rivers and Streams

Lotic systems present more problems for stocking programs than do reservoirs due to their harsher and more fluctuating environment. In the Pacific Northwest, various populations and species of salmon have important adaptations to location and timing of spawning, including the behavior of young. Thus, subpopulations are different in their development of coadapted gene complexes. It is now clear that introductions of hatchery-reared salmon can cause outbreeding depression through the loss of locally adapted gene complexes. Warmwater streams are far less likely to be the focus of stocking programs than are coldwater streams. However, largemouth bass and other warmwater sport fish that are adapted to one area often are captured and used as broodstock for other areas. Also, illegal and accidental stockings via bait buckets or escapement from reservoirs have resulted in the introduction of alien stream adapted forms, such as channel catfish, smallmouth bass, and white sucker, which have been stocked into nonnative streams on purpose or by accident. Regulation of warmwater streams has encouraged hybrid introgression by changing the timing, duration, and extent of flood flows used for spawning and rearing. Examples: After closure of dams, there is a general period of filling in which water levels in downstream reaches is very low. I believe that such conditions caused a hybrid swarm below Flaming Gorge Dam on the Green River, and for many years, hybrids of native bluehead, flannelmouth, and razorback sucker were routinely captured. However, crosses with introduced species also occur, an example given by invasion of the white sucker, ostensibly by bait bucket. Inhabiting the cooler upper tributaries, the white sucker has crossed with native bluehead and flannelmouth suckers. These hybrids are easy to detect, in part due to the great differences in mouth structure and scale sizes. The sensitivity of riverine fishes to habitat change and invasive species is finally being learned. Management agencies are beginning to realize that protection of wild stocks and maintenance of their habitat is the best path to take in managing stream populations (White et al. 1995).

Estuarine and Marine Systems

Hybridization of fishes in the marine and estuarine environment includes numerous species of gadoids, pleuronectids, and clupeids. Clupeids that form hybrids include two species of menhaden (Yellowfin, *B. smithi*, and Atlantic, *B. tyrannus*) (Gardner 1997), the most abundant fishes in the Atlantic coastal zone. In addition to salmon, several other fishes are propagated in hatcheries for supplementing or enhancing depleted populations of marine sport fish. Some of these are striped bass, red drum, spotted sea bass, white sea bass, and striped mullet. Gardner (1997) reported that hybridization in the sea is not rare, but less is known about the extent or the problems compared with freshwater systems.

Issues

Until now, we have focused on extinctions caused by agents of the "evil quartet," namely habitat alteration and adverse effects due to native fish encounters with humans and introduced species. However, as Rhymer and Simberloff (1996) point out, there is another type of extinction referred to as "genetic extinction" that can occur by hybridization and introgression that can occur between separate native fish populations, and also between native and introduced fishes. Hybridization requires energy and reduces survival, while introgression amounts to a gradual replacement of genes and gene complexes of the native fish with those of introduced ones, especially if the introduced stock is larger. The latter process has been likened to a gradual "invasion of the body snatchers," and its insidious nature has made this seriousness of the problem underestimated (Carlton and Ruiz 2005, p. 141).

Hybrid fishes are the result of the cross fertilization of eggs of one genetically distinct population with those of another, thus it can be intraspecific, interspecific, or intergeneric. Hybridization is best thought of as an extreme case of outbreeding, and it can have either beneficial or harmful effects. Production of hybrid fishes between fish subspecies, species, and genera is not uncommon in nature, and well-known examples of natural hybrids occur in several families (Bond 1996; Gardner 1997), including freshwater (e.g., cyprinids, catostomids, salmonids, and centrarchids) and marine fishes (e.g., gadoids, pleuronectids, and clupeids). Some hybrids are reproductively fertile and have produced new species, but the main effect on fish populations has been harmful.

If hybridization produces infertile offspring, there is no chance of *hybrid introgression*, a flow of genes from one population to another population. However, production of sterile offspring has a cost: It reduces fitness, requires energy expenditure, and results in forgone reproduction—wasted energy and reproductive effort. In case of introgression, the outbreeding depression that occurs can be deadly because transfer of genes can result in the loss of locally adapted gene complexes affecting adaptations such as mating behavior, timing of migration and spawning, habitat selection, and environmental tolerance.

Presence of fish hybrids has been detected for a long time. Carl Hubbs (1955) conducted an extensive survey of naturally occurring fish hybrids of North America, and Frank Schwartz (1981) did a global accounting of both artificial and naturally occurring fish hybrids, producing 3759 references (including a few of mine). Those studies make it clear that hybridization among fishes is not a rare event. However, the extent of hybridization varies according to the environment. Fish hybridization appears to be correlated with environmental disturbances, thus temperate freshwater fish hybrids appear to be common, while tropical marine do not (Campton 1987).

The causes of hybridization between two distinct populations in nature are usually related to a breakdown in reproductive barriers. These breakdowns are related to behavior choices such as timing (zeitgebers) with respect to flow, temperature, and photoperiod; selection and location of spawning or other types of habitat; and aggression. Habitat alteration and introduction of fish from other populations (including hatcheries) are responsible for most cases of hybridization today.

WATER POLLUTION

Effects

Lakes

By definition, lakes are standing bodies of water, and in most natural lakes and some reservoirs, water residence time is measured in years. Lakes also may have stratified water levels that do not mix all year. In more eutrophic lakes, there also may be a lack of oxygen in the lower stratified level. For these reasons, lakes are more threatened by pollution. For lakes and reservoirs located in highly populated areas there also are problems with inputs of domestic wastes from runoff and sewage. Shallow lakes and ponds in such areas may be dominated by emergent vegetation, algae, duckweed, and other submerged or floating plants. Even in large lakes, excess nutrient input can cause an overload in which decomposing masses of algae and plant material use up the oxygen and the bottom becomes devoid of oxygen (anoxic). In addition to inputs of contaminants from the drainage basin, very large lakes also can be contaminated by airborne particles, and recreational use of lakes can result in substantial inputs of hydrocarbons from boat traffic. Examples: The Laurentian Great Lakes have long been contaminated with a plethora of contaminants, including Polychlorinated biphenyls (PCBs) and mercury. Although conditions in the Great Lakes have improved, fishes of all sorts are heavily contaminated. In some locations, pregnant women and children are advised to consume no fatty fish such as salmonids, and men should eat fish no more than once a week.

Rivers and Streams

In contrast, flowing streams are flushing water downstream and residence time is measured in hours or days instead of years. This aids stream recovery from inputs of pollution. Also, streams seldom have problems associated with low oxygen levels. Domestic and industrial polluters understand and take advantage of this attribute of streams to get rid of wastes, thus even with stream renewal properties, large fish kills can occur due to spills. Throughout the world, large rivers are polluted to the point that drinking water has become scarce and waterborne diseases are common. Fishes in agricultural area can be exposed to a variety of pesticides and herbicides that can negatively affect them. Although organochlorides used in farming can be lethal to fish, I have seen several canisters of parathion (used on tobacco) disposed of by dumping them into streams in North Carolina.

Estuaries

In temperate zones, estuaries are typically areas of high nutrient content, principally nitrogen and phosphorus, which can be elevated due to point source and nonpoint source inputs in sewage, agricultural runoff, and even transported by air. Associated with high concentration of nutrients are blooms of different algae, including introduced dinoflagellates, which can produce toxic conditions such as red tide and other harmful blooms. North Carolina estuaries of Albemarle and Pamlico sounds have had major fish kills due to various algae, including the dinoflagellate *Pfiesteria piscicida* (Helfman 2007). Over a billion fish (mostly menhaden) were killed in the 1980–1990s. Estuaries also are contaminated by pesticides, herbicides, industrial wastes, heavy metals, and other substances. It is in estuaries that young forage fishes such as menhaden can assimilate heavy metals such as lead and mercury. As these young fish mature and move offshore, they promote bioaccumulation up the food chain to larger and larger fish. Larger tuna (albacore) are loaded with mercury, about four times that of light tuna (yellowfin, etc.), which means that a human should restrict intake of albacore and eat light chunk tuna no more than about twice a week to be safe. Methylmercury is not too good for fish either, causing liver disease, impaired swimming and feeding, and, over time, death.

Marine Systems

Oceans are the sink for all contaminants. Fortunately, marine life is very resilient, and many pollutants, such as oil, can be degraded. With this logic, unknown tons of garbage and even more sewage have been dumped into oceans. Thus, much of the coastal zone in the world is polluted. Beach closings have become common in many areas, principally due to high levels of bacteria present from incompletely treated sewage. Oil pollution also is a serious problem due to runoff from land, tanker discharges, and losses from drilling rigs. In the open ocean, bioaccumulation of toxic compounds continues in large predators like swordfish and tunas.

Issues

Water can be polluted with many different substances by introducing unnatural substances into it, or by increasing the concentration of natural substances so they become harmful. Humans have disposed of thousands of chemicals into public waters, including contaminants of all kinds. Two major groups of chemicals are nutrients readily taken up by plants, and contaminants, which can be toxic if found in high enough concentration. Also, water can be polluted by living organisms, including pathogens, which cause disease, or other organisms that may produce a toxic waste product such as "red tide." Water also can be polluted with solid debris of all sorts, especially plastics, most of which float at the surface. Substances and organisms also may change the quality of water by affecting its pH (to be acidic or basic), amount of dissolved oxygen, temperature, degree of eutrophication, and other factors.

Virtually all surface waters have been polluted with nutrients and other chemicals found in agricultural runoff, sometimes called "nutrient enrichment" because of the high content of plant nutrients nitrogen and phosphorous, and by municipal releases of sewage and other wastewater, called "cultural eutrophication." Nutrients placed into receiving waters can affect the number and kinds of organisms that live there, can cause blooms of algae, some of which can be poisonous, and result in oxygen deficiencies.

Water may be contaminated with both organic and inorganic substances, which may be natural or *xenobiotic* (not natural). Organic pollutants include household chemicals such as soaps, detergents, disinfectants, herbicides, insecticides; household food wastes, including vinegar, spices, fats and oils; fuels and lubricants, solvents, cosmetics, medicines, and more. Inorganic compounds can include fertilizers mentioned above, industrial wastes that can be acidic or basic, heavy metals, ammonia, and more.

All of the aforementioned pollutants are transported in water, which in most cases winds up in estuaries and finally into the oceans. Nutrients affect the numbers and kinds of organisms present, by stimulating plant growth and promoting eutrophication, which may require so much oxygen that anoxic conditions occur. Toxic contaminants such as heavy metals or fat-soluble compounds can kill or sicken organisms and also can be concentrated in tissues as organisms are eaten and biomass passes up food chains. Medicines passed in human wastes, such as birth control pills and tranquilizers can affect sexual development and reproduction in fishes and other organisms. Humans are susceptible to contaminants from eating fish, and the U.S. Environmental Protection Agency (EPA) provides advisory service (at www.epa.us.gov) for which fish are safe and what quantities are safe for consumption over time. Primary contaminants of fish that affect humans include mercury, PCBs, and pesticides.

In a 4-year study, National Study of Chemical Residues in Lake Fish Tissue, EPA has found that mercury concentrations exceeding standards for consumption were found in 49% of lakes and reservoirs in the country, and PCBs and dioxins also were widespread. The most significant source of mercury is a result of burning coal. EPA also is conducting stream surveys for contaminants and the results are to be available in 2011. Information can be obtained at www.epa.gov/waterscience/fishstudy/.

This section would not be complete without some reference to the British Petroleum oil spill in the Gulf of Mexico, which was the largest accidental oil spill in history. Coverage of the disaster was provided by the Encyclopedia of Earth (2010) at www.eoearth.org. Also, the October 2010 issue of *National Geographic* dedicated three major articles and a supplement to the disaster, which included an oil platform and well located about 50 miles southeast of the Mississippi River delta. The following account is taken from those sources: On April 20, 2010, a large gas bubble entered into the drill casing of the deep Macondo well, accessed by a pipeline located about 5000 ft below the water surface and extending another 13,000 ft below the sea flow. An explosive blowout occurred as the gas spread upward, destroying the offshore oil platform Deepwater Horizon and killing 11 persons. The shattered and twisted wreck of the $560 million platform sank to the ocean floor on April 22. After leaking oil from the ruptured pipe for 89 days, the oil flow was stopped on July 15. In all, about 4.9 million barrels of oil were placed into the open waters of the Gulf of Mexico. Evidently, safeguards were unable to shut off a rupture in the pipe. The spill dwarfs that of the Exxon Valdez (only 262 thousand barrels) and resulted in a presidential proclamation that placed a moratorium on deepwater drilling until facts were compiled and causes were understood. In addition to the oil, dispersants also were used to break up oil at the surface and sink it. These dispersants are toxic to marine organisms. The U.S. Commerce Secretary declared a fishing disaster due the adverse economic impacts. Environmental concerns extend for at least decades. Damages to the marine and estuarine systems are mind-boggling and include adverse impacts on plankton, wetlands, fish, and other vertebrates, including marine mammals, birds, turtles, and so on. British Petroleum, in addition to cleanup costs, has pledged $500 million just to study the effects of the spill. It is too soon, and not enough is known to even guess at the damage to estuarine and marine systems, which will take at least decades to assess. At stake is one of the biggest industries in the Gulf region, amounting to annual commercial fisheries of 1.27 billion pounds valued at $660 million. The largest commercial catches are shrimp and red snapper. Also, there is a recreational fishery of 3.2 million people. Especially vulnerable to the oil are oysters, shrimp, and blue crab. Although bacteria can and do consume oil, it is a long process and uses up oxygen, which can result in anoxic (dead) zones on the bottom. Wetlands, marshes, and swamps are of concern because of their high productivity and the vulnerability of birds, turtles, and other animals. Eggs and larvae of fishes are highly sensitive to environmental contaminants such as the oil, dispersants used to treat it, and anoxia. In all, a loss of productivity with impacts on the entire food web is expected in the Gulf of Mexico, including effects on everything from plankton to whales. Effects of the spill can be expected to continue for at least a decade or more.

ARE ALL SUSPECTS GUILTY?

Once again, we invoke the "Murder on the Orient Express Hypothesis," this time for the five causes we have evaluated as anthropogenic agents responsible for the decline, endangerment, and extinction of fishes. In concept, the hypothesis seems to fit, because overall, every suspect is an accomplice and all suspects are guilty. Sometimes, not all five factors will fit, but in very large systems, there is a good chance that they will. This is quite interesting, because we came to the same conclusion about natural extinctions in chapter 7.

The reader is cautioned to consider all five causative factors as potential suspects, and always to look for synergy. For example, fish cannot exist without habitat (which includes water!) and food. In freshwater systems, massive habitat destruction and alteration such as converting a river into a reservoir, thus fragmenting habitat and blocking movements, can leave little or no place for native riverine fish to reproduce or forage. Stocking the new reservoir with "preadapted" hardy and aggressive nonnatives can take away what little habitat there is left for riverine fishes and further limit food. If the introduced fish is predaceous on one life stage of the natives, and present in large numbers, survival may continue for individuals, but fitness approaches zero. Even worse, contaminants can produce stress, limit physiological tolerance, and increase vulnerability to other factors, such as invasive species.

In the marine environment, population sizes of a target fish might be reduced to low levels by fishing, then habitat is degraded and prey reduced by trawling over and over again. More stress is added via contaminants or lowering oxygen concentration, and then other predaceous species move in to the disturbed habitat. This is a case of ecosystem overfishing, and it seems very doubtful that the target fish population would be able to tolerate such conditions for very long.

The key to recovery of declining fishes is provision of native habitat and natural conditions of flow and other requirements. Restoration should include the following: removal or extermination of invasive species; regulation of fishing to reduce effort and use of gear that does not destroy habitat or kill prey species; redirection of hatchery augmentation programs to reduce hybridization between stocked and wild fish and to avoid introgression; and reduction of stress on fish populations by providing good habitat conditions for all of the life history stages, including high-quality water. Unfortunately, in our thirst for oil, unpredictable disasters like the 2010 Gulf of Mexico oil spill can be expected in the future as deepwater drilling continues.

CASE STUDY: CHESAPEAKE BAY—AN ECOLOGICAL DISASTER

Background

Chesapeake Bay, the largest tidal estuary in the United States, is a drowned river valley network that includes about 150 streams and rivers that merge as they drain 65,000 square miles of six coastal states and the District of Columbia into the Atlantic Ocean (Figure 31.5). The largest rivers are the Susquehanna, James, and Potomac, which penetrate 190 miles inland. It also is the most productive estuary in the United States—its rich waters support hundreds of fish and shellfish species, of which its historically abundant oysters, striped bass, and blue crab are very desirable as food for humans. The early colonists at Jamestown (1607) lauded its majesty in many letters to England, in which the bay was compared with Eden (Roberts 2007). Now we are using Chesapeake Bay as an example of things gone wrong on an ecosystem scale. These unnatural changes are due to multiple stresses induced by humans, resulting in frightening impacts. The Chesapeake Bay Ecological Foundation, a nongovernmental support group sponsoring restoration of the bay, has considered the situation to be "an undeclared ecological disaster" (Price 2010). Excellent accounts of change in the bay have been written by Franklin (2007) and Roberts (2007).

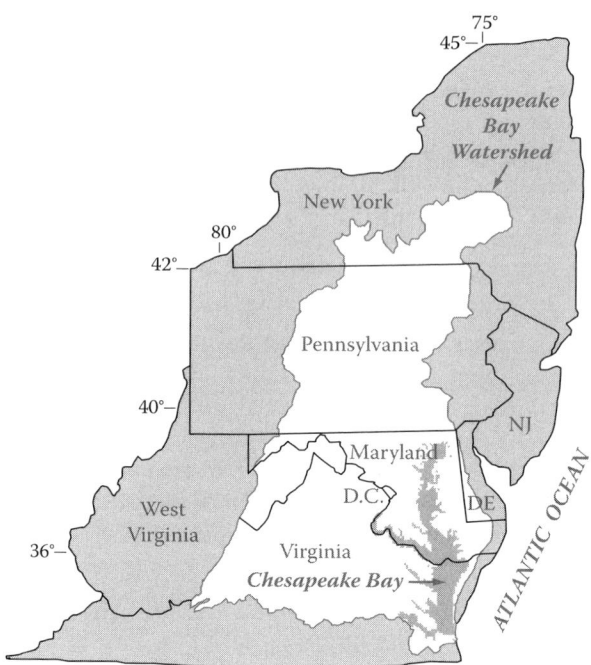

Figure 31.5 Chesapeake Bay basin. (Courtesy of USGS, http://commons.wikimedia/wiki/File:Chesapeake_Bay_Watershed.gif.)

Historically, Chesapeake Bay as we know it did not exist until after the close of the last ice age (Pleistocene) when melting glaciers produced a sea level rise of approximately 100 m, flooding its river valleys. The resultant estuary was rich in food and supported huge populations of fishes. Hildebrand and Schroeder (1928), in their monumental coverage of Chesapeake Bay fishes, provided a comprehensive account of fishes and fisheries of the bay and adjacent areas. They reported capture of 202 species, noting that fishes in the bay were almost exclusively transient, with commercial species leaving for deeper offshore waters in the fall and returning in the spring. However, the authors did little to illuminate the bay as a functioning ecosystem, instead focusing on species accounts and commercial fisheries.

By 1920, some species had declined significantly as indicated by commercial fishing records (Hildebrand and Schroeder 1928). Sturgeons, sought for their caviar and flesh, went from a catch of 814,000 lb in 1890 to 22,000 lb in 1920, thereafter never recovering. Fishing records also showed declining stocks of shad, herrings, bluefish, sheepshead, spot, butterfly fish, kingfish, and sea trout, while landings of croaker and summer flounder increased (from fishing records covering the period 1908–1920). Anadromous fishes remained very abundant in the bay, and the most abundant food fishes were (1) river herrings, (2) croakers, (3) American shad, (4) sea trout, and (5) striped bass. Also, the menhaden industry was very important and a catch of over 350,000 lb was reported for 1920. The industry supported 18 reduction factories, employing 900 workers and 42 steam vessels with 1500 persons.

Oyster Reefs

The estuary also was shallow and inhabited by large populations of the American oyster *Crassostrea virginica*, a reef building bivalve mollusk that grows in size from 2 to 6 in. in length. This oyster lives in the subtidal zone in most of its range, but also inhabits the intertidal zone in the southern Atlantic (Burrell 1986). Huge oyster reefs or beds can be built by oysters because they cement their shells to rocks or other hard substances, then once started, build a reef by attaching to the shells of other oysters, thus perpetuating their own habitat. Oysters reproduce by releasing gametes that unite to make planktonic larvae. These larvae are mobile, free-living organisms that find a likely spot and attach (as spates) there for life. When early colonists arrived in the bay area, they found millions of oysters living in extensive oyster reefs, in some places extending up to 150 km into the shallow bay. Oyster reefs also are prime habitats for many fishes.

The American oyster is a filter-feeding animal and it pumps about 5 L of water an hour through its body, removing and consuming plankton and detritus. Oyster beds are mostly found lining tidal creeks where they filter vast amounts of water entering the estuary from the river drainage basins. They provide an ecosystem service by removing organic matter, nutrients and other particles from the water, assimilating it into oyster tissue (including their larvae) and releasing the concentrated organic matter into the benthic substrate as feces and pseudofeces (called bentho-pelagic coupling). Packaged into larger particles, the feces are consumed by worms and other benthos. In filtering water, oysters also reduce turbidity and this encourages the growth of sea grasses, which stabilize bottoms, take up nutrients, and provide food and cover for other organisms as well.

Oyster reefs can be robust and modify flow and sedimentation patterns in the estuary, resulting in different bottom types. Also, they are used as refuge by many organisms, and some species also use the reefs as feeding or nesting habitat, including finfish. Coen et al. (1999) considered the oyster reef as an essential fish habitat, listing seven permanent residents, five facultative residents, and 67 transient fishes associated with the reefs. Oysters provide food for other organisms directly as adults and planktonic larvae. Also, they convert plankton and detritus into larger food pellets that are consumed by benthic organisms, they clarify the water and thus encourage growth of sea grasses, and they provide habitat for fish, crabs, shrimp, barnacles, and other invertebrates.

Oysters are good to eat, and were highly sought by colonists. At first, humans did little damage, but by the early 1800s, the fishery greatly expanded, and by the end of that century, millions of bushels of oysters were shipped annually from the bay. At first, humans gathered oysters by hand, then by tongs, and finally by use of dredges towed by boats. By 1870, a thousand dredge boats sought oysters in the bay (Roberts 2007). The steel dredges tore off the top layers of the reefs and ground them down, layer by layer, removing the shell without replacement (Breitburg and Riedel 2005). In time, the fisheries began to collapse, and by the early 1900s, wild oysters were scarce. At that time, efforts were made to cultivate oysters by private parties. The bay continued to produce oysters, but in 1957, the disease MSX (*Hapalosporidium nelsoni*) was introduced in Asian oysters. This disease kills over 90% of American oysters in 2–3 years (Burrell 1986). Although some oysters are produced commercially in Chesapeake Bay each year, the annual commercial output has dropped from a peak of about 15 million bushels to only 80 thousand (Price 2010).

As humans depleted the oysters, they also began to change the landscape. Heavily wooded with a great diversity of deciduous trees when first settled, by the close of the 1800s, over 80% was deforested and farmed (Roberts 2007). This greatly increased soil erosion. As a result, river channels filled with silt and debris began to fill the bay. Increased agriculture and development of inorganic fertilizers resulted in nutrient loading and increased people resulted in increasing discharges of sewage. By the 1930s, anoxic conditions were reported (Franklin 2007). With so few oysters to filter the water, organic matter (including detritus and dead plankton) began to build up because it could not be decomposed fast enough. Meanwhile, more people inhabited the bay and drainage basin (presently about 15 million persons), increasing inputs of sewage and industrial waste. Increases in agriculture, chicken and hog farms, and lawns resulted in increased runoff of plant nutrients. By the 1960s, increasing eutrophication and the increase in phytoplankton shaded the bottom. This resulted in the loss of about one-half of the sea grass beds (Roberts 2007) that provided food and shelter to a host of organisms. Dead zones soon appeared due to low or no oxygen. Life in the bay decreased in the benthic zone and became more restricted to the pelagic zone. Presently about

300 million pounds of nitrogen enter Chesapeake Bay each year due to point source discharge, runoff, and air pollution (Chesapeake Bay Foundation 2010).

Loss of Planktivores

Historically, there were two major planktivores in Chesapeake Bay: the American or eastern oyster, a benthic mollusk, and the Atlantic menhaden (Figure 31.6), a pelagic fish. Oysters utilized areas where flowing water assisted them by moving food past their filtering devices. However, there were vast areas of open water in the bay that was too deep and slow moving for efficient consumption by oysters. These open waters required a pelagic planktivore, and this role was filled by billions of Atlantic menhaden (*Brevoortia tyranus*). This estuarine-dependent clupeid (adults about 37 cm in length) has long, slender, and numerous gill rakers that it uses to filter phytoplankton, small zooplankton, and detritus from the water column, which it converts into rich oily flesh. The flesh is sought by larger fish species, such as bluefish and striped bass.

The menhaden are sensitive to low-oxygen conditions, and some fish kills have occurred, but in general, the fish overcame the eutrophic water conditions of the 1960s and with a great supply of food, multiplied in numbers. Historically, the young menhaden feed in estuaries in summer and grow quickly, then leave for deeper water in fall. The adult fish are consumed by larger predaceous fishes in the coastal and offshore zones of the Atlantic Ocean. However, menhaden is a highly desirable commercial fish, and large numbers are captured in the bay. These fish are rendered by heating, producing oil, fish meal, and fertilizer (bones). About 150,000 metric tons of menhaden are captured annually in Chesapeake Bay. However, the fate of the bay, larger commercial and sport fish, and many of its inhabitants seems to be linked with the ability of millions of menhaden to recycle phytoplankton (Price 2010).

Ecological Disaster

By the 1970s, a disaster was evident. Chesapeake Bay was exhibiting regular toxic plankton blooms and low oxygen levels were causing mass fish kills. It was during this same time that striped bass numbers fell, reaching low numbers in the 1980s. Striped bass (Figure 31.7) is a very important sport fish and management action resulted in a moratorium on fishing. It worked and resulted in a great increase of the fish by 2000. But then, fishermen began catching diseased striped bass, infected with sores and in poor condition. Many of these fish were very thin, and obviously, their diet was inadequate. It was decided that poor nutrition was due to a lack of their favorite food: menhaden. It seems that the striped bass historically consumed fish (primarily menhaden) as 99% of their diet, but these fish were eating invertebrates, mostly young blue crabs (review by Franklin 2007). This change in diet from red, oily, high-energy menhaden to the white, low-nutritional flesh

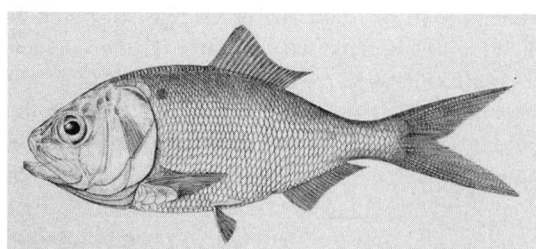

Figure 31.6 Atlantic menhaden: the most important fish? (Courtesy of the National Oceanic and Atmospheric Administration, http://commons.wikimedia/wiki/File:Brevoortia_tyrannus1jpg.)

Figure 31.7 Striped bass: an important sport fish. (Courtesy of USFWS, http://commons.wikimedia/wiki/File:Striped_bass_FWS_1.jpg.)

of invertebrates was not healthy for the fish. At the same time, it was noted that the abundance of other clupeids (shad and herring) already in low supply, had declined. It was presumed that the stripers and bluefish were eating them as the best substitute for menhaden.

The troubles of Chesapeake Bay were not restricted to loss of menhaden or stripers. Humans now got directly involved in the bay's problems. In 1997, about two dozen fishermen and water quality workers got sick with rashes and suffered memory loss (Roberts 2007). The cause was determined to be a result of exposure to a toxic algae *Pfiesteria piscida*, a microorganism (dinoflagellate) that produces a harmful toxin under conditions of nutrient loading. Other outbreaks have occurred since then.

Other changes have been noticed in the bay, including increased abundance of comb jellies (ctenophores) which feed on zooplankton (Franklin 2007). These organisms worsen the eutrophication problem by consuming the zooplankton predators that forage on phytoplankton.

Humans have interfered with the Chesapeake Bay system in numerous ways, producing a monstrous condition from which no easy solution is possible. However, there is great hope that the bay will improve if nutrient levels can be reduced and a moratorium is placed on commercial fishing in the bay for menhaden. It is believed that millions of menhaden will be able to reduce the eutrophication problem by recycling the algae and detritus into fish flesh and then removing the biomass to the ocean or to larger fish that are captured and removed from the bay. The malnutrition and obvious starvation of striped bass also appears related to overfishing of their preferred prey. It seems logical to use menhaden in restoring the bay to a more healthy condition. From a socioeconomic perspective, this would seem to be more desirable to decision makers to have menhaden clean up our mess and also feed other fish, such as striped bass, than to reduce the fish to oil and fish meal. Hopefully, a program can be developed to protect the menhaden while providing some relief to the present commercial fishing interest. To that end, a new initiative has been launched by presidential executive order to provide a comprehensive package of federal actions to aid in restoration of the bay. This new strategy includes provisions for close collaboration with citizens and state agencies. (Interested parties can obtain information at www.epa.gov or at www.chesapeakebay.net/.)

Things are looking up for bay restoration. The EPA has recently requested a 25% reduction in nitrogen and phosphorous, and a 16% reduction in sediment (USEPA 2010). Furthermore, in May 2010, based on a stock assessment that showed overfishing and poor recruitment, the Atlantic States Marine Fisheries Commission directed the menhaden technical to commit to consider alternate goals for population management (Blankenship 2010). Hopefully, this will mean a reduced commercial harvest of menhaden.

SUMMARY

Five causes linked to the decline and endangerment of fishes and discussed for lakes, streams, estuaries, and marine systems. These causes were (1) physical habitat alteration (destruction, alteration, and replacement of habitats), (2) competition and predation from introduced species, (3) hybridization (with or without genetic introgression), (4) chemical alteration, contamination, or pollution of water, and (5) overfishing. In general, these causes were more applicable to some aquatic systems than others were; however, all appeared pertinent to the problem. Logical causes for extinction varied between aquatic habitats. Habitat alteration has been dominant with streams, competition and predation fit best with lakes, hybridization applies almost exclusively to freshwater systems, and chemical alteration seemed a best fit in lakes and estuaries. Overfishing now occurs mostly in a few large lakes and in marine systems. The Chesapeake Bay study illustrates the main points of this chapter and demonstrates how major ecosystems are becoming ecological catastrophes due solely to human induced changes. Catastrophes associated with deepwater oil extraction can be expected to continue as oil becomes more sought in the future.

CHAPTER **32**

Changes and the Future

INTRODUCTION

In chapters 2 and 7, we spent some time puzzling over the evolution, radiations, and extinctions of fishes, drawing some lessons from the past. One lesson applies to the very nature of the planet we live on: if Earth is stable, then stability certainly has wide limits. But there is a great problem in extending the lessons learned from the past into the future. That problem is global anthropogenic climate change. In finishing this book, I felt the necessity to extend our studies to global changes that may affect fish distribution and abundance in the twenty-first century. Most of the anticipated changes are familiar because they have occurred in the past. However, the difference is likely to be the speed and extent of change due to human influence.

It is anticipated that the twenty-first century will be characterized by global changes resulting in increasing socioeconomic hardships on human civilization. Hardships will be caused by three pervasive and synergistic factors: global climate change (GCC), declining oil and gas supplies, and an increasing human population. A detailed discussion of those hardships is outside the scope of this book. But within the scope is concern about increasing environmental impacts on fish populations. Not only are direct effects anticipated as a result of the three factors above (i.e., effects and issues due to the "five causes" identified in the last chapter), but socioeconomic pressures associated with increasing costs of living will result in less concern about maintaining environmental quality in the twenty-first century (e.g., see discussion by Czúcz et al. 2010). Thus, there is the likelihood of increasing environmental impacts, but less concern about them.

There is great concern about the effects of GCC among ecologists, but there also has been some confusion in the media about whether it even exists. In view of the wide variation in opinion about the matter, I felt it necessary to be somewhat philosophical in considering what the issues are and why there has been confusion, the validity of GCC science, and what the direct and indirect effects might be. Then we can apply what we know about GCC to evaluate likely changes on aquatic systems and the ecology and conservation of fishes. Finally, this chapter closes with the identification of fish-related issues that need to be addressed and thoughts about the future of fisheries management.

INTERESTING TIMES

The English phrase "May you live in interesting times," possibly a translation of a Chinese proverb, is attributed to Sir Austen Chamberlain in 1936. First published in 1939 (Shapiro 2006), it was purported to be a curse given to an enemy! Curse or not, we are living in increasingly interesting times, in no small measure due to our understanding about changing conditions on the planet we live on. The world is also getting more "interesting" as human population numbers soar to dizzying heights!

We live in an age of science and technology in which basic science is used to draw conclusions about nature and applied science (technology) uses the conclusions to aid humans in some way. Science is meant to be nonbiased (objective), and conclusions are drawn from evidence (Agin 2006). Hypotheses are tested, and a null hypothesis is accepted or not, based on statistical testing. Scientific results are announced to a community, which then attempts to redo experiments and validate the conclusions. However, it seems that scientists are not the only humans to practicing the trade.

To put it bluntly, humans are conducting some dangerous and rather poorly planned experiments with Planet Earth. These "experiments" are recycling stored fossil fuels back into the active carbon pool (atmosphere), homogenizing plant and animal communities, simplification of ecosystems by reducing habitat diversity, making and spreading poisons, increasing extinction rates, and changing climate. Let us consider climate. The object of one experiment could be put in question: *can we raise the average world temperature by increasing methane gas or carbon dioxide (CO_2) levels in the atmosphere?* The answer to this question is found in several sources (presented in more detail later): *increased greenhouse gases, including methane and CO_2, amplify the warming effect.* However, it is not that simple, because raising world temperature could result in additional effects. Put in the form of other questions, we might ask the following: Will a rise of temperature be associated with an ice-free Arctic Ocean in summer? And if so, does this happen before the Greenland ice sheet slumps into the ocean? Does the melting of the polar ice caps cause sea level rise? Could all that meltwater shut off the Gulf Stream? If so, would that increase hypoxia on the ocean floor and cause England to have a climate more like Iceland? The questions are pertinent (even if unscientific), because all of these effects have been postulated. There are many more interesting questions to ask, but the level of complexity is rapidly increasing, so we will stop at these.

Let us assume that a group of scientists proposes an interesting experiment. They wish to double the atmospheric CO_2 levels and see if green plants grow better. They would want to obtain some financial support so they approach the United Nations. They are told there are some legal technicalities, so they go to the International Court of Justice (World Court) to get their blessing. They are amazed to be thrown out of the court and threatened with internment in a facility for the mentally deranged! No one is willing to take such a risk on the only inhabited planet we have. Unfortunately, that experiment is already in progress, but there is no target level of CO_2, nor, it seems, do we have any way of controlling the amount entering the atmosphere. We have an experiment that has no bounds and no control that is already in progress. In addition, there are some pretty horrid side effects.

Ecologists, as scientists, must consider what the results might be if the planet warms, because of the potential for disaster(s). Some environmentalists have gone so far as to accept GCC as proof of the death of nature, because all ecosystems and habitats on the planet have been affected by humans to some degree (e.g., McKibben 1989; Wapner 2010). However, there are some people that for economic, political, religious, or other reasons reject notions that GCC is occurring and that humans have had any influence on climate. Some detractors believe there is nothing to worry about, because serious effects of change are so far in the future that they will not be affected. Others simply believe that GCC is a myth. But the number of doubters seems to be dwindling as publications appear.

Agin (2006) contrasted concepts of science, bad science, and junk science. Bad science is science gone wrong, with improper methods or controls, producing conclusions that are not real, improper interpretation, and overextension of results. On the other hand, junk science is corrupt science, clouding reality with fear, myth, or deception for some conflicting purpose. Junk science violates the scientific principles of objectivity, statistical interpretation, and replication. What are the reasons for junk science? It is used to discredit unfavorable results and things you do not want to hear. Motivation can be induced by politics, religion, industry, government, fraud, and ignorance.

Alexander (2010) reviewed three new books that discuss various aspects of climate-change denial: The first book, *Climate Cover-up: The Crusade to Deny Global Warming*, written by Hoggan and Littlemore (2010), found that some deniers, funded by industry, have worked hard to establish and to spread confusion over issues. The second book, *The Climate War: True Believers, Power Brokers, and the Fight to Save the Earth*, by Pooley (2010), took a balanced approach in examining both sides of the issue, but found that climate legislation has been effectively blocked in Congress by lobbying of senators and representatives. The third book, *Merchants of Doubt: How a Handful of Scientists Obscured the Truth on Issues from Tobacco Smoke to Global Warming*, by Oreskes and Conway (2010), exposed a maverick group of scientists for hire, challenging GCC and other issues for industry.

Scientists are stunned that others would distort and ignore scientific data and conclusions. Instead, they had rather believe that their opponents are overcome by misinformation or just ignorant (Mooney 2010). In truth, it is political conviction, not education that seems to be driving whether you believe in GCC or not. Alexander (2010) argues that an easy way to fill television space is to allow advocates to "debate" charged issues and simply present the opinions of both sides on equal footing. However, pitting opinion and value judgment (usually about politics and religion) against science may be entertaining, but it is inappropriate journalism and solves no problems.

Here is a word about doubters: It is perfectly normal to doubt. Virtually no one believes everything in the newspaper or on the television because mistakes are made, and not everyone is trustworthy. Also, the human mind interprets what we see and hear based on perceptions. GCC is something we have not experienced and may not understand, especially for a person who lacks knowledge about climate, geological time scales, and past global events. Perhaps the concept of GCC is just too much for some individuals. In the best-selling book *Future Shock*, author Alvin Toffler (1970) envisioned that "information overload" due to technological and social change would produce "shattering stress and disorientation." Future shock is brought on by too much change in too little time. Humans tend to expect the usual, and interpret the world based on what we have experienced in our lives.

Another author, Benjamin Franklin, understood this problem and wrote "Soliloquy of a venerable ephemera who had lived 420 minutes," which appeared in the December 4, 1735, edition of the *Pennsylvania Gazette* in an essay "Of human vanity." The story centers on an old mayfly who was convinced that the world would surely end because the sun was falling from the sky (setting). In all of his great age, he had never known of the sun to do that. The adult mayfly felt he was qualified to make such a statement because he had lived the great age of 7 h (Sparks 1836, pp. 178–179). We do not want to think like this mayfly, so we will review some basics about climate on Planet Earth and learn a little about the drivers of climate and how fishes are being affected.

GLOBAL CLIMATE

Changes in Progress

Temperature, wind, moisture, and precipitation characterize the *climate* at a given place on the surface of the Earth. Climate is generated by *weather*, or the motion of air that results from uneven heating. It is the heating that is the key to weather and thus climate.

Our only supply of heat comes from radiant energy transmitted from the sun to the Earth, and this supply is expressed as the *solar constant* (it fluctuates around 2 cal/m^2/min). Absorption in the atmosphere and reflectance from clouds and the surface deprives the Earth from receiving about 55% of incoming radiation, leaving about 45% to strike the surface. Atmospheric absorption (mostly by water vapor and CO_2) is continuous and is reradiated to space and also back to the planet. In visualizing this, the atmosphere acts much like a blanket to retain heat in the same way that a greenhouse keeps plants warm. This phenomenon was noticed at least as early as 1824 (by Joseph Fourier) and testing was done to explore the *greenhouse effect* by Arrhenus in 1896. The greenhouse effect is very important for life on the planet, because without it the surface temperature would fall about 33°C (to −15°C) (Serreze 2010), and make life as we know it impossible. This effect is one of the principles necessary to understand global climate and it is well founded and accepted. What is not so accepted is how increases and decreases in greenhouse gases might change the average global temperature.

Naturally occurring greenhouse gases (in order of importance) are water vapor, carbon dioxide, methane, and ozone. As the industrial revolution continued, higher levels of carbon dioxide (a 36% increase), and methane (148%) were measured in the atmosphere, along with chlorofluorocarbon and nitrous oxide. Carbon dioxide values from ice cores studied by the Scripps CO_2 Program show that the atmosphere over the past 400,000 years has ranged from about 200 ppm during period of glaciation and about 280 ppm during warmer interglacials. Present levels of CO_2 are 386 ppm recorded at the Mauna Loa observatory (Figure 32.1), where the concentration has risen at about 2 ppm/year. The present concentration greatly exceeds the expected average, and this level of CO_2 has not been exceeded for about 20 million years (University Corporation for Atmospheric Research 2009).

Present atmospheric CO_2 levels are about 30% greater than 150 years ago (ya). The higher levels of CO_2 and other greenhouse gases are produced by human-mediated recycling of buried fossil fuels back into the atmosphere. Human-produced gases are absorbed by plants (25%) and the oceans (25%), and 50% remains in the atmosphere for up to 100 years or longer. The increase in greenhouse gases has been implicated in the rise of surface temperature by about 0.75°C during the twentieth century, and projected to increase another 1.1°C–6.4°C by the close of the twenty-first century, when atmospheric CO_2 levels are expected to rise to 720 ppm (Serreze 2010).

Global warming is defined as an average temperature increase for the planet. However, temperature changes on the globe vary, mostly due to albedo and the angle of the sun with respect to a given surface area. *Albedo* is the percentage of solar radiation that is reflected back from the surface of the Earth. Albedo is low for dark substances and high for light ones. Albedo for water is low, about 2% for direct rays, but it is very high for low-angle rays (the well-known glare effect), whereas snow and ice varies from 45% to 90%. Texture also is important for albedo, and forests in summer have a very low albedo (5%), while grasslands reflect more (30%).

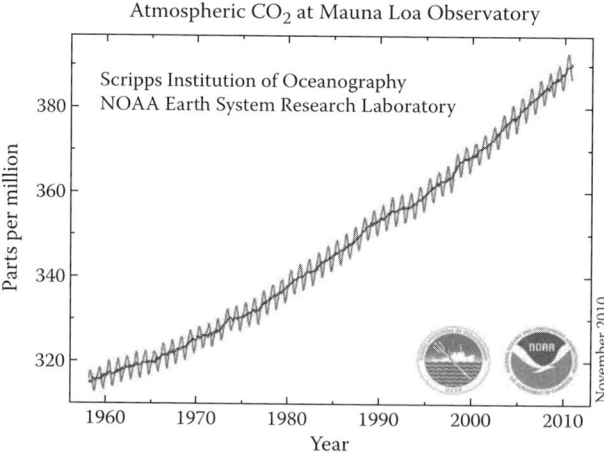

Figure 32.1 Global CO_2 concentration. (Courtesy of the National Oceanic and Atmospheric Administration, http://search.usa.gov/search/images?query=atmospheric+co2+at+mauna+loa&locl=m=false and commit=search.)

GCC has been widely debated and publicized (e.g., Dell'Amore 2009). Extensive research has been done and is in progress by hundreds of organizations and government teams around the world. The United Nations involves about 1500 experts (including almost all of the climatologists on the Earth). It has three working groups of about 50 scientists (each of whom also involve support workers and reviewers), each on the Intergovernmental Panel on Climate Change (IPCC 2007), which published its most recent findings in 2007. It will review and publish again (fifth report) in 2014. Also, FAO (Food and Agriculture Organization of the United Nations) has released its own report (Cochrane et al. 2009). Those findings are supported by the American Association for the Advancement of Science, the United States National Academy of Science, about 40 other scientific groups, and thousands of other scientists. It is pertinent here to review the scientific findings of the IPCC, and I will use the Working Group 1 *Summary for Policymakers* document to provide the key conclusions (percentages in parentheses added; note that in IPCC statements, "most" means >50%, "likely" means at least a 66% likelihood, and "very likely" means at least a 90% likelihood):

- Warming of the climate system is unequivocal.
- Most (>50%) of the observed increase in globally averaged temperatures since the mid-twentieth century is very likely (>90%) due to the observed increase in anthropogenic (human) greenhouse gas concentrations.
- Anthropogenic warming and sea level rise would continue for centuries due to the timescales associated with climate processes and feedbacks, even if greenhouse gas concentrations were to be stabilized, although the likely amount of temperature and sea level rise varies greatly depending on the fossil intensity of human activity during the next century.
- The probability that this is caused by natural climatic processes alone is less than 5%.
- World temperatures could rise by between 1.1°C and 6.4°C (2°F to 11.5°F) during the twenty-first century.
  - *Sea levels* will probably rise by 18 to 59 cm (7.08 to 23.22 in.).
  - *There is* a >90% confidence level that there will be more frequent warm spells, heat waves, and heavy rainfall.
  - *There is* a >66% confidence level that there will be an increase in droughts, tropical cyclones, and extreme high tides.
- Both past and future anthropogenic carbon dioxide emissions will continue to contribute to warming and sea level rise for more than a millennium.
- Global atmospheric concentrations of carbon dioxide, methane, and nitrous oxide have increased markedly as a result of human activities since 1750 and now far exceed preindustrial values over the past 650,000 years.

In the United States, Congress passed the Global Change Research Act of 1990 (PL 101-606), and thirteen departments and agencies participate in the United States Global Change Research Program "to understand, assess, predict, and respond to human-induced and natural processes of global change." The program has provided their key findings on the U.S. Global Change Web site: www.globalchange.gov/publications/reports/scientific-assessments/us-impacts/key-findings. The consensus is that global warming is real and primarily human induced, climate change is in progress and will increase, water resources will be stressed via drought, crops and livestock production will be challenged, coastal areas are at risk due to sea level rise, and GCC will interact with and exacerbate many social and environmental stresses.

The preceding casts little doubt that GCC is real and warming of the planet is increasing. Although there are uncertainties in what the effect will be on the planet or on the biodiversity and ecology of fishes, there are some major impacts anticipated on the world as we know it. Also, there is uncertainty about the speed at which some of these impacts may occur. For example, polar regions are experiencing differential warming due to the *snow and ice—albedo* effect, which is producing a *positive feedback*. As sea ice melts, it leaves open water, which has a very low albedo; hence, it warms more than the ice. As more ice melts, the exposed water speeds up the melting process. In a similar fashion, the melting of snow over tundra results in a darker and more irregular surface that has a lower albedo than the snow. As more tundra is exposed, the faster the melting will occur. Both of these examples show the effect of positive feedback. Unlike our home thermostat, there are no set point controls in nature, and homeostasis is replaced by homeorhesis (pulsating conditions).

Warming/Cooling: Evidence and Tipping Points

The present understanding of GCC has it as a very gradual process in which global temperatures will increase worldwide. Notice that I use global climate *change* rather than *warming*, because effects of GCC will vary in space and time. Predicting what fish habitat will be like in the future is not a simple matter of adding the average predicted temperature to every system, and then upping the value every decade or so. In reality, places on Earth will vary in combinations of hot/cold, wet/dry, and windy/calm conditions. Also, there is no reason to think that the process of change will be gradual. There is ample fossil and other evidence that the global climate has experienced major shifts in only decades (Gagosian 2003). The driving force in warming the Earth's atmosphere is the greenhouse effect; however, it is also possible that warming in one area will be accompanied by cooling in another. The *Ocean Conveyor* provides a sobering example.

The equatorial oceans are very hot and experience a high evaporation rate. This results in very warm, salty water in the Caribbean Sea. The Caribbean also piles up and moves north in the Gulf Stream due to the currents generated by the rotational power of the Earth and the Coriolis effect. This water flows along the east coast of the United States until it meets the cold Labrador Current that is moving south from the Arctic. The very warm water gives up heat to the atmosphere and it also is cooled by North Atlantic water. The cold salty water thus becomes very dense and sinks, leaving the warmed, less salty surface water to proceed northeast, warming the atmosphere as it travels. Two things happen: (1) The Gulf Stream water acts as a huge heat pump warming the atmosphere as it travels north. The eastern North Atlantic warms on average by 3°C–5°C, and western Europe can gain more. (2) The cold, salty water sinks to great depths and joins the deep part of the Ocean Conveyor, which is the deep thermohaline flow of the global ocean (Figure 32.2). The Atlantic portion of the Ocean Conveyor (i.e., the Atlantic meridional overturning circulation) is an important source of oxygen to the ocean floor, an important source of heat and moisture to Europe and the North Atlantic, and used by fishes in their reproductive and seasonal migrations.

There is ample evidence that the Ocean Conveyor has been disrupted in the past to the point that it has slowed and even shut down on several occasions (reviewed by Pearce 2007). What would cause this? As Arctic sea ice and polar icecaps melt, the freshwater and diluted seawater floats. Also, more sea exposed will result in more evaporation and increased precipitation. Both of these processes will make the surface waters more buoyant, slowing down or shutting off the flow of dense salty water to depths, thus interfering with circulation of the conveyor. This will make the North Atlantic colder and reduce fish distribution and production. Also, the lack of atmospheric warming would result in less moisture for Europe, producing droughts and affecting freshwater fish there.

The Ocean Conveyor above is considered a tipping element because it is a "large-scale component of the Earth ... that may pass a tipping point" (Lenton et al. 2008). A tipping point is analogous to tipping over a boat. The boat may sway side to side as it moves through the water, but there is a limit at which the boat will fill with water and sink. A tipping point is "a critical threshold at which a tiny perturbation can qualitatively alter the state or development of a system" (Lenton et al. 2008). It is thought that a tipping point for the Ocean Conveyor might be only decades away, and then GCC would be responsible for plunging eastern North America and western Europe into a deep freeze (Barry 2004). The tipping point in question might also be considered as a switch. Is the switch activated according to whether the water freezes before it sinks or sinks before it freezes? Fresher very

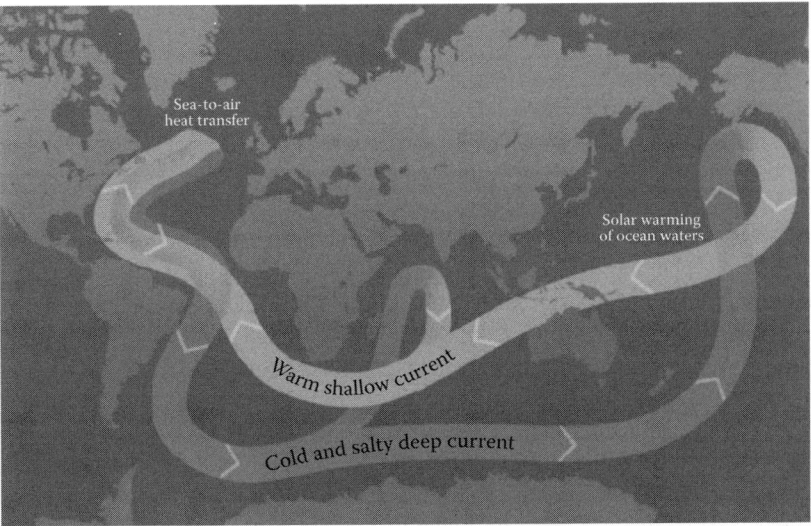

Figure 32.2 Ocean Conveyor. (Courtesy of the U.S. Global Change Research Program, http://commons.wikimedia/wiki/File:Ocean_circulation_conveyor_belt.jpg.)

cold water freezes, floats on top, and shuts down the conveyor. Cold (but warmer) salty water sinks and runs the conveyor. Let us check the notion with the climatic record.

Arctic ice cores from Greenland and other data provide a record of climatic changes beginning with the last ice age, which was in full stage about 18,000 ya. At 16,000 ya, there was strong warming, but ice sheets were reforming at 15,000 ya. This cold was soon to be followed by intense warming in 14,500 ya that produced a 65-ft sea level rise from melting ice caps. However, the cold gave way to warming 13,000 ya, which changed again to cold conditions 12,800 ya. The preceding record is like a wild roller coaster ride of ups and downs with no intermediate condition. Clearly, this looks like a switch affecting Ocean Conveyor for which there is some support. The cold period of 12,800 ya has been related to immense flooding of Pleistocene Lake Agassiz into the North Atlantic. The cold freshwater must have stopped the sinking of dense water and shut down the conveyor. It took another 1300 years for the conveyor to return and the climate to warm again, and then 8200 ya, the ice returned, lasting about 350 years, when warm conditions returned and persisted to the present (reviewed by Pearce 2007).

There are other tipping elements associated with GCC. One of these is the Greenland ice sheet. Instead of the ice melting and running off on the surface, it seems that surface meltwater accumulates in supraglacial lakes that empty with gravity flow down through the glacier all the way to the bed. There it tends to lubricate the base of the glacier and facilitates its movement (Das et al. 2008). The movement increases with warming and fracture propagation. So instead of melting in thousands of years, the ice sheet could slump into the sea much faster once a tipping point is reached—perhaps in decades. If the 2-mile-deep glacier enters the sea by melting or ice flow, it could raise the global sea level by about 6–7 m (20–23 ft), cooling and reducing productivity in the North Atlantic. Another possibility of concern for fisheries is the El Niño–Southern Oscillation (ENSO), and it is feared that with GCC, the warm oceanic phase could become the average condition, greatly reducing coastal upwellings along the western coastlines of North and South America and reducing productivity.

GCC: EFFECTS ON FISH AND HABITAT

Direct and Indirect Effects

Although the effects of GCC are complex and thus difficult to predict, scientists are extremely concerned about an increasing loss of biodiversity. Many studies are being generated on how to reduce the impending impacts, and general "calls to arms" are appearing (e.g., see Hunter et al. 2010) in an effort to encourage more participation from conservation biologists.

As previously stated, humans have caused an increased loading of greenhouse gases, especially CO_2, into the atmosphere. This has amplified the greenhouse effect, which causes warming of the planet surface, affecting both terrestrial and aquatic systems, such as the melting of polar ice sheets. The magnitude of warming is important in how habitat and organisms will be affected. A relatively high degree of warming will affect the abiotic and biotic components of an organism's habitat and physiological effects of warming will affect the ability of organisms to compete, survive, and reproduce.

Another direct effect is *ocean acidification*, which occurs when CO_2 is taken up by seawater. Acidification can reduce the concentration of, or even eliminate carbonate compounds needed by shellfish and invertebrates to make their shells and skeletons. Warming and acidification are a deadly combination for some organisms, especially coral reefs.

Indirect effects extend beyond the warming effect of CO_2 and other greenhouse gases. They include effects associated with changing climate, such as a rise in sea level, changes in precipitation patterns, changes in ecosystem food webs, and extirpation or extinction of organisms. Indirect effects on aquatic habitat also may be linked to changes in the terrestrial environment, including different runoff patterns, erosion, nutrients, habitat availability, and so on. As discussed above, indirect effects can be catastrophic, such as discussed for the Ocean Conveyor.

Ecology and Fish Production

The American Fisheries Society is the advocate for all branches of fisheries science, promoting conservation, development, and wise use. The society has drafted a new policy statement on climate change and fisheries available at (www.fisheries.org/afs). Acting in the interest of freshwater and marine fisheries, the society believes that global warming will cause problems for fish and fisheries. It calls for an immediate reduction in greenhouse gas emissions and implementation of adaptive management to develop new policies.

Freshwater fishes are usually classified as coldwater, coolwater, or warmwater based on physiology and preferences. It is fairly easy to see that coldwater fish in temperate zones would be limited by a warming climate, while warmwater fish might expand their range upward. However, in the tropics, warming might be extreme (too hot and/or low oxygen) and cause problems even for warmwater species.

Warming is expected to have a major effect on coldwater streams in the United States. Most headwater streams are already occupied by salmonids, and there is little unoccupied habitat left that meets year round requirements for them. It is assumed that the warmwater zone will extend further upslope and the coldwater zone will contract because once the top of the mountain is reached there is no habitat left. In most cases, this means coldwater fishes will be further limited in carrying capacity than at present. As a result, trout and salmon are expected to be lost from 18% to 38% of their present habitat by 2090 (National Resources Defense Council 2002) and trout habitat in the Rocky Mountain region may decrease 50% by the end of the century (Kinsella et al. 2008).

The general response to GCC is a polar migration of the temperate zone, which will extend the warmwater portion of streams and subject them to different hydrologic conditions, especially changes in runoff patterns that guide migrating fishes (e.g., by zeitgebers). Also, warmer climate will place more demands on water use for agriculture and other human purposes, resulting in a loss of diversity in fish populations and possibly increased extinctions.

Freshwater lakes are highly vulnerable to climate change because of differences in the amount and timing of precipitation, runoff, and evaporation. In northern climates, the duration of ice cover will be shortened, allowing for more growth of aquatic plants (Barange and Perry 2009). There will be more winter flooding and a general shifting in timing of abiotic variables. Large reservoirs with high surface-to-volume relationships, and high discharge, such as Lake Powell in the southwestern United States may become run-of-river impoundments. Other reservoirs supplying irrigation and waters may be lowered by drought, to the detriment of fish populations.

Lakes of high latitude or altitude will sustain warmer water temperatures and shorter periods of ice cover. Also, more evaporation will lower water levels. Those conditions should result in an increase in productivity. Deep tropical lakes will be affected differently, because increased warming is not of itself enough to increase productivity and could result in the opposite effect, likely related to less inflow and nutrient supply (Barange and Perry 2009). GCC has already affected African great lakes. Warming in Lake Tanganyika has resulted in an increase in surface temperature, which increased water column stability. This, and a decrease in wind velocity, has resulted in less nutrient upwelling and a decrease in primary productivity, which translated into a loss in fish yields of about 30% (O'Reilly et al. 2003). In another lake study, Winder and Schindler (2004) reported that warmer water disrupted the trophic linkage between phytoplankton and zooplankton due to early spring warming, which now occurs earlier (>20 d). As a result, Daphnia populations have declined due to a mismatch with the diatom bloom, causing potential problems for upper trophic levels.

River basins in general are expected to suffer loss of biodiversity as a result in warming and reduced water availability. Peak flows are expected to occur earlier (to late winter) and lower, hotter water conditions will occur. Xenopoulos et al. (2005) considered fish diversity in over 300 river basins worldwide. These authors found that river basins with fish data (n > 130) would lose 80% of their water, mostly due to withdrawals. About half of these rivers would lose about 10% of their fishes, but the predictions given are considered underestimated because other factors such as invasive species were not considered.

Marine systems also should experience the contraction of coldwater species toward the poles and an expansion of warmwater species. Pelagic fish communities would increase vertical habitat use in response to epipelagic (surface) warming. Nutrient limitations and increased stratification in equatorial seas could lead to a decrease in productivity, while polar areas should experience an increase (Barange and Perry 2009). Oceanic migrations, especially related to seasonal migration are expected to be earlier according to the degree of warming. Tropical fish communities are likely to decline in abundance, distribution, and diversity as warmwater is heated further and physiological tolerances are stressed.

Shallow estuarine and inshore systems used as nursery areas for fish will be greatly affected by sea level rise and increased warming. It is anticipated that a continuing slow rise in sea level will permit expansion of these shallow systems inshore. However, many shoreline areas have been filled and bulkheaded, restricting fish access, and in other areas, shoreline erosion has resulted in high banks.

The Atlantic croaker (*Micropogonias undulates*) (Figure 32.3) fishery has recently been evaluated under a climate-change scenario. This fishery is presently an $8-million industry along the mid-Atlantic coast of the United States. Using temperature forecasts from 14 general circulation models under three carbon dioxide scenarios, Hare et al. (2010) developed a coupled climate–population model to predict effects of continued fishing and climate change on the croaker stock. Their model also is based on likely effects of temperature on overwinter juvenile survival in estuaries. At present levels of fishing, they predicted a 60%–100% increase in spawning biomass and a shift in population distribution 50–100 km north by 2100.

Figure 32.3 Atlantic croaker is one fish expected to benefit from Global Climate Change. Image retrieved from http://nefsc.noaa.gov/press-release/2010.

The Atlantic croaker predictions are based on effects of temperature on shallow estuarine nursery areas, and we could anticipate that deeper marine systems would not be as affected by GCC. However, the distribution of about two-thirds of fishes in the North Sea has changed in mean latitude, depth, or both in the past 25 years (Perry et al. 2005). These fish stocks are presumably following plankton blooms, which are occurring earlier and extending further north. Warmer conditions would be accompanied by differences in deep currents, and possible hypoxia.

Coral reef systems are extremely sensitive to the adverse effects of GCC, specifically the direct effects of warming and ocean acidification. Many coral reefs around the world already have been lost. For example, corals were stressed by warm water in the eastern Caribbean in 2005, a condition called bleaching, in which the corals lose their algae (zooxanthellae) and turn white. The corals can recover from this, however, once stressed they are vulnerable to disease infection (e.g., white plague), which resulted in the loss of 60% of the reefs in the Virgin Islands and 53% of reefs in Puerto Rico's La Parguera Natural Marine Reserve (Encyclopedia of Earth 2009). Increasing atmospheric CO_2 will make matters worse by further heating, but concomitant ocean acidification will deprive coral reefs, mollusks, and invertebrates the carbonates needed to build their calcareous skeletons and shells (Fabry et al. 2008). As the corals go, also the habitat for the coral reef fish community will go as well.

Invasive Species

Invasive fishes are mostly hardy and aggressive warmwater species. It is probable that GCC will benefit these species by enhancing their competitive and predatory ability. This will place many native fishes at a disadvantage. Rahel and Olden (2008) regarded climate change and invasive species as forces acting in synergy to displace native fishes. In this case, warmer water, less ice cover, changes in flow regimes, increased water development, and other changes are advantegeous to the invasion and proliferation of nonnative fishes, because these changes also place the native fish at a disadvantage. For example of this, I refer to the channel catfish study done in critical habitat for endangered Colorado River fishes (chapter 25). We found that growth rate was so slow in the upper Green and Yampa rivers, that channel catfish were about 12 years old (392 mm) before they became large enough to be piscivorous. In warmer climates, the catfish might reach that size in only 4 years. These findings indicated that warming the river would favor catfish to the detriment of native species. Another nonnative example is provided by the humpback chub population in the Little Colorado River, Grand Canyon. Cold discharge from Glen Canyon dam supports a trout fishery, and the cold water has effectively isolated this population in the tributary stream. Although the chub is not able to colonize the mainstream river, cold water has discouraged invasion by large piscivorous fishes that occur in warmer downstream reaches (e.g., striped bass, flathead catfish). GCC may warm the water so much that warmwater fishes might invade upstream. If so, protective isolation of the endangered humpback chub might end.

FISH AND FISHERIES IN THE FUTURE: BAD NEWS AND GOOD PROSPECTS

Historically, inland fish management has protected, regulated, supplemented, and propagated populations of "valuable" fish species for aesthetic, recreational, or commercial use by humans. Past management practices have too often focused on commercial or sport fisheries and even then considered one species at a time. There has been little understanding or concern about natural fish communities or ecosystem ramifications. In freshwater systems, humans "controlled" fish populations in ponds and small lakes and then moved on to "tame" even large river systems. Construction of large reservoirs produced new artificial systems. Little was known about successful management of fisheries in them, and efforts were often by trial and error. As more reservoirs were built, sometimes one after another, the natural riverine system was lost and native fish populations declined. This decline was not entirely due to dams and reservoirs; it also was related to invasive nonnative fish introductions. In the case of diadromous fishes, their populations universally declined due to barriers created by dams and agricultural diversions, sometimes even when attempts were made to incorporate passageways for them. In the recent past, reservoirs were considered so economically necessary that few if any provisions were made for maintaining remnants of the natural system. Once completed, we are told to live with them. But reservoirs are not permanent. They fill up with sediment, and otherwise outlive their intended purposes. Due to the high cost of removal, some dams remain a useless barrier. These old dams, some of which continue to block fish migrations, are now being breached to restore fish movements and migrations. More removals are contemplated.

Meanwhile, highly valuable coastal fisheries have undergone a protracted period of decline, including Pacific salmonids in the northwestern United States, and Atlantic salmon and anadromous clupeids in the northeastern United States. In the arid southwestern United States, native freshwater fish communities have declined at an alarming rate. If present trends continue for freshwaters of the United States, there will be many more fish, including evolutionarily significant units (ESUs) that will be endangered and perhaps even extinct in the foreseeable future.

As pointed out earlier, no endangered fish has ever been recovered to the point that it could be delisted, even though the reasons for the endangered status of listed freshwater fishes are known. The environmental movement of the 1960–1970 period resulted in more environmental-friendly development due to the requirement for all major federal actions to have public participation under NEPA. The public was mostly concerned about economics and unnecessary habitat alteration. Public interactions included hearings and comments on proposed actions, which can result in environmental assessments and impact statements. Under the Endangered Species Act of 1973 (ESA), all actions, including listings, critical habitat, notices of a change in status, and recovery plans, also require public involvement—and can address the biodiversity crisis. States were brought in as cooperators under NEPA, receiving matching funds under Section 6. However, states soon followed with protective laws, and major nonfederal actions in most states that affect water or biodiversity now require review by state agencies. In addition, states developed their own list of species of special concern. The role of state fish and game organizations as only purveyors of sport fishing and hunting has been changing.

There are similarities in the management of fishes and fisheries among the states. In most cases, fish and game agencies have been in existence for a very long time and their mandate was, for the most part to regulate fishing and hunting of game species. To this day, most state wildlife agencies receive the bulk of their funding from sales of fishing and hunting licenses—western states may receive over 90% of their funds in that manner. In most cases, state wildlife agencies are administered by a board of commissioners who are appointed to their positions, and these boards make final decisions about natural resource management and conservation (Scalet et al. 1996). In some cases, the boards are merely advisory and can be overruled by the governor. A standard complaint among professional biologists is that the commissions tend to make decisions based solely on socioeconomic perspectives and do not appreciate the scientific aspects of management.

Fish and wildlife have been separated as either "game" or "nongame" by most states. Consequently, "game" received the bulk of the support via fishing and hunting interests, and federal aid programs for fish and wildlife restoration. The "nongame" component was less well supported by income tax check-offs (donations) and Section 6 of the ESA. The vast majority of state agencies continue to be dominated by sport fish and game management interests. Even so, state agencies have expanded into nontraditional areas involving nongame species and they are engaged in research and management activities for supporting natural communities. This has been a bumpy road, and conflicts arise between state game programs and U.S. agencies over resource management issues, including highly publicized recovery efforts (e.g., gray wolves, Canada lynx, and stocking and control of nonnative fishes).

It has been evident that with the increase in human populations, freshwater systems cannot support heavy exploitation for very long. Thus, except for large lakes, rivers and coastal systems, states do not allow commercial fishing operations. Instead, states have continued with their primary mission of regulating recreational fishing. Their interests are in promoting recreational use and setting regulations such as dates and times for fishing, types of gear allowed, creel limits, and size restrictions. Fisheries management in many states primarily involved stocking of hatchery fishes, many times for put-and-take fishing. Thus, state fisheries biologists were primarily hatchery oriented.

States will have a valuable role in defining the future of freshwater fish management according to the roles they establish with other agencies. State contributions will include assisting with recovery of threatened and endangered fishes, reducing the spread of nonnative fishes, such as regulating the baitfish and aquarium industries, and by stopping additional introductions. Nonnative fish control and eradication (e.g., snakeheads, Asian carp) is already being implemented by state fishery agencies. States also maintain hatcheries, and make supplemental stockings. Most states are proactive and have stopped the practice of introducing nonnative fishes. State hatcheries operations are being subjected to scrutiny and rigorous testing to maintain genetic diversity and to avoid spreading diseases.

Freshwater fisheries are increasingly being affected by pollution. Although many fishermen wish to consume their catch, in many areas they are advised not to do so. Mercury in fish now exceeds Environmental Protection Agency standards in many lakes and reservoirs in the United States, and other contaminants such as PCBs and dioxins are widespread in fishes. States will no doubt be active in this area in the future.

Most states also are involved in restoration of fisheries through various federal aid programs and also novel ideas such as the "adopt a herring" initiative in the northeastern United States. States also are interested in restoration of large rivers, and especially those that have the potential for freshwater and anadromous fisheries.

In marine systems, the focus is on commercially important fisheries. Unfortunately, many of these have been exploited to the point of recruitment overfishing—but even worse, some of those stocks have suffered a great loss of habitat due to fishing methods and gear. Their "management" consisting solely of modeling yields of various stocks and overfishing related to advances in fishing technology, unreported bycatch, bykill, ghost fishing, and other poor management practices. In international seas, exploitation has even extended to long-lived and slow-growing fishes, which will have a difficult time recovering from disturbance and habitat alteration, if this is even possible. Such fisheries problems are now being recognized by laypersons, such as Greenberg (2010) who identified problems and provided suggestions for managing salmon, tuna, sea bass, and cod as food for future generations.

Marine fisheries management in the United States is changing for the best. The FCMA mandates sustainability and provides a chance to make some real progress in restoring and conserving commercial stocks. Unfortunately, some of the regional councils that make the decisions have been dominated by socioeconomic rather than conservation interests. This seems to be changing with more public involvement. Certainly, there are hundreds of capable fishery scientists that are ready to do the work needed to bring the fisheries back. I have been encouraged by recent publications and especially by the attitudes of contributors to Beamish and Rothschild (2009).

WELCOME TO THE TWENTY-FIRST CENTURY

We now live in a time of sophisticated computer technology and virtual reality. Although new inventions can assist us in conservation, there remains a need for knowledge and wisdom in designing, interpreting, and implementing new conservation practices. There is great danger of continuing on the same old historic path of habitat alteration, technoarrogance, and denial.

As pessimistic as I have to be, there are signs of encouragement. For one thing, the average citizen is far better educated and concerned about environmental issues than ever before. In addition, there appears to be a new awareness and good intentions on the part of fishery management agencies and fishery scientists. And certainly, there seems to be increasing cooperation at the national level to approach natural resource management with resolve. If we can have the freedom to develop and practice good science, the wisdom to recognize that more attention to ecological principles is a necessity, and the ability to form multidisciplinary teams in problem solving, then we have a chance to improve our present situation.

Developing a new perspective and a unified approach may be easier for marine fisheries management than for management of inland fishes. Regional councils under the FCMA are under one federal agency, whereas there are 50 states involved in management of freshwater fisheries, including game and nongame species. Guidance in developing programs differs between the federal and state programs and more coordination and cooperation will be required in the future. In addition, states also would benefit from developing aggressive regional networks of specialists to work on emerging problems among states and other agencies. Professional societies such as the American Fisheries Society have played a key role in management by identifying and clarifying national and local concerns through their national, regional, and local chapters.

The American Fisheries Society has communicated with federal agencies, congressional committees, and others involved with groups that are engaged in policy setting for fisheries (Rassam 2009). The society has raised "issues of concern" and gone on record to recommend goals for a new U.S. fisheries policy. The society also has expressed concerns for the future of fisheries professionals in government service: professionals should be able to express their scientific views and to participate in professional societies.

The society believes that the goal of U.S. fisheries policy should be "to enhance the sustainability of existing stocks while rebuilding troubled stocks when feasible." However, the society has gone further and believes that the following issues of concern must be considered in the formulation of a national policy (information in parentheses added):

- Commercial and sport harvests (overharvesting)
- Habitat degradation (fish cannot live without habitat)
- GCC (will aid invasive species and result in communities lost)
- Invasive species (a threat to biodiversity)
- Declines and extinctions (homogenization of fish communities, diversity loss)
- Ecosystem alterations (cumulative impacts disrupt stability, services)

Another professional organization, the American Institute of Fishery Research Biologists, sponsored a symposium that has been published in book form with the assistance of a host of important fishery-related agencies and organizations: *The Future of Fisheries Science in North America* (2009). The symposium and book attempt to identify needs in the future, and the volume editor (Noakes 2009) places the effort in context by highlighting three pervasive concepts: the need to consider biodiversity, ecosystem management, and climate change in fisheries management. It is pertinent to note that of the 33 chapters of the book, the term *ecosystem* appeared in more than 1/3 of titles, closely followed by the need to improve stock assessment and modeling.

In summarizing the symposium, Rothschild and Beamish (2009) stated that fishery science needs to have a better understanding of the biology of major species, and fishery management must consider biodiversity and an ecosystem approach to management (EAM) of commercial stocks. In order to accomplish this, research needs to do the following:

- Develop new models and new technologies to obtain knowledge needed for an ecosystem approach, and to improve stock assessment.
- Delineate stocks by genetics.
- Apply knowledge from other fields such as engineering and operations research.
- Understand the relationship between climate and stock variability.

Inclusion of biodiversity and EAM in fishery management will require new objectives: (1) managing ecosystems, (2) managing habitat, (3) stopping overfishing, (4) using a precautionary approach, and (4) rebuilding stocks. All of these objectives are going to be difficult to implement. But they are revolutionary. In short, this means that fisheries management in the United States is in transition to a "regional, ecosystem-based stewardship." A common theme is the need to enhance existing theories with environmental data and to develop new ones. What appears to be new is the recognition that climate and ecosystem changes have to be considered in every aspect of fisheries management. In addition, there is a need to understand human impacts of fishes and fisheries. How is this to be done? It will require an interdisciplinary team approach with, as a minimum, oceanographers, fishery managers, fishery scientists, and fish ecologists. One suggestion in expediting all of these changes is the formation of a North American version of the International Council for Exploration of the Sea.

How can we prepare for implementing new approaches and answering questions such as the following: What are the major changes in climate that we can expect in this century, and how will ecosystems be affected? What are major issues and potential solutions for the biodiversity crisis? In this context, it is alarming that Berkson et al. (2009) reported that only 7% of U.S. universities offered all of the nine courses (including fish ecology) considered essential to the education of a fishery assessment scientist and suggested that there may be a shortage of scientists with adequate training to work in fisheries population dynamics and modeling.

Fisheries scientists need a good background in basic sciences, math, statistics, and modeling, which is essential for stock assessment work. However, I also believe that some fishery biologists at the journeyman level (M.S.) and even above presently do not have a good understanding about the ecology of fishes, especially behavioral ecology, to appreciate the complexity of these organisms. As for fish ecologists, we never learn enough because the field is broad and changing, and like the Red Queen, we have to keep moving! I sincerely hope that this volume will be of some help to fisheries biologists, ecologists, and enthusiasts who are intrigued with fish—the ultimate aquatic animal and the ancestral vertebrate.

SUMMARY

Humans are conducting a dangerous experiment on Earth by liberating vast amounts of greenhouse gases into the atmosphere. The direct effect of these gases, principally CO_2, is GCC, whose initial effect is global warming—resulting in melting of polar ice (ice sheets, sea ice, and glaciers), and ocean acidification. Indirect effects could be rise in sea level, disruption of the Ocean Conveyor, ENSO persistence, and other effects of GCC, many of which can be viewed as potentially catastrophic. Most of these effects presumably will be felt by the end of the twenty-first century, but some may occur as early as decades if tipping points are reached. Temperate fish communities are already feeling the effects of warming, which results in contraction of coldwater systems and dispersion or expansion of warmwater ones. Tropical fish communities are likely to decline in abundance, distribution, and diversity as warmwater is heated further and physiological tolerances are stressed. Warming is expected to aid invasion of nonnative fishes, most of which are hardy and warmwater tolerant. Recreational and commercial fisheries management will require new approaches to protect biodiversity, reduce habitat alteration, and to implement an ecosystem approach. However, human overpopulation, declining oil and gas supplies, and other problems may act to de-emphasize conservation programs. State agencies will play a decisive role in protecting freshwater systems and federal agencies will determine the fate of marine commercial fisheries.

APPENDIX
A Guide to Major Fish Groups

Fish Diversity Exercise: Major Groups of Extant Fishes

INTRODUCTION

With about 32,500 extant fishes in 515 families (Nelson 2006), it is exceedingly difficult, if not impossible, to learn all or even most of them and clearly that is not our task. However, ecologists and conservationists who work with fish need to have an appreciation of fish diversity in morphology, adaptation, and behavior. If we do not have the skill to identify fishes, we cannot develop such an appreciation. This skill requires, at a minimum, that we know how to use field guides and other sources to help us, and that we look at a fish with basic knowledge of known relationships and adaptations. This requires some knowledge of scientific terminology, gross morphology (qualitative features), and standard meristics (quantitative counts and measures). The first step is to learn how to recognize fishes that are well known and important in an ecological, commercial, or recreational sense. This we will do by using features of a fish that are easy to see and adaptations that are evident. The approach is that of a prehistoric saga, the fauna changing as groups appear and disappear in time.

This exercise is most helpful when specimens are available to examine, and for those who are not familiar with fish taxonomy, it is useful to complete the exercise prior to reading chapters 4–6. In those chapters, the goal is to appreciate the great diversity in fishes, to develop an understanding of factors that have produced the present fauna, and to consider ecological attributes of species and fish communities. On the other hand, here we simply identify major groups and discuss some important species.

After doing the exercise, a trip to a fishing dock or fish market to view the catch, or to a museum or sporting goods store to see mounted fish can be helpful. But viewing live fish in action is even better, and an aquarium visit or viewing a video, such as "Eyewitness Video: Fish" (BBC), will augment and reinforce what has been learned.

Almost all fish look like fish, and just about everyone can correctly identify an animal to be a fish if it is one. Furthermore, once a few characters are pointed out, it will be easy to separate fishes into groups. This ability is related to *gestalt*, a mental process in which we have the innate ability to perceive an organism as more than a collection of parts and to recognize differences in organisms even though we may never have seen them before. In this exercise, success is related to an ability to identify separate structures and to note when differences occur in those structures.

A fish can be conveniently described as "an aquatic vertebrate with gills that persist throughout life and limbs in shape of fins." This definition of a fish requires that we know what vertebrates are and that we are able to identify gills and fins. As presented in the early chapters of this book, we characterize fish as true vertebrates that have altered inherited structures from nonvertebrate chordates (such as the notochord, gill slits, dorsal nerve cord, and muscle myomeres) and derived new ones (such as a bony cranium, jaws, paired fins, etc.). The development of gill filaments for extracting oxygen from water, paired fins for aiding in movement, and jaws for accomplishing a variety of tasks aided the diversification of fishes. However, even these basic structures are not the same in all fish. In fact, some fishes will appear more worm- or eel-like compared to popular notions about how a fish is supposed to appear. Also, there are surprises when we discover what the most recently evolved groups look like.

INSTRUCTIONS

In this exercise, we recognize six major groups of living fishes, which may contain several families each, and some of the representative species they contain. Their group separation can be based on a few characters shown in Table A1 (other differences presented in chapters 4–6). For example, some ancestral forms can be lumped as agnathans because they have no jaws, leaving the great remainder of fishes and all vertebrates as animals with jaws. In this remainder, we can separate the sharks and their relatives, whose supporting skeleton is based on cartilage, from all other fishes that have bony skeletons, and so forth. Such distinguishing characteristics will be identified and discussed as the groups emerge. A cautionary note: because such a long time has passed during the evolution of fishes, some characteristics may have appeared in a species, but then were lost. The use of lower and higher teleosts as groups is done purely for simplicity and convenience.

Detailed drawings and pictures, some with various parts labeled, will be used to allow comparisons to be made with other fish groups and species. To aid in learning general nomenclature, anatomical positions and axes are provided in Figure A1. It is also most helpful to have specimens in hand or at least on display for discovering what some characteristics actually look or feel like. In viewing an unknown fish, one should pay close attention to the following:

(1) General body shape. For example, is a fish mostly torpedo-shaped or different, such as flattened from top to bottom (dorsoventrally), side-to-side (laterally), or snake-like?
(2) The location and shape of appendages: where are the fins, are they in the midline, or are there also paired fins that are similar to our arms and legs? Where are the paired fins located and how are they attached to the body?
(3) Presence of jaws and the appearance of teeth.
(4) Type of body covering: does it have scales? If so, are they like sharp teeth (placoid scales of sharks), like bony platelike scales of gar, or more like shingles or finger nails (bony ridge scales of teleosts) in gross appearance? If not, are there bony plates like sturgeons or smooth skin as in many catfish?
(5) Considering the gill openings: are the slits visible or covered? Note the gill arches (Figure A2) that support gill filaments (for obtaining oxygen) and the number and length of gill rakers (used to retain food). Are there many long rakers for retaining small food such as zooplankton, or are there less, but heavy rakers for retaining larger prey such as fish?
(6) The kind of caudal fin present, if any, and look for the tail vertebrae. Do they extend into the upper lobe of the caudal fin or terminate in midline?
(7) Are there other structures such as spines, barbels, or sucking discs?

The need to ask these questions will be apparent as we consider groups of fishes.

One may ask why we study some fishes that we may never see. Truly if we studied the fishes according to number of species and groups, we would look at teleosts and little else (e.g., see numbers of extant fishes presented in Figure 4.1). To do so would hamper our understanding of paleoecology and tell us little about phylogeny of fishes. So in this section, as in the rest of the text, we look at diversity from an evolutionary prospective.

Table A1 Very Simplified Way of Dividing Living Fishes into Six Groups

Group	Jaws	Skeleton of Bone	Ray Fins	Homocercal Tail Fin	Spines in Fins, etc.
Agnathans	No	No	No	No	No
Sharks, rays	Yes	No	No	No	No
Lobefin fish	Yes	Yes	No	No	No
Ancestral ray fins	Yes	Some bone	Yes	No	No
Lower teleosts	Yes	Yes	Yes	Yes	No
Higher teleosts	Yes	Yes	Yes	Yes	Yes, may be lost

APPENDIX

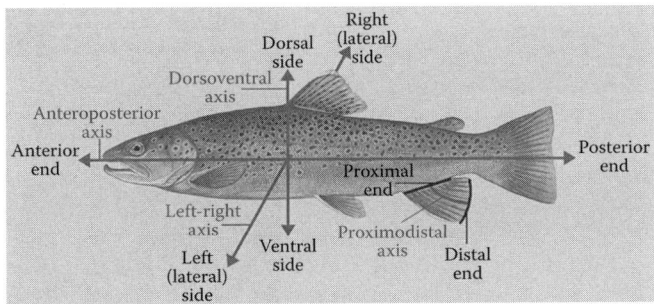

Figure A1 Anatomical features, positions, and axes shown for a brown trout. (Courtesy of http://en.wikipedia.org/wiki/File:Anatomical_Directions_and_Axes.JPG; Raver, D., U.S. Fish and Wildlife Service (USFWS), http://commons.wikimedia.org/wiki/File:Bachforelle_ Zeichnung.jpg.)

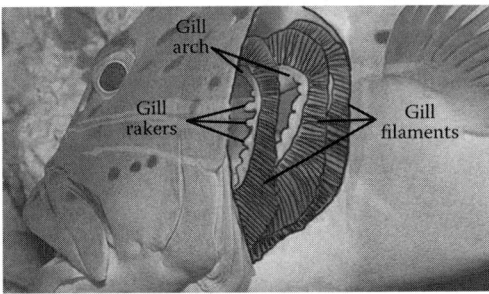

Figure A2 Gill arches, filaments, and rakers of a fish. (Courtesy of Richardson, M., http://commons.wikimedia.org/wiki/File:Fish_Gills_ Labeled.jpg.)

AGNATHANS

Seven superclasses have been identified for the first major radiation of fishes, a group commonly called Agnatha (i.e., no jaws; Nelson 2006); however, only two of these groups have living representatives, the hagfishes and lampreys. Judging by persistence over hundreds of millions of years, both of these strange animals are successful. What are their attributes? Both appear to be large worms with naked, scale-less bodies. To the uninformed they are sometimes called eels, and especially the hagfishes which are commonly called "slime eels." However, eels have jaws (derived from gill arches), but hagfish and lampreys do not have jaws. Also, they both lack paired fins, but that is where the similarity ends (Figure A3).

Hagfish are relatively simple organisms. They are blind scavengers and feed on dead animals by extending and covering food with a toothy tongue, which is then retracted to compress and tear off pieces. Hagfishes differ conspicuously from lampreys. They lack a separate dorsal fin—what appears to be a dorsal fin is part of the caudal fin. The hagfishes have mucous cells that produce a thick slime; they have sensory projections called barbels around the mouth (which is not a circular disc), and degenerate eyes. There are many other, less conspicuous differences between the two taxa that are discussed in chapter 4.

Lampreys are true vertebrates and presumably descendants of the extinct ostracoderms. They can be identified as the only living agnathan with a dorsal fin, a disc-like mouth, and functional eye. They lack the slime cells and barbels of the hagfish, have teeth on the oral disc as well as tongue, and have a larval stage. The sea lamprey is a parasite on teleost fishes.

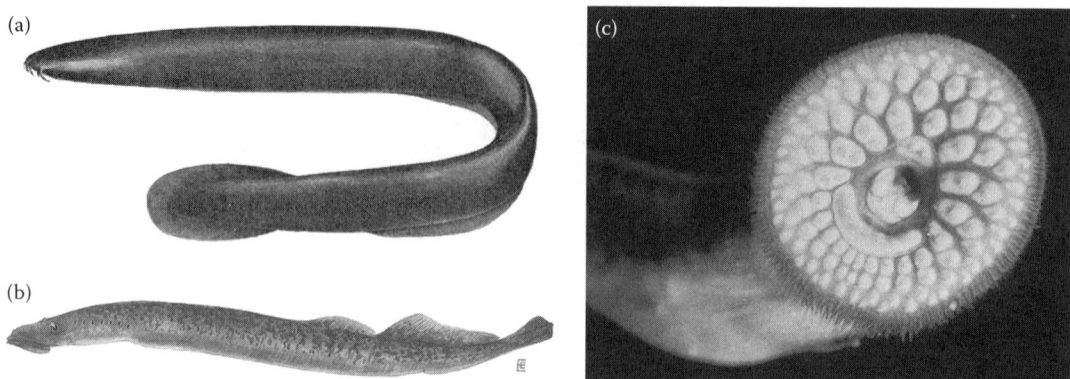

Figure A3 Agnathans: (a) the hagfish (*Eptatretus*) and (b) sea lamprey (*Petromyzon)* with its oral disc (c). (Courtesy of USFWS, Bulletin of the United States Fish Commission 26:1906, http://www.archive.org/details/bulletinofunited261906unit, restored image retrieved from http://commons.wikimedia.org/wiki/File:Eptatretpus_deani.jpg; Edmonson, E., and Chrisp, H., New York State Department of Environmental Conservation, http://www.ny,gov/animals/52634.html; USFWS digital photo (http://www.fws.gov/digitalmedia), http://commons.wikimedia.org/wiki/File:Petromyson-marinus1_FWS.jpg.)

GNATHOSTOMES (JAWS IN THE MOUTH)

All of the remaining extant fishes (and tetrapods) have jaws derived from modified gill arches and paired limbs. An ancient and distinct class of jawed fishes includes the sharks and their relatives (Class Chondrichthyes), which are subdivided into two groups, the ratfishes (or chimeras) and the elasmobranchs (sharks and rays; Figure A4). Because sharks and rays are shown in movies, television, and aquaria, they need little description. With that in mind, chimeras visually stand apart from others in their class. They have a cover over the gill slits, and they also have rodent-like teeth, pointed tail, and a dorsal poisonous spine. The elasmobranchs are also rather easy to identify (Figure A5) because of their separate platelike gills and large pectoral fins that may be united with the body, as in the rays. They are covered with a tough skin that has small, pointed, toothlike

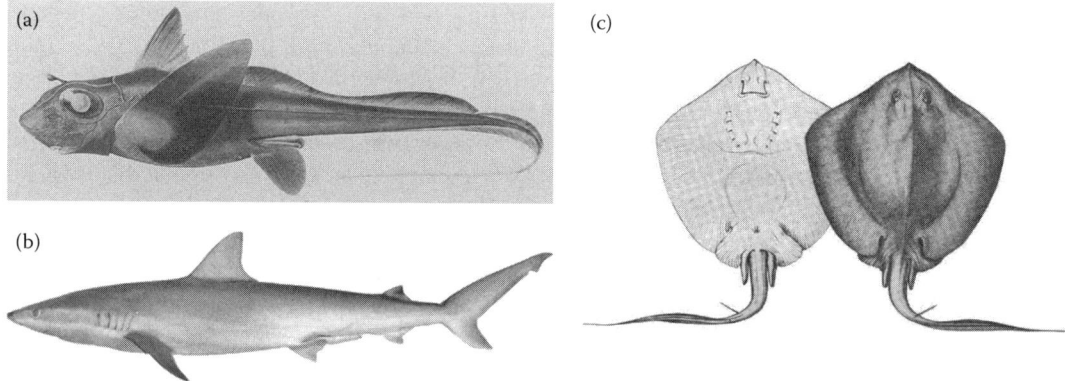

Figure A4 Chondrichthyans: (a) chimera (*Hydrolagus*), (b) blacknose shark (*Carcharhinus*), and (c) Say's stingray (*Dasyatis*). (Courtesy of Murray and Hort, 1912, http://commons.wikimedia.org/wiki/File:Hydrolagus_mirabilis.jpg; Hoffmayer, E., Iglesias, S., and McAuley, R., National Oceanic and Atmospheric Administration (USNOAA; http://www.nmfs.noaa.gov,sfa/hms/sharks/2008), http://commons.wikimedia.org/wiki/File:Carcharhinus_acronotus.jpg; New Jersey State Museum (1907 Annual Report), http://commons.wikimedia.org/wiki/File:Dasyatis_say_njsm.jpg.)

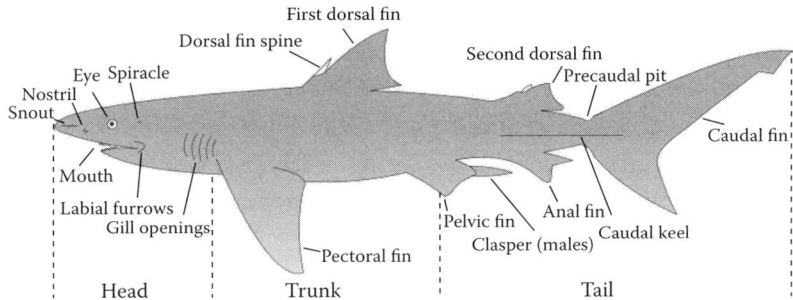

Figure A5 Shark morphology, labeled. (Courtesy of Huh, C., http://commons.wikimedia.org/wiki/File:Parts_of_a_shark.svg.)

(placoid) scales. Some forms retain an opening from the remains of a gill arch called a spiracle. It is used as an inlet for water to flow over the gills, especially in rays. Rays have large flexible pectoral fins that are used like wings in swimming. Their flattened bodies are ideal for hiding in bottom substrates. In general, sharks have pointed rostrums, inferior mouths, and uneven caudal fin lobes that give them a distinctive look. The caudal fin is heterocercal (uneven) because the vertebrae extend into and enlarge the upper lobe of the fin. However, there are some sharks that have large pectoral fins and look like rays, and some rays with small fins that look like sharks. General practice is to consider the location of the gill slits. If they open on the dorsal side of the animal, it is considered to be a ray or skate. If they open on the lateral side, consider it a shark.

A new group of fishes emerged to share aquatic systems with the sharks. This group (Teleostomes = whole mouth) improved upon the simple jaw and mouth of the sharks, and many other new characteristics emerged as well. This group was ancestral to three classes of vertebrates, of which only the rayfin and lobefin lines remain alive today.

The lobefin fishes (Class Sarcopterygii = flesh + fin) are easy to identify because their paired fins are appendages with fleshy bases (as in an ear lobe). In the lobefins, the bony rays that support the fins emerge from the fleshy appendage and not directly from the body. Three major groups survive: the lungfishes, the coelacanths (Figure A6), and the tetrapods. The fishes have a strange caudal fin that is confluent in lungfishes but with three lobes in coelacanths. Coelacanths are marine and can also be separated from the lungfishes by the presence of two dorsal fins, an anal fin, and many other, less conspicuous features. Lungfishes are freshwater lobefin fishes with a reduced number of unpaired fins and functional lungs.

The rayfin fishes (Class Actinopterygii = ray + fin) are in the main line of fish evolution. Their fin rays are generated in the body and not from the base of the fins. Over millions of years, the

Figure A6 (a) Coelacanth (*Latimeria*) and (b) Australian lungfish (*Neoceratodus*). (Courtesy of HTO, http://commons.wikimedia.org/wiki/File:Latimeria_model_01,JPG; Cada, R.N., FishBase.org, http://commons.wikimedia.org/wiki/File:Nefor_u0_gif.)

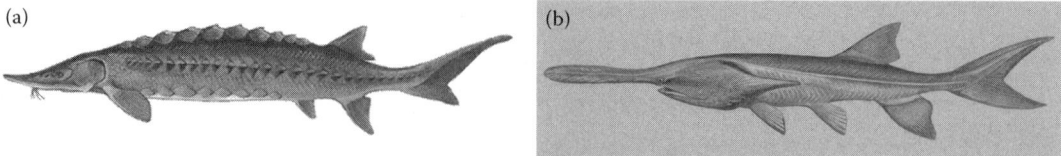

Figure A7 (a) Atlantic sturgeon (*Acipenser*) and (b) paddlefish (*Polyodon*). (Courtesy of Raver, D., USFWS.)

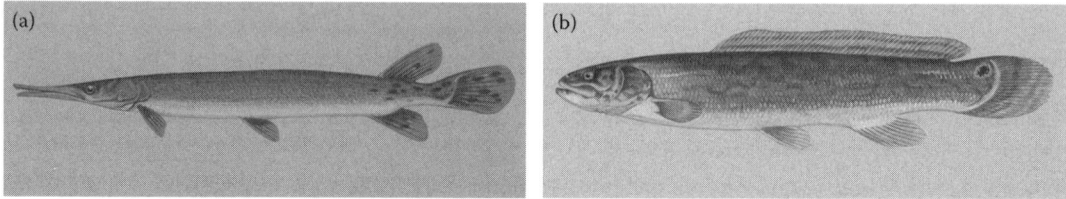

Figure A8 (a) Spotted gar (*Lepisosteus*) and (b) bowfin (*Amia*). (Courtesy of Raver, D., USFWS.)

scales have changed in emerging fishes from platelike ganoid scales to the familiar flexible bony ridge scales. Also, rayfins have at least one branchiostegal ray, which is used to seal the opercle against the body, thus aiding the mouth forming suction. Relict (nonteleost) members of this class include the sturgeon and paddlefish, which retain the heterocercal caudal fin and are cartilaginous (hence the Subclass Chondrostei, cartilage + bone). Both are easy to identify and separate due to the five rows of bony scutes on the sturgeons and the long paddle-shaped rostrum of the two remaining paddlefish species (Figure A7). However, early naturalists did confuse them with sharks.

A more derived group of relict rayfin fishes are gars and bowfins. These fish are called Neopterygii (new + fins) because each ray in the fin is supported by a separate element in the body, instead of sprouting several rays from one element. Also, they have a distinctive caudal fin that separates them from all living fish: It is an abbreviated heterocercal form in which only a few vertebral elements move into the upper lobe. Both gar and bowfin are fish of warm sluggish water, and they have vascularized swim bladders that allow them to breathe air. They are easily separated: Gars are long fish with a thin, narrow snout full of teeth, and are also heavily armored with a continuous cover of platelike ganoid scales. Bowfins are more robust in appearance, much shorter than gars, have large smooth (cycloid) scales, and a large reptile-like head (Figure A8).

LOWER TELEOSTS

Teleost fishes, which comprise 96% of all living fishes, have a two-lobed caudal fin that has a symmetrical appearance (homocercal) on the outside due to new bones that support it internally. Early teleosts, however, are different in appearance compared with more recently derived forms (e.g., see Figure 1.1). Thus, early groups of fishes are referred to as lower teleosts (i.e., found lower, thus earlier in geological record).

An early teleost offshoot, the Family Clupeidae contains several first-level consumers, and they are important forage fishes: They include herrings, shads, sardines, menhaden, and others. The most valuable group of commercially fished marine species, they have laterally compressed bodies, one soft-rayed dorsal fin, smooth cycloid scales, a distinctive keeled saw-toothed arrangement of scales along the ventral side (belly), and abdominal pelvic fins (Figure A9).

Another early offshoot is the Otophysii (ear + bladder), a large group of freshwater fishes that have small bones along the vertebrae connecting the swim bladder with the inner ear, thus

Figure A9 Clupeids; blueback herring (*Alosa*). (Courtesy of Raver, D., USFWS.)

amplifying sound. Included are minnows, suckers, catfishes, and characins. Minnows (Cyprinidae) comprise the largest freshwater fish family (2420 species). Minnows and closely related suckers (Catostomidae) are similar to the clupeids, but more diversified. Although lacking jaw teeth, they have pharyngeal teeth (located on the last gill arch) that aid in processing food. They also have open swim bladders connected to the esophagus and have cycloid scales, and their pelvic fins are abdominal (Figure A10). Suckers have thick fleshy lips for sucking up detritus and sometimes scraping mouth parts to remove algae, etc., from rocks (Figure A11). North American freshwater catfishes (ictalurids) are part of this group and very easy to identify because of the prominent (cat-like) barbels, stout spine-like locking rays in the dorsal and pectoral fins, and the absence of scales. They also have a small fatty adipose fin located between the dorsal and caudal fins (Figure A12). Marine catfishes are also common along the southeastern coast of the United States.

The first two groups mentioned here have strayed from the main evolutionary line leading to the higher teleosts. However, the families Salmonidae and Esocidae are considered to represent prototypes or first models of the higher teleosts. An important recreational and commercially exploited group, familiar salmonids include whitefish, salmon, trout, and char. All have jaw teeth, adipose fin, and small scales. In the United States, trout are avidly sought by anglers. In the western United States, the major native fish groups are cutthroat and rainbow trouts (Figure A13a). They are derived from Pacific salmon and have dark spots on a lighter body and caudal fin. In the eastern and northern United States, native trouts are actually chars, very cold hardy fish such as lake trout and brook trout. The chars have white edges to the ventral fins and vermicular shading on the darker body (Figure A13b). A common nonnative trout, widely introduced, is the European brown trout and it is related to the Atlantic salmon (i.e., *Salmo*). It has red spots with halos and dark spots on the light body, but no spots on the fins. Trouts may be the only fish at higher elevations, and their elongated, somewhat laterally compressed body, small scales, adipose fin, and bright colors and spots make it hard to mistake them for another fish.

Figure A10 Cyprinids: (a) common carp (Cyprinus): 1 = barbel, 2 = opercle, 3 = pectoral fin, 4 = pelvic fin, 5 = anal fin, 6 = dorsal fin, 7 = homocercal caudal fin. (b) Creek chub. (Courtesy of Raver, D., USFWS, http://commons.wikimedia.org/wiki/File:Common_carp_tagged.png.)

Figure A11 Catostomids (North American suckers): white sucker (catostomidae). (Courtesy of Edmonson, E., and Chrisp, H., *1927–1940 New York Biological Survey*, New York State Dept Environmental Conservation.)

Figure A12 Ictalurids: channel catfish (*Ictaluris punctatus*). (Courtesy of Raver, D., USFWS.)

Figure A13 (a) Rainbow trout (*Oncorhynchus*) and (b) brook trout (*Salvelinus*). (Courtesy of Raver, D., USFWS.)

Figure A14 Northern pike (*Esox lucius*). (Courtesy of Knepp, T., USFWS, http://commons.wikimedia.org/wiki/File:Esox_lucius1.jpg.)

APPENDIX 471

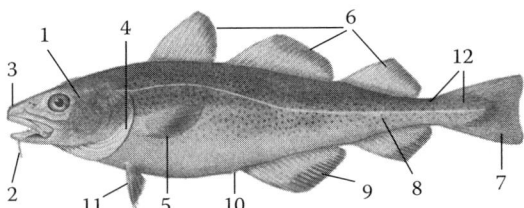

Figure A15 Atlantic cod (*Gadus morhua*): 1 = opercle, 2 = barbel, 3 = snout, 4 = branchiostegal rays, 5 = pectoral fin, 6 = dorsal fin, 7 = homocercal caudal fin, 8 = lateral line, 9 = anal fin, 10 = anus, 11 = pelvic fin, 12 = caudal fin base. (Courtesy of USNOAA, http://commons.wikimedia.org/wiki/File:Atlantic_cod.jpg.)

Fishes in the pike family all appear similar. Northern pike, muskellunge, and pickerels have elongated bodies, a duck-bill-shaped rostrum, a large mouth with many very sharp teeth, and a posterior position of the dorsal fin (Figure A14).

Codfishes, haddock, pollock, hakes, burbot, cusk, etc. (Family Gadidae) are large predators and quite good to eat. The cods are very closely related to higher teleosts. They are long fish with tapering bodies and subdivided dorsal and anal fins. Also note the forward position of the pelvic fin and the presence of numerous branchiostegal rays (Figure A15). Benthic feeders, cods, and their relatives are often collectively called "groundfish." They have been commercially fished for centuries.

HIGHER TELEOSTS

Great diversity is built around central themes of higher teleosts. Although body shapes vary considerably, the theme is evident in the Perciformes, the largest order of vertebrates: new ctenoid scales, which have little sharp projections and feel rough to the touch, a closed air bladder, spines in the dorsal and anal fins, a protrudable mouth, teeth on the jaws and in the mouth, forward movement of the pelvic fins, and dorsal movement of the pectoral fins.

Good examples of higher teleosts are the laterally compressed sunfishes (also called bream) and freshwater basses of the family Centrarchidae because they have all of the characteristics named above. Examples of recreational fishes are bluegill and redear (shellcracker) sunfishes, crappie, rock bass, and largemouth and smallmouth basses (Figure A16).

The Family Percidae, however, is the ideal higher teleost group and truly a spiny-rayed fish. In addition to the preceding characteristics, fishes in this family have two dorsal fins: the first is spinous, and spines are present in pelvic fins, which are thoracic or jugal and lateral. Examples are yellow perch and darters (Figure A17).

Figure A16 Centrarchids: (a) largemouth bass (*Micropterus salmoides*) and (b) bluegill sunfish (*Lepomis macrochirus*). (Courtesy of Raver, D., USFWS.)

Figure A17 Percids: (a) yellow perch (*Perca flavescens*) and (b) candy darter (*Etheostoma osburni*). (Courtesy of Raver, D., USFWS; Courtesy of Burkhead, N., USNOAA, http://fl.biology.usgs.gov/pics/nativefish/nativefish.html.)

The diversity in higher teleosts is amazing due to the conversion of some structures into others, such as the sucking disc of remoras, converted from the dorsal fin. Highly evolved and swift marine predators are all higher teleosts such as jacks, mackerels, and, of course, tuna and billfishes (Figure A18). In case of fast swimming tunas and marlins, the fins can be retracted into grooves and recesses on the body in order to reduce drag, and special finlets near the tail serve to break up viscous laminar flow.

If the apex of fish evolution can be considered the most recently derived forms, two groups are candidates, and they are truly bizarre: the flatfishes (flounders, sole, etc.) in which one eye rotates from one side of the head to the other, and the tetraodonts (they have four teeth), which include puffers and molas that almost defy description (Figures A19 and A20). It seems that the flatfishes have taken up an occupation that competes with rays, and at least some of the tetraodonts emulate pelagic forms of the missing ostracoderms, with a few added tricks such as body inflation and poison in the tissues.

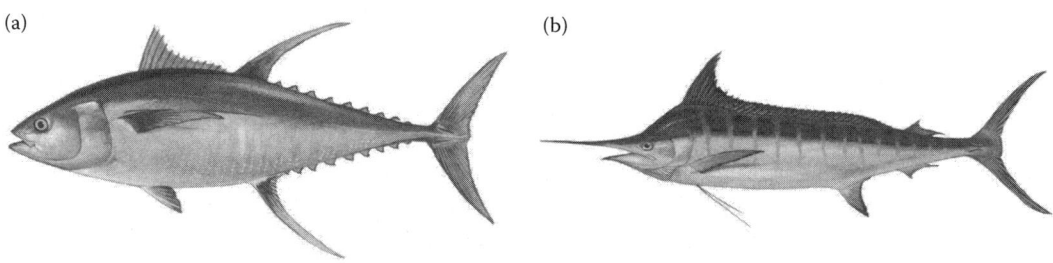

Figure A18 (a) Yellowfin tuna (*Thunnas albacores*) and (b) blue marlin (*Makaira nigricans*). (Courtesy of Raver, D., and NC Division of Marine Fisheries.)

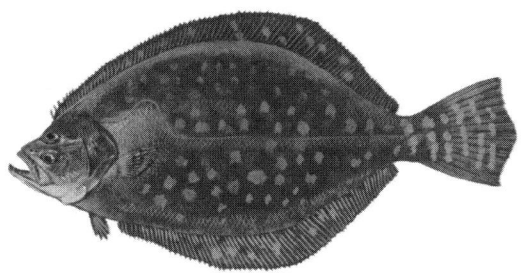

Figure A19 Flatfishes: southern flounder (*Paralichthys lethostigma*). (Courtesy of Raver, D., and NC Division of Marine Fisheries.)

Figure A20 Tetraodonts: puffer (*Sphoeroides nephelus*). (Courtesy of Raver, D., and NC Division of Marine Fisheries.)

The preceding discussion emphasizes well-known fishes and places them in a logical phylogenetic order. We will build on this coverage during the remainder of the text. Some readers may not be familiar with all of the terms used here, although many terms are defined in the text. Most or hopefully all troublesome words should also be defined in the glossary section.

Glossary

abyss: the oceans at great depths; marine habitat at 4000 to 6000 m.
adipose fin: a small fleshy median fin.
aerobic: oxygen is present for use in metabolism.
agnatha: lacking jaws.
alarm substance: see *shreckstoff*.
albedo: the percentage of solar radiation reflected from a place on Earth.
alien: not native or indigenous to an region, continent, or planet.
Allee effect: population growth rate is highest at intermediate densities.
altricial: eggs, embryos, and larvae with a small yolk supply and juveniles that are not are not fully formed at swim-up.
allochthonous: substances (e.g., organic matter) imported into a system.
ambient energy hypothesis: favorable ambient conditions (e.g., tropics) result in higher density and greater species diversity.
ampullae of Lorenzini: organs that allow electroreception in elasmobranchs.
anaerobic: oxygen is not available for metabolism.
anadromous: migrating from marine to freshwater for spawning.
anal fin: a ventral unpaired median fin located just anterior of the caudal fin.
anchoveta: a common name for the Peruvian anchovy (*E. ringens*).
anoxia: no oxygen present.
anthropogenic: human produced (e.g., effects on organisms and systems).
applied ecology: study of anthropogenic effects on nature.
aphotic: without light.
autochthonous: substances (e.g., organic matter) produced within a system.
autotrophic: producing organic matter and energy from inorganic substances.
bathyal: deep sea below the edge of the continental shelf.
bathypelagic: open-water marine aphotic zone at depths of 1000 to 4000 m.
biodiversity: species richness, e.g., the number of native species per area.
biogeography, first principle of (Buffon's law): different regions have different species even if the climate is similar.
benthic: the bottom or substrate of aquatic habitat.
branchiostegal rays: rays that strengthen gill membranes and enhance suction feeding.
bycatch and bykill: harvesting or indirectly killing nontarget species or nontarget sizes.
catadromous: migrating from freshwater to oceans for spawning.
caudal fin: unpaired posterior (tail) fin of a fish.
chessboard model: mass extinctions not only remove players but can destroy the board; these catastrophic events can disrupt evolutionary patterns in addition to species loss (Erwin 2006, p. 219).
chordate: belonging to an animal phylum that includes all deuterostomes, i.e., having a hollow dorsal nerve cord, notochord, pharyngeal slits, and shaped muscle myomeres.
chloride cells (α, β types): cells of gills that are used for osmoregulation in teleost fishes.
claspers: intromittant (for insertion) sexual organs of male chondrichthians.
coextinction: when the extinction of one species results in the extinction of one or more additional species, potentially due to mutualism.
compensation zone or depth: a place where energy derived from photosynthesis is equal to energy spent by respiration.
constant extinction, law of (Van Valen's law): the probability of a taxa surviving a mass extinction is not related to its longevity.

consumptive use: resource use that results in direct effects to the resource, such as adverse impact or loss.
Coriolis effect: effect of Earth's rotation in deflecting air and ocean currents to the right in the northern hemisphere or to the left in the southern, resulting in gyres. Effect is maximum at poles, 0 at equator.
cranium: part of the skull covering the brain, usually of cartilage or bone in vertebrates.
critical period concept (attributed to J. Hjort): a period of high mortality in ontogeny, especially at first exogenous feeding.
cropper: animals that eat smaller living and dead organisms.
ctenoid scale: a bony-ridge scale with sharp projections on the exposed surface that is common in higher teleosts.
cultivation hypothesis (Walters and Kitchell 2001): young of large predators are protected by their parents, which also prey on intermediate-sized predators, thus cultivating the environment for their young.
current: flow of water, air, etc., measured in centimeters, meters, or feet per second.
cutaneous: referring to the skin.
cycloid scale: a smooth, bony-ridge scale that is common in lower teleosts.
decomposer: an organism that feeds on deposits of dead organic matter.
decomposition: breakdown of complex organic matter into simpler compounds by abiotic and biotic processes.
deep sea: oceanic systems below the continental shelf.
deep sea as a desert hypothesis: Sanders 1968, Dayton and Hessler 1972.
demersal: living on or close to the bottom.
depensation: a reduced population density results in lower per capita growth.
dermal: the living part of the skin and associated structures.
detritus: bits and pieces of dead and decomposing organic matter.
deuterostome: an organism in which the blastopore become the anus instead of the mouth in the blastula stage of ontogeny.
diadromous: migrating between fresh and saltwater.
diphycercal: a bi-lobed caudal fin present in coelacanths.
dispersal: a one-way, directed movement.
disturbance concept (Grime 1973): disturbance can reduce interspecific competition while encouraging species diversity (also see *food cropper hypothesis*). It is now thought that an intermediate level of disturbance is best because both r- and K-selected species can coexist.
Dollo's law: evolution is not reversible.
dorsal fin: the principal unpaired median dorsal fin in fishes.
drag: resistance of an object to movement through a liquid due to viscosity of the liquid and the shape of the object.
drift: objects or organisms transported in stream flow.
ecological equivalent: organisms living in different ecosystems that have similar adaptations.
ecology: study of: the structure and function of nature, the relationships among organisms and between organisms and their environment, and factors influencing the distribution and abundance of organisms.
ecosystem simplification: loss of ecosystem complexity due to human-induced effects.
ectotherm: organism whose body temperature is controlled externally.
ecotourism: outdoor recreation/education using an ecosystem perspective.
eddy: water flowing in a circular rotation or countercurrent flow.
endotherm: organism that controls body temperature internally.
elasmobranchs: sharks and rays.

ENSO (El Niño—southern oscillation): phenomenon resulting in movement of warm surface waters eastward in the Pacific that overlie and reduce coastal upwellings along the west coast of North and South America.
epifauna: organisms that live in the substrate.
epipelagic: the upper, aphotic zone of the sea.
esca: a lure such as in anglerfish.
estuary: a semi-enclosed aquatic system in which seawater is measurably diluted by inputs of freshwater.
eutrophic: rich food.
euryhaline: tolerant of a broad range of salinity.
extant: living.
extinction coefficient: ratio of the light intensity at a depth relative to that at the surface.
fecundity: related to the number of offspring produced, i.e., *fecund* would mean that many offspring are produced by an individual or species.
fish: an aquatic vertebrate with limbs shaped like fins and gills that persist throughout life.
fish lice: ectoparasitic copepods of fishes (Order Lernaeoptodida and Caligoidida).
fishery: a continuing exploitation of fish by humans for a specific fish or location.
fitness: the degree of success in producing viable progeny; the genetic contribution of an organism to the next generation.
flood pulse concept: the stream floodplain is an integral part of the lotic ecosystem and connected by energy flow during seasonal flooding.
food chain: the transfer of energy by consumption of organic matter in a number of steps; there is a grazing and a detrital chain.
food cropper hypothesis: especially in the deep sea, organisms consume anything that is small enough; thus, specialized species are rare. Consumers are scarce, competition is low. Diversity, once developed, is maintained. (Also see *disturbance hypothesis*.)
food web: a system of interacting food chains.
FPOM: fine particulate organic matter, with particle size < 1 mm.
foraging: behavior related to obtaining food for consumption.
fundamental niche: niche space reflective of the sum total of adaptations of a species unconstrained by interspecific competition.
GAS (Selye): general adaptation syndrome associated with stress.
gas bladder: a.k.a. swim bladder. Used as hydrostatic organ and also for obtaining oxygen (open type).
genotype: the total genetic constitution of an individual or population.
geographic area hypothesis (attributed to Terbough 1973): species in the tropics have large areas of suitable habitat, resulting in large organisms and a low rate of extinction.
geologic time: a system of time based on geologic events (stratigraphy) and includes eons, eras, and periods.
gill: organ used to obtain oxygen and to regulate salt balance in fishes.
gill filaments: portion of the gill that functions in the exchange of gas and ions.
gill raker: portion of the gill that prevents food from escaping and protects the gill filaments.
Gondwanaland (Gonwana): a supercontinent of the southern hemisphere that resulted from the breakup of Pangea (ca. 180 mya).
grazer: an organism (herbivore) that consumes green plants and does not necessarily kill its prey. Also used for some carnivores that "graze" algae or sessile invertebrates, such as rays that consume the siphon tubes of molluscs.
greenhouse effect: warming of an area by re-radiation.
guyot: an isolated underwater mountain with a flat top/a flat seamount.

gyre: circular rotation of water in ocean basins partially as a result of the Coriolis effect.
habitat: the space occupied by an organism including abiotic and biotic components.
hadal: a part of the ocean that lies below 6000 m.
hemoglobin: a respiratory pigment that aids in the transport of oxygen and carbon dioxide in the blood.
herbivore: an organism that feeds on green plants.
heterocercal: the caudal fin of teleosts, which appears symmetrical on the outside.
heterotrophic: a system in which most of the carbon is obtained from consumption of other consumers and not from photosynthetic organisms.
homeorhesis: the attribute of systems to maintain conditions in a varying state of flux, or dynamic equilibrium, in the absence of set-point controls.
homeostasis: the tendency of organisms and organs to resist change and to maintain a stable condition, usually around a set point controlled by negative feedback.
home range: an area inhabited by an organism and familiar to it, but not defended.
homing: the ability of an organism to remember and locate a specific site, as in homing to natal spawning areas.
HSOB: home site olfactory bouquet, a mix of organic chemicals detected by olfaction and presumably useful in establishing an imprint or long-term memory.
Hypertonic: exceeding some concentration.
hypocercal: a type of caudal fin that has a descending shape.
hypural plate: bones of the caudal fin of teleost fishes that form the basis for a homocercal tail.
hydrothermal vents: venting of water at the ocean floor caused by water seeping down and contacting magma.
ichthyology: the study of fish, especially of taxonomy and phylogenetic systematic analysis.
illicium: a line used as a lure.
imprinting: production of a lasting memory of an object or set of conditions that can be recalled later in life by a stimulus such as sight, odor, or hormones.
infauna: organisms living in the substrate.
intermittent stream: a stream in which surface water is not continuous.
iteroparous: capable of spawning repeatedly.
keystone species: a species important for maintaining ecosystem integrity.
lacustrine: lake-like.
larva(e) (fry): the earliest free-living stage of an organism, usually following the embryo stage, in which the individual has not yet developed to the point of being a little adult, or juvenile.
Laurasia: the northern hemispheric land mass that resulted from the breakup of the supercontinent Pangea.
Leibig's law of the minimum: the metabolite that is in the least supply will restrict the number and growth of organisms.
litter: coarse particulate organic matter, especially leaves, needles, and woody debris.
littoral: intertidal.
lobefin fish: group of fishes (Sarcopterygii) with fleshy bases to their fins.
marine snow: sinking particulate matter derived from organisms.
mass extinction: occurs when loss of species or families follows some catastrophic event.
match–mismatch hypothesis: posits that organisms must produce offspring at the time and location needed for their optimum survival, usually to match the production of larval food.
maximum sustained yield (MSY): the concept that largest sustainable yields can be harvested if fishing yields do not exceed one half of the carrying capacity.
median fins: the unpaired fins lying in the midline of a fish (e.g., dorsal and anal).
mesopelagic: twilight zone, light too low to support autotrophy.
microcosm: early term for an ecosystem, e.g., the lake as a microcosm.

millions of eggs hypothesis: the notion that fish stocks can be heavily fished because the numbers taken can be rapidly replaced due to the large number of eggs produced by marine fishes.
murder on the Orient Express hypothesis: a case in which all suspects are guilty; refers to multiple causes, especially in catastrophic changes such as mass extinctions.
myomeres: segmented blocks of muscles; in fishes these can be "V," "W," or more complex in shape.
natural: an ecosystem characterized by no or few artificial structures.
nekton: an organisms living in the water column.
neritic province: the marine system overlying the continental shelf.
neuston: organisms living at the water surface.
niche (ecological niche): the role of an organism in nature, produced by the sum total of its adaptations. A niche includes both fundamental and realized niche. Niches may change with ontogeny.
notochord: a supportive rod of connective tissue along the dorsal side of chordates that is used to support muscles in coordinated swimming.
ocean conveyor: a system of ocean currents that forms a loop of flowing, oxygenated water at the surface and great depths due to thermohaline (water density changes with heat and salt content) circulation.
oceanic province: open marine realm offshore of the continental shelf.
oligotrophic: a low-food situation.
ontogeny: the development of an organism from egg to adult.
ontogenetic niche: the fundamental niche of an organism which has changed characters due to the process of developing to an adult.
operculum: a structure that covers and protects the gills in fish.
ostracoderms: a general term for four extinct classes of armored fishes lacking jaws derived from gill arches, pelvic fins, and other characters of jawed fishes.
otoliths: calcareous stones that occupy space in the membranous sacs of the inner ear of fishes and aid in orientation to gravity.
overfishing: occurs when fishing removes individuals faster than recruitment can replace them; when harvest exceeds maximum sustained yield or other target.
oviparous: bearing eggs.
ovoviviparous: bearing young produced from eggs.
Pangaea: supercontinent that formed about 250 mya.
Patchiness: nonuniform density due to clumps or patches.
pectoral fins: paired fins homologous to arms.
pelvic fin: paired fins homologous with legs.
pelagic realm: the open-water zones of the ocean and lakes.
Phanerozoic: an eon characterized by abundant fossils, including the Paleozoic, Mesozoic, and Cenozoic eras.
phenotype: the morphological, physiological, or behavioral characteristics of an organism due to the interaction of its genotype with environmental factors or constraints.
physoclistus: type of gas bladder that is closed, with the gas regulated only by the blood.
phenotypic plasticity: the flexibility of an organism in adapting its phenotype to environmental conditions.
physostomus: gas bladder is open to esophagus and filled by gulping air.
plankton: small aquatic organisms that have little or no locomotion ability and primarily are at drift with the current.
potamodramous: migration in freshwater.
potamon: pertaining to a river.

preadapted: an organism that is adapted (native) to a system that is similar to the one in which it is being placed.
predation: consumption of living organisms smaller than the predator.
primary fish group: a taxa that cannot live in seawater for very long.
primary productivity: the production of biomass by photosynthetic and chemosynthetic organisms.
pristine: a place in nature that is pure and unspoiled.
pseudoextinction: a result of passing genes along to another group; phyletic transformation of one species that develops into another.
purse seine: a type of net used to catch schooling pelagic fishes by encircling them. It has a drawstring that purses the net and prevents escape by diving.
ram ventilation: movement through water acts as a ram to circulate water through open gills, as in elasmobranches.
rayfin fish: a fish whose fin supports are within the body and not on the bases of the fin.
realized niche: the niche of an organism in which the fundamental niche is constrained by competing species.
recruitment overfishing: harvesting fish that have not lived long enough to reproduce.
red tide: reddish surface water due to an infestation of small organisms, usually dinoflagellates, whose effects include toxins and lowering of oxygen.
Red Queen hypothesis (Van Valen 1973): evolutionary changes in a lineage are likely related to keeping up with environmental change in the present rather than preparing for new environments in the future. The Red Queen in Louis Carrol's novel (*Through the Looking Glass*) had to keep running just to stay in the same place. A concept very close to Romer's Rule (Allaby 2005).
rhithron: the upper reaches of a coldwater stream where water temperatures range from about 4–20°C.
river continuum concept: posits that river systems are connected from origin to end in a predictable manner.
Romer's rule (attributed to A.S. Romer by Hockett and Ascher 1964): changes in organisms are generally conservative and aid in maintaining a traditional way of life in an altered environment. Red Queen Hypothesis is similar (Allaby 2005).
salt marsh: a shallow wetland influenced by diluted salt water and dominated by *Spartina* and other adapted plants.
saltation, theory of: the concept that ontogeny in some organisms occurs in a series of major steps, usually due to morphological changes.
sea floor spreading/subduction: movement of deep sea floor as plates from ocean ridges to their disappearance under the continents.
sea grass: not true grasses, but flowering plants adapted to underwater life, such as eelgrass and turtlegrass.
sea mounts: underwater volcanic "islands" or mountain tops.
semicircular canals: part of the inner ear in vertebrates, used for balance and maintaining body position.
Shelford's law of tolerance: existence and success of an organism depends on its range of tolerance to environmental factors. The distribution of an organism is likely constrained by the environmental variable for which it has the most narrow tolerance range (Allaby 2005, p. 395).
shreckstoff: "scary stuff" can cause fright reaction in fishes.
spiracle: paired water intake organs of elasmobranchs; aids hiding in benthic rays.
stability–time hypothesis: persistence and adaptation of organisms are favored by system stability over long time.

stable ocean hypothesis (Lasker): marine fish recruitment is enhanced by stable ocean conditions, especially calm winds.
symbiosis: two species that live in the same general location.
territory: an area of home range that is defended.
thermal stratification: the layering of water or air due to density effects and a lack of mixing.
tide: movement of large bodies of water due to the gravitational effects of the moon and sun.
tipping point: a threshold where a small disturbance can alter the state or development of a system.
trawl: type of fishing gear in which a huge net with weights is dragged along the bottom.
trophic level: relative position in a food chain based on feeding as a primary producer or a consumer.
tubercles: breeding structures of fish that appear as bumps.
turbidity: decreased clarity of water due to suspended materials.
ventral: a position on the vertebrate body that is on the abdomen or chest; opposite of dorsal.
vicariance biogeography: a state of geographic distribution in which organisms did not actively transport themselves.
viscosity: the property of a liquid to resist movement through it.
viviparous: bearing live young.
Weberian apparatus: a set of small bones that enhance hearing in Ostariophysian fishes.
wave: energy transmission in water by the familiar patterns of waves, which move in circular, orbital motion.
wilderness: a natural system that is unsettled by humans.
xenobiotic: a substance that is not produced by nature.
zeitgeber: "time-giving"; an environmental signal.

Literature Cited

Able, K.W., and M.P. Fahay. 1998. *The first year in the life of estuarine fishes in the Mid Atlantic Bight.* Rutgers University Press, New Brunswick, NJ.

Able, K.W., and M.P. Fahay. 2010. *Ecology of estuarine fishes: Temperate waters of the western North Atlantic.* The Johns Hopkins University Press, Baltimore, MD.

Adams, S.M. 1976. Feeding ecology of eelgrass fish communities. *Transactions of the American Fisheries Society* 105:514–519.

Adams, S.M. (Ed.). 1990. *Biological indicators of stress in fish.* American Fisheries Society Symposium 8, Bethesda, MD.

Adams, S.M., and J.E. Breck. 1990. Bioenergetics. Pages 389–415. In: C.B. Schreck and P.B. Moyle (Eds.). *Methods for fish biology.* American Fisheries Society, Bethesda, MD.

Adler, H.E. 1975. *Fish behavior: Why fish do what they do.* TFH publications, Neptune City, NJ.

Agin, D. 2006. *Junk science: How politicians, corporations, and other hucksters betray us.* St. Martin's Press, New York.

Ahlberg, P.E. 1972. Coelacanth fins and evolution. *Nature* 358:459.

Ahlstrom, E.H., J.L. Butler, and B.Y. Sumida. 1976. Pelagic stromateoid fishes (Pisces: Perciformes) of the eastern Pacific: Kinds, distributions, and early life histories and observations on five of these from the northwest Atlantic. *Bulletin of Marine Science* 26:285–402.

Alexander, C. 2010. Beyond a reasonable doubt: Three new books dissect the anatomy of climate-change denial. *Conservation* 11(3):43–46.

Alfieri, M.S., and L.A. Dugatkin. 2006. Cooperation and cognition in fishes. Pages 203–222. In: C. Brown, K. Laland, and J. Krause (Eds.). *Fish cognition and behavior.* Blackwell Publishing Ltd., UK.

Alò, D., and T.F. Turner. 2005. Effects of habitat fragmentation on effective population size in the endangered Rio Grande silvery minnow. *Conservation Biology* 2005:1138–1148.

Allaby, M. 2005. *A dictionary of ecology.* 3rd Ed. Oxford University Press, New York.

Allan, J.D. 1995. *Stream ecology: Structure and function of running waters.* Chapman and Hall, New York.

Allee, W.C. 1931. *Animal aggregations, a study in general sociology.* University of Chicago Press, Chicago.

Allee, W.C. 1941. Integration of problems concerning protozoan populations with those of general biology. *American Naturalist* 75:473–487.

Allee, W.C., A.E. Emerson, O. Park, T. Park, and K.P. Schmidt. 1949. *Principles of animal ecology.* W.B. Saunders Co., Philadelphia, PA.

Allen, E.J. 1914. Fluctuations in the yield of sea fisheries. *Nature* 93:672–673.

Allen, L.G., and J.N. Cross. 2006. Surface waters. Pages 320–341. In: L.G. Allen, D.J. Pondella II, and M.H. Horn (Eds.). *The ecology of marine fishes: California and adjacent waters.* University of California Press, Berkeley.

Allsopp, M., R. Page, P. Johnston, and D. Santillo. 2007. Oceans in peril: Protecting marine biodiversity. *Worldwatch Report* 174. Worldwatch Institute, Washington, DC.

Alpert, P. 1995. Incarnating ecosystem management. *Conservation Biology* 9:952–955.

Anderson, J.T. 1988. A review of size dependent survival during pre-recruit stages of fishes in relation to recruitment. *Journal of North West Atlantic Fisheries* 8:55–66.

Anderson, R.O. 1986. Symposium wrap-up. Pages 114–116. In: J.G. Dillard, L. K. Graham, and T.R. Russell (Eds.). *The paddlefish: status, management, and propagation.* Special Publication 7, North Central Division, American Fisheries Society, Bethesda, MD. (NB: Oral presentation written by editors from notes).

Angel, M.V. 1997. What is the deep sea? Pages 2–41. In: D.J. Randall and A.P Farrell (Eds.). *Deep-sea fishes.* Academic Press, New York.

Armantrout, N.B. 1998. *Glossary of aquatic habitat inventory terminology.* Western Division, American Fishery Society, Bethesda, MD.

Armour, C.L., and J.G. Taylor. 1991. Evaluation of the Instream Flow Incremental Methodology by U.S. Fish and Wildlife Service field users. *Fisheries* 16(5):36–43.

Austen, P.J., and R.W. Langton. 1999. The effects of fishing on fish habitat. Pages 150–187. In: L.R. Genaka (Ed.). *Fish habitat: Essential fish habitat and rehabilitation.* American Fisheries Society Symposium 22, Bethesda, MD.

Axelrod, H.R., and W.E. Burgess. 1986. *African cichlids of Lakes Malawi and Tanganyika.* 11th Ed. T.F.H. Publications, Neptune City, NJ.

Baker, R.R. 1978. *The evolutionary ecology of animal migration.* Holmes and Meier Publishers, Inc., New York.

Baker, R.R. 1982. *Migration: Paths through time and space.* Holmes and Meier Publishers, Inc., New York.

Balon, E.K. 1979. The theory of saltation and its application in the ontogeny of fishes: Steps and thresholds. *Environmental Biology of Fishes* 4:97–101.

Balon, E.K. 1984. Patterns in the evolution of reproductive styles in fishes. Pages 35–53. In: G.W. Potts and R.J. Wootton (Eds.). *Fish reproduction: Strategies and tactics.* Academic Press, New York.

Balon, E.K. 1985. *Early life histories of fishes: New developmental, ecological and evolutionary perspectives.* Dr. W. Junk, Kluwer Academic Publishers, Boston.

Balon, E.K., M.N. Bruton, and H. Fricke. 1988. A fiftieth anniversary reflection on the living coelacanth, *Latimeria chalumnae*: Some new interpretations of its natural history and conservation status. *Environmental Biology of Fishes* 23:241–280.

Barange, M., and R.I. Perry. 2009. Physical and ecological impacts of climate change relevant to marine and inland capture fisheries and aquaculture. Pages 7–106. In: K. Cochrane, C. De Young, D. Soto, and T. Bahri (Eds.). Climate change implications for fisheries and aquaculture: Overview of current scientific knowledge. *FAO Fisheries and Aquaculture Technical Paper* 530. Food and Agricultural Organization of the United Nations, Rome.

Barcott, B. 2010. The gulf of oil: Forlorn in the bayou. *National Geographic* 218(4):62–74.

Barlow, G.W. 2000. *The cichlid fishes: Nature's grand experiment in evolution.* Perseus Publishing, Cambridge, MA.

Barnett, M.A. 1984. Mesopelagic fish zoogeography in the central tropical and subtropical Pacific Ocean: Species composition and structure at representative locations in three ecosystems. *Marine Biology* 82(2):199–208.

Barnosky, A.D., N. Matzke, S. Tomiya, G.O.U. Wogan, B. Swartz, T.B. Quental, C. Marshall, J.L. McGuire, E.L. Lindsey, K.C. McGuire, B. Mersey, and E.A. Ferrer. 2011. Has the Earth's sixth mass extinction already arrived? *Nature* 471:51–57 (March 3, 2011).

Barrett P.J., J. Wullschleger, and J. Sjoberg. 2008. *2008 Annual report on recovery and management actions for the Devils Hole pupfish (Cyprinodon diabolis).* U.S. National Park Service, Death Valley National Park, Death Valley, CA.

Barry, P.L. 2004. A chilling possibility: By disturbing a massive ocean current melting arctic sea ice might trigger colder weather in Europe and North America. NASA Science: Science News. March 3, 2004. Access at: http://science.nasa.gov/science-news/science-at-nasa/2004/05mar_arctic/ (accessed on June 24, 2011).

Bateson, P., and R.A. Hinde. 1987. Developmental changes in sensitivity to experience. Pages 19–34 In: M.H. Bornstein (Ed.). *Sensitive periods in development: Interdisciplinary perspectives.* Lawrence Erlbaum Associates, Hillsdale, NY.

Baum, J.K., R.A. Myers, D.G. Kehler, B. Worm, S.J. Harley, and P.A. Doherty. 2003. Collapse and conservation of shark populations in the Northwest Atlantic. *Science* 299:389–392.

Bayley, P.B., and H.W. Li. 1996. Riverine fishes. Pages 92–122. In: G. Petts and P. Calow (Eds.). *River biota: Diversity and dynamics.* Blackwell Science, Cambridge, MA.

BBC/Lionheart and DK Vision. 1994. Eyewitness: Fish. VHS. DK Productions, Dorling Kindersley, Inc., New York.

Beamesderfer, R.C., and B.E. Rieman. 1991. Abundance and distribution of northern squawfish, walleyes, and smallmouth bass in John Day Reservoir, Columbia River. *Transactions of the American Fisheries Society* 120:439–447.

Beamish, R.J., and B.J. Rothschild (Eds.). 2009. *The future of fisheries science in North America.* Fish & Fisheries Series 31. Springer Science + Business Media B.V., Dordrecht, Netherlands.

Bean, M.J. 1983. *The evolution of national wildlife law.* Praeger Publishing Division of Greenwood Press, Westport, CT.

Becker, L., R.J. Poreda, A.R. Basu, K.O. Pope, T. M. Harrison, C.N. Nicholson, and R. Lasky. 2004. Bedout: A possible end Permian impact crater offshore of northwestern Australia. *Science* 34(5676):1469–1476.

Beckman, W.C. 1952. Guide to the fishes of Colorado. Leaflet 11, University of Colorado Museum, Boulder.

Begon, M., C.R. Townsend, and J.L. Harper. 2006. *Ecology: From individuals to ecosystems.* 4th Edition. Blackwell Publishing, Malden, MA.

Behnke, R.J. 1992. *Native trout of western North America*. Monograph 6, American Fisheries Society, Bethesda, MD.

Behnke, R.J. 2002. *Trout and salmon of North America*. The Free Press, New York.

Belding, D.L. 1921. *A report upon the alewife fisheries of Massachusetts*. Contribution 11, Division of Marine Fisheries, Massachusetts Department of Natural Resources, Boston.

Belkin, D. 2009. States cast for way to stop carp. *The Wall Street Journal* (Friday December 4, 2009):A3.

Bell, G. 1982. *The masterpiece of nature, the evolution and genetics of sexuality*. University of California Press, Berkeley.

Belt, D. 1992. Russia's Lake Baikal: The world's great lake. *National Geographic* (June):2–36.

Benaka, L.R. (Ed.). 1999. *Fish habitat: Essential fish habitat and rehabilitation*. Proceedings of the Sea Grant Symposium on Fish Habitat. American Fisheries Society, Bethesda, MD.

Benard, M.F. 2004. Predator-introduced phenotypic plasticity in organisms with complex life histories. *Annual Review of Ecology, Evolution and Systematics* 35:651–673.

Bennett, G.W. 1971. *Management of lakes and ponds*, 2nd Ed. Van Nostrand Reinhold Company, New York.

Benton, M.J. 2003. *When life nearly died: The greatest mass extinction of all time*. Thames & Hudson, Ltd., London.

Berec, L., E. Angulo, and F. Couchamp. 2006. Multiple Allee effects and population management. *Trends in Ecology and Evolution* 22(4):185–191.

Berkson, J., K.M. Hunt, J.C. Whitehead, D.J. Murie, T.J. Kwak, J. Boreman. 2009. Is there a shortage of fisheries stock assessment scientists? *Fisheries* 34(5):217–219.

Berra, T.M. 2007. *Freshwater Fish Distribution*. The University of Chicago Press, Chicago, IL.

Bettoli, P., J.A. Kerns, and G.D. Scholten. 2009. Status of paddlefish in the United States. Pages 23–38. In: C.P. Paukert and G.D. Scholten (Eds.). *Paddlefish management, propagation and conservation in the 21st Century: Building from 20 years of research and management*. American Fisheries Society Symposium 66, Bethesda, MD.

Bigelow, H.G., and W.C. Schroeder. 1953. *Fishes of the Gulf of Maine*. Fishery Bulletin 74. U.S. Fish and Wildlife Service, Washington, DC.

Biro, P.B., and J.R. Post. 2008. Rapid depletion of genotypes with fast growth and bold personalities from harvested fish populations. *Proceedings of the National Academy of Sciences of the United States of America* 105(8):2919–2922.

Blankenship, C. 2010. Menhaden study suggests overfishing may have taken place: Fishery agency will consider new management options. *Briefs, The Newsletter of the American Institute of Fisheries Research Biologists* 39(3)8–9.

Blaxter, J.H.S. (Ed.). 1974. *The early life history of fish*. Springer-Verlag, New York.

Blaxter, J.H.S. 1988. Pattern and variety in development. Pages 1–58, In: W/S/Hoar and D.J. Randall (Eds.). *Fish physiology*. Volume XIa. Academic Press, San Diego, CA.

Blaxter, J.H.S., and G. Hempel. 1963. The influence of egg size on herring larvae (*Clupea harengus* L.). *ICES Journal of Marine Science* 28(2):211–240.

Bond, C.E. 1979. *Biology of fishes*. W.B. Saunders, Philadelphia, PA.

Bond, C.E. 1996. *Biology of fishes*. 2nd Ed. Saunders College Publishing, New York.

Bone, Q., and N.B. Marshall. 1982. *Biology of fishes*. Blackie Academic and Professional, Glasgow, Scotland.

Bone, Q., and R.H. Moore. 2008. *Biology of fishes*. 3rd Ed. Taylor & Francis Group, New York.

Bone, Q., N.B. Marshall, and J.H.S. Blaxter.1995. *Biology of fishes*. 2nd Ed. Chapman and Hall, New York.

Boreman, J., G.S. Nakashima, J.A. Wilson, and R.L. Kendall, Eds. 1997. *Northwest Atlantic groundfish: Perspectives on a fishery collapse*. American Fisheries Society, Bethesda, MD.

Borror, D.J. 1960. *Dictionary of word roots and combining forms*. Mayfield Publishing, Palo Alto, CA.

Boucher, D.H. 1992. Mutualism and cooperation. Pages 208–211. In: E.F. Keller and E.A. Lloyd (Eds.). *Keywords in evolutionary biology*. Harvard University Press, Cambridge, MA.

Bourne, J.D., Jr. 2010. The gulf of oil: The deep dilemma. *National Geographic* 218(4):40–53.

Bower, A., C. Reedy, and J. Yelin-Kefer. 2001. Consensus versus conservation in the Upper Colorado River Basin recovery implementation program. *Conservation Biology* 15:1001–1007.

Bowles, E.C. 1995. Supplementation: Panacea or curse for the recovery of declining fish stocks? *American Fisheries Society Symposium* 15:277–283.

Boxall, B. 2009. Utility agrees to remove Klamath dams. *The Seattle Times*. Sunday, October 4, 2009. (http://seattletimes.nwsource.com/html/localnews/).

Bradford, M.J., and G. Cabana. 1997. Interannual variability in stage-specific survival rates and the causes of recruitment variation. Pages 469–493. In: R.C. Chambers and E.A. Trippel (Eds.). *Early life history and recruitment in fish populations*. Fish and Fisheries Series 21. Chapman and Hall, New York.

Brannon, E.L., D.F. Amend, M.A. Cronin, et al. 2004. The controversy about salmon hatcheries. *Fisheries* 29(9):12–31.

Brannon, E.L., T.P. Quinn, G.L. Lucchetti, and B.D. Ross. 1981. Compass orientation of sockeye sac fry from a complex river system. *Canadian Journal of Zoology* 59:1548–1553.

Breder, C.M., Jr., and D.E. Rosen. 1966. *Modes of reproduction in fishes*. American Museum of Natural History, T.F.H. Publications, Jersey City, NJ.

Breederland, B., and J. Daniels. 2008. Where botulism is killing waterfowl. *Refuge Update* 5(3):5. US Fish and Wildlife Service, Washington, DC.

Breitburg, D.L., and G.F. Riedel. 2005. Multiple stressors in marine systems. Pages 167–182. In: E.A. Norse and L.B. Crowder (Eds.). *Marine conservation biology: The science of maintaining the sea's biodiversity*. Island Press, Washington, DC.

Brett, J.R., and T.D.D. Groves. 1979. Physiological energetics. Pages 279–352. In: W.S. Hoar, D.J. Randall, and J.R. Brett (Eds.). *Fish physiology, Volume 8*. Academic Press, New York.

Bridges, C.R. 1993. Ecophysiology of intertidal fish. Pages 375–400. In: J.C. Rankin and F.B. Jensen (Eds.). *Fish ecophysiology*. Fish and Fisheries Series 9, Chapman and Hall, New York.

Briggs, J.C. 1986. Introduction to the zoogeography of North American fishes. Pages 1–16. In: C.H. Hocutt, and E.O. Wiley (Eds.). *The zoogeography of North American Freshwater fishes*. John Wiley & Sons, New York.

Briggs, J.C. 2008. The North Atlantic Ocean: Need for proactive management. *Fisheries* 33(4):180–185.

Briggs, J.C., B.W. Bowen, and M.A. Rex. 2004. Marine biogeography. Pages 233–237. In: M.V. Lomolino and L.R. Heaney (Eds.). *Frontiers of biogeography: New directions in the geography of nature*. Sinaueer Associates, Sunderland, MA.

Brooks, D.R. 2004. Reticulations in historical biogeogrphy: The triumph of time over space in evolution. Pages 125–244. In: Lomolino, M.V. and L.R. Heaney (Eds.). *Frontiers of biogeography: New directions in the geography of nature*. Sinauer Associates, Sunderland, MA.

Brooks, J.L., and S.I. Dodson. 1965. Predation, body size, and composition of plankton. *Science* 53:119–126.

Brophy, D., B.S. Danilowicz, and P.A. King. 2006. Spawning season fidelity in sympatric populations of Atlantic herring (Clupea harengus). *Canadian Journal of Fisheries and Aquatic Sciences* 63:607–616.

Browman, H.I. 1989. Embryology, ethology and ecology of ontogenetic critical periods in fish. *Brain, Behavior and Evolution* 34:5–12.

Brown, C. 2004. Not just a pretty face. *New Scientist Issue* 2451:42 (June 12, 2004). www.newscientist.com (accessed October 1, 2010).

Brown, C., and K.N. Laland. 2003. Social learning in fishes: A review. *Fish and Fisheries* 4(3):280–288.

Brown, C., K. Laland, and J. Krause. 2006. *Fish cognition and behavior*. Blackwell Publishing, Ltd., Oxford, UK.

Brown, F.A., Jr., J.W. Hastings, and J.D. Palmer (Eds.). 1970. *The biological clock: Two views*. Academic Press, New York.

Brown, F.A., Jr. 1970. Hypothesis of environmental timing of the clock. Pages 13–60. In: F.A. Brown, J.W. Hastings, and J.D. Palmer (Eds.). *The biological clock: Two views*. Academic Press, New York.

Brown, J.A. 1993. Endocrine response to environmental pollutants. Pages 277–296. In: J.C. Rankin and F.B. Jensen (Eds.). *Fish ecophysiology*. Fish and Fisheries Series 9. Chapman and Hall, New York.

Brown, L.R., and P.B. Moyle. 1981. The impact of squawfish on salmonid populations: A review. *North American Journal of Fisheries Management* 1:104–111.

Brown, L.R., and P.B. Moyle. 2005. Native fishes of the Sacramenso-San Joaquin drainage, California: A history of decline. Pages 75–98. In: J.N. Rinne, R.M. Hughes, and B. Calamusso (Eds.). *Historical changes in large river fish assemblages of the Americas*. American Fisheries Society Symposium 45, Bethesda, MD.

Brown, R.S., W.A. Hubert, and S.F. Daly. 2011. A primer on winter, ice, and fish: What fisheries biologists should know about winter ice processes and stream-dwelling fish. *Fisheries* 36(1):8–26.

Bryan, J. 2011. Conserving a long-lived leviathon: Lake sturgeon. *Reflections on Fisheries Conservation* 3(3):24–27. U.S. Fish and Wildlife Service, Washington, DC.

Bshary, R. 2006. Machiavellian intelligence in fishes. Pages 223–242. In: C. Brown, K. Laland, and J. Krause (Eds.). *Fish cognition and behavior*. Blackwell Publishing Ltd., UK.

LITERATURE CITED

Burian, R.M. 1992. Adaptation: Historical perspectives. Pages 7–12. In: E.F. Keller and E.A. Lloyd (Eds.). *Keywords in evolutionary biology*. Harvard University Press, Cambridge, MA.

Burgess, R.F. 1970. *The sharks*. Doubleday & Company, Garden City, NY.

Burrell, V.G., Jr. 1986. American oyster. Species profiles: Life history and environmental requirements of coastal fishes and invertebrates (South Atlantic). *U.S. Fish and Wildlife Service Biological Report* 82(11.57).

Busack, C.R., and K.P. Currens. 1995. Genetic risks and hazards in hatchery operations: Fundamental concepts and issues. *American Fisheries Society Symposium* 15:71–80.

Busacker, G.P., I.R. Adelman, and E.M. Goolish. 1990. Growth. Pages 363–387. In: C.B., Schreck and P.B. Moyle (Eds.). *Methods for fish biology*. American Fisheries Society, Bethesda, MD.

Butchart, S.H.M., M. Walpole, B. Collen, et al. 2010. Global biodiversity: Indicators of recent declines. *Science* 328(5982):1164–1168.

Bye, V.J. 1984. The role of environmental factors in the timing of reproductive cycles. Pages 187–206. In: G.W. Potts and R.J. Wootton (Eds.). *Fish reproduction: Strategies and tactics*. Academic Press, New York.

Cairns, J., Jr. 1990. The emergence of global environmental awareness. *Journal of Environmental Science (China)* 2:1–18.

Campbell, N.A. 1993. *Biology*. 3rd Ed. Benjamin Cummings Publishing Company, New York.

Campbell, R. 2010. Snakehead eradication proving tough in Arkansas. *Eddies: Reflections on Fisheries Conservation* 3(2):7. U.S. Fish and Wildlife Service, Washington, DC.

Campton, D.E. 1987. Natural hybridization and introgression in fishes: Methods of detection and genetic interpretations. Pages 161–192. In: N. Ryman and F. Utter (Eds.). *Population genetics & fishery management*. Washington Sea Grant Program, University of Washington Press, Seattle.

Campton, D.E. 1995. Genetic effects of hatchery fish on wild populations of Pacific salmon and steelhead: What do we really know? Pages 337–353, In: H.L. Schramm, Jr., and R.G. Piper (Eds.). *Uses and effects of cultured fishes in aquatic ecosystems*. American Fisheries Society Symposium 15, Bethesda, MD.

Cantrell, M.A., and A.K. Hill. 2009. How dammed is your watershed: First approximation of an index to relative dammed-ness of US. Watersheds. *American Fisheries Society Symposium* 69:919–921.

Carlson, D.M., and P.S. Bonislawsky. 1981. The paddlefish (*Polydon spathula*) fisheries of the midwestern United States. *Fisheries* 6(2):12–22.

Carlton, J. 1998. Apostrophe to the ocean. *Conservation Biology* 12:1165–1168.

Carlton, J.T., and G.M. Ruiz. 2005. The magnitude and consequence of bioinvasions in marine ecosystems: Implications for conservation biology. Pages 123–148. In: E.A. Norse, and L.G. Crowder (Eds.). *Marine conservation biology: The science of maintaining the sea's biodiversity*. Island Press, Washington, DC.

Carpenter, A.R. 1996. Microcosm experiments have limited relevance for community and ecosystem ecology. *Ecology* 77:677–680.

Carpenter, J. 2005. Competition for food between an introduced crayfish and two fishes endemic to the Colorado River basin. *Environmental Biology of Fishes* 72:335–342.

Carpenter, J. and G. Mueller. 2008. Small nonnative fishes as predators of larval razorback suckers. *Southwest Naturalist* 53:236–242.

Carpenter, S.R., J.F. Kitchell, and J.R. Hodgson. 1985. Cascading trophic interactions and lake productivity. *Bioscience* 35:634–639.

Carson, R. 1962. *Silent spring*. Houghton Mifflin, Boston.

Castagna, J. 2010. Vision unveiled for "Mosaic of Habitats" in the New York/New Jersey estuary. *Fisheries* 35(5):213–214, 250.

Castro, P., and M. E. Huber. 2007. *Marine biology*. 6th Ed. McGraw Hill, New York.

Casazza, T.L., and S.W. Ross. 2008. Fishes associated with pelagic *Sargassum* and open water lacking *Sargassum* in the Gulf Stream off North Carolina. *Fisheries Bulletin* 106:348–363.

Chambers, R.C., and E.A. Trippel (Eds.). 1997. *Early life history and recruitment in fish populations*. Fish and Fisheries Series 21. Chapman and Hall, New York.

Charnov, E.L. 1976. Optimal foraging: The marginal value theorem. *Theoretical Population Biology* 9:129–136.

Chen, J. 2005. Across the apocalypse on horseback: Imperfect legal responses to biodiversity loss. *Washington University Journal of Law and Policy* 17(12):13–35.

Chesapeake Bay Foundation. 2010. State of the bay 2010. Available on line at http://www.cbf.org/Document.Doc?id=596 (12/2010) Accessed March 30, 2011.

Chipps, S.R., and D.H. Wahl. 2008. Bioenergetics modeling in the 21st century: Reviewing new insights and constraints. *Transactions of the American Fisheries Society* 137:298–313.

Claessen, D.C., and U. Dieckmann. 2002. Ontogenetic niche shifts and evolutionary branching in size-structured populations. *Evolutionary Ecology Research* 4:189–217.

Cochrane, K., C. De Young, D. Soto, and T. Bahri (Eds.). 2009. Climate change implications for fisheries and aquaculture: Overview of current scientific knowledge. *FAO Fisheries and Aquaculture Technical Paper* 530. Food and Agricultural Organization of the United Nations, Rome.

Coen, L.D., M.W. Luckenbach, and D.L. Breitburg. 1999. The role of oyster reefs as essential fish habitat: A review of current knowledge and some new perspectives. Pages 438–454. In: L.R. Genaka (Ed.). *Fish Habitat: Essential fish habitat and rehabilitation*. American Fisheries Society Symposium 22, Bethesda, MD.

Cohen, D.M. 1970. How many recent fishes are there? *Proceedings of the California Academy of Sciences* 38:341–346.

Colinvaux, P. 1993. *Ecology 2*. John Wiley & Sons, Inc., New York.

Colewell, R.K. 1992. Niche: A bifurcation in the conceptual lineage of the term. Pages 241–248. In: E.F. Keller and E.A. Lloyd (Eds.). *Keywords in evolutionary biology*. Harvard University Press, Cambridge, MA.

Collier, M., R.H. Webb, and J.C. Schmidt. 2000. *Dams and rivers: A primer on the downstream effects of dams*. U.S. Geological Survey Circular 1126. Branch of Information Services, Denver, CO.

Connell, J.H. 1961. The influence of competition and other factors on the distribution of the barnacle *Cthalamus stellatus*. *Ecology* 42:710–723.

Cooke, G.D., E.B. Welch, S.A. Peterson, and S.A. Nichols. 2005. *Restoration and management of lakes and reservoirs*. 3rd Ed. Taylor & Francis Group/CRC Press, Boca Raton, FL.

Cooper, J.C., and P.J. Hirsch. 1982. The role of chemoreception in salmonid homing. Pages 343–362. In: T.J. Hara (Ed.). *Chemoreception in fishes. Developments in aquaculture and Fisheries Science*, 8. Elsevier Scientific Publishing/Company. New York.

Copeland, B.J., H.T. Odum, and F.N. Mosely. 1969. Migrating subsystems. Pages 1354–1385. In: H.T. Odum, B.J. Copeland, and E.A. McMahan (Eds.). *Coastal ecological systems of the United States. Volume 2. A report to the Federal Water Pollution Control Administration*. Institute of Marine Sciences, University of North Carolina Press. Chapel Hill.

Courtenay, W.R., Jr. 1995. The case for caution with fish introductions. *American Fisheries Society Symposium* 15:413–424.

Courtenay, W.R., Jr., B.B. Collette, T.E. Essington, R. Hilborn, J.W. Orr, D. Pauly, J.E. Randall, and W.F. Smith-Vaniz. 2009. Risks of introductions of marine fishes: Reply to Briggs. *Fisheries* 34(4):181–186.

Cowan, J.H., Jr., K.A. Rose, and E.D. Houde. 1997. Size-based foraging success and vulnerability to predation: selection of survivors in individual-based models of larval fish populations. Pages 457–386. In: R.C. Chambers and E.A. Trippel (Eds.). *Early life history and recruitment in fish populations*. Fish and Fisheries Series 21. Chapman and Hall, New York.

Crecco, V.A., and T. Savoy. 1987. Review of recruitment mechanisms of the American shad: The critical period and match/mismatch hypotheses reexamined. Pages 455–568. In: M.J. Dadswell, R.J. Klauda, C.M. Moffitt, R.L. Saunders, R.A. Rulifson, and J.E. Cooper (Eds). *Common strategies of anadromous and catadromous fishes*. American Fisheries Society Symposium 1, Bethesda, MD.

Crowder, L.B. 1984. Character displacement and habitat shift in a native cisco in southeastern Lake Michigan: Evidence for competition? *Copeia* (1984):878–883.

Crowder, L.B. 2005. Back to the future in marine conservation. Pages 19–29. In: E.A. Norse and L.G. Crowder (Eds.). *Marine conservation biology: The science of maintaining the sea's biodiversity*. Island Press, Washington, DC.

Curtis, B. 1949. *The life story of the fish: His manners and morals*. Harcourt, Brace and Company, New York.

Cushing, D. 1990. Plankton production and year-class strength in fish populations: An update of the match/mismatch hypothesis. *Advances in Marine Biology* 26:250–293.

Cushing, D. 1995. *Population production and regulation in the sea: A fisheries perspective*. Cambridge University Press, New York.

Czúcz, B., J.P. Gathman, and G.R. McPherson. 2010. The impending peak and decline of petroleum production: An underestimated challenge for conservation of ecological integrity. *Conservation Biology* 24(2):948–956.

Dadswell, M.J., R.J. Klauda, C.M. Moffitt, R.L. Saunders, R.A. Rulifson and J.E. Cooper (Eds.). 1987. *Common strategies of anadromous and catadromous fishes*. American Fisheries Society Symposium 1, Bethesda, MD.

LITERATURE CITED

Daniel, H., and F. Minot. 1954. *The inexhaustible sea*. Dodd, Mead, New York.

Darlington, P.J., Jr. 1957. *Zoogeography: The geographical distribution of animals*. John Wiley and Sons, New York.

Darwin, C., and P. Horan. 1979. *The origin of species*. (Reprint of the 1859 J. Murray edition: On the origin of species by means of natural selection. A new forward added). Gramercy Books, an imprint of Random House Value Publishing, New York.

Das, S.B., I. Joughin, M.D. Behn, I.M. Howat, M.A. King, D. Lizarralde, and M.P. Bhatia. 2008. Fracture propagation to the base of the Greenland ice sheet during supraglacial lake drainage. *Science* 320(5877):778–781.

Day, D. 1981. *Vanished species*. Gallery Books, an imprint of W.H. Smith Publishers, New York.

Dayton, P., and R.H. Hessler. 1972. The role of biological disturbance in maintaining diversity in the deep sea. *Deep-Sea Research* 19:19–208.

Daw, T., W.N. Adger, K. Brown, and M.-C. Badjeck. 2009. Climate change and capture fisheries: Potential impacts, adaptation and mitigation. Pages 107–150. In: K. Cochrane, C. De Young, D. Soto, and T. Bahri (Eds.). *Climate change implications for fisheries and aquaculture: Overview of current scientific knowledge*. FAO Fisheries and Aquaculture Technical Paper 530. Food and Agricultural Organization of the United Nations, Rome.

Deacon, J.E., G. Kobetich, J.D. Williams, and S. Contrreras. 1979. Fishes of North America, endangered, threatened, or of special concern. *Fisheries* 4(2):29–44.

Deacon, J.E., and C.D. Williams. 1991. Ash Meadows and the legacy of the Devils Hole pupfish. Pages 69–91. In: W.L. Minckley and J.E. Deacon (Eds.). *Battle against extinction: Native fish management in the American west*. University of Arizona Press, Tucson.

DeAlteris, J.T., and D.L. Morse. 1997. Fishing gear management. Pages 167–175. In: J. Boreman, G.S. Nakashima, J.A. Wilson, and R.L. Kendall, (Eds.). *Northwest Atlantic groundfish: Perspectives on a fishery collapse*. American Fisheries Society, Bethesda, MD.

Dean, B. 1895. *Fishes, living and fossil: An outline of their forms and probable relationships*. Columbia University Biological Series III. MacMillan and Company, New York.

Dell'Amore, C. 2009. Photos: Five Global Warming 'tipping points,' *National Geographic News*, February 18, 2009, accessed February 1, 2010, http://news.nationalgeographic.com/news/2009/03/photogalleries/tipping-points-climate-change/.

Dempsey, D. 2004. *On the Brink: The Great Lakes in the 21st Century*. Michigan State University Press, East Lansing.

Dence, W.A. 1948. Life history, ecology and habits of the dwarf darter Catostomus commersonni utawana Mather at the Huntington Wildlife Station, Rooseovelt. *Wildlife Bulletin* 8(4):221–242.

DeVries, R.D., R.A. Stein, and P.L. Chesson. 1989. Sunfish foraging among patches: The patch departure decision. *Animal Behaviour* 37:455–464.

Diamond, J. 1989. Overview of recent extinctions. Pages 37–41. In: E. Western and M. Pearl (Eds.). *Conservation for the Twenty-First Century*, Oxford University Press, New York.

Diana, J.S. 1995. *Biology and ecology of fishes*. Cooper Publishing Group, Carmel, IN.

Diana, J.S. 2004. *Biology and ecology of fishes*. 2nd Ed. Biological Sciences Press, Cooper Publishing Group, Traverse City, MI.

Dionne, M., F.T. Short, and D.M. Burdick. 1999. Fish utilization of restored, created, and reference salt marsh habitat in the Gulf of Maine. Pages 384–404. In: L.R. Benaka (Ed.). *Fish habitat: Essential fish habitat and rehabilitation*. Proceedings of the Sea Grant symposium on fish habitat. American Fisheries Society, Bethesda, MD.

Ditty, J.G., L.A. Fuiman and R.F. Shaw. 2003. Characterizing natural intervals of development in the early life of fishes: an example using blennies (Teleostomi: Blenniidae). Pages 405–418. In: H.I Browman and A.B. Skiftesvik (Eds.). *The big fish bang: Proceedings of the 26th Annual Larval Fish Conference*. Institute of Marine Research, Bergen, Norway.

Dodson, J.J. 1997. Fish migration: an evolutionary perspective. Pages 10–36. In: Godin, J.-G.L. (Ed.). *Behavioural ecology of teleost fishes*. Oxford University Press, Inc., New York.

Doppelt, B., M. Scurlock, C. Grissell, and J. Karr. 1993. *Entering the watershed: A new approach to save America's river ecosystems*. The Pacific Rivers Council. Island Press, Washington, DC.

Døving, K.B., and A.D. Kasumyan. 2008. Chemoregulation. Pages 331–394. In: R.N. Finn and B.G. Kapoor (Eds.). *Fish larval physiology*. Science publishers, Enfield, NH.

Drenner, R.W., and A. Mazumder. 1999. Microcosm experiments have limited relevance for community and ecosystem ecology: Comment. *Ecology* 80:1081–1085.

Drinkwater, K.F., and D.G. Mountain. 1997. Climate and oceanogrphy. Pages 3–25 In: J. Boreman, B.S. Nakashima, J.A. Wilson, and R.L. Kendall (Eds.). *Northwest Atlantic groundfish: Perspectives on a fishery collapse*. American Fisheries Society, Bethesda, MD.

Drucker, P.F. 1973. *Management: Tasks, responsibilities, practices*. Harper & Row, New York.

Dulvy, N.K., Y. Sadovy, and J.O. Reynolds. 2003. Extinction vulnerability in marine populations. *Fish and Fisheries* 4(1):25–64.

Echelle, A.A., E.V. Carson, A.F. Echelle, R.A. Van Den Bussche, T.E. Dowling, and A. Meyer. 2005. Historical biogeography of the new-world pupfish genus *Cyprinodon (Telesostei: Cyprinodontidae) Copeia* 2005:320–339.

Edwards, R.T. 1998. The hyporheic zone. Pages 399–429. In: R.J. Naiman and R.E. Bilby (Eds.). *River ecology and management: Lessons from the Pacific coastal ecoregion*. Springer Verlag, New York.

Edwards, T.C., Jr. 1989. The Wildlife Society and the Society for Conservation Biology: Strange but unwilling bedfellows. *Wildlife Society Bulletin* 17:340–343.

Ehrenfeld, D.W. 1981. *The arrogance of humanism*. Oxford University Press, New York.

Ehrlich, P., and A. Ehrlich. 1981. *Extinction: The causes and consequences of the disappearance of species*. Ballantine Books, New York.

Ellner, S.P., J. Fieberg, D. Ludwig, and C. Wilcox. 2002. Precision of population viability analysis. *Conservation Biology* 16:258–261.

Elton, C. 1927. *Animal Ecology*. The Macmillan Company, New York.

Elwood, J.W., J.D. Newbold, R.V. O'Neill, and W.V. Winkle. 1983. Resource spiraling: An operational paradigm for analyzing lotic ecosystems. Pages 3–27. In: T.D. Fontaine and S.M. Bartell (Eds.). *Dynamics of Lotic Ecosystems*. Ann Arbor Science Publishers, Ann Arbor, MI.

Encyclopedia of Earth. 2009. Global warming takes a toll on coral reefs. (http://www.eoearth.org/article/Global_warming_takes_a_toll_on_cora...) updated March 21, 2009, accessed March 30, 2009.

Encyclopedia of Earth. 2010. Deepwater Horizon oil spill. Revised 10/15/2010 (http://www.eoearth.org/article/Deepwater_Horizon_oil_spill?topic=50359) Accessed June 24, 2011.

Engel, J., and R. Kvitek. 1998. Effects of otter trawling on a benthic community in Monterey Bay National Marine Sanctuary. *Conservation Biology* 12:1204–1214.

Erdmann, M.V., R.L. Caldwell, and M.K. Moosa. 1998. Indonesian "King of the Sea" discovered. *Nature* 395:335.

Erwin, D.H. 2006. *Extinction: How life on Earth nearly ended 250 million years ago*. Princeton University Press, Princeton, NJ.

Etnier, D.A., and W.C. Starnes. 1993. *The fishes of Tennessee*. University of Tennessee Press, Knoxville.

Fabry, V.J., B.A. Seibel, R.A. Feely, and J.C. Orr. 2008. Impacts of ocean acidification on marine fauna and ecosystem processes. *ICES Journal of Marine Science: Journal du Conseil* 65(3):414–432.

Fausch, K.D., and T.R. Cummings. 1986. Effects of brook trout competition on threatened greenback cutthroat trout. Final report to the US, Natonal Park Service, Rocky Mountain Regional Office, Denver, CO.

Feyrer, F. and M.P. Healey. 2003. Fish community structure and environmental correlates in the highly altered southern Sacramento–San Joaquin Delta. *Environmental Biology of Fishes* 66(2):123–132.

Ffolliott, P.F., L.A. Bojorquez-Tapia, and M. Hernandez-Narvaez. 2001. *Natural resources management: A primer*. Iowa State University Press, Ames.

Finn, R.N., and B.G. Kapoor (Eds.). 2008. *Fish larval physiology*. Science Publishers, Enfield, NH.

Finney, S. 2011. Keeping Asian carp out of the Great Lakes. *Eddies: Reflections on Fisheries Conservation* 3(3):32–33. U.S. Fish and Wildlife Service, Washington, DC.

Fish, M.P. 1932. Contributions to the early life histories of sixty-two species of fishes from Lake Erie and its tributary waters. *Bulletin of the U.S. Bureau of Fisheries* 47:293–398.

Fitzgerald, G.J., and R.J. Wootton. 1986. Behavioural ecology of sticklebacks. Pages 409–458. In: T.J. Pritcher (Ed.). *The behavior of teleost fishes*. The Johns Hopkins University Press, Baltimore, MD.

Flagg, T.A., F.W. Waknitz. D.J. Maynard, G.B. Milner, and C.V.W. Mahnken. 1995. The effect of hatcheries on native coho salmon populations in the lower Columbia River. Pages 366–375. In: H.L. Schramm, Jr., and R.G. Piper. *Uses and effects of cultured fishes in aquatic ecosystems*. American Fisheries Society Symposium 15, Bethesda, MD.

Flickinger, S.A., F.J. Bulow, and D.W. Willis. 1999. Small Impoundments. Pages 561–588. In: C.C. Kohler and W.A. Hubert (Eds.). *Inland fisheries management in North America,* 2nd Edition. American Fisheries Society, Bethesda, MD.

Forbes, S.A. 1887. The lake as a micocosm. *Bulletin of the Scientific Association* (Peoria, Illinois) 1887:77–87.

Foster, N.R. 1985. Lake trout reproductive behavior: influence of chemosensory cues from young-of-year reproductive by-products. *Transactions of the American Fisheries Society* 114:794–803.

Frank, K.T., B. Petrie, J.S. Choi, and W.C. Leggett. 2005. Trophic cascades in a formerly cod-dominated ecosystem. *Science* 308:1621–1623.

Frankel, G. S., and D.L. Gunn. 1961. *The orientation of animals: Kineses, taxes and compass reactions.* (An expanded version of the first edition published in 1940 by Oxford University Press.) Dover Publications, Inc., New York.

Frankel, O.H., and M.E. Soulé. 1981. *Conservation and evolution.* Cambridge University Press, New York.

Franklin, A., A. Haro, and T. Castro-Santos. 2009. An evaluation of nature-like fishways for passage of anadromous alewife. Pages 907–909. In: A.J. Haro, K.L. Smith, R.A. Rulifson, C.M. Moffitt, R.J. Klauda, M.J. Dadswell, R.A. Cunjak, J.E. Cooper, K.L. Beal, and T.S. Avery (Eds.). *Challenges for diadromous fishes in a dynamic global environment.* American Fisheries Society Symposium 69, Bethesda, MD.

Franklin, H.B. 2007. *The most important fish in the sea: Menhaden and America.* Island Press, Washington, DC.

Franklin, I.R. 1980. Evolutionary change in small populations. Pages 135–149. In: M.E. Soulé and B.A. Wilcox (Eds.). *Conservation Biology: An evolutionary-ecological perspective.* Sinauer Associates, Sunderland, MA.

Frazier, L.N. 2008. Sea-cage aquaculture, sea lice, and declines of wild fish. *Conservation Biology* 23(3):599–607.

Frimpong, E.A., and P.L. Angermeier. 2009. Fish traits: A database of ecological and life-history traits of freshwater fishes of the United States. *Fisheries* 34(10):487–495 (obtain database at www.cnr.vt.edu/fisheries/fishtraits/).

Fry, F.E. 1971. The effect of environmental factors on the physiology of fish. Pages 1–98. In W.S. Hoar and D.J. Randall (Eds.). *Fish physiology. Volume VI.* Academic Press, New York.

Fuiman, L.A., and D.M. Higgs. 1997. Ontogeny, growth and the recruitment process. Pages 225–249. In: R.C. Chambers and E.A. Trippel (Eds.). *Early life history and recruitment in fish populations.* Fish and Fisheries Series 21. Chapman and Hall, New York.

Fuiman, L.A. and R.G. Werner (Eds.). 2002. Fisheries science: The unique contribution of early life history stages. Blackwell, Malden, MA.

Fuller, P.M., L.G. Nico, and J.D. Williams. 1999. *Nonindigenous fishes introduced into inland waters of the United States.* American Fisheries Society Special Publication 27, Bethesda, MD.

Funk, J.L. 1970. Warm-water streams. Pages 141–152. In: N.G. Benson (Ed.). *A century of fisheries in North America.* American Fisheries Society, Washington, DC.

Futuyma, D.J. 1998. Evolutionary biology. 3rd Edition. Sinauer Associates. Sunderland, MA.

Gadgil, M., and W.H. Gossert. 1970. Life historical consequences of natural selection. *American Naturalist* 104:1–24.

Gagosian, R.B. 2003. Abrupt Climate Change: Should We Be Worried?, Woods Hole Oceanographic Institute, Woods Hole, MA, accessed February 4, 2010, http://www.whoi.edu/page.do?.

Ganong, W.F. 1963. *Review of medical physiology.* Lange Medical Publishers, Los Altos, CA.

Gardiner, H. and P. Herlinger. 2006. Incredible journey of the greenback cutthroats. (DVD). Borderlines Product.

Gardner, J.P.A. 1997. Hybridization and speciation in the sea. Pages 1–79. In: J.H.S. Blaxter and A.J. Southwood. *Advances in Marine Biology*, Volume 31. Academic Press, San Diego, CA.

Gartner, J.V., Jr., R.E. Crabtree, and K.J. Sulak. 1997. Feeding at depth. Pages 115–193. In: D.J. Randall and A. P. Farrell (Eds.). *Deep-sea fishes.* Academic Press, San Diego, CA.

Gates, F.C. 1942. The bogs of northern lower Michigan. *Ecological Monographs* 12:213–254.

Gause, G.F. 1934. Experimental analysis of Vito Volterra's mathematical theory of the struggle for existence. *Science* 79:16–17.

Gende, S.M., R.T. Edwards, M.F. Willson, and M.S. Wipfli. 2002. Pacific salmon in aquatic and terrestrial ecosystems. *BioScience* 52(10): 917–928.

Gengerke, T.W. 1986. Distribution and abundance of paddlefish in the United States. Pages 22–35. In: J.G. Dillard, L. K. Graham, and T.R. Russell (Eds.). *The paddlefish: status, management, and propagation.* Special Publication 7, North Central Division, American Fisheries Society, Bethesda, MD.

George, A.L., B.R. Kuhajda, J.D. Williams, M.A. Cantrell, P.L. Rakes, and J.R. Shute. 2009. Guidelines for propagation and translocation for freshwater fish conservation. *Fisheries* 34(11):529–545.

Gerken, J.E., and C.P. Paukert. 2009. Threats to paddlefish habitat: Implications for conservation. Pages 173–183. In: C.P. Paukert and G.D. Scholten (Eds.). *Paddlefish management, propagation and conservation in the 21st Century: Building from 20 years of research and management.* American Fisheries Society Symposium 66, Bethesda, MD.

Gerking, S.D. 1959. The restricted movement of fish populations. *Biological Reviews* 34(2):221–242.

Gerking, S.D. 1994. *Feeding ecology of fish.* Academic Press, Inc., San Diego, CA.

Gerlach, J.M., and A.W. Kahnle. 1981. Larval fish drift in a warmwater stream. Pages 154–162. In: L.A. Krumholz (Ed.). Proceedings of the warmwater streams symposium. American Fisheries Society, Bethesda, Md.

Gessel, M.H., J.G. Williams D.A. Brege, R.F. Krcma, and D.R. Chambers. 1991. Juvenile salmonid guidance at the Bonneville Dam Second Powerhouse, Columbia River, 1983–1989. *North American Journal of Fisheries Management* 11:400–412.

Gibson, R.N. 1986. Intertidal teleosts: Life in a fluctuating environment. Pages 338–408. In: T.J. Pitcher (Ed.). *The behavior of teleost fishes.* The John Hopkins University Press, Baltimore, MD.

Gilliam, J.F., and D.F. Fraser. 1987. Habitat selection under predation hazard: Test of a model with foraging minnows. *Ecology* 68:1856–1862.

Glebe, B.D., and W.C. Leggett. 1981. Latitudinal differences in energy allocation and use during the freshwater migrations of American shad (*Alosa sapidissima*) and their life history consequences. *Canadian Journal of Fisheries and Aquatic Sciences* 38:806–820.

Glude, J.B. 1971. Pacific northwest fisheries. Pages 234–257. In: S. Shapiro (Ed.). *Our changing fisheries.* U.S. Department of Commerce, National Marine Fisheries Service, Washington, DC.

Godin, J.-G. L. (Ed.). 1997. *Behavioural ecology of teleost fishes.* Oxford University Press, Inc., New York.

Godwin, C.H., and R.A. Rulifson. 2002. Use of water control structures for passing fish and invertebrates into Lake Mattamuskeet, NC. Federal Aid Report 1228-40181-97-G022 to the U.S. Fish and Wildlife Service, Raleigh, NC.

Goldschmidt, T. 1996. *Darwin's dream pond: Drama in Lake Victoria.* The MIT Press, Cambridge, MA.

Goldstein, R.J. 1986. *Coastal fishing in the Carolinas, from surf, pier and jetty.* 3rd Ed. John F. Blair, Publisher, Winston Salem, NC.

Good, T.P., T.J. Beechie, P. McElhany, M.M. McClure, and M.H. Ruckelshaus. 2007. Recovery planning for Endangered Species Act–listed Pacific salmon: Using science to inform goals and strategies. *Fisheries* 32(9):426–440.

Goode, G.B., and T.H. Bean. 1896. *Oceanic Ichthyology, a treatise on the deep-sea and pelagic fishes of the world, based chiefly upon the collections made by the steamers Blake, Albatross, and Fish Hawk in the Northwestern Atlantic.* Special Bulletin of the U.S. National Museum, Washington, DC.

Gorbunova, N.N., and N. Ya. Lipskaya. 1975. Feeding of larvae of the blue marlin *Makaira nigricans* (Pisces, Istiophoridae). *Journal of Ichthyology* 15:95–101.

Gorr, T., and T. Kleinschmidt. 1993. Evolutionary relationships of the coelacanth. *American Scientist* 81: 72–82.

Gosline, W.A. 1965. Teleostean phylogeny. *Copeia* 1965:186–194.

Gould, S.J. 1977. *Ontogeny and phylogeny.* The Belknap Press of Harvard University, Cambridge, MA.

Gould, S.J. 1993. *Eight little piggies: Reflections in natural history.* W.W. Norton and Company, New York.

Govoni, J.J., and R.B. Follower, Jr. 2008. Buoyancy. Pages 495–522. In: R.N. Finn and B.J Kapoor (Eds.). *Fish larval physiology.* Science Publishers, Enfield, NH.

Grady, J.M., and B.S. Elkington 2009. Establishing and maintaining paddlefish populations by stocking. Pages 385–396. In: C.PO. Paukert and G.D. Scholten (Eds.). *Paddlefish management and conservation in the 21st Century.* American Fisheries Society Symposium 66. Bethesda, MD.

Graham, L.K., E.J. Hamilton, T.R. Russell, and C.E. Hicks. 1986. Culture of paddlefish—A review of methods. Pages 78–94. In: J.G. Dillard, L.K. Graham, and T.R. Russell (Eds.). *The paddlefish: status, management, and propagation.* Special Publication 7, North Central Division, American Fisheries Society, Bethesda, MD.

LITERATURE CITED

Gray, G.G. 1993. *Wildlife and people: The human dimensions of wildlife ecology*. University of Illinois Press, Urbana.

Greenberg, P. 2010. *Four fish: The future of the last wild food*. The Penguin Press, New York.

Griesemer, J.R. 1992. Niche: Historical perspectives. Pages 231–240. In: E.F. Keller and E.A. Lloyd (Eds.). Keywords in evolutionary biology. Harvard University Press, Cambridge, MA.

Griffith, J.S. 1999. Coldwater streams. Pages 481–504 (Chapter 18) In: C.C. Kohler and W.A. Hubert (Eds.). *Fisheries Management in North America*. 2nd Ed. American Fisheries Society, Bethesda, MD.

Griffiths, S.W., and J.D. Armstrong. 2002. Kin-based territory overlap and food sharing in Atlantic salmon juveniles. *Journal of Animal Ecology* 71(3):480–486.

Grime, J.P. 1977. Evidence for the existence of three primary strategies in plants and its relevance to ecological and evolutionary theory. *American Naturalist* 111:1169–1194.

Grinnell, J. 1904. The origin and distribution of the chesnut-backed chickadee *The Auk* 21:364–379.

Grinnell, J. 1917. The niche-relationship of the California thresher. *Auk* 34:427–433.

Gross, G.M., and E. Gross. 1995. *Oceanography: A view of Earth*. Prentice Hall, Upper Saddle River, NJ.

Gross, M.R. 1984. Sunfish, salmon, and the evolution of alternative reproductive strategies and tactics in fishes. Pages 55–75. In: G.W. Potts and R.J. Wootton (Eds.). *Fish reproduction: Strategies and tactics*. Academic Press, New York.

Gudjónsson, S., and D.L. Scarnecchia. 2009. Even the evil need a place to live: Wild salmon, salmon farming, and zoning the Icelandic coastline. *Fisheries* 34(10): 477–486.

Hall, T.E. 2008. Pattern formation. Pages 3–25. In: R.N. Finn and B.G. Kapoor (Eds.). *Fish larval physiology*. Science Publishers, Enfield, NH.

Hamilton, B.T., S.E. Moore, T.B. Williams, N. Darby, and M.R. Vinson. 2009. Comparative effects of rotenone and antimycin on macroinvertebrate diversity in two streams in Great Basin National Park, Nevada. *North American Journal of Fisheries Management* 29:1620–1635.

Hansen, K.A., and C.P. Paukert. 2009. Current management of paddlefish sport fisheries. Pages 277–290. In: C.P. Paukert and G.D. Scholten (Eds.). *Paddlefish management, propagation and conservation in the 21st Century: Building from 20 years of research and management*. American Fisheries Society Symposium 66, Bethesda, MD.

Hansen, M.J., D. Boisclair, S.G. Grandt, W.W. Hewett, J.F. Kitchell, M.C. Lucas, and J.J. Ney. 1993. Applications of bioenergetic models to fish ecology and management: Where do we go from here? *Transactions of the American Fisheries Society* 122:1019–1030.

Hara T.J. (Ed.). 1982. *Chemoreception in fishes*. Developments in Aquaculture and Fisheries Science, 8. Elsevier Scientific Publishiing Company, New York.

Harden-Jones, F.R.H. 1968. *Fish migration*. Edward Arnold, London.

Harden-Jones, F.R.H. 1981. Fish migration: Strategy and tactics. Pages 139–165. In: D.J. Aidley (Ed.). *Animal migration*. Seminar Series 13, Society for Experimental Biology. Cambridge University Press, New York.

Hardin, G. 1968. The tragedy of the commons. *Science* 162:1243–1248.

Hardy, J.D., Jr., G.E. Drewry, R.A. Fritzsche, G.D. Johnson, P.W. Jones, and F.D. Martin. 1978. *Development of fishes of the Mid-Atlantic Bight, an atlas of egg, larval and juvenile stages*. U.S. Fish and Wildlife Service Publication FWS/OBS 78/12:1–6.

Hare, J.A. and R.K. Cowen. 1997. Size, growth, development, and survival of the planktonic larvae of *Pomatomus saltatrix* (Pisces:Pomatomidae). *Ecology* 78:2415–2431.

Hare, J.A., M.A. Alexander, M.J. Fogarty, E.H. Williams, and J.D. Scott. 2010. Forcasting the dynamics of a coastal fishery species using a coupled climate-population model. Ecological Applications 20:452–464.

Harrison, I.J., and M.J. Stiassny. 1999. The quiet crises: a preliminary listing of the freshwater fishes of the world that are extinct or "missing in action." Pages 271–332. In: R.D.E. MacPhee (Ed.). *Extinctions in near time: causes, contexts, and consequences*. Kluwer Academic/Plenum Publishing, New York.

Hart, P.J.B. 1986. Foraging in teleost fishes. Pages 211–235. In: T.J. Pitcher (Ed.). *The behavior of teleost fishes*. The Johns Hopkins University Press, Baltimore, MD.

Hart, P.J.B. 1997. Foraging tactics. Pages 104–133. In: J.J. Godin (Ed.). *Behavioural ecology of teleost fishes*. Oxford University Press, New York.

Hartley, P.H.T. 1948. Food and feeding relationships in a community of fresh water fishes. *Journal of Animal Ecology* 17:1–14.

Hartman, K.J., and J.F. Kitchell. 2008. Bioenergetics modeling: Progress since the 1992 Symposium. *Transactions of the American Fisheries Society* 137: 216–223.

Harvey, B.C. 1987. Susceptability of young-of-the-year fishes to downstream displacement by flooding. *Transactions of the American Fisheries Society* 116:851–855.

Harvey, B.C., and M.A. Wilzbach. 2010. Carcass addition does not enhance juvenile salmonid biomass, growth, or retention in six Northwestern California streams. *North American Journal of Fisheries Management* 30:1445–1451.

Hasler, A.D. 1971. Orientation and fish migration. Pages 429–510. In: W.W. Hoar and D.J. Randall (Eds.). *Fish physiology*, Volume VI., Academic Press New York.

Hasler, A.D., and A.T. Scholz. 1983. *Olfactory imprinting and homing in salmon.* Zoophysiology 14. Springer-Verlag, New York.

Hasler, A.D., and W.J. Wisby. 1951. Descrimination of stream odors by fishes and relation to parent stream behavior. *American Naturalist* 85:223–238.

Hastings, J.W. 1970. Cellular-biochemical clock hypothesis. Pages 63–91. In: F.A. Brown, J.W. Hastings, and J.D. Palmer. *The biological clock: Two views.* Academic Press, New York.

Havey, K.S. 1961. Restoration of anadromous alewives at Long Pond, Maine. *Transactions of the American Fisheries Society* 90:281–286.

Hawkins, L.A., A.E. Magurran, and J.D. Armstrong. 2004a. Innate predator recognition in newly-hatched Atlantic salmon. *Behavior* 141:1249–1262.

Hawkins, L.A., A.E. Magurran, and J.D. Armstrong. 2007. Innate abilities to distinguish between predator species and cue concentration in Atlantic salmon. *Animal Behavior* 73:1051–1057.

Hawkins, L.A., A.E. Magurran, and J.D. Armstrong. 2008. Ontogenetic learning of predator recognition in hatchery-reared Atlantic salmon, Salmo salar. *Animal Behavior* 75:1663–1671.

Hawkins, L.A., H.M. Tyus, W.L. Minckley, and D.L. Schultz. 2004b. Comparison of four techniques for aging adult Colorado pikeminnow, Ptychocheilus lucius. *Southwestern Naturalist* 49:203–208.

Hay, J. 1959. *The Run.* Doubleday, Garden City, NY.

Hayes, D.B., W.W. Taylor, and P.A. Soranno. 1999. Natural lakes and large impoundments. Pages 589–622. In: C.C. Kohler and W.A. Hubert (Eds.). *Inland fisheries management in North America*, 2nd edition. American Fisheries Society, Bethesda, MD.

Hazon, N., and R.J. Belment. 1997. Endocrinology. Pages 441–464. In: D.H. Evans (Ed.). *The physiology of fishes.* 2nd Edition. CRC Press, Boca Raton, FL.

Heath, A.G. 1990. Summary and perspectives. *American Fisheries Society Symposium* 8:181–191.

Heath, D.D., J.W. Heath, C.A. Bryden, R.M. Johnson, and CW. Fox. 2003. Rapid evolution of egg size in captive salmon. *Science* 299:1738–1740.

Hedgecock, D. 1994. Does variance in reproductive success limit effective population sizes of marine organisms? Pages 122–134. In: A.R. Beaumont (Ed.). *Genetics and evolution of aquatic organisms?* Chapman Hall, London.

Helfman, G.S. 2007. *Fish conservation: A guide to understanding and restoring global aquatic biodiversity and fishery resources.* Island Press, Washington, DC.

Helfman, G.S., B.B. Collette, D.E. Facey, and B.W. Bowen. 2009. *The diversity of fishes: Biology, evolution, and ecology.* 2nd Edition. Wiley-Blackwell, John Wiley & Sons, Hoboken, NJ.

Helfman, G.S., D.E. Facey, L.S. Hales, Jr., and E.L. Bozeman, Jr. 1987. Reproductive ecology of the American eel. Pages 42–56. In: M.J. Dadswell, R.J. Klauda, C.M. Moffitt, R.L. Saunders, R.A. Rulifson and J.E. Cooper (Eds.). *Common strategies of anadromous and catadromous fishes.* American Fisheries Society Symposium 1, Bethesda, MD.

Heppell, S.S., S.A. Heppell, A.J. Read, and L.B. Crowder. 2005. Effects of fishing on long-lived marine organisms. Pages 211–231. In: E.A. Norse, and L.G. Crowder (Eds.). *Marine conservation biology: The science of maintaining the sea's biodiversity.* Island Press, Washington, DC.

Herke, W.H., and G.D. Rogers. 1999. Maintenance of the estuarine environment. Pages 321–342. In: C.C. Kohler and W.A. Hubert (Eds.). *Inland fisheries management in North America*, 2nd Ed. American Fisheries Society, Bethesda, MD.

Hilborn, R. 2005. Are sustainable fisheries achievable? Pages 247–259. In: E.A. Norse and L.B. Crowder (Eds.). *Marine conservation biology: The science of maintaining the seas's biodiversity.* Island Press, Washington, DC.

Hildebrand, S.F., and W.C. Schroeder. 1928. *Fishes of Chesapeake Bay.* Document 1024, Bulletin of the Bureau of Fisheries XLII, part 1, Washington, DC. (Reprinted in 1972 for the Smithsonian Institution by T.F.H. Publications, Neptune, NJ).

Hinckley, A.D. 1976. *Applied ecology: A nontechnical approach.* Macmillian Publishing Co., New York.

Hockett, C.F., and R. Asher. 1964. The human revolution. *American Scientist* 52:70–92.

Hocutt, C.H., and E.O. Wiley (Eds.). 1986. *The zoogeography of North American Freshwater fishes.* John Wiley & Sons, New York.

Hoese, H.D., and R.H. Moore. 1998. *Fishes of the Gulf of Mexico.* 2nd Ed. Texas A&M University Press, College Station.

Hoffman, G.L. 1999. *Parasites of North American freshwater fishes.* 2nd Ed. Cornell University Press, Ithaca, NY.

Hoggan, J., and R. Littlemore. 2010. *Climate cover-up: The crusade to deny global warming.* Greystone Books, New York.

Holopainen, J., J. Aho, M. Vornanen, and H. Huuskonen. 2005. Phenotypic plasticity and predator effects on morphology and physiology of Crucian carp in nature and the laboratory. *Journal of Fish Biology* 50:781–798.

Horn, M.H., and K.L.M. Martin. 2006. Rocky intertidal zone. Pages 205–226. In: L.G. Allen, D.J. Pondella III, and M.H. Horn (Eds.). *The ecology of marine fishes: California and adjacent waters.* University of California Press, Berkeley.

Horn, M.H., L.G. Allen, and R.N. Lea. 2006. Biogeography. Pages 3–25. In: L.G. Allen, D.J. Pondella III, and M.H. Horn (Eds.). *The ecology of marine fishes: California and adjacent waters.* University of California Press, Berkeley.

Horn, M.H., K.L.M. Martin, and M.A. Chotkowski (Eds.). 1999. *Intertidal fishes: Life in two worlds.* Academic Press Inc., New York.

Houde, E.D. 1987. Fish early life dynamics and recruitment variability. Pages 17–29. In: Hoyt, R.D. (Ed.). *10th Annual Larval Fish Conference.* American Fisheries Society Symposium 2, Bethesda, MD.

Houde, E.D. 1997. Patterns and consequences of selective processes in teleost early life histories. Pages 173–196. In: R.C. Chambers and E.A. Trippel (Eds.). *Early life history and recruitment in fish populations.* Fish and Fisheries Series 21. Chapman and Hall, New York.

Hubbs, C.L. 1943. Terminology of early life stages of fish. *Copeia* 1943:260.

Hubbs, C.L. 1955. Hybridization between fish species in nature. *Systematic Zoology* 4:1–20.

Hubbs, C.L., K.F. Lagler, and G.R. Smith. 2004. *Fishes of the Great Lakes Region.* Revised Edition. University of Michigan Press, Ann Arbor.

Hughes, R.M., and J.M. Omernik. 1981. Use and misuse of the terms watershed and stream order. Pages 320–326. In: Krumholtz, L.A. (Ed). *The warmwater streams symposium.* Southern Division, American Fisheries Society, Bethesda, MD.

Hughes, R.M., J.R. Rinne, and B. Calamusso. 2005. Introduction to historical changes in large river fish assemblages of the Americas. Pages 1–12. In: J.N. Rinne, R.M. Hughes, and B. Calamuusso (Eds.). *Historical changes in large river fish assemblages of the Americas.* American Fisheries Society Symposium 45, Bethesda, MD.

Hughes, R.N. 1997. Diet selection. Pages 134–162. In: Godin, J.-G.J. (Ed.). *Behavioural ecology of teleost fishes.* Oxford University Press, New York.

Hume, S., A. Morton, B.C. Keller, R.M. Leslie, O. Langer, and D. Staniford. 2004. *A stain upon the sea: West coast salmon farming.* Harbour Publishing Company, Ltd., Madeira Park, BC.

Humphries, C.J., and M.C. Ebach. 2004. Biogeography on a dynamic earth. Pages 67–86. In: Lomolino, M.V. and L.R. Heaney (Eds). *Frontiers of biogeography: New directions in the geography of nature.* Sinauer Associates, Sunderland, MA.

Hunt, D.M., J. Fitzgibbon, S. J. Slobodyanyuk, J.K. Bowmaker, and K.S. Dulai. 1997. Molecular evolution of the cottoid fish endemic to Lake Baikal deduced from nuclear DNA evidence. *Molecular Phylogenetics and Evolution* 8:415–422.

Hunt, R.L. 1993. *Trout stream therapy.* The University of Wisconsin Press, Madison.

Hunter, J.R. 1972. Swimming and feeding behavior of larval anchovy, *Engraulis mordax*, larvae. *U.S. Fishery Bulletin* 70:821–838.

Hunter, J.R. 1981. Feeding ecology and predation of marine fish larvae. Pages 33–79. In: R. Lasker (Ed.). *Marine fish larvae: Morphology, ecology, and relation to fisheries.* University of Washington Press, Seattle.

Hunter, J.R. and K.M. Coyne. 1982. The Onset of Schooling in Northern Anchovy Larvae *Engraulis mordax*, CalCOFI Report 23: 246–251. Accessed March 12, 2011, http://swfsc.noaa.gov/publications/cr/1982/8217.

Hunter, M., Jr., E. Dinerstein, J. Hockstra, and D. Lindenmayer. 2010. A call to action for conserving biological diversity in the face of climate change. *Conservation Biology* 24(5):1169–1171.

Hutchings, J.A. 1997. Life history responses to environmental variability in early life. Pages 139–168. In: R.C. Chambers and E.A. Trippel (Eds.). *Early life history and recruitment in fish populations*. Fish and Fisheries Series 21. Chapman and Hall, New York.

Hutchings, J.A., and J.D. Reynolds. 2004. Marine fish population collapses: Consequences for recovery and extinction risk. *Bioscience* 54:297–309.

Hutchinson, G.E. 1957. Concluding remarks. *Cold Spring Harbor Symposium on Quantitative Biology* 22: 415–427.

Hutchinson, G.E. 1978. *An introduction to population ecology*. Yale University Press, New Haven, CT.

Hutchinson, G.E. 1979. *The kindly fruits of the earth: Recollections of an embryo ecologist*. Yale University Press, New Haven, CT.

Hutchinson, G.E., and H. Löffler. 1956. The thermal classification of lakes. *Proceedings of the National Academy of Sciences of the United States of America* 42:84–86.

Hynes, H.B.N.1970. *The ecology of running waters*. University of Toronto Press, Canada.

Idyll, C.P. 1971. *Abyss: The deep sea and the creatures that live in it*. Revised Edition. Thomas Y. Crowell Company, New York.

Idaho Chapter of the American Fisheries Society (Idaho-AFS). 1995. Why isn't science saving salmon? *Fisheries* 20(9):4–9.

Ilves, K.L., and D.J. Randall. 2007. Why have primitive fishes survived? Pages 516–536. In: D.J. Mckenzie. A.P. Farrell. and C.J. Brauner (Eds.). *Primitive fishes*. Fish Physiology Series 26. Elsevier (Associated Press), New York.

Intergovernmental Panel on Climate Change (IPCC). 2007. Summary for policymakers. In *Climate Change 2007: The Physical Science Basis. Contribution of Working Group I to the Fourth Assessment Report of the IPCC*, ed. S. Solomon et. al., Cambridge University Press, Cambridge, UK, and New York, http://ipcc-wg1.ucar.edu/wg1/report/argwg1.

International Union for the Conservation of Nature and Natural Resources (IUCN). 2009. *The IUCN red list of threatened species, 2009 update*. www.iucnredlist.org/.

Irving D.B., and T. Modde. 2000. Home-range fidelity and use of historic habitat by adult Colorado pikeminnow (*Ptychocheilus lucius*) in the White River, Colorado and Utah. *Western North American Naturalist* 60(1):16–25.

Ivlev, V.W. 1961. *Experimental ecology of the feeding of fishes*. Yale University Press, New Haven, CT.

Jackson, J.B.C. 2008. Ecological extinction and evolution in the brave new ocean. *Proceedings of the National Academy of Sciences of the United States of America*: 105 (Supplement 1):11458–11465.

Jackson, R.W. 1980. *The fish of fossil lake: the story of Fossil Butte National Monument*. Dinosaur Nature Association, Dinosaur National Park, Jensen, UT.

Jacobs, S. and M. Jepson. 2009. Creating a community context for the fishery stock sustainability index. *Fisheries* 34(5):228–232.

James, N. 2002. *Lake Victoria cichlids: Practical fishkeeping*. Ringpress Books, Surry, UK.

Janvier, P. 2007. Living primitive fishes from deep time. Pages 1–53. In: D.J. Mckenzie. A.P. Farrell. and C.J. Brauner (Eds.). *Primitive fishes*. Fish Physiology Series 26. Elsevier (Associated Press), New York.

Jearld, A., Jr. 1983. Age and growth. Pages 301–324. In: L.A. Nielsen and D.L. Johnson (Eds.) *Fisheries techniques*. American Fisheries Society, Bethesda, MD.

Jelks, H.L., S.J. Walsh, N.M. Burkhead, S. Contreras-Balderas, E. Diaz-Pardo, D.A. Hendrickson, J. Lyons, N.E. Mandrak, F. McCormick, J.S. Nelson, S.P. Platania, B.A. Porter, C.G. Renaud, J.J. Schmitter-Soto, E.G. Taylor, and M.L. Warren, Jr. 2008. Conservation status of imperiled North American freshwater and diadromous fishes. *Fisheries* 33(8):372–407.

Jenkins, R.M. 1970. Reservoir fish management. Pages 173–182. In: N.G. Benson (Ed.). *A century of fisheries management in North America*. Special Publication 7, American Fisheries Society, Bethesda, MD.

Jennings, C.A., and S.J. Zigler. 2009. Biology and life history of paddlefish in North America: An update. Pages 1–22. In: C.P. Paukert and G.D. Scholten (Eds.). *Paddlefish management, propagation and conservation in the 21st Century: Building from 20 years of research and management*. American Fisheries Society Symposium 66, Bethesda, MD.

Jennings, M.J., and D.P. Philipp. 1994. Biotic and abiotic factors affecting survival of early life history intervals of a stream-dwelling sunfish. *Environmental Biology of Fishes* 39:153–159.

Jennings, S., M.J. Kaiser, and J.D. Reynolds. 2001. *Marine fisheries ecology*. Blackwell Scientific, Malden, MA.

Jobling, M. 1993. Bioenergetics; feed intake and energy partitioning. Pages 1–44. In: J.D. Rankin and F.B. Jensen (Eds.). *Fish ecophysiology*. Fish and Fisheries Series 9. Chapman and Hall, New York.

Jobling, M. 1995. *Environmental biology of fishes*. Fish and Fisheries Series 16. Chapman and Hall, New York.

Johnson, B.L., W.B. Richardson, and T.J. Naimo. 1995. Past, present, and future concepts in large river ecology. *Bioscience* 45:134–141.

Johnson, D. 2007. *Fish of Colorado field guide*. Adventure Publications, Inc., Cambridge, MN.

Johnson, D.R., H.M. Perry, J. Lyczkowski-Schultz, and D. Hanisko. 2009. Red snapper larval transport in the northern Gulf of Mexico. *Transactions of the American Fisheries Society* 138:458–470.

Johnson, J.E., M.G. Pardew, and M.M. Lyttle. 1993. Predation recognition and avoidance by larval razorback sucker and northern hog sucker. *Transactions of the American Fisheries Society* 122:1139–1145.

Johnson, P.B. 1978. *Behavioral mechanisms of upstream migration and home stream selection in coho salmon (Oncorhynchus kisutch)*. Ph.D. Thesis, University of Wisconsin, Madison.

Jones, M., A. Laurila, N. Peuhkuri, J. Piironen, and T. Seppä. 2003. Timing an ontogenetic niche shift: responses of emerging salmon alevins to chemical cues from predators and competitors. *Oikos* 102:155–163.

Jordan, D.S., and B.W. Evermann. 1900. *The fishes of North and Middle America: A descriptive catalogue of the species of fish-like vertebrates found in the waters of North America, north of the Isthmus of Panama*. Bulletin 47 Part IV of the United States National Museum. Government Printing Office, Washington, DC. (Reprinted in 1963 for the Smithsonian Institution by T.F.H. Publications, Inc., Jersey City, NJ.)

Jordan, D.S., and B.W. Evermann. 1902. *American food and game fishes. A popular account of all the species found in America north of the Equator, with keys for ready identification, life histories and methods of capture*. Doubleday, New York.

Jordan, D.S., and B.W. Evermann. 1923. *American food and game fishes. A popular account of all the species found in America north of the Equator, with keys for ready identification, life histories and methods of capture*. Doubleday, Page and Company. Revised edition. (Last revision of the original 1902 edition. Reprinted 1969 by Dover Publications, Inc., New York.)

Jude, D.J., and J. Leach. 1999. Great Lakes Fisheries. Pages 623–664. In: C.C. Kohler and W.A. Hubert (Eds.). *Fisheries Management in North America*, 2nd Edition. American Fisheries Society, Bethesda, MD.

Jumper, G.Y., Jr., and R.C. Baird. 1991. Location by olfaction: A model and application to the mating problem in the deep-sea hatchetfish *Argyropelecus hemigynmnus*. *The American Naturalist* 138:1431–1458.

Junk, W.J., P.B. Bayley, and R.E. Sparks. 1989. The flood pulse concept in river floodplain systems. Pages 110–127. In: D.P. Dodge (Ed.). *Proceedings of the international large rivers symposium*. Canadian Special Publication in Fisheries and Aquatic Sciences 106, Ottawa.

Kaiser, M.J. 1998. Significance of bottom fishing disturbance. *Conservation Biology* 12:1230–1235.

Kalff, J. 2002. *Limnology: Inland Water Ecosystems*. Prentice Hall, Upper Saddle River, NJ.

Kamler, E. 1992. *Early life history of fish: An energetics approach*. Fish and Fisheries Series 4. Chapman and Hall, London.

Kaneko, T., and J. Hiroi. 2008. Osmo- and ionoregulation. Pages 163–184. In: R.N. Finn and B.G. Kapoor (Eds.). *Fish larval physiology*. Science Publishers, Enfield, NH.

Kaplan, I.C., and P. Levin. 2009. Ecosystem-based management of what? An emerging approach for balancing conflicting objectives in marine resource management. Pages 77–95. In: R.J. Beamish and B.J. Rothschild (Eds.). *The future of fisheries science in North America*. Springer Science and Business Media B.V.

Kaplan, R.H., and W.S. Cooper. 1984. The evolution of developmental plasticity in reproductive characteristics: An application of the "Adaptive Coin-Flipping" principle. *The American Naturalist* 123:393–410.

Kardong, K.V. 1998. *Vertebrates: Comparative anatomy, function, evolution*, 2nd Edition. McGraw-Hill, Boston, MA.

Karp, C.A., and H.M. Tyus. 1990. Behavior and interspecific interactions of Colorado squawfish *Ptychocheilus lucius* and five other fish species. *Copeia* 1990:25–34.

Kaufman, L. 1992. Catastrophic changes in species-rich freshwater ecosystems: The lessons of Lake Victoria. *Bioscience* 42:846–858.

Keenleyside, M.H.A. 1979. *Diversity and adaptation in fish behaviour*. Springer-Verlag, New York.

Keller, E.F. 1992. Competition: Current uses. Pages 68–73. In: E.F. Keller and E.A. Lloyd (Eds.). *Keywords in evolutionary biology*. Harvard University Press, Cambridge, MA.

Kelsh, R.N., and D.M. Parichy. 2008. Pigmentation. Pages 27–49. In: R.N. Finn and B.G. Kapoor (Eds.). *Fish larval physiology*. Science Publishers, Enfield, NH.

Kendall, A.W., Jr., E.H. Ahlstrom, and H.G. Moser. 1984. Early life history stages of fishes and their characters. Pages 11–22. In: H.G. Moser, W.J. Richards, D.M. Cohen, M.P. Fahay, A.W. Kendall, Jr., and S. L. Richardson (Eds.). *Ontogeny and systematics of fishes*. American Society of Ichthyologists and Herpetologists. Special Publication 1, Washington, DC.

Kendall, R.J. (Ed.). 1978. *Selected coolwater fishes of North America*. Special Publication 11, American Fisheries Society, Washington, DC.

Kimmel, B.L., O.T. Lind, and L.J. Paulson. 1990. Reservoir primary production. Pages 133–194. In: K.W. Thornton, B.L. Kimmel, and F.E. Payne (Eds.). *Reservoir limnology: Ecological Perspectives*. John Wiley & Sons, New York.

Kinsella, S., T. Spencer, and B. Farling. 2002. Trout in trouble: The impact of global warming on trout in the interior west. Natural Resources Defense Council, Issue Paper, New York. (www.nrdc.org/globalwarming/trout/trout.pdf; July 23, 2008, accessed June 24, 2011).

Kishida, O., G.C. Trussell, and K. Nishimura. 2007. Geographic variation in predator-induced defense and its genetic basis. *Ecology* 88:1948–1954.

Kitchell, J.F. 1983. Energetics. Pages 312–338. In: P. Webb and D. Weihs (Eds.) *Fish biomechanics*. Praeger, New York.

Kleerekoper, H. 1982. The role of olfaction in the orientation of fishes. Pages 201–225. In: T.J. Hara (Ed.). *Chemoreception in fishes*. Developments in aquaculture and Fisheries Science, 8. Elsevier Scientific Publishing Company. New York.

Koel, T.M., P.E. Bigelow, P.D. Doepke, B.D. Ertel, and D.L. Mahony. 2005. Nonnative lake trout result in Yellowstone cutthroat trout decline and impacts to bears and anglers. *Fisheries* 30(11):10–14.

Koh, P.K., R.R. Dunn, N.S. Sodhi, R.K. Colwell, H.C. Proctor and V.S. Smith. 2004. Species coextinctions and the biodiversity crisis. *Science* 305(5690):1632–1634.

Kohler, C.C., and J.J. Ney. 1982. A comparison of methods for quantitative analysis of feeding selection of fishes. *Environmental Biology of Fishes* 7(4)363–368.

Kolar, C.S., W.R. Courtenay, Jr., and L.G. Nico. 2010. Managing undesired and invading fishes. Pages 218–260. In: W.A. Hubert and M.J. Quist (Eds.). *Inland fisheries management in North America*. 3rd Edition. American Fisheries Society, Bethesda, MD.

Kontula, T., S.F. Kirilchik, and R. Väinölä. 2003. Endemic diversification of the monophyletic cottoid fish species flock in Lake Baikal explored with mtDNA sequencing. *Molecular Phylogenetics and Evolution* 27:143–155.

Kottelat, M., R. Britz, T.H. Hui, and K.-E. Witte. 2006. Paedocypris, a new genus of southeast Asian cyprinid fish with a remarkable sexual dimorphism, comprises the world's smallest vertebrate. *Proceedings of the Royal Society B: Biological Sciences* 273:895–899.

Krebs, C.J. 1972. *Ecology: The experimental analysis of distribution and abundance*. Harper & Row Publishers, New York.

Krebs, C.J. 2008. *The ecological world view*. University of California Press, Berkeley.

Kriegler, E., J.W. Hall, H. Held, R. Dawson, and H.J. Schellnhuber. 2009. Imprecise probability assessment of tipping points in the climate system. *Proceedings of the National Academy of Sciences of the United States of America (PNAS)* 106(13):5041–5046.

Krohne, D.T. 2001. *General ecology*. 2nd Ed. Brooks/Cole, Pacific Grove, CA.

Krueger, C.C., and D.D. Decker. 1999. The process of fisheries management. Pages 31–60. In: C.C. Kohler and W.A. Hubert (Eds.). *Inland fisheries management in North America*. 2nd Ed. American Fisheries Society, Bethesda, MD.

Kurlansky, M. 1998. *Cod: A biography of the fish that changed the world*. Penguin/Putnam, Inc. New York.

Kyle, H.M. 1926. *The biology of fishes*. The Macmillan Company, New York.

Lackey, R.T. 2009. Salmon decline in western North America: Historical context. In: J. Siry (Ed.). Salmon Decline. Encyclopedia of Earth http://www.eoearth.org/article/salmon_decline_in_western_north_america/.

Lackey, R.T., and L.A. Nielsen (Eds.). 1980. *Fisheries management*. John Wiley & Sons, New York.

Lagler, K.F. 1952. *Freshwater fishery biology*. W.C. Brown, Dubuque, Iowa.

Lagler, K.F. 1956. *Freshwater fishery biology*. 2nd Edition. W.C. Brown, Dubuque, Iowa.

Lagler, K.F., J.E. Bardach, and R.R. Miller. 1962. *Ichthyology*. John Wiley & Sons, New York.

Lande, R. 1995. Mutation and conservation. *Conservation Biology* 9:782–791.

Lanham, U. 1962. *The fishes*. Columbia University Press. New York.

La Rivers, I. 1994. *Fish and fisheries of Nevada*. Revised edition of the 1962 publication. University of Nevada Press, Reno.

Larkin, P.A. 1980. Objectives of management. Pages 245–262. In: R.T. Lackey and L.A. Nielsen (Eds.). 1980. *Fisheries management*. John Wiley & Sons, New York.

Lasker, R. 1975. Field criteria for survival of anchovy larvae: the relation between inshore chlorophyll maximum layers and successful first feeding. *United States Fishery Bulletin* 73:453–462.

Lasker, R. 1981. The role of a stable ocean in larval fish survival and subsequent recruitment. Pages 80–86. In: R. Lasker (Ed.). *Marine fish larvae: Morphology, ecology, and relation to fisheries*. Washington Sea Grant Program, University of Washington Press, Seattle.

Lasker, R. 1987. Use of fish eggs and larvae in probing some major problems in fisheries and aquaculture. Pages 1–16. In: Hoyt, R.D. (Ed.). *10th Annual Larval Fish Conference*. American Fisheries Society Symposium 2, Bethesda, MD.

Lassuy, D.R. 1995. Introduced species as a factor in extinction and endangerment of native fish species. Pages 391–396. In: H.L. Schramm, Jr., and R.G. Piper (Eds.). *Uses and effects of cultured fishes in aquatic ecosystems*. American Fisheries Society Symposium 15, Bethesda, MD.

Lawrence, E. 1989. *Henderson's dictionary of biological terms*. 10th Edition. John Wiley & Sons, New York.

Laycock, G. 1966. The alien animal. 1970 Printing, Ballantine Books edition, New York.

Leary, R.F., F.W. Allendorf, and G.K. Sage. 1995. Hybridization and introgression between introduced and native fish. Pages 91–101. In: H.L. Schramm, Jr., and R.G. Piper (Eds.). *Uses and effects of cultured fishes in aquatic ecosystems*. American Fisheries Society Symposium 15, Bethesda, MD.

Leet, W.S., and F.K. Cramer. 1971. Pacific southwest fisheries. Pages 218–233. In: S. Shapiro (Ed.). *Our Changing Fisheries*. U. S. Department of Commerce, Washington, DC.

Leggett, W.C., and J.E. Carscadden. 1978. Latitudinal variation in reproductive characteristics of American shad (*Alosa sapidissima*): evidence for population specific life history strategies in fish. *Journal of the Fisheries Research Board of Canada* 35:1469–1478.

Legler, N.D., T.B. Johnson, D.D. Heath, and S.A. Ludsin. 2010. Water temperature and prey size effects on the rate of digestion of larval and early juvenile fish. *Transactions of the American Fisheries Society* 139:868–875.

Leis, J.M. 1991. The pelagic stage of reef fishes: The larval biology of coral reef fishes. Pages 183–230. In: Sale, P.F. (Ed.). *The ecology of fishes on coral reefs*. Academic Press, New York.

Lema, S.C. 2008. The phenotypic plasticity of Death Valley's pupfish. *The American Scientist* 96:28–36.

Lenton, T.M., H. Held, E. Kriegler, J.W. Hall, W. Lucht, S. Rahmstorf, and H.J. Schellnhuber. 2008. Tipping elements in the Earth's climate. *Proceedings of the National Academy of Sciences of the United States of America (PNAS)* 105(6):1786–1793.

Leopold, A. 1949. *A Sand County almanac. and sketches here and there*. Oxford University Press, London.

Leopold, L.B. 1994. *A view of the river*. Harvard University Press, Cambridge, MA.

Levin, P.S., E.E. Holmes, K.R. Piner, and C.J. Harvey. 2006. Shifts in a Pacific Ocean fish assemblage: The potential influence of exploitation. *Conservation Biology* 20:1181–1190.

Lewis, W.M., Jr. 1983. A revised classification of lakes based on mixing. *Canadian Journal of Fisheries and Aquatic Sciences* 40:1779–1787.

Lewis, W.M., Jr. 2001. *Wetlands explained: Wetland science, policy and politics in America*. Oxford University Press, New York.

Li, H.W., and P.B. Moyle. 1999. Management of introduced fishes. Pages 345–374. In: C.C. Kohler and W.A. Hubert (Eds.). *Inland fisheries management in North America*. 2nd Edition. American Fisheries Society, Bethesda, MD.

Liem, K.F. 1980. Acquisition of energy by teleosts: Adaptive mechanisms and evolutionary patterns. Pages 299–334. In: M.A. Ali (Ed.). *Environmental physiology of fishes*. Plenum Press, New York.

Limburg, K.E., and J.R. Walden. 2009. Dramatic declines in North Atlantic diadromous fishes. *Bioscience* 59(11):955–965.

Lindemann, R.L. 1942. The trophic-dynamic aspect of ecology. *Ecology* 23:399–418.

Lindsey, C.C. 1981. Stocks are chameleons: Plasticity in gill rakers of coregonid fishes. *Canadian Journal of Fisheries and Aquatic Sciences* 38:1497–1506.

Liston, C., C.A. Karp, L. Hess, and S. Hiebert. 1994. Predator removal activities and intake channel studies, 1991–1992. *Tracy Fish Collection Facility Studies, Volume 1*. U.S. Bureau of Reclamation, Denver, CO.

Litvak, M.K., and W.C. Leggett. 1992. Age and size-selective predation on larval fishes: the bigger-is-better hypothesis revisited. *Marine Ecology Progress Series* 81:13–24.

Loesch, J.G. 1987. Overview of life history aspects of anadromous alewife and blueback herring in freshwater habitats. Pages 89–103. In: M.J. Dadswell, R.J. Klauda, C.M. Moffitt, R.L. Saunders, R.A. Rulifson and J.E. Cooper (Eds.). *Common strategies of anadromous and catadromous fishes.* American Fisheries Society Symposium 1, Bethesda, MD.

Long, J.A. 1995. *The rise of fishes: 500 million years of evolution.* John Hopkins University Press, Baltimore.

Long, J.A., K. Trinajstic, G.C. Young, and T. Senden. 2008. Live birth in Devonian. *Nature* 453:650–652.

Love, M.S. 2006. Subsistence, commercial and recreational fisheries. Pages 567–594. In: L.G. Allen, D.J. Pondella II, and M.H. Horn (Eds.). *The ecology of marine fishes: California and adjacent waters.* University of California Press, Berkeley.

Lucas, M.C., and E. Baras. 2001. *Migration of freshwater fishes.* Blackwell Science Ltd., Malden, MA.

Ludwig, D., R. Hilborn, and C. Walters. 1993. Uncertainty, resource exploitation, and conservation: Lessons from history. Policy Forum. *Science* 260:5104:17,36.

MacArthur, R.H. 1968. The theory of the niche. Pages 159–176. In: R.C. Lowontin (Ed.). *Population biology and evolution.* Syracuse University Press, New York.

MacArthur, R.H. 1972. *Geographical ecology: Patterns in the distribution of species.* Osford University Press, New York.

MacArthur, R.H., and E.O. Wilson. 1967. *The theory of island biogeography.* Princeton University Press, Princeton, NJ.

Maclean, N.F. 1989. *A river runs through it.* University of Chicago Press, Illinois.

Maddux, H.R., W.R. Noonan, L.A. Fitzpatrick, and H.M. Tyus. 1994. Endangered and threatened wildlife and plants: Determination of critical habitat for four Colorado River endangered fishes, Final Rule. Fish and Wildlife Service, 50 CFR Part 17. *Federal Register* 599540:13374–13400.

Madenjian, C.P., R. O'Gorman, D.B. Bunnell, R.L. Argyle, E.F. Roseman, D.M. Warner, J.D. Stockwell and M.A. Stapanian. 2008. Adverse effects of alewives on Laurentian Great Lakes fish communities. North American Journal of Fisheries Management 28:263–282.

Magnuson, J.J. 1988. Two worlds for fish recruitment: Lakes and oceans. Pages 1–6. In: R.D. Hoyt (Ed.). 11th Annual Larval Fish Conference. American Fisheries Society Special Publication 5, Bethesda, MD.

Magnússon, K.G. 1995. An overview of multispecies VGA—theory and applications. *Reviews in Fish Biology and Fisheries* 5:195–212.

Maisey, J.G. 1996. *Discovering Fossil Fishes.* Westview Press, Boulder, CO.

Mansueti. A.J., and J.D. Hardy, Jr. 1967. *Development of fishes of the Chesapeake Bay region. Volume I.* Port City Press, Baltimore, MD.

Margalef, R.1968. *Perspectives in ecological theory.* University of Chicago Press, Chicago, IL.

Markle, D.F., S.A. Reithel, J. Crandall, T. Wood, T. Tyler, M. Terwilliger, and D.C. Simon. 2009. Larval fish transport and retention and the importance of location for juvenile fish recruitment in upper Klamath Lake, Oregon. *Transactions of the American Fisheries Society* 138:328–347.

Marshall, N.B. 1966. *The Life of Fishes.* (Fifth printing 1976). The Universe Natural History Series. Universe Books, New York.

Marshall, N.B. 1984. Progenetic tendencies in deep-sea fishes. Pages 91–102. In: G.W. Potts and R.J. Wootton (Eds.). *Fish reproduction: Strategies and tactics.* Academic Press, New York.

Martin, D. 1995. The collapse of the northern cod stocks. *Fisheries* 20(5):5–8.

Martin, M.L. 2005. Avoiding extinction: An evaluation of short-term recovery recommendations for the endangered Rio Grande silvery minnow. Masters Thesis, Environmental Policy and Management, University College, University of Denver. (DU electronic capstones, theses & dissertations at: https://ectd.du.edu).

Martin, R.A. 2004. *Missing links: Evolutionary concepts & traditions through time.* Jones & Bartlett Publishers, Sudbury, MA.

Masaitis, V.L. 2002. Middle Devonian Kaluga impact crater, Russia. *Deep Sea Research Part II: Topical Studies in Oceanography* 49:1157–1169.

Mathis, A., D.P. Chivers, and R.J.E. Smith. 1996. Cultural transmission of predator recognition in fishes: Intraspecific and interspecific learning. *Animal Behaviour* 51:185–200.

Matthews, W.J. 1998. *Patterns in freshwater fish ecology.* Chapman and Hall, New York.

Mayer, E. 1964. *Systematics and the origin of species from the viewpoint of a zoologist.* Dover Publications, New York. (Reprint of the original 1942 Columbia University Press edition with added preface).

McCleave, J.D., R.C. Kleckner, and M. Castonguay. 1987. Reproductive sympatry of American and European eels and implications for migration and taxonomy. Pages 286–297. In: M.J. Dadswell, R.J. Klauda, C.M. Moffitt, R.L. Saunders, R.A. Rulifson and J.E. Cooper (Eds.). *Common strategies of anadromous and catadromous fishes*. American Fisheries Society Symposium 1, Bethesda, MD.

McCutchan, J.H., Jr., and W.M. Lewis, Jr. 2002. Relative importance of carbon sources for macroinvertebrates in a Rocky Mountain stream. *Limnology and Oceanography* 47:742–752.

McDowall, R.M. 1987. The occurrence and distribution of diadromy among fishes. Pages 1–13. In: M.J. Dadswell, R.J. Klauda, C.M. Moffitt, R.L. Saunders, R.A. Rulifson and J.E. Cooper (Eds.). *Common strategies of anadromous and catadromous fishes*. American Fisheries Society Symposium 1, Bethesda, MD.

McHugh, J.L. 1980. Coastal fisheries. Pages 323–346 (Chapter 14). In: R.T. Lackey and L.A. Nielsen (Eds.). *Fisheries Management*. Halsted Press, New York.

Mckenzie, D.J., A.P. Farrell. and C.J. Brauner. 2007. *Primitive fishes*. Fish Physiology Series 26. Elsevier (Associated Press), New York.

McKeown, B. A. 1984. *Fish migration*. Timber Press, Portland, OR.

McKibben, B. 1989. *The end of nature*. Random House, New York.

McKinney, M.L. 1997. Extinction vulnerability and selectivity: Combining ecological and paleontological views. *Annual Review of Ecology and Systematics* 28:495–516.

McKinney, M.L., and J.L. Lockwood. 1999. Biotic homogenization: a few winners replacing many losers in the next mass extinction. *Trends in Ecology and Evolution* 14(11):450–453.

McLusky, D.S. 1989. *The estuarine ecosystem*. 2nd Edition. Chapman and Hall, New York.

McMahon, T.E., and D.H. Bennett. 1996. Walleye and northern pike: Boost or bane to northwest fisheries. *Fisheries* 21(8):6–13.

McMichael, G.A., M.B. Eppard, T.J. Carlson, J.S. Carter, B.D. Ebberts, R.S. Brown, M. Weiland, G.R. Ploskey, R.A. Harnish, and Z.D. Deng. 2010. The juvenile salmon acoustic telemetry system: A new tool. *Fisheries* 35(1):9–22.

McQuinn, I.H. 1997. Metapopulations and the Atlantic herring. *Reviews in Fish Biology and Fisheries* 7:297–329.

Meffe, G.K., C.R. Carroll, and contributors. 1997. *Principles of conservation biology*. Sinauer Associates, Sunderland, MA.

Metcalf, J.L., V.L. Pritchard, S.M. Silvestri, J.B. Jenkins, J.S. Wood, D.E. Cowley, R.P. Evans, D.K. Shiozawa, and A.P. Martin. 2007. Across the great divide: genetic forensics reveals misidentification of endangered cutthroat trout populations. *Molecular Ecology* 16:4445–4454.

Michael, S.W. 1993. *Reef sharks and rays of the world: A guide to their identification, behavior, and ecology*. Sea Challengers, Monterey, CA.

Milinski, M. 1994. Long-term memory for food patches and implications for ideal free distribution in sticklebacks. *Ecology* 75:1150–1156.

Mille et une Productions. 2004. *Darwin's nightmare*. DVD. Uni France Films, Paris. Distributed by Image Entertainment, Chatsworth, CA.

Miller, R.R. 1946. *Gila cypha*, a remarkable new species of cyprinid fish from the Colorado River in Grand Canyon, Arizona. *Journal of the Washington Academy of Science* 36:409–415.

Miller, R.R. 1981. Coevolution of deserts and pupfishes (Genus Cyprinodon) in the American Southwest. Pages 39–94. In: R.J. Naiman and D.L. Soltz (Eds.). *Fishes in North American deserts*. John Wiley & Sons, New York.

Miller, R.R., J.D. Williams, and J.E. Williams. 1989. Extinctions of North American Fishes during the past century. *Fisheries* 14(6):22–38.

Minckley, W.L. 1982. Trophic interrerlations among introduced fishes in the lower Colorado River, southwestern United States. *California Fish and Game* 68(2):78–89.

Minello, T.J. 1999. Nekton densities in shallow estuarine habitats of Texas and Louisiana and the identification of essential fish habitat. Pages 43–75. In: L.R. Benaka (Ed.). *Fish habitat: Essential fish habitat and rehabilitation*. Proceedings of the Sea Grant symposium on fish habitat. American Fisheries Society, Bethesda, MD.

Mitchell, K. 2000. The greenbacks are doing swimmingly. *Coloradoan* 4(4):2. (Alumni Association of the University of Colorado, Boulder).

Modde, T. 1980. State stocking policies for small warmwater impoundments. *Fisheries* 5(5):13–17.

Montgomery, D.R. 2003. *King of fish: The thousand year run of salmon.* Westview Press, Boulder, CO.

Montgomery, J., and N. Pankhurst. 1997. Sensory physiology. Pages 325–349. In: D.J. Randall and A.P. Farrell (Eds.). 1997. *Deep-sea fishes.* Fish Physiology Series 16. Academic Press, New York.

Mooney, C. 2010. We have met the enemy and it isn't ignorance. *Conservation* 11(3):48.

Moore, B., and M. Moore. 1997. *Dictionary of Latin and Greek origins: A comprehensive guide to the classical origin of English words.* Barnes and Noble Books, New York.

Moser, H.G. 1981. Morphological and functional aspects of marineish larvae. Pages 89–131. In: R. Lasker (Ed.). *Marine fish larvae: Morphology, ecology, and relation to fisheries.* Washington Sea Grant Program, University of Washington Press, Seattle.

Moser, H.G. 1996. *The early stages of fishes in the California Current Region.* California Cooperative Oceanic Fisheries Investigations. Atlas No. 33. Allen Press, Lawrence, KS.

Moser, H.G., and W. Watson. 2006. Ichthyoplankton. Pages 269–319. In: L.G. Allen, D.J. Pondella II, and M.H. Horn (Eds.). *The ecology of marine fishes: California and adjacent waters.* University of California Press, Berkeley.

Moulton, M.P., and J. Sanderson. 1997. *Wildlife issues in a changing world.* St. Lucie Press, Delray Beach, FL.

Moyle, P.B. 1993. *Fish: An Enthusiasts Guide.* University of California Press, Berkeley and Los Angeles.

Moyle, P.B. 2002. *Inland fishes of California.* University of California Press, Berkeley and Las Angeles.

Moyle, P.B., and B. Hebold. 1987. Life history patterns and community structure in stream fishes of western North America: Comparisons with eastern North America and Europe. Pages 25–32. In: W.J. Matthews and D.C. Heins (Eds.). *Community and evolutionary ecology of North American stream fishes.* University of Oklahoma Press, Norman.

Moyle, P.B., and J.J. Cech, Jr. 1982. *Fishes: An introduction to ichthyology.* Prentice-Hall, Inc., Englewood, NJ.

Moyle, P.B., and J.J. Cech, Jr. 2004. *Fishes: An introduction to ichthyology.* 5th edition. Prentice-Hall, Inc. Upper Saddle River, NJ.

Moyle, P.B., H.W. Li, and B.A. Barton. 1986. The Frankenstein effect: Impact of introduced fishes on native fishes in North America. Pages 412–426. In: R.H. Stroud (Ed.). *Fish culture in fisheries management.* Fish Culture and Fisheries Management Section of the American Fisheries Society, Bethesda, MD.

Mueller, G.A., and P.C. Marsh. 2002. Lost, a desert river and its native fishes: A historical perspective of the lower Colorado River. Information and Technology Report No. USGS/BRD/ITR-2002-0010. U.S. Government Printing Office, Denver, CO.

Mueller, G.A., P.C. Marsh, and W.L. Minckley. 2005. A legacy of change: The lower Colorado River, Arizona—California—Nevada, USA, and Sonora—Baja California Norte, Mexico. Pages 139–156. In: J.N. Rinne, R.M. Hughes, and B. Calamusso (Eds.). *Historical changes in large river fish assemblages of the Americas.* American Fisheries Society Symposium 45, Bethesda, MD.

Müller, U.K. 2008. Swimming and muscle. Pages 523–549. In: R.N. Finn and B.G. Kapoor (Eds.). *Fish Larval Physiology.* Science Publishers, Enfkeld, NH.

Mullin, M.M. 1993. *Webs & scales: Physical and ecological processes in marine fish recruitment.* Washington Sea Grant Program, University of Washington Press, Seattle.

Murawski, S.A., J.J. Maquire, R.K. Mayo, and F.M. Serchuk. 1997. Groundfish stocks and the fishing industry. Pages 27–70. In: J. Boreman, B.S. Nakashima, J.A. Wilson, and R.L. Kendall (Eds.). *Northwest Atlantic groundfish: Perspectives on a fishery collapse.* American Fisheries Society, Bethesda, MD.

Myers, R.A., and C.A. Ottensmeyer. 2005. Extinction risk in marine species. Pages 58–79. In: E.A. Norse, and L.G. Crowder (Eds.). *Marine conservation biology: The science of maintaining the sea's biodiversity.* Island Press, Washington, DC.

Myers, R.A., J. Baum, T.A. Sheperd, S.P. Powers, and C.H. Peterson. 2007. Cascading effects of the loss of apex predatory sharks from the coastal ocean. *Science* 315:1846–1850.

Myklebust, J. 2006. Status review of the greenback cutthroat trout. Masters Thesis, Environmental Policy and Management, University College, University of Denver. (DU electronic capstones, theses & dissertations at: https://ectd.du.edu).

Nash, R. 1982. *Wilderness and the American mind.* 3rd edition. Yale University Press, New Haven, CT.

Nash, R.D.M., A.H. Valencia, and A.J. Gerren. 2006. The origin of Fulton's condition factor—setting the record straight. *Fisheries* 31(5):236–238.

National Marine Fisheries Service. 2009. Marine and Anadromous Fish: Status of Fish Species. Office of Protected Resources. Accessed January 15, 2010, http://www.nmfs.noaa.gov/pr/species/fish.

National Research Council (NRC). 1996. *Upstream: Salmon and society in the Pacific Northwest*. National Academy Press, Washington, DC.
Natural Resources Defense Council. 2002. Global Warming Threatens Cold-Water Fish. May 21, 2002. Accessed June 24, 2011, http://www.nrdc.org/globalwarming/ntrout.asp.
Nehlsen, W., J.E. Williamson, and J.A. Lichatowich. 1991. Pacific salmon at the crossroads: Stocks at risk from California, Oregon, Idaho, and Washington. *Fisheries* 16(2):4–21.
Nelson, J.S. 2006. *Fishes of the World*. 4th edition. John Wiley & Sons, Inc. Hoboken, NJ.
Nelson, J.S., E.J. Crossman, H. Expinosa-Pérez, L.T. Findley, C.R. Gilbert, R.N. Lea, and J.D. Williams. 2004. *Common and scientific names of fishes from the United States, Canada, and Mexico*. 6th Edition. American Fisheries Society Special Publication 29. Bethesda, MD.
Nelson, K., and M. Soulé. 1987. Genetical conservation of exploited fishes. Pages 345–368. In: N.Ryman and F. Utter (Eds.). *Population genetics & fisheries Management*. Washington Sea Grant Program, University of Washington Press, Seattle.
Nicholson, A.J. 1954. An outline of the dynamics of animal populations. *Australian Journal of Zoology* 2:9–65.
Nielsen, L.A. 1999. History of inland fisheries management in North America. Pages 3–30. In: C.C. Kohler and W.A. Hubert (Eds.). *Inland fisheries management in North America*. 2nd edition. American Fisheries Society, Bethesda, MD.
Nielson, A. 1950. The torrential invertebrate fauna. *Oikos* 2:176–196.
Nikolskii, G.V. 1961. *Special ichthyology (Chastnaya ikhtiologfya)*. Translated from Russian. 2nd edition published in 1954 by the USSR. Published for the National Science Foundation and Smithsonian Institution by the Israel Program for Scientific Translations, Jerusalem. U.S. Department of Commerce, Office of Technical Services, Washington, DC.
Nikolsky, G.V. 1963. *The Ecology of Fishes*. Academic Press, Inc., London. (1978 Reprint by T.F.H. Publications, Inc, Neptune, NJ).
Noakes, D.L.G. 1986. The genetic basis of fish behaviour. Pages 3–22. In: T.J. Pritcher (Ed.). *The behavior of teleost fishes*. The Johns Hopkins University Press, Baltimore, MD.
Noakes, D.L.G. 2009. Volume foreword. Pages ix–xi, In: R.J. Beamish and B.J. Rothschild (Eds.). *The future of fisheries science in North America*. Fish and Fisheries Series 31. Springer Science + Business Media B.V. Dordrecht, Netherlands.
Nobel, R.L. 1980. Management of lakes, reservoirs and ponds. Pages 265–295. In: R.T. Lackey and L.A. Nielsen (Ed.). *Fisheries management*. John Wiley & Sons, New York.
Norse, E.A., L.B. Crowder. K. Gjerde, D. Hyrenbach, C.M. Roberts, C. Safina, and M.E. Soulé. 2005. Place-based ecosystem management in the open ocean. Pages 302–327. In: E.A. Norse, and L.G. Crowder (Eds.). *Marine conservation biology: The science of maintaining the sea's biodiversity*. Island Press, Washington, DC.
Noss, R.F. 1997. Hierarchial indicators for monitoring changes in biodiversity. Essay 4.A, Pages 88–89. In: G.K. Meffe, C.R. Carroll, and contributors. 1997. *Principles of conservation biology*. Sinauer Associates, Sunderland, MA
Novales Flamarique, I., G.A. Mueller, C.L. Cheng, and C.R. Figiel. 2007. Communication using eye roll reflective signaling. *Proceedings of the Royal Society B*: Biological Sciences (2007) 274:877–882.
Nybakken, J.W. 1997. *Marine biology: An ecological approach*. Addison-Wesley Educational Publishers, Addison Wesley Longman, Reading, MA.
Odling-Smee, L., S.D. Simpson, and V.A. Braithwaite. 2006. The role of learning in fish orientation. Pages 119–138. In: C. Brown, K. Laland, and J. Krause (Eds.). *Fish cognition and behavior*. Blackwell Publishing, Ltd., Oxford, UK.
Odum, E.P. 1959. *Fundamentals of ecology*. 2nd Edition. W.B. Saunders Company, Philadelphia, PA.
Odum, E.P. 1971. *Fundamentals of ecology*. 3rd Edition. W.B. Saunders Company, Philadelphia, PA.
Odum, E.P. 1997. *Ecology: A bridge between science and technology*. Sinauer Associates, Sunderland, MA.
Odum, H.T., and E.C. Odum. 1976. *Energy basis for man and nature*. McGraw Hill, New York.
Odum, H.T., B.J. Copeland, and E.A. McMahan. 1969. *Coastal ecological systems of the United States*. (3 vols). The Conservation Foundation and Office of Coastal Environment, U.S. National Oceanic and Atmospheric Administration. Washington, DC.
Olden J.D., N.L. Poff, and K.R. Bestgen. 2006. Life-history strategies predict fish invasions and extirpations in the Colorado River basin. *Ecological Monographs* 76(1):25–40.

Ono, R.D., J.D. Williams, and A. Wagner. 1983. *Vanishing fishes of North America*. Stone Wall Press, Washington, DC.

O'Reilly, C.M., S.R. Alin, P.-D. Plisnier, A.S. Cohen, and B.A. McKee. 2003. Climate change decreases aquatic ecosystem productivity of Lake Tanganyika, Africa. *Nature* 424(6950):766–768.

Oreskes, N., and E.M. Conway. 2010. Merchants of doubt: How a handful of scientists obscured the truth on issues from tobacco smoke to global warming. Bloomsbury Publishing, New York.

Orth, D.J., and R.J. White. 1999. Stream habitat management. Pages 249–284. In: C.C. Kohler and W.A. Hubert (Eds.). *Inland fisheries management in North America*, 2nd edition. American Fisheries Society, Bethesda, MD.

Osborne, M.J., M.A. Benivides, and T.F. Turner. 2005. Genetic heterogeneity among pelagic egg samples and variance in reproductive success in an endangered freshwater fish, *Hybognathus amarus* (Cyprinidae). *Environmental Biology of Fishes* 73:(4):463–472.

Osmundson, D.B., R.J. Ryel, and T.E. Mourning. 1997. Growth and survival of Colorado squawfish in the upper Colorado River. *Transactions of the American Fisheries Society* 126:687–698.

Pacheco, A.L., and G.C. Grant. 1965. Seasonal occurrence of juvenile menhaden and other small fishes in a tributary creek of Indian River, Delaware, 1957–1958. Part 1, Studies of the early life history of Atlantic menhaden in estuarine nurseries. Special Scientific Report—Fisheries 504, U.S. Fish and Wildlife Service, Washington, DC.

Palmer, J.D. 1970. Introduction to biological rhythms and clocks. Pages 3–12. In: F.A. Brown, Jr., J.W. Hastings, and J.D. Palmer (Eds.). *The biological clock: Two views*. Academic Press, New York.

Pankhurst, P.M. 2008. Mechanoreception. Pages 305–329. In: R.N. Finn and B.G. Kapoor (Eds.). *Fish larval physiology*. Science Publishers, Enfield, NH.

Park, T. 1948. Experimental studies of interspecies competition: 1. competition between populations of flour beetles, *Trilobium confusum* Duval nd *Trilobium castaneum* Herbst. *Ecological Monographs* 18:265–308.

Park, T. 1954. Experimental studies of interspecific competition II. Temperature, humidity, and competition in two species of Tribolium. *Physiological Zoology* 27:177–238.

Park, Y.-S., J. Chang, S. Lek, W. Cao, and S. Brasse. 2003. Conservation strategies for endemic fish species threatened by the Three Gorges Dam. *Conservation Biology* 17:1748–1758.

Parker, G.A. 2000. Scramble in behavior and ecology. *Philosophical Transactions of the Royal Society of London B* 355:1637–1645.

Parsons, G.R., E.R. Hoffmayer, J.M. Hendon, and W.V. Bet-Sayad. 2008. A review of shark reproductive ecology: Life history and evolutionary implications. Pages 435–469. In: M.J. Rocha, A. Arukwe, and B.G. Kapoor (Eds.). *Fish reproduction*. Science Publishers, Enfield, NH.

Patzner, R.A. 2008. Reproductive strategies of fish. Pages 311–350. In: M.J. Rocha, A. Arukwe, and B.G. Kapoor (Eds.). *Fish reproduction*. Science Publishers, Enfield, NH.

Paukert, C.T., and G.D. Scholten (Eds.). 2009a. *Paddlefish management, propagation, and conservation in the 21st Century: Building from 20 years of research and management*. American Fisheries Society Symposium 66, Bethesda, MD.

Paukert, C.T., and G.D. Scholten. 2009b. Where we are today in paddlefish conservation and management. Pages 441–443. In: C.P. Paukert and G.D. Scholten (Eds.). *Paddlefish management, propagation and conservation in the 21st Century: Building from 20 years of research and management*. American Fisheries Society Symposium 66, Bethesda, MD.

Paulsen, C.M., and T.R. Fisher. 2001. Statistical relationship between parr-to-smolt survival of Snake River spring–summer chinook salmon and indices of land use. *Transactions of the American Fisheries Society* 130:347–358.

Pauly, E. 1989. An eponym for Reuben Lasker. *United States Fishery Bulletin* 87:383–384.

Pearce, F. 2007. *With speed and violence: Why scientists fear tipping points in climate change*. Beacon Press, Boston, MA.

Pearsons, T.N. 2008. Misconception, reality and uncertainty about ecological interactions and risks between hatchery and wild salmonids. *Fisheries* 33:278–290.

Pearsons, T.N. 2010. Operating hatcheries within an ecosystem context using the Adaptive Stocking Concept. *Fisheries* 35(1):23–31.

Pelster, B. 1997. Buoyancy at depth. Pages 195–237. In: D.J. Randall and A P. Farrell (Eds.). *Deep-sea fishes*. Academic Press, San Diego, CA.

Perry, A.L., P.J. Low, J.R. Ellis, and J.D. Reynolds. 2005. Climate change and distribution shifts in marine fishes. *Science* 308(5730):1912–1915.

Peterman, R.M., and M.J. Bradford. 1987. Wind speed and mortality rate of a marine fish, the northern anchovy *Engraulis mordax*. Science 235(4786):354–356.

Petersen, J.H., and J.F. Kitchell. 2001. Climate regimes and water temperature changes in the Columbia River: Bioenergetic implications for predators of juvenile salmon. *Canadian Journal of Fisheries and Aquatic Sciences* 58:1831–1841.

Peterson, W.T., and F.B. Schwing. 2003. A new climate regime in northeast pacific ecosystems. *Geophysical Research Letters* 30(17):1896.

Pflieger, W.L. 1997. *The fishes of Missouri*. Missouri Department of Conservation, Jefferson City.

Phillips, J., and S.J. Gould. 1860. *Life on Earth: Its origin and succession*. Ayer Publishing (1980 reprint of the original 1860 edition published by MacMillian, Cambridge).

Pietsch, T.W. 1976. Dimorphism, parasitism and sex: Reproductive strategies among deepsea ceratoid anglerfishes. *Copeia* 1976:781–793.

Pietsch, T.W. 2005. Dimorphism, parasitism and sex revisited: node of reproduction among deepsea ceratoid anglerfishes (Teleostei:Lophiiformes). *Ichthyol Res* 52:207–236.

Pimm, S.L. 1996. Lessons from a kill. *Biodiversity and Conservation* 5(4):1059–1067.

Pimm, S.L., and P. Raven. 2000. Biodiversity: Extinction by the numbers. *Nature* 403:843–845 (24 Feb 2000).

Pimm, S.L., G.J. Russell, J.L. Gittleman, and T.M. Brooks. 1995. The future of biodiversity. *Science* 269:347–350.

Pine, W.E., III, S.J.D. Martell, C.J. Walters, and J.F. Kitchell. 2009. Counterintuitive responses of fish populations to management actions: Some common causes and implications for predictions based on ecosystem modeling. *Fisheries* 34(4):165–180.

Piper, R.G., I.B. McElwain, L.E. Orme, J.A. McCraren, L.G. Fowler, and J.R. Leonard. 1982. *Fish hatchery management*. U.S. Fish and Wildlife Service, Washington, DC.

Pitcher, T.J. 1986. Functions of shoaling behaviour in teleosts. Pages 294–337. In: T.J. Pitcher (Ed.). *The behavior of teleost fishes*. Johns Hopkins University Press, Baltimore, MD.

Pitcher, T.J. , and P.J.B. Hart. 1982. *Fisheries ecology*. AVI Publishing Company, Westport, CT.

Pooley, E. 2010. The climate war: True believers, power brokers, and the fight to save the earth. Hyperion Books, New York.

Pope, J.G., J.G. Sheperd, and J. Webb. 1994. Successful surf-riding on size spectra: The secret of survival in the sea. *Philosophical Transactions of the Royal Society of London B* 343:41–49.

Portz, D.E. 1999. Fish humps in two Colorado River fishes: An induced morphological response to cyprinid predation? Submitted to the Department of Environmental, Population, and Organismic Biology, University of Colorado at Boulder, as a requirement for the MA degree.

Portz, D.E., and H.M. Tyus. 2004. Fish humps in two Colorado River fishes: An induiced morphological response to cyprinid predation? *Environmental Biology of Fishes* 71:233–245.

Post, G. 1987. *Textbook of fish health*. T.F.H. Publications, Inc., Neptune City, NJ.

Potts, G.W., and R.J. Wootton (Eds.).1984. *Fish reproduction: Strategies and tactics*. Academic Press, New York.

Pough, F.H., C.M. Janis, and J.B. Heiser. 2005. *Vertebrate life*. 7th edition. Pearson Prentice Hall, Upper Saddle River, NJ.

Povilitis. A., and K. Suckling. 2010. Addressing climate change threats for endangered species in U.S. recovery plans. *Conservation Biology* 24(2):372–376.

Power, D.M., N. Silva, and M.S. Campinho. 2008. Smoltification. Pages 639–681. In: R.N. Finn and B.G. Kapoor (Eds.). *Fish larval physiology*. Science Publishers, Enfield, NH.

Power, M.E., D. Tilman, J.A. Estes, B.A. Menge, W.J. Bond, L.S. Mills, B. Daily, J.C. Castilla, J. Lubchenco, and R.T. Paine. 1996. Challenges in the quest for keystones. *BioScience* 46(8):609–620.

Preston, J.L. 1978. Communication systems and social interactions in a goby-shrimp symbiosis. *Animal behavior* 26:791–802.

Price, J. 2010. Chesapeake Bay: An undeclared ecological disaster. Chesapeake Bay Ecological Foundation, Inc., Easton. MD, http://www.chesbay.org/contact.

Pritchard, D.W. 1967. What is an estuary: physical viewpoint. Pages 3–5. In: G.H. Lauff (Ed.). *Estuaries*. Publication 83. American Association for the Advancement of Science. Washington, DC.

Prochazka, K. 1996. Seasonal patterns in a temperate intertidal fish community on the west coast of South Africa. *Environmental Biology of Fishes* 45(2):133–140.

Prugh, L.R., C.J. Stoner, C.W. Epps, W.T. Bean, W.J. Ripple, A.S. Laliberte, and J.S. Brashares. 2009. The rise of the mesopredator. *Bioscience* 59(9):779–791.

Pyke, G.H. 1984. Optimal foraging theory: A critical review. *Annual Review of Ecology and Systematics* 15:523–575.

Quinn, J.W. 2009. Harvest of paddlefish in North America. Pages 203–221. In: C.P. Paukert and G.D. Scholten (Eds.). *Paddlefish management, propagation and conservation in the 21st Century: Building from 20 years of research and management.* American Fisheries Society Symposium 66, Bethesda, MD.

Radonski, G.C., and A.J. Loftus. 1995. Fish genetics, fish hatcheries, wild fish, and other fables. Pages 1–4. In: H.L Schramm, Jr., and R.G. Piper (Eds.). *Uses and effects of cultured fishes in aquatic ecosystems.* American Fisheries Society Symposium 15, Bethesda, MD.

Rahel, F.J., and J.D. Olden. 2008. Assessing the effects of climate change on aquatic invasive species. *Conservation Biology* 22:521–533.

Randall, D.J., and A.P. Farrell (Eds.). 1997. *Deep-sea fishes.* Fish Physiology Series 16. Academic Press, New York.

Raney, E.C., and D.A. Webster. 1942. The spring migration of the common white sucker, *Catostomous c. commersonni* (Lacèpéd), in Skaneateles Lake Inlet, New York. *Copeia* 1942:139–148.

Rasmussen, J.L. (Chairman).1997. Introduction of aquatic species. Pages 33–35. In: *Resource policy handbook.* 1st Ed. Resource Policy Committee, American Fisheries Society, Bethesda, MD.

Rassam, G. 2009. Column: Director's Line: Issues of concern to AFS. *Fisheries* 34(4):187.

Rassam, G. 2011. American Fisheries Society Issues: Policy statement on climate change. *Fisheries* 36(2):86.

Raup, D.M. 1991. *Extinction: Bad Genes or Bad Luck?* W.W. Norton & Company, New York.

Raymond, H.L. 1979. Effects of dams and impoundments on migrations of juvenile chinook salmon and steelhead from the Snake River, 1966–1975. *Transactions of the American Fisheries Society* 108:505–529.

Reed, D.H., J.J. O'Grady, B.W. Brook, J.D. Ballou, and R. Frankham. 2003. Estimates of minimum viable population sizes for vertebrates and factors influencing those estimates. *Biological Conservation* 113: 23–34.

Reed, J.M., L.S. Mills, J.B. Dunning, Jr., E.S. Menges, K.S. McKelvey, R. Frye, S.R. Beissinger, M-C. Anstett, and P. Miller. 1999. Emerging issues in population viability analysis. *Conservation Biology* 16:7–19.

Reeves, G.H., R.H. Everest, and J.D. Hall. 1987. Interactions between the redside shiner (*Richardsonius balteatus*) and steelhead trout (*Salmo gairdneri*) in western Oregon: the influence of water temperature. *Canadian Journal of Fisheries and Aquatic Sciences* 44:1603–1613.

Regetz, J. 2003. Landscape-level constraints on recruitment of chinook salmon (*Oncorhynchus tshawytscha*) in the Columbia River basin, USA. *Aquatic Conservation: Marine and Freshwater Ecosystems* 13:35–49 (Wiley Interscience).

Reid, G.K. 1961. *Ecology of inland waters and estuaries.* Reinhold Publishing Corporation. New York.

Reiser, A., C.G. Hudson, and S.E. Roady. 2005. The role of legal regimes in marine conservation. Pages 362–374. In: E.A. Norse and L.B. Crowder (Eds.). *Marine conservation biology: The science of maintaining the seas's biodiversity.* Island Press, Washington, DC.

Rhymer, J.M., and D. Simberloff. 1996. Extinction by hybridization and intergression. *Annual Review of Ecology and Systematics* 27:83–109.

Rice, J.A., L.B. Crowder, and E.A. Marshall. 1997. Predation on juvenile fishes: dynamic interactions between size-structured predators and prey. Pages 333–355. In: R.C. Chambers and E.A. Trippel (Eds.). *Early life history and recruitment in fish populations.* Fish and Fisheries Series 21. Chapman and Hall, New York.

Ricker, W.E., 1975. *Computation and interpretation of biological statistics of fish populations.* Bulletin 191, Fisheries Research Board of Canada. Canadian Government Publishing Centre, Ottawa.

Richter, B.D., D.P. Braun, M.A. Mendelson, and L.L. Master 1997. Threats to imperiled freshwater fauna. *Conservation Biology* 11:1081–1093.

Ricklefs, W.C. 1990. *Ecology.* 3rd Ed. W.H. Freeman and Company, New York.

Riddle, B.R., and D.J. Hafner. 2004. The past and future roles of phylogeography. Pages 93–110. In: M.V. Lomolino and L.R. Heaney (Eds.). *Frontiers of biogeography: New directions in the geography of nature.* Sinauer Associates, Sunderland, MA.

Rijnsdorp, A.D., M.A. Peck, G.H. Englehard, C. Mollmann, and J.K. Pinnegar. 2009. Resolving the effect of climate change on fish populations. *ICES Journal of Marine Science: Journal du Conseil* 66(7):1570–1583.

Rinne, J.N., R.M. Hughes, and B. Calamusso (Eds.). 2005. *Historical changes in large river fish assemblages of the Americas.* American Fisheries Society Symposium 45, Bethesda, MD.

Roberts, C. 2007. *The unnatural history of the sea.* Island Press/Shearwater Books. Washington, DC.

Robins, C.R., and G.C. Ray. 1986. *A field guide to Atlantic Coast Fishes: North America.* Houghton Mifflin Company, Boston.

Robinson, B.W., A.J. Januszkiewicz, and J.C. Koblitz. 2008. Survival benefits and divergence of predator-induced behavior between pumpkinseed sunfish ecomophs. *Behavioral Ecology* 19:263–271.

Robinson, B.W., D.S. Wilson, and A.S. Margosian. 2000. A pluralistic analysis of character release in pumpkinseed sunfish (*Lepomis gibbosus*). *Ecology* 81:2799–2812.

Robison, H.W. 1986. Zoogeographic implications of the Mississippi River basin. Pages 267–285. In: C.H. Hocutt and E.O. Wiley (Eds.). *The zoogeography of North American Freshwater fishes.* John Wiley & Sons, New York.

Robison, H.W., and T.M. Buchanan. 1988. *Fishes of Arkansas.* University of Arkansas Press, Fayetteville.

Rocha, M.J., A. Arukwe, and B.G. Kapoor (Eds.). Fish reproduction. Science Publishing, Inc., Enfield, NH.

Romer, A.S. 1966. *Vertebrate paleontology,* 3rd Edition. University of Chicago Press, Chicago, IL.

Rose, G.A. 2004. Reconciling overfishing and climate change with stock dynamics of Atlantic cod (Gadus morhua) over 500 years. *Canadian Journal of Fisheries and Aquatic Sciences* 61:1553–1557.

Rosen, R.A., and D.C. Hales. 1981. Feeding of paddlefish, *Polyodon spathula. Copeia* 1981:441–455.

Rosenberg, A.A., W.J. Bolster, K.E. Alexander, W.G. Leavenworth, A.B. Cooper, and M.G. McKenzie. 2005. The history of ocean resources: modeling cod biomass using historical records. *Frontiers in Ecology and Environment* 3(2):84–90.

Ross, M.R. 1997. *Fisheries conservation and management.* Prentice Hall, Upper Saddle River, NJ.

Rothschild, B.J., A.F. Sharov, and M. Lambert. 1997. Single-species and multispecies management. Pages 141–152. In: J. Boreman, G.S. Nakashima, J.A. Wilson, and R.L. Kendall (Eds.). *Northwest Atlantic groundfish: Perspectives on a fishery collapse.* American Fisheries Society, Bethesda, MD.

Rothschild, B.J., and R.J. Beamish. 2009. On the future of fisheries science. Pages 1–11. In: R.J. Beamish and B.J. Rothschild (Eds.). *The future of fisheries science in North America.* Fish and fisheries series, Volume 31. Springer Science + Business Media B.V. Dordrecht, Netherlands.

Rowe, L., and D. Ludwig. 1991. Size and timing of metamorphosis in complex life cycles: Time constraints and variation. *Ecology* 72:413–427.

Rowe, S., J. Hutchings, D. Bekkevold, and A. Rakitin. 2004. Depensation, probability of fertilization, and the mating system of Atlantic cod (*Gadus morhua* L.). *ICES Journal of Marine Science* 6i(7):1144–1150.

Ruelle, R. and P.S. Hudson. 1977. Paddlefish (*Polyodon spathula*): growth and food of young of the year and a suggested technique for measuring length. *Transactions of the American Fisheries Society* 94:160–168.

Ruiz, G.M., P. Fofonoff, A.H. Hines, and E.D. Grosholz. 1999. Non-indigenous species as stressors in estuarine and marine communities: Assessing invasion impact and interactions. *Limnology and Oceanography* 44(3):950–972.

Rulifson, R.A., and B.L. Wall. 2006. Fish and blue crab passage through water control structures of a coastal bay lake. *North American Journal of Fisheries Management* 26:317–326.

Runkel, R.L. 2007. Toward a transportation based analysis of nutrient spiraling and uptake in streams. *Limnology and Oceanography: Methods* (5):50–62.

Russell, T.R. 1986. Biology and life history of the paddlefish—A review. Pages 2–21. In: J.G. Dillard, L. K. Graham, and T.R. Russell (Eds.). *The paddlefish: status, management, and propagation.* Special Publication 7, North Central Division, American Fisheries Society, Bethesda, MD.

Ruttner, F. 1963. *Fundamentals of limnology,* 3rd Edition. (Translated by D.G. Frey and F.E.J. Fry, 1973 printing). University of Toronto Press.

Ruzycki, J.R., D.A. Beauchamp, and D.L.Yule. 2003. Effects of introduced lake trout on native cutthroat trout in Yellowstone Lake. *Ecological Applications* 13:23–37.

Sale, P.F. 1991. Introduction. Pages 3–15. In: P.F. Sale (Ed.). *The ecology of fishes on coral reefs.* Academic Press, New York.

Salomon, A.K., S.K. Gaichas, N.T. Shears, J.E. Smith, E.M.P. Madini, and S.D. Gaines. 2010. Key features and context-dependence of fishery-induced trophic cascades. *Conservation Biology* 24(2):382–394.

Sanders, H.L. 1968. Marine benthic diversity, a competitive study. *American Naturalist* 102:243–282.

Sanderson, B.L., K.A. Barnas, and A.M. Warago Rub. 2009. Nonindigenous species of the Pacific Northwest: An overlooked risk to endangered salmon? *BioScience* 59(3):245–256.

Sanderson, E.W. 2006. How many animals do we want to save? The many ways of setting population target levels for conservation. *Bioscience* 56(11):911–922.

Sargent, R.C., and M.R. Gross. 1993. William's principle: An explanation of parental care in teleost fishes. Pages 333–362. In: T.J. Pitcher (Ed.). *Behavior of teleost fishes*, 2nd Edition. Chapman and Hall, New York.

Sassa, C., K. Kawaguchi, and K. Taki. 2007. Larval fish assemblages in the Kuroshio-Oyashio Transition Region of the western North Pacific. *Marine Biology* 150:1403–1415.

Saunders, R., M.A. Hachey, and C.W. Fay. 2006. Maine's diadromous fish community: Past, present and implications for Atlantic salmon recovery. *Fisheries* 31:537–546.

Scalet, C.G., L.D. Flake, and D.W. Willis. 1996. *Introduction to wildlife and fisheries: An integrated approach.* W.H. Freeman, New York.

Scarce, R. 2000. *Fishy business: Salmon, biology, and the social construction of nature.* Temple University Press, Philadelphia.

Schaffer, B., and M. Willilams. 1977. Relationships of fossil and living elasmobranchs. *American Zoology* 17:293–302.

Scheidegger, K.J., and M.B. Bain. 1995. Larval fish distribution and microhabitat use in free-flowing and regulated rivers. *Copeia* 1995:125–135.

Schellart, N.A.M., and R.J. Wubbels. 1997. The auditory and mechanosensory lateral line system. Pages 283–312, In: D.H. Evans (Ed.). *The physiology of fishes*, 2nd Edition. CRC Press, Boca Raton, FL.

Schieber, J., and D.J. Over. 2005. Sedimentary fill of the late Devonian Flynn Creek crater: A hard target marine impact. Pages 51–69. In: D.J. Over, J.R. Morrow, and P.B. Wignall (Eds.). *Understanding late Devonian and Permian–Triassic biotic and climatic events: Towards an Integrated approach.*, Elsevier Publications.

Schindler, D.E., X. Fleishman, N.J. Mantua, B. Riddell, M. Riddell, M. Ruckelshaus, J. Seeb, and M. Webster. 2008. Climate change, ecosystem impacts, and management of Pacific salmon. *Fisheries* 33(10):502–506.

Schlosser, I.J. 1991. Stream fish ecology: A landscape perspective. *Bioscience* 41(10):704–711.

Schoener, T.W. 1985. Some comments on Connell's and my reviews of field experiments in interspecific competition. *American Naturalist* 125:730–740.

Schoener, T.W. 1989. The ecological niche. Pages 79–113. In: J.M. Cherrett (Ed.). *Ecological concepts: The contribution of ecology to an understanding of the natural world.* British Ecological Society. Blackwell Scientific Publications, Cambridge, MA.

Schoener, T.W. 2009. Ecological niche. Pages 3–13. In: S.A. Levin (Ed.). *The Princeton Guide to Ecology.* Princeton University Press, Princeton, NJ.

Schooley, J.D., A.P. Karam, B.R. Kesner, P.C. Marsh, C.A. Pacey, and D.J. Thornbrugh. 2008. Detection of larval remains after consumption by fishes. *Transactions of the American Fisheries Society* 137:1044–1049.

Schramm, C.H., Jr., and W.A. Huber. 2005. Ecosystem management. Pages 111–123. In: C.C. Kohler and W.A.Huber (Eds.). *Inland fisheries management in North America*, 2nd Edition. American Fisheries Society, Bethesda, MD.

Schwartz, F.J. 1981. *World literature to fish hybrids, with an analysis by family, species and hybrid.* NOAA Technical Report NMFS SSRF-750, U.S. Department of Commerce (NTIS-PB82-169715).

Schwartz, F.J. and J. Tyler. 1970. *Marine fishes common to North Carolina.* North Carolina Department of Conservation and Development. Raleigh.

Scotese, C.R. 2004. Cenozoic and Mesozoic paleogeography: Changing terrestrial biogeographic pathways. Pages 9–26. In: M.V. Lomolino and L.R. Heaney (Eds.). *Frontiers of biogeography: New directions in the geography of nature.* Sinauer Associates, Sunderland, MA.

Selye, H. 1950. Stress and the general adaptation syndrome. *British Medical Journal* 1950: 1383–1392.

Selye, H. 1973. The evolution of the stress concept. *American Scientist* 61:692–699.

Serreze, M.C. 2010. Understanding recent climate change. *Conservation Biology* 24:10–17.

Sette, O.E. 1971. Resources and their environment. Pages 20–39. In: S. Shapiro (Ed.). *Our changing fisheries.* U.S. Department of Commerce. U.S. Government Printing Office, Washington, DC

Shaffer, M.L. 1981. Minimum population sizes for species conservation. *Bioscience* 31(2):131–134.

Shapiro, F.R. 2006. The Yale book of quotations. Yale University Press, New Haven, CT.

Shapiro, S. (editor). 1971. *Our changing fisheries.* U.S. Department of Commerce. U.S. Government Printing Office, Washington, DC.

Sheehan, R.J., and J. L. Rasmussen. 1999. Large rivers. Pages 529–559. In: C.C. Kohler and W.A. Hubert (Eds.). *Inland fisheries management in North America.* 2nd Edition. American Fisheries Society, Bethesda, MD.

Shelford, V.E. 1911. Physiological animal geography. *Journal of Morphology* 22:551–618.
Shu, D.-G., et al. 1999. Lower cambrian vertebrates from south China. *Nature* 402:42–46.
Sideleva, V. G. 2003. *The endemic fishes of Lake Baikal*. Backhuys Publishers, The Netherlands.
Sigfried, S., D. Croft, S. Hiebert, and M. Bowen. 2000. Continuous monitoring of fish eggs and larvae at the Tracy Fish Collection Facility, Tracy, California, February–June 1994. Volume 6. *Tracy Fish Collection Facility Studies*. U.S. Bureau of Reclamation, Denver.
Sigler, W.F., and J.W. Sigler. 1987. *Fishes of the Great Basin: A natural history*. University of Nevada Press, Reno.
Simon, T.P., and E.B. Emery. 1995. Modification and assessment of an index of biotic integrity to quantify water resource quality in great rivers. *Regulated Rivers Research & Management* 11:283–298.
Sinclair, M. 1988. *Marine populations: An essay on population regulation and speciation*. Washington Sea Grant Program. University of Washington Press, Seattle.
Sinclair, M. 1997. Prologue—Recruitment in fish populations: the paradigm shift generated by ICES Committee A. Pages 1–30. In: R.C. Chambers and E.A. Trippel (Eds.). *Early life history and recruitment in fish populations*. Fish and Fisheries Series 21. Chapman and Hall, New York.
Sinclair, M. 2009. Herring and the ICES: A historical sketch of a few ideas and their linkages. *ICES Journal of Marine Science* 66 (advanced online access April 30, 2009, at http://icesjms.oxfordjournals.org .retrieved 7/28/09).
Sinclair, A.F., and S.A. Murawski. 1997. Why have groundfish stocks declined? Pages 71–93. In: J. Boreman, B.S. Nakashima, J.A. Wilson, and R.L. Kendall (Eds.). *Northwest Atlantic groundfish: Perspectives on a fishery collapse*. American Fisheries Society, Bethesda, MD.
Sinitsyna, T. 2008. Baikal without waste water. Opinion and Analysis Section, Future Media—International forum. Russian News and Information Agency Novosti, August 21, 2008. Available on line at http://en15,rian.ru/analysis/20080821/116181368.html. Accessed June 13 2011.
Slobodkin, L.B. 1961. *Growth and regulation of animal populations*. Holt, Rinehart and Winston. New York.
Smith, C.L. (Ed) 1988. *Fisheries research in the Hudson River*. Hudson River Environmental Society. State University of New York Press, Albany.
Smith, G.R. 1981a. Late Cenozoic freshwater fishes of North America. *Annual Review of Ecology and Systematics* 12:163–193.
Smith, H.M. 1907. *The fisheries of North Carolina*. Vol. II. North Carolina Geological and Economic Survey, Raleigh.
Smith, L.S. 1982. *Introduction to fish physiology*. T.F.H. Publications Incorporated, Neptune, NJ.
Smith, P.E. 1981b. Fisheries on coastal pelagic schooling fish. Pages 1–32. In: R. Lasker (Ed.). *Marine fish larvae: Morphology, ecology, and relation to fisheries*. Washington Sea Grant Program, University of Washington Press, Seattle.
Smith, R.J.F. 1985. *The control of fish migration*. Zoophysiology 17. Springer- Verlag, New York.
Smith, R.L., and T.M. Smith. 2001. *Ecology and Field Biology*. 6th Ed. Benjamin Cummings, San Francisco.
Smith, S.H. 1970. Trends in fishery management of the Great Lakes. Pages 107–114. In: N.G. Benson (Ed.). *A century of fisheries in North America*. American Fisheries Society, Washington, DC.
Smith, S.H. 1973. Application of theory and research in fishery management of the Laurentian Great Lakes. *Transactions of the American Fisheries Society* 102:156–163.
Smith, T.B., and S. Skúlason. 1996. Evolutionary significance of resource polymorphisms in fishes, amphibians and birds. *Annual Review of Ecology and Systematics* 27:111–133.
Snieszko, S.F. 1974. The effect of environmental stress on outbreaks of infectious diseases of fishes. *Journal of Fish Biology* 6:197–208.
Snyder, D.E. 1976. Terminologies for intervals of larval fish development. Pages 41–60. In: J. Boreman (Ed.). *Great lakes fish egg and larvae identification*. U.S. Fish and Wildlife Service Report No. FWS/OBS-76/23, Ann Arbor, MI.
Snyder, D.E. 1983. Fish eggs and larvae. Pages 165–197. In: L.A. Nielsen and D.L. Johnson (Eds.). *Fisheries techniques*. American Fisheries Society, Bethesda, MD.
Snyder, D.E. 1990. Fish larvae—ecologically distinct organisms. Pages 20–23. In: M.B. Bain (Ed.). *Ecology and assessment of warmwater streams*. U.S. Fish and Wildlife Service Biological Report 90(5).
Snyder, D.E., and R.T. Muth. 1988. Description and identification of June, Utah, and mountain sucker larvae and early juveniles. State of Utah, Natural Resource Publication 88-8, Salt Lake City.
Snyder, D.E., and R.T. Muth. 1990. Description and identification of razorback, white, Utah, bluhead and mountain sucker larvae and early juveniles. Colorado Division of Wildlife, Technical Bulletin 38, Denver.

Snyder, N.F.R. 1994. The California condor recovery program: Problems in organization and execution. Pages 183–204. In: T.W. Clark, R.P. Reading, and A.L. Clarke (Eds.). *Endangered species recovery: Finding the lessons improving the process.* Island Press, Washington, DC.

Soltz, D.L., and R.J. Naiman. 1978. *The natural history of native fishes in the Death Valley system.* Natural History Museum of Los Angeles County, Science Series 30:1–76. California.

Solemdal, P. 1997. Epilog—The three cavaliers: a discussion from the golden age of Norwegian marine research. Pages 551–565. In: R.C. Chambers and E.A. Trippel. Early life history and recruitment in fish populations. Fish and Fisheries Series 21. Chapman and Hall, New York.

Soulé, M.E. 1980. Thresholds for survival: Maintaining fitness and evolutionary potential. Pages 151–169 (Chapter 9) In: M.E. Soulé and B.A. Wilcox (Eds.). *Conservation Biology: An evolutionary-ecological perspective.* Sinauer Associates, Sunderland, MA.

Southall, P.D., and W.A. Hubert. 1984. Habitat use by adult paddlefish in the upper Mississippi River. *Transactions of the American Fisheries Society* 113:125–131.

Sparks, J. 1836. *The works of Benjamin Franklin, containing several political and historic tracts. Volume II.* Hilliard Gray and Company, Boston.

Sparrowe, R.D. 1986. Threats to paddlefish habitat. Pages 36–45. In: J.G. Dillard, L. K. Graham, and T.R. Russell (Eds). *The paddlefish: status, management, and propagation.* Special Publication 7, North Central Division, American Fisheries Society, Bethesda, MD.

Springer, C. 2011. An unnatural history: Controlling the parasitic sea lamprey in the Great Lakes. *Eddies: Reflections on Fisheries Conservation* 3(3):28–31. U.S. Fish and Wildlife Service, Washington, DC.

Springer, V.G., and J.P. Gold. 1989. *Sharks in question*: The Smithsonian answer book. Smithsonian Institution Press, Washington, DC.

Stacey, N.E. 1984. Control of the timing of ovulation by exogenous and endogenous factors. Pages 207–222. In: G.W. Potts and R.J. Wootton (Eds.). *Fish reproduction: Strategies and tactics.* Academic Press, New York.

St. Louis, V.L., C.A. Kelly, E. Duchemin, et al. 2000. Reservoir surfaces as sources of greenhouse gases to the atmosphere: A global estimate. *BioScience* 50:766–775.

Stanley, T.R., Jr. 1995. Ecosystem management and the arrogance of humanism. *Conservation Biology* 9:255–262.

Steffansson, S.D., B.Th. Björnsson, L. O.E. Ebbesson, and S.D. McCormick. 2008. Smoltification. Pages 639–681. In. R.N. Finn and B.G. Kapoor (Eds.). *Fish larval physiology.* Science Publishers, Enfield, NH.

Stephens, P.A., W.J. Sutherland, and R.P. Freckletoon. 1999. What is the Alee effect? *Oikos* 87(1):185–190.

Stilling, P. D. 1999. *Ecology: Theories and application*, 3rd Edition. Prentice-Hall, Inc. Upper Saddle River, NJ.

Stockwell, J.D., M.P. Ebener, J.A. Black, O.T. Gorman, T.R. Hrabik, R.E. Kinnunen, W.P. Mattes, J.K. Oyadomari, S.T. Schram, D.R.Schreiner, M.J. Seider, S.P. Sitar, and D.L. Yule. 2009. A synthesis of cisco recovery in Lake Superior: Implications for native fish rehabilitation in the Laurentian Great Lakes. *North American Journal of Fisheries Management* 29:626–652.

Strahler, A.N. 1957. Quantitative analysis of watershed geomorphology. *Transactions of the American Geophysical Union* 38:913–920.

Strecker, A.L., P.M. Campbell, and J.D. Olden. 2011. The aquarium trade as an invasion pathway in the Pacific northwest. *Fisheries* 35(2):74–85.

Sumich, J.L., and J.F. Morrissey. 2004. *Introduction to the biology of marine life.* 8th edition. Jones and Bartlett Publishers, Sudbury, MA.

Summerfelt, R.C. 1999. Lake and reservoir habitat management. Pages 285–320. In: C.C. Kohler and W.A. Hubert (Eds.). *Inland fisheries management in North America*, 2nd Ed. American Fisheries Society, Bethesda, MD.

Suntsov, A.V., E.A. Widder, and T.T. Sutton. 2008. Bioluminescence. Pages 51–90. In. R.N. Finn and B.G. Kapoor (Eds.). *Fish larval physiology.* Science Publishers, Enfield, NH.

Swingle, H.S. 1956. Determination of balance in farm fish ponds. *North American Wildlife Conference* 20(1955):298–318.

Swingle, H.S. 1970. History of warmwater pond culture in the United States. Pages 95–105. In: N.G. Benson (Ed.). *A century of fisheries in North America.* Special Publication No. 7, American Fisheries Society, Washington, DC.

Tansley, A.G. 1935. The use and abuse of vegetational concepts and terms. *Ecology* 16:284–307.

Taylor, J.N., W.R. Courtenay, Jr., and J.A. McCain. 1984. Known impacts of exotic fishes in the continental United States. Pages 322–373. In: W.R. Courtenay, Jr., and J.R. Stauffer, Jr. (Eds.). *Distribution, biology, and management of exotic fishes.* The Johns Hopkins University Press, Baltimore, MD.

Taylor, M.I., and M.E. Knight. 2008. Mating systems. Pages 277–310. In: M.J. Rocha, A. Arukwe, and B.G. Kapoor (Eds.). *Fish reproduction.* Science Publishing, Inc., Enfield, NH.

Tear, T.H., J.M. Scott, P.H. Hayward, and B. Griffith. 1995. Recovery plans and the Endangered Species Act: Are criticisms supported by data? *Conservation Biology* 9(1):182–195.

Temple, S.A. 1990. The nasty necessity: Eradicating exotics. *Conservation Biology* 4(2):113–115.

Terborgh, J., and B. Winter. 1980. Some causes of extinction. Pages 119–133. In: M.E. Soulé and B.A. Wilcox (Eds.). *Conservation Biology: An evolutionary-ecological perspective.* Sinauer Associates, Sunderland, MA.

Thomas, C.D. 1990. What do real population dynamics tell us about minimum viable population sizes? *Conservation Biology* 4:324–327.

Thomson, K.S. 1991. *Living fossil: The story of the coelacanth.* W.W. Norton & Company, New York.

Thornton, K.W., B.L. Kimmel, and F.E. Payne (Eds.). *Reservoir limnology: Ecological Perspectives.* John Wiley & Sons, New York.

Thunberg, B.E. 1971. Olfaction in parent stream selection by the alewife (*Alosa pseudoharengus*). *Animal Behavior* 19:217–225.

Thurman, H.J., and A.P. Trujillo. 2002. *Essentials of oceanography*, 7th Edition. Prentice Hall, Upper Saddle Creek, NJ

Toffler, A. 1970. *Future shock.* Random House, New York.

Tolkien, J.R.R. 1965. *The hobbit or there and back again.* Ballentine Books, New York (First published in 1937 by George Allen & Unwin, UK).

Tollrian, R. and C.D. Harvell. 1998. *The ecology and evolution of inducible defenses.* Princeton University Press, Princeton, NJ.

Traill, L.W. C.J. Bradshaw, and B.W. Brook. 2007. Minimum viable population size: A meta-analysis of 30 years of published estimates. *Biological Conservation* 139:159–186.

Trammel, M.A., E.P. Bergersen, and P.J. Martinez. 1993. Evaluation of Colorado squawfish stocking in a mainstream impoundment on the white River. *Southwestern Naturalist* 38:362–369.

Trandahl. A. 1978. Preface. Pages IX–X. In: R.J. Kendall (Ed.). *Selected coolwater fishes of North America.* Special Publication 11, American Fisheries Society, Washington, DC.

Tsukamoto, K., M.M. Miller, A. Kotake, J. Aoyama, and K. Uchida. 2009. The origin of fish migration: The random escapement hypothesis. Pages 45–61. In: A.J. Haro, K.L. Smith, R.A. Rulifson, C.M. Moffitt, R.J. Klauda, M.J. Dadswell, R.A. Cunjak, J.E. Cooper, K.L. Beal, and T.S. Avery (Eds.). *Challenges for diadromous fishes in a dynamic global environment.* American Fisheries Society Symposium 69, Bethesda, MD.

Tsukamoto, K., T. Otake, N. Mochioka, T. Lee, H. Fricke, T. Inagaki, J. Aoyama, S. Ishikawa, S. Kimura, M.J. Miller, H. Hasumoto, M. Oya, and Y. Suzuki. 2004. Sea mounts, new moon and eel spawning: The search for the spawning site of the Japanese eel. *Environmental Biology of Fishes* 66:221–229.

Turner, J.R.G., and B.A. Hawkins. 2004. The global diversity gradient. Pages 171–190. In: M.V. Lomolino and L.R. Heaney (Eds.). *Frontiers of biogeography: New directions in the geography of nature.* Sinauer Associates, Sunderland, MA.

Tyus, H.M. 1971. Population size, harvest and movements of alewives, *Alosa pseudoharengus* (Wilson) during spawning migrations to Lake Mattamuskeet, North Carolina. Ph.D. Thesis. Department of Zoology, North Carolina State University, Raleigh.

Tyus, H.M. 1974. Movements and spawning of anadromous alewives *Alosa pseudoharengus* (Wilson) at Lake Mattamuskeet, North Carolina. *Transactions of the American Fisheries Society* 103:392–396.

Tyus, H.M. 1990a. Effects of altered stream flows on fisheries resources. *Fisheries* 15(3):18–20.

Tyus, H.M. 1990b. Potamodromy and reproduction of Colorado squawfish in the Green River basin, Colorado and Utah. Transactions of the American Fisheries Society 119:1035–1047.

Tyus, H.M. 1991. Movements and habitat use of young Colorado squawfish, *Ptychocheilus lucius*, in the Green River of Utah. *Journal of Freshwater Ecology* 6:43–51.

Tyus, H.M. and C.A. Karp. 1989. Habitat use and stream flow needs of rare and endangered fishes, Yampa River, Colorado. *U.S. Fish and Wildlife Service Biological Report* 89(14):1–27. Washington, DC.

Tyus, H.M., and J.F. Saunders, III. 2000. Nonnative fish control and endangered fish recovery: Lessons from the Colorado River. *Fisheries* 25(9):17–24.

Tyus, H.M., and N.J. Nikirk. 1990. Abundance, growth, and diet of channel catfish Ictalurus punctatus in the Green and Yampa rivers, Colorado and Utah. *Southwestern Naturalist* 35:233–244.

Tyus, H.M., and W.L. Minckley. 1988. Migrating Mormon crickets, *Anabrus simplex* (Orthoptera:Tettigoniidae), as food for stream fishes. *Great Basin Naturalist* 48:25–30.

Tyus, H.M., C.A. Brown, and J.F. Saunders, III. 2000. Movements of young Colorado pikeminnow and razorback sucker in response to water flow and light level. *Journal of Freshwater Ecology* 15:525–535.

Underhill, J.C. 1986. The fish fauna of the Laurentian Great Lakes, the St. Lawrence lowlands, Newfoundland, and Labrador. Pages 106–186. In: C.H. Hocutt and E.O. Wiley (Eds.). *Zoogeography of North American freshwater fishes*. John Wiley & Sons, New York.

University Corporation for Atmospheric Research. 2009. Climate Change: Frequently Asked Questions. Boulder, CO. Accessed February 2, 2010, http://www.ucar.edu/news/features/climatechange/faqs.jsp.

Unkenholz, D.G. 1986. Effects of dams and other habitat alterations on paddlefish sport fisheries. Pages 54–61. In: J.G. Dilliard, L.K. Graham, and T. Russell (Eds.). *The paddlefish: status, management, and propagation*. American Fisheries Society Special Publication 7, Bethesda, MD.

U.S. Bureau of Reclamation (USBR). 2001. Tracy fish facility improvement program. *Final environmental assessment and initial study*. USBR Technical Service Center, Denver Federal Center, Denver, CO.

U.S. Environmental Protection Agency. 2008. Great Lakes Factsheet No. 1. Accessed June 13, 2011, http://www.epa.gov/glnpo/atlas/gl-fact1.html.

U.S. Environmental Protection Agency. 2010. Draft TMDL for Chesapeake Bay, http://www.epa.gov/chesapeakebaytmdl/.

U.S. Fish and Wildlife Service (FWS). 1998. *Greenback cutthroat trout recovery plan*. Prepared by the Greenback Cutthroat Trout Recovery Team for Region 6, U.S. Fish and Wildlife Service, Denver, CO.

U.S. Fish and Wildlife Service (FWS). 2008. Recovering Threatened and Endangered Species: A Report to Congress, Fiscal Year 2005–2006, http://www.fws.gov/endangered/recovery/reports_to_congress/2005-6/summary/.

U.S. Fish and Wildlife Service (FWS). 2009a. List of Threatened and Endangered Species, http://ecos.fws.gov/tess_public/TESSBoxscore.

U.S. Fish and Wildlife Service (FWS). 2009b. List of Threatened and Endangered Species: Listed Fishes, http://ecos.fws.gov/tess_public/speciesReport.

U.S. Government Accountability (GAO). 2002. *Columbia River basin salmon and steelhead: Federal agencies' recovery responsibilities, expenditures and actions*. Report No. GAO-02-612, July 26, 2002, Washington, DC.

U.S. Congress Office of Technology Assistance (OTA). 1993. *Harmful nonindigenous species in the United States*. Publications OTA-F-565 (report) and OTA-566 (summary). U.S. Superintendent of Documents, Washington, DC.

Valdez, R.A., and R.T. Muth. 2005. Ecology and conservation of native fishes in the upper Colorado River basin. Pages 157–204. In: J.N. Rinne, R.M. Hughes, and B. Calamusso (Eds.). *Historical changes in large river fish assemblages of the Americas*. American Fisheries Society Symposium 45, Bethesda, MD.

Van Den Avyle, M.J., and R.S. Hayward. 1999. Dynamics of exploited fish populations. Pages 127–166. In: C.C. Kohler and W.A. Hubert (Eds.). *Inland fisheries management in North America*, 2nd Edition. American Fisheries Society, Bethesda, MD.

Vannote, R.L., G.W. Minshall, K.W. Cummins, J.R. Sedell, and C.E. Cushing. 1980. The river continuum concept. *Canadian Journal of Fisheries and Aquatic Sciences* 37:130–137.

Van Valen, L. 1973. A new evolutionary law. *Evolutionary Theory* 1:1–30.

Van Winkle, W., P. Anders, D.H. Secor, and D. Dizon. 2002. *Biology, management and protection of North American sturgeon*. American Fisheries Society Symposium 28, Bethesda, MD.

Varley, J.D., and P. Schullery. 1998. *Yellowstone fishes: Ecology, history, and angling in the park*. Stackpole Books, Mechanicsburg, PS.

Vigg, S., T.P. Poe, L.A. Pendergast, and H.C. Hansel. 1991. Rates of consumption of juvenile salmonids and alternative prey fish by northern squawfish, walleyes, smallmouth bass, and channel catfish in John Day Reservoir, Columbia River. *Transactions of the American Fisheries Society* 120:421–438.

Voight, W. 1948. *The road to survival*. William Sloan, New York.

Vøllestad, L.A., K. Varreng, and A.B.S. Poléo. 2004. Body depth variation in crucian carp, *Carassius carassius*: an experimental individual-based study. *Ecology of Freshwater Fish* 13:197–202.

von der Emde, G. 1997. Electroreception. Pages 313–343. In: D.H. Evans (Ed.). *The physiology of fishes*. 2nd Edition, CRC Press, LLC, Boca Raton, FL.

Walburg, C.H. 1971. Loss of young fish in reservoir discharge and year class survival, Lewis and Clark Lake, Missouri River. Pages 441–448. In: G.E. Hall (Ed.). *Reservoir fisheries and limnology*, American Fisheries Society Special Publication 8, Bethesda MD.

Walker, J.D., and J.W. Geissman. 2009. 2009 GSA geologic time scale. *GSA Today* 19(4):60–61. Also available at http://www.geosociety.org/science/timescale/.

Wallace, A.S. 1876. *The geographical distribution of animals with a study of the relations of living and extinct faunas as elucidating the past changes of the earth's surface.* In two volumes. Harper and Brothers, New York.

Walters, C., and J.F. Kitchell. 2001. Cultivation/depensation effects on juvenile survival and recruitment: implications for the theory of fishing. *Canadian Journal of Fisheries and Aquatic Science* 58:39–50.

Wapner, P. 2010. *Living through the end of nature: The future of American environmentalism.* The MIT Press, Cambridge, MA.

Ward, J.V. 1992. A mountain river. Pages 493–510. In: P. Calow and G.E. Petts. (Eds.). *The rivers handbook: Hydrological and ecological principles*. Volume One. Blackwell Scientific Publications, London.

Ward, J.V. 1994. Ecology of alpine streams. *Freshwater Biology* 32:277–294. Also published online May 30, 2006, by Wiley Interscience at http://www.eawag.ch/publications/eawagnews/www_en54/.

Ward, J.V., and B.C. Kondratieff. 1992. *An illustrated guide to the mountain stream insects of Colorado*. University Press of Colorado, Niwot.

Ward, J. V., and J.A. Stanford. 1983. The serial discontinuity concept of lotic ecosystems. Pages 29–42. In: T.D. Fontaine and S.M. Bartell (Eds.). *Dynamics of Lotic Ecosystems*. Ann Arbor Science Publishers, Ann Arbor, MI.

Warren, C.E., and G.E. Davis. 1967. Laboratory studies on the feeding, bioenergetics, and feeding of fish. Pages 175–214. In: S.D. Gerking (Ed.). *The biological basis of freshwater fish production.* Wiley, New York.

Washington Department of Fish and Wildlife and Oregon Department of Fish and Wildlife. 2002. Status Report: Columbia River Fish Runs and Fisheries, http://wdfw.wa.gov/fish/columbia/2000_status_report_text.pdf.

Waters, J.M., J.M. Epifanio, T. Gunter, and B.L. Brown. 2000. Homing behavior facilitates subtle genetic differentiation among river populations of *Alosa sapidissima*. *Journal of Fish Biology* 56:622–636.

Watling, L. 2005. The global destruction of habitat by mobile fishing gear. Pages 198–210. In: E.A. Norse, and L.G. Crowder (Eds.). *Marine conservation biology: The science of maintaining the sea's biodiversity.* Island Press, Washington, DC.

Watling, L., and E.A. Norse. 1998. Disturbance of the seabed by mobile fishing gear: A comparison to forest clearcutting. *Conservation Biology* 12:1180–1197.

Watson, W., and H.J. Walker, Jr. 2004. The world's smallest vertebrate, *Schindleria brevipinguis*, a new paedomorphic species in the family Schindleriidae. (Perciformes: Gobiodei). *Records of the Australian Museum* 56(2):139–142.

Webb, P.W. 1978. Partitioning of energy into metabolism and growth. Pages 184–214. In: S.D. Gerking (Ed.). *Ecology of freshwater fish production.* John Wiley & Sons, New York.

Webb, P.W. 1997. Swimming. Pages 3–24. In: D.H. Evans (Ed.). *The Physiology of Fishes*, 2nd Edition. CRC Press, Boca Raton.

Webster, J.R. 1975. Analysis of potassium and calcium dynamics in stream ecosystems on three Appalacian watersheds of contrasting vegetation. Ph.D. Thesis, University of Georgia, Athens.

Webster, J.R., and B.C. Patten. 1979. Effects of watershed perturbation on stream potassium and calcium dynamics. *Ecological Monographs* 49:51–72.

Wedemeyer, G.A., B.A. Barton, and D.J. McLeay. 1990. Stress and acclimation. Pages 451–489. In: C.B. Schreck and P.B. Moyle (Eds.). *Methods for fish biology*. American Fisheries Society, Bethesda, MD.

Wegener, A.L. 1966. *The origins of continents and oceans*, translated from the 4th German edition by J. Biram. Dover Publications, Mineolta, NY.

Welch, P.S. 1952. *Limnology*, 2nd Edition. McGraw Hill Book Company, New York.

Werner, E.E., and D.J. Hall. 1974. Optimal foraging and the size selection of prey by the bluegill sunfish (*Lepomis macrochirus*). *Ecology* 55:1042–1052.

Werner, E.E., and D.J. Hall. 1976. Niche shift in sunfishes: Experimental evidence and significance. *Science* 191:404–406.

Werner, E.E., and D.J. Hall. 1977. Competition and habitat shiftin two sunfishes (Centrarchidae). *Ecology* 58:867–876.

Werner, E.E., and J.F. Gilliam. 1984. The ontogenetic niche and species interactions in size-structured populations. *Annual Review of Ecology and Systematics* 15:393–425.

Werner, R.G. 1979. Homing mechanism of spawning white suckers in Wolf Lake New York. *New York Fish and Game Journal* 26:48–58.

Wesche, T.A., and D.J. Isaak. 1999. Watershed management and land-use practices. Pages 217–248. In: C.C. Kohler and W.A. Hubert (Eds.). *Inland fisheries management in North America*, 2nd Edition. American Fisheries Society, Bethesda, MD.

West-Eberhard, M.J. 1992. Adaptation: Current usages. Pages 13–18. In: E.F. Kerrer and E.A. Lloyd (Eds.). *Keywords in evolutionary biology*. Harvard University Press, Cambridge, MA.

Westrum, R. 1994. An organizational perspective: Designing recovery teams from the inside out. Pages 327–349. In: T.W. Clark, R.P. Reading, and A.L. Clarke, (Eds.). *Endangered species recovery: Finding the lessons improving the process*. Island Press, Washington, DC.

Wetzel, R.G. 1975. Limnology. W.B. Saunders, Philadelphia, PA.

Wetzel, R.G. 1990. Reservoir ecosystems: Conclusions and speculations. Pages 227–238. In: K.W. Thornton, B.L. Kimmel, and F.E. Payne (Eds.). *Reservoir limnology: Ecological Perspectives.* John Wiley & Sons, New York.

Wetzel, R.G. 2001. *Limnology: Lake and River Ecosystems*, 3rd Edition. Academic Press, New York.

Wheeler, J.P., and G.H. Winters. 1984. Homing of Atlantic herring (*Clupea harengus harengus*) in Newfoundland waters as indicated by tagging data. *Canadian Journal of Fisheries and Aquatic Sciences* 41:108–117.

White, R., B., Mefford and C. Liston. 2000. Evaluation of mitten crab exclusion technology at the Tracy Fish Collection Facility, California. Volume 14. *Tracy Fish Collection Facility Studies.* U.S. Bureau of Reclamation, Denver.

White, R.J., J.R. Karr, and W. Nehlsen. 1995. Better roles for fish stocking in aquatic resource management. Pages 527–547. In: H.L. Schramm, Jr., and R.G. Piper. *Uses and effects of cultured fishes in aquatic ecosystems*. American Fisheries Society Symposium 15, Bethesda, MD.

Wikipedia. 2010. Global warming. Revised February 1, 2010. Accessed February 1, 2010, http://en.wikipedia.org/wiki/Global_warming/.

Wildhaber, M.L., R.F. Green,, and L.B. Crowder. 1994. Bluegill continuously update patch giving- up time based on foraging experience. *Animal Behaviour* 47:501–513.

Wiley, M.L., and B.B. Collette. 1970. Breeding tubercles and contact organs in fishes, their occurrence, structure, and significance. *Bulletin of the American Museum of Natural History* 143(3):143–216.

Wilhere, G.F. 2008. The how-much-is-enough myth. *Conservation Biology* 22(3):514–517.

Williams, G.C. 1966. Natural selection, the costs of reproduction, and a refinement of Lack's principle. *American Naturalist* 100:687–690.

Williams, J.D., and G.K. Meffe. 1998. Non-indigenous species. Pages 117–129. In: M.J. Mac, P.A. Opler, C.E. P. Haecker, and P.J. Doran. (Eds.). *Status and trends of the nation's biological resources.* U.S. Geological Survey, Washington, DC.

Williams, J.D., J.E. Johnson, D.A. Hendrickson, S. Contreras-Balderas, J.D. Williams, M. Navarrl-Mendoza, D.E. McAllister, and J.E. Deacon. 1989. Fishes of North America, endangered, threatened or of special concern 1989. *Fisheries* 14(6):2–20.

Williams, J.G. 2010. Sampling for environmental flow assessments. *Fisheries* 35(9): 434–443.

Williams, J.G., S.G. Smith, and W.D. Muir. 2001. Survival estimates for downstream migrant yearling juvenile salmonids through the Snake and Columbia rivers hydropower system, 1966–1980 and 1993–1999. *North American Journal of Fisheries Management* 21:310–317.

Williams, M.G. 1992. Species: Current usages. Pages 318–323. In: E.F. Kerrer and E.A. Lloyd (Eds.). *Keywords in evolutionary biology*. Harvard University Press, Cambridge, MA.

Willis, B. 1993. Paddlefish: An American Treasure (DVD). Earthwave Society Inc., Fort Worth, TX, http://www.earthwave.org/paddlefish.htm.

Wilson, E.O. 2002. *The future of life*. Alfred A. Knopf, New York.

Wilson, R., and J.Q. Wilson. 1992. *Pisces guide to watching fishes: Understanding coral reef fish behavior*. Gulf Publishing Company, Houston, TX.

Wiltzius, W.J. 1985. *Fish culture and stocking in Colorado, 1873–1978*. Division Report 12, Colorado Division of Wildlife, Denver.

Windell, J.T. 1978. Estimating food consumption rates in fish populations. Pages 227–254. In: *Methods for assessment of fish production in fresh waters*. IBP Handbook No. 3., 3rd Edition. Blackwell Scientific Publications, Oxford, GB.

Winder, M., and D.E. Schindler. 2004. Climate change uncouples trophic interactions in an aquatic ecosystem. *Ecology* 85(8):2100–2106.

Winemiller, K.O., and K.A. Rose. 1992. Patterns of life-history diversification in North American fishes: Implications for population regulation. *Canadian Journal of Fisheries and Aquatic Sciences* 49: 2196–2216.

Winger, P.V. 1981. Physical and chemical characteristics of warmwater streams: A review. Pages 32–44 In: L.A. Krumholz (Ed.). *Warmwater streams symposium: A national symposium on fisheries aspects of warmwater streams*. Southern Division of the American Fisheries Society, Bethesda, MD.

Wipfli, M.S., J.P. Hudson. J.P. Caduette, and D.T. Chaloner. 2003. Marine subsidies in freshwater ecosystems: salmon carcasses increase the growth rates of stream–resident salmonids. *Transactions of the American Fisheries Society* 49:2196–2216.

Woldt, A. 2010. Asian carp caught above electric barrier. *Eddies: Reflections on Fisheries Conservation* 3(2):6. U.S. Fish and Wildlife Service, Washington, DC.

Woodward, G.M. 1956. *Commercial fisheries of North Carolina: An economic analysis*. University of North Carolina Press, Chapel Hill.

Wootton, R.J. 1990. *Ecology of teleost fishes*. Fish and fisheries series 1, Chapman and Hall, New York.

Worm, B., E.B. Barbler, N. Beaumont, et al. 2006. Impacts of biodiversity loss on ocean ecosystem services. *Science* 314(5800):787–790.

Wydoski, R.S., and R.R. Whitney. 2003. *Inland fishes of Washington*. University of Washington Press, Seattle.

Wydoski, R.S., and R.W. Wiley. 1999. Management of undesirable fish species. Pages 403–430. In: C.C. Kohler and W.A. Hubert. *Inland fisheries management in North America*. 2nd Ed. American Fisheries Society, Bethesda, MD.

www.fws.gov/nevada/protected_species/fish/dhp (accessed June 14, 2011).

Xenopoulos, M.A., D.M. Lodge, J. Alcamo, M. Märker, K. Schulze, and D.P. Van Vuuren. 2005. Scenarios of freshwater fish extinctions from climate change and water withdrawal. *Global Change Biology* 11:1557–1564.

Zhu, M., and X.-B. Yu. 2002. A primitive fish close to the common ancestor of tetrapods and lungfishes. *Nature* 410:81–84.

Zimmerman, M.P. 1999. Food habits of smallmouth bass, walleyes, and northern pikeminnow in the lower Columbia River basin during outmigration of juvenile anadromous salmonids. *Transactions of the American Fisheries Society* 128:1036–1054.

Zimmerman, M.S., and C.C. Krueger. 2009. An ecosystem perspective on re-establishing native deepwater fishes in the Laurentian Great Lakes. *North American Journal of Fisheries Management* 29:1352–1371.

Plate 1 Some salmonid fishes of North American coldwater streams and lakes: (a) western Yellowstone cutthroat trout, (b) Pacific salmon (e.g., sockeye), (c) eastern brook trout, (d) Atlantic salmon, (e) Arctic char, (f) grayling, and (g) introduced brown trout. (h) Central image: opening day of trout season in New York. ((a) Courtesy of USDA-Forest Service, http://commons.wikimedia.org/wiki/File:Yellowstone_Cutthroat_Trout.jpg. (b, d) From Knepp, T., U.S. Fish and Wildlife Service (USFWS), http://commons.wikimedia.org/wiki/File:Oncorhynchus_nerka.jpg and http://commons.wikimedia.org/wiki/File:Salmo_salar.jpg. With permission. (c, g) Courtesy of Raver, D., Freshwater Fish Collection, USFWS. With permission. (e) Courtesy of Gaither, J., Digital Library of USFWS, http://images.fws.gov/, http://commons.wikimedia.org/wiki/File:Arctic_Char.jpg. (f) Courtesy of U.S. Bureau of Fisheries (1909), retouched by National Oceanic and Atmospheric Administration (NOAA), http://www.photolib.noaa.gov/htmls/fish3019.htm, http://commons.wikimedia.org/wiki/File:Arctic_grayling.jpg.)

Plate 2 North American fishes of coolwater streams and lakes include (a) mountain whitefish (central image), (b) mountain sucker, (c) northern pike, (d) burbot, (e) smallmouth bass, (f) yellow perch, (g) sauger, and (h) slimy sculpin. ((a) Courtesy of Wooster, H. M., http://commons.wikimedia.org/wiki/File:Prosopium_williamsoni.jpg. (b) Courtesy of Kredit, H., U.S. National Park Service USNPS, http://www.nps.gov/features/yell/slidefile/fishherps/fish/Images/00889.jpg. (c) Courtesy of Knepp, T., USFWS, http://images.fws.gov/, http://commons.wikimedia.org/wiki/File:Esox_lucius1.jpg. (d) Courtesy of Great Lakes Research Laboratory, NOAA, http://commons.wikimedia.org/wiki/File:Lota_lota_GLERL_1.jpg. (e, f, g) From Raver, D., USFWS, usa.gov. With permission. (h) By Edmonson, E., and Chrisp, H., Courtesy of New York State Department of Environmental Conservation (NYSDEC), http://www.ny.gov/animals/52634.html. With permission.)

Plate 3 Fishes of the Laurentian Great Lakes include native (a) lake sturgeon, (b) blue pike (extinct), (c) lake herring (a cisco), (d) lake whitefish, (e) lake trout, (f) walleye, centrarchid basses and sunfishes—e.g., (g) pumpkinseed—and nonnatives such as (h) sea lamprey and (i) alewife. (j) Central image: Great Lakes system. ((a–d, h) By Edmondson, E., and Chrisp, H., Courtesy of NYSDEC, http://www.ny.gov/animals/52634.html. With permission. (e) Courtesy of USFWS, http:images.fws.gov/, http://commons.wikimedia.org/wiki/File:Fisher_holding_Lake_trout.jpg. (f, g, i) From Raver, D., FW fish collection, USFWS. With permission. (j) Courtesy of Detroit District, U.S. Army Corps of Engineers, http://commons.wikimedia.org/wiki/File:Great_Lakes_2.PNG.)

Plate 4 Fishes of North American warmwater streams and natural lakes include native (a) largemouth bass; (b) sunfishes (e.g., black crappie); (c) channel catfish; (d) pickerels, e.g., chain pickerel; minnows, e.g., (e) redbelly dace and (f) Colorado pikeminnow; (g) suckers, e.g., shortnose redhorse; (h) darters (colorful candy darter shown); and numerous nonnatives. (i) Central image: largemouth bass in nature. ((a, b, d) From Raver, D., USFWS, FW fish collection. With permission. (c, e, g) By Edmonson, E., and Chrisp, H., Courtesy of New York Biological Survey, NYSDEC. (h) Courtesy of Burkhead, N., U.S. Geological Survey, http://fl.biology.usgs.gov.pics/nativefish/nativefish.html. (i) Courtesy of Knepp, T., USFWS, http://commons.wikimedia.org/wiki/File:Micropterus_salmoides_2.jpg.)

Plate 5 The Missouri–Mississippi river system has an old and diverse fish fauna, including (a) alligator gar, (b) shovelnose sturgeon, (c) paddlefish, (d) bowfin, (e) flathead catfish, (f) blue catfish, (g) minnows (e.g., fathead minnow), (h) buffalo (e.g., smallmouth buffalo), and (i) endangered pallid sturgeon (center). ((a, c–h) From Raver, D., USFWS, FW fish collection. With permission. (b) Courtesy of Bighorn Canyon National Recreation Area, USNPS, http://www.nps.gov/bica/naturescience/checklist-of-bighorn-canyon-fish.htm. (i) Courtesy of USFWS Digital Library, http://www.fws.gov/digitalmedia/cdm4/results.php?.)

Plate 6 Some sport and food fishes stocked in North American impoundments and reservoirs include (a) largemouth bass, (b) bluegill, (c) striped bass, (d) white bass, (e) white crappie, (f) channel catfish, (g) Western salmonids (e.g., rainbow trout), (h) tilapia, and (i) lake trout (center; shown with FWS biologist). ((a–g) Courtesy Raver, D., USFWS. With permission. (i) Courtesy of Hanson, D., USFWS.)

Plate 7 Eastern North American has an extensive network of coastal streams, ponds, and lowlands with a fish fauna that includes (a) longnose gar, (b) bowfin, (c) redfin pickerel, (d) white catfish, (e) sunfishes (e.g., mud sunfish), (f) anadromous clupeids (American shad), and (g) nonnative common carp held by D. Stephenson. (h) Central image: freshwater marsh and ponds. ((a–e) Courtesy Raver, D., USFWS. With permission.)

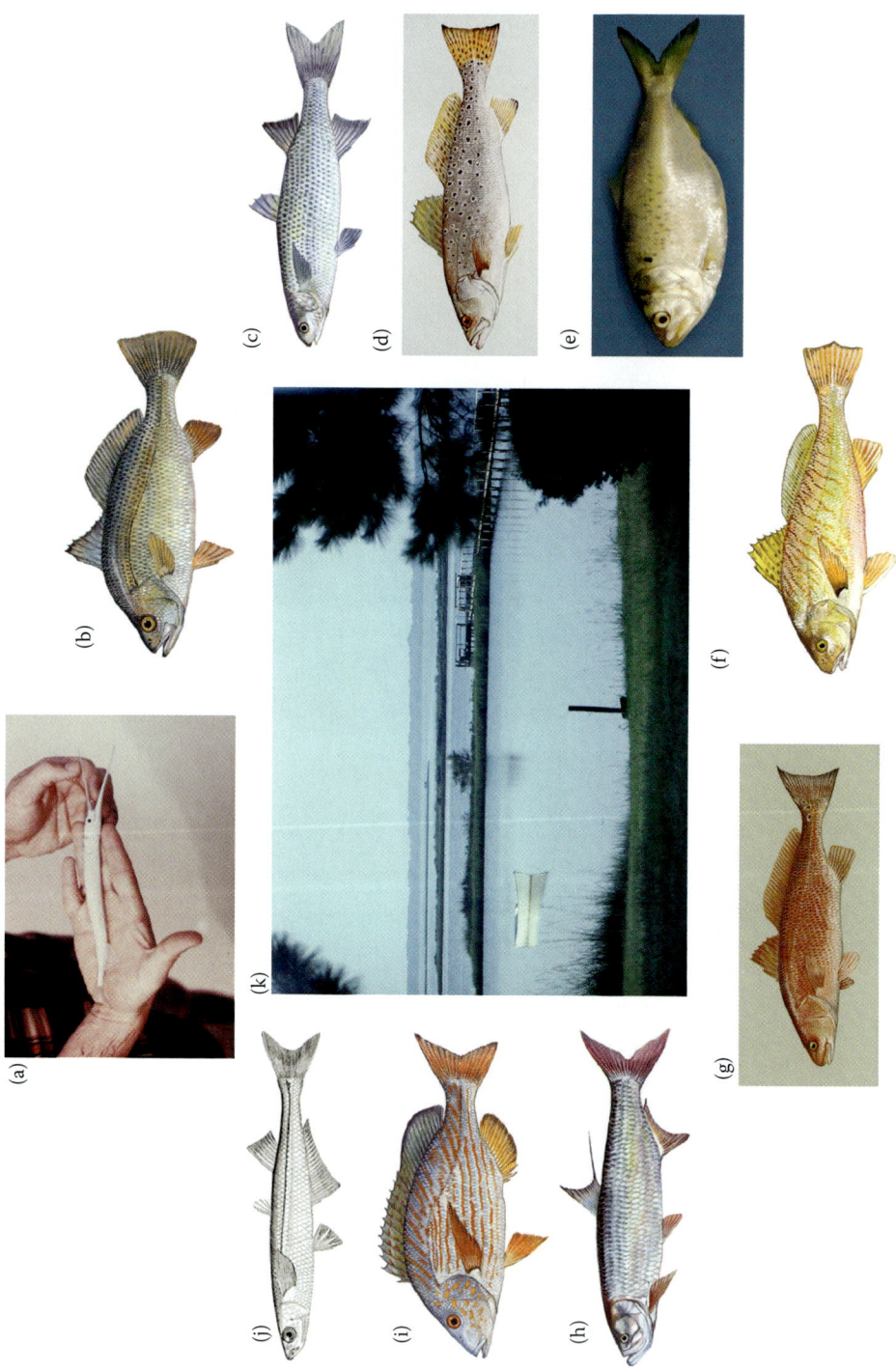

Plate 8 Fishes of North American Southeastern estuaries are diverse and include (a) needlefish (Atlantic), (b) silver perch, (c) mullet (e.g., striped), (d) sea trout (spotted), (e) Gulf menhaden, (f) croaker, (g) drum (red), (h) tarpon, (i) pigfish, and (j) silversides. (k) Central image: Bogue Sound, North Carolina. ((b–d, f–i) From Raver, D., Fishfinder Collection, North Carolina Division of Marine Fisheries (NCDMF). With permission. (e) Courtesy of Bournje, J., http://commons.wikimedia.org/wiki/File:B.patronus.JPG. (j) Courtesy of NOAA, http://www.nefsc.noaa.gov/lineart/atlantic%20silverside.jpg, http://commons.wikimedia.org/wiki/File:Atlantic_silverside.jpg.)

Plate 9 Some subtidal fishes of inshore marine beaches and bays: (a) bluefish, (b) northern kingfish (whiting), (c) puffer, (d) spot, (e) southern stingray, (f) big skate, (g) flounder (summer), (h) pinfish, and (i) sheepshead. (j) Central image: beach seining. ((a–d, g–i) From Raver, D., Courtesy of the NCDMF. (e, f) Courtesy of NOAA http://www.photolib.noaa.gov/htmls/fish4182.htm and http://www.photolib.noaa.gov/htmls/fish4021.htm.)

Plate 10 Some epipelagic offshore (neritic) fishes: (a) Pacific herring, (b) Pacific sardine, (c) northern anchovy, (d) yellowtail, (e) king mackerel, (f) bonito (oceanic), (g) blue marlin, (h) mako shark, and (i) giant manta (center). ((a) Courtesy of Islandwood.org, http://commons.wikimedia.org/wiki/File:Clupea_pallasii.jpeg. (b, i) Courtesy of NOAA, http://www.nmfs.noaa.gov/fishwatch/species/sardine.htm and http://Sanctuaries.noaa.gov/sos2006/flowergarden.html. (e, g, h) From Raver, D., Fishfinder collection, NCDMF. With permission.)

Plate 11 Selected benthic (neritic) fishes: (a) Atlantic cod, (b) haddock, (c) rockfish (vermillion), (d) sea bass (black), (e) red snapper, (f) red grouper, and (g) angel shark (center). ((a, b) Courtesy of Bloch, M. E., et al., http://gallica.bnf.fr/ark:/12148/btv1b2300245v.planchecontact, fish #63 and 62, http://commons.wikimedia.org/wiki/File:Gadus_morhua2.jpg and http://commons.wikimedia.org/wiki/File:Melanogrammus_aeglefinus1.jpg. (c) Courtesy of Simon, V., NOAA Photo Library, http://www.photolib.noaa.gov/htmla/fish4070.htm, http://commons.wikimedia.org/wiki/File:Sebastes_miniatus_NOAA.jpg. (d, e) From Raver, D., Fishfinder Collection, NCDMF. With permission. (f) Courtesy of NOAA, http://www.noaa.gov/, http://www.photolib.noaa.gov/fish/fish3131.htm, http://commons.wikimedia.org/wiki/File:Epinephelus_morio.jpg. (g) Courtesy of Chess, T., NOAA, http://swfsc.noaa.gov/textblock.asps?ParentMenuId=123&id=985, http://commons.wikimedia.org/wiki/File:Squatina_californica.jpg.)

Plate 12 Examples of deep-sea fishes include mesopelagic, bathypelagic, and benthopelagic fishes such as (a) lantern fishes (*Myctophum*), (b) bristlemouths (*Sigmops*), (c) hatchetfishes (*Sternoptyx*), (d) fangtooths (*Anoplogaster*), (e) swallower (*Saccopharynx*), (f) anglerfishes (*Melanocetus*), and (g) softhead, and (h) ghostly grenadiers (center). ((a, b, c) Courtesy of Kissling, E., 1911 Monacan Expedition (Résultats des campagnes scientifiques accomplies sur son yacht par Albert Ier, prince souverain de Monaco), http://www.archive.org/stream/rsultatsdescam35albe/rsultatsdescam35albe_djvu.txt, http://commons.wikimedia.org/wiki/+File:Myctopohum_punctatum1.jpg (for a), File:Sigmops_bathypilus.jpg (for b), and File:Sternoptyx_diaphana1.jpg (for c). (d) Courtesy of Brauer, A., 1898 German Expedition (Brauer, *Die Tiefsee-Fische.I.Systematischer Teil*), Fishbase.org, http://commons.wikimedia.org/wiki/File:Anoplogaster_cornuta_Brauer.jpg. (e, f) Courtesy of Gunther, A., 1887; (e) http://www.us.archive.org/GnuBook/?id=reportonscientif22grea#676, http://commons.wikimedia.org/wiki/File:Saccopharynx.JPG; (f) http://www.archive.org/details/reportondeepseaf00gn, http://commons.wikimedia.org/wiki/File:Melanocetus_murrayi_(Murrays_abyssal_anglerfish).jpg. (g) Courtesy of Noble, B., NOAA, http:www.photolib.noaa.gov.bigs/fish4320.jpg, http://commons.wikimedia.org/wiki/File:Softhead_grenadier_(Malacocephalus_laevis).jpg. (h) Courtesy of Bari, M., NOAA, http://www.mbnms-simon.org/other/photos/photo-info.php?photoID=420, http://commons.wikimedia.org/wiki/File:Coryphaenoides_leptolepis_1.jpg.)

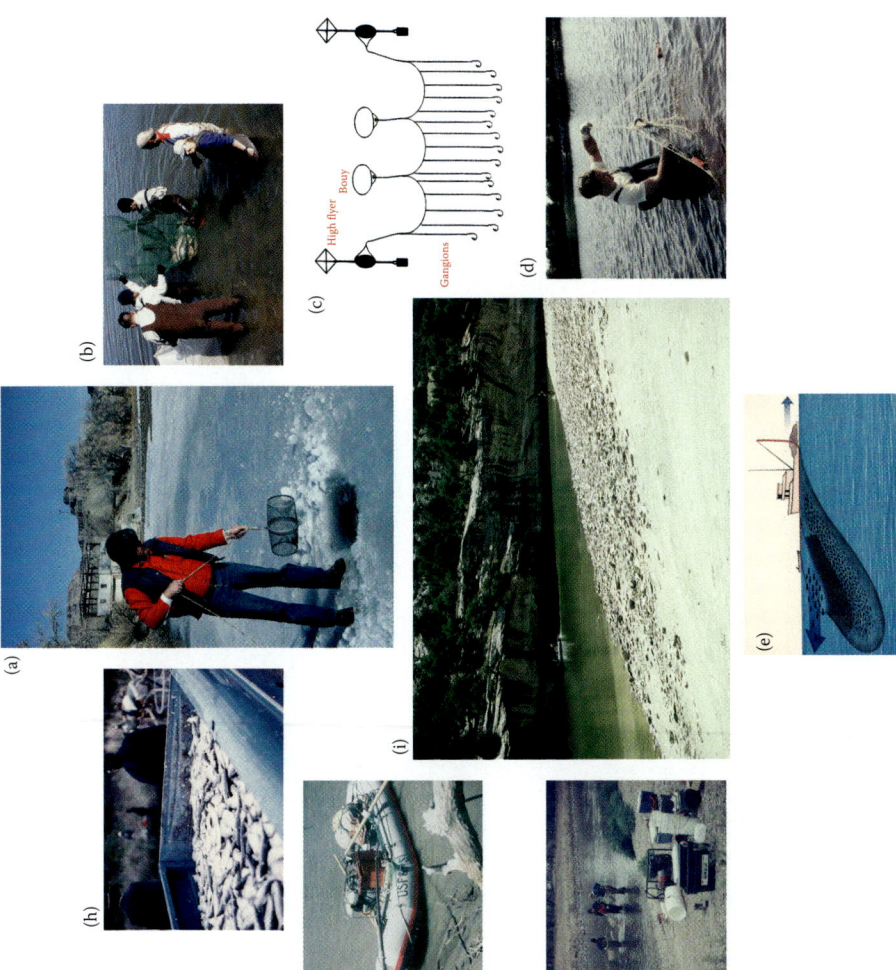

Plate 13 Fish-sampling techniques. Passive collecting gear includes (a) baited minnow traps, (b) hoop (fyke) nets, (c) baited hooks used alone or in a long-line array, and (d) entanglement nets (gill or trammel net). Active methods include purse seining (e), trolling, trawling, haul seining, or electrofishing using a generator and variable voltage pulsator. Electrofishing can be done with the generator on the bank (f) or in a boat (g). Although primitive, even dip nets can be effective if the fish are in migration (h). Some passive gear (e.g., a gill net) also can be used actively (i) (center). (c) Courtesy of NOAA, Apex Predator Investigation, NE Fisheries Science Center, http://na.nefsc.noaa.gov/sharks/img/survey5.jpg. (e) Courtesy of NOAA, NOAA Photo Library, http://www.photolib.noaa.gov/bigs/fish0830.jpg.)

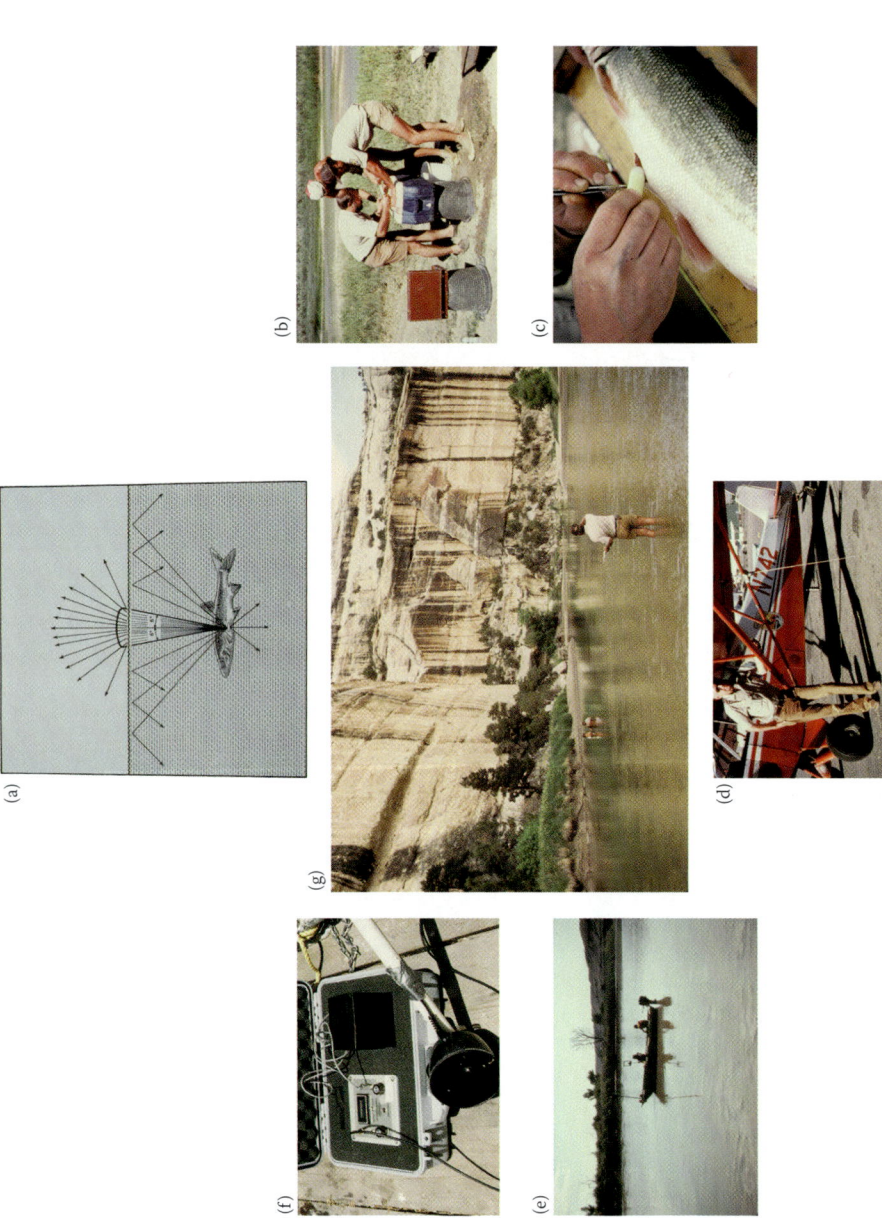

Plate 14 Biomonitoring fish with telemetry: (a) In freshwater, radio waves from transmitting fish travel in air. To minimize handling and holding stress, researchers should be prepared to anesthetize (b), implant (c), and release fish at capture site. Fish tracking can be done by airplane (d), but habitat use data requires a boat (e). Ultrasonics requires a hydrophone (f) that detects sound waves in water to determine underwater position of a fish. Antennas locate the fish from radio waves at the water–air interface. (g) Central image: determining habitat parameters at triangulated fish locations. (All of the persons shown here are USFWS personnel.)

Plate 15 Fish studies occur in a changing environment. Natural changes include *seasonal flows*, illustrated here as habitat change in Colorado pikeminnow spawning grounds of the Green River from June (a) to August (b) (note reference bush on river right). Fish also move to different habitats, even going airborne (c), when a grass carp escapes capture. Anthropogenic changes may be *direct*, such as this (d) endangered fish killed in a national park, or a radio-tagged fish (e) that was being prepared for eating. More commonly, they are *indirect*, such as habitat loss by stream channelization (f), and blockage and habitat change by dams. Predation, and competition by introduced fish such as this northern pike (held by J. Beard) (g), and a deadly meal of introduced channel catfish (h), shown stuck tightly in the throat of a Colorado pikeminnow.

Plate 16 People and fish-related moments: (a) Walt Courtenay works on a snakehead. (b) H. S. Swingle talks to visitors about pond culture at Auburn University. (c) F. E. (Gene) Hester wets a hook. N.C. State students aboard the old RV Eastward (d) bringing in the long line, with (e) W. W. Hassler at the gaff. (f) Former graduate students Chuck Manooch, John Merriner, and the author mend a net at Montego Bay. (g) W. L. Minckley with a razorback at Lake Mohave. (h) Hatchery manager Roger Hamman and Don Portz discuss endangered fishes at Dexter NFH. (i) John Hawkins works a fyke net on the Yampa River. (j) National Park Ranger Myron Chase with a razorback sucker. (k) An Ecology of Fishes field trip. (l) an endangered Colorado pikeminnow with a meal of large sucker. (m) Caryl Williams, W. L. Minckley, and the author electrofishing in Lodore Canyon. (a) Courtesy of Doug Finger, The Gainesville Sun; (g) and (m) courtesy of Walter Courtenay. With permission.

Taxonomic Index*

PREVERTEBRATE CHORDATES AND FISHES

A

Acanthodii, 29f, 39, 40, 40f
Actinopterygii, 29f, 39, 43–46
Agnatha (superclass), 14, 28, 29f, 30–31, 507, 508f
agnathans
    early fishes *Myllokunmingia*, 14, 14f
    and *Haikouichthys*, 14, 14f
Amphioxus (lancelets), Branchiostomatidae
    Branchiostoma, 12f, 14
Anchovies (Engraulidae)
    northern (*Engraulis mordax*), 52f , 355
    Peruvian anchoveta (*E. ringens*), 209f, 209–211
Angelfishes, butterfly fishes, Chaetodontidae, 193–194
Anglerfishes (lophiiformes), 56f
    black sea devils, Melanocetidae, 225f
    Cryptosaras couesii, 226f
    frogfishes, Antennariidae, 225
    goosefishes, Lophiidae, 225, 310f, 311
    Photocorynus, 6, 7f
    sea devils, Ceratiidae, 224, 224t, 225f
Asian carps, Cyprinidae, 432–433, 433f
    bighead (*Hypophthalmichthys nobilis*), 432–433
    black (*Mylopharyngodon piceus*), 133
    grass carp (*Ctenopharyngodon idella*), 249
    silver (*H. Molitrix*), 433f

B

Barracuda, Sphyraenidae, 195, 433f
Basking shark (*Cetorhinus maximus*), 6
Barracudinas, Paralepididae, 219t
Bass, freshwater, Centrarchidae, Largemouth bass (*Micropterus salmoides*), 6f, 147, 148f, 283
    Smallmouth bass (*M. dolomieu*), 417–418
    and rock bass, *Amboplites rupestris*, 106
Bass, marine. Serranidae: sea bass and groupers. Black sea bass (*Centropristis striatus*), 122, 296
    Red grouper (*Epinephelus morio*), Plate 11f
    Giant grouper (*Epinephelus lanceolatus*), 295f
    gag (*Mycteroperca*), 191
    kelp bass (*Paralabrax clathratus*), 192
Bass, temperate. Moronidae (brackish, freshwater, and marine)
    striped bass (*Morone saxatilis*), 445f
    white bass (*M. chrysops*), 353
    white perch, (*M. americanus*), 164
Bigscale fishes, Melamphaidae, 219t
Billfishes, Istiophoridae: blue marlin, 360, 360f
Blennies, Labrisomidae, 167, 180, 193
Bluefish, Pomatomidae, 195f, 174, 195, 389
Bonytongues, Osteoglossidae, 50, 51f
    arapaima, (*Arapaima gigas*) (Pirarucú), 51, 51f
    arawana, (*O. bicirrhosum*) (silver arawana, monkeyfish) and, 51, 310
    fossil bonytongue, (*Phareodus*), 67f, 68
Bowfin, Amiidae:(*Amia calva*), 46, 46f, 47
Bristlemouths, Gonostomatidae, 216f

C

Cardinalfish, Apogonidae, 406
Catfishes, NA freshwater, Ictaluridae—bullheads, (*Ameiurus spp*), 54f, 54
    blue, (*Ictalurus furcatus*) Plate 5f
    channel (*Ictalurus punctatus*), 307, 307f
    flathead (Pylodictis), 54, 321, 429, 456
Catfishes, marine, Ariidaehardhead, (*Ariopsis felis*) 54
    characins, Characidae; e.g., piranhas, 53–54
    gafftopsail, (*Bagre marinus*), 54
    and pacu, 309, 308
Chimeras (ratfishes) Chimaeriformes, 35f, 34, 37
Chondrostei. *See* sturgeons and paddlefishes
Cichlids, Cichlidae
    Haplochromis, 135
    H. latifasciata, 135f
    Oreochromis, 138
    Tilapia, 135
Clupeids. *See* herrings and relatives
Cladoselache, 35, 36f
Cobia, Rachycentridae: (*Rachycentron canadum*), 195
Cods and relatives, Gadidae: Atlantic cod, (*Gadus morhua*), 55, 56f, 197, 197f
    and cusk, (*Brosme*), 56f
    haddock (*Melanogrammus aeglefinus*), 55, 199
    hake (Phycidae, *Urophycis*), 55, 199
Coelacanths, Latimeria, 41–42, 41f, 42f
Conodonts, as early agnathans, 31–32, 31f
Cusk eels, Ophidiidae, 220
Cyprinidontidae, toothed carps. *See* Devils Hole pupfish

D

Damselfishes, Pomacentridae, 192, 193, 296
Darters, Etheostoma. *See* Percidae
Delta smelt– (*Hypomesus transpacificus*), 396, 398
Devils Hole pupfish, (*Cyprinodon diabolis*), 151–152, 152f
Dolphinfish, Coryphaenidae: (AKA dorado, mahimahi) (*Coryphaena hippurus*), 195, 205, 219t
Dragonfish, Stomiidae, 218, 219t
Drums and croakers, kingfish, Sciaenidae
    Atlantic croaker (*Micropogonias undulates*), 455, 456, 456f
    northern kingfish (*Menticirrhus saxatilis*), 195, 195f

---

* Page numbers followed by f and t indicate figures and tables, respectively.
Common and scientific names (Nelson et al. 2004 and Nelson 2006)

Drums and croakers, kingfish, Sciaenidae *(continued)*
    red drum *(Sciaenops ocellatus)*, Plate 8g, 164f, 165
    silver perch *(Bairdiella chrysoura)*, Plate 8b, 191
    spot *(Leiostomus xanthurus)*, 164
    spotted sea trout *(Cynoscion nebulosis)*, 164, 165, 166, 197
Dunkeosteus (Placodermi) 33, 33f, 34f

E

Eels, Anguillidae: American eel, *(A. rostrado)*, 51–52, 51f, 343f
Elasmobranch. *See* sharks and rays

F

Fangtooth, Anoplogastridae, 219t
Filefishes, Monacanthidae, 191
Flatfishes
    Atlantic, Hippoglossus, 358f
    California, *(Paralichthys)*, 57f, 195
    flounders, soles, halibuts. *(O. Pleuronectiformes)*
        Southern flounder *(Paralichthyes)*, 58, 165, 166
    halibut, 57f, 358f
    hogchoker, Achiridae (North American soles) *(Trinectes maculates)*, 195
    summer flounder *(Paralichthyes)*, 442
Flying fish, Exocoetidae: bandwing, 207, 208f
Furu: haplochromine cichlids of Lake Victoria, 135

G

Gar, Lepisosteidae
    alligator *(Atractosteus)*, 46
    long nose *(L. Osseus)*, 46f, 169
    spotted *(L. oculatus)*, 510f
Gobi, Gobiidae, 165, 167, 306
    round gobi, Neogobius, and tubenose gobi, Proterorhinus, 133
Grenadiers, Macrouridae
    ghostly grenadier, Plate 12h
    giant grenadier, 220, 220f
    softhead grenadier, Plate 12g
Groupers and sea basses, Serranidae. *See* bass, marine
Gulper and pelican eels, Eurypharyngidae
    pelican, 218, 219f, 219t
Grunts, Haemulidae
    pigfish *(Orthopristis chrysoptera)*, Plate 8i, 191

H

Hagfishes (Mixini), 28, 30, 30t, 30f
Hatchetfishes, Sternoptychidae, 217f
halibut, Paralichthyidae (sand flounders). *See* flatfishes
Herrings and relatives, Clupeidae
    alewife, *(Alosa pseudoharengus)*, 130, 131f
    American shad, *(A. sapidissima)*, 250
    Atlantic herring, *(Clupea harengus)*, larva, 354f
    blueback herring *(Alosa aestivalis)*, 52
    Menhadens, *(Brevoortia spp)*, 444, 444f
    Sardines: Pacific, 52, 52f, 210, 210f
Humpback chub, *(Gila cypha)*, 276, 318f. *See also* minnows

J

Jacks, Carangidae
    amberjack *(Seriola)*, 195, 207
    jack cravelle *(Caranx)*, 195
    Pompanos *(Trachinotus)*, 195
    yellowtail *(S. lalandi)*, 195, 196f

K

Kelp bass. *See* Serranidae
Killifishes, Fundulidae
    killifishes, 56, 121, 164, 166
    and mummichog, *(Fundulus heteroclitus)*, 165
    topminnows, 150
Kingfish (aka whiting). *See* drums

L

Lake sturgeon *(Acipenser fulvescens)*. *See* sturgeons
Lamprey, sea, Petromyzonidae, *(Petromyzon marinus)*, 30f, 31f, 130, 130f, 133
Lancetfish, Alepisauridae, 216, 217f
Lanternfishes, Myctophoridae, 216, 216f
Ling cod, Hexagrammidae, *(Ophiodon elongates)*, 195
Lizardfish (Synodontidae), 195
Lungfishes (Dipnoi), 42–43, 43f

M

Mackerels and relatives, Scombridae
    bluefin, *(T. thunnus)*, 206f, 207
    king mackerel, *(Scomberomorus cavalla)*, 311, 514
    little tunny, *(E. alletteratus)*, 195
    Pacific bonito, *(S. chiliensis)*, 195
    skipjack *(K. pelamis)*, 208, 250f, 251
    Spanish mackerel, *(S. maculates)*, 166, 195
    yellowfin tuna, *(T. albacores)*, 195, 237, 439
Menhaden. *See* herrings
Myllokunmingia, 14f, 28
Minnows (Cyprinidae), 53–54, 53f
    Chubs, *(Gila spp.)*– bonytail, 319, 426
        and humpback, 109, 276, 318f
    creek chub, *(Semotilus atromaculatus)*, 318
    fathead minnow, *(Pimephales promelas)*, 53f, 200, 299
    golden shiner, *(Notemigonus crysoleucas)*, 6f
    pikeminnows, Colorado *(Ptychocheilus lucius)* and northern *(P. oregonensis)*
    red shiner, *(Cyprinella lutrensis)*, 299, 300
    redside shiner, *(Richardsonius balteatus)*, 299
    splittail, *(Pogonichthys macrolepidotus)*, 396
Mudskippers (e.g. *Periophthalmus*), Gobiidae, 181
Mullet, Mugilidae, striped, *(Mugil cephalus)*, 166, 195, 306f
Mummichug *(Fundulus heteroclites)*, 165

TAXONOMIC INDEX

N

Needlefish, Belonidae
    Atlantic, (*Strongylura marinus*), 195
Neopterygii. *See* gar and bowfin
Nile perch, (*Lates nilotica*), 136–138, 137f
Northern pikeminnow. *See* minnows

O

Ocean sunfish, Molidae, (*Mola mola*), 58f, 207
Ostariophysi, 53, 73
Osteolepidiformes, 41
Orange roughy, (*Hoplostethus atlanticus*), 224f, 406
Osteichthyes, p., 39
Ostracoderms (anaspids, cephalaspids, pteraspids, thelodonts), 32f

P

Pacific sardine, (*Sardinops sagax*). *See under* herrings
Pacu, (*Myleus pacu*), 308, 309f
Paddlefish, American, (*Polyodon spathula*), 45f, 109, 109f
Parrotfishes, Scaridae, 193, 194
Patagonian toothfish, (*Dissotichus eleginoides*) (Chilean seabass), 224, 224f, 406
Pelagic cods, Melanonidae, 56
Perch and darters, Percidae
    and blue pike (extinct), 128, 129
    candy darter (*Etheostoma*), 514f
    sauger, (*Sander canadense*), 57f, 73
    walleye (*S. vitreus*), 73, 128
    yellow perch (*Perca flavescens*), 73, 124
Peruvian anchoveta. *See* anchovies
Pinfish, porgies, Sparidae
    pinfish (*Lagodon rhomboides*), 165
Pikaia, 14, 14f
Pikes and pickerals, Esocidae
    chain pickerel (*E. niger*), Plate 4d
    northern pike (*Esox lucius*), 55, 55f
    and redfin pickerel (*E. Americana*), Plate 7e
Pipefishes, Syngnathidae, 191, 207, 405
Placoderms, *Dunkeostomus*, 33–34, 33f, 34f
    Materpisces attenboroughi, 34
Pompano. *See* jacks
Pinfish and porgies, Sparidae
    pinfish (*Lagodon rhombides*), 165
    sheepshead, 165
Puffers, Tetraodontidae, 58f, 241, 309
Pupfishes, Cyprinodontidae
    Devils hole, (*Cyprinodon diabolis*), 151–152, 151f

R

Rainbow trout (*Oncorhyncus mykiss*). *See* trouts
Rays and skates
    Rajidae:big skate (*Raja binoculata*), Plate 9f
    giant manta, Mobulidae, 183
    and little skate (*Leucoraja erinacea*), 36f
    Rajiidae
    southern stingray Dasyatidae, 195
Remoras, Echeneidae, 219t, 234, 272
Rockfishes, Scorpaenidae, 192, 195, 406
Roundtail chub (*Gila robusta*). *See* cyprinidae

S

Sabertooth, Evermannellidae, 219f, 219t
salmonids, *See* trouts, salmon and relatives
Sarcopterygii, 29f, 40–43, 63, 509
Sargassumfish (*Histrio histrio*), Antennariidae, 207
sculpins, Cottidae, 73, 94, 95, 97t, 104, 123, 124, 128, 133
    Mottled sculpin, (*Cottus bairdi*), 330
    Slimy sculpin, (*C. cognatus*), 95f
Sea devils, Ceratiidae, 225
Sea lamprey, Petromyzontidae, 130, 133, 139, 272, 311, 344, 381, 387, 429, 434
    Petromyzon marinus, 30f, 31f
Sea robins, Triglidae, 195
Sea trouts. *See* drums
Sharks, Selachii (Nelson 2006), 33f
    Angel, Squatinidae, Plate 11g
    mako, Lamnidae Plate 10h
        blacknose, Carcharhinidae, 508f
    whale, Rhincodontidae, 6f
Spiny dogfish, Squalidae, 197, 287, 402
silversides, Atherinopsidae, 164, 191, 195
    Atlantic, (*Menidia menidia*), 165
snakeheads, Channidae, 432, 433, 458
    Channa spp., 433f
Snappers, Lutjanidae, 57, 59, 194
    red snapper (*Lutjanus*), 356, 440
    vermillian snapper, Rhomboplites
Snipe eels, Nemichthyidae, 219t
Spiny eels, Notacanthidae, 219t
Squirrelfishes, Holocentridae, 194
Sturgeons, Acipensaridae, 44–46, 404, 435, 442
    and pallid sturgeon (*S. albus*), 44, 435
    shovelnose (*Scaphirhynchus platorynchus*), Plate 5b
    white sturgeon, (*A. transmontanus*), 195, 397f
Suckers, Catostomidae, 53–54, 54f, 95, 97t, 304, 308
    bluehead (*C. discobolus*), 308f, 437
    mountain (*C. platyrhynchus*), 54f, 95, 308
    razorback (*Xyrauchen texanus*), 318f
    redhorse (*Moxostoma*), 106
    white (*C. commersoni*), 54f, 277, 296, 437
Sunfishes, Centrarchidae
    black (*Pomoxis nigromaculatus*) and white (*P. annularis*) crappie, 149
    bluegill (*Lepomis macrochirus*), 121, 283–284, 386f
    bluespotted (*Enneacanthus obesus*), 57f
    green (*L. cyanellus*), 135
    mud (*Acantharcus pomotis*), Plate 7e
    P. annularis, Plates 4b, 6e
    pumpkinseed (*L. gibbosus*), 268–269, 321
    rock bass (*Ambloplites rupestris*), 106, 513
Surfperches, Embiotocidae, 192
Surgeonfishes, rabbitfiishes, Acanthuridae, 193

Swallower, Saccopharyngidae, 218
Swordfish, Xiphiidae (*Xiphias gladius*), 211, 335, 439

T

Tarpon, Megalopidae, 51
   Megalops atlanticus, Plate 8h
Teleost, lower, upper, characteristics of, 50–51
Teleostomes, 39
Temperate basses, Moronidae
   striped bass, (*M. saxatilis*), 444
   white, (*M. chrysops*), 121
   white perch, (*M. Americana*), 164
Tetrapods, 29f, 41
Tilapia. *See* cichlids, Plate 6h
Toadfishes, Batrachoididae, Gulf toadfish (*Opsanus beta*), 195
Triggerfishes, Balistidae, 58
trouts
   salmon, char and relatives (Salmoniformes, Salmonidae), 95, 96, 101, 102, 123, 332, 511
      Arctic char, (*Salvelinus alpines*), 54, 95, 96
   Atlantic salmon (*Salmo salar*), 55, 95–96, 131f
   Brook trout (*S. fontinalis*), 96f, 101
   ciscos (*Coregonus*), 54, 131f, 133
   coho (*O. kisutch*), 96, 132, 341
   European brown trout (*Salmo trutta*), 95–96, 96f, 511
   grayling (*Thymallus arcticus*), Thamallus, 54, 121
   greenback cutthroat (*O. clarki stomias*), 100f, 100–102, 406
   lake trout, (*S. namaycush*), 99, 99f, 128
   Pacific salmon: Chinook (*Oncorhyncus tshawytscha*), 55f, 96, 413–415
   pink (*O. gorbuscha*), 55f, 132, 342
   rainbow (*O. mykiss*), 96f, 101, 512f
   sockeye (*O. nerka*), 96, 338
   western interior—cutthroat (*O. clarki*), 96
   whitefishes (*Prosopium*), 54, 121, 124, 128

W

Whalefish, Cetomimidae, 219t
Wrasses, parrotfish, Labridae, 192, 193, 270

INVERTEBRATES

Aquatic invertebrates, 94, 307, 309, 312
black fly, Simulidae, 94
blue crab (*Callinectes sapidus*), 444–445
caddis fly, Chironomidae, 94
Dobson fly, Megaloptera, (hellgrammite), 94
Erypterids; giant water scorpions, 15
oysters, Ostreidae, 166–167, 443
quagga mussel (*Dreissena bugensis*), 133
stone fly, Plecoptera, 94
Zebra mussel, *Dreissena polymorpha*, 132, 132f, 133

TETRAPODS

alligator, *Alligator spp*, 34, 46
beaver, (*Castor Canadensis*), 141
Caspian tern, 418
Cormorant, 210, 399, 418
grizzly bear (*Ursus arctos*), 99, 429
Tetrapods, 29f, 41

Subject Index*

A

Abyss, 218–219
Adaptive management. *See also* Tracy case study
   concept, 393
Allee effect, 280, 281, 287
Amensalism, 272
Adaptations
   behavioral, 289–291
   of elasmobranchs, 35–37
   physiological, 233–234, 235–240
   of rayfin fishes, 509–510
Acclimation, acclimatization, 231, 257, 264
Anabolism
   compared to catabolism, 247, 248f, 252
African Great Lakes
   characteristics, 134–135, 134t
   fishes of, 134–138
   formation, 134
Age and growth
   determination, 254–256, 254t, 256t
   significance, 258–260, 259t, 260f
Agnatha. *See also* Ostracoderms
   hagfish, 28–31, 29f, 30t, 30f, 31f
   lamprey, 28–31, 29f, 30t, 30f, 31f
   prevertebrate chordates, 14
Anthropogenic effects. *See also* Case study on Columbia River and Chesapeake Bay
   in altering habitat, 378
   chemical alteration/pollution, 424
   as a way of classifying ecosystems, 378
   defined, 378
   degree of, 378
   and disrupting communities, 426
   effects due to introducing nonnative species, 430–432
   fragmenting habitat, 276
   and overharvest, 424
Apex predator
   as keystone species, 403
   and trophic cascades, 305
Applied ecology
   defined, 377
Aquatic systems. *See* lotic, lentic, reservoirs, estuaries, marine, neritic, oceanic
Archea
   productivity of, 203
Ash Meadows. *See also* case study
   and pupfish, 150, 152
Autotrophy
   in lakes, 120
   in oceans, 204
   in streams, 97t

B

Behavior
   cognition, 292–293
   communication, 293–294
   foraging and predator avoidance, 313–315
   innate and learned, 289–291
   intelligence, 296–297
   migratory, 335–336
   reproductive, 327–330
   social, 293–296
   spacing, 294–295
Behavioral ecology
   and fitness, 297
   behavior, 289
   in energy allocation, 327–330
   in life histories, 324–327
   as a response to environmental factors, 289
Benthic
   Fishes, 6
   food webs in stream, 94, 102
   habitat in oceans, 221–222, 222t
   oceans, 178, 179, 186
Biodiversity
   a basis for conservation biology, 409–410
   biodiversity crisis, 402–406
   endangered species, 406–409
   illustrated by a baseball team of catchers, 402
   maintaining, 402
   as part of the planetary life support system, 401–402
   and species richness, 71
Bioenergetics, 251
   energy budgets, 247–250
   energy allocation in carnivorous fish/herbivorous fish, 248–250
Biorhythm, 291
   biological clock, 291
   circadian rhythm, 292
Biogeography
   ecological, 78
   historical, 78
Blue planet
   nature of, 11
   surface water, 17, 22
Branchiostegal rays
   appearance, 44, 46
   function, 39, 43
Buoyancy
   hydrodynamic and static lift, 223, 238
   significance, 219, 223
   ways to attain, 223

---

* Page numbers followed by f and t indicate figures and tables, respectively.
  For major topics, chapter noted in parentheses.

C

Canadian fisheries
    for Atlantic Cod, 196, 197f
Carbon dioxide. *See also* global climate change
    atmospheric concentration increasing, 450f, 451
    in ocean acidification, 454
Carrying capacity
    abiotic and biotic influences on, 288
    defined, 285
Case studies
    African Great Lakes, 134
    aging hard structures, 258
    alewife migration, 168
    behavioral interactions, 298
    deepsea anglerfish, 224
    Devils Hole pupfish, 150
    fish migration, 246
    fossil Lake, 66
    greenback cutthroat trout, 100
    Lake Baikal, 123
    management and fish salvage in California, 394
    movement and transport of fish larvae, 370
    NW Atlantic groundfishes, 196
    paddlefish, 109
    Peruvian anchoveta, 209
    predator-prey interactions, 318
    saving salmon, 413
    sharks, 183
    timing of spawning, 332
    troubles in Chesapeake Bay, 441–445
Caudal fin
    evolution of, 43
    function, 233
Character displacement
    an indication of competition, 271
Chessboard model *See* ecological concepts
Chesapeake Bay
    historic fisheries, 435, 436
    overfishing, 424, 428, 434
    pollution of, 432, 438
    Setting, 428
Codland
    importance of cod to Europeans, 196
Coldwater streams and fishes
    Definition, 93–96
    food webs, 97, 122
    fishes of, 95–96
Colorado River
    abiotic and biotic alterations in, 109
    fishes of, 370–373
Columbia River
    case study of, 413
    can science save the salmon, 421
    decline, 413, 414
    historic importance for salmon, 413–415
Commensalism, 272
Competition
    competitive exclusion, 265–268
    detecting in fishes, 271
    exploitative, 269, 270, 273
    interference, 269, 271, 273
    interspecific, 270
    intraspecific, 269
    preemptive, 270
    significance, 266–273
    and trophic adaptability, 271, 311
Conservation
    50/500 "rule", 409–410
    emerging (conservation biology), 378, 406
    Evolutionarily Significant Unit (ESU), 411
    fish declines. *See* Five causes
    Minimum Viable Population (MVP) size, 408
    Population viability analysis, 410
    traditional (wise use), 393, 415
    vulnerability and extinction vortex, 407–409
Continents as zoogeographic regions
    work of A.R. Wallace, 71
Continental Drift
    Concept, 73
    Historic, 78
    role in vicariance biogeography, 77–78
Coriolis effect, 76, 208
Countercurrent exchange
    of heat, 235–237
    oxygen, 237–238
Courtenay, W.P., Jr., 432, 490
Critical period, 365–374, 498
Cultivation Hypothesis, 287, 329, 362, 498
Cushing, D.H., 283, 367–368

D

Dams and fish, 146
Death of nature concept, 448
Decomposition, 119–120, 498
    fishes as decomposers, 53
    role in aquatic systems, 121
Deep scattering layer, 216
Deep sea (abyssal) fishes, 219. *See also* Table 19.2
Devils hole. *See* case study of Devils Hole pupfish
Devonian Period
    dates of, 61–63
    extinctions in, 63
    significance in fish evolution, 61–68
Diadromy. *See* migrations
Diseases
    ich, 284
    and stress, 258
    whirling disease, 272, 436
Diversity of fishes, 27, 39, 49
Dollo's Law. *See* ecological concepts

E

Ecology
    applied, 3–4
    definition, 4

evolutionary, 3–4
of fishes, 5
history, 3
Ecological concepts (hypotheses, laws, etc) related to
 EVOLUTION
  Chessboard model, 66
  Constant extinction, 66
  Dollo's law, 66
  Murder on the Orient Express, 66
  Red Queen, 66
  specialization and extinction vulnerability, 66
  Williams' Principle, 330
 ZOOGEOGRAPHY
  First principle of biogeography, 80
  hypotheses
   ambient energy, 80
   dispersal, 80
   disturbance, 80
   evolutionary speed, 80
   geographic area, 80
   interspecific interactions, 80
   productivity, 80
   vicariance, 80
 LOTIC SYSTEMS
  flood pulse, 90, 91f
  river continuum, 89–90, 89t, 90f
  serial discontinuity, 91
 LENTIC SYSTEMS
  lakes as microcosms, 122
  succession and evolution of lakes, 122
  trophic state and food web models, 122
 ESTUARIES
  critical salinity, 167
  migrating subsystems, 167
  as nurseries, 167
  as nutrient traps, 167
  Remanes curve, 167
 MARINE SYSTEMS
  critical period,
  interconnection, 183
  match/mismatch hypothesis,
  ocean conveyer, 183
  seafloor spreading, 183
  stable ocean, 183
  vertical zonation, 183
 ANTHROPOGENIC EFFECTS
  Fisheries effects
   Cultivation, 287
   density compensation and depensation, 287
   fisheries yield models, 287–288
   Lee's phenomenon, 286
   Overfishing, 285
   sustained yield, 286–287
  General effects
   ecosystem simplification, 91
   global climate change (GCC), 451
   global warming, 450
   ozone depletion, 450
   recycling of carbon, 450

Ecophysiology
 elements of, 231
 of fishes, 232–233
Ecosystem management, 392
 arrogance, 394
 concept, 392
 considered as techno-arrogance, 379, 385–389, 394–400
 relate to ecosystem approach to management, 392–394
 related to the Arrogance of Humanism, 394
 usefulness, 382
Endangered fish recovery, 433
 as amended, 413
 case study of Pacific salmon decline in the Columbia River, 413–415
 Endangered Species Act of 1973, 428, 457
 Endangered species defined, 423, 433–434
 factors in, 423–424
 Listing factors, 412
 poor success, 428
 species problem, 410–413
Endangered species (29–31)
 Causes of recent (anthropogenic) endangerment, 353, 424
 direct and indirect causes, 423, 424
 estuaries, 435
 the "evil quartet", 437
 five causes discussed for lakes, 424, 429, 434, 436, 438
 and marine systems, 435
 more than one cause, 423–440
 with pertinent issues identified, 435–436
 streams, 435
ENSO (El Niño southern oscillation), 453
 effect on fisheries, 454
 El Niño, 453
 on GCC, 454
 La Niña, 177, 195, 209
Essential fish habitat (EFH), 167, 391
 and fisheries conservation and management, 390
Estuaries
 formation, 158f
 as nutrient traps, 167
 types of, 158
 use by fishes, 163
Equatorial divergences, 206, 211
Evolution (Part II)
 of fishes, 506, 514
 General, 4
Evolutionarily significant units (ESU)
 as Endangered Species, 457
Exploitation of wildlife
 by early European invaders (e.g., fate of heath hen, passenger pigeon, lake sturgeon) in westward expansion (e.g. bison, pronghorn, native cutthroat trouts), 380
 by Native Americans (e.g. Pleistocene overkill), 379
 of North America, 378–381
Extinctions of species
 causes of, 65
 coextinction, 423–426

Extinctions of species (*continued*)
 concept of First Strike and problems of small
  populations, 407–410
 extinction patterns, 71
 extinctions of North American freshwater fishes in
  20th Century, 78
 extinct vs functionally and arguably extinct, 406
 of fishes, 63
 freshwater fishes, 71, 76
 Lazarus species, 406
 marine fishes, 181–183
 pseudoextinction, 64
Extinctions, mass
 of acanthodians, 39–40, 40f, 46
 anthropogenic factors, 428
 The Big Five, 403
 causes, 408
 events, 402–403
 ostracoderms, 49, 59
 placoderms, 29f–30, 32–37
 the present (Holocene) extinction, 403
Extinction rate
 baseline, 62
 Devonian, 29, 32–37
 during Ordovician, 61–62
 geologic vs. recent, 61–66
 K-T and Holocene, 65, 75, 403
 Permian, 62–68, 74f

F

Feeding
 competition and resource use, 311, 318
 ram, 241
 by suction, 43–44
 trophic adaptability, 311
Farm ponds and fish management, 147–148, 384, 385
Faunal groups, 194
 of beaches, 179
 of bottoms, 179, 191, 192
 of shorelines, 120, 157, 180
Fins
 caudal, 50, 59
 evolution of, 39
 median, 27, 59, 500
 paired, 27, 40, 43–46
 types, 27, 50
Fish communities
 bathypelagic, 203, 213–214, 219–222, 497
 benthopelagic, 219
 deserts, 150
 epipelagic, 203–212
 estuaries, 157–159, 163
 of Freshwater. *See* streams
 groundwater, 18, 141, 149, 155
 lakes, 18, 67, 120, 134
 marine. *See* Inshore
 mesopelagic, 214, 215–219, 222
 neritic, 189
 rivers, 18, 338, 385

Fish culture. *See* hatcheries
Fish management
 American Institute of Fishery Research Biologists
  identified needs, 459
 in estuaries, 388, 425
 in freshwater, 383, 385
 improving: American Fisheries Society issues of
  concern, 383f, 386–387, 391, 404–405, 432
 marine systems, 388–389, 426, 430, 435, 437
 ponds, 384–386
 reservoirs, 429
 streams, 387
Fish salvage. *See* case study
Fitness
 definition, 231
 maximizing, 231, 290, 314
Fish
 definition, 5
 diversity, 27
 extinctions, 61
 occurrence, 12
Fisheries
 collapse, 201, 210, 287
 exploitation, 285
Fish kills. *See also* Chesapeake Bay
 due to anoxia= winter kill and summer kill, 438–440
Fish populations
 carrying capacity, 282, 284
 regulation of, 280
Fish salvage. *See* case study
Fisheries management
 early practices result in over fishing, 390
 habitats, 377, 384
 and humans, 382
 recent practice conceived as three overlapping circles,
  383f
 requires knowledge of ecology, 393
 as a scientific approach, 384
Food webs
 of benthic, 85
 pelagic systems, 17, 166
Foraging
 behavior, 307, 313
 changes in energy – time allocation, 254f
 foraging theory, 313
 give-up time, 314, 317
 optimum foraging, 313
 patch depression, 314, 317
 patch model, 314, 316, 322
 predator ability, 313, 315
 predator model, 314, 316, 322
 prey defenses, 315
 risk sensitive, 314, 318
Forbs, S.A., 122. *See also* microcosm
Fossil Butte
 fishes of, 67–68
 setting, 63
 significance, 61
Fossils and geologic time, 29f, 66
 paleoecology, 267

SUBJECT INDEX

Future of fisheries
    marine fishes under the FCMA, 458
    role of states in freshwater, 457–458
Future shock, 449

G

Gape
    predator limitations, 234, 241
    prey response, 314
General adaptation syndrome, 257. *See also* stress
Ghost fishing
    effects of, 184
    types of lost gear, 184
Gills
    in osmoregulation, 239
    role in oxygen extraction, 237
Global climate change
    associated effects on, 454
        coldwater, 454, 455
        invasive species, 456
        lakes, 455
        marine fishes, 454–456
        temperate zone, 454
        tropical systems, 455
        warmwater, 454, 455
    causes, 447
    present evidence for, 452
Global climate change denial, 448
    bad science and junk science, 448
Glycogen
    role in swimming and metabolism, 236
Greenhouse effect
    Gases, 448
Greenland ice sheet, 453
Goldschmidt, T.
    and Lake Victoria, 135
Gondwanaland
    breakup of, 42, 81
    continents of, 73–77
Gravity
    avoiding, 21–22
    as a control on speed, 21
Great Basin and fishes, 115
Great Lakes. *See* Laurentian Great Lakes
Great Salt Lake, 115
Grinnell, J., 266–268
Gulf of Maine, 207, 354
Gulf Stream
    importance, 176–178
    fishes of, 194, 205–211
Guyots, 221
Groundwater
    as fish habitat, 149, 150
    in Devil's Hole, 150

H

Hatcheries
    fish, 416
    harmful effects of, 432
    seen as a panacea/failure in Pacific northwest, 416
    use in fish farming, 438
Habitat alterations
    constraints on fish, 435
Hemoglobin
    oxygen binding, 237
Hjort, Johan, 366–367
Homeorhesis
    and regulation of ecosystems, 11–12
Homeostasis
    and regulation of physiology, 252
Holarctic
    characteristics of, 77
Homing
    olfactory hypothesis, 339–340
        imprinting, 341–344
        navigation, 338
        orientation, 338, 339
        piloting, 338
    parent(home) stream "theory", 236–237
Hutchinson, G.E., 365
    Hutchinsonian (ecological) niche, 267, 273
    n-dimensional hypervolume of resource axes, 267–268, 273. *See also* niche
Hybridization, 424
Hydrothermal vents, 221
Hypotheses. *See* ecological concepts

I

Ice
    anchor, 93–97
    cover and lake circulation, 119
    frazil, 93–97
    structure, 18, 21–22
Ichthyology
    defined, 4
    relationship to ecology, 4
Insects, stream, 94, 97
    collectors, 89, 90f, 94
    scrapers, 90, 94
    shredders, 89, 90f, 94
Iteroparity, 330
IUCN Red Book, 334, 403–406

J

Japan current, 176, 177f
Jaws
    evolution of, 14
    pharyngeal jaws, 135
    usefulness, 32

K

Kelp forests, 192
Kitchell, J.F., 201, 250, 251, 254, 287, 306, 329, 418

L

Lactic acid, 236
Lakes
    characteristics, 117
    fishes, 123
    formation, 115–116
    function, 120
    Laurentian and African Great Lakes, 126, 134–135
    stratification, 118, 118f, 122
    temperate vs. tropical lakes, 125
    types, 116–117
    zones of, 120, 124
Lake Baikal. *See* case studies
Lakes Victoria, Tanganyka, Malawi. *See* case studies
Lanham, Ural, 49, 238
Larvae
    types of
        altricial, 355
        leptocephalus, 357
        precocial, 355
    transport of, 355
Larval fish
    as ecologically distinct organisms, 355
    fitness and altricial larva, 363
    metamorphosis, 356
    niche concept and larvae, 365
    and salutatory ontogeny, 373
    taxonomy of, 356
Lasker, R., 367, 368f
Laurasia
    and breakup, 81
    conditions on, 71–80
    Formation, 74
Laurentian Great Lakes
    Erie, 128–132
    history of change. *See* case studies
    Huron, 127–132
    Ontario, 127f, 128f, 127–131
    Michigan, 127–132
    Superior, 128–132
Law
    of Constant Extinction, 65, 66
    Shelford's law, etc. *See* ecological concepts
Laws(statutes) = Endangered Species Act of 1973 et seq, 398
    Fisheries Conservation and Management Act of 1976 (AKA Magnuson Act), 390
    National Environmental Policy Act of 1969, 353
lentic systems
    definitions, 115
    structure and function, 120
    temperate vs tropical lakes, 125
Lewis, W.M., Jr., 149
Lindeman, R.L., 303
Locomotion, movement, 233
Life history patterns, traits, 324
    r and K selection, 331
    semelparity and iteroparity, 330
    tradeoffs, 328

Light
    properties in water, 17
Lotic systems
    coldwater streams, 93
    general, 83
    warmwater streams, 103
Luciferin
    and light production, 223

M

Management
    fisheries practices, 387
    future trends, 457
    need for change, 391–394
    sustainability, 391
Marine systems, 181–182, 183
    critical period, 365–367
    match/mismatch hypothesis, 367–369
Metabolism, 232
    aerobic and anaerobic, 236
    standard, 248
Migration
    of alewife, 337
    anguillid eels, 343–344
    Atlantic herring, 345
    defined, 335
    diadromy, 340–341
    oceanadromy, 340
    Pacific salmon, 341
    pikeminnow, 341
    potamodromy, 341
    See also homing
Minckley, W.L., 429
Mississippi River, 109
Missouri River, 108
Muscle
    red, 236
    warm, 235–237
    white, 236
Mutualism, 270, 296
MVP (minimum viable population)
    with changing environment, 408
    factors influencing, 408
    of wild populations, 409–410
Myers, George, 72–73

N

Narragansett Bay, 166, 435
Natural systems
    characteristics of (e.g., diversity, stability, redundancy), 402
Neritic Province, 189–201
    function, 190
    productivity, 190
Nerve cord, 12, 12f, 13
Newfoundland, 189, 196, 197–198, 199
    groundfishes. *See* codland
    and historic fisheries, 196–198

SUBJECT INDEX

Niche, 265–269
  concepts of, 265
    fundamental, 265, 267
    niche shifts, 268–269
    ontogenetic change in, 269
    realized, 265, 267
  population/persistence, 268
  recess/role, 268
  resource/utilization niches, 268
    "vacant niches", 268
  types of niche, 265, 268
Nonnative fishes, 71, 79, 108, 125, 126, 133, 147, 276, 299, 300–301
  effects on native aquatic systems, 429–430
Northwest Atlantic groundfishery, 196–201
  extent and collapse of, 200–201
  lack of recovery. *See* case studies
Notochord, 12, 12f, 13

O

Oceanadromy, 344–345
Ocean conveyor. *See* waves and currents
Oceanic systems
  fishes, 205–208, 216–218
  realms and provinces, 203–204, 213–215
  zones, 204–208, 215–218
Ontogeny, 357–359
  and phylogeny in fishes, 507–515. *See also* niche
Optimal foraging theory, 313–314, 314t
  basic prey model, 314, 322
  patch model, 314, 322
Organic matter
  allochthonous, 93, 94, 104, 105
  autochthonous, 93, 94, 117
  CPOM, 93–94, 104
  detritus, 84, 88, 94, 105, 179
  DOM, 94f, 105
  FPOM, 94, 104, 105
  litter, 84, 88, 89, 94, 94f, 104
  woody debris, 94, 104
Otter trawl, 198f, 198–199
Otoliths, 39, 42, 43, 259t
Overfishing, 110, 285–288, 366, 390, 391, 392
Oxygen (20)
  from air, 237–238
  extracting from water, 237–238
Oyster reefs, 166–167, 442–444

P

Parasites, of fish, 311
  brood parasitism, 272
  sea lamprey, 30f, 31f, 130, 130f, 133, 311
  sea lice, 272, 417
  tapeworms, 272
  trematodes, 272
Pelagic fishes, 204, 205
  food webs, 17, 166
  epipelagic, 205–208, 206f

Pelagic Realm, 190
  bathypelagic, 203, 218–219, 219f, 222t
  benthopelagic, 219–221
  epipelagic, 190, 203–208, 222t
  mesopelagic, 190, 203, 215–218, 222t
PHABSIM (Physical habitat simulation model)
  appropriate application of, 428
  concept, 428
Pharynx, 12f, 13
Photopores, 216, 217f, 222, 223
Planet Earth, 61, 79, 402
  unintentional experiments on, 448
Pleistocene glaciations, 78–79
Plymouth Colony fish law(15), 168, 380
Polar migration hypothesis, 366
Pollution
  effects on fishes, 438–440
Population ecology, 275–288
  dynamics of population growth, 278–280
    biotic potential, 279
    environmental resistance, 279
    logistic model, 279
  regulation, 280–284
    density dependence and independence, 283–284
    equilibrium, 283–284
  size of
    at MVP, 409–410
  viability
    age structure and age classes, 277
    sex ratios of, 277
    survival rates, 278
Potamodromy, 341
Predation, predation risk, 370
Predators
  ambush, 46, 311
  pursuit, 311
Primary division of freshwater fishes, 73
Principle, of constant proportions, 21
Pseudoextinction, 64
Purse seine, 209–210

R

Radiant energy
  and albedo, 450
  solar constant, 449
r and K selection, 331–332
Red Queen Hypothesis. *See* ecological concepts
Relict bony fishes, 44–46
Reservoirs and impoundments, 147–149
  farm pond management, 147–148
  reservoir fishes, 147–149
Rivers
  Colorado, 53, 100, 101, 109, 234, 272, 276, 298, 318, 346, 350, 354, 370, 381, 425, 426, 427f, 429, 435, 456
  Mississippi, 76, 79, 84, 105, 106, 109, 110, 111–112, 126, 127f, 133, 158, 381, 440
  Missouri, 45, 79, 108, 110, 146, 149, 381, 387, 425, 435

Rivers (*continued*)
    Snake, 146, 414f, 418, 419–421
    Yangtze, 85t, 109
Reproduction
    alternative breeding strategies, 326–327
    bearers of eggs or young
        oviparous, 324, 327, 355
        ovoviviparous, 324, 355
        viviparous, 324, 355
    cost of, 330
    courtship, 329
    mate selection, 326
    migration, 329
    secondary sex characters, 329
    spawning, 327, 329
Reproductive ecology, 323–334
    energy allocation for prespawning, 329
    reproductive allocation of energy to parents, 329–330
    reproductive life history strategies, 330–331
        iteroparity, 330–331
        semelparity, 330
    spawning and post spawning, 327
    synchronous spawning, 362
    timing and environmental cues, 324–325
    young, 327–329
Rete mirable (wonder net), 238
River estuary, 157–162
    Chesapeake, 159
    Hudson, 159
Romer, A.S., 33, 66

S

Sacramento and San Juaquin rivers, 108, 394, 430
Sargasso Sea, 207, 343–344
San Francisco Bay, 158, 430
Scales
    of fish, 43, 45f
    of geologic time, 122, 138, 267
Schmidt, Johannes, 343, 366
Sea mounts, 221
Selye, Hans, 257
Semelparity, 330
Sex, 325–326
    sex dominance, 326
    sexual selection of mates, 326
Sharks
    as predators, 40
    swimming, 43
    vulnerability, 183–186
Spawning. *See* reproduction
Species concepts
    Nominalist and Pluralist viewpoints, 411
Speciation
    microallopatric, 135
    species flocks, 123, 134, 135
Smokers. *See* hydrothermal vents
St. Lawrence River, 129, 130, 198, 199
Streams. *See* coldwater and warmwater streams

Stream transport
    Cpm, 371–373
    of larval fish, 339, 370–373
    paddlefish, 111
Stress, 258
Swimming
    lift based, 233
    resistance based, 233
Swingle, H.S., 146, 147
Symbiosis, species interactions, 270

T

Tolkien, J.R.R., 5
Teeth, 242
    jaw, 242
    pharyngeal, 242
Teleosts, 49–59, 510–515
Territories, 295
Tetrapods, 41
Tipping points, 452–453
Tolerance, 264
    Shelfords law of, 264
Tragedy of the Commons, 286, 392, 407
Trash fish, 53, 103, 129, 304, 418, 434
Trawls. *See* Otter trawl
Trophic concepts, 303–321
    adaptability, 311–312
    autotrophs, 303
    carnivores, 309–311
    cascades, 304–306
    detritivores, 306
    herbivores, 303, 307–308
    pyramid, 303–305
Truman, Harry S., 150, 152, 390

U

Upwelling, 208–209
    coastal, 208
    equatorial, 208
U.S. Army Corps of Engineers (USACE), 159, 414, 420
U.S. Supreme Court, 152, 152
User, of natural resources, 378
    concept of ecotourism, 378
    consumptive and nonconsumptive use, 378, 392
    direct use, 378
    indirect, 378

V

Vicariance biogeography, 77–78

W

Wallace, A.R., 71, 75, 76, 80
Warmwater streams and rivers
    description, 103
    fishes of, 103–112
    structure and function, 103–106

Water
    density, 19–20
    molecules, 18–19
    properties, 17–20
Water storage
    groundwater, 115, 149–150
    in reservoirs, 141–149
Waves and currents
    gyres, 176–177
    thermohaline flow, 177–178, 214, 452
    tides, 159–160, 162
    upwellings, 177, 204, 208–209
Wegener, Alfred, 73, 78
Whirling disease, 99, 272
Wilderness
    perception by invading Europeans, 377, 380
Winberg equation, 247–248
Woods Hole, 207

Z

Zoogeography of fishes, 71–80
    concept, 79–80
Zoogeographic Regions
    Australian, 77
    Ethiopian, 75–76
    Nearctic, 76–77
    Neotropical, 76
    Oriental, 76
    Palearctric, 76